Introduction to the Theory and Applications of Functional Differential Equations

Mathematics and Its Applications

Volume 463

Introduction to the Theory and Applications of Functional Differential Equations

by

V. Kolmanovskii

Moscow State University of Electronics and Mathematics
and Space Research Institute (IKI) of the
Russian Academy of Sciences,
Moscow, Russia

and

A. Myshkis

Moscow State University of
Communications (MIIT),
Moscow, Russia

SPRINGER-SCIENCE+BUSINESS MEDIA, B.V.

A C.I.P. Catalogue record for this book is available from the Library of Congress.

ISBN 978-90-481-5148-6 ISBN 978-94-017-1965-0 (eBook)
DOI 10.1007/978-94-017-1965-0

Printed on acid-free paper

Dedicated to the memory of
the outstanding mathematician,
our friend and teacher
Professor Mark Alexandrovich Krasnoselskii
1920 – 1997

Contents

Part V.
CONTROL AND ESTIMATION IN HEREDITARY SYSTEMS

PREFACE

At the beginning of this century Émil Picard wrote:

"Les équations différentielles de la mécanique classique sont telles qu'il en résulte que le mouvement est déterminé par la simple connaissance des positions et des vitesses, c'est-à-dire par l'état à un instant donné et à l'instant infiniment voison.

Les états antérieurs n'y intervenant pas, l'hérédité y est un vain mot. L'application de ces équations où le passé ne se distingue pas de l'avenir, où les mouvements sont de nature réversible, sont donc inapplicables aux êtres vivants".

"Nous pouvons rêver d'équations fonctionnelles plus compliquées que les équations classiques parce qu'elles renfermeront en outre des intégrales prises entre un temps passé très éloigné et le temps actuel, qui apporteront la part de l'hérédité". (See "La mathématique dans ses rapports avec la physique, Actes du IVe congrès international des Mathématiciens, Rome, 1908.)

Many years have passed since this publication. These years have seen substantial progress in many aspects of Functional Differential Equations (FDEs). A distinguishing feature of the FDEs under consideration is that the evolution rate of the processes described by such equations depends on the past history. The discipline of FDEs has grown tremendously, and publication of literature has increased perhaps twofold over publication in the previous decade. Several new scientific journals have been introduced to absorb this increased productivity. These journals reflect the broadening interests of scientists, with ever greater attention being paid to applications.

In the middle 1980s, we wrote a book on applications of FDEs entitled "Applied theory of functional differential equations", published in the beginning 1992 by Kluwer Academic Publishers.

The first edition of the book have been used numerous time, mostly in teaching FDEs to students of applied mathematics and science, where the purpose was to provide an understanding of the FDEs techniques as well as practice in using them. The book should prove useful for professional applied mathematicians and scientists because of its emphasis on analysis of mathematical models using FDEs and their detailed investigation. Students, colleagues and other users have provided us with many constructive comments and proposals to make the future edition both better medium for prospective users of FDEs and better for classroom instruction or self-study.

At the time of writing of the first edition, the field of FDEs is making significant breakthroughs in its practice which are no longer only a specialist's field. The number of FDEs used in the sciences and in applied areas has been increasing tremendously, where they continue to play a crucial role. Applications range from physiology and enzyme kinetics to whaling control and food webs, from neural networks to lazer optics, and from studies of engines to the theory of business cycles. Moreover as our activities get more and more automated, the importance of taking into consideration delay becomes evident as increased numbers of large industrial equipment such as engines, blowers, fans, transformers, and compressors are in use.

The present book is intended as a replacement of the previous one, and is by no means a revision of it. We tried to write a new book for those wishing to learn about FDEs with little or no prior knowledge and lead them from the simplest to the most complicated subjects in simple and understandable way. Yet we never forget the readers with a main interest in applications.

A considerable amount of thought, feedback, and effort have gone into the writing and production of the new book which represents a wide–ranging interdisciplinary study of FDEs.

In spite of the general topic remains the same, the overlap between two books has turned out to be very small indeed. But the intended readership is precisely the same, namely postgraduate and advanced undergraduate students and various users of FDEs. Moreover we note that the book will serve as a basic course in FDEs theory for students of applied mathematics, engineering and physical sciences.

Emphasis in the subject of FDEs has changed substantially during last years. New methods in stability theory, optimal control, optimal estimation, periodic solutions were being proposed. Along with this huge variety of new applications has emerged.

Yet, many recent results are inaccessible and have never brought together. For this reason, in this book we have tried to reflect those changes and to bring coverage of literature up to date. So some parts of the old book were dropped and replaced by newer material, to the current situation, that updates the subject relative to the latest developments. Problems of implementation of general methods are considered in much more detail. Great care is done to include topics that can be used in everyday practice and make the book more effective as a textbook.

Examples are interspersed with the text at the appropriate sections. In the main, they are intended to illustrate and to extend the text.

Studying a smaller number of methods in greater depth tends to raise the level of difficulty. In order to avoid this we are giving sometimes non–rigourous accounts where appropriate. If rigour is essential, we have quoted theorems, and supplied proofs especially if they are constructive.

Every efforts were done to explain basic ideas with the minimal mathematical equipment and to appeal to ones intuition along with ones mathematical skills.

The approach of the book is similar to that of the classical applied mathematics, whose genuine interest in real world problems does not detract from his delight in watching mathematics at work.

Keeping in mind a special applied flavour of the book we tried sometimes avoid the strict axiomatic approach, but there has been no dilution of the standard of mathematical argument. Learning to follow and construct a logical sequence of ideas is one of the most important feature of this book and at the same time to convey some notion of the connection between a mathematical model and the real world.

We hope that the book provide a readable account of the newest development of FDEs theory.

The preparation of the new book presented an opportunity to improve the old one by rewriting many sections, by eliminating errors and misprints, by strengthening the part devoted to theoretical developments of FDEs and their applications.

In particular we have included new material on:

- phenomena modelled by FDEs,
- proofs of all main theorems,
- boundary value problems,
- formal procedure for Liapunov functionals construction for different classes of FDEs,
- non autonomous equations of first and second orders,
- periodic solutions of FDEs,
- periodicity in the theory of stochastic FDEs,
- minimax filtering and estimation,
- model reference adaptive control,
- regularity properties of solutions of stochastic FDEs,

to name a few.

Most of the material included in the book is acceptable to any reader with a reasonable background in real analysis, and some acquaintance with differential equations and probability theory.

There are several sections which require more sophisticated knowledge. In such cases adequate references are provided. A detailed table of contents should enable a reader to fairly quickly navigate through the book in searching for desired topics.

Chapters 1,2 are rather elementary, providing an easy introduction to the subject. They contain introductory sections on a number of topics. Although some readers would like to jump right over this elementary chapter, they might find it convenient to refer back to it to put later parts of the book into context. Detailed description of some new phenomena modelled by FDEs is given here.

The main purpose of Ch. 2 is to describe various ways in which FDEs arise as models in various applications. One of the most important tasks in the analysis of systems is mathematical modeling of the systems. Modeling of nonlinear systems is usually quite complex. Therefore one often has the task of determining not only how to accurately describe a system mathematically, but, more importantly, how to make proper assumptions and approximations, whenever necessary, so that the system may be adequately characterized by a simple mathematical model.

We are sure the topics featured in this chapter will convince the reader that FDEs constitute a very useful tool in contemporary mathematical modeling theory. Moreover, it is shown that sometimes delays do really exist in the laws of Nature (out there waiting to be discovered), and sometimes they are enacted by ourselves to guide us through the jungle of experience. It is becoming increasingly clear that FDEs allow us to improve essentially the clarity of mathematically complex models. As a result they become understandable to those who are not directly involved with modelling, thus enhancing the usefulness of these models.

We also demonstrate that ignoring effects that are adequately represented by functional differential equations is not an alternative because it can lead to wrong (and potentially disastrous) consequences in terms of stability analysis and control design.

For example, engine cycle delays and transport delays are, typically, included in mean–value models of automotive internal combustion engines. These delays are time–variable because they depend on the engine speed that can vary. The implications of these delays for stability analysis and control synthesis of advanced technology internal combustion engines can be described for several problems such as the idle speed control problem and the closed loop air–fuel ratio control problem.

Modern general theory of FDEs is considered in Ch. 3, where all main theorems are equipped with full proofs which are quite instructive and are thus outlined in detail. Various techniques for qualitative investigation have been considered at appropriate places. Theoretical developments and mathematical treatment of the principles involved are included as needed for understanding of the techniques presented.

More intensive treatment of these subjects should be obtained, if desired, from many references available.

Stability theory is treated in chapters 4–10. The most important methods and techniques used in applications (e.g., Laplace transformation method, direct Liapunov method, comparison theorems, etc.) are included in these chapters. In particular, formal procedure to construct Liapunov functionals for various classes of FDEs is proposed and illustrated by numerous examples as well as applications of differential inequalities for stability investigation. New stability conditions are given for particular FDEs.

Chapters 11–13 are devoted to periodicity of FDEs and boundary value problems. Problems of existence of periodic solutions for retarded, neutral and stochastic FDEs both linear and quasilinear are considered.

Chapters 14–16 are connected with optimal control and estimate problems under incomplete state information. Minimax estimation problems and model reference adaptive control are considered in chapters 18 and 19. In particular, it is shown that purposeful introduction of delay into system can essentially improve its qualitative and quantitative characteristics and many phenomena, such as periodicity, oscillation, instability, etc., can be explained in terms of delay.

Chapters of the book are relatively independent of each other. It means that they can be read in any order. At the same time a beginner is encountered to first read chapter 1 which is a simplified introduction in the subject.

Finally, the book contains an extensive bibliography of the subject, while it cannot be claimed to be complete. There many references to journal papers, aimed at indicating the origination of the results, or contain supplementary material directly related to the text.

Two numbers are used for theorems and formulas inside a single chapter, and three numbers when referring to other chapters.

Preparation of the book was supported by Russian Fund for Fundamental Researches; INTAS; Institut franco–russe A. M. Liapunov d'informatique et de mathematiques appliquées; International Soros Science Education Program (ISSEP).

Readers having comments and suggestions are invited to send these to us. Of course, full acknowledgments will be made.

We are grateful to many our students, associates and colleagues for their quite fruitful discussions and many useful suggestions for improvements, problem statements, consultation, and numerous changes made throughout the text to improve readability and presentation. But our decision was not to name them individually as an indication of how many collaborators we have had.

Part I

MODELLING BY FUNCTIONAL DIFFERENTIAL EQUATIONS

In this part we discuss the possibility of modeling real phenomena by FDEs.

Many actual systems have the property of *aftereffect*, i.e. the future states depend not only on the present, but also on the past history. Afereffect is believed to occur in mechanics, control theory, physics, chemistry, biology, medicine, economics, information theory, etc. This wide appearance of aftereffect is reason to regard it as a universal property of the surrounding world.

Many papers have been devoted to the description and investigation of real phenomena taking into account aftereffect, see, e.g. [1,7,9,14,19,21,26, 28, 35, 39, 48, 54, 55, 65, 80, 88, 92, 93, 105, 107, 116, 120, 122, 125, 128, 146, 155, 156, 160, 161, 167, 173, 182, 183, 185, 187, 200, 201, 206–209, 214, 217–220, 222, 225, 227, 229, 233, 235, 240–245, 247, 248, 250–252, 271, 272, 277, 293, 296, 299, 312, 314, 315, 332, 333, 341, 342, 355, 363, 365, 371, 377, 387, 398, 400, 422, 427, 435, 441, 442, 450, 454, 456, 459–462, 468, 480, 487, 499, 500, 502, 505–509, 511, 512, 514, 541]. Moreover during last years some surveys and books have been published in which various theoretical and applied aspects of time delay systems were considered. They provide a good motivation to study time delay systems.

In this part we give some models with delay, in particular because the only way to learn how to model is to exhibit a great variety of problems and a corresponding variety of models describing them.

CHAPTER 1

THEORETICAL PRELIMINARIES

This chapter covers some theoretical backgrounds of the equations used in the book without concentration on mathematical rigour. General classification of the equations and some properties of their solutions will be discussed.

1. FUNCTIONAL DIFFERENTIAL EQUATIONS (FDEs)

1.1. *SOME CLASSES OF FDES*

We consider equations with an unknown scalar or vector function depending on a continuous argument t, which may be treated as time. The equations may be scalar or vector equations, and should have the same dimension as the unknown function. Unless stated otherwise, all variables under consideration are real.

As it well known, an *ordinary differential equation* (ODE) is an equation connecting the values of an unknown function and some of its derivatives for one and the same argument value. For example, the equation

$$F(t, x, dx/dt, d^2x/dt^2) = 0$$

may be written as $F(t, x(t), \dot{x}(t), \ddot{x}(t)) = 0$, where dots indicate derivatives: $\dot{x}(t) = dx(t)/dt$.

A functional equation (FE) is an equation involving an unknown function for different argument values. The equations $x(2t) + 2x(3t) = 1$, $x(t) = t^2x(t+1) - [x(t-2)]^2$, $x(x(t)) = x(t) + 1$, etc., are examples of FEs. The differences between the argument values of an unknown function and t in a FE are called *argument deviations.* If all argument deviations are constant (as in the second example above), then the FE is called a *difference equation.*

Above we gave some examples of FEs with *discrete* (or *concentrated*) *argument deviations.* By increasing in the equation the number of summands and simultaneously decreasing the differences between neighbouring argument values, one naturally arrives at FEs with *continuous* (or *distributed*) and *mixed* (both continuous and discrete) *argument deviations.* These are called *integral* and *integral–functional equations* (in particular, *integral–difference equations*). However, it is meaningless to give these classes in detail, since they only serve as orientation in investigations.

Combining the notions of differential and functional equations, we obtain the notion of *functional differential equation* (FDE), or, equivalently, *differential equation with deviating argument.* Thus, this is an equation connecting the unknown function and some of its derivatives for, in general, different argument values. Here also the argument values can be discrete, continuous or mixed. Correspondingly one introduce the notions of *differential–difference equation* (DDE), *integro–differential equation* (IDE), etc.

A FDE is called *periodic* with *period* $T > 0$ (briefly, T–periodic) if it is invariant under the change $t \mapsto t + T$. In more detail this means that if we replace t by $s + T$, set $x(t) = y(s)$, and then replace the letters s, y by t, x, then we should obtain the initial equation. For example, contrary to its appearance, the equation $\dot{x}(t) = x(\sin t)$ is not periodic. A FDE is called *autonomous* if it is invariant under the change $t \mapsto t + T$ for all $T \in \mathbf{R}$.

The *order* of a FDE is the order of the highest derivative of the unknown function entering in the equation. So, a FE may be regarded as a FDE of order zero. Hence the notion of FDE generalizes all equations of mathematical analysis for function of a continuous argument. A similar assertion holds for function depending on several arguments. Therefore the creation of a sufficiently substantional theory of FDEs is possible only for certain reasonably restricted classes of FDEs. Most of these classes are chosen guided by applications.

The naturally defined notions of functional differential inequality or inclusion, and stochastic FDE, which are close to that of FDE, will be discussed later.

1.2. *SOLUTION CONCEPT FOR A FDE*

The study of FDEs is closely connected with a refinement of the solution concept for a FDE. This refinement should either be stated explicitly, or be implicitly implied. By an example we make clear the difference in solution concept between FDEs and ODEs.

The FDE

$$\dot{x}(t) = kx(t - h), \quad k, h = \text{const}, \quad k \neq 0, \tag{1.1}$$

is a ODE for $h = 0$. Suppose $h > 0$. We would like to investigate the solution x of this equation on an interval $\alpha \leq t \leq \beta$, i.e. $x : [\alpha, \beta] \to \mathbf{R}$, where $\beta > \alpha + h$. Then for $\alpha \leq t < \alpha + h$ we have $t - h < \alpha$, and so the righthand side of this equation is not defined. This obstacle may be overcome in two ways:

1) We consider a solution x of (1.1) for $\alpha \leq t \leq \beta$ so that (1.1) holds for all $t \in [\alpha, \beta]$, but for $t-h < \alpha$ we replace $x(t-h)$ on the righthand side of (1.1) by the values of some *initial function* ϕ defined on $\alpha - h \leq t < \beta$. Hence we assume by definition that in (1.1),

$$x(t-h) = \phi(t-h) \quad \text{for } t-h < \alpha.$$

This initial function is either given, or constructed from additional conditions.

2) We interpret a solution x of (1.1) on $[\alpha, \beta]$ as a function defined on this interval, but satisfying (1.1) only for $\alpha + h \leq t \leq \beta$.

Of course, in both approaches we must also indicate, as for ODEs, the class of functions from which a solution is chosen and in what sense the equation should be satisfied: everywhere, almost everywhere, etc. More detailed assertions will be given in Chapter 3.

In essence, both approaches are equivalent. Transition from the first to the second is done by extending the solution x by the initial function ϕ. Conversely, for a solution $x : [\alpha, \beta] \to \mathbf{R}$ in the sense of the second definition the restriction $x|_{[\alpha+h,\beta]}$ is a solution in the first sense for the initial function $\phi = x|_{[\alpha,\alpha+h)}$. So the choice of one of these approaches over the other is more connected with the particular point of view: either the initial function is an independent object or it is part of the solution (physically, it is the previous history of the process). During the creation of the general theory the first approach prevailed, but nowadays the second is used more widely. Depending on convenience, we will use both approaches.

The necessary to make more precise the solution concept arises especially when we would like to change variables in FDEs. E.g., if in (1.1) we replace $x(t-h) = y(t)$, we obtain the new FDE $\dot{y}(t+h) = ky(t)$. Considering the solution of this equation on the interval $[\gamma, \delta]$, then under the first approach we must prescribe a *finite function* $\tilde{\phi} : (\delta, \delta + h] \to \mathbf{R}$, and under the second approach the equation must be satisfied for $\gamma \leq t \leq \delta - h$. The solution of this problem under the second approach, as it will be shown in Subs. 3.1.5, has essentially other properties than that of with the initial function for Eq. (1.1).

It can be shown that for different, still seemingly natural, refinements of the solution concept for formally one and the same problem, these solutions may prove vastly different. Consider the example with initial function (i.e. we use the first approach)

$$\dot{x}(t) = \dot{x}(t-1), \quad 0 \leq t < \infty; \quad x(t-1) = \phi(t-1) \equiv 0, \quad t-1 < 0. \quad (1.2)$$

We also give the *initial value* $x(0) = 1$. One of the common refinements of the solution concept gives, when applied to (1.2), that the solution is a locally absolutely continuous function $x : [0, \infty) \to \mathbf{R}$ satisfying the equation for almost all t. It is clear that the function $t \mapsto x_1(t) \equiv 1$ is such.

There is also another widespread refinement of the solution concept: a solution is a locally integrable function $x : [0, \infty) \to \mathbf{R}$, right continuous at $t = 0$ (this is important in connection with the initial value) that, when extended to $[-1, \infty)$ by the initial function ϕ, satisfies (1.2) on $(0, \infty)$ as a generalized function. (Note that here it looks like transition from the first to the second approach is performed.) This allows us to integrate both parts of the equation, giving

$$x(t) = x(t-1) + C, \quad 0 \le t < \infty, \quad C = \text{const},$$

where $C = 1$ by the initial condition. Hence, setting successively $t \in [j, j+1)$, $j = 0, 1, \ldots$, we obtain the solution $t \mapsto x_2(t) = [t] + 1$ (here $[t]$ is the integer part of t). Clearly, x_1 and x_2 differ.

1.3. *FDE WITH RETARDATIONS*

As will be shown below (in Chapter 2), such equations arise when modelling biological, physical, etc., processes whose rate of change of state at any moment of time t is determined not only by the present state, but also by past states.

First we restrict ourselves to the case when there are finitely many argument deviations whose dependence on t is known. Then we obtain the equation of general form

$$x^{(m)}(t) = f(t, x^{(m_1)}(t - h_1(t)), \ldots, x^{(m_k)}(t - h_k(t))). \tag{1.3}$$

Here $x(t) \in \mathbf{R}^n$, all $m_i \ge 0, h_i \ge 0$, i.e. all argument deviations are nonnegative. ($\mathbf{R}^n$ is the real euclidean space of n–dimensional column vectors, with norm $|\cdot|$.) This property constitutes the definition of *functional differential equation with retardations*. In (1.3) the function f and *delays* h_i are given, and x is the unknown function of t. More precise assumptions and solution concepts will be given in Chapter 3.

According to a now universally accepted proposal of G. Kamenskii, (1.3) is called a *functional differential equation of retarded type*, or *retarded functional differential equation* (RDE), if $\max\{m_1, \ldots, m_k\} < m$; a *functional differential equation of neutral type* (NDE) if $\max\{m_1, \ldots, m_k\} = m$; and a *functional differential equation of advanced type* (ADE) if $\max\{m_1, \ldots, m_k\} > m$. In particular, RDE can be characterized as follows: it is an equation in which the value of the higher derivative for any values of the arguments is defined by the values of the lower derivatives depending on the lesser values of the argument (in nonstrict sense).

Sometimes FDEs with discrete retardations are called as *equations with time lag*, and equations with continuous retardations as *equations with aftereffect*.

Experience in mathematical modelling has shown that the evolution equations of actual processes with retardations are almost exclusively RDEs and NDEs. On the other hand, the investigation of various problems for these equations has revealed that RDEs and NDEs (in contrast to ADEs) have many "nice" mathematical properties, which will be discussed in Chapter 3.

As for ODEs, we can transform the equation (1.3) to a first order vector equation of the higher dimension by taking as new unknown functions the lower derivatives of x. Preserving the notation x for the new unknown function and f for the new righthand side, we can write an RDE as

$$\dot{x}(t) = f(t, x(t - h_1(t)), \dots, x(t - h_k(t))), \tag{1.4}$$

and an NDE as

$$\dot{x}(t) = f(t, x(t-h_1(t)), \dots, x(t-h_k(t)), \dot{x}(t-g_1(t)), \dots, \dot{x}(t-g_l(t))). \tag{1.5}$$

Note that any FDE is equivalent to a *hybrid system* of ODEs and functional equations, in particular, difference equations. For example, (1.5) is equivalent to the following hybrid system:

$$\left.\begin{aligned} &\dot{x}(t) = y(t), \\ &y(t) = f(t, x(t - h_1(t)), \dots, x(t - h_k(t)), y(t - g_1(t)), \dots, y(t - g_l(t))). \end{aligned}\right\}$$

More complicated types of retardation have also been considered, and also have real applications. For example, the delay may depend on the unknown solution, and have the form $h_i(t, x(t))$. Such delays are sometimes termed *autoregulative*.

Similarly, FDEs with retardation and continuous or mixed argument deviations may have the form of an IDE or a functional integro–differential equation of some structure of *Volterra type*. The latter means that for any t under consideration, the integrals depend on the values of the unknown function $x(\cdot)$ on some interval of $(-\infty, t]$. For example, the continuous analog of (1.4) could be the Volterra type IDE

$$\dot{x}(t) = f\left(t, \int_{t-h(t)}^{t} K(t, \theta, x(\theta))d\theta\right), \quad h(t) \geq 0. \tag{1.6}$$

Such equations were introduced by V. Volterra to investigate actual hereditary phenomena (whence their name).

We now consider some classes of equations of type (1.6) which can be naturally extended to other FDEs with retardation. If $h(t)$ takes finite values only, then (1.6) has *finite aftereffect*; if $h(t) \equiv \infty$, it has *infinite aftereffect*. If $\sup h(t) < \infty$, then it has *bounded aftereffect*, and it has *unbounded aftereffect* in the opposite case. Sometimes the aftereffect time is interpreted as $\sup h(t)$; then finiteness and infiniteness of aftereffect means boundedness, respectively unboudedness, of it. However, in some cases our more detailed classification is useful.

If $t - h(t) \to \infty$ as $t \to \infty$, then (1.6) is said to have the property of *completely forgetting the past*. This means that the values of the solution x on any finite t-interval do not influence the righthand side of the equation for sufficiently large t. In other words, the rate of change of the state at any moment is determined by the states of the process at preceding moments which are not too remote. Sometimes such equations are said to have *fading memory*. If $t - h(t) \not\to \infty$ as $t \to \infty$, then either the past is not forgotten (in this case talks about *residual phenomena*), or is asymptotically forgotten.

Sometimes Volterra type IDEs can be reduced to an (in some sense) equivalent system of ODEs, which occasionally find application. We explain for the following equation with exponentially fading memory:

$$\dot{x}(t) = f\left(t, \int_{t_0}^{t} x(\theta)e^{-k(t-\theta)}d\theta\right), \quad t \geq t_0.$$

Put

$$y(t) := \int_{t_0}^{t} x(\theta)e^{-k(t-\theta)}d\theta.$$

Differentiating this equality, we obtain

$$\dot{y}(t) = x(t) - k\int_{t_0}^{t} x(\theta)e^{-k(t-\theta)}d\theta.$$

Hence the initial equation is equivalent to the system of ODEs

$$\dot{x}(t) = f(t, y(t)), \quad \dot{y}(t) = x(t) - ky(t),$$

with the additional condition $y(t_0) = 0$.

This equivalence is likely to be surprising, because IDEs describe processes whose rate is determined by all previous states, while ODEs describe processes only determined by the current states. It is the particular way the past influences the presence, as if by accumulation, which explains this phenomenon. This feature can be understood by the example of the transition.

$$\left(x(t) = \int_{t_0}^{t} f(\theta)d\theta \right) \Leftrightarrow (\dot{x}(t) = f(t), \quad x(t_0) = 0).$$

Similar transitions from IDEs into ODEs can always be realized if the past influences the present only by integrals of the form

$$\int_{t_0}^{t} x(\theta)K(t-\theta)d\theta,$$

where K satisfies a linear homogeneous ODE with constant coefficients. In other words, K should be a *quasipolynomial.*

Equations (1.4) and (1.6) are the most widely used types of nonlinear RDEs. As already mentioned, delays and equations with aftereffect may have a more complicated structure, but they all have the *Volterra property:* the rate of evolution of the process depends on the past and the present, but not on the future. Keeping this property in mind, we can write (following N. Krasovskii) the general RDE in the form

$$\dot{x}(t) = f(t, x_t). \tag{1.7}$$

Here $x(t) \in \mathbf{R}^n$ and x_t (for a given t) is the function defined by

$$x_t(\theta) := x(t+\theta), \quad \theta \in J_t \subseteq (-\infty, 0],$$

where J_t is a given interval $[-h(t), -g(t)]$ or $(-\infty, -g(t)]$. The transition from x to x_t for $J = [-h, 0]$ is shown in Fig. 1.1.1. Note that x_t may be treated as the fragment of the function x at the left of the point t, observed from this point. The righthand side of (1.7) is a function of t, and a functional of x_t, i.e. to any function $\psi : J_t \to \mathbf{R}^n$ in some class of functions corresponds a vector $f(t, \psi) \in \mathbf{R}^n$.

Equations (1.4) and (1.6) are particular cases of (1.7). Namely, we obtain (1.4) for

$$f(t, \psi) := f(t, \psi(-h_1(t)), \ldots, \psi(-h_k(t))),$$

and (1.6) for

$$f(t, \psi) := f\left(t, \int_{-h(t)}^{0} K(t, t+\theta, \psi(\theta))d\theta \right).$$

Similarly, the general NDE can be written as

$$\dot{x}(t) = f(t, x_t, \dot{x}_t), \tag{1.8}$$

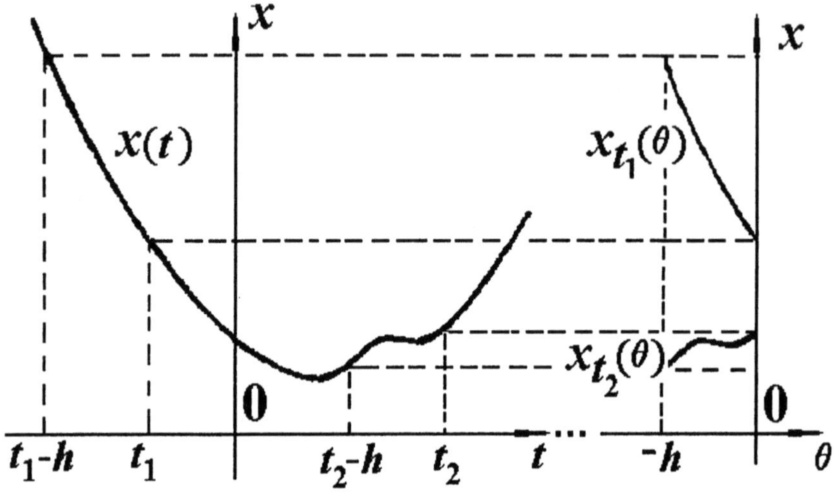

Figure 1.1.1. Geometrical interpretation of the transition from x to x_t

where $\dot{x}_t$ denotes $(\dot{x})_t$, or, equivalently, $(x_t)^\bullet$.

Note that if J_t does not reduce to a point, then formally the righthand side of (1.8) can be written as $\tilde{f}(t, x_t)$, because by giving a function we also give its derivative. Of course, this does not mean that there are no principal distinctions between RDEs and NDEs, since the conditions natural for RDEs are usually not satisfied by $\tilde{f}$. Roughly speaking, for variable t the righthand side of an RDE must define a bounded operator on an appropriate space, and the righthand side of a "true" NDE, an unbounded operator.

In recent years a majority authors preferred general form of NDEs (following J.Hale):

$$[x(t) - g(t, x_t)]^\bullet = f(t, x_t). \tag{1.9}$$

In general, (1.8) and (1.9) cannot be reduced to each other. However, principal distinctions between these forms of NDEs are usually absent in simple cases, provided the solution concept is reasonably modified. It may occur, however, that after natural corrections we obtain quite different results for each of these forms of writing NDEs. (This has been demonstrated using equation (1.2): the second solution concept given there may be immediately obtained if the equation under consideration is written as $[x(t) - x(t-1)]^\bullet = 0$.)

1.4. *A LITTLE BIT OF PHILOSOPHY*

How can the states of a process at previous moments of time influence directly the present state of evolution of this process? It is well known that at each moment the object "knows" only the situation at that moment, and hence can react only to this. Even if we take into consideration the memory of materials or human beings, we should be able to consider only the memory state at the current moment. We discuss this problem using a simple example.

Imagine a showering person wishing to achieve the desired value T_d of water temperature T by rotation of the mixer handle for cold and hot water (Fig. 1.1.2). Assume that the change ΔT in water temperature at the mixer output is proportional to the angle $\Delta\alpha$ of rotation of the mixer, with coefficient k. (We could give a more realistic dependence of ΔT on $\Delta\alpha$, but this is not important here.) Let $T_m(t)$ denote the water temperature in the mixer output, and h the constant time needed by the water to go from the mixer output to the tip of the person's head. Assume that the rate of rotation of the handle is proportional to the deviation in water temperature from T_d perceived by the person, with coefficient κ. This κ depends on the person's temperament, and is larger for an energetic person than it is for a phlegmatic one.

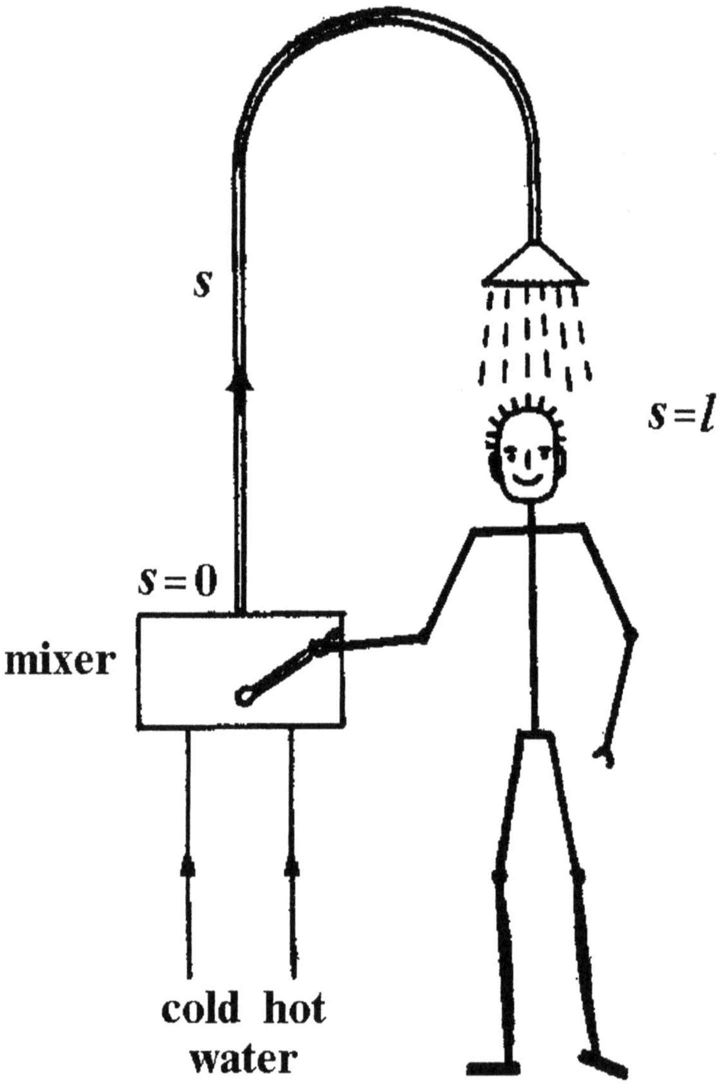

Figure 1.1.2. Qualitative model of water temperature regulation by a showering person

Because at time t the person feels the water temperature leaving the mixer at time $t-h$, we find $\dot{\alpha}(t) = -\kappa[T_m(t-h) - T_d]$. This implies an equation for the temperature T_m:

$$\dot{T}_m(t) = -k\kappa[T_m(t-h) - T_d]. \tag{1.10}$$

This is a typical RDE. The properties of its solution, with obvious interpretations, will be described in Subs. 3.6.3.

We now consider this phenomenon in more detail. We introduce a coordinate s, measuring the length from the mixer output ($s=0$) to the person's head ($s=l$). Assuming the velocity of water flow to be constant (equal to v), we find $h = l/v$. Let $T(t,s)$ be the water temperature at point s at time t, and assume that the temperature of small portions of water does not change during flow. Then we obtain the following equation for the evolution of the temperature:

$$\frac{dT}{dt} = 0 \Leftrightarrow \frac{\partial T(t,s)}{\partial t} + v\frac{\partial T(t,s)}{\partial s} = 0, \quad 0 \le s \le l. \tag{1.11}$$

The regulation (or control) condition becomes the boundary condition

$$\frac{\partial T(t,0)}{\partial t} = -k\kappa[T(t,l) - T_d]. \tag{1.12}$$

We see that neither (1.11) nor the boundary condition (1.12) involves delays: at any moment t the system reacts only on the situation at this moment. Equation (1.10) can be obtained from this model by the simplification excluding s. In fact, the general solution of (1.11) is $T(t,s) = f(vt-s)$. Putting $T(t,0) = T_m(t)$, we find $T_m(t) = f(vt), T(t,l) = f(v(t-l/v)) = T_m(t-h)$. Using (1.12) this implies (1.10).

This example is apparently generic. In other cases it is usually not difficult too show that the appearance of delay in a differential equation is the result of some essential simplification of the model. A detailed analysis of the reasons leading to the delay leads to an evolution equation (most often involving partial derivatives) all terms of which relate to the same moment. Excluding all dependencies that knowledge seems redundant (in the example above, full knowledge of the function $(t,s) \mapsto T(t,s)$), we obtain an RDE. So the processes (or systems) with aftereffect are the processes (or systems) for which a change of the basic variables can be done naturally using a simplified model founded on direct influence of the past into the future.

We note one peculiarity of the boundary condition (1.12) that seems physically absurd. According to this condition, transmission of a signal from $s=l$ to $s=0$ happens instantly. Of course, in practice this is not true, i.e. we have a simplifying assumption not impairing the adequacy of the solution. The transmission of a signal through the brain and neural chains requires some time, and is more complicated than temperature transmission, although it is possible to propose an evolution model without delay also in this case. However, the use of the delay equation (1.10) may be advantageous, since there is no need to analyze details of the transmission, and it allows us to add a term to the delay h to make up for the time of signal transmission. Like κ, this term may be regarded as an empirical constant depending on the person's temperament.

In conclusion we would like to pay attention to the fact that the results obtained on the basis of this model are only qualitative, because of the approximate nature of the hypothesis used when deriving (1.10) and the low reliability of the empirical constants. However, this model is especially useful for its transparent imitation of other models of the same kind.

CHAPTER 2
MODELS

1. VISCOELASTICITY

In mechanics, models with aftereffect are used to describe stress–strain states of materials, for example of wood, polymers, plastics, ice, etc. The theory of elastic aftereffect for rigid bodies was proposed by Boltzmann and developed by Volterra. The latter called it the hereditary theory of elasticity (see the references in [19, 43, 116–118, 156]). In modern literature the phrase "hereditary elasticity" is usually replaced by *viscoelasticity.* The theory of viscoelasticity lies at the foundation of designing many construction elements that have to function under complicated exploiting conditions.

From the point of view of the theory of FDEs, viscoelastic structures can be described by equations with unbounded distributed aftereffect. This point of view was systematically used in the books [19, 156.3].

Consider the equations of linear approximation of the transversal motion of a rectilinear hinged–hinged homogeneous viscoelastic bar subjected to an axial load P (Fig. 2.1.1). The viscosity effect relaxes the bending stresses in the bar. Formally this effect is equivalent to the appearance of a counteracting bending moment. Assume that the unit bending moment applied to a bar element during time interval $(t, t+dt)$ gives at each moment $t+\tau > t$ rise to the counteracting bending moment $r(t,\tau)dt$. (Here $r \geq 0$ is called the *relaxation function;* the dependence of r on t is connected with a possible aging of the material.) If the process starts at time $t = t_0 \geq -\infty$, we obtain the equation of motion

$$\left.\begin{aligned} &\rho\frac{\partial^2}{\partial t^2}y(t,x) = -\frac{\partial^2}{\partial x^2}[Py(t,x) + M(t,x)], \\ &M(t,x) = EI\frac{\partial^2}{\partial x^2}\left[y(t,x) - \int_{t_0}^{t} r(\tau, t-\tau)y(\tau,x)d\tau\right], \\ &t \geq t_0, \quad 0 \leq x \leq l. \end{aligned}\right\}$$

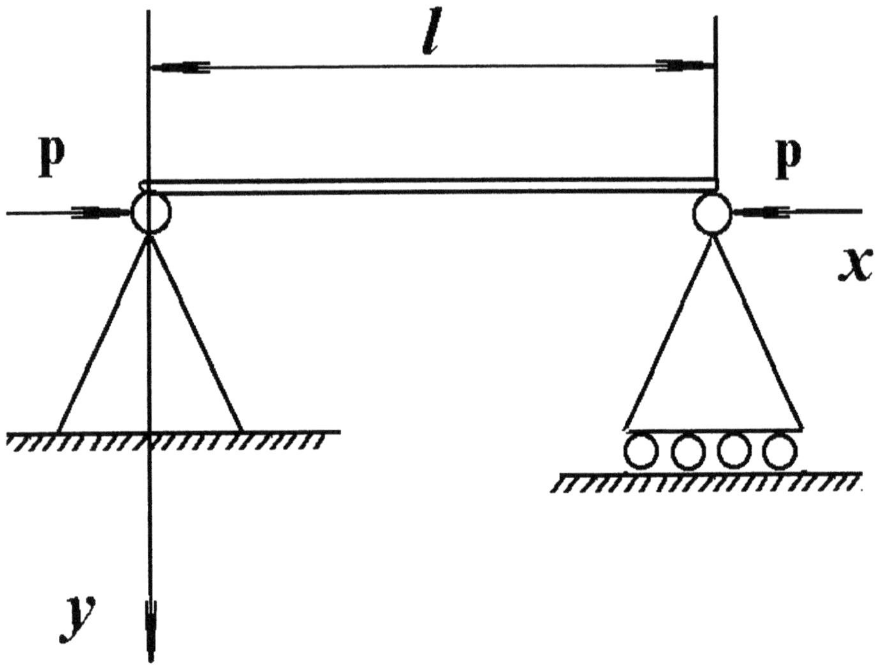

Figure 2.1.1. Hinged-hinged squeezed bar

Here ρ is the linear density of the bar material, I the moment of inertia of a cross section of the bar, E the elasticity modulus, and l the length of the bar. The boundary conditions have the form

$$y(t,0) = y(t,l) = 0, \quad \left.\frac{\partial^2 y(t,x)}{\partial x^2}\right|_{x=0} = \left.\frac{\partial^2 y(t,x)}{\partial x^2}\right|_{x=l} = 0.$$

The same type of equations arise when studying the motion of viscoelastic bodies.

Consider a one-dimensional viscoelastic body which moves longitudinally and occupies initially the interval $[0,l] \subset \mathbf{R}$. Let $x + u(x,t)$ be the position of the material particle with Lagrange coordinate x at time t. Then the equation of motion is

$$\rho u_{tt}(t,x) = \sigma_x(t,x) + f(t,x).$$

Here, ρ is the density, σ the Piola–Kirchhoff stress, and f the force on the body. The stress has the form

$$\sigma(t,x) = P(u_x(t,x)) - \int_0^\infty a(t-\tau)q(u_x(\tau,x))d\tau.$$

In this equation, the first term, P, describes the instantaneous response of the material, and the second term depends on the history of displacement in $(-\infty,t] \times [0,l]$. To solve this equation we have to prescribe: the history of displacement $u(t,x)$ for $t \in (-\infty,t_0]$, $x \in [0,l]$; the initial velocity $u_t(t_0,x)$; boundary data, e.g., $u(t,0)$ and $u(t,l)$.

2. AFTEREFFECT IN MECHANICS

Many problems whose models involve equations with unbounded distributed aftereffect appear when investigating the motion of bodies with regard to their interaction with the environment. Such problems are called *aeroautoelasticity problems*, or *problems with viscous aftereffect.* In these problems it is essential that the disturbances (or viscous waves) created by the moving body in the environment influence the body during a long period of time. For this reason, at any moment the reaction on the body from the environment does not depend only on the velocity of the body at this moment, but also on the previous history of the motion. Both stochastic and partial differential equations are used to model such motion of a body.

2.1. *MOTION OF A PARTICLE IN A LIQUID*

Consider a Brownian particle moving in an unbounded medium. Let $x_1(t)$ be the coordinate and $x_2(t)$ the velocity of the particle. The equations of motion are:

$$\dot{x}_1(t) = x_2(t), \quad m dx_2(t) = f(t)dt + \sigma d\xi(t),$$

where m is the mass of the particle and $\sigma d\xi(t)$ the random action of the medium on the particle. The function f represents the systematic action of the medium on the particle. According to Boussinesq it equals

$$f(t) = -hx_2(t) - m_1\dot{x}_2(t)\sigma\left(\frac{1}{\pi}hm_1\right)^{1/2}\int_0^\infty \dot{x}_2(t-s^2)ds.$$

Here, $-hx_2$ is the Stokes friction force, m_1 the apparent additional mass (equal to half the mass of the material of the medium ousted by the body), and the integral term describes the viscous hydrodynamic aftereffect. Note that in this case the Brownian motion is described by a stochastic NDE.

One also uses neutral partial differential equations to describe the motion of a small particle in a liquid. An example is the following equation in dimensionless variables:

$$K\left(1+\frac{\kappa}{2}\right)\frac{d\eta}{d\tau} = \frac{3}{2}K\kappa\frac{d\zeta}{d\tau} - Gj - (\eta-\zeta) -$$

$$-3\left(\frac{K\kappa}{2\pi}\right)^{1/2}\int_{-\infty}^{\tau}\left(\frac{d\eta}{d\tau}-\frac{d\zeta}{d\tau}\right)\Big|_{\tau=\tau'}(\tau-\tau')^{-1/2}d\tau' + \frac{3}{2}\frac{K\kappa}{R}\frac{L^2}{V}\mathrm{grad}\,\mathrm{div}u,$$

where

$$\eta := \frac{v}{V}, \quad \zeta := \frac{u}{V}, \quad \tau := \frac{Vt}{L}, \quad \kappa := \frac{d_1}{d_2},$$

$$G := \frac{2}{9}(d_2-d_1)\frac{ga^2}{\mu V}, \quad K := \frac{2}{9}\frac{Va^2d_2}{L\mu}, \quad R := 2\frac{LVd_1}{\mu}.$$

Here, a and d_2 are the characteristic size and density of the particle, d_1 and μ are the density and dynamic viscosity of the liquid, g is the gravity acceleration, j is a vertical unit vector, r is the radius vector of the particle, $v(t)$ and $u(r,t)$ are the velocities of the particle and the liquid, respectively, L and V are the characteristic linear and velocity scales of the flow, K and R are the Stokes and Reynolds numbers. One can see that for a given function ζ there arises a Volterra type IDE with infinite delay with respect to η.

2.2. *CONTROLLED MOTION OF A RIGID BODY*

Equations with unbounded delay appear when modelling the motion of mechanical systems with regulators depending on the whole previous history. For example, consider the mechanical system representing a set of interacting rigid bodies. The system is assumed to be controlled by the standard proportional integral (PI) or proportional integro–differential (PID) regulators.

Let $q \in \mathbf{R}^n$ be the vector of generalized coordinates of the system, $p = H(q)\dot{q}$ the generalized impulse, and $T = (1/2)p'H^{-1}(q)p$ the kinetic energy. Here, $H(q)$ is a positive definite bounded symmetric inertia matrix, and the prime denotes transposition. Assume that besides the control u there are no generalized forces acting upon the system. Then in the canonical variables q, p the equations of motion of this system have the Hamiltonian form

$$\dot{p}(t) = -\frac{\partial L}{\partial q} + u, \quad \dot{q}(T) = H^{-1}(q)p, \quad t \geq t_0, \tag{2.1}$$

where $(q,p) \mapsto L(q,p)$ is the Lagrangian.

The control u can be given by PI or PID regulators, so that

$$u(t) = Bq(t) + \int_{t_0}^{t} g(t-\tau)\dot{q}(\tau)d\tau \tag{2.2}$$

or

$$u(t) = A\dot{q}(t) + \int_{t_0}^{t} g(t-\tau)q(\tau)d\tau. \tag{2.3}$$

Here A and B are given matrices, and g is a scalar function which is absolutely integrable over $[0, \infty)$. Note that (2.1) is an ODE if the dependence of u on t is given in advance, or if $u = u(t, q, p)$, while it becomes an FDE if u is given by (2.2) or (2.3). The technical applications of the system (2.1)–(2.3) are likely to be related to robot manipulations. A general approach to designing controllers of the types (2.2) or (2.3) has been proposed. It has been proved that the resulting closed loop system is asymptotically stable, and sufficient conditions for stability have been obtained [12].

There are studies devoted to systems with intentional delays in the controller's design. It has been shown that the cost functional may be improved by a judicious use of time–delay actions. Moreover, a controller with time delays can eliminate overshoot and quench the oscillation, yielding a smooth and fast transient response. For certain systems, the proportional minus delay controller may replace the proportional plus derivative regulator (PD regulator).

One has proposed a controller for general multivariate nonlinear systems in which distributed delays are included in the feedback loop. As a result, the closed feedback system can be described by either RDE or NDEs. For example, consider the plant with transfer function $s \mapsto W(s) := K_0(Ts)^{-1}\exp(-s)$, where K_0 is the amplification, T a time constant, and the delay of the plant equals 1. The plant is regulated by a P or PD type controller. The corresponding single–loop feedback system is shown in Fig. 2.2.1. If the regulator transfer function $s \mapsto W_1(s) :\equiv K_1$ is constant, then the equation of the closed system takes the form

$$T\dot{x}(t) + x(t) + Kx(t-1) = K_0 y(t-1), \quad K := K_0 K_1.$$

For a PD regulator with transfer function $s \mapsto W_1(s) := K_1(1+sD)$ (where D is a constant), the closed system is described by the NDE

$$T\dot{x}(t) + KD\dot{x}(t-1) + x(t) + Kx(t-1) = K_0 y(t-1).$$

Controllers with delay (e.g., those of Smith and Reswick) are often used to obtain plant output without delay [289.1, 289.3].

2.3. *MODELS OF POLYMER CRYSTALLIZATION*

Consider the process of injection moulding, used in the manufacture of polymers. A main feature of this process is dependence of the characteristics of the final product on the whole history of the evolution of temperature and pressure.

Let $W(t)$ be the fraction of crystals at time t, and let W_0 be its maximum. Then we obtain for the function $t \mapsto \delta(t) := 1 - W_0^{-1}W(t)$ (see [14])

$$-W_0\dot{\delta}(t) = \frac{4}{3}\pi r^3 \nu f_1(\delta(t)) +$$

$$+4\pi\rho\nu f_2(\delta(t)) \int_0^t f_1(\delta(\tau)) \left[r + \int_\tau^t f_2(\delta(s))ds \right]^2 d\tau,$$

$$t \geq 0, \quad \delta(0) = 1.$$

Here, r is the radius of the nucleus, the known functions f_1 and f_2 are such that

$$f_i(0) = 0, \quad f_i(1) = 1, \quad \dot{f}_i(t) \geq 0 \quad (i = 1, 2),$$

and ν, ρ are given positive constants.

2.4. *STRETCHING OF A POLYMER FILAMENT*

The process of stretching, and subsequent elastic recovery, of a homogeneous filament of a polymeric liquid can be described by the equation [338]

$$\dot{y}(t) = \int_{-\infty}^t K(t-\tau)[y^3(t)y^{-2}(\tau) - y(\tau)]d\tau.$$

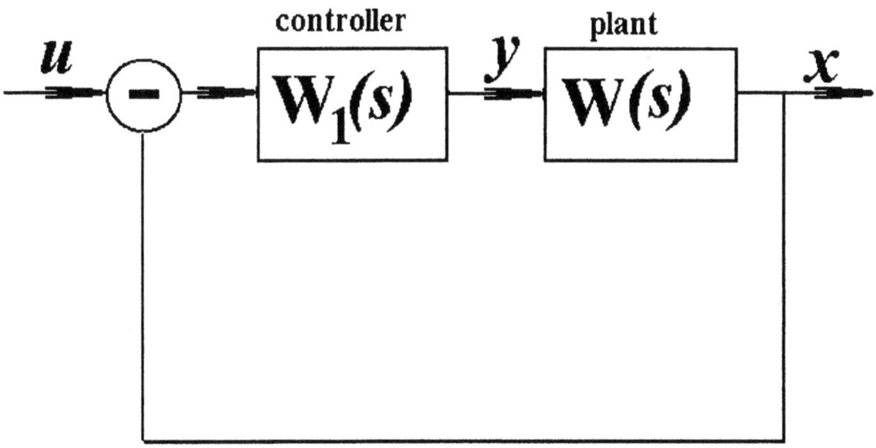

Figure 2.2.1. Single-loop feedback control system

Here, the kernel K represents the memory, and $y(t)$ is elastic strain.

3. HEREDITARY PHENOMENA IN PHYSICS

The description of certain phenomena in physics has to take into account that the rate of propogation is finite [122, 155.1, 250, 480, 509, 516, 541].

3.1. *DYNAMICS OF OSCILLATORS*

Oscillation in a vacuum tube can be described by the following equation in dimensionless variables [370, 408]:

$$\ddot{x}(t) + 2r\dot{x}(t) + \omega^2 x(t) + 2q\dot{x}(t-1) = \epsilon\dot{x}^3(t-1).$$

In this equation, time delay is due to the fact that the time necessary for electrons to pass from the cathode to the anode in the tube is finite. The same equation has been used in the theory of selfexcited oscillations and in the theory of stabilization of ships [370, 408].

The interaction of an oscillator with an energy source, taking into account delays, has been modelled by coupled equations:

$$\left.\begin{aligned} &\ddot{y}(t) + [\omega^2 + \epsilon\lambda\sin\phi(t-h_1)]y(t) = -\epsilon[\delta\dot{y}(t) + \gamma y^3(t-h_2)], \\ &\ddot{\phi}(t) = \epsilon[L(\dot{\phi}(t)) - H\dot{\phi}(t) - a_1 y^2(t-h_3)\cos\phi(t) + \\ &\qquad -a_2\sin\phi(t) - a_3\cos\phi(t)]. \end{aligned}\right\}$$

The dynamics of an autogenerator with delay and second–order filter was described in [220] by the equation

$$\ddot{x}(t) + 2\delta\dot{x}(t) + x(t) = f(x(t-h)).$$

These equations are examples of systems of coupled oscillators which are widely used in electronics, chemistry and ecology (see the references in [207, 289.3, 301, 380]).

3.2. *RELATIVISTIC DYNAMICS*

The problem of the relativistic motion of a particle of mass m in a central field has been investigated in [57].

Let r be the vector joining the particle with an immobile center, and let $\tau(r)$ be the interaction delay. Then the equation of motion takes the form [287.1]

$$m\ddot{r}(t) = -kr(t-\tau(r))|r(t-\tau(r))|^{-3}.$$

It has been proved that circular spherical orbits exist if and only if the radii satisfy a certain quantization condition coinciding with the Bohr condition.

We will now consider the motion of two, or more, charged particles under the following assumptions:

1) the motion of each particle lies in the electromagnetic field(s) of the other(s);

2) the fundamental laws of physics are symmetric with respect to time reversal.

Note that electromagnetic fields propagate with finite velocity. Hence the first assumption means that the equations of the motion of a particle must contain time delays depending on the trajectories of the particles. However, the second assumption implies that the modelling equations should involve advance terms too. Thus the model is described by FDEs involving delayed and advanced terms [250, 516, 541].

We describe the equations of motion of two identical, charged particles moving symmetrically about the point $x = 0$ on the x–axis. The coordinate of one particle is $x(t) > 0$, so the coordinate of the second is $-x(t)$. The motion of the particle at $x(t)$ is in the fields of the other particle, one field emitted at an earlier time $t - r(t)$ and one a later time $t + q(t)$. Then the equations of motions of motion take the form [250].

$$\left.\begin{aligned} \dot{x}(t) &= cv(t), \\ \frac{\dot{v}(t)}{[1 - v^2(t)]^{3/2}} &= \frac{k}{r^2}\frac{1 - v(t - r(t))}{1 + v(t - r(t))} + \frac{k}{q^2}\frac{1 + v(t + r(t))}{1 - v(t + r(t))}, \\ cr(t) := x(t) + x(t - r(t)), &\quad cq(t) := x(t) + x(t + q(t)). \end{aligned}\right\}$$

Here, c is the velocity of light, the constant $k > 0$ depends on the rest masses and charges of the particles, and $|v(t)| < 1$.

3.3. *NUCLEAR REACTORS*

FDEs are widely used to model the dynamics of nuclear reactors. The physical reasons for the appearance of delay are various: transportation delays caused by finiteness of time of heat transport along different elements of the circulation contours; warming up time of the reactor; snapping time of the control system; etc.

In [209] the following models were used:

$$\left.\begin{aligned} \dot{x}(t) &= [ax(t) + by(t - h)][1 + x(t)], \\ \dot{y}(t) &= x(t) - y(t); \end{aligned}\right\}$$

$$\left.\begin{aligned}\dot{x}(t) &= [\phi(x(t-h_1)) + \psi(y(t-h_2))][1+x(t)],\\ \dot{y}(t) &= x(t) - y(t);\end{aligned}\right\}$$

$$\left.\begin{aligned}\dot{x}(t) &= [a_1\theta_1(t) + a_2\theta_2(t)][1+x(t)] - a_3[x(t)-y(t)],\\ \dot{y}(t) &= a_4[x(t)-y(t)],\\ \gamma\dot{\theta}_1(t) &= (1-a)x(t) - b[\theta_1(t)-\theta_2(t)],\\ \dot{\theta}_2(t) &= \alpha\theta_2(t-h) - \theta_2(t) + ax(t) + b[\theta_1(t)-\theta_2(t)].\end{aligned}\right\} \quad (3.1)$$

Here, $x(t)$ is the relative change of neutron density, $y(t), \theta_1(t), \theta_2(t)$ are proportional to the relative change in temperature of the reactor, fuel and de–acceleration device, respectively. The first two models do not take into account the delayed neutrons, but (3.1) does. In (3.1), the delay h is the time of liquid fuel transportation along a circulation contour.

3.4. *DISTRIBUTED NETWORKS (LONG LINE WITH TUNNEL DIODE)*

A tunnel diod is a two–electrode device on the basis of semiconducting crystals having a very narrow potential barrier hindering the motion of electrons. Such diodes are widely used in high–frequency amplifiers of electric oscillations, and in other devices.

Given a homogeneous linear conductor with distributed parameters and without energy loss, at the end $x = 0$ of which there is an external source of constant voltage E, while the other end $x = l$ is grounded by means of a tunnel diod. The current $i(t,x)$ and voltage $v(t,x)$ being the functions of t, x satisfy the system of telegraph equations

$$L\frac{\partial i}{\partial t} + \frac{\partial v}{\partial x} = 0, \quad C\frac{\partial v}{\partial t} + \frac{\partial i}{\partial x} = 0. \quad (3.2)$$

Here L and C are the inductivity and capacity of the conductor per unit length. The boundary conditions may be taken as

$$(v + R_0 i)|_{x=0} = E, \quad \left[i - C_l\frac{\partial v}{\partial t} - f(v)\right]\Bigg|_{x=l} = 0, \quad (3.3)$$

where R_0 is the resistance at the input, C_l is the capacity at the output, and $f(v)$ is the current–voltage characteristic of the diode. This empirical characteristic is such that $f(0) = 0$, and further has a very steep maximum, followed by a slanting positive minimum, after which the function increases (see Fig. 2.3.1).

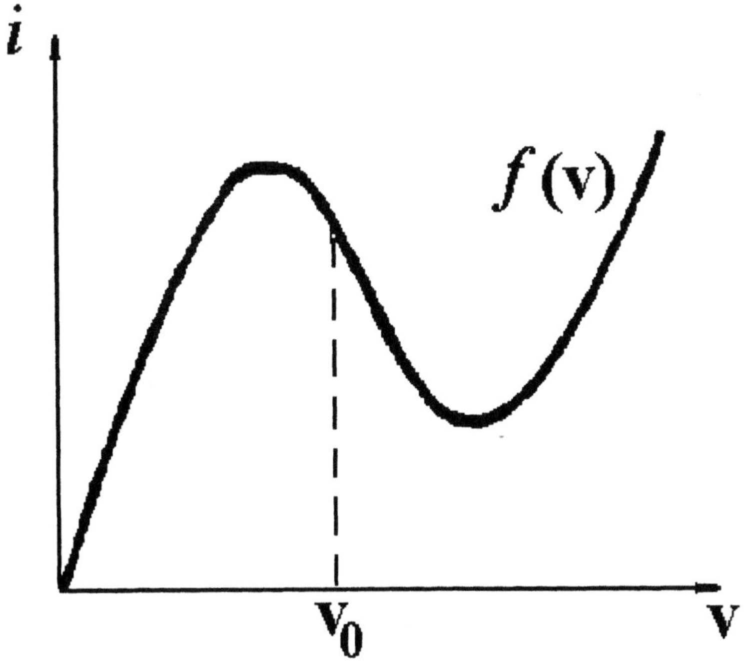

Figure 2.3.1. Current-voltage characteristic

The system of equations

$$v + R_0 i = E, \quad i - f(v) = 0$$

defines the possible stationary values $v = v_0, i = i_0$. Assume that the point (v_0, i_0) lies near the highest point of the graph $i = f(v)$, at the right of it. Then for the working part of the characteristic we can take the relation

$$f(v) = f(v_0) - a(v - v_0) - p(v - v_0)^2, \quad a, p = \text{const} > 0.$$

Using d'Alembert's formula, for the solution of (3.2) we obtain

$$v - v_0 = \phi_1(t - bx) + \phi_2(t + bx),$$

$$i - i_0 = Z^{-1}[\phi_1(t - bx) - \phi_2(t + bx)],$$

$$b := \sqrt{LC}, \quad Z := \sqrt{LC^{-1}}.$$

This and the first condition of (3.3) imply that $\phi_1(t) = (R_0 - Z)(R_0 + Z)^{-1}\phi_2(t)$. The second condition in (3.3) gives

$$\dot{x}(\tau) - K\dot{x}(\tau - h) + (1 - \alpha)x(\tau) + K(1 + \alpha)x(\tau - h) = [x(\tau) - Kx(\tau - h)]^2. \quad (3.4)$$

Here, all quantities are dimensionless:

$$\tau := (C_l Z)^{-1} t, \; x(\tau) := pZ\phi_2(t), \; h := 2lCC_l^{-1},$$

$$K := (Z - R_0)(Z + R_0)^{-1}, \; \alpha := aZ.$$

Note also that under some conditions this circuit can be reduced to that of a continuous nonlinear difference equation which can lead to space–time chaos. [1]

3.5. *HEAT FLOW IN MATERIALS WITH MEMORY*

This type of hereditary phenomenon can also be modelled using FDEs [339, 446].

For example, in [339] the following partial differential equations are used for describing heat flow in materials with memory:

$$\frac{\partial u(t,x)}{\partial t} = (\Delta + c)\int_{-\infty}^{t} k(t - s)u(s,x)ds + k_0 u(t,x) + f(t,x), \quad t \geq t_0,$$

[1] A. N. Sharkovsky, P. H. Deregel, L. O. Chua, Dry turbulence and period–adding phenomena from A 1–D map with time delay. Nonlinear science, ser. B., Vol. 8, World Scientific, 1995.

where Δ is the Laplace operator.

Note that to describe the behaviour of nonconducting material dielectrics with memory, as well as some geophysical phenomena, we can also use functional partial differential equations with unbounded delay.

To model certain materials with memory, in [356] the authors used stochastic FDEs.

3.6. *MODELS OF LASERS*

FDEs are widely used to model the dynamic properties of lasers. E.g., the following equations were introduced in [468]:

$$\left.\begin{aligned} \dot{x}_1(t) &= vx_1(t)[x_2(t) - 1 - m - \alpha m x_1(t-h)] + vU_0, \\ \dot{x}_2(t) &= K_0 - K(t)[x_1(t) + 1], \end{aligned}\right\}$$

where $x_1(t)$ is the radiation density and $x_2(t)$ the amplification coefficient. The other parameters are constants depending on the properties of the laser.

3.7. *NEURAL NETWORK*

In 1984, J.J. Hopfield has introduced a model for a network of n saturating voltage amplifiers (neurons). This model consists of the system of ODEs and describes the evolution of the voltage on the input of each neuron. It was assumed that updating and propagation occur instantaneously. However, it was implicitly observed and explicitly pointed out by C. M. Marcus and R. M. Westervelt in [357] that in both real neural networks and their hardware implementations, neurons do not respond and communicate instantaneously and time lags always exist. Later the processing time in each neuron was incorporated into equations of Hopfield.

As a result RDEs model of a neural network of n neurons was obtained (see, e.g., [37, 357, 508]) in the form

$$C_i \frac{d}{ds} u_i(s) = -\frac{1}{R_i} u_i(s) + \sum_{j=1}^{n} T_{ij} f_j(u_j(s - \tau_{ij}^*)), \quad 1 \le i \le n,$$

where $u_i(s)$ represents the voltage on the input of the ith neurons, C_i is the input capacitance of the cell memberance of the ith neuron, T_{ij} is the synapse efficacy between neuron i and j with $T_{ij} = R_{ij}^{-1}$ when the noninverting output of neuron j is connected to the input of neuron i through a resistance R_{ij} and $T_{ij} = R_{ij}^{-1}$ when the inverting output of neuron j is connected to the input of neuron i through a resistance R_{ij}, R_i is the total parallel transmembrance resistance of the ith neuron, and

$$\frac{1}{R_i} = \frac{1}{\rho_i} + \sum_{j=1}^{n} \frac{1}{R_{ij}}, \quad 1 \le i \le n,$$

with ρ_i denoting the input resistance of neuron i corresponding to the connection to the outside of the network, and the transfer function $f_i(u)$ is called the input–output relation which is nonlinear and sigmoidal, saturating at ± 1 with the maximal slope at $u = 0$. It was demonstrated that delay may cause the sustained oscillations in the neural network.

4. MODELS WITH DELAYS IN TECHNICAL PROBLEMS

It is believed that the appearance of technical problems involving different delays was at the beginning of the 20th century. For example, we may mention problems of the influence of hydroshocks on the oscillation of turbines [503], the feedback system for a hydroelectric power station [58], ship stabilization [289.1], time delays in feedback systems [15], control processes for pressure, heat transfer regulation, etc. [201]. Moreover delayed feedback can be deliberately introduced into the system to improve its qualitative characteristics [292.8].

In this section we elucidate only several applications of FDEs in technical problems.

4.1. *INFEED GRINDING AND CUTTING*

To describe processes of infeed grinding and cutting, many mathematical models with delay were proposed [167, 367, 484].

Consider the simplified cutting scheme in Fig. 2.4.1, where m is the mass of the cutter, α the viscosity coefficient, and c the elasticity coefficient. Let v_0 be the velocity of shaving relative to the cutter, $f(v_0)$ the friction coefficient, d_0 the desirable cutting depth, and $x(t)$ the position of the cutter. For the deviation $z(t) = x(t) - x_0$ from the static rest position $x_0 = f(v_0)qd_0c^{-1}$, where the coefficient q depends on the geometry of the cutter and the width of the cut layer, we obtain

$$m\ddot{z}(t) + \alpha\dot{z}(t) + cz(t) + f(v_0)q[z(t) - z(t-h)] =$$

$$= -q[f(v_0 + \dot{z}(t)) - f(v_0)][d_0 + z(t) - z(t-h)].$$

In this model, the time delay h arises due to the dependence of the cutting process on the surface state at the previous rotation. A detailed stability analysis of this model was given in [484].

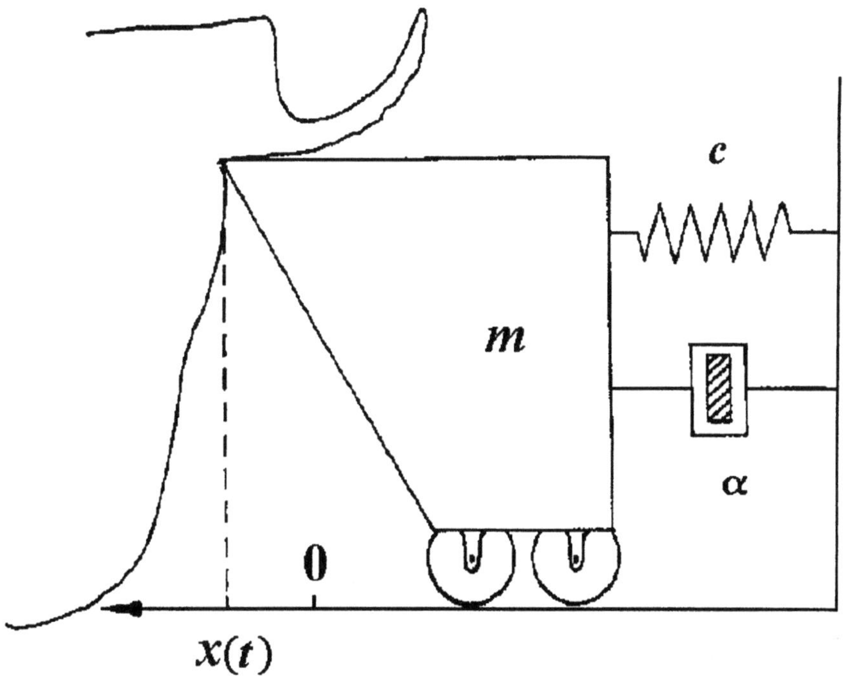

Figure 2.4.1. A simplified cutting scheme

One of the most important problems connected with cutting is oscillation of the cutter. Such oscillation can be described by an RDE with delays depending on the unknown solution [484]:

$$\left.\begin{aligned}&\ddot{x}(t)+\alpha_1\dot{x}_1(t)+\alpha_2 x(t)+\alpha_3 v x(t-h_1(\dot{x}(t),\dot{y}(t),v))=0,\\ &\ddot{y}(t)+\beta_1\dot{y}(t)+\beta_2 y(t)+\beta_3 x(t-h_2(\dot{y}(t),v))=0.\end{aligned}\right\}$$

Here α_i and β_i are constants, v is the rate of cutting, and the delays h_i are given by

$$h_1(\dot{x},\dot{y},v)=\frac{\alpha_4}{v+\dot{y}}+\frac{\alpha_5}{v+\dot{x}+\dot{y}},\quad h_2(\dot{y},v)=\frac{\alpha_4}{v+\dot{y}}.$$

4.2. *TECHNOLOGICAL DELAY*

Physical and chemical processes in reactors are characterized by their complexity. Changes in the amount of liquid entering the system happen to cause a change in the amount of liquid leaving the system only during a time h. In reality there is also a certain time needed to mix the liquids in the vessel, for chemical reaction, and for transportation of liquid from one part the reactor to another.

The feedback control loops in integrated communication and control systems are subject to network–induced delays, in addition to the delays incurred in digital sampling and data processing [345].

Such type of delay is called *technological delay*, and occurs if it is necessary to take into account the finiteness of the time needed to complete a technological process.

E.g., processes in an absorbing column with recycling have been described by the following equation [289.3] (after transition to dimensionless variables and substitution of concrete numerical values for the parameters):

$$\dot{x}(t)=3.2[-x(t)+x(t-h)]+u(t-0.625h),$$

where $x(t)$ is a value of circulating mixture and $u(t)$ a control.

Another type of delay is connected with energy or signal transmission. Hereditary phenomena are especially important when controlling objects with high velocity (e.g., airplanes, rockets), and creating long distance control devices. Delays may occur in the automatic regulation landing system of an airplane, because of the finiteness of time of the propulsion reaction on a deviation of the control lever of the engine. It is also essential to take into account delays when treating the control of aerodynamic rudder servomechanisms.

4.3. *CAR CHASING*

(See [44, 82, 163, 244].) Consider $n \geq 2$ identical cars, one following the other without passing possibility. It is assumed that acceleration at time t of each i–th car is proportional to the relative velocity of i–th and $(i+1)$–th cars at time $t-h$, and inversely proportional to the distance between them at this moment $t-h$, where h is the driver reaction time. Let m be the mass of a car, $x_i(t)(i=1,\ldots,n)$ the position of car i at time t ($t \mapsto x_n(t)$ is a given function). Then the system of dynamic equations of motion is:

$$m\ddot{x}_i(t)[x_i(t-h) - x_{i+1}(t-h)] = \alpha[\dot{x}_i(t-h) - \dot{x}_{i+1}(t-h)]$$

$$(i = 1, \ldots, n-1),$$

where α is the (constant) sensitivity coefficient of the driver.

4.4. *SHIP COURSE STABILIZATION*

The model of ship course stabilization under conditions of uncertainty may be described by the equation

$$I\ddot{x}(t) + H\dot{x}(t) = -K\psi(t) + b_0\dot{\xi}_0(t), \quad x(t_0) = x_0, \quad \dot{x}_0(t_0) = 0,$$

where $x(t)$ is the angle of deviation from the course, $\psi(t)$ the turning angle of the rudder, and $\xi_0(t)$ stochastic disturbances. It is assumed that the process $t \mapsto x(t)$ is unobservable. The process $t \mapsto y(t)$ of observations satisfies the equation

$$\dot{y}(t) = x(t-h) + b_1\dot{\xi}_1(t), \quad x(s) = 0, \quad s < t_0, \quad y(t_0) = 0.$$

Here, $h \geq 0$ is the information delay and $\xi_1(t)$ the noise in the observation channel. The rudder angle $\psi(t)$ is controlled by the automatic helmsman rule:

$$T\dot{\psi}(t) + \psi(t) = \beta\dot{y}(t) + \alpha y(t), \quad \psi(t_0) = 0.$$

In these equations, $I, H, K, b_0, b_1, T, \alpha, \beta$ are certain nonnegative constants.

This model has been investigated in [289.3].

4.5. *PROCESS OF COMBUSTION IN SMALL ROCKETS*

The model of fuel combustion in small rockets is described by the equations [428]:

$$\left.\begin{aligned} \dot{x}_1(t) + (1-\gamma)x_1(t) + \gamma x_1(t-\tau) - x_2(t-\tau) &= f_1(t), \\ \dot{x}_2(t) + \alpha x_1(t) + \alpha k x_2(t) &= f_2(t). \end{aligned}\right\}$$

Here x_1 is a pressure in the combustion chamber, x_2 – is a consumption of the fuel, positive constants γ, τ, α, k are defined by the characteristics of the rocket, f_1 and f_2 are regulation inputs.

4.6. *DELAY-DIFFERENTIAL EQUATIONS IN ENGINEERING APPLICATIONS*[1]

4.6.1 Background. RDEs can be used to model a great variety of engineering processes and their importance for system analysis and feedback control design can be hardly underestimated. The use of delays for engineering process modelling can be a very effective tool in the following situations:

- A high-dimensional process model without delays can be often approximated by a lower-dimensional model with delay. This lower-dimensional model with delay often turns out to be more amenable to analysis and control design than the original, higher dimensional model without delay.
- Engineering processes often include subsystems that are most appropriately modelled by delays. Familiar examples include transport delays due to finite speed of gas propogating through a long pipe, process delays due to residence time before a chemical reaction starts or communication delays due to finite speed of radio signal propogating through a long distance in space.
- In computer-controlled systems, computational delays due to finite-time required to execute embedded control algorithms may be significant. Although modern computers are becoming faster, the number of different tasks a typical automotive or aerospace on-board computer is required to handle increases as well. As a result, non-negligible computational delays often do exist.

In what follows we illustrate some of these ways in which delays can be introduced in engineering process models. At last we discuss the importance and performance limitations imposed by delays for stability and feedback control system design.

4.6.2 Approximating higher order model by lower order models plus delay. Consider a delay-free single input single output tenth order system with the transfer function

$$G(s) = \frac{1}{(s+1)^{10}}. \tag{4.1}$$

The response of this system to a step input and a zero initial condition is shown in Fig. 2.4.2. Clearly, this system can be mistaken for a first or

[1]This subsection is written jointly with I.V. Kolmanovsky, Ford Research Laboratories.

second order system with a time-delay. This approach of approximating higher order models with lower order models plus time delays has been extensively used in process control industry. Reference [239] argues that the use of delays in model approximation is generic in a sense that model reduction of high order linear systems with significant excess of poles over zeros must include delays to match the phase lag of the original system. For example, the three term, first-order plus time delay (FOPTD) has been used very extensively to approximate models of higher order industrial processes, see e.g. [275] and references therein. It is given by the transfer function

$$\frac{R_{ss}e^{-hs}}{\tau s+1}.$$

The constants R_{ss}, τ and h are determined from the step response of the original system with zero initial condition. The steady-state gain, R_{ss}, is equal to the final value of the output. A tangent line is drawn to the point of maximum slope and intersects the time axis at $t = h$ and reaches the height R_{ss} at time $t = h + \tau$. For the system (4.1), we estimate $R_{ss} = 1$, $h = 7$ and $\tau = 5$. The resulting approximation is indicated in Fig. 2.4.3 by the dashed line. Fig. 2.4.4 shows the approximation by a second order system with delay. Once the parameters of the FOPTD model have been identified, simple proportional+integral (PI) controllers can be tuned using conventional rules, see e.g. [275] and references therein. The normalized dead-time is given as h/τ and the processes with h/τ larger than 0.5 are generally considered difficult to control using feedback only and a feedforward controller may be needed.

We next consider a more complex example of a two input two output reactive ion etching system which is used in semiconductor manufacturing. Reactive ion etching is a critical technology for VLSI circuit fabrication. In a reactive ion etcher gases such as CF_4, CCl_4, O_2 are ionized and plasma consisting of ions, free radicals and electrons is formed. The ions bombard the surface of the wafer thereby etching the surface physically. Free radicals diffuse to the surface of the wafer where they react with the exposed material thereby etching the surface chemically. The quality of the etched wafer surface is extremely important for successful integrated circuit manufacturing. The processes involved in the etching are fairly complex and can be modelled by high order systems. Simple models can be developed as well. For example, reference [423] identifies a locally valid model with

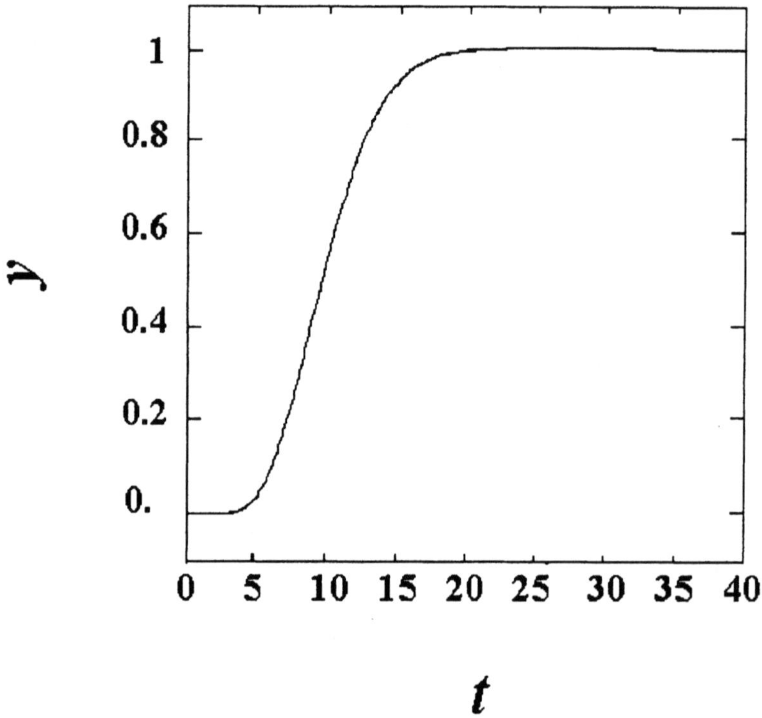

Figure 2.4.2. A tenth order system without delay can be mistaken for a lower order system with delay

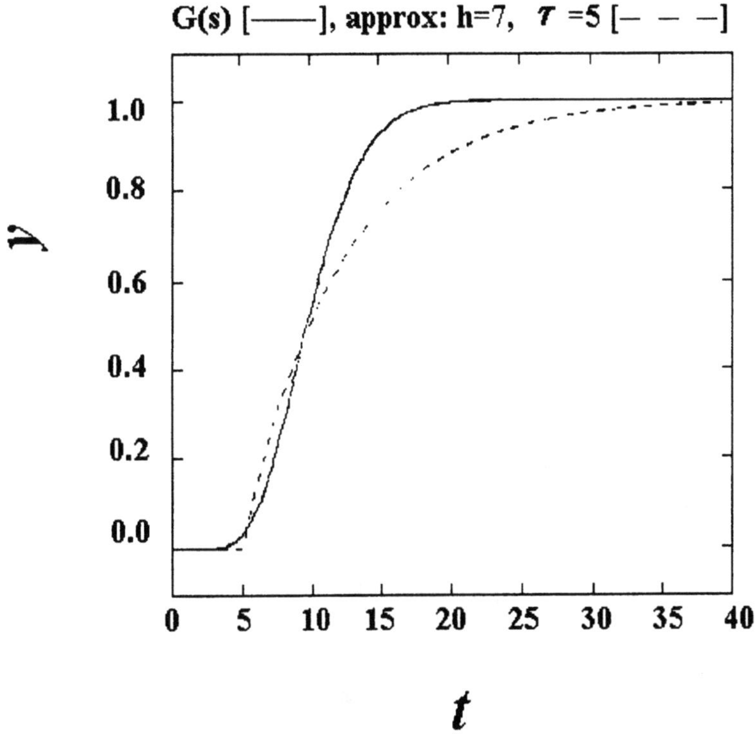

Figure 2.4.3. Approximating the tenth order system in Fig. 2.4.2 by the first order system with delay. Step response of the tenth order system is shown by the solid line. Step response of the first order system with delay is shown by the dashed line

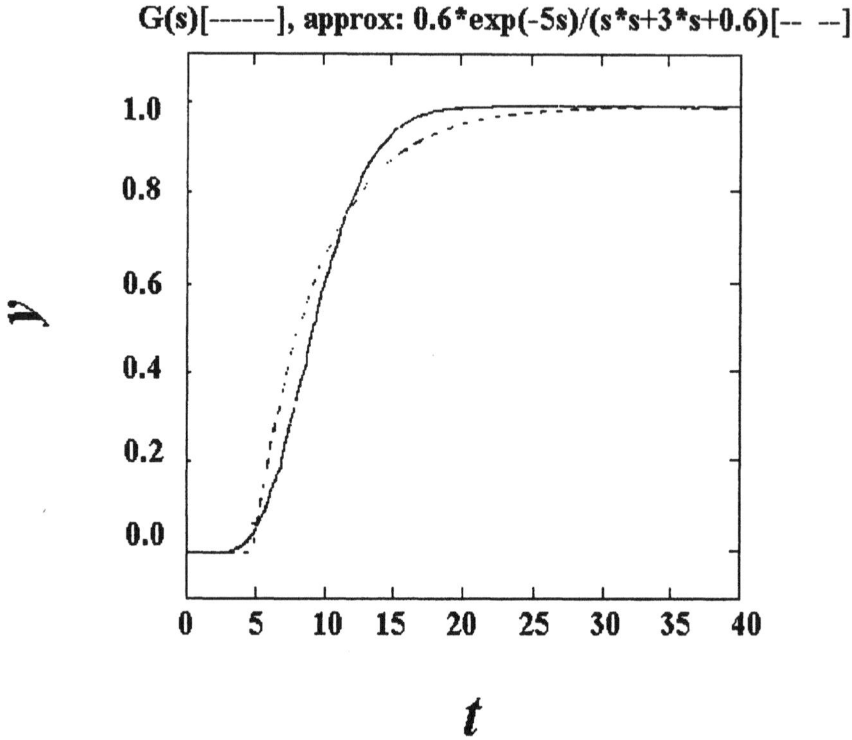

Figure 2.4.4. Approximating the tenth order system in Figure 2.4.2 by a second order system with delay. Step response of the tenth order system is shown by the solid line. Step response of the first order system with delay is shown by the dashed line.

delay that has the transfer function matrix

$$\begin{pmatrix} V_{bias} \\ (F) \end{pmatrix} = \begin{pmatrix} \dfrac{2.41s+9.75}{s^2+4s+0.7}e^{-\frac{1}{2}s} & \dfrac{0.28}{s+0.97} \\ \dfrac{-0.17}{s+0.24}e^{-\frac{1}{2}s} & \dfrac{0.17s+0.7}{s^2+15s+26.7} \end{pmatrix} \begin{pmatrix} throttle \\ power \end{pmatrix}.$$

The outputs are: the dc-bias voltage, V_{bias}, that develops between the electrodes in the etcher and which determines mechanical energy of impinging ions; and $[F]$ which represents the absolute concentration of the fluorine involved in chemical etching. The inputs are: position of the throttle located between the etcher chamber and a cryo pump which is used to regulate pressure in the etcher chamber; and the power applied across the two electrodes by a RF (13.56 MHz) power source. Based on this simple model with delay, a controller was designed in [423] that experimentally demonstrated stability and good performance.

The final example has to do with evaluation of flying qualities of military aircraft. Flying qualities give an indication of whether the response to pilot input will be acceptable. Flying qualities are evaluated by approximating the actual high order response of the aircraft from pilot inputs to outputs of interests by low order equivalent system (LOES). The parameters if the LOES must lie within certain bounds for the closed loop aircraft to have adequate flying qualities. For the lateral-directional axes of aircraft LOES from roll rate command to roll rate is [467]:

$$\frac{\dot{\mu}}{\dot{\mu}_c} = \frac{K_\mu(s^2+2\zeta\omega s+\omega^2)}{(T_R s+1)(T_S s+1)(s^2+2\zeta_D\omega_D s+\omega_D^2)}e^{-s\tau}.$$

The equivalent time delay τ represents neglected higher order modes in the lower order approximation. The time delay is caused primarily by cascading numerous dynamic elements that form the control system. This is the undesirable dead-time perceived by the pilot in the initial aircraft response to a command and it is the critical flying qualities factor [461]. Requirements for the parameters of the LOES are found in the military specification, MIL-STD-1797A. In particular, the equivalent time delay in the LOES must be less than 0.15 seconds.

4.6.3 Delays as subsystem models. Here we describe in more details three situations where delays can be introduced as subsystem models: transport delays, process delays and communication delays. We emphasize that this list is hardly exhaustive and examples of delays not covered by this classification are easy to find. For example, in control of complex chemical processes the operator may take a sample and analyze it in the lab for a few hours before deciding on a corrective control action. This type of measurement delay is perhaps most similar to transport delay discussed below.

4.6.3.1 Transport delays. A classical example of a transport delay occurs in rolling mills. In a rolling mill, the incoming steel strip is rolled to a desired thickness when it is passed through a pair of rollers, see Fig. 2.4.5. The distance between the rollers and, hence, the thickness of the rolled strip, can be adjusted by an electric motor. The thickness of the strip is measured by a thickness sensor located downstream of the rollers. The transport delay is due to the time it takes the strip to move from the rollers to the sensor. While a variety of models for rolling mills has been developed in the literature[1], we consider a simple, conceptual model [208]. The strip after the rollers is assumed to have thickness x, width w and velocity v. The distance between the rollers and the thickness sensor is d. Thickness sensor measurements are used to control the distance u between the rollers. The delay $\tau(t)$ between the rollers and the thickness sensor is given by

$$d := \int_{t-\tau(t)}^{t} v(s)ds.$$

By differentiating this expression we obtain:

$$v(t) - v(t-\tau(t))(1-\dot{\tau}(t)) = 0.$$

Assuming that the width of the strip does not change, mass conservation provides

$$x(t)v(t) = \text{constant}.$$

Hence,

$$\frac{1}{x(t)} - \frac{1-\dot{\tau}(t)}{x(t-\tau(t))} = 0.$$

Suppose that u is generated by an integral controller with gain k_i so that

$$u(t) = k_i z(t), \quad \dot{z}(t) = x(t-\tau(t)) - x_d,$$

where x_d is the desired thickness of the strip. Then, we obtain the following system of delay-differential equations governing controlled process operation:

$$\begin{aligned} \dot{\tau}(t) &= \frac{-x(t-\tau(t)) + x(t)}{x(t)}, \\ \dot{x}(t) &= k_i(x(t-\tau(t)) - x_d). \end{aligned}$$

[1] See e.g. [441] for more advanced models.

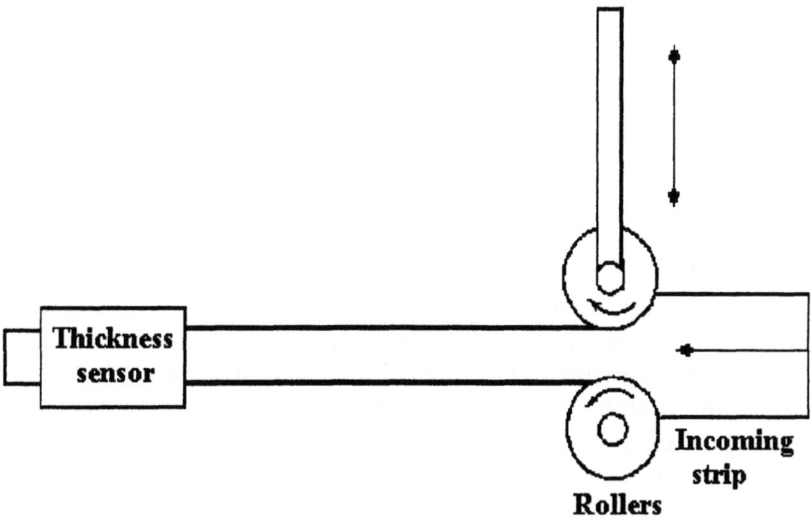

Figure 2.4.5. Metal rolling system

Rolling mill is not the only example in metal industry with significant delays. Reference [260] describes another example, production of galvanized sheet steel. Steel strip, of order 1m wide by 1mm thick, is preheated and passed at speeds of order 100 m/min through a bath of liquid zinc at temperature of order 500°C. A zinc film is entrained onto the strip as it emerges from the bath, see Fig. 2.4.6. The amount of zinc deposition is regulated by an air-knife which consists of a pair of nozzles that blow air at the strip forcing return flow of liquid zinc from the strip into the bath. The pressure in the nozzles and the distance between the nozzles and the strip can be controlled (higher pressure or smaller distance result in less deposit). The film solidifies whilst the strip runs vertically upwards for a distance of order 50m to the first set of rollers which guide the finished product to a delivery section. To provide acceptable quality of the final product tight control of deposited zinc mass is required. The feedback measurements are provided by X-ray fluorescent gauges located on each side of the moving strip. The gauges scan back and forth across the strip giving measurements of deposited mass. The gauges cannot operate at temperatures of molten zinc and must therefore be mounted a significant distance L downstream from the air-knife, where the strip is cooler. Each measurement is thus subject to a delay L/v (of order 1 minute), where v is the velocity of the strip as it moves vertically upward. I In fact, the delay is so large that the rest of the dynamics can be modelled by steady-state maps, see [260].

As the previous two examples illustrate processes in metal industry can have substantial delays. In particular, in FOPTD approximation of these processes,

$$\frac{R_{ss}e^{-hs}}{\tau s+1},$$

the normalized dead-time, h/τ, is a measure of difficulty in controlling the process with a feedback. The processes with normalized dead-time larger than 0.5 are generally considered difficult. Reference [275] point out that in metal industry the normalized dead-time is larger than in many other industrial processes (up to 10). Nevertheless, delays found in many other engineering systems are still substantial. We review some of these examples next.

The operation of a paper machine is illustrated in Fig. 2.4.7. The thick stock valve and the filler valve control the flows of the thick stock (cellulose fiber mixed with water) and filler. The filler is added to improve opacity and homogenuity of the paper. Filler and thick stock are mixed together and diluted with white water (recirculated water from wire section containing both fiber and filler) in the white water section. In the headbox the stock is distributed across the wire. In the wire section the sheet is formed when the stock is dewatered. Most of the water falls through the wire and is recycled

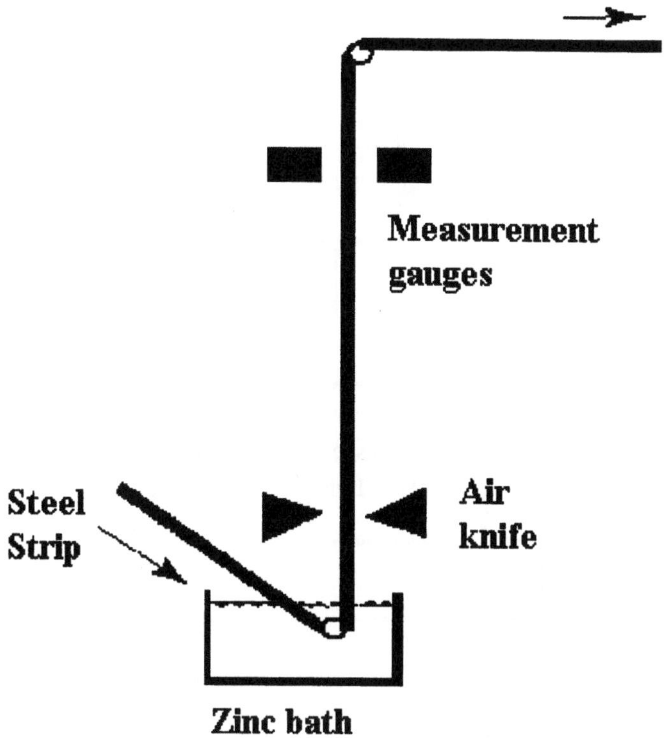

Figure 2.4.6. Production of galvanized sheet steel

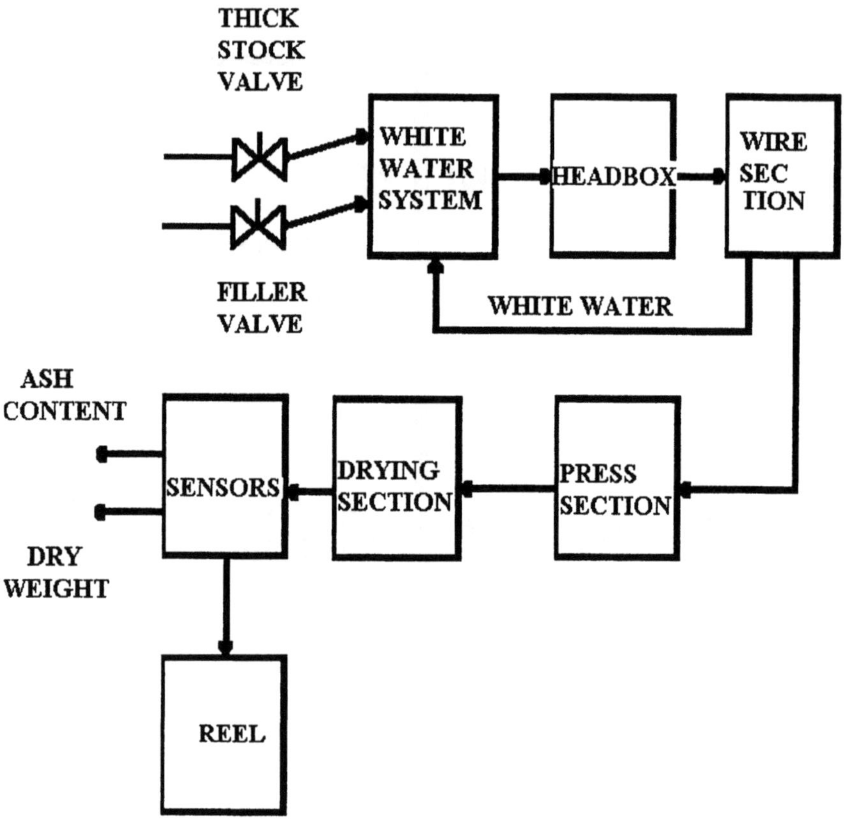

Figure 2.4.7. Paper machine operation

to the white water system. The paper web passes through the press section where further dewatering takes place. In the drying section the paper is dryed. Near the reel scanning sensors measure paper dry weight and ash content (amount of filler). A more detailed description of the paper machine operation can be found e.g. in [256]. It is clear that multiple transport delays are present in this system. For example, the pipes can be modelled by pure transport delays. The paper machine operation is modelled in [256] by its transfer function matrix,

$$G(s) = \begin{pmatrix} \dfrac{0.214e^{-68s}}{125s+1} & \dfrac{-0.192e^{-68s}}{17s+1} \\[2ex] \dfrac{0.153e^{-68s}}{125s+1} & \dfrac{0.93e^{-68s}}{17s+1} \end{pmatrix},$$

where the inputs are filler valve and thick stock valve and the outputs are ash content and dry weight. In this case the delays account not only for transport delays in the system but also for neglected higher order modes and delay time required for scanning sensors to move across the paper web.

Our next example has to do with operation of an internal combustion engine. A schematic diagram of a four-stroke port fuel injected spark ignition gasoline engine is shown in Fig. 2.4.8. Driver's pedal is mechanically linked to the throttle which controls the flow of air into the intake manifold. The intake manifold branches out into four runners leading to individual cylinders of the engine. The amount of fuel is injected by electronic injectors located in the intake manifold runners close to the intake valves to match the estimated by on-board computer amout of air so that the air-to-fuel ratio is close to stoicheometric. The mixture of air and fuel is inducted in a cylinder of the engine on the induction stroke; then it is compressed on the compression stroke; then, spark is used to start the combustion and expanding gases push the piston thereby generating torque on the engine crankshaft; and, finally, the exhaust gases are pushed out of the exhaust manifold on the exhaust stroke. In a four-cylinder engine at any given time there is one cylinder that undergoes induction stroke, one cylinder that undergoes compression stroke, one cylinder that undergoes power stroke and one cylinder that undergoes exhaust stroke. The intake and exhaust valves for each cylinder open and close to admit the mixture in the cylinders or to let the exhaust gases out of the cylinders. The opening and closing of the valves is accomplished via a mechanical linkage of the crankshaft with the camshaft. Spark timing (i.e. time when the spark fires) can be adjusted for each cylinder. The spark timing is usually measured in degrees of crankshaft revolution between the position when the spark is fired and the position corresponding to the top dead center of the cylinder.

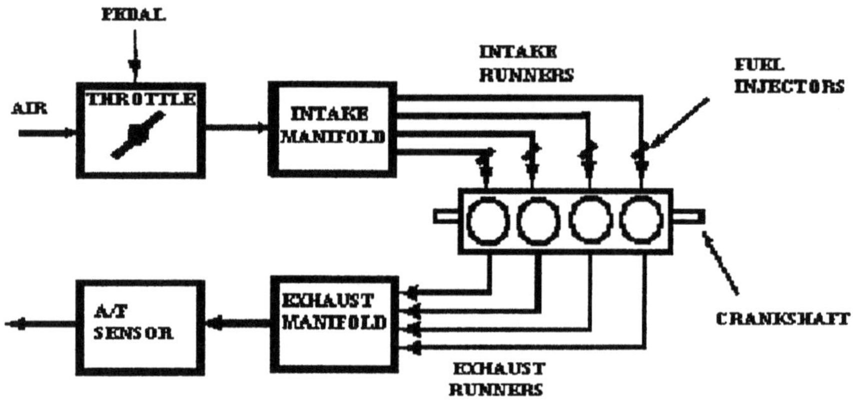

Figure 2.4.8. Schematic diagram of a four-stroke gasoline port-injected spark ignition interval combustion engine

A model for a spark-ignition engine has been reported in [419]. It is based on the mean-value assumption: the values of variables representing engine behavior are averaged over an engine cycle. For example, the variations in intake manifold pressure due to intake valve opening and closing are not captured but are averaged out. The resulting mean-value model is limited to the low frequency range. For completeness, we present a derivation for the model. The dynamics of the intake manifold pressure are obtained by differentiating the ideal gas law:

$$\dot{m} = \frac{\dot{p}V}{RT},$$

where m is the mass of air in the intake manifold in kg, p is the pressure in intake manifold in kPa, T is the intake manifold temperature in K which is assumed constant and $R = 0.2897$ kJ/kg is the gas constant Using mass conservation,

$$\dot{m} = W_{throttle} - W_{engine},$$

so that

$$\dot{p} = \frac{RT}{V}(W_{throttle} - W_{engine}), \tag{4.2}$$

where $W_{throttle}$ is the flow into the intake manifold through the throttle body while W_{engine} is the flow out of the intake manifold into the engine. This equation is valid during normal driving. When the engine is idling (driver's foot is off the pedal), $W_{throttle} = 0$ and the air-by-pass valve provides air to the engine. Thus during idle equation (4.2) is modified by replacing $W_{throttle}$ with $W_{air-by-pass}$. The flow through the throttle body is calculated from an orifice equation of the form

$$W_{throttle} = f(\theta)g(p),$$

where θ is the throttle position and the flow is measured in kg/sec. A simplified model for g is of the form:

$$\begin{aligned} g(p) &= 1, \quad p < p_a/2, \\ g(p) &= \frac{2}{p_a}\sqrt{pp_a - p^2}, \quad p \geq p_a/2, \end{aligned}$$

where p_a is the atmospheric pressure. The flow through the air-by-pass valve is calculated using a similar expression. Reference[1] reports f for a throttle on a four-cylinder spark-ignition engine:

$$f(\theta) = (2.821 - 0.05231\theta + 0.10299\theta^2 - 0.00063\theta^3)/1000,$$

[1]Crossley, P.R., Cook, J.A., "A nonlinear engine model for drivetrian system development," *IEE Conference Publication Number 332*, Vol. 2, pp. 921–923, 1991.

where θ is the throttle angle in degrees. The intake flow into the engine is calculated as a nonlinear function of engine speed $w(t)$ and intake manifold pressure. In addition the following expression for a four-cylinder spark-ignition engine is reported

$$W_{engine} = (-0.366 + 0.08979\omega(p/100) - 0.0337\omega(p/100)^2 +$$

$$+0.0001\omega^2(p/100))/1000,$$

where ω is engine speed in rad/sec. Crankshaft dynamics are described by

$$J_e\dot{\omega} = (T_e - T_{load}),$$

where J_e is the crankshaft moment of inertia (measured in kg/m^2), T_e is the brake torque generated by the engine (measured in N m) and T_{load} is the external load torque on the crankshaft (measured in N m). The brake torque generated by the engine is calculated as

$$T_e(t) = f_T(\phi(t - \sigma(t)), W_{engine}(t - \sigma(t)), \omega(t), \delta(t)),$$

where ϕ is the air-to-fuel ratio and $\delta(t)$ is the spark-timing (measured in degrees). The time delay, $\sigma(t)$, is referred to as the induction-to-power stroke delay and its value in seconds is dependent on engine speed as

$$\sigma(t) = \frac{2\pi}{\omega(t)}.$$

This delay is introduced because the torque value is determined by the air flow and fuel flow values at the beginning of the induction stroke while the actual torque generation takes place during the power stroke. The induction-to-power delay is important for low engine speed operating region, in particular, for analysis of stability of idle speed control loop. Reference mentioned above reports a torque map for a four-cylinder spark ignition engine:

$$\begin{aligned} T_e &= -181.3 + 379.36m_a + 21.91\phi - 0.85\phi^2 + \\ &+ 0.26\delta - 0.028\delta^2 + 0.027\omega - 0.000107\omega^2 + \\ &+ 0.00048\omega\delta + 2.55\delta m_a - 0.05\delta^2 m_a, \end{aligned}$$

where m_a is the mass of air in g inducted in one cylinder during the induction stroke,

$$m_a(t) = \frac{1000\pi}{\omega(t)} W_{engine}(t),$$

and δ is the spark timing advance in degrees measured after top dead center. For a vehicle employing an automatic transmission, the external load torque consists of the load applied by the torque converter plus external torque disturbances which may arise because of auxiliary loads imposed on the engine (engagement of the air conditioner compressor, for example). The torque from the converter is generally specified as the square of the ratio of engine speed to a converter input capacity factor, K_i. The external torque load can then be expressed as

$$T_{load} = (\omega(t)/K_i)^2 + T_d \; '$$

where K_i is a constant and T_d is the torque disturbance. For emission control, tight air-to-fuel ratio regulation is required. The air-to-fuel ratio is measured by a sensor located on the engine exhaust side. The measurement of the air-to-fuel ratio is delayed because of the time delay from induction stroke to exhaust stroke combined with gas transport delay from exhaust valve to the sensor location:

$$\phi_{meas}(t) = \phi(t - h(t)), \quad h(t) = \frac{2\pi}{\omega(t)} + \frac{L}{v}.$$

Here L is the distance to the sensor, while v is the gas velocity. The delay $h(t)$ depends on engine speed and is typically of order of $8\pi/\omega(t)$ seconds or two engine cycles. This delay in measurements of the air-to-fuel ratio is important in the design of the air-to-fuel ratio regulation loop because it sets a limit on how fast the feedback can be made. In typical production vehicles an additional complication exists: a less expensive switching EGO sensor is used that can only indicate if $\phi_{meas}(t) > \phi_s$ or $\phi_{meas}(t) < \phi_s$, where ϕ_s is the desired stoicheometric air-to-fuel ratio, $\phi_s = 14.6$; the actual value of $\phi_{meas}(t)$ is unknown.

Next example has to do with operation of a commercial boiler. Reference[2] develops a model for boiler No. 2 at Abbott Power Plant in Champaign, Illinois, see Fig. 2.4.9. In a boiler, the steam pressure at the outlet of the drum (header pressure), the level of the water in the drum and the oxygen level in the flue gas must be maintained at their specified levels despite variations in steam flow rate in the header, fluctuations in the heating value of the fuel, and other commonly present disturbances such as variations in ambient temperature and various leaks. The control system is capable of actuating air, fuel and feedwater flow rates using measurements of steam pressure, actual fuel flow rate, drum water level, actual feedwater flow rate, steam flow rate, steam temperature, and oxygen level in the flue gas. The

[2] Pellegrinetti, G., and Bentsman, J., "Nonlinear control oriented boiler modeling – A benchmark problem for controller design," *IEEE Transactions on Control Systems Technology,* Vol. 4, pp. 57–64m, 1996.

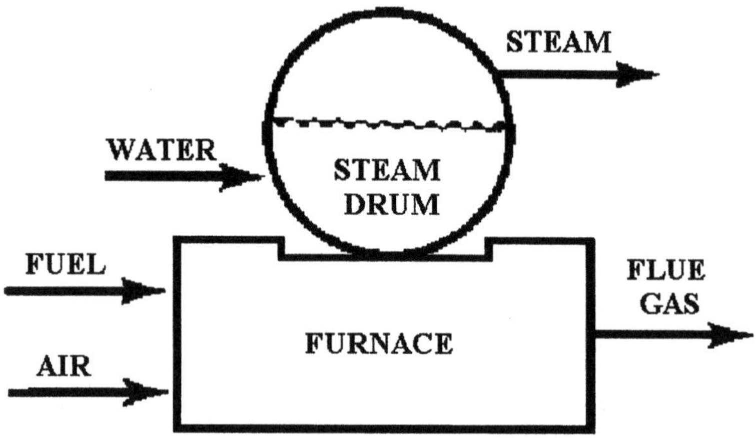

Figure 2.4.9. Schematic representation of a commercial boiler

model developed in [2] p. 55 has the form of the system of a stochastic FDEs:

$$\begin{aligned}
\dot{x}_1(t) &= c_{11}x_4(t)x_1^{9/8}(t) + c_{12}u_1(t-\tau_1) - c_{13}u_3(t-\tau_3),\\
\dot{x}_2(t) &= c_{21}x_2(t) + \frac{c_{22}u_2(t-\tau_2) - c_{23}u_1(t-\tau_1) - c_{24}u_1(t-\tau_1)x_2(t)}{c_{25}u_2(t-\tau_2) - c_{26}u_1(t-\tau_1)},\\
\dot{x}_3(t) &= -c_{31}x_1(t) - c_{32}x_4(t)x_1(t) + c_{33}u_3(t-\tau_3),\\
\dot{x}_4(t) &= -c_{41}x_4(t) + c_{42}u_1(t-\tau_1) + c_{43} + n_5(t),\\
y_1(t) &= c_{51}x_1(t-\tau_4) + n_1(t),\\
y_2(t) &= c_{61}x_2(t-\tau_5) + n_2(t),\\
y_3(t) &= c_{70}x_1(t-\tau_6) + c_{71}x_3(t-\tau_6) + c_{72}x_4(t-\tau_6)x_1(t-\tau_6) +\\
&+ c_{73}u_3(t-\tau_3-\tau_6) + c_{74}u_1(t-\tau_1-\tau_6) +\\
&+ \frac{(c_{75}x_1(t-\tau_6) + c_{76})(1 - c_{77}x_3(t-\tau_6))}{x_3(t-\tau_6)(x_1(t-\tau_6) + c_{78})} + c_{79} + n_3(t),\\
y_4(t) &= (c_{81}x_4(t-\tau_7) + c_{82})x_1(t-\tau_7) + n_4(t).
\end{aligned}$$

Here x_1 is the drum pressure state; y_1 is the measured drum pressure; y_2 and x_2 are the measured excess oxygen level and its state, respectively, (percent); x_3 is the system fluid density; y_3 is the drum water level (in); y_4 is the steam flow rate; u_1, u_2, u_3 are, respectively, the fuel, air, and feed water flow rate inputs which take values between 0 and 1; x_4 is the exogenous variable related to the load disturbance intensity (0-1); and the variables n_i are the outputs of the first order noise models driven by zero mean, unit variance white noise:

$$n_1 = \frac{0.75s + 0.1}{s + 0.001}w_1, \quad n_2 = \frac{0.019s + 0.001}{s + 0.024}w_2,$$

$$n_3 = \frac{0.105s + 0.038}{s + 0.010}w_3, \quad n_4 = \frac{0.01s + 0.0001}{s + 0.001}w_4,$$

$$n_5 = \frac{0.003s + 0.003}{s + 0.0075}w_5 + w_d,$$

where n_i, $i = 1, \cdots, 5$, is the colored noise, w_i is the unit variance white noise, and w_d is the deterministic part of the disturbance that defines the load level. The time delays, τ_i, include measurement instrumentation delays, process delays and transport delays that are, in general, time-varying because they depend on travel distances, gas flow rates, etc. The value of the coefficients for this model can be found in [2] p. 55.

Our final example is the control of pH in a mixing tank [536]. An influent mixture enters a mixing tank along with an acid or base reagent. After a time delay the dynamic pH electrode senses the pH of the mixture. Based on this measurement the controller adjusts the base and acid readgent flows to maintain a desired pH. The delay can be viewed as a transport delay caused by reagents and influent propogating through the large volume of the tank with a finite speed to the sensor location. Let $x_1 = [OH^-]$ be the concentration of base in tank, $x_2 = [H^+]$ be the concentration of acid in tank. Using the notation of [536], the mass balances of the ionic components results in

$$\dot{x}_1(t+\tau) = [ccb \cdot u_1(t) + cbr \cdot qa(t) - (qa(t) + u_1(t) + u_2(t)) \cdot x_1(t+\tau)]/vol,$$

$$\dot{x}_2(t+\tau) = [cca \cdot u_2(t) + car \cdot qa(t) - (qa(t) + u_1(t) + u_2(t)) \cdot x_2(t+\tau)]/vol,$$

where $ccb = 20$ moles/liter is the concentration of the base reagent, cbr is the disturbance in influent base concentration (mole/liter), $qa = 100$ liters/min is the influent flow rate, u_1 is the input flow rate of base reagent (control), u_2 is the input flow rate of acid reagent, $vol = 1000$ liters is the volume of the mixing tank, $cca = 20$ moles/liter is concentration of acid reagent, car is the disturbance in influent acid concentration measured in moles/liter and τ is time-varying depending on the flow rates, with the nominal value of about 0.52 min. The pH electrode dynamics are modelled as first order system with different time constants for increasing and decreasing pH:

$$\dot{x}_3 = (-\ln x_2(t) - x_3(t))/\delta,$$

where

$$\delta = \begin{cases} \delta_1, & -\ln x_2(t) \text{ increasing} \\ \delta_2, & -\ln x_2(t) \text{ decreasing,} \end{cases}$$

where $\delta_1 = 0.1$ minute, $\delta_2 = 0.2$ minute.

4.6.3.2 Process delays. Some processes include delays that are of somewhat different physical character than transport delays or communication delays. We refer to these delays as process delays.

We first note that in the pH control example of the previous subsection, we interpreted the delay as a transport delay. However, this delay can also be viewed as a process delay because the molecules have to remain in the tank for a certain period of time for the chemical reactions to go to completion [438]. Therefore, the tank has to have a sufficiently large volume to allow for adequate residence time for the molecules.

Another example concerns combustion in diesel engines [233]. In a direct injection diesel engine fuel is injected directly in the cylinder which contains air heated up on the compression stroke. The fuel droplets are heated up and then vaporize. Mixing of the vaporized fuel and air takes place and the mixture starts to burn by self-ignition. The time it takes for physical and chemical transformations to the fuel and air mixture to occur before combustion starts is known as the ignition delay and it is one of the most important characteristics of diesel engine operation, affecting noise, torque generation and emissions.

The final example is the car driver's model [317]. The car driver generates steering commands to the front wheels based on the perceived heading and lane position errors. Because humans cannot react to events instantaneously, the driver's model must include a time delay, τ. Another major component is a lead or anticipation term, $(T_L s+1)$, that the driver adopts to counteract the lag in vehicle lateral response. The model in [317] is given by a transfer function,

$$\delta_{sw} = -K_{\Psi}(T_L s + 1)e^{-\tau s}\left(K_y(y - y_d) + \Psi\right),$$

where δ_{sw} is the magnitude of the steering wheel correction applied by the driver, $y - y_d$ is the vehicle lateral offset in meters, Ψ is the vehicle heading angle in degrees, τ is measured in seconds, and $K_y = 0.061$, $K_{\Psi} = 15.91$, $\tau = 0.13, T_L = 0.09$. This transfer function is improper and has to be augmented with full vehicle dynamics model to be used for simulation and analysis. The vehicle dynamics model determines vehicle heading, yaw rate, lateral displacement and sideslip angle based on the steering wheel correction applied by the driver, see e.g. [317].

4.6.3.3 Communication delays. Familiar examples of communication delays include signal transmission delays imposed by limits on the speed of light or sound. For satellites in low Earth orbit, round trip signal delays are minimally 0.4 seconds; for spacecraft near Moon these delays are of order of 3 seconds. Reference [450] and other articles in that Special Issue of IEEE Transactions on Robotics and Automation provide an overview of issues that arise when controlling robots remotely in space. In deep ocean operation, acoustic telemetry is often employed. Comminication with an underwater vehicle at depth of 1700 meters imposes a 2 second round-trip delay because the speed of sound in water is about $1700 m/s$. Some underwater vehicles are steered by dragging them with a cable, causing time delays of ten or more minutes. Similar delays may occur for tethered satellites in orbit.

As a simple, conceptual example consider a satellite in an orbit around the Earth [208]. Let x_1, x_2, x_3 be the coordinates of satellite center of mass, $x = (x_1, x_2, x_3)'$, r be the radius of the orbit, $r = \sqrt{x_1^2 + x_2^2 + x_3^2}$, m be the mass of the satellite, u_1, u_2, u_3 be the control forces that can be applied to the satellite by thruster jets along each coordinate direction, and $u = (u_1, u_2, u_3)'$. Then,

$$\ddot{x}_1(t) = -\frac{k x_1(t)}{r^3(t)} + \frac{u_1(t)}{m},$$

$$\ddot{x}_2(t) = -\frac{kx_2(t)}{r^3(t)} + \frac{u_2(t)}{m},$$

$$\ddot{x}_3(t) = -\frac{kx_3(t)}{r^3(t)} + \frac{u_3(t)}{m}.$$

If the satellite is controlled from the Earth, the control inputs u_1, u_2, u_3 are generated from the telemetric data of satellite coordinates and velocities

$$u = R(x, \dot{x}).$$

Because of the communication delay between the Earth and the satellite, the satellite receives commands with a delay and based on delayed values of its coordinates and velocities so that

$$u(t) = R(x(t - 2\tau), \dot{x}(t - 2\tau)).$$

In this equation τ is the value of the signal transmission delay between the Earth and the satellite.

4.6.4 Performance and stability limitations imposed by delays on feedback control systems. Delays impose severe limitations on the operation of feedback control systems. The issues are best understood by considering a specific example. The system is

$$\dot{x} = u + w,$$

where x is a scalar state, u is a scalar control input and w is a disturbance. The system can be stabilized to a desired equilibrium r by a proportional feedback on x:

$$u(t) = -k(x(t) - r).$$

Let

$$V(x) = \frac{1}{2}(x - r)^2.$$

Then the derivative of V along the trajectories of the closed loop system is of the form

$$\dot{V}(x) = -k(x - r)^2 + (x - r)w.$$

Suppose w is bounded in an interval $[-1, 1]$ but is otherwise unknown. Then, for $|x - r| > \frac{1}{k}$, $\dot{V}(x) < 0$. This implies that every trajectory of the closed loop system asymptotically converges to the set $\Lambda(k) = [r - \frac{1}{k}, r + \frac{1}{k}]$. By increasing the gain k better and better performance is obtained because the speed with which the trajectories converge to the set $\Lambda(k)$ increases and the set $\Lambda(k)$ shrinks.

It is not possible to increase gain without limit to obtain better and better performance when delays are present. For example, suppose the control is delayed so that

$$u(t) = -k(x(t-\tau) - r(t-\tau)),$$

and, for simplicity, suppose $w \equiv 0$, $r(t) = 0$ for $t > 0$, and $x(t) = 1$, $r(t) = 1$ for $t \in [-\tau, 0]$. The dash-dotted line in Fig. 2.4.10 shows the well-behaved response of the system with $\tau = 0, k = -2$. The solid line shows the response for $\tau = 1, k = -2$. This response in unstable. If we lower the gain k to 1 we obtain the response shown by the dashed line: it is stable but is very oscillatory.

There is a simple, intuitive reason for loss of stability. The feedback $u = -k(x(t) - r)$ is designed to decrease system energy V and can be viewed as damping. But because of the delays $u(t)$ may not change sign when $x(t) - r$ changes sign and, as a result, energy is pumped into the system instead of being taken away from the system. This "delay-caused" energy addition may outweigh the dissipation if the value of k is sufficiently large relative to the value of the delay. The rigorous stability analysis of systems with delay is a subject of great importance and it is treated in part 3 of the book.

5. AFTEREFFECT IN BIOLOGY

Since the time of V. Volterra, FDEs are widely used to model biological processes. The transmission of control signals in biological systems is related to such long processes as birth, growth (development), and death. Because of this, the evolution of biological systems depends in an essential way on the whole previous history, and can be modelled by FDEs. Moreover, using FDEs allows us to take into account various insecure factors such as: inhomogeneity of the species ages; finite lifetime and interaction time; discrete seasons of development; finite acceptance time for external signals and finite time for elaborating counteractions; pollution effects, resulting in additional mortality with time delay; spatial environmental heterogeneity; etc. The importance of aftereffects in population dynamics, and the new effects stipulated by it, have been indicated in [207, 301, 502].

5.1. *EVOLUTION EQUATIONS OF A SINGLE SPECIES*

To describe species populations struggling for a common food, numerous models with retardation have been proposed. One of the simplest such models of intraspecific struggle for. the case of limited self-renewing food

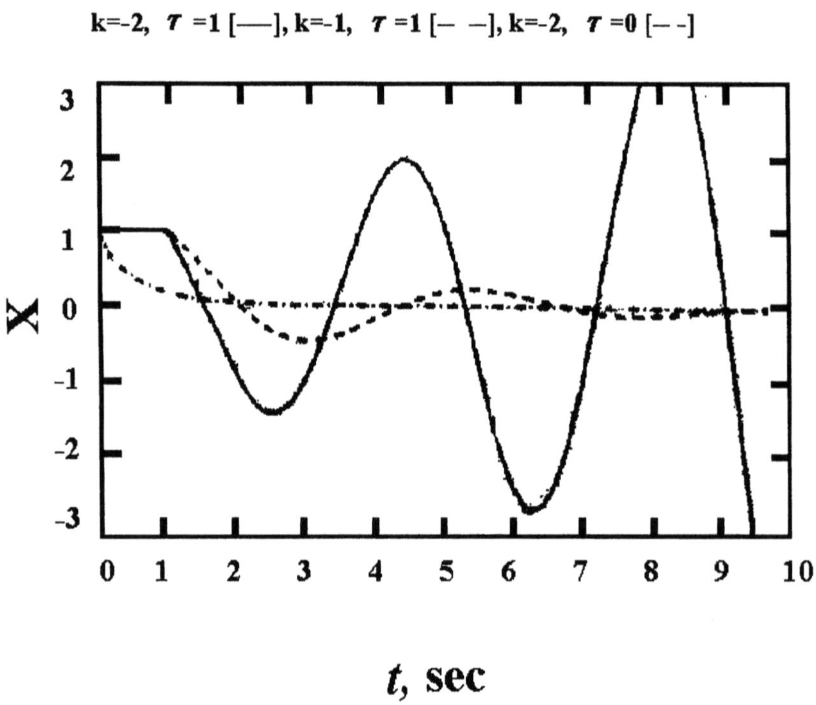

Figure 2.4.10. Instability caused by high-gain feedback and a delay

resources, is the logistic model [207]

$$\dot{x}(t) = \gamma[1 - K^{-1}x(t-h)]x(t). \tag{5.1}$$

Here $x(t)$ is the population number (number of individuals in the population), regarded as a continuous quantity, and h is the production time of food resources (which is sometimes interpreted as the average age of producers). For $h = 0$ we obtain the well–known logistic equation; taking into account a delay $h > 0$ means that the food resources at time t are determined by the population number at time $t - h$. The constant γ is related to the reproduction of species, and represents the difference between birth and death rates. Usually, γ is called the *Maltus coefficient* of linear growth. It is the inverse of the time of increase of $N(t)$ at e–time under the condition that $N(t)$ is sufficiently small. The constant K is the average population number, and is related to the ability of the environment to sustain the population. Models like (5.1) do not take into consideration such phenomena as immigration, emigration, age heterogeneity, interactions with other species, etc. At the same time, (5.1) can be used to study hatching periods, pregnancy duration, egg–laying, etc.

Putting $x(t) = K[1 + y(t/h)]$, we obtain the following equation for y:

$$\dot{y}(t) = -\gamma h y(t-1)[1 + y(t)].$$

This equation is also encountered in number theory [527]. Other RDEs arising in number theory have been studied in [95, 218, 515].

The model (5.1) has been generalized in various directions, taking into account migration, inhomogeneity and stochastic properties of the environment, diffusion, incubation, etc. Moreover, although discrete delays do occur in certain ecological models, more often the delays are distributed, and so the present state of a time–varying system is affected by all past values [420].

E.g., the following model has been proposed in [128], in which the whole past history is needed, and therefore the relation to be satisfied involves an integral of the unknown function:

$$\dot{x}(t) = \gamma\left[1 - \int_0^\infty x(t-s)dK(s)\right]x(t).$$

Here the kernel $K(s)$ is a function of bounded variation on $[0, \infty)$; it can be interpreted as the residual intensity of population. It is interesting to note that the same type of equation arises in nuclear reactor theory [209].

If the growth coefficient γ, or some environmental characteristics, are unknown, then the stochastic equation

$$dx(t) = \gamma\left[1 - \int_0^\infty x(t-s)dK(s)\right] x(t)dt + x(t)d\xi(t)$$

can be used for the population number, where $t \mapsto \xi(t)$ is stochastic process. If $t \mapsto \xi(t)$ is a Wiener process, this equation is an Itô's stochastic differential equation with aftereffect. To model Volterra population processes in a stochastic environment two types of such equations have been investigated in [228]:

$$dx(t) = x(t)a_1(x_t)dt + a_2(x(t))d\xi(t),$$

$$dx(t) = \gamma\left[1 - \int_0^\infty K(s)x(t-s)ds\right] x(t)dt + a_2(x(t))d\xi(t). \tag{5.2}$$

Another possibility to model population processes is to use NDEs. E.g., the following generalization of (5.1) has been proposed in [207, 301]:

$$\dot{x}(t) = \gamma\{1 - K^{-1}[x(t-h) - c\dot{x}(t-h)]\}x(t),$$

where γ, K, c, h are positive constants.

Environmental inhomogeneity leads to population diffusion, which may be taken into account by appropriately modifying (5.1). E.g., it has been shown in [302] that the population density $N(t,x)$ at time t and space point x (here we consider the one–dimensional case) satisfies the equation

$$\frac{\partial N(t,x)}{\partial t} = D\frac{\partial^2 N(t,x)}{\partial x^2} + \gamma[1 - K^{-1}N(t-h,x)]N(t,x),$$

where D is the diffusion coefficient, and h, K are positive constants. This equation has been studied in [366].

Along with equations involving time delays, equations involving shifts of spatial variables are also used (see, e.g., [329]). So in [329] the following model has been considered:

$$\frac{\partial N(t,x)}{\partial t} = D\frac{\partial^2 N(t,x)}{\partial x^2} + \gamma\left[1 - \int_{-\infty}^{\infty} K(y)N(t,x-y)dy\right] N(t,x).$$

Note that this FDE does not involve aftereffect; it is close to the equations known as *equation of Barbashin type*.

A drop in birth rate caused by accumulation of metabolic products can be expressed by a power of the integral term in the logistic equation (5.1):

$$\dot{x}(t) = \gamma \left\{ 1 - x(t) - \left[\int_0^\infty x(t-s)dK(s) \right]^j \right\} x(t), \quad j \geq 0.$$

For $j = 2$ this model has been investigated in [128].

A deficiency of the models above is that they do not incorporate any age heterogeneity. The widely–used MacKendrich – Von Foerster model takes this kind of inhomogeneity into account. The delay generalization of this model can be done as follows.

Let $\rho(t,\tau)$ be the population density at time t for age τ. This means that at time t the total number of population members of ages τ to $\tau + \Delta\tau$ is equal to $\rho(t,\tau)\Delta\tau + o(\Delta\tau)$. Let $a(\tau)$ and $b(\tau)$ be the birth and death rates. Then the conservation law gives

$$\left(\frac{\partial}{\partial t} + \frac{\partial}{\partial \tau} \right) \rho(t,\tau) = -b(t)\rho(t,\tau).$$

The initial condition is

$$\rho(0,\tau) = f(\tau),$$

where $f(\tau)$ is a given age distribution at time $t = 0$. The boundary condition is due to the birth rate with delay $h > 0$ (pregnancy duration):

$$\rho(t,0) = \int_0^\infty a(\tau)\rho(t-h,\tau)d\tau.$$

5.2. *INTERACTION OF TWO SPECIES*

The logistic growth model (5.1) has been generalized to the case of two species interacting in a restricted way. Suppose there are two populations, prey and predator (or host and parasite), with numbers x_1 and x_2, respectively. Usually it is assumed that the decay mechanism for each population can be expressed by reduction of its decay rate by a factor proportional to the size of the other population. Predator–prey interactions abound in the biological world, and are one of the most important topics in theoretical ecology. The study of predation has a long history, beginning with the work of Lotka and Volterra and continuing to be of interest today. In most ecological models the growth rate of a species does not only depend on the instantaneous population size, but also on the past history of the population. E.g., in the predator–prey model the loss of prey by predation will affect the growth rate of predators at a future time.

The first predator–prey model with aftereffect was proposed by Volterra [502]. A coupled system for two interacting species has been described in [502]:

$$\left.\begin{aligned} \dot{x}_1(t) &= a_1 x_1(t) - a_2 x_1(t) x_2(t) - a_3 x_1^2(t), \\ \dot{x}_2(t) &= -a_4 x_2(t) + a_5 x_1(t-h) x_2(t-h). \end{aligned}\right\} \tag{5.3}$$

Here, $a_i > 0$ are constants, and the delay $h > 0$ is the average time between death of a prey and birth of a subsequent number of predators.

The model (5.3) has been modified to take into account predator–prey contacts at previous moments, environmental inhomogeneity, and rate of interspecial growth and death, competition, age structure, etc. In such cases we often obtain equations with distributed unbounded delay. E.g., the following predator–prey model has been studied in [207]:

$$\left.\begin{aligned} \dot{x}_1(t) &= [a_1 - a_2 x_2(t) - a_3 \int_0^\infty x_1(t-s) dG_1(s)] x_1(t), \\ \dot{x}_2(t) &= -[a_4 - a_5 \int_0^\infty x_1(t-s) dG_2(s)] x_2(t). \end{aligned}\right\}$$

In these equations, the functions G_i characterize the influence of the previous prey population on the present predator–prey state.

5.3. *POPULATION DYNAMICS MODEL OF N INTERACTING SPECIES*

Consider the population dynamics model in which it is assumed that the rate of change of each species depends on the past history as well as the dispersion rates. The latter are caused by environmental inhomogeneity [128]. There is no immigration or emigration. Let $u_i(t,x)$ be the population density of species i at time t and location $x \in D \subset \mathbf{R}^n$, $i = 1, \dots, N$. Let d_i measure the dispersion rate; it may depend on $t, x, u_i(t,x)$. Then we obtain the system of RDEs [301, 302]

$$\frac{\partial u_i(t,x)}{\partial t} = \sum_{k=1}^n \frac{\partial}{\partial x_k}\left[d_i \frac{\partial u_i(t,x)}{\partial x_k}\right] + b_i(u_i(t,x)) \int_0^\infty \sum_{j=1}^N u_j(t-s,x) dK_{ij}(s),$$

$$(t,x) \in [t_0, \infty) \times D, \quad i = 1, \dots, N,$$

with boundary conditions

$$\frac{\partial}{\partial \nu} u_i(t,x) = 0 \text{ on } [t_0, \infty) \times \partial D,$$

and initial conditions

$$u_i(t,x) = \phi_i(t,x) \quad \text{on} \quad (-\infty, t_0] \times D.$$

Here, $\partial/\partial\nu$ is the outward normal derivative, the functions b_i are strictly increasing with $b_i(0) = 0$, and K_{ij} are functions of bounded variation.

The model of N interacting species in a random medium has been considered [49, 307], using the Itô's equation

$$dx(t) = [Ax(t) + B(x_t)]dt + C(x_t)d\xi(t),$$

where A is a constant matrix and B, C are bounded linear functionals.

5.4. *COEXISTENCE OF COMPETITIVE MICRO-ORGANISMS*

The following model of competing micro-organisms surviving on a single nutrient and with delays in birth and death processes has been described in [173, 186–188]:

$$\left.\begin{aligned} \dot{x}_0(t) &= 1 - x_0(t) - x_1(t)f_1(x_0(t)) - x_2(t)f_2(x_0(t)), \\ \dot{x}_1(t) &= [f_1(x_0(t-h_1)) - 1]x_1(t), \\ \dot{x}_2(t) &= [f_2(x_0(t-h_2)) - 1]x_2(t). \end{aligned}\right\}$$

Here x_0 is the nutrient concentration, x_1, x_2 are the concentrations of competing micro-organisms, $h_i > 0$ are (constant) delays, and $f_i(0) = 0$, $\dot{f}_i(x) > 0$ for $x > 0$.

The stability of a periodic solution of this model has been studied in [186].

5.5. *CONTROL PROBLEMS IN ECOLOGY*

In each of the above mentioned models we can introduced a control $u(t)$, taking into account some purposeful action of various factors on the system. For example, consider the action of pesticides, and assume these act only on preys. Using the model (5.3), this action can be described by the equations

$$\left.\begin{aligned} \dot{x}_1(t) &= [a_1 - a_2x_2(t) - a_3x_1(t) - b_1(t)u(t)]x_1(t), \\ \dot{x}_2(t) &= -a_4x_2(t) + a_5x_1(t-h)x_2(t-h), \quad b_1 \geq 0. \end{aligned}\right\}$$

If pesticides act only on predators, then a control appears at the second equation (5.3):

$$\left.\begin{aligned} \dot{x}_1(t) &= [a_1 - a_2x_2(t) - a_3x_1(t)]x_1(t), \\ \dot{x}_2(t) &= -a_4x_2(t) + a_5x_1(t-h)x_2(t-h) - b_2(t)x_2(t)u(t), \quad b_2(t) \geq 0. \end{aligned}\right\}$$

If pesticides act upon both species, then a control appears in both equations in (5.3).

Finally, it is also possible that pesticides have, next to an instantaneous effect, a prolonged effect by means of a residuum $R(t)$. Then the behaviour of the predator–prey system can be modelled by the equations

$$\left.\begin{aligned} \dot{x}_1(t) &= [a_1 - a_2 x_2(t) - a_3 x_1(t) - b_1 u(t) - a_6 R(t)] x_1(t), \\ \dot{x}_2(t) &= [-a_4 - b_2(t) u(t) - a_7 R(t)] x_2(t) + a_5 x_1(t-h) x_2(t-h), \\ \dot{R}(t) &= -a_8 R(t) + u(t), \quad a_i \geq 0, \quad b_i(t) \geq 0. \end{aligned}\right\}$$

For the model of the production of white blood cells, optimal control problems for a Lotka – Volterra system have been considered in [284.2] and [79].

5.6. *CONTROL PROBLEMS IN MICROBIOLOGY (CHEMOSTAT MODELS)*

Certain delay models are used to control processes of microbiological growth of cells and production of a useful product. We will consider one of them, describing the continuous reproduction of micro–organisms, the production of ferments, the degradation of wastes, etc. The process here discussed is as follows (Fig. 2.5.1).

Bacteria are introduced into a vessel with an entrance for nourishing substances and an entrance for extraction of resulting products. The bacteria consume the nourishing substances, reproduct, and produce at a certain moment of time some quantity of the resulting product.

Dynamics of microbial competition can be described in the framework of the chemostat theory [462]. It allows us to take into account such complicated phenomena as interactions with more trophic levels, spatial heterogene ity, size–structured species, diffusion etc. The chemostat is a piece of laboratory apparatus used to culture microorganisms.

Under some conditions the process can be described by the bilinear delay model [284.2]

$$\left.\begin{aligned} \dot{x}_1(t) &= \gamma(t) x_1(t) - u(t) x_1(t) - \beta x_1(t-h), \\ \dot{x}_2(t) &= \gamma(t) \alpha^{-1} x_1(t) - u(t) x_2(t) + b u(t). \end{aligned}\right\} \tag{5.4}$$

The first equation is the balance equation of biological substrate, the second equation characterizes the production of resulting mass by the bacteria. Here,

$x_1(t)$ is the volume of microbiological substance;
$x_2(t)$ is the volume of the resulting product;

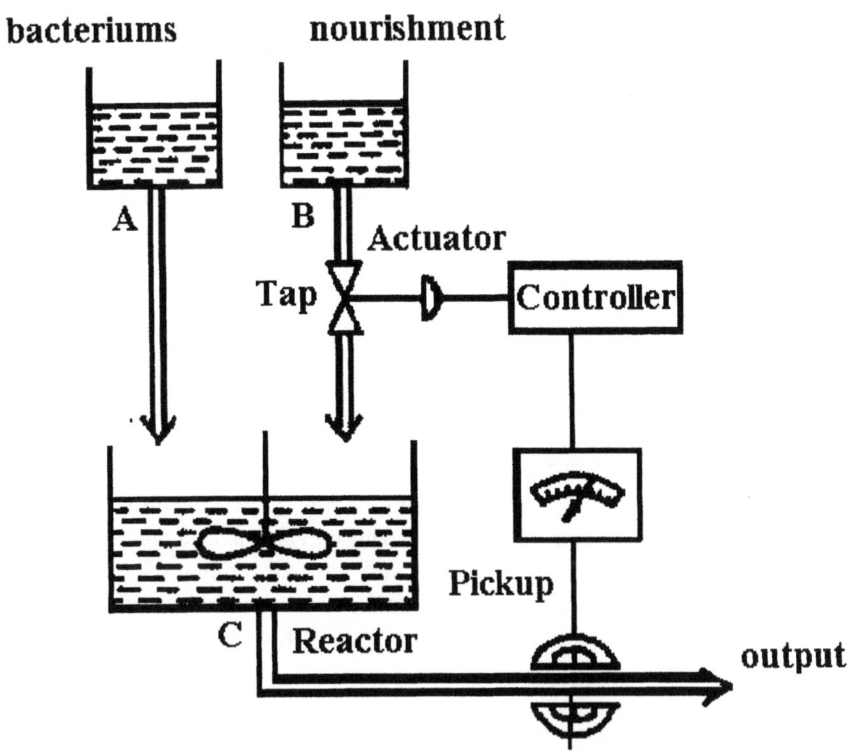

Figure 2.5.1. Biological reactor

$u(t)$ is the volume of nourishing environment in the vessel;

$\gamma(t)$ is the rate of biological growth;

$x_1(t-h)$ accounts for the loss of bacteria of great vitality during a finite time h;

β and b are constants in the model;

α is the rate of growth of the useful product.

Models like (5.4) can be used in the process of biological clearance of sewage, when dirty water goes into a vessel with active substances that come to react with the contaminator. The rates of supply of dirty water and active substances are controlled. As a result of the reaction we obtain clean water, and a residue of biopolymers. The values of the delays in this model depend on the rate of mixing, temperature, density etc.

5.7. *NICHOLSON BLOWFLIES MODEL*

To describe the dynamics of Nicholson's blowflies, the following equation has been used in [227]:

$$\dot{x}(t) = -ax(t) + bx(t-h)e^{-\gamma x(t-h)}, \quad t \geq t_0.$$

Here, $x(t)$ is the size of the population at time t, b is the maximum daily production rate of eggs per capita, γ^{-1} is the size at which the production rate of the population is maximal, a is the daily adult death rate per capita, and h is the lifetime.

Some of Nicholson's experiments on the Australian sheep blowfly praude examples of cycles with a period comparable in magnitude to the generation time. The mechanism which causes these cycles is assumed to be the delay between the change in fecundity and the subsequent change in the rate of recruitment to the adult population. A number of models have been developed, all showing cycling behaviour as a result of this delay [227, 399, 500].

5.8. *HELICAL MOVEMENT OF TIPS OF GROWING PLANTS*

The moving of the tip of a growing plant (especially of the sunflowers) has been modelled in [303, 337, 466] by the equation

$$\ddot{x}(t) + \alpha\dot{x}(t) + \beta f(x(t-h)) = 0.$$

Here α, β, h are positive constants, and $xf(x) > 0$, if $x \neq 0$.

5.9. *GRAZING SYSTEM*

The degradation of a grass biomass in a forested grassland is a consequence of many factors, e.g. overgrazing by ungulates, cattle migrating from plains into the grassland during the draught period, cutting of grass for food, fodder and fuel, etc. Amount of grass biomass at a given instant depends upon its growth grazing rate, both in the past and present. Dynamics of grass biomass and cattle population can be described using the delay concept of predator–prey models. Let $G(t)$ and $U(t)$ be positive functions representing the density of grass biomass and cattle population. Then [26, 310, 422]

$$\frac{dG}{dt} = Gg(G,U),$$

$$\frac{dU}{dt} = U\left[-\gamma - f(U) + \beta\int_{-\infty}^{t} w(t-s)h(G(s))ds\right],$$

where $g(G,U)$ denotes the specific growth rate of grass biomass; $\gamma > 0$ is the natural death rate of the cattle population; $f(U)$ is the self–inhibition rate of the cattle population; $h(G)$ is the grass response to cattle; $w(t)$ is a nonnegative continuous function called the weight function (or kernel); $\beta > 0$ is the delay coefficient.

6. AFTEREFFECT IN MEDICINE

6.1. *MATHEMATICAL MODELS OF THE SUGAR QUANTITY IN BLOOD*

FDEs can be efficiently used to describe various processes in living organisms. Thus in [126] propagation of electrical pulses on a ring of cardiac tissue was modelled by a single integral–delay equation. Various hereditary models have been proposed in[1] to describe the functioning of thyroid gland the system of maintaining the sugar level in blood, and blood production. Certain parameters in these models can be regulated (temperature, diet, drugs, etc.). E.g. the control model for the sugar level in blood has the form[1]

$$\left.\begin{aligned}
\dot{x}_1(t) &= a_1\{a_2x_4(t) + a_3[a_2x_4(t) - a_4x_2(t)] - a_5x_1(t-h)\}x_1(t),\\
\dot{x}_2(t) &= a_6\{a_2x_4(t) + b_1u(t) - a_7[a_2x_4(t) - a_5x_1(t)] - a_4x_2(t)\}x_2(t),\\
\dot{x}_3(t) &= a_8\{a_5x_1(t) + b_2u(t) + a_9[a_5x_1(t) - a_4x_2(t)] - a_{10}x_3(t)\}x_3(t),\\
\dot{x}_4(t) &= a_{11}\{1 + u(t) + a_{12}[1 - a_4x_2(t)] - a_2x_4(t)\}x_4(t).
\end{aligned}\right\}$$

Here, $x_1(t)$ is the amount of insulin produced by the pancreas, $x_2(t)$ is the amount of active insulin in the blood, $x_3(t)$ is the total amount of insulin

[1]Shvitra, D.I., Physiology systems dynamics, Mikslas, Vilnius, 1989

in the blood, $x_4(t)$ is the amount of sugar (all at time t); a_2, a_4, a_5, a_{10} are the averages of these amounts; the delay h characterizes the finite time needed for production of insulin, and a_1 is the rate of insulin production; a_6, a_8, a_{11} reflect the increase of insulin, total amount of insulin and sugar in the blood; finally, $b_1 \geq 0, b_2 \geq 0$, a_3, a_7, a_9, a_{12} are feedback coefficients. The control $u(t)$ is fulfilled by choice of a diet, and may affect the amount of sugar in the blood.

We also mention the Mackay – Glass model of blood production:

$$\dot{x}(t) = \frac{ax(t-h)}{1 - x^2(t-h)} - bx(t).$$

A numerical treatment of this equation has been given in [176], in which also the appearance of chaotic attractors has been shown. In [222], an analysis of the blood control mechanism is based on the equation

$$\dot{x}(t) = \frac{kx^m(t-h)}{1 + k_0 x^n(t-h)} - k_1 x(t).$$

6.2. *MODEL OF ARTERIAL BLOOD PRESSURE REGULATION*

In [206], FDEs have been used to describe arterial pressure control and mechanisms for the origin of hypertension. In it, for the first time these phenomena were treated as resulting from the stress–strained vascular state, measured by numerous receptor systems. It proved possible to completely re–evaluate well–known experimental and clinical observations, and to design an orderly concept of them. The cardiac vacular system is presented in Fig. 2.6.1, where A and B are arterial and venous vessels connected with each other, R is the peripheral resistance, Q is the rate of blood flow from A to B, and Q_h is the productivity of the heart. The rate of the incoming liquid is Q_i, that of the outgoing Q_o.

Let $V_a(t)$ and $V_b(t)$ be the stressed arterial and venous volume. Then

$$\begin{aligned} \dot{V}_a(t) &= k_1 k_b V_b(t) - k_4 k_a V_a(t) + \\ &+ f(t) - [k_a V_a(t) - k_b V_b(t)] \cdot [k_2 - k_3 V_a(t-h)]^{-1}, \end{aligned} \tag{6.1}$$

$$\dot{V}_b(t) = Q_i + [k_a V_a(t) - k_b V_b(t)] \cdot [k_2 - k_3 V_a(t-h)]^{-1} - k_1 k_b V_b(t). \tag{6.2}$$

The constants k_i depend on the characteristics of the cardiac vascular system, and h is the delay of baroreceptors. The function f in (6.1) is equal to $t \mapsto V_a(t - h_1)$ (where h_1 is the delay in the aldosterone circuit) if the aldosterone circuit is taken into account, and $f = 0$ otherwise.

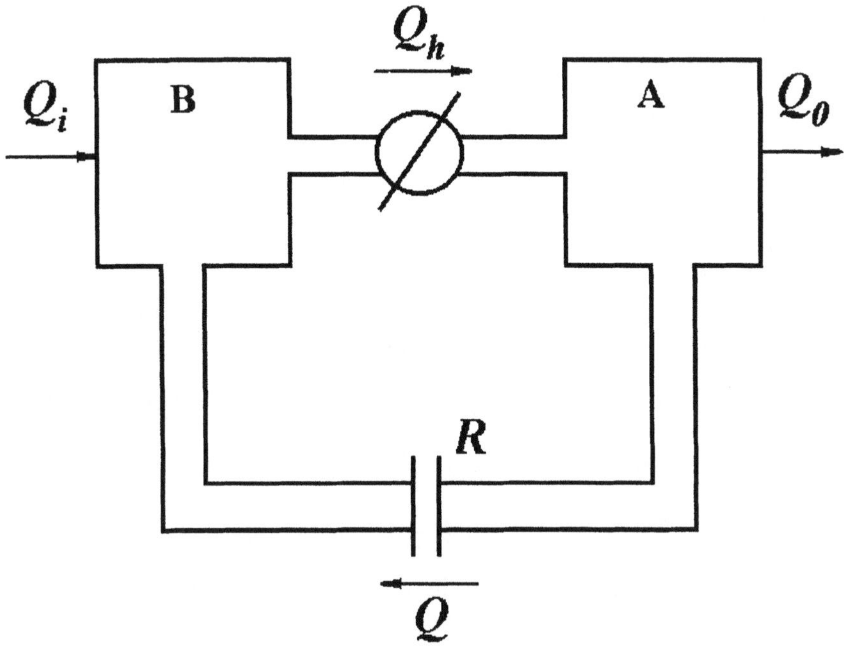

Figure 2.6.1. Scheme of the cardiac vasular system

More detailed models of arterial blood pressure regulation, allowing one to take into account features neglected in (6.1) and (6.2), have been considered in [206]. However, these models contain a larger number of parameters to be identified, and hence in actual applications their significance is less.

6.3. *CANCER CHEMOTHERAPY*

There are some papers devoted to the effect of antitumor drugs on the kinetics of the cell cycle. In them it is assumed that the drugs act instantaneously, and without aftereffect. However, in reality, to determine the time–varying effects of drugs on the cell cycle one has to take into account that the action of drugs has finite duration.

We will consider the interaction process between drugs and tumor cells under the following assumptions:

1) the entire cell population grows at a constant rate α;

2) the time for a cell cycle remains constant.

It is known [51] that the cell cycle consists of three basic phases, and that the number of cancer cells doubles every two months, on the average. The dynamics of the three phases of the cell cycle can be represented by the following bilinear system of delay differential equations:

$$\left.\begin{aligned}
\dot{x}_1(t) &= \alpha[a_3 + u_3(t-h_3)]x_3(t-h_3) - [a_1 + u_1(t-h_1)]x_1(t-h_1),\\
\dot{x}_2(t) &= [a_1 + u_1(t-h_1)]x_1(t-h_1) - [a_2 + u_2(t-h_2)]x_2(t-h_2),\\
\dot{x}_3(t) &= [a_2 + u_2(t-h_2)]x_2(t-h_2) - [a_3 + u_3(t-h_3)]x_3(t-h_3).
\end{aligned}\right\}$$

Here, $x_i(t)$ is the number of cells at time t, the function $t \mapsto a_i + u_i(t)$ is the amount of cells changing at phase i. The effect of drugs at time t is represented by controls $u_i(t)$ (one for each phase), and α is a constant of the cell cycle. For a model without delay (i.e., $h = 0$) α =1.6 (see [51]).

6.4. *MATHEMATICAL MODELS OF LEARNING*

The following model has been proposed to describe the behaviour of the central nervous system in a learning process [452]

$$\dot{x}(t) = K[x(t) - x(t-1)][N - x(t)], \tag{6.3}$$

$$x(t) = 0 \quad (-1 \le t < 0), \quad x(0) = x_0. \tag{6.4}$$

Here, K and N are positive constants, $0 < x_0 < N$. Another learning model has been proposed in [216.2, 216.3], using a system of nonlinear RDEs with discrete delays.

6.5. *MATHEMATICAL MODELS IN IMMUNOLOGY AND EPIDEMIOLOGY*

FDEs can be efficiently used in the simulation of immune and epidemic process. Here we will describe the model of virus disease [355].

Let $V(t)$ be the number of viruses, $C(t)$ the number of plasmacytes producing antibodies, $m(t)$ the relative amount of damages tissue, and $F(t)$ the amount of antibodies. Then the following system of FDEs holds:

$$\left.\begin{aligned} \dot{V}(t) &= [\beta - \gamma F(t)]V(t), \\ \dot{C}(t) &= \alpha\xi(m(t))F(t-h)V(t-h) - \mu_C[C(t) - C_0], \\ \dot{F}(t) &= \rho C(t) - \eta\gamma F(t)V(t) - \mu_f F(t), \\ \dot{m}(t) &= \delta V(t) - \mu_m m(t). \end{aligned}\right\} \tag{6.5}$$

Here, C_0 is a constant level of plasmacytes, the delay h is the time elapsing from the beginning of the stimulation of the lymphocytes to the beginning of the production of antibody mass, and $\xi(\cdot)$ is a nonincreasing function. All other coefficients in (6.5) are constants.

The following RDE with delay depending on the unknown function was used [120] to model epidemics:

$$\dot{x}(t) = ax(t - h(t, x(t))).$$

To model infection diseases, the following equation was used in [315]:

$$\dot{x}(t) = \int_0^t K(t-\tau)\dot{x}(\tau)d\tau - f(t).$$

Mathematical models of infectious diseases, antivirial immune and antibacterial response applied to the study of hepatitis B, influenza A, infectious bacterial pneumonia, and mixed infections were developed in [355]. These mathematical models are formulated as systems of FDEs. More complicated model consisting of 10 FDEs to describe the basic immune processes operating during acute viral infection with interactions was considered in [54, 55].

6.6. *MODEL OF THE HUMAN IMMUNODEFICIENCY VIRUS (HIV) EPIDEMIC*

To study the dynamics of HIV, which is the etiological agent for AIDS (acquired immunodeficiency syndrome), the following model has been proposed in [92]:

$$\left.\begin{aligned}
\dot{S}(t) &= \Lambda - B(t) - S(t),\\
\left(\frac{\partial}{\partial t} + \frac{\partial}{\partial \tau}\right) i(t,\tau) &= -[1+\alpha(\tau)]i(t,\tau),\\
i(t,0) = B(t) &= S(t)C(T(t))\frac{W(t)}{T(t)},\\
T = I + S, \quad I(t) &= \int_0^\infty i(t,\tau)d\tau,\\
W(t) &= \int_0^\infty \lambda(\tau)i(t,\tau)d\tau,\\
\dot{A}(t) &= \int_0^\infty \alpha(\tau)i(t,\tau)d\tau - (1+\nu)A(t).
\end{aligned}\right\} \tag{6.6}$$

This model was created to mimic the dynamic of HIV in a homogeneously mixed (male) homosexual population. In this population, three groups are distinguished: S (uninfected, but susceptible), I (HIV infected), and A (fully developed AIDS disease). It is supposed that members of I and S are sexually active and choose their partner at random, while members of A are sexually inactive. Also, the following notation is used: t is time, τ is time elapsed from the moment of infection, Λ is the (constant) rate of growth of the sexually active population, $B(t)$ is the number of new cases of infection per unit time, $i(t,\tau)$ is the infection–age density, so that the chance that a randomly selected partner was infected τ moments ago (has infection age τ) is $i(t,\tau)/[I(t)+S(t)]$, the function $\tau \mapsto \alpha(\tau)$ is the rate with which the infected persons with infection age τ stop being sexually active due to the disease, $C(T(t))$ is the average number of sexual contacts an average person has in unit time, ν is the average death rate of a person with AIDS, and $\tau \mapsto \lambda(\tau)$ is a given nonnegative function.

Another model for HIV in a homogeneously mixed single–gender group with distributed waiting times has been described in [91–93], using equations with distributed delay:

$$\left.\begin{aligned} \dot{S}(t) &= \Lambda - B(t) - \mu S(t), \\ I(t) &= I_0(t) + \int_0^t B(s)e^{-\mu(t-s)}P(t-s)ds, \\ A(t) &= A_0(t) + A_1 e^{-\varepsilon t} + \int_0^t d\tau \int_0^\tau B(s)e^{-\mu(t-s)}[-\dot{P}(\tau - s)e^{-\varepsilon(t-\tau)}]ds. \end{aligned}\right\} \tag{6.7}$$

Here the functions $I_0(\cdot), A_0(\cdot)$ and constants A_1, ε are given; μ is the rate with which sexual activity is stopped; $P(s)$ is the proportion of individuals infected at time t that are still alive at time $s+t$. The other notation in (6.7) is as in (6.6).

In [273] the authors suggest a model for describing the risk of infection by HIV as a function of time and an individual's history of sexually transmitted disease. The equation describing the epidemic progress is:

$$\frac{\partial H(t,k)}{\partial t} = \int_0^\infty \beta(C,k)P(C,k)\frac{N(k)-H(t,k)}{N(k)} \cdot \frac{H(t,C)}{N(C)}dC - \mu H(t,k).$$

It is an integro–differential equation of Barbashin type without deviation of time argument. Here $H(t,k)dk$ is the number of people with rate between k and $k+dk$ which are infected with HIV at time t; $\beta(C,k)$ is the probability that partnership between an uninfected person of rate k and an infected person of rate C results in transmission of the infection; $P(C,k)dCdk$ is the number of partnerships formed in unit time between people with rate between C and $C+dC$ and those with rate between k and $k+dk$. The number of people with rate between k and $k+dk$ is

$$N(k)dk = \frac{1}{k}\int_0^\infty P(C,k)dCdk.$$

Hence, $\int_0^\infty P(C,k)dC$ is the total number of partnerships formed by people with rate between k and $k+dk$. The constant μ is the rate with which sexual activity is stopped.

6.7. *MODEL OF SURVIVAL OF RED BLOOD CELLS*

A model for the survival of red blood cells in an animal has described in [511] by the equation

$$\dot{x}(t) = -ax(t) + be^{-\gamma x(t-h)}, \quad t \geq t_0,$$

where $x(t)$ is the number of red blood cells at time t, a is the probability of death of a red blood cell, $b, \gamma > 0$ are constants related to the production of red blood cells per unit time, and the delay $h > 0$ is the time required to produce a red blood cells.

Other model for concentration $x(t)$ of blood cells at time t was proposed in [358]

$$\dot{x}(t) = -ax(t) + f(x(t-h)), \quad f(x) = \frac{\alpha x}{1 + x^\beta}.$$

In this model it is assumed that the cells are lost from the circulation at a rate a, the flux f of the cells into the circulation from the stem cell compartment depends on $x(t-h)$ at a time $t-h$, and α, β are some positive constants.

6.8. *VISION PROCESS IN THE COMPOUND EYE*

This process has been modelled in [235] by

$$\dot{x}(t) + qx(t) + p\max\{x(\tau(t)), c\} = 0,$$

where $c < 0, p, q$ are constants.

6.9. *HUMAN RESPIRATORY SYSTEM*

Various mathematical models of the human respiratory system have been developed since 1940 to study a wide range of features of this complicated object. The corresponding phenomena referred to as periodic breathing (including Cheyne Stokes respiration and apneustic breathing) are assumed to be caused by delay in the feedback signals. This hypothesis has been studied beginning with the paper [214]. In [277] the model was proposed consisting of 5 nonlinear RDEs with discrete delays. The following types of delay are taken into account in [277]: lung to brain delay; lung to tissue transport delay; venous side transport delay from tissue to lung; lung to carotid artery delay.

6.10. *REGULATION OF GLUCOSE-INSULIN SYSTEM*

The known model of glucose–insulin system describes glucose concentration changes in blood as depending from spontaneous, insulin–independent net glucose tissue uptake, from insulin–dependent net glucose tissue uptake and from constant baseline liver glucose production. The term "net glucose uptake" indicates that changes in tissue glucose uptake and in liver glucose delivery are considered together.

Insulin plasma concentration changes are considered to depend from a spontaneous constant–rate decay, due to insulin catabolism, and from pancreatic insulin secretion. The delay term refers to the pancreatic secretion of insulin: effective pancreatic secretion (after the liver first–pass effect) at time t is considered to be proportional to the average value of glucose concentration in the b_5 minutes preceding time t.

The RDEs of the model have the form

$$\left.\begin{aligned} \frac{dG}{dt} &= -b_1 G(t) - b_4 I(t) G(t) + b_7, \\ G(t) &\equiv G_b, \quad \forall t \in [-b_5, 0), \quad G(0) = G_b + b_0, \\ \frac{dI}{dt} &= -b_2 I(t) + \frac{b_6}{b_5} \int_{t-b_5}^{t} G(s)ds, \quad I(0) = I_b + b_3 b_0, \end{aligned}\right\}$$

where G is the glucose plasma concentration; G_b is the basal (pre injection) plasma glucose concentration; I is the insulin plasma concentration; I_b is the basal (pre injection) insulin plasma comentation; b_i are some constants.

6.11. *A DISEASE TRANSMISSION MODEL*

Let a total variable population size be divided into disjoint classes of individuals who are susceptible, exposed (i.e. infected but not yet infective), and recovered with temporary immunity; with sizes denoted by $S(t), E(t)$, $I(t)$ and $R(t)$, respectively.

The equations for the model take the forms for $t > \overline{\tau}$ (see [120])

$$\left.\begin{aligned}
\dot{S}(t) &= bN(t) - dS(t) - \frac{\lambda S(t)I(t)}{N(t)} + \gamma I(t-\tau)e^{-d\tau}, \\
\dot{I}(t) &= \frac{\lambda S(t-\omega)I(t-\omega)}{N(t-\omega)}e^{-d\omega} - (\varepsilon + \gamma + d)I(t), \\
\dot{E}(t) &= \frac{\lambda S(t)I(t)}{N(t)} - \frac{\lambda S(t-\omega)I(t-\omega)}{N(t-\omega)}e^{-d\omega} - dE(t), \\
\dot{R}(t) &= \gamma I(t) - \gamma I(t-\tau)e^{-d\tau} - dR(t).
\end{aligned}\right\}$$

Here the parameter $b > 0$ is the birth rate constant (all newborns are assumed susceptible and vertical transmission is ignored), $d > 0$ is the natural death rate constant, $\varepsilon \geq 0$ is the disease related death rate constant in the infective class, and $\gamma \geq 0$ is the rate constant for recovery.

The force of infection is assumed to be of the standard type, namely $\lambda I/N$, with $\lambda > 0$ the effective per capita contact rate constant of infective individuals. Here λ is the product of the average number of contacts of an individual per unit time and the probability of transmitting the disease during one contact by an infective. Thus individuals leave the susceptible class at a rate $\lambda SI/N$. There is a constant period $\omega \geq 0$ of latency and a constant period $\tau \geq 0$ of temporary immunity.

The disease is assumed to have been in the population for at least a time $\overline{\tau} = \max\{\omega, \tau\}$, so that initial perturbations have died out.

7. AFTEREFFECT IN ECONOMY AND OTHER SCIENCES

Delay models are frequently encountered in economics and management science. To explain for price fluctuations, commodity supply fluctuations, trade cycles, etc., many authors use delay between beginning of a production decision and the delivery of goods to the market [39, 195, 312, 476]

7.1. *OPTIMAL SKILL WITH RETARDED CONTROLS*

In certain papers labour training, which is a necessary requirement for employment, is treated by the use of models with retarded controls. In such models, delays may occur because of differences between labour demanded and offered.

Let $x_1(t)$ be the fraction of labour force in training, $x_2(t)$ the fraction of unskilled labour force, and $x_3(t)$ the fraction of skilled labour force (all at time t). Then these variables can be described by the equations [443, 444]

$$\left.\begin{aligned} \dot{x}_1(t) &= \beta[-x_1(t) - c + H(u(t)) + \\ &+ \int_{-\infty}^{t} e^{-\beta(t-\tau)} h(\max\{t-\tau, u(\tau)\})\psi_1(t-\tau, u(\tau))d\tau], \\ \dot{x}_2(t) &= \beta[1 - H(u(t)) - x_2(t)], \\ x_3(t) &= 1 - x_1(t) - x_2(t). \end{aligned}\right\}$$

Here, β is a positive constant, and $h(x), x \geq 0$, is a density function such that

$$H(u) = \int_0^u h(s)ds, \quad H(\infty) = 1.$$

$H(u)$ is the proportion of new laborers that have been trained for at most u years (have untrainability index at most u), and $\psi_1(x,y) = 1$ if $x > y$ and $\psi_1(x,y) = 0$ if $x < 0$. The control action $u(t)$ is interpreted as the duration of time such that at time t any new laborer with untrainability index at most $u(t)$ will enter training. The restrictions on $x_i(t)$ and $u(t)$ are as follows:

$$x_1(t) \geq 0, \quad x_2(t) \geq 0, \quad x_1(t) + x_2(t) \leq 1, \quad u(t) \geq 0.$$

Finally, a cost functional, depending on $x_i(t)$, should represent the future output.

7.2. *OPTIMAL ADVERTISING POLICIES*

Several delay models have been proposed in which advertising expenditures are regarded as an investment in a goodwill stock $x(t)$.

Let $u(t)$ be the amount of advertising expenditures, and $t \mapsto f(t)$ a density function. Then [405]

$$\dot{x}(t) = g(u(t), x(t)) - \int_{-\infty}^{t} f(t-\tau)g(u(\tau), x(\tau))d\tau,$$

where g is the goodwill production function. The cost functional is the net profit stream, and the aim is to maximize it:

$$\int_0^T e^{-rt}R(u(t), x(t))dt + e^{-rT}S(x(T)) \to \max, \quad u \geq 0,$$

where $r > 0$ is a (constant) amount of discount, R is the amount of profit, and S is a given function.

7.3. *COMMODITY PRICE FLUCTUATIONS*

Consider a single commodity market. Let $x(t)$ be the percentual deviation of the commodity price from equilibrium value. The following equation has been proposed in [237]:

$$\dot{x}(t) = -ax(t) - b\int_0^\infty g(s)x(t-s)ds,$$

where a is proportional to the demand and b is proportional to the supply.

To explain the 4–year cycle in pork price, the following linear RDE with discrete constant delay h has been used in [312]:

$$\dot{x}(t) = k[x(t-h) - x_0].$$

A more complicated model of price fluctuation has been investigated in [39]. Let $x_1(t)$ be the current market price of the commodity, and $x_2(t)$ the current demanded price. Then [39]

$$\left.\begin{aligned} \dot{x}_1(t) &= [f_1(x_2(t)) - f_2(x_1(t-h(x_1(t))))]x_1(t), \\ x_2(t) &= \int_{-\infty}^{t} K(t-s)x_1(s)ds. \end{aligned}\right\}$$

Here the kernel K is related to previous market prices, and the functions f_1, f_2 denote the demand and supply of the commodity.

7.4. *MODEL OF THE FISHING PROCESS*

Let $P(t)$ be the average population of the resource, and $E(t)$ the fishing level. Then [297]

$$\left.\begin{aligned} \dot{P}(t) &= P(t)F(P(t)) - qH(E(t),P(t)), \\ \dot{E}(t) &= k[qH(E(t-rE(t)),P(t-rE(t))) - cE(t-rE(t))]. \end{aligned}\right\}$$

Here, $H(E,P)$ is the harvesting strategy, $F(\cdot)$ the relative growth rate, q the catchability coefficient, and c the cost per unit effort. All constants q, k, c, r are positive.

7.5. *RIVER POLLUTION CONTROL*

Let $z(t)$ and $q(t)$ be the concentrations per unit volume of biological oxygen demand (BOD) and dissolved oxygen (DO), respectively, at time t. It is assumed that the flow rate is constant, water is well mixed, and there exists $\tau > 0$ such that BOD and OD concentrations entering at time t are equal to the corresponding concentrations τ time units ago.

Using mass balance concentration, the following equations have been derived [319]:

$$\left.\begin{aligned} \dot{z}(t) &= -k_1(t)z(t)+ \\ &+v^{-1}[Q_1(m+u_1(t))+Qz(t-\tau)-(Q+Q_1)z(t)]+v_1(t), \\ \dot{q}(t) &= -k_3(t)z(t)+k_2(t)[q_0-q(t)]+ \\ &+v^{-1}[Qq(t-\tau)-(Q+Q_1)q(t)]+u_2(t)+v_2(t). \end{aligned}\right\}$$

Here, $k_i(\cdot)$ denote the BOD decay rate, the BO re–aereation rate, and the BOD deoxygenation rate; q_0 is the DO saturation concentration; Q and Q_1 are the stream flow rate and the effluent flow rate; v is the constant volume of water under consideration; $u_i(\cdot)$ are controls; $v_i(\cdot)$ are random disturbances affecting the rates of change of BOD and DO; and m is a constant.

7.6. *CONTROL OF FINANCIAL MANAGEMENT*

Consider a system in which the money can be used in two different ways: part of them can be reserved in a bank account and can be taken at any time, the other part can be used to buy the bonds in order to obtain more profit. Model of this system governed by two ODEs is given in [444]. But the important feature of this phenomenon is a delay which must be incorporated in the model of [444]. The appearance of this delay reflects that the profit can not be obtained instantaneously but after some time h. Denote $x_1(t)$ a current deposit with the interest ratio a_1 and $x_2(t)$ the total amount of bonds with the interest ratio a_2. Further, let $b \in (0,1)$ be a ratio fee to be paid to bonddealers and $u(t)$ be a buing or selling ratio. The sign of $u(t)$ corresponds buying ($u(t) > 0$) and selling ($u(t) < 0$). It is also assumed that the ratio $u(t)$ of buying or selling can not exceed a quantity $u_0 \geq |u(t)|$. As a result we have a model

$$\left.\begin{aligned} \dot{x}_1(t) &= a_1x_1(t-h)-u(t)-b|u(t)|+f(t), \\ \dot{x}_2(t) &= a_2x_2(t-h)+u(t), \end{aligned}\right\}$$

where $f(t)$ is a rate of consumption. The cost functional to be maximized is a total sum of money at the moment T, i.e. $x_1(T) + x_2(T) \to \max$ for all $u : |u(t)| \leq u_0$.

Part II

THEORETICAL BACKGROUND OF FUNCTIONAL DIFFERENTIAL EQUATIONS

CHAPTER 3

GENERAL THEORY

1. INTRODUCTION. METHODS OF STEPS

1.1. *NOTATIONS*

Let K be a subset of a set M, and f a function on M. Then $f|_K$ is a restriction of f to K.

If $x \in \mathbf{R}^n$, then $|x|$ is the Euclidean norm of x. For $f : M \to \mathbf{R}^n$, we set, by definition,

$$\| f \| := \sup_{t \in M} |f(t)|;$$

the symbol := means "equality by definition". Let us denote by $\mathbf{R}^{n\times n}$ the set of real square $n \times n$ matrices. For $A \in \mathbf{R}^{n\times n}$ we let $|A|$ be the operator norm of A:

$$|A| := \max_{x\in\mathbf{R}^n, |x|=1} |Ax|.$$

Any connected subset of $\mathbf{R}$ consisting of more than one point is called an *interval*. In what follows an interval is denoted by J, and $\mathbf{R}_+ := [0, \infty)$.

For given J and n the set of all continuous functions $F : J \to R^n$ is denoted by $C(J, \mathbf{R}^n)$. This is a linear space with respect to the usual operations over functions. If J is a closed interval then $C(J, \mathbf{R}^n)$ is a Banach space with respect to the norm $\| \cdot \|$. If the value of n is clear from the context, we write $C(J)$ or CJ.

For a measurable function $f : J \to \mathrm{R}^n$ we put

$$\| f \|_p := \left[\int_J |f(t)|^p dt \right]^{1/p} \quad (1 \le p < \infty), \quad \| f \|_\infty := \operatorname{vrai}\sup_{t\in J} |f(t)|$$

$$:= \inf_M \sup_{t\in M} |f(t)|.$$

Here $M \subset J$ is an arbitrary set with the meas $(J \backslash M) = 0$. For the functions f defined almost everywhere on J one must take such M where f is defined.

For given J, n, p $(1 \le p \le \infty)$, the set of all measurable functions $f : J \to R^n$ with $\| f|_{J_1} \|_p < \infty$ for any closed interval $J_1 \subseteq J$ is denoted

$L_p(J, R^n), L_p(J), L_pJ$. It is a linear space, and if J is closed interval then it is the Banach space with respect to the norm $\|\cdot\|_p$. In addition, it is assumed that the functions being distinguished only on the subset of J with the zero measure represent the same element in L_pJ.

The linear space of all continuously differentiable functions $f : J \to R^n$ is denoted $C^1(J, R^n)$ (or $C^1(J), C^1J$). If J is a closed interval then it is a Banach space with respect to the norm

$$\| f \|_{C^1} := \| f \| + \| \dot{f} \| .$$

The spaces $C^k(J, \mathbf{R}^n)$ $(k = 2, 3, \ldots)$ are similarly defined.

The linear space of all functions $f : J \to \mathbf{R}^n$ satisfying Lipschitz condition on every closed interval $J_1 \subseteq J$ is denoted $\mathrm{Lip}(J, \mathbf{R}^n)$ (or $\mathrm{Lip}(J)$, $\mathrm{Lip}J$). If J is a closed interval, then it is a Banach space with respect to the norm

$$\| f \|_{\mathrm{Lip}} := \| f \| + \| \dot{f} \|_\infty .$$

1.2. *CAUCHY PROBLEM FOR FDES*

The *Cauchy problem* (also called *initial problem* or *basic initial problem*) for a first order FDE is to find the solution of this equation subjected to a given initial function and initial value. Consider, for example, RDE (1.1.7) with finite aftereffect, and let, for some $t_0 \in \mathbf{R}$, the function $(t, u) \mapsto f(t, u)$ be defined for all $t \in [t_0, \infty), u \in C(J_t)$, where $J_t = [-h(t), -g(t)] \subset (-\infty, 0]$. The point t_0 will be called the *initial point* for the solution. Assume for the simplicity that

$$\bar{t}_0 := \inf_{t \geq t_0} [t - h(t)] > -\infty.$$

(In application we usually get $\bar{t}_0 = t_0 - h(t_0)$.) The *initial function* ϕ for Eq. (1.1.7) with a prescribed initial point t_0 is given on the *initial interval* $[\bar{t}_0, t_0)$. If $\bar{t}_0 = t_0$ then the initial interval $[\bar{t}_0, t_0)$ is empty and the initial function should not be prescribed. (This is a *Cauchy problem without previous history.*) Also the *initial value* $x(t_0)$ of the solution must be given.

It is essential that the solution $x(t)$ must be constructed in the direction of increasing t, i.e. on some interval J_x with the left end $t_0 \in J_x$. Here, x is considered as a prolongation of the initial function, that is $x(t+\theta) := \phi(t+\theta)$ for $t+\theta < t_0$. Usually the initial value of the solution is included in the initial function, i.e. the latter is assumed to be prescribed on the interval $[\bar{t}_0, t_0]$ and $\phi(t_0) = x(t_0)$. We shall also do the same without further mentioning this.

In reality, t_0 does not coincide always with the moment of the beginning of the process under investigation. Generally speaking, the hereditary differential law under consideration turns out to be applicable only from some stage onwards in the evolution of the process, and t_0 cannot be taken before this, but may be taken at a later stage.

The Cauchy problem for NDE (1.1.8), (1.1.9) is analogously stated. We shall discuss some further refinements in Sec. 3.

1.3. *STEPS METHOD FOR FDES OF RETARDED TYPE (RDES)*

For discrete delays, very often (practically always) to solve the Cauchy problem on finite interval (practically not too large) one can apply a steps method. Let us explain this method using the scalar equation with one discrete delay

$$\dot{x}(t) = f(t, x(t), x(t-h)), \quad t \geq t_0, \quad h = \text{const} > 0. \tag{1.1}$$

Let $f : [t_0, \infty) \times \mathbf{R}^2 \to \mathbf{R}$ be a continuous function satisfying a Lipschitz condition with respect to the second argument. For Eq. (1.1) the initial function ϕ must be assigned on the interval $[t_0 - h, t_0]$ and we suppose it to be continuous.

If $t \in [t_0, t_0 + h]$ (it is the first step), then $t - h \in [t_0 - h, t_0]$. Therefore RDE (1.1) converts into ODE

$$\dot{x}(t) = f(t, x(t), \phi(t-h)), \quad t_0 \leq t \leq t_0 + h.$$

Solving this equation under the initial value $x(t_0) = \phi(t_0)$, we get the desired solution on the interval $[t_0, t_0 + h]$. In particular, the value $x(t_0 + h)$ is defined also. If now, $t_0 + h \leq t \leq t_0 + 2h$ (it is the second step), then $t - h \in [t_0, t_0 + h]$, and so $x(t - h)$ is known from the first step. Hence Eq. (1.1) for $t_0 + h \leq t \leq t_0 + 2h$ once again converts into ODE, which with the known initial value $x(t_0 + h)$ defines the solution x on this interval. After that we shall consider the interval $[t_0 + 2h, t_0 + 3h]$, etc. In this way the solution can be obtained for arbitrarily large t (theoretically, for the whole semiaxis $[t_0, \infty)$).

For the simplest case when the righthand side of Eq. (1.1) does not contain $x(t)$, at each step the solution is reduced to the integration of a given function.

An interesting problem is the problem about the degree of smoothness of the constructed solution. Let for simplicity the function f has continious derivatives of all orders. Then it follows from Eq. (1.1) that $x \in C^1[t_0, \infty)$. But even for an arbitrarily smooth function ϕ we get in general case

$$\dot{x}(t_0^+)(= f(t_0, \phi(t_0), \phi(t_0 - h))) \neq \dot{\phi}(t_0).$$

It means that if the function x is considered as a prolongation of the function ϕ, then the composite function get a break for $t = t_0$ (a jump of the first derivative). By virtue of Eq. (1.1) from $x \in C^1[t_0, \infty)$ we get $\dot{x}|_{[t_0+h,\infty)} \in C^1[t_0 + h, \infty)$, i.e. $x|_{[t_0+h,\infty)} \in C^2[t_0 + h, \infty)$ (the space C^2 is introduced in much the same way as we introduced C^1). But even for arbitrarily smooth ϕ, we get in general $\ddot{x}(t_0 + h^+) \neq \ddot{x}(t_0 + h^-)$. Further, using Eq. (1.1), we derive that $x|_{[t_0+2h,\infty)} \in C^3[t_0 + 2h, \infty)$ and so on.

This property of *solutions smoothing* for the increasing t is a characteristic feature of RDE. Along with some another properties it makes RDEs to resemble partial differential equations of parabolic type. However, the reasons for this resemblance are not entirely clear.

For some very scarce cases the steps method leads us to the formulas for the exact solution of a problem. Thus, for Eq. (1.1.1) being considered on an interval $[t_0, \infty)$ with $h > 0$ and with the initial function $t \mapsto \phi(t) \equiv \phi_0 = \text{const}$ $(t_0 - h \leq t \leq t_0)$ this method gives

$$x(t) = \phi_0 \sum_{j=0}^{\infty} \frac{h^j}{j!} [t - t_0 - (j-1)h]^j H[t - t_0 - (j-1)h], \quad t_0 \leq t < \infty,$$

where H is a *Heaviside function* $(H(t) := (1+\text{sgn}t)/2)$. Here the summation is in fact taken over finite number of terms and the greater t the greater number of these terms. For $t_0 - h < t \leq t_0$, the formula gives the initial function.

Denoting the righthand side of this formula by $\phi_0 X(t - t_0; k, h)$ we can deduce the expression for the solution of Eq. (1.1.1) with arbitrary continuous initial function $\phi : [t_0 - h, t_0] \to \mathbf{R}$ as follows

$$x(t) = X(t - t_0 - h; k, h)\phi(t_0) + k \int_{t_0-h}^{t_0} X(t - \tau - 2h; k, h)\phi(\tau)d\tau$$

$$(t_0 \leq t \leq \infty)$$

where $X(\theta; k, h) = 0$ for $\theta < -h$. If $\phi \in C^1[-h, 0]$, then integrating by parts we have

$$x(t) = X(t - t_0; k, h)\phi(t_0 - h) + \int_{t_0-h}^{t_0} X(t - \tau - h; k, h)\dot{\phi}(\tau)d\tau.$$

Return to Eq. (1.1). Because at every step it is converted to an ODE, all methods for ODE's numerical solution (Runge – Kutta method etc.) can be applied to Eq. (1.1). In this case it is necessary to store the values of the solutions at the present step, since they are used at the next step. If the number of steps is not too large, this storage (filling the memory of a computer) can be avoided by a simple method proposed, apparently, by R. Bellman.

Assume, for simplicity, that $t_0 = 0$ and designate

$$x_j(t) = x(t + (j-1)h),\ 0 \le t \le h,\ j = 1, 2, \ldots,\ x_0(t) = \phi(t-h).$$

Suppose we would like to perform N steps. Then the solution of Eq. (1.1) is reduced to solving a sequence of N systems of ODEs, the j-th of which has the form

$$\dot{x}_i(t) = f(t + (i-1)h,\quad x_i(t), x_{i-1}(t)),\quad 0 \le t \le h,\quad i = 1, \ldots, j$$

with initial conditions

$$x_i(0) = x_{i-1}(h),\quad i = 1, \ldots, j$$

which are taken from the solution of the $(j-1)$th system (if $j = 1$ then $x_1(0) = \phi(0)$). So, only these initial conditions must be stored. Having reached $j = N$ we got the solution x on the whole interval $[0, Nh]$.

A similar steps method is applied to the more general equation (1.1.4) for continuous positive functions h_i. However, one of them may be identically zero. Assume for simplicity that all functions $t \mapsto t - h_i(t)$ tend to infinity as $t \to \infty$. Then successive steps are performed on the intervals $[t_0, t_1]$, $[t_1, t_2]$,..., where

$$t_{j+1} = \inf\{t : \max_i [t - h_i(t)] > t_j\},\quad j = 0, 1, \ldots$$

where the index i with $h_i(t) \equiv 0$ is not taken into consideration. At every step RDE is converted to an ODE with regard to all previous steps. The totally of all steps leads, under natural assumptions about f and ϕ, to the solution construction on semiaxis $[t_0, \infty)$, provided that this solution does not tend to infinity for a finite t. The latter is possible if the Lipschitz condition, mentioned above for Eq. (1.1), will be only locally fulfilled.

More general RDEs for which steps method can be applied are

$$\dot{x}(t) = f(t, x(t), x_t). \tag{1.2}$$

The meaning of x_t is the same as in Eq. (1.1.7), where the function g defining the right end of the interval J_t is continuous and positive.

1.4. *STEPS METHOD FOR FDES OF NEUTRAL TYPE (NDES)*

Steps method for NDEs can be applied in much the same way as for RDEs. Let for example the defining scalar equation be

$$\dot{x}(t) = f(t, x(t), x(t-h), \dot{x}(t-h_1)),\quad t_0 \le t < \infty;\ h, h_1 = \text{const} > 0.$$

Sufficient conditions to apply steps method for this equation are the following: the continuous function $f : [t_0, \infty) \times \mathbf{R}^3 \to \mathbf{R}$ satisfies the Lipschitz condition with respect to the second argument, and initial function $\phi \in C^1[t_0 - \max\{h, h_1\}, t_0]$. The scheme of steps method realization is the same as for Eq. (1.1), but the value of steps $\Delta t = \min\{h, h_1\}$.

Usually, when applying steps method to Eq.(1.1.5) one assumes that the zero delay is admitted in the unknown function, but not in this derivative, i. e. the equation is solved with respect to $\dot{x}(t)$.

The essential feature of solutions of the Cauchy problem for NDEs is the absence of the smoothing property mentioned in Subs. 1.3. The reason of this is clear: at every step the solution of a "proper" NDE (this remark must be done because formally every RDE is at the same time a NDE) must first be differentiated and be only then integrated.

To explain it consider the simple scalar problem

$$\dot{x}(t) = \dot{x}(t-1), \quad 0 \le t < \infty; \quad x(t) = \phi(t), \; -1 \le t \le 0$$

where $\phi \in C^1[-1, 0]$ is a given initial function. The solution is

$$x(t) = \phi(t-j) + j[\phi(0) - \phi(-1)], \quad j-1 \le t \le j, \quad j = 1, 2, \ldots .$$

Hence if $\dot{\phi}(-1) \neq \dot{\phi}(0)$ then $\dot{x}(t)$ has jumps at $t = 0, 1, \ldots$. If $\dot{\phi}(-1) = \dot{\phi}(0)$ but $\phi \in C^2[-1, 0]$ and $\ddot{\phi}(-1) \neq \ddot{\phi}(0)$ then $\dot{x}$ is continuous and $\ddot{x}(t)$ has jumps for $t = 0, 1, \ldots$, etc.

This example is generic: if the right hand side of NDE (1.1.8) and the initial function ϕ are arbitrarily smooth, then the continuous solution get, generally speaking, jumps of derivative for arbitrarily large t. The jumps would be absent if the initial function ϕ satisfies the necessary and sufficient *sewing condition*

$$\dot{\phi}(t_0) = f(t_0, \phi_{t_0}, \dot{\phi}_{t_0}).$$

Considered property along with certain another ones makes NDEs to resemble partial differential equations of hyperbolic types.

1.5. *PROBLEM FOR A PROCESS WITH AFTEREFFECT RENEWAL*

Some obstacles to solve differential equations with advanced argument (ADE) can be illustrated by the steps method. We have already mentioned that until now such equations have not appeared in applications. However, from the applied point of view, it is natural to consider the problem of renewal of a process with aftereffect given a final observation, which is equivalent to the problem of forward prolongation of solutions of ADEs.

Consider, for example, the simplest equation

$$\dot{x}(t) = x(t-1), \quad -\infty < t \le 1.$$

Given the final values $x(t) = \psi(t)$, $0 \leq t \leq 1$, where $\psi \in C^\infty[0,1]$. The problem is to renew $x|_{t<0}$. At the first step let us set $t \in [0,1)$. Then

$$x(t) = \dot{\psi}(t+1), \quad -1 \leq t < 0.$$

Hence, if $\psi(0) \neq \dot{\psi}(1)$, then x has a jump at $t = 0$

$$x(0^+) - x(0^-) = \psi(0) - \dot{\psi}(1).$$

Using distributions, subsequent steps give a formal solution of the problem

$$x(t) = \sum_{j=1}^{\infty} \sum_{i=1}^{j} [\psi^{(i-1)}(0) - \psi^{(i)}(1)] \delta^{(j-i)}(t+j) + y(t),$$

where

$$y(t) = \psi^{(j)}(t+j), \quad -j \leq t < -j+1, \quad j = 1, 2, \dots$$

and δ is the delta-function.

Consequently, in order to avoid the delta-summands one must impose the conditions

$$\psi^{(i-1)}(0) = \psi^{(i)}(1), \quad i = 1, 2, \dots$$

The number of these conditions increases with the growth of the time-interval on which the solution is to be renewed. But even if these conditions do hold, the renewal operator of a hereditary process with final observation is an unbounded one already at the first step, since it is the differentiation operator (a function can be small but with an arbitrarily large derivative). So the renewal problem described above is ill-posed, like many other inverse problems, and for its solution corresponding regularization methods must be applied.

In relation to the similar problem for NDE, the result depends on whether the equation would be "rear defined". Thus, the inverse problem for the equation

$$\dot{x}(t) = \dot{x}(t-1) + x(t-2)$$

is an ill-posed one, but for the equation

$$\dot{x}(t) = x(t-1) + \dot{x}(t-2)$$

it is well-posed, as can be proved by the steps method. The latter equation is an example of a *bilateral NDEs* (by Kamenskii's classification) which preserve their type under time inversion. Remark that the general theory of these equations appears to be rich in content has not been developed yet.

2. CAUCHY PROBLEM FOR RDEs

2.1. *BASIC SOLVABILITY THEOREM*

Here we give a widely-used version of a solvability theorem for the initial value problem for RDEs, assuming that the solution belongs to C^1 (i. e. the solution is a continuously differentable function of t) and a time of possible aftereffect is constant and finite. So we consider the problem

$$\dot{x}(t) = f(t, x_t); \quad x_t(\theta) := x(t+\theta), \quad -h \leq \theta \leq 0, \tag{2.1}$$

$$x_{t_0} = \phi, \tag{2.2}$$

where $h = \text{const} \in [0, \infty)$, $x(t) \in \mathbf{R}^n$, $t_0 \in \mathbf{R}$, $\phi : [-h, 0] \to \mathbf{R}^n$, $n \geq 1$. We emphasize that, in contrast to §1, the initial function ϕ is now assumed to be defined at the "canonical" interval $[-h, 0]$ instead of on $[t_0 - h, t_0]$, i. e. the solution $t \mapsto x(t)$ $(t \geq t_0)$ is a prolongation of the function $t \mapsto \phi(t - t_0)$ $(t_0 - h \leq t \leq t_0)$.

A function $x \in C^1(J_x)$ is said to be a *solution* of the problem (2.1), (2.2) on an interval J_x with left end $t_0 \in J_x$ if it satisfies Eq. (2.1) complemented by (2.2) at all points J_x, (i.e. $x_t(t+\theta) := \phi(t + \theta - t_0)$ for $t + \theta < t_0$) and $x(t_0) = \phi(0)$. Designate $D_h := [t_0, \infty) \times C[-h, 0]$.

Theorem 2.1 Let $\phi \in C[-h, 0]$ and a functional $f : D_h \to \mathbf{R}^n$ be continuous and satisfy in some neighbourhood of any point $(t, u) \in D_h$ a Lipschitz condition in the second argument (with Lipschitz constant depending, in general case on this point). Then there is $t_\phi (= t_{\phi, t_0, f}) \in (t_0, \infty]$ such that:

a) there exists a solution x of Pr. (2.1), (2.2) on the inerval $[t_0, t_\phi)$;

b) at any interval $[t_0, t_1] \subset [t_0, t_\phi)$ this solution is unique (more exactly, any two solutions of this problem coincide on the common interval of their existence, i.e. one of this solution is a prolongation of the other solution);

c) if $t_\phi < \infty$, then $x(t)$ has no finite limit as $t \to t_\phi^-$, i. e. $[t_0, t_\phi)$ is the maximal interval of solution's existence of Pr. (2.1), (2.2);

d) the solution x depends continuously on f, ϕ.

It means that for any $t_1 \in (t_0, t_\phi), \varepsilon > 0$ there exists $\delta > 0$ such that if f, ϕ in (2.1), (2.2) are replaced by $\overline{f}, \overline{\phi}$ where

$$\| \overline{\phi} - \phi \| < \delta, |\overline{f}(t, \psi) - f(t, \psi)| < \delta, \text{if} \quad t_0 \leq t \leq t_1, \quad \| \psi - x_t \| < \varepsilon \tag{2.3}$$

and $\overline{f}, \overline{\phi}$ satisfy the same restrictions as f, ϕ then for the solution $\overline{x}$ of the transformed problem with maximal interval of solution's existance $[t_0, t_{\overline{\phi}})$ we get

$$t_{\overline{\phi}} > t_1, \quad |\overline{x}(t) - x(t)| < \varepsilon, \quad t_0 \leq t \leq t_1. \tag{2.4}$$

(Taking $t - t_0$ as a new independent variable we can investigate the solution dependence on t_0 in much the same way as its dependence on f. This remark applies equally to subsequent solvability theorems.)

Proof of Thm. 2.1. Problem (2.1), (2.2) is equivalent to the problem of existence of continuous solution of the integro–functional equation

$$x(t) = \phi(0) + \int_{t_0}^{t} f(\tau, x_\tau)d\tau, \quad t \in J_x. \tag{2.5}$$

Let us consider the operator, defined by the righthand side of (2.5), in the metric space $\{x \in C([t_0, t_0 + \alpha], [\phi(0) - \beta, \phi(0) + \beta]), \, x(t_0) = \phi(0)\}$, where $\alpha, \beta > 0$ are sufficiently small, and apply contraction mapping principle. As a result we obtain existence and uniqueness of the solution on each sufficiently small interval $[t_0, t_0 + \alpha]$. From here it follows uniqueness of the solution on each interval where this solution does exist. Really, if there exists two such solutions x^1 and x^2, then, shifting the initial point from t_0 into the point $\inf\{t > t_0 : x^1(t) \neq x^2(t)\}$, we should obtain the contradiction. Now, uniting all the intervals $[t_0, t_1]$ $\{t_0 < t_1 < \infty\}$ on which the solution exists, we can obtain the maximal interval of the existence of the solution. This interval is open from the right, i. e. it has a form $[t_0, t_\phi)$, $t_0 < t_\phi \leq \infty$. Hence, the assertions *a)* and *b)* are proven.

Assume now that $t_\phi < \infty$ and there exists $x(t_\phi) := \lim x(t)$ as $t \to t_\phi^-$. Then the function x complemented by $x(t_\phi)$ is continuous on $[t_0, t_\phi]$. From here and continuity of the function f it follows that if equation (2.5) is valid for $t \in [t_0, t_\phi)$, then it should be also valid for $t = t_\phi$. Hence, the solution exists on $[t_0, t_\phi]$, but it contradicts the definition of t_ϕ. The assertion *c)* is proven.

For the proof of *d)* assume that $t_1 \in (t_0, t_\phi)$ and $\varepsilon > 0$ are fixed, and $\varepsilon > 0$ is so small that the function $(t, \psi) \mapsto f(t, \psi)$ is bounded and satisfies Lipschitz condition with respect to ψ in the strip $t_0 \leq t \leq t_1$, $\| \psi - x_t \| < \varepsilon$ (such $\varepsilon > 0$ can be chosen by virtue of the finite covering theorem). Prove now, that if $\delta > 0$ is sufficiently small and estimates (2.3) are valid, then the equality $|\overline{x}(t) - x(t)| = \varepsilon$ is impossible for $t \in [t_0, t_1]$. From here it follows that $\dot{\overline{x}}(t)$ is bounded and therefore, due to *c)* both relations (2.4) hold.

Assume the opposite, i. e. that $\delta > 0$ with the desired properties does not exist. Then there exist sequences $\delta^k \to 0$ $(\forall \delta^k \in (0, \varepsilon))$, $\overline{f}^k, \overline{\phi}^k$, satisfying (2.3) (where $\overline{f}, \overline{\phi}, \delta$ are replaced by $\overline{f}^k, \overline{\phi}^k, \delta^k$) such that for some $t^k \in (t_0, t_1]$ the solutions $\overline{x}^k$ of the suitable problem (2.1), (2.2) satisfy the relations

$$|\overline{x}^k(t) - x(t)| < \varepsilon \quad (t_0 \leq t < t^k), \; |\overline{x}^k(t^k) - x(t^k)| = \varepsilon. \tag{2.6}$$

But the set of the functions $\overline{x}^k : [t_0, t^k] \to \mathbf{R}^n$ is uniformly bounded and equicontinuous. Therefore, without loss of generality after passing to the subsequence we can assume that $t^k \to \bar{t} > t_0$ and $\bar{x}^k \to \bar{x}$ as $k \to \infty$ uniformly on each interval $[t_0, \tilde{t}] \subset [t_0, \bar{t})$. In addition, the function $\bar{x}$ is uniformly continuous on $[t_0, \bar{t})$ and therefore can be continuously prolonged on $[t_0, \bar{t}]$ with $|\bar{x}(\bar{t}) - x(\bar{t})| = \lim |\bar{x}^k(t^k) - x(t^k)| = \varepsilon$.

The function $\bar{x}^k$ is a solution of the problem

$$\bar{x}^k(t) = \bar{\phi}^k(0) + \int_{t_0}^{t} \bar{f}^k(\tau, \bar{x}^k_\tau)d\tau, \quad t_0 \leq t \leq t^k,$$

$$\bar{x}^k(t) = \bar{\phi}^k(t - t_0), \quad t_0 - h \leq t \leq t_0.$$

Pass here to the limit as $k \to \infty$ on any above mentioned interval $[t_0, \tilde{t}]$, using the estimate

$$\left| \int_{t_0}^{t} \bar{f}^k(\tau, \bar{x}^k_\tau)d\tau - \int_{t_0}^{t} f(\tau, \bar{x}_\tau)d\tau \right| \leq$$
$$\leq \int_{t_0}^{\tilde{t}} \left|\bar{f}^k(\tau, \bar{x}^k_\tau)d\tau - f(\tau, \bar{x}_\tau)\right| d\tau + \int_{t_0}^{\tilde{t}} \left|f(\tau, \bar{x}^k_\tau) - f(\tau, \bar{x}_\tau)\right| d\tau. \tag{2.7}$$

Both terms at the right-hand side of (2.7) tend to zero as $k \to \infty$: the former because of the uniform convergence $\bar{f}^k$ to f in the strip $t_0 \leq t \leq t_1$, $\| \psi - x_t \| < \varepsilon$, the latter due to the Lipschitz condition for f with respect to the second argument in the same strip.

So, we obtain that $\bar{x}$ satisfies equation (2.5) on $[t_0, \tilde{t}]$ with initial condition (2.2). Hence, by *c)* we have $\bar{x}(t) \equiv x(t)$ for $t_0 \leq t \leq \tilde{t}$. Since $\tilde{t} \in (t_0, \bar{t})$ is arbitrary, we get $\bar{x}(\bar{t}) = x(\bar{t})$ which is impossible. Thm. 2.1 is proven.□

2.2. *VARIANTS*

2.2.1 Application of Thm. 2.1 at each step in the steps method (Subs. 1.3) proves that the first three conclusions in this theorem are valid if the Lipschitz condition holds in the following weaker form: in some neighbbourhood of any point of D_h the inequality $|f(t, \psi_1) - f(t, \psi_2)| \leq L \| \psi_1 - \psi_2 \|$ holds for any $\psi_1, \psi_2 \in C[-h, 0]$ such that $\psi_1(\theta) \equiv \psi_2(\theta)$, $-h \leq \theta \leq -\varepsilon$. Here L and $\varepsilon \in (0, h]$ may depend on the chosen point of D_h.

If $f(t, x_t) = f(t, x(t), x(t - h_1(t)), \ldots, x(t - h_j(t))$, then this remark covers all arguments of the form $x(t - h_i(t))$ where h_i is continuous and (strictly) positive: for these arguments Lipschitz condition is not required.
2.2.2 Let a functional f be defined and satisfy the above formulated conditi-

ons only on some set $D \subset D_h$, open in D_h. Of course, one must require $(t_0, \phi) \in D$. In this case the first, second and fourth assertions of Thm. 2.1 remain valid. The third assertion contains also the posible case: $t < \infty$, $x(t)$ has a finite limit as $t \to t_\phi^-$ and

$$\lim_{t \to t_\phi^-} d((t, x_t), \partial D) = 0$$

where d means a distance and ∂D is a boundary of D. Hence for this case the maximal interval of existence of Pr. (2.1), (2.2) solution is also $[t_0, t_\phi)$.
2.2.3 The third assertion of Thm. 2.1 may be sharpened under some additional not very restrictive requirements. Thus if a functional f is bounded on every bound subset of D_h (it is always fulfilled in real examples under Thm. 2.1 conditions), then $x(t)$ is unbounded as $t \to t_\phi^-$ and $t_\phi < \infty$. Moreover, there are important cases in which $|x(t)| \to \infty$. (It is known that for ODEs, i. e. for $h = 0$, all solutions that take a prolongation only on finite t-interval possess this property.) In particular, this is a case for which the steps method is applicable in the form described in Subs. 1.3, i.e. Eq. (2.1) takes the form (1.2) and the additional restrictions concerning (1.2) are met. Examples for which under Thm. 2.1 conditions the following relations hold

$$\limsup |x(t)| = \infty, \quad \liminf |x(t)| < \infty, \quad \text{as} \quad t \to t_\phi^- < \infty$$

do not appear to have significance in applications.

A solution x of Pr. (2.1), (2.2) with $|x(t)| \to \infty$ as $t \to t_\phi^- < \infty$, characterizes a *peaking regime*. Really, such a *blowing up* solution means either a catastrophic evolution of the process or the necessary to change in principle from some time onwards the mathematical model of the phenomenon.
2.2.4 What are the additional assumptions such that under Thm. 2.1 conditions would be $t_\phi = \infty$? In particular, $t_\phi = \infty$ if the functional f is *sublinear*, i.e.

$$|f(t, \psi)| \leq A(t) \parallel \psi \parallel + B(t), \quad \forall (t, \psi) \in D_h \tag{2.8}$$

where $A, B \in C[t_0, \infty)$. The inequality (2.8) takes place if f satisfies on every bounded set $M \subset D_h$ the Lipschitz condition with respect to the second argument with Lipschitz constant, depending only on $\sup_{(t,u) \in M} t$.

Rather more general than (2.8) is *Nagumo type condition*

$$|f(t, \psi)| \leq \Phi(t, \parallel \psi \parallel), \quad \forall (t, \psi) \in D_h \tag{2.9}$$

where $\Phi : [t_0, \infty) \times [0, \infty) \to (0, \infty)$ is continuous, nondecreasing on t and

$$\int_0^\infty \frac{ds}{\Phi(t,s)} = \infty, \quad t_0 \le t < \infty.$$

Condition (2.9) is sufficient for $t_\phi = \infty$.

More refined conditions for unbounded prolongation of solutions of (2.1), (2.2), taking into account the directions of the field f, are similar to those of stability theory, using Liapunov functions or functionals. Consider a simple example of this kind utilizing Razumikhin type stability theorem (see Ch. 7). Let there exist a function $V : \mathbf{R}^n \to \mathbf{R}, V \in C^1$ such that $V(x) \to \infty$ as $|x| \to \infty$ and for all sufficiently large $r > 0$

$$\sup_{t_0 \le t \le r, \|\psi\| \le r} |f(t, \psi)| < \infty$$

whereas if $\psi(0) = x, V(x) = r, V(\psi(\theta)) < r, \quad -h \le \theta < 0$ then

$$f'(t, \psi) \cdot \mathrm{grad} V(x) < 0$$

(here the prime is a transposition). Then $t_\phi = \infty$ for any solution of (2.1), (2.2).

2.2.5 Thm. 2.1 and its variants *2.2.1* – *2.2.4* can be generalized directly to RDEs with finite variable aftereffect which may even be unbounded as $t \to \infty$; more precisely, to Eqs. (1.1.7) if $g, h \in C[t_0, \infty)$. In fact, having defined $h_1 = \max_{t_0 \le t \le t_1} h(t)$ for a fixed $t_1 \in (t_0, \infty)$ the functional $f(t, \cdot)$ is expected to be defined on $C[-h_1, 0]$ for every $t \in [t_0, t_1]$ by the relation

$$f(t, \psi) = f(t, \psi|_{[-h(t), -g(t)]}), \quad \forall \psi \in C[-h_1, 0].$$

In addition, the conditions of the theorem (and its variants) must be reformulated for this extended functional. Since t_1 is arbitrary, all assertions in the theorem are met. Here the initial function must be defined on the union of all intervals

$$[t - t_0 - h(t), t - t_0 - g(t)] \bigcap (-\infty, 0], \quad t_0 \le t < \infty.$$

2.3. *SEMIGROUP RELATION*

Let us term $x(t; t_0, \phi)$ be the solution of Pr. (2.1), (2.2) under Thm. 2.1 conditions. Assume that for any initial point $t = t_1 \ge t_0$ and any initial function $\phi \in C[-h, 0]$ the solution of the corresponding Cauchy problem can be prolonged on the semiaxis $[t_1, \infty)$. Then from Thm. 2.1 it follows that $x(\cdot;\cdot,\cdot)$ is continuous and satisfies the *semigroup relation*

$$x(t;t_2,x_{t_2}(\cdot;t_1,\phi)) = x(t;t_1,\phi), \quad t \geq t_2 \geq t_1 \geq t_0, \quad \phi \in C[-h,0]$$

obtained by *translation of the initial point* to t_2. This operation, encountered often in the theory of equations with aftereffect, consists of application of the following simple argument for the problem of the form (2.1), (2.2). The solution $x : [t_0, t_\phi) \to \mathbf{R}^n$ restricted to any interval $[t_1, t_\phi)$ $(t_0 < t_1 < t_\phi)$ is also a solution of Eq. (2.1) with initial point $t = t_1$ and initial function $\theta \mapsto x(t_1 + \theta), \quad -h \leq \theta \leq 0$, prolonged by ϕ if $t_1 < t_0 + h$.

If Eq. (2.1) is *autonomous,* i.e. has the form

$$\dot{x}(t) = f(x_t)$$

and the functional f satisfies the Th. 2.1 conditions, then by virtue of this theorem, a solution of the Cauchy problem can be represented in the form $t \mapsto x(t-t_0;\phi) := x(t-t_0;0,\phi)$. Moreover, if every solution can be infinitely extended, i. e. all $t_\phi = \infty$, then the semigroup relation takes the form

$$x(\sigma_2; x_{\sigma_1}(\cdot;\phi)) = x(\sigma_1 + \sigma_2;\phi), \quad \sigma_1, \sigma_2 \geq 0, \quad \phi \in C[-h,0].$$

These relations may be interpreted in a way proposed by Krasovskii, that has the fundamental significance for FDEs now. Namely, assume that to Pr. (2.1), (2.2) solution corresponds not only its *graph* in $\mathbf{R} \times \mathbf{R}^n$ but also the *integral curve* in $\mathbf{R} \times C[-h,0]$ representing the set of all pairs (t, x_t) with $t \in [t_0, t_\phi)$.

Consideration of the integral curve is convenient, because for a given Eq. (2.1) it is completely determined by its initial point (t_0, ϕ). Hence the equation of this curve can be taken as

$$x_t = T(t,t_0)\phi, \quad t_0 \leq t < t_\phi$$

where $T(t,t_0) : C[-h,0] \to C[-h,0]$ is a *shift operator* from t_0 to t, corresponding to Eq. (2.1). Analogously the operator $T(t,t_1)$ is defined, where $t \geq t_1 \geq t_0$. It is clear that $T(t,t) = I_{C[-h,0]}$ (the identity operator)

Under the assumption made at the beginning of this subsection the operator $T(t,t_1)$ is defined and continuous for all $t_1 \geq t_0$, $t \geq t_1$. Moreover, if F is bounded on any bounded subset of D_h, then $T(t,t_1)$ for $t \geq t_1 + h$ is completely continuous, i.e. it is continuous and maps any bounded set of $C[-h,0]$ into a compact set.

Semigroup relation in terms of the shift operator takes the form

$$T(t,t_2)T(t_2,t_1) = T(t,t_1), \quad t \geq t_2 \geq t_1 \geq t_0.$$

For autonomous equations we get $T(t, t_0) = T(t - t_0, 0)$ and the righthand side may be designated by $T(t - t_0)$ (shift operator over time $t - t_0$). In this case semigroup relation is

$$T(\sigma_2)T(\sigma_1) = T(\sigma_1 + \sigma_2) \quad (\sigma_1, \sigma_2 \geq 0).$$

2.4. *ABSOLUTELY CONTINUOUS SOLUTIONS*

Having weakend the requirements in the definition of Eq. (2.1) solution, one can extend the class of equations for which Cauchy problem is solvable. One of the most widespread extension is as follows.

We say that x is an *absolutely continuous solution* (an AC solution) of (2.1), (2.2) if x is defined on some interval J_x with left end $t_0 \in J_x, x$ is absolutely continuous in some vicinity of any point of J_x, satisfies Eq.(2.1) (complemented by (2.2)) for almost all $t \in J_x$ and equals $\phi(0)$ for $t = t_0$. Further, we say that a functional $(t, \psi) \mapsto f(t, \psi)$ satisfies the *modified Carathéodory condition* on D_h if it is continuous in ψ for almost all fixed $t \in [t_0, \infty)$, measurable in t for each fixed $\psi \in C[-h, 0]$ and if for some neighbbourhood of any point of D_h there exists a Lebescue integrable function $\zeta : [t_0, \infty) \to [0, \infty)$ (generally speaking, depending upon this neighbourhood) such that

$$|f(t, \psi)| \leq \zeta(t), \quad |f(t, \psi_1) - f(t, \psi_2)| \leq \zeta(t) \parallel \psi_1 - \psi_2 \parallel . \qquad (2.10)$$

Inequality (2.10) must be valid for almost all t in the above mentioned neighbourhood.

The modified Carathéodory condition is especially useful because it covers linear RDEs with discontinuous but integrable coefficients.

Theorem 2.2 Let a functional f satisfy on D_h the modified Carathéodory condition, and $\phi \in C[-h, 0]$. Then there is a $t_\phi \in (t_0, \infty]$ such that

a) on interval $[t_0, t_\phi)$ there is AC solution x of Pr.(2.1), (2.2);

b) on any inerval $[t_0, t_1] \subset [t_0, t_\phi)$ this AC solution is unique;

c) if $t_\phi < \infty$, then $x(t)$ has no finite limit as $t \to t_\phi^-$;

d) AC solution x continuously depends on f, ϕ.

It means that for any $t_1 \in (t_0, t_\phi)$, $\varepsilon > 0$ there exists a $\delta > 0$ such that if f, ϕ in Pr. (2.1), (2.2) are replaced by $\overline{f}, \overline{\phi}$, where $\overline{f}, \overline{\phi}$ satisfy the same requirements as f, ϕ and

$$\int_{t_0}^{t_1} \sup_{\|u - x_t\| < \varepsilon} |\overline{f}(t, u) - f(t, u)| dt < \delta, \quad \| \overline{\phi} - \phi \| < \delta$$

then for the AC solution $\overline{x}$ of the changed problem with the maximal interval of existence $[t_0, \overline{t}_{\overline{\phi}})$ we get

$$\overline{t}_{\overline{\phi}} > t_1, \quad |\overline{x}(t) - x(t)| < \varepsilon, \quad t_0 \leq t \leq t_1.$$

The proof of Thm. 2.2 is just the same as the proof of Thm. 2.1.

The variants of these assertions given in Subs. 2.2 remain valid for AC solutions, except for the last statement in *2.2.4*. Also, all considerations connected with semigroup relation are valid.

Note that an ordinary solution of Eq. (2.1) is at the same time an AC solution. But, if F and ϕ are continuous, then the AC solution of Pr. (2.1), (2.2) coincides with the ordinary one. Hence in this case a relaxation of the requirements does not lead to new solutions.

A further generalization lies in considering discontinuous initial function ϕ. Let, for example, $\phi : [-h, 0] \to \mathbf{R}^n$ be a fixed function. Designate by M_ϕ the set of all functions $y : [t_0 - h, \infty) \to \mathbf{R}^n$ such that $y_{t_0} = \phi, \quad y|_{[t_0, \infty)} \in C[t_0, \infty)$ and let D_ϕ be the set of all pairs $(t, y_t) : t_0 \leq t < \infty, \quad y \in M_\phi$, equipped with the distance function

$$d((t_1, \psi_1), (t_2, \psi_2)) = |t_1 - t_2| + \| \overline{\psi}_2 - \overline{\psi}_1 \| .$$

Here $\overline{\psi}_i = \psi_i$ for $t_i \geq t_0 + h$, and if $t_0 \leq t_i < t_0 + h$ then

$$\overline{\psi}_i(\theta) = \begin{cases} \psi_i(\theta), & t_0 - t_i \leq \theta \leq 0, \\ \\ \psi_i(t_0 - t_i), & -h \leq \theta \leq t_0 - t_i, \end{cases}$$

Assume that $f : D_\phi \to \mathbf{R}^n$ and for any $y \in M_\phi$ the function $t \mapsto f(t, y_t)$ is defined for almost all $t \in [t_0, \infty)$ and belongs to $L_1[t_0, \infty)$. Let, moreover, the functional f be continuous in ψ for almost all t and conditions (2.10) are fulfilled in a neighbbourhood of any point of D_ϕ. Under these assumptions all assertions of Subs. 2.4 hold for AC solutions of (2.1), (2.2), except the assertion that the solution depends continuously on the initial function which we have taken fixed. (It is easy to overcome this obstacle, but we prefer not to do it here.)

In fact, the generalization just mentioned is useful (and can be used) if, in particular, the function ϕ is uniformly continuous on $[-h, 0)$ but $\phi(0^-) \neq \phi(0)$.

2.5. *RDES WITH INFINITE DELAY*

Another generalization of Thm. 2.1 is to consider RDEs with infinite delay

$$\dot{x}(t) = f(t, x_t), \quad t \geq t_0; \quad x_t(\theta) := x(t + \theta), \quad -\infty < \theta \leq 0, \qquad (2.11)$$

$$x_{t_0} = \phi : (-\infty, 0] \to \mathbf{R}^n. \tag{2.12}$$

Here, imposing some restrictions, we must regulate the behaviour of the considered functions as $t \to -\infty$, by indicating a complete metric space $K \subseteq C(-\infty, 0]$ containing the parts x_t of the Cauchy problem solution. Let us introduce also the set $\widetilde{K}$ of all functions $y \in C(\mathbf{R})$ such that $y|_{(-\infty,0]} \in K$. Assume that $\widetilde{K}$ is invariant under time shifts (in t), and that the metric $d(= d_K)$ on K satisfies the conditions:

if $y \in \widetilde{K}$, then the map $(t \mapsto y_t) : \mathbf{R} \to K$ is continuous;

if $y, y_i \in \widetilde{K}$ $(i = 1, 2, \ldots)$, $-\infty < t_1 < t_2 < \infty$, $d((y_i)_{t_1}, y_{t_1}) \to 0$, $\| (y_i - y)|_{[t_1,t_2]} \| \to 0$ as $i \to \infty$, then $\max\limits_{t_1 \le t \le t_2} d((y_i)_t, y_t) \to 0$ as $i \to \infty$.

Consider some examples of K:

1) The Banach space of functions $\psi \in C(-\infty, 0]$ such that $e^{pt}\psi(t)$ is bounded and uniformly continuous, with the norm

$$\| \psi \| := \sup_{-\infty < t \le 0} e^{pt} |\psi(t)|.$$

Here, $p \in \mathbf{R}$ is fixed;

2) The Banach space of the same functions as in 1), with $p = 0$ for which

$$\| \psi \| := \sup_{-\infty < t \le 0} |\psi(t)| + \int_{-\infty}^{0} |\psi(t)| dt < \infty.$$

3) The space $C(-\infty, 0]$ equipped with the metric

$$d(u_1, u_2) = \sum_{i=1}^{\infty} \frac{1}{2^i} \frac{\| (u_1 - u_2)|_{[-i,0]} \|}{1 + \| (u_1 - u_2)|_{[-i,0]} \|}$$

and so on.

Theorem 2.3 Let $\phi \in K$, the functional $f : D_\infty (:= [t_0, \infty) \times K) \to \mathbf{R}^n$ be continuous and for any $(t, \psi) \in D_\infty$ there are $\varepsilon, L > 0$ for which the inequality $|f(t + \tau, \psi_1) - f(t + \tau, \psi_2)| \le L \| \psi_1 - \psi_2 \|$ is valid if $\tau \in (0, \varepsilon)$, $\psi_i \in K$, $(\psi_i)_{-\tau} = \psi$, $\max\limits_{-\tau \le \theta \le 0} |\psi_i(\theta) - \psi_i(0)| < \varepsilon$, $i = 1, 2$. Then there is a $t_\phi \in (t_0, \infty]$ such that:

a) there exists a solution x of Pr. (2.11), (2.12) on interval $[t_0, t_\phi)$;

b) on any interval $[t_0, t_1] \subset [t_0, t_\phi)$ this solution is unique;

c) if $t_\phi < \infty$, then $x(t)$ has no finite limit as $t \to t_\phi^-$;

d) the solution x depends continuously on the functional f and the function ϕ.

It means that for any $t_1 \in (t_0, t_\phi)$, $\varepsilon > 0$ there is a $\delta > 0$ such that if f, ϕ in Pr. (2.11), (2.12) are replaced by $\overline{f}, \overline{\phi}$, where $\overline{f}, \overline{\phi}$ satisfy the same restrictions as do f, ϕ, and

$$|\overline{f}(t,\psi) - f(t,\psi)| < \delta, \text{ for } t_0 \le t \le t_1, \ d(\psi, x_t) < \varepsilon, \ d(\overline{\phi}, \phi) < \delta$$

then for the solution $\overline{x}$ of the changed problem with the maximal interval of existance $[t_0, t_{\overline{\phi}})$ we get

$$t_{\overline{\phi}} > t_1, \ |\overline{x}(t) - x(t)| < \varepsilon, \ t_0 \le t \le t_1.$$

Proof of Thm. 2.3. The conditions, imposed on K, imply that if $\phi \in K$, then for any function $x : [t_0, t_x) \to \mathbf{R}^n$, $t_0 < t_x < \infty$, which is a continuous prolongation for $t \mapsto \phi(t - t_0)$, $-\infty < t \le t_0$, the function x_t, $t_0 \le t < t_x$, belongs to K and depends continuously on t. Therefore, Pr. (2.11), (2.12) as previously is equivalent to the problem of existence of the continuous solution of the integro–functional equation

$$\left.\begin{aligned} x(t) &= \phi(0) + \int_{t_0}^{t} f(\tau, x_\tau)d\tau, \quad t_0 \le t < t_\phi, \\ x_t(\theta) &:= x(t+\theta), \quad -\infty < \theta \le 0. \end{aligned}\right\} \tag{2.13}$$

Local existence of the solution and its nonlocal uniqueness follow from the Lipschitz condition. From here as in the proof of Thm. 2.1 it follows the assertions *a), b), c)*. For the proof of *d)* assume the opposite. Then for any $t_1 \in (t_0, t_\phi)$ and sufficiently small $\varepsilon > 0$ we can construct the sequences $\{\overline{F}^k\}$ and $\{\overline{\phi}^k\}$ such that

$$\sup\{|\overline{f}^k(t,\psi) - f(t,\psi)| : t_0 \le t \le t_1, d(\psi, X_t) < \varepsilon\} \to 0,$$

$$d(\overline{\phi}^k, \phi) \to 0 \text{ as } k \to \infty,$$

but for the sequence $\overline{x}^k : [t_0, t^k] \to \mathbf{R}^n$ of solutions of integro–functional equations (2.13), where f and ϕ are replaced by $\overline{f}_k$ and $\overline{\phi}^k$, relations (2.6) hold. In addition, as in the proof of Thm. 2.1, we can suppose that $t^k \to \overline{t} \in (t_0, t_1]$ and $x^k \to \overline{x}$ as $k \to \infty$ uniformly on each interval $[t_0, \tilde{t}] \subset [t_0, \overline{t})$. Passing to the limit is fulfilled as $k \to \infty$ in these equations according to the scheme (2.7) and leads to the contradiction. Note that the second term in the righthand side tends to zero because of Lipschitz condition in the neighbourhood of the point $(t, \psi) = (\tilde{t}, \overline{x}_{\tilde{t}})$ and the second condition for the metric of K. Thm. 2.3 is proven. □

It is the virtual indefiniteness of $f(t, x_t)$ that compels us to consider Pr. (2.11), (2.12) with an infinite delay on any finite time interval $[t_0, t_1]$ for a fixed initial function. If the behaviour of $\phi(t)$ as $t \to -\infty$ implies that f is definite, then we can immediately reduce the considered problem to a similar problem with finite delay. In fact, for any pair $(t, \widetilde{\psi})$, where $t_0 \le t < \infty$, $\widetilde{\psi} \in C[t_0 - t, 0]$, $\widetilde{\psi}(t_0 - t) = \psi(0)$, we can define

$$\widetilde{f}(t, \widetilde{\psi}) := f(t, \psi), \quad \psi(\theta) := \begin{cases} \widetilde{\psi}(\theta), & t_0 - t \le \theta \le 0, \\ \phi(\theta - t_0 + t), & -\infty < \theta \le t_0 - t. \end{cases}$$

Using these notations, one can conclude that Pr. (2.6), (2.7) is equivalent to the Cauchy problem for the equation $\dot{x}(t) = \widetilde{f}(t, x_t)$ with finite delay $t - t_0$ (though unbounded as $t \to \infty$) and single–point initial condition $x(t_0) = \phi(0)$.

This remark makes it possible to generalize the assertions of Subs. 2.2, 2.3 to the equations with an unbounded delay. It is especially relevant for the Cauchy problem for the Volterra integro–differential equations, e.g.

$$\dot{x}(t) = \int_{-\infty}^{t} f(t, \tau, x(t), x(\tau))d\tau, \quad t \ge 0, \quad x(t) = \phi(t), \quad -\infty < t \le 0$$

because this equation is equivalent to the equation with a finite aftereffect

$$\dot{x}(t) = \int_{-\infty}^{0} f(t, \tau, \phi(t), \phi(\tau))d\tau + \int_{0}^{t} f(t, \tau, x(t), x(\tau))d\tau, \quad t \ge 0, \ x(0) = \phi(0),$$

under the condition that the first integral in the righthand side is convergent.

2.6. *FEATURES OF THE CAUCHY PROBLEM FOR RDES*

Consider some distinguishing features of the solvability theory of the Cauchy problem for RDEs in contrast to those of for ODEs. Two general distinctions are believed to take place, and the other distinctions can either be reduced to these or have a more special nature. The two main ones are:

1) the necessity to pose not only the initial value of the solution, but also the initial function. If this function is not fixed it leads to an infinite dimensional manifold of solutions;

2) the existance of the Cauchy problem solution which has the usual properties only forward in time.

Remark, however, that although in general the dimension of the solutions manifold is infinite, there exists an important case in which it is n-dimensional where n is the solution dimension. The Cauchy problem without previous history possesses such property. For Eq. (1.1.7) with initial point t_0 it means that $h(t) \leq t - t_0$ for all $t \geq t_0$, i.e. the initial interval is empty and instead of the initial function one must pose only an initial value $x(t_0)$. (The initial interval is said as well to degenerate into a point, and therefore only $\phi(0)$ must be posed.) For example, aftereffect may be realized as an integral in an unknown function (which has to be determined) between the limits t_0 and t. Note that if for such an equation is considered the Cauchy problem for $t = \bar{t}_0 > t_0$ then, generally speaking, the solutions manifold appears to be infinite dimensional. It once more proves the impossibility to extend the solution for decreasing t.

It may be formally assumed that the initial condition for Cauchy problem without previous history set up at $t = t_0$ is defined by the function $\phi : [-h, 0] \to \mathbf{R}^n$ for any $h > 0$ because such a function also defines $\phi(0) = x(t_0)$. This method, sometimes being used for uniformity, does not certainly convert the dimension of solutions manifold into infinite because $\phi(\theta)$ for $-h \leq \theta < 0$ is not utilized, and may be ignored. Such *ignoring of nonutilized* (for example, *too early*) *values of an initial function* is widely used in the theory of FDEs. Thus for variable delays the initial function may be set up for all $t \leq t_0$, but actually only its values influencing on the solution must be taken into consideration.

Because of an initial function is set on the inerval with length equal to the time aftereffect, it may occur that two solutions coincide on the interval of a smaller length, and then become different. This effect is connected with the difference between the notions of "graph" and "integral curve" for RDE solutions, introduced in Subs. 2.3.

The one-sided forward extendibility of the Cauchy problem solutions makes it possible that the solutions *stick* together. In contrast to ODEs, this interesting phenomenon may occur for RDEs with an arbitrarily smooth righthand side, and consists of the fact that solutions which are originally different may coincide beginning out of some time moment. This phenomenon along with the ill–posedness of problem for processes with aftereffect renewal (Subs. 1.5) induce in particular that the shift operators for RDEs, considered in Subs. 2.3, form a semigroup, but not a group.

Example 2.1 The Cauchy problem for the eqution

$$\dot{x}(t) = x\left(t - \frac{t^2}{1+t^2}\right), \quad n = 1$$

with the initial point $t_0 = 0$ has no previous history. Hence, if any $t_0 < 0$ is assumed to be an initial point, then two arbitrary solutions x_1 and x_2 with $x_1(0) = x_2(0)$ coincide (stick together) for all $t \geq 0$. Thus, the infinite dimensional manifold of solutions after "passage" through the point $t = 0$ has become 1-dimensional!

Example 2.2 Let be given scalar equtions

$$\dot{x}_1(t) = 0, \quad \dot{x}_i(t) = x_{i-1}(t-1), \quad i = 2, \ldots, n, \quad 0 \leq t < \infty \tag{2.14}$$

with given continuous initial functions

$$t \mapsto x_i(t) = \phi_i(t), \quad -1 \leq t \leq 0, \quad i = 1, \ldots, n.$$

By integrating these equations we successively obtain

$$x_i(t) = a_{i0}t^{i-1} + a_{i1}t^{i-2} + \ldots + a_{i,i-1}, \quad i - 1 \leq t < \infty, \quad i = 1, \ldots, n.$$

Recurrence relations for the coefficients a_{is} are following from the identities

$$a_{i0}(i-1)t^{i-2} + a_{i1}(i-2)t^{i-3} + \ldots + a_{i,i-2} =$$

$$= a_{i-1,0}(t-1)^{i-2} + a_{i-1,1}(t-1)^{i-3} + \ldots + a_{i-1,i-2}, \quad i = 2, \ldots, n.$$

Each of these relations gives the possibility to find $a_{i,0}, \ldots, a_{i,i-2}$ if $a_{i-1,0}, \ldots, a_{i-1,i-2}$ are known. The value of $a_{i,i-1}$ is expressed through the functions $\phi_1, \ldots, \phi_i$ by the formulas omitted here. So the infinite dimensional manifold of Eq. (2.14) solutions on the interval $[0, n-1]$ becomes n-dimensional for $t \geq n - 1$. We can accept $a_{10}, a_{21}, \ldots, a_{n,n-1}$ as parameters of this restricted solution.

Example 2.3 Interesting example of a nonlinear scalar autonomous equation with solutions sticking together was considered by H. Hausrath ([238], §3.2).

$$\dot{x}(t) = \beta[\max_{-1 \leq \theta \leq 0} |x(t+\theta)| - x(t)|, \quad 0 \leq t < \infty$$

where $\beta = \text{const} > 0$. (This RDEs referred as *differential equation with maxima*, i.e. FDE that involves maximum or minimum of the unknown function.) It is clear that any solution is nondecreasing function for $t \geq 0$, and that any nonnegative constant is a solution. Hence, for $\phi(0) \geq 0$ the solutions stick together with a constant solution, from some $t \in [0, 1)$ onwards.

For $\phi(0) < 0$ the possible behaviour of solutions depends on whether $\beta > \beta_0$, or $\beta \leq \beta_0$ where $\beta_0 \approx 0.278465$ is a root of the equation $\beta e^{1+\beta} = 1$. If $\beta > \beta_0$, then every solution sticks together with a constant, from some moment onwards which is not greater than some $t_\beta > 0$, where $t_\beta \to \infty$ as $\beta \to \beta_0^+$. But, if $\beta \leq \beta_0$, then there are both solutions sticking together with constant and solutions without this property. Moreover, for every such β there are solutions for which the time when sticking starts it as much as desirable.

3. CAUCHY PROBLEM FOR NDEs

3.1. *SMOOTH SOLUTIONS*

Consider the following problem:

$$\dot{x}(t) = f(t, x_t, \dot{x}_t), \quad t \geq t_0, \quad x_t(\theta) = x(t+\theta), \quad -h \leq \theta \leq 0, \tag{3.1}$$

$$x_{t_0} = \phi, \quad \dot{x}_{t_0} = \dot{\phi}, \tag{3.2}$$

where, like in Subs. 2.1, the time $h > 0$ of possible aftereffect is assumed to be constant and finite. As we have just noted in Sec. 1.4, for an arbitrarily smooth functional f and initial function ϕ the solution x may fail to be continuously differentiable if the sewing condition is not fulfilled. It leads to some complications in the NDEs theory, because in this case the application of the space $C[0,h]$ is, generally speaking, unsufficient.

If it is assumed that the sewing condition is valid, then the statement about Pr. (3.1), (3.2) solution is similar to Thm. 2.1. In this case the notion of a solution is the same as in Subs. 2.1.

Theorem 3.1 Let $f : E_h(:= [t_0, \infty) \times (C[-h,0])^2) \to \mathbf{R}^n$ be a continuous functional and in some neighbbourhood of any point of E_h it satisfies the condition

$$|f(t, \psi_1, \chi_1) - f(t, \psi_2, \chi_2)| \leq L \parallel \psi_1 - \psi_2 \parallel + l \parallel \chi_1 - \chi_2 \parallel \tag{3.3}$$

with constants $L \in [0, \infty)$, $l \in [0,1)$ (which may depend on the point). Assume also that $\phi \in C^1[0,h]$ and the sewing condition

$$\dot{\phi}(0) = f(t_0, \phi, \dot{\phi})$$

is fulfilled. Then there is a $t_\phi \in (t_0, \infty]$ such that

a) there exists a solution x of Pr. (3.1), (3.2) on the interval $[t_0, t_\phi)$;

b) on any interval $[t_0, t_1] \subset [t_0, t_\phi]$ this solution is unique;

c) if $t_\phi < \infty$, then $\dot{x}(t)$ has not a finite limit as $t \to t_\phi^-$;

d) the solution x and $\dot{x}$ depend continuously on f, ϕ in the stated below sense.

Let for any fixed $t_1 \in (t_0, t_\phi)$ and $\varepsilon > 0$ be prescribed a set U of perturbed functionals $((t, \psi, \chi) \mapsto \overline{f}(t, \psi, \chi)) : E_h \to \mathbf{R}^n)$ equicontinuous in the strip

$$\Pi_{t_1, \varepsilon} : t_0 \leq t \leq t_1, \quad \parallel \psi - x_t \parallel < \varepsilon, \quad \parallel \chi - \dot{x}_t \parallel < \varepsilon$$

and each functional $\overline{f}$ satisfies in this strip the Lipschitz condition (3.3) with the constants $L_{\overline{f}} \geq 0$, $l_{\overline{f}} \in [0,1)$.Then there is a $\delta > 0$ such that if

f, ϕ in (3.1), (3.2) are replaced by $\overline{f} \in U, \overline{\phi} \in C^1[-h,0]$ with

$$\left. \begin{array}{c} |\overline{f}(t,\psi,\chi) - f(t,\psi,\chi)| < \delta \text{ in } \Pi_{t_1,\varepsilon}, \; \| \overline{\phi} - \phi \| < \delta, \\ \| \dot{\overline{\phi}} - \dot{\phi} \| < \delta, \;\; \dot{\overline{\phi}}(0) = \overline{f}(t_0, \overline{\phi}, \dot{\overline{\phi}}), \end{array} \right\} \tag{3.4}$$

then for the solution $\overline{x}$ of the transformed problem with the maximal interval of existence $[t_0, t_{\overline{\phi}})$ we get

$$t_{\overline{\phi}} > t_1, \;\; |\overline{x}(t) - x(t)| < \varepsilon, \;\; |\dot{\overline{x}}(t) - \dot{x}(t)| < \varepsilon, \;\; t_0 \le t \le t_1.$$

Proof of Thm. 3.1. Let us denote $\dot{x}(t) = y(t)$ and integrate both parts of Eq. (3.1). We obtain that Pr. (3.1), (3.2) is equivalent to the problem of existence of a continuous solution of the equation

$$x(t) = \phi(0) + \int_{t_0}^{t} f(\tau, x_\tau, y_\tau) d\tau, \;\; y(t) = f(t, x_t, y_t), \;\; t_0 \le t < t_x$$

with the initial condition (3.2), where $\dot{x}$ is replaced by y. Consequently, we must find a fixed point of the mapping of the space $\{\psi, \chi \in C[t_0, t] : \psi(t_0) = \phi(0), \; \chi(t_0) = \dot{\phi}(0^-); t = \text{const} > t_0\}$ into itself

$$(\psi, \chi) \mapsto (\tau \mapsto \phi(0) + \int_{t_0}^{\tau} f(\tau_1, \psi_{\tau_1}, \chi_{\tau_1}) d\tau_1, \; \tau \mapsto f(\tau, \psi_\tau, \chi_\tau)) =:$$

$$=: (A^1(\psi,\chi), A^2(\psi,\chi)) =: A(\psi,\chi),$$

under corresponding initial condition.

Applying the Lipschitz condition in a neighbourhood of the point $(t_0, \phi, \dot{\phi})$, we obtain in this neighbourhood the following estimates for $t > t_0$

$$\| A^1(\psi^2,\chi^2) - A^1(\psi^1,\chi^1) \| \le L(t-t_0) \| \psi^2 - \psi^1 \| + l(t-t_0) \| \chi^2 - \chi^1 \|,$$

$$\| A^2(\psi^2,\chi^2) - A^2(\psi^1,\chi^1) \| \le L \| \psi^2 - \psi^1 \| + l \| \chi^2 - \chi^1 \| .$$

Let us introduce for a fixed $t > t_0$ a norm

$$|||(\psi,\chi)||| := \| \psi \| + (t-t_0)^{1/2} \| \chi \|,$$

then we have that

$$|||A(\psi^2,\chi^2) - A(\psi^1,\chi^1)||| \le [L(t-t_0) + L(t-t_0)^{1/2}] \| \psi^2 - \psi^1 \|$$

$$+ [l(t-t_0) + l(t-t_0)^{1/2}] \| \chi^2 - \chi^1 \| \le k |||(\psi^2,\chi^2) - (\psi^1,\chi^1)|||,$$

where $k := [(t-t_0)^{1/2}+1]\max\{L(t-t_0)^{1/2}, l\}$. Hence, if $t-t_0$ is sufficiently small, then $k < 1$, i. e. the map A is a contraction. Put $\psi^0(\tau) \equiv \phi(0)$, $\chi^0(\tau) \equiv \dot{\phi}(0)$. Then the points (t, ψ^0, χ^0) and $(t, A(\psi^0, \chi^0))$ are arbitrarily close to $(t_0, \phi, \dot{\phi})$ if $t-t_0$ is sufficiently small. Denote

$$p := |||A(\psi^0, \chi^0) - (\psi^0, \chi^0)|||, \quad r := p(1-k)^{-1}.$$

For sufficiently small p and $|||(\psi, \chi) - (\psi^0, \chi^0)||| \leq r$ we obtain

$$|||A(\psi, \chi) - (\psi^0, \chi^0)||| \leq |||A(\psi, \chi) - A(\psi^0, \chi^0)||| - |||A(\psi^0, \chi^0)$$

$$-(\psi^0, \chi^0)||| \leq kr - p = r.$$

Hence, A maps the closed set $|||(\psi, \chi) - (\psi^0, \chi^0)||| \leq r$ into itself. From here it follows the local existence and uniqueness of the solution of Pr. (3.1), (3.2).

Further application of the standard scheme implies the assertions *a)* and *b)*. If $\dot{x}(t)$ has a finite limit as $t \to t_\phi^-$, then $x(t)$ has also a finite limit as $t \to t_\phi^-$, and the solution can be prolonged for $t > t_\phi^-$. It means that the assertion *c)* holds.

For the proof of *d)* prove firstly, that if $\varepsilon > 0$ is sufficiently small and the solution $\overline{x}$ of the perturbed problem (3.1), (3.2) is defined on the interval $[t_0, \overline{t})$, where $\overline{t} < t_\phi$ and

$$|\overline{x}(t) - x(t)| < \varepsilon, \quad |\dot{\overline{x}}(t) - \dot{x}(t)| < \varepsilon, \quad t_0 \leq t < \overline{t},$$

then $\dot{\overline{x}}$ is uniformly continuous on $[t_0, \overline{t})$. (So, $\overline{x}$ can be prolonged as a solution of the same problem for $t \geq \overline{t}$.) In fact, as in the proof of Thm. 2.1, we have that for $t_0 \leq t \leq \overline{t}$, $\| \psi - x_t \| < \varepsilon$, $\| \chi - \dot{x}_t \| < \varepsilon$ the function $\overline{f}$ is continuous and satisfies the similar Lipschitz condition as f does satisfy. Therefore, for $t_0 \leq t_1 < t_2 < \overline{t}$ we get

$$|\dot{\overline{x}}(t_2) - \dot{\overline{x}}(t_1)| \leq |\overline{f}(t_2, \overline{x}_{t_2}, \dot{\overline{x}}_{t_2}) - \overline{f}(t_1, \overline{x}_{t_2}, \dot{\overline{x}}_{t_2})| +$$

$$+|\overline{\phi}(t_1, \overline{x}_{t_2}, \dot{\overline{x}}_{t_2}) - \overline{f}(t_1, \overline{x}_{t_1}, \dot{\overline{x}}_{t_1})| \leq \tag{3.5}$$

$$\leq \overline{\omega}(t_2 - t_1) + \overline{L} \| \overline{x}_{t_2} - \overline{x}_{t_1} \| + \overline{l} \| \dot{\overline{x}}_{t_2} - \dot{\overline{x}}_{t_1} \|,$$

where $\overline{L} := L_{\overline{f}}, \overline{l} := l_{\overline{f}}$ and $\overline{\omega}$ is a modulus of continuity of the function $\overline{f}$ in mentioned domain. If $t_0 - h \leq t_1 < t_2 \leq t_0$, then taking into account the initial condition for $\overline{x}$, we must instead of (3.5) write $|\dot{\overline{x}}(t_2) - \dot{\overline{x}}(t_1)| \leq \overline{\omega}_1(t_2 - t_1)$, where $\overline{\omega}_1$ is a modulus of continuity of $\dot{\overline{\phi}}$. If $t_0 - h \leq t_1 < t_0 < t_2 < \overline{t}$, then

$$|\dot{\overline{x}}(t_2) - \dot{\overline{x}}(t_1)| \leq |\dot{\overline{x}}(t_2) - \dot{\overline{x}}(t_0)| + |\dot{\overline{x}}(t_0) - \dot{\overline{x}}(t_1)| \leq$$

$$\leq \overline{\omega}(t_2 - t_0) + \overline{L} \| x_{t_2} - \overline{\phi} \| + \overline{l} \| \dot{\overline{x}}_{t_2} - \dot{\overline{\phi}} \| + \overline{\omega}_1(t_0 - t_1).$$

Further, we have

$$\| \overline{x}(t_2) - \overline{x}(t_1) \| = \max_{-h\leq\theta\leq 0} |\overline{x}(t_2+\theta) - \overline{x}(t_1+\theta| =$$

$$= \max_{-h\leq\theta\leq 0} \Big| \int_{t_1+\theta}^{t_2+\theta} \overline{F}(\tau, \overline{x}_\tau, \dot{\overline{x}}_\tau) d\tau \Big|, \tag{3.6}$$

where for $\tau < t_0$ the integrand must be replaced on $\dot{\overline{\phi}}(\tau - t_0)$. Consequently, this integrand is bounded for $\tau \geq t_0$ and $\tau < t_0$. From here and (3.6) it follows that the middle summand at the righthand side of estimate (3.5) has a form $O(t_2 - t_1)$.

Denote for $q \in (0, \overline{t} - t_0)$

$$\alpha(q) := \sup\{|\dot{\overline{x}}(y+r) - \dot{\overline{x}}(t)| : 0 \leq r \leq q, \quad t_0 - h \leq t < \overline{t} - r\}.$$

For a given q there exist the sequences $\{r_k\}$ and $\{t_k\}$, belonging to the above mentioned intervals, such that

$$\alpha(q) = \lim_{k\to\infty} |\dot{\overline{x}}(t_k + r_k) - \dot{\overline{x}}(t_k)|.$$

And so, by (3.5) and further estimates of $|\dot{\overline{x}}(t_2) - \dot{\overline{x}}(t_1)|$ we obtain, that for some $C > 0$

$$\alpha(q) \leq \liminf_{k\to\infty} [\overline{\omega}(r_k) + \overline{\omega}_1(r_k) + \overline{L}Cr_k + \overline{l}\alpha(r_k)] \leq$$

$$\leq \overline{\omega}(q) + \overline{\omega}_1(q) + \overline{L}Cq + \overline{l}\alpha(q).$$

From here it follows that

$$\alpha(q) \leq (1 - \overline{l})^{-1}[\overline{\omega}(q) + \overline{\omega}_1(q) + \overline{L}Cq] \to 0 \quad \text{as} \quad q \to 0^+.$$

Hence, uniform continuity of the function $\dot{\overline{x}}$ on $[t_0, \overline{t})$ is proven.

Assume now that assertion *d)* is not true. Then, similar to the proof of Thm. 2.1, we obtain for fixed $t_1 \in (t_0, t_\phi)$ and $\varepsilon > 0$ the sequences $\{\overline{f}^k\}$ and $\{\overline{\phi}^k\}$, satisfying (3.4) with $\delta^k (\to 0^+)$ instead of δ, and also other conditions of Thm. 3.1. In addition, we have for the corresponding solutions $\overline{x}^k : [t_0, t^k] \to \mathbf{R}^n$ of perturbed Pr. (3.1), (3.2)

$$t^k \to \overline{t} > t_0, \quad |\overline{x}^k(t) - x(t)| < \varepsilon, \quad |\dot{\overline{x}}^k(t) - \dot{x}(t)| < \varepsilon, \quad t_0 \leq t < t^k,$$

$$\max\{|\overline{x}^k(t) - x(t^k)|, \quad |\dot{\overline{x}}^k(t^k) - \dot{x}(t^k)|\} = \varepsilon.$$

Uniform boundedness of the functions $\overline{f} \in U$ in $\Pi_{t_1,\varepsilon}$ imply that $\bar{t} > t_0$. Moreover, the sequence $\{\overline{x}^k\}$ converges uniformly on each interval $[t_0, \tilde{t}] \subset [t_0, \bar{t})$ to some function $\overline{x} \in C[t_0, \bar{t}]$.

Without loss of generality we can assume that $t_1 - t_0 < 1$. (In fact, the validity of assertion *d)* follows from here for arbitrary $t_1 > t_0$ by the translation of the initial point using a finite number of steps.) But if $t_1 - t_0 < 1$ and $|\dot{\overline{x}}^k(t) - \dot{x}(t)| < \varepsilon$ $(t_0 \leq t < t^k)$, then for all sufficiently large k we have $|\overline{x}^k(t^k) - x(t^k)| < \varepsilon$. Hence $|\dot{\overline{x}}^k(t^k) - \dot{x}(t^k)| = \varepsilon$.

Further, from the uniform local Lipschitz condition for $\{\overline{f}^k\}$ and the uniform local estimates from above of the modulus of continuity of $\{\overline{f}^k\}$ (similar to the uniform continuity of the function $\dot{\overline{x}}$ proved above) it follows equicontinuity of the set of the functions $\{\dot{\overline{x}}^k\}$. Equicontinuity of the sequence $\{\dot{\underline{\phi}}^k\}$ follows from its uniform convergence. Passing to the subsequence, we can assume that $\{\dot{\overline{x}}^k\}$ tends uniformly on each interval $[t_0, \tilde{t}] \subset [t_0, \bar{t})$ to a continuous function, which must coincide with $\dot{\overline{x}}$, and also, from our mentioned argument it follows the existence of this derivative and its uniform continuity on $[t_0, \bar{t})$. Hence, $\overline{x} \in C^1[t_0, \bar{t}]$ and $|\dot{\overline{x}}(\bar{t}) - \dot{x}(\bar{t})| = \varepsilon$. But passing to the limit in perturbed problem (3.1), (3.2) as $k \to \infty$, we obtain that $\overline{x}$ is a solution of the nonperturbed problem. Therefore, by *b)* we have $\overline{x}(t) = x(t)$ $(t_0 \leq t \leq \bar{t})$ which contradicts to the previous equality. Thm. 3.1 is proven. □

Variants of assertions of Thm. 3.1, like those considered in Subs. 2.2, the consideration in Subs. 2.3 and an analog of Thm. 2.3 remain valid. In particular, note the consequence from variant *2.2.1*: for NDEs (1.1.5) the Lipschitz condition with respect to the functional arguments with positive delays is not required.

Remark 3.1. Lipschitz condition (3.3) can be easily generalized as follows: for any point of E_h there are constants $L \in [0, \infty)$, $l \in [0, 1)$, and $g \in [0, h]$ such that in some its neighbbourhood the inequality

$$|f(t, \psi_1, \chi_1) - f(t, \psi_2, \chi_2)| \leq L(\| \psi_1 - \psi_2 \| + \| (\chi_1 - \chi_2)|_{[-L,-g]} \| +$$

$$+ l \, \| (\chi_1 - \chi_2)|_{[-g,0]} \|$$

hold. In detailed formulation of the property *d)* the constant g must be the same for all points from $\Pi_{t_1,\varepsilon}$. The same remarks are valid for Thms.3.2-3.5.

3.2. *NDES WITH A FUNCTIONAL OF INTEGRAL TYPE*

If the sewing condition (3.3) is not fulfilled, then, generally speaking, a solution has no continuous derivative, and certainly, it has no a two-sided derivative for $t = t_0$. It, of course, influences on the assertions related to the solution of Pr. (3.1), (3.2).

The simpler case is NDEs with a *functional* $f(t, x, \dot{x}_t)$ *of the integral type* in $\dot{x}$. It means that the third argument of f may be considered as an element of $L_\infty[-h, 0]$. But in this case one must keep in mind the following feature of the norm $\| \cdot \|_\infty$: if a function $u \in L_\infty[t_1, t_2]$, $(t_2 > t_1 + h)$ has an unremovable discontinuity for some $t_3 \in [t_1, t_2]$, then function $t \mapsto u_t$, considered as a map of the interval $[t_1 + h, t_2]$ into Banach space $L_\infty[-h, 0]$, will be discontinuous for every $t \in [\max\{t_3, t_1 + h\},\ \min\{t_2, t_3 + h\}]$.

Therefore, for functions belonging to $L_\infty[-h, 0]$ along with the norm $\| \cdot \|_\infty$ the norm $\| \cdot \|_1$ will be used. The convenience of the latter is determined by the following simple lemma:

Lemma 3.1 If $u \in L_1[-h-\varepsilon, \varepsilon]$ *(in particular, if* $u \in L_\infty[-h-\varepsilon, \varepsilon]$*), where* $\varepsilon > 0$, *then* $\| u_s - u_0 \|_1 \to 0$ *as* $s \to 0$.

The proof is immediately obtained by approximation in $L_1[-h-\varepsilon, \varepsilon]$ of the function u by continuous functions.□

Application of the two norms may be avoided if, for example, the functional $(t, \psi, \chi) \mapsto f(t, \psi, \chi)$ is defined for all $\chi \in L_1[-h, 0]$. But for a nonlinear functional f it restricts essentialy its growth in χ that is not desirable in general existence theorem for the Cauchy problem.

We shall say that a function x is a solution of (3.1), (3.2) on some inerval J_x containing its left end t_0 if x is of Lipschitz class and satisfies Eq. (3.1) complemented by (3.2) almost everywhere on J_x, and $x(t_0) = \phi(0)$. For any $N > 0$ designate $L_1^N[-h, 0]$ the metric space of functions $\psi \in L_\infty[-h, 0]$ such that $\| \psi \|_\infty \leq N$, equipped with the metric $d(\psi_1, \psi_2) = \| \psi_1 - \psi_2 \|_1$.

Theorem 3.2 Let the functional $(t, \psi, \chi) \mapsto f(t, \psi, \chi)$ *in Eq. (3.1) be defined on* $[t_0, \infty) \times C[-h, 0] \times L_\infty[-h, 0]$ *and be continuous on* $[t_0, \infty) \times C[-h, 0] \times L_1^N[-h, 0]$ *for any* $N > 0$. *Assume that* $\phi \in \mathrm{Lip}[-h, 0]$, *and for any* $t^* \in [t_0, \infty)$ *and any bounded set* $K \subset C[-h, 0] \times L_\infty[-h, 0]$ *there exist* $L \geq 0$, $l \in [0, 1)$, $g \in (0, h)$ *for which*

$$|f(t, \psi_1, \chi_1) - f(t, \psi_2, \chi_2)| \leq L \| \psi_1 - \psi_2 \| + l \| \chi_1 - \chi_2 \|_\infty \tag{3.7}$$

if $t \in [\max\{t_0, t^* - g\}, t^* + g]$, $(\psi_i, \chi_i) \in K$, $i = 1, 2$. *Then there exists a* $t_\phi \in (t_0, \infty]$ *such that*

a) there exists a solution x *of Pr. (3.1), (3.2) on* $[t_0, t_\phi)$;

b) on any interval $[t_0, t_1] \subset [t_0, t_\phi)$ *this solution is unique;*

c) $x \in C^1[t_0, t_\phi)$ *and Eq. (3.1) is satisfied for all* $t \in [t_0, t_\phi)$;

d) if $t_\phi < \infty$ *then* $\| \dot{x}_t \|_\infty$ *is unbounded as* $t \to t_\phi^-$;

e) the solution x *and* $\dot{x}$ *continuously depend on* f, ϕ *in the stated below sense.*

Let for any fixed $t_1 \in (t_0, t_\phi)$ and $\varepsilon > 0$ be given a set U of perturbed functionals $((t,\psi,\chi) \mapsto \overline{f}(t,\psi,\chi))$ of the same type as in Thm. 3.2 which are equicontinuous in the strip

$$\Pi^1_{t_1,\varepsilon} : t_0 \le t \le t_1, \quad \| \psi - x_t \| < \varepsilon, \| \chi - \dot{x}_t \|_{\mathrm{Lip}} < \varepsilon$$

and satisfy the Lipschitz condition (3.7) (with their own Lipschitz constants). Then there is a $\delta > 0$ such that if f, ϕ in (3.1), (3.2) are replaced on $\overline{f} \in U, \overline{\phi} \in \mathrm{Lip}[-h,0]$ and

$$|\overline{f}(t,\psi,\chi) - f(t,\psi,\chi)| < \delta \ \text{ in } \ \Pi^1_{t_1,\varepsilon}, \ \| \phi - \overline{\phi} \|_{\mathrm{Lip}} < \delta \tag{3.8}$$

then for the solution $\overline{x}$ of the transformed problem with the maximal interval of existence $[t_0, \overline{t}_{\overline{\phi}})$ we get

$$\overline{t}_{\overline{\phi}} > t_1, \quad |\overline{x}(t) - x(t)| < \varepsilon, \quad t_0 \le t \le t_1,$$

$$|\dot{\overline{x}}(t) - \dot{x}(t)| < \varepsilon, \quad t_0 + \varepsilon \le t \le t_1.$$

Proof of Thm. 3.2. First of all, by Lemma 3.1 we obtain that if $x \in \mathrm{Lip}[t_0, t_x)$ and $x(t_0) = \phi(0)$, then the function $t \mapsto f(t, x_t, \dot{x}_t)$ $(t_0 \le t < t_x)$ with initial condition (3.2) is continuous. From here it follows that under condition of Thm. 3.2 Pr. (3.1), (3.2) is equivalent to the problem of existence of the Lipschitzian solution of the equation

$$x(t) = \phi(0) + \int_{t_0}^{t} f(\tau, x_\tau, \dot{x}_\tau) d\tau, \quad t_0 \le t < t_x,$$

with initial condition (3.2), and also such a solution belongs $C^1[t_0, t_x)$. Hence, *c)* follows from *a)*.

Putting in the condition of Thm. 3.2 $t^* = t_0$, we can prove (just in the same way as in the proof of Thm. 3.1) the local existence and uniquiness of the solution of Pr. (3.1), (3.2). After that we can prolong this solution on a maximal possible interval using the standard procedure and obtain assertions *a)* – *c)*.

To prove *d)* it is sufficient to note that if $\| \dot{x}_t \|$ is bounded as $t \to t_\phi^-$, then $x(t)$ has a finite limit. If we put $x(t_\phi) := x(t_\phi^-)$ and transfer the initial point in t_ϕ, then we should be able to prolong the solution beyond t_ϕ. But this contradicts to the definition $[t_0, t_\phi)$ as a maximum interval of existence of the solution.

To prove *e)* assume that *e)* is incorrect for some $t_1 \in (t_0, t_\phi)$. Let us denote $t^* \in [t_0, t_\phi)$ as an infimum of such t_1 and take some sufficiently small $\varepsilon > 0$. Similar to the proof of Thm.3.1 we can assume that $t^* - t_0 < 1$.

Then for each $k \in \mathbf{N}$ there are $t_1^k, t^k \in (t_0, t_1^k]$ and δ_k such that $t_1^k \to t^{*+}$ and $\delta_k \to 0^+$ as $k \to \infty$, and the solution $x^k : [t_0, t^k] \to \mathbf{R}^n$ of perturbed problem (3.1), (3.2), satisfying estimates (3.8) with δ_k instead of δ, has the properties:

$$|\overline{x}^k(t) - x(t)| < \varepsilon, \quad t_0 \le t \le t^k,$$

$$|\dot{\overline{x}}^k(t) - \dot{x}(t)| < \varepsilon, \quad t_0 \le t < t^k,$$

$$|\dot{\overline{x}}^k(t^k) - \dot{x}(t^k)| = \varepsilon.$$

In addition, we can suppose that $t^k \to \overline{t}$ as $k \to \infty$, and $\overline{x}^k \to \overline{x}$ uniformly on each interval $[t_0, \tilde{t}] \subset [t_0, \overline{t})$.

By definition of t^* we have $\overline{t} = t^*$. Estimating $|\overline{f}^k(t+\tau, \overline{x}_{t+\tau}^k, \dot{\overline{x}}_{t+\tau}^k) - \overline{f}^k(t, \overline{x}_t^k, \dot{\overline{x}}_t^k)|$ under conditons of Thm. 3.2, we obtain (as in the proof of Thm. 3.1) that the sequence $\{\dot{\overline{x}}^k\}$ is equicontinuous. Consequently this sequence can be considered as an uniformly convergent sequence to $\dot{\overline{x}}$ on each above mentioned interval $[t_0, \tilde{t}]$. Hence, $|\dot{\overline{x}}(\overline{t}^-) - \dot{x}(\overline{t})| = \varepsilon$. But as previously we obtain that $\overline{x}$ is a solution of Pr. (3.1), (3.2), and so $\dot{\overline{x}}(\overline{t}^-) = \dot{x}(\overline{t})$ by *b)*. This contradiction finishes the proof of Thm. 3.2. □

Variants of assertions of Thm. 3.2 and other assertions, like those of considered in Subs. 2.2, 2.3 and Thm. 2.3, remain in force.

So, for a functional of integral type in the derivative of the unknown function the extension of the solutions class actually does not happen in comparison with Thm. 3.1 in spite of the essential extension of the class of initial functions and failure of the sewing condition: the solutions remain smooth. From here it follows that equations of the considered type and with sufficiently smooth righthand side have the property of solutions smoothing that was above mentioned in Subs. 1.3. This property relates such equations with RDEs. Moreover, in some cases they can immediately be reduced to RDEs. For example, if aftereffect is realized in the form of the integral $\int_{t-h}^{t} K(t,s)\dot{x}(s)ds$ with $K \in C^1$, then this reduction can be done by integration by parts.

3.3. *APPLICATION OF THE STEPS METHOD*

For NDEs with discrete delays in the derivative of the unknown function some obstacle arises, because a functional depending on discrete values of a function cannot be considered as given on a space like L_∞ where the function is defined only almost everywhere. There are different ways to overcome this obstacle. Thus, the initial function ϕ may be approximated by a sequence of functions ϕ_i that satisfy the conditions of Thm. 3.1, by virtue of which there exists a sequence of solutions x_i. If f is such that the limit of the sequence x_i exists, then this limit is called a generalized solution of the original problem. But this way may lead to cumbersome formulations, that will be omitted here.

Another way that is likely to lead to the results which may be sufficient for applications, is to restrict the class of equations under investigation, so as to have a possibility to apply the steps method with respect to discrete delays in the derivative of the unknown function.

Let for instance the functional f in Eq. (3.1) take the form (compare with Eq. (1.1.5)):

$$f(t, x_t, \dot{x}_t) =$$

$$= f(t, x(t-h_1(t)), \ldots, \dot{x}(t-h_k(t)), \ldots, \dot{x}(t-g_1(t)), \dot{x}(t-g_l(t)) + \quad (3.9)$$

$$+f_1(t, x_t, \dot{x}_t).$$

A solution of Pr. (3.1), (3.2) is defined in much the same way as in Subs.3.2. Using the steps method we are able to obtain the following theorem.

Theorem 3.3 Assume that:

1) all functions $h_i : [t_0, \infty) \to [0, \infty), g_j : [t_0, \infty) \to (0, \infty)$ are continuous, and every function $t \to t - g_j(t)$ may take on any interval $[t_0, t_1]$ an arbitrary value at only finitely many points;

2) the function $f : [t_0, \infty) \times \mathbf{R}^{nk+nl} \to \mathbf{R}^n$ is continuous and satisfies the local Lipschitz condition in those arguments $x(t - h_i(t))$ for which the function h_i can take the value zero;

3) the functional f_1 satisfies Thm. 3.2 conditions;

4) the initial function ϕ is piecewise smooth (i.e. ϕ is continuous and $\dot{\phi}$ is continuous on the initial interval except at a set of discontinuity points of the first kind not having limit points).

Then for the solution x of Pr. (3.1), (3.2) are valid the assertions of Thm. 3.2 with the following modifications: a solution $x : [t_0, t_\phi) \to \mathbf{R}^n$ is a piecewise smooth function and Eq. (3.1) is satisfied for all $t \in [t_0, t_\phi)$ except on a set without limit points; the solution x depends, together with $\dot{x}$, continuously on f_1, ϕ (as in Thm.3.2), and also on f, h_i in the sense of the uniform metric.

Proof of Thm. 3.3 is reduced to the sequential application of Thm. 3.2 on the inervals $[t_0, t_1]$, $[t_1, t_2]$,.... Here $t_0 < t_1 < t_2 \ldots$ is a sequence of points which put together a set $M_\phi \subset [t_0, \infty)$, where the solution can have breakpoints. This set is defined as the minimal set containing t_0, and such that if $t \in [t_0, \infty)$ and for some $j, t - g_j(t)$ either belongs to M_ϕ or is a discontinuity point of $t \mapsto \phi(t - t_0)(-\infty < t \leq t_0)$, then $t \in M_\phi$. The set discontinuity point of $t \mapsto \phi(t-t_0)(-\infty < t \leq t_0)$, then $t \in M_\phi$. The set M_ϕ has no limit points, and the derivative of a solution of Pr. (3.1), (3.2) is continuous at each point of $[t_0, t_\phi) \setminus M_\phi$.□

Also, the results similar to those of Thm. 2.3 and Subs. 2.2, 2.3 will be valid.

Concerning the dependence of the solution on the functions g_i, remark that variation of g_i generally speaking induces a shift of the discontinuity points of $\dot{x}$. Hence, although x depends continuously on the functions g_i (in the uniform metric, on any interval $[t_0, t_1] \subset [t_0, t_\phi)$ as in Thm. 3.2) this in general is not the case for $\dot{x}$ near these discontinuity points.

The following statement takes place: for any $\varepsilon > 0$ the values $\dot{x}(t), t \in [t_0, t_1] \setminus U(M_\phi, \varepsilon)$ (where $U(M_\phi, \varepsilon)$ is the ε-neighbourhood of the set M_ϕ), vary uniformly by an arbitrarily small amount under a uniformly sufficiently small variation of all functions g_j. If the set $U(M_\phi, \varepsilon)$ is not eliminated, then the smallness of $\dot{x}$ alteration under a such variation of g_j only be asserted in the metric of $L_1^N[t_0, t_1]$ for sufficiently large N.

The same results can be proved for some cases when discrete and integral terms in (3.1) are connected more closely than for the functional (3.9).

3.4. *TRANSITION TO THE OPERATOR EQUATION*

There is one more way to overcome the obstacle arising in formulation of the general solvability thorems for Cauchy problem (3.1), (3.2), if according to the sense of the problem must be $\dot{x}_t \in L_\infty$ but the functional $(t, u, v) \mapsto f(t, u, v)$ cannot be considered to be defined on the functions $v \in L_\infty$.

Remark that essentially the same obstacle appears for ODEs with integrable terms. Consider, for instance, the simplest scalar equation

$$\dot{x}(t) = f(t)x(t) \tag{3.10}$$

with an integrable function $f : [0,1] \to \mathbf{R}$. If f is considered as an element of $L_1[0,1]$, then one must have in mind that every such element is, strictly speaking, not the unique concrete function, but a class of all integrable functions differing from each other on arbitrary set of measure zero. It is no sense under this interpretation to say that Eq. (3.10) is satisfied for a given t. However, it is more appropriate to consider both sides of (3.10) "in the large", as elements of $L_1[0,1]$, because under this inerpretation uncertainty and even distinction between the functions on a measure zero set are not essential. In other words, relation (3.10) is interpreted as an equality of operators values of x, mapping the set of absolutely continuous functions on $[0,1]$ into $L_1[0,1]$, but not as equality of functionals values for every or almost every $t \in [0,1]$.

Such approach is widely used for general FDEs at last years. It consists of interpretation the equation as the equality of values of the operators defined by both sides of the considered equations in an appropriate space. In particular, this approach allows us to investigate NDEs like (3.9) with discrete delays in the derivative of unknown function, because with regard to the initial function the righthand side defines an operator acting from the space of continuously differentiable functions $x : [t_0, \infty) \to \mathbf{R}^n$ into $L_1[t_0, \infty)$. But, in the general case one must be careful, because of the appearance of composite functions. Thus for the arguments $t - g_j(t)$ of $\dot{x}$ one must assume that the function $t \mapsto t - g_j(t)$ cannot map a set of positive measure to a set of measure zero. (If the function g_j is analytic or piecewise analytic as usually encounters in application, then the last condition means that the function $t \mapsto t - g_j(t)$ has no intervals of constancy.) Moreover, if $x \in \mathrm{Lip}[t_0, \infty)$ is not required, then some essential restrictions on f must be imposed, which we shall not concern here.

Consider a variant of the solvability theorem for the Cauchy problem for NDEs under the operator approach. Represent the equation in the form

$$Dx = \mathcal{F}x^{\phi}. \tag{3.11}$$

Here, D is the differentiation operator, mapping the set $\mathrm{Lip}[t_0, \infty)$ into $L_\infty[t_0, \infty)$. The operator $\mathcal{F}$ being defined by the right side of the FDE must map the set $\mathrm{Lip}[t_0 - h, \infty)$ into $L_\infty[t_0, \infty)$ and should be a *Volterra operator*, i.e. for any $t_1 \in (t_0, \infty)$ the equality $u_1(t) \equiv u_2(t)$, $(t_0 - h \le t \le t_1)$, $(u_1, u_2 \in \mathrm{Lip}[t_0 - h, \infty))$ implies $(\mathcal{F}u_1)(t) = (\mathcal{F}u_2)(t)$ for almost all $t \in [t_0, t_1]$. The function x^ϕ is the prolongation of a function $x : [t_0, T) \to \mathbf{R}^n$ (where, *apriori* T is free) by the values $\phi(t)$ $(t_0 - h \le t < t_0)$ to the inerval $[t_0 - h, T]$. It is assumed that $x \in \mathrm{Lip}[t_0, T)$, $\phi \in \mathrm{Lip}[t_0 - h, t_0]$, $x(t_0) = \phi(t_0)$.

In contrast to (3.1) righhand side of Eq. (3.11) does not contain $\dot{x}$ directly. But (3.6) includes NDEs, due to the indication of the spaces, to and from which the operator $\mathcal{F}$ acts. Remark that the operator $\mathcal{F}$ being a Volterra one, acts also in the spaces $\mathrm{Lip}[t_0 - h, t_1]$ for any $t_1 \in (t_0, \infty)$, mapping it into the space $L_\infty[t_0, t_1]$. Hence, Eq. (3.11) may be considered on any interval containing its left end t_0 and the solution of Eq. (3.11) on each such interval is always the solution of this equation on any smaller interval without the left end t_0.

For the equation under consideration it is natural to define the transferring the initial point t_0, using the following scheme. Let a solution $(x^\phi)^1$ of Eq. (3.11) be given on an interval $[t_0, t_1]$. Let us introduce the operator $\mathcal{F}_1 : P_1\mathcal{F}$, where P_1 is an operator of restriction on $[t_1, \infty)$; then $\mathcal{F}_1 : \mathrm{Lip}[t_0 - h, \infty) \to L_\infty[t_1, \infty)$. Also, introduce the initial function $\phi_1 : [t_0 - h, t_1] \to \mathbf{R}^n$, which is a prolongation of ϕ by $(x^\phi)^1$. (In fact, if the memory is bounded by h then we have $\mathcal{F}_1 : \mathrm{Lip}[t_1 - h, \infty) \to L_\infty[t_1, \infty)$, and the initial function is a restriction ϕ_1 on $[t_1 - h, t_1]$.) After transferring the initial point into t_1 Eq. (3.11) can be written in the form $D_1 x = \mathcal{F}_1 x^{\phi_1}$, where D_1 is an operator of differentiaton on $[t_1,\infty)$. Function x^{ϕ_1}, defined on some interval with the left end t_1, is a solution of this equation if and only if the function x^{ϕ}, "composed" from $(x^{\phi})^1$ and x^{ϕ_1}, is a solution of (3.11).

It is natural to say that the operator $\mathcal{F}$ possesses a delay less or equal h, if from $u_1, u_2 \in \text{Lip}[t_0-h,\infty)$, $u_1(t) \equiv u_2(t)(\forall t \in [t_1-h,t_1]$, $t_1 \geq t_0$ it follows that$(\mathcal{F}u_1)(t) = (\mathcal{F}u)(t)$ for almost all $t \in [t_1,\infty)$. Here we do not need in such assumption, i.e. delay in Eq. (3.11) can be unbounded.

A generalization of Thm. 3.2 to possible discrete delays in the argument of $\dot{x}$ is as follows.

Theorem 3.4 Assume that all the above mentioned conditions for Eq. (3.11) are fulfilled and for any $t_1 \in (t_0,\infty)$ *the operator* $\mathcal{F}:\text{Lip}[t_0-h,t_1] \to L_\infty[t_0,t_1]$ *is continuous. Further suppose that for any* $t^* \in [t_0,\infty)$ *and* $M>0$ *there exist* $L>0, l\in[0,1), g\in(0,h)$ *such that if* $t \in [\max\{t_0, t^*-g\}, t^*+g], u_i \in \text{Lip}[t_0-h,t], \| u_i \|_{\text{Lip}} < M (i=1,2)$ *then*

$$\| \mathcal{F}u_1 - \mathcal{F}u_2 \| \leq L \| u_1 - u_2 \| + l \| \dot{u}_1 - \dot{u}_2 \| . \tag{3.12}$$

Then there is a $t_\phi \in (t_0,\infty]$ *such that*

a) there exists a solution x *of Eq. (3.11) on the interval* $[t_0,t_\phi)$ *(initial condition (3.2) has been included in this equation);*

b)on any interval $[t_0,t_1] \subset [t_0,t_\phi)$ *this solution is unique;*

c) if $t_\phi < \infty$ *then* $\operatorname*{vrai\ sup}_{t_0 \leq t < t_\phi} |\dot{x}(t)| = \infty$*;*

d) the solution x *and* $\dot{x}$ *depend continuously upon* $\mathcal{F}$ *and* ϕ *in the stated below sense.*

Let for any fixed $t_1 \in (t_0,t_\phi)$ and $\varepsilon>0$ be prescribed a set U of perturbed continuous operators

$$(u \to \overline{\mathcal{F}}u) : \text{Lip}[t_0-h,t_1] \to L_\infty[t_0,t_1]$$

which are equicontinuous for $u \in \Pi^2_{t_1,\varepsilon} := \{u \in \text{Lip}[t_0-h,t_1] : \| u - x^{\phi}_{[t_0-h,t_1]} \| < \varepsilon\}$ and satisfy the Lipschitz condition (3.12) (with their own Lipschitz constants). Then there is a $\delta>0$ such that if $\mathcal{F},\phi$ in (3.11) are replaced on $\overline{\mathcal{F}} \in U, \overline{\phi} \in \text{Lip}[t_0-h,t_0]$ with

$$\| \overline{\mathcal{F}}u - \mathcal{F}u \|_\infty < \delta \text{ in } \Pi^2_{t_1,\ \varepsilon}, \quad \| \overline{\phi} - \phi \|_{\text{Lip}} < \delta$$

then for the solution $\overline{x}$ of the transformed problem with the maximal interval of existence $[t_0, t_{\overline{\phi}})$ we get

$$t_{\overline{\phi}} > t_1, \quad \| (\overline{x} - x)|_{[t_0,t_1]} \|_{\text{Lip}} < \varepsilon.$$

The proof of Thm. 3.4 is similar to the proof of Thm. 3.2.

Variants of these assertions similar to those investigated in Subs. 2.3 and Thm. 2.3 remain also in force.

3.5. *HALE'S FORM OF NDES*

Consider NDEs in the form proposed by J. Hale

$$[x(t) - g(t, x_t)]^{\bullet} = f(t, x_t) \tag{3.13}$$

(see Eq. (1.1.9)) with the same initial condition (3.2). If the functional g is differentiable, then (3.13) can be reduced to (3.1). But, in general such differentiability is not assumed, and moreover the formulations of solvability theorems for Eq.(3.13) are simpler than for (3.1). Therefore the above mentioned reduction will not be fulfilled.

Note that the converse reduction from (3.1) to (3.13), is not possible, in general. Necessary condition imposed on Eq. (3.1) for this reduction is *quasilinearity*, i.e. linearity with respect to $\dot{x}$. The following examples show, however, that it is not sufficient:

$$\dot{x}(t) = x(t-1)\dot{x}(t-1), \quad \dot{x}(t) = x(t-1)\dot{x}(t-2)$$

The former can be brought to the form (3.13), the latter – not. Nevertheless, many NDEs (including those important for applications), and even whole classes of them can be brought to the form (3.13). Moreover, such form of the equation should follow from the actual sense of the problem. This and the simpler formulation of the thorems are responsible for the fact that the form (3.13) has been widely used during recent years. In the theory of NDEs the form (3.13) takes about the same place as does the divergent form (or the form with divergent principal part) in the theory of partial differential equations.

Let us assume that the functionals f, g in Eq. (3.13) are defined on $[t_0, \infty) \times C[-h, 0]$ $(0 < h < \infty)$, and $\phi \in C[-h, 0]$. A solution of Pr. (3.13), (2.2) is called a continuous function x on some interval J_x with the left end $t_0 \in J_x$ for which $x(t_0) = \phi(0)$ and also the function $(t \mapsto x(t) - g(t, (x^{\phi})_t)) \in C^1(J_x)$ must satisfy Eq. (3.13) for all $t \in J_x$. The function x^{ϕ} is the prolongation of a function x on the interval $[t_0 - h, t_0]$ by the values $\phi(t - t_0)$. Note that differentiability of x is not presupposed. Hence the notion of solution here is similar to that of the quasisolutions in the theory of ODEs.

Theorem 3.5 Let the functionals $f, g : [t_0, \infty) \times C[-h, 0] \to \mathbf{R}^n$ be continuous and for any $t_1 \in [t_0, \infty)$ and bounded set $K \subset C[-h, 0]$ there are $L \geq 0$ $l \in$

$[0,1)$, $g \in (0,h)$ *such that if* $t \in [\max\{t_0, t_1 - g\}, t_1 + g]$, $(u_1, u_2) \in K$, *then*

$$\left. \begin{aligned} |g(t,u_1) - g(t,u_2)| &\le l \parallel u_1 - u_2 \parallel, \\ |f(t,u_1) - f(t,u_2)| &\le L \parallel u_1 - u_2 \parallel . \end{aligned} \right\} \tag{3.14}$$

Then there is a $t_\phi \in (t_0, \infty]$ *such that*

a) there exists a solution x *of Pr. (3.13), (2.2) on the interval* $[t_0, t_\phi)$*;*

b) on any interval $[t_0, t_1] \subset [t_0, t_\phi)$ *this solution is unique;*

c) if $t_\phi < \infty$, *then* $x(t)$ *does not have a finite limit as* $t \to t_\phi^-$*;*

d) the solution x *depends continuously on* f, g, ϕ *in the stated below sense.*

Let for any fixed $t_1 \in (t_0, t_\phi)$ and $\varepsilon > 0$ be prescribed a set $U_f(U_g)$ of perturbed continuous functionals $((t,\psi) \to \overline{f}(t,\psi))$ (correspondingly $((t,\psi) \to \overline{g}(t,\psi)))$: $[t_0, \infty) \times C[-h,0] \to \mathbf{R}^n$ which are equicontinuous in the strip $\Pi^3_{t_1,\varepsilon} : t_0 \le t \le t_1$, $\parallel \psi - x_t \parallel < \varepsilon$ and satisfy the Lipschitz condition (3.14) with their own Lipschitz constants. Then there is a $\delta > 0$ such that if f, g, ϕ in (3.13), (2.2) are replaced by $\overline{f} \in U_f, \overline{g} \in U_g$, $\overline{\phi} \in C[-h,0)$ with

$$|\overline{f}(t,\psi) - f(t,\psi)| < \delta, \ |\overline{g}(t,\psi) - g(t,\psi)| < \delta \ \text{ in } \ \Pi^3_{t_1,\varepsilon}, \ \parallel \overline{\phi} - \phi \parallel < \delta$$

then for the solution $\overline{x}$ of the transformed problem with maximal interval of existence $[t_0, t_{\overline{\phi}})$ we have

$$t_{\overline{\phi}} > t_1, \ |\overline{x}(t) - x(t)| < \varepsilon, \ t_0 \le t \le t_1.$$

Variants of these assertions like to those in Subs. 2.2, 2.3 and Thms. 2.2, 2.3 are also valid. In particular, if it is known that f is bounded on each bounded subset of $[t_0, \infty) \times C[-h,0]$, then in *c)* above, "does not have a finite limite" can be replaced by "is unbounded".

The proof of Thm. 3.5 is a simplified version of the proof of Thm. 2.2. In fact, it is sufficiently to note that a solution of (3.13), (2.2) is equivalent to existence of the continuous solution of the equation

$$x(t) - g(t, x_t) = \phi(0) + \int_{t_0}^{t} f(\tau, x_\tau) d\tau, \quad t_0 \le t < t_x$$

with initial condition (2.2).

4. DIFFERENTIAL INCLUSIONS OF RETARDED TYPE (RDIs)

4.1. *INTRODUCTION*

The general form of retarded type differential inclusion (RDI) of the first order is

$$\dot{x}(t) \in \Phi(t, x_t). \tag{4.1}$$

The meaning of x_t is just the same as earlier. $\Phi(t, x_t)$ is bounded subset in $\mathbf{R}^n$, depending on t and x_t. Detailed assumptions about Φ will be given later. Let us suppose for definiteness that $x_t : [-h, 0] \to \mathbf{R}^n$.

Let us list some cases of natural appearance of RDIs.

A. Let the initial mathematical model have the form of RDE (2.1), but with some uncertainty in the righthand side, which is known incompletely. It is this uncertainty that converts RDE (2.1) into RDI (4.1). The exact solution of RDE (e.g.,under a given initial condition) is principally impossible, but guaranteed estimates of it can be obtained by the investigation of RDI.

B. Consider the following control problem with aftereffect in the phase coordinates

$$\dot{x}(t) = F(t, x_t, u(t)), \quad u(t) \in U(t) \tag{4.2}$$

where $U(t)$ is a given set depending on t. Put for every $v \in C[-h, 0]$

$$\Phi(t, v) = F(t, v, U(t)) : \{F(t, v, u) : \; u \in U(t)\}.$$

Then we can transform (4.2) into (4.1) and investigation of the latter gives us information about the possibilities of control.

C. Let the initial mathematical model be described by a *functional differential inequality of retarded type*

$$\Psi(t, \dot{x}(t), x_t) \leq 0. \tag{4.3}$$

Obviously, (4.3) can be written in the form (4.1), so (4.1) include also this class of inequalities. Remark, however, that conditions on Φ formulated below restrict essentially the class of inequalities (4.3) to which RDIs theory can be applied. E. g., these conditions eliminate the inequalities linear in $\dot{x}(t)$.

D. RDIs (4.1) would appear if one wants to define in a more correct manner the notion of solution of an RDE (2.1) with a righthand side which is discontinuous in the second (functional) argument. This will be discussed later in Subs. 4.4. (Note that integrable discontinuties in the first argument are covered by Caratheodory conditions.)

There are also other cases of natural appearance of RDIs (4.1) and among them in applications, for example, to mechanical systems with forced vibration and combined dry and viscous damping, theory of plasticity, electrical circuits etc.

4.2. *MULTIMAPS*

A *multimap* Φ from a set X into a set Y is a correspondence between these sets associating with each element $x \in X$ a subset $\Phi(x) \subseteq Y$, called the image of x (under Φ). Let us recall some notions related to multimaps used below.

Let $K(\mathbf{R}^n)$ be the set of all compact (i.e. bounded and closed) nonempty subsets of $\mathbf{R}^n$. It is always assumed that $\Phi(x) \in K(\mathbf{R}^n)$ for all $x \in X$. For a metric space X such a map Φ is called *upper semicontinuous* if for any $x_0 \in X, \varepsilon > 0$ there is $\delta > 0$ such that from $x \in U(x_0, \delta)$ it follows $\Phi(x) \subset U(F(x_0), \varepsilon)$. (Here $U(\alpha, \eta)$ or $U(A, \eta)$ are the η–neighbourhood of the point α, and set A respectively.) In the particular case of a simple–valued Φ this definition turns into the usual definition of continuity.

By equivalent definition "of Heine" a map $\Phi : X \to K(\mathbf{R}^n)$ is upper semicontinuous if it is locally bounded and from $x_i \to \overline{x}$ $(i \to \infty), y_i \in \Phi(x_i), y_i \to \overline{y}$ it follows $\overline{y} \in \Phi(\overline{x})$. (Here $\Phi(\overline{x})$ may contain "redundant" points that are nonlimit for sequences $y_i \in \Phi(x_i)$. For example, the multi-valued function $\Phi : \mathbf{R} \to K(\mathbf{R})$ defined by the formulas

$$\Phi(t) = \begin{cases} \{0\}, & t \neq 0, \\ [0,1], & t = 0, \end{cases}$$

is upper semicontinuous.)

A multimap $\Phi : X \to K(\mathbf{R}), X \subseteq R$ is called measurable if the set X is measurable, $\Phi(x)$ is defined for almost all $x \in X$, and for any $\varepsilon > 0$ there exists a measurable set $X_\varepsilon \subseteq X$ such that $\text{meas}(X \setminus X_\varepsilon) < \varepsilon$ and $\Phi|_{X_\varepsilon}$ is upper semicontinuous. If F is a single–valued map, then this definition becomes the well–known measurability definition of Luzin.

We say that a multimap $(t, v) \mapsto \Phi(t, v), \Phi : [t_0, \infty) \times C[-h, 0] \to K(\mathbf{R}^n)$ satisfies the *Carathéodory condition* if: for almost all $t \in [t_0, \infty)$ a map $\Phi(t, \cdot) : C[-h, 0] \to K(\mathbf{R}^n)$ is upper semicontinuous; for any $v \in C[-h, 0]$ the map $\Phi(\cdot, v) : [t_0, \infty) \to K(\mathbf{R}^n)$ is measurable; on any bounded set $Q \subset [t_0, \infty) \times C[-h, 0]$ all values $|\Phi(t, v)|$ are not greater than $f_Q(t)$, where f_Q is an integrable function depending on Q.

(In Subs. 2.4 the modified Carathéodory condition contains also a Lipschitz type inequality that was sufficient for the solution uniqueness. Here that is not the case since for RDIs the uniqueness problem does not arise usually.)

4.3. *SOLVABILITY OF THE CAUCHY PROBLEM FOR RDIS*

Designate by $Kv(\mathbf{R}^n)$ the family of all convex set in $K(\mathbf{R}^n)$. Assume in this subsection that $\Phi : [t_0, \infty) \times C[-h, 0] \to Kv(\mathbf{R}^n)$ and the initial condition (the same as (2.2)) is

$$x_{t_0} = \phi. \tag{4.4}$$

A solution of Pr. (4.1), (4.4) is a function x, defined on some interval J_x with the left end $t_0 \in J_x$, which is absolutely continuous on every closed subinterval of J_x, satisfying (4.4) for $t = t_0$ and (4.1) for almost all $t \in J_x$ (where x has to be extended by the function $t \mapsto \phi(t - t_0)$ to the inerval $[t_0 - h, t_0))$.

Theorem 4.1 Let in Eq. (4.1) the map $\Phi : [t_0, \infty) \times C[-h, 0] \to Kv(\mathbf{R}^n)$ *be upper semicontinuous and bounded on each bounded subset from* $[t_0, \infty) \times C[-h, 0]$, *and* $\phi \in C[-h, 0]$. *Then there is* $t_\phi \in [t_0, \infty)$ *such that*

a) there exists at least one solution of Pr. (4.1),(4.2) on the interval $[t_0, t_\phi)$;

b) for any $t_1 \in (t_0, t_\phi)$ *any solution of this problem defined on* $[t_0, t_1]$ *or* $[t_0, t_1)$ *can be extended to a solution defined on* $[t_0, t_\phi)$;

c) if $t_\phi < \infty$, *then on* $[t_0, t_\phi)$ *there is at least one solution of this problem that is unbounded as* $t \to t_\phi^-$ *(so* t_ϕ *is a maximal value for which both previous statements a), b) are valid);*

d) for any $t_1 \in (t_0, t_\phi)$ *the set all solutions of Pr. (4.1), (4.4) on* $[t_0, t_1]$, *considered as a subset of the space* $C[t_0, t_1]$, *is compact and connected.* (A closed set is said to be *connected* if it cannot be represented as a sum of two nonempty, nonintersecting closed subsets.)*;*

e) for any $t_1 \in (t_0, t_\phi)$ *the solutions set of Pr. (4.1), (4.4) on the interval* $[t_0, t_1]$ *depends upper semicontinuously on the righthand side* Φ *and initial function* ϕ.

This means that for any $\varepsilon > 0$ there exists $\delta > 0$ with the following property. Let $\overline{\Phi}$ and $\overline{\phi}$ satisfying the same general requirement as do Φ, ϕ with $\| \overline{\phi} - \phi \| < \delta$, and for any solution x of the original problem on $[t_0, t_1]$ and any $t \in [t_0, t_1], v \in C[-h, 0]$ such that $\| v - x_t \| < \varepsilon$ the inlclusion $\overline{\Phi}(t, v) \subset U(\Phi(t, v), \delta)$ takes place. Then for the modified in such a way problem with the maximal interval of the solutions existence $[t_0, \bar{t}_{\overline{\phi}})$ we get $\bar{t}_{\overline{\phi}} > t_1$, and for any solution $\overline{x}$ of the modified problem on $[t_0, t_1]$ there is a solution x of the original problem on $[t_0, t_1]$ for which

$$\| \overline{x}(t) - x(t) \| < \varepsilon, \ t_0 \le t \le t_1.$$

Semicontinuity of the solutions set dependence of the Cauchy problem (4.1), (4.4) on Φ and ϕ means, roughly speaking, that this set can not be essentially widen, but can be made essentially smaller, by slight variation of Φ, ϕ.

Thm. 4.1 can also be applied to Pr. (2.1), (2.2) if in Thm. 2.1 the Lipschitz condition on F, sufficient for the solution uniqueness, fails to be valid.

Proof of Thm. 4.1. Construct in a usual way the sequence of the Euler refined broken lines $x = x^k(t) \in \mathbf{R}^n$. Then we obtain that the sequence $\{x^k\}$

is equicontinuous on a sufficiently small interval $[t_0, \bar{t}]$. Hence, going over to a subsequence, we can assume that $\{x^k\}$ tends uniformly to the function $\bar{x} : [t_0, \bar{t}] \to \mathbf{R}^n$. For each $t \in [t_0, \bar{t}]$ and $\varepsilon > 0$, using upper semicontinuity of the map Φ in the $(t, \bar{x}_t)$ $(\bar{x}_t(\theta) := \phi(t - t_0 + \theta)$, if $t + \theta \leq t_0)$ and convexity of the set $U(\Phi(t, \bar{x}_t, \varepsilon)$, we have (if k is sufficiently large and $|t_1 - t| > 0$ is sufficiently small)

$$\{[x^k(t_1) - x^k(t)](t_1 - t)^{-1}\} \in U(\Phi(t, \bar{x}_t), \varepsilon).$$

If $k \to \infty$ then we get that the similar inclusion is valid for $\bar{x}$ instead of x^k. After that, passing to the limit as $\varepsilon \to 0^+$, we prove that all limit values of the quotient $[\bar{x}(t_1) - \bar{x}(t)]/(t_1 - t)$, $t_1 \to t$, belongs to $\Phi(t, \bar{x}_t)$. But the function $\bar{x}$ satisfies the Lipschitz condition. Therefore, $\bar{x}$ is a solution of Eq. (4.1).

The solution $\bar{x}$ can be extended in a usual way. As a result we obtain the solution $x : [t_0, t_x)$ of Pr. (4.1), (4.4) such that if $t_x < \infty$ then x is not uniformly continuous on $[t_0, t_x)$ (otherwise there is $x(t_x^-)$ and further extension would be possible). Denote by t_ϕ the infimum of t_x fo all solutions of this problem. From the boundedness Φ in the neighbbourhood of (t_0, ϕ) it follows that $t_0 < t_\phi$. Hence, the assertions *a)* and *b)* follow from the definition of t_ϕ.

Let us prove *c)*. If *c)* is untrue, then by the definition of t_ϕ, there is a sequence of the solutions $x^k : [t_0, t_k) \to \mathbf{R}^n$ of Pr. (4.1), (4.4) such that $t_k \to t_\phi^+$ as $k \to \infty$, and $\sup |x^k(t)| = \infty$ for $t_0 \leq t < t_k$. Let $\bar{t} \in (t_0, t_\phi]$ be a supremum of those $t \in (t_0, t_\phi]$, for which the sequence $\{x^k\}$ is bounded on $[t_0, t]$. Without loss of generality, passing to the subsequence, we can assume that every subsequence of the sequence $\{x^k\}$ is unbounded on every interval $[t_0, t)$ $(\bar{t} < t \leq t_\phi)$. Further, applying the Cantor diagonal process we can assume that this sequence tends uniformly to some function $\bar{x} : [t_0, \bar{t}) \to \mathbf{R}^n$ on each interval $[t_0, t] \subset [t_0, \bar{t})$. The function $\bar{x}$ is also a solution of Pr. (4.1), (4.4), and by *b)* and our assumption on *c)* it is bounded. Therefore, $\bar{x}$ can be extended on $[t_0, \bar{t}]$. Hence, $\dot{x}^k(t)$ are uniformly bounded in sufficiently small neighbbourhood of the point $\bar{t}$. But this contradicts to the definition of $\bar{t}$, if $\bar{t} < t_\phi$, and to the way of construction of the sequence $\{x^k\}$, if $\bar{t} = t_\phi$. So, *c)* is proven.

The assertion *e)* is proven by contradiction, as previously, using the fact that the set of solutions of the inclusions of the form (4.1) with uniformly bounded righthand sides is equicontinuous.

Let us prove *d)*. The boundedness of the set $M \subset C[t_0, t_1]$ of all the solutions of (4.1), (4.4) on the interval $[t_0, t_1]$ for a given $t_1 \in (t_0, t_\phi)$ is proven by contradiction as previously. From here it follows that the set of these solutions is equicontinuous, and hence by virtue of its completeness, this set is a compact.

Prove at last connectedness of M. Assume the opposite, i.e. $M = M_1 \bigcup M_2, M_1 \bigcap M_2 = \emptyset$, where M_1 and M_2 are nonempty closed sets. Let $\Phi^\varepsilon(t,\psi)$ be the closure of the set $U(\Phi(t,\psi),\varepsilon)$ with $t \in [t_0,\infty), \psi \in C[-h,0]$. This set $\Phi^\varepsilon(t,\psi)$ is convex, and the map $(t,\psi) \mapsto \Phi^\varepsilon(t,\psi)$ is upper semicontinuous and bounded on every bounded set. Using just the same arguments as in the first item of this proof, we obtain that if $\varepsilon_k \to 0^+$ and x^k is a solution of Pr. (4.1), (4.4) with Φ^{ε_k} instead of Φ on $[t_0,t_1]$, where x^k tends uniformly to the function $\overline{x}$, then $\overline{x}$ is also a solution of Pr. (4.1), (4.4).

Denote by M^ε the family of solutions of Pr. $(4.1)^\varepsilon$, (4.4) (with Φ^ε instead of Φ). It is clear that $M \subset M^\varepsilon$, and from *e)* it follows that

$$\forall\eta > 0, \exists\varepsilon_0 > 0 : \forall\varepsilon \in (0,\varepsilon_0) \Rightarrow M^\varepsilon \subset U(M,\eta).$$

Put $M_i^\varepsilon := M^\varepsilon \bigcap U(M_i,\alpha)(i = 1,2)$, where α is the forth of the shortest distance between M_1 and M_2, and $M^\varepsilon \subset U(M,\alpha)$. So, M^ε is a sum of the closed, nonempty, disjoint sets M_1^ε and M_2^ε. Further, let us associate to each solution $x : [t_0,t_1] \to \mathbf{R}^n$ of Pr. (4.1), (4.4) a function $\widetilde{x} : [t_0,t_1] \to \mathbf{R}^n$, which coincides with x in the points $t = t^k := t + k\delta_N$ $(k = 0,\dots,N; \delta_N := (t_1 - t_0)/N)$ and is linear between them. Assume that N is so large that each $\widetilde{x} \in M^\varepsilon$, and each Euler broken line with the step δ_N for the Pr. $(4.1)^\varepsilon$, (4.4) is distant from some function belonging M^ε less than α. Existence of such N can be easily proven by contradiction. Then each of these Euler broken lines belongs to one of the disjoint sets $U(M_i^\varepsilon,2\alpha)(i = 1,2)$. But $x \in M_i \Rightarrow \widetilde{x} \in M_i^\varepsilon$, and each function $\widetilde{x}$ is the Euler broken line for itself. Hence, the compact set of the Euler broken lines with the step δ_N, and chosen ε and N, is disconnected.

Let us show that this conclusion is impossible. In fact, denote by $E_k \subset C[t_0,t^k := t_0 + k\delta_N]$ $(k = 0,\dots,N)$ the family of all Euler broken lines with the step δ_N for Pr. $(4.1)^\varepsilon$, (4.4) on $[t_0,t^k]$. The set E_0 consisting of a unique point is a connected set. The set E_N is disconnected. Hence, there is $k < N$ such that E_k is connected, and E_{k+1} can be represented as a union of the compact, nonempty, disjoint sets D_1 and D_2. Let $f_k : E_{k+1} \to E_k$ be a map of the restrictions of the functions from $[t_0,t^{k+1}]$ into $[t_0,t^k]$. Remind that E_k is connected set, and $f_k(D_i)$ are nonempty and closed sets. Hence, there are the broken lines $s_1 \in D_1$ and $s_2 \in D_2$ for which $f_k(s_1) = f_k(s_2)$, i.e. they coincide for $t_0 \le t \le t^k$. But the $\Phi^\varepsilon(t^k,(s_1)_{t^k})$ is a convex set. Therefore, the "points" s_1 and s_2 can be connected by a "segment", belonging entirely to E_{k+1} that contradicts to the definition of D_i. Consequently, M is a connected set. Thm. 4.1 is proven.□

The assertions and the proof of Thm. 4.1 does not change if instead of the boundedness of the map Φ on every bounded subset of the set $[t_0,\infty) \times C[-h,0]$ we assume that for any such subset there is an integrable function $t \mapsto g(t) \ge 0$ such that $|\Phi(t,\psi)| \le g(t)$ everywhere in M. Further generalizations of this theorem are also possible, in particular, for the case when the map $\Phi : [t_0, \infty) \times C[-h, 0] \to Kv(\mathbf{R}^n)$ satisfies the Carathéodory condition (see Subs. 4.2).

4.4. *GENERALIZED SOLUTIONS OF RDES AND RDIS*

As mentioned in Subs. 4.1, differential inclusions (DIs) arise when we try to give a more precise definition of solution of a differential equation with discontinuous righthand side. E.g., consider the scalar ODE (without after-effect) of motion of a body with vanishingly small mass under switching external and dissipative forces:

$$k\dot{x} = F_0 - (\mathrm{sgn}\, x)F_1, \tag{4.5}$$

where $F_1 > F_0 > 0, \quad k > 0$.

In Fig. 3.4.1 the corresponding direction field in the (x, t)-plane is shown for $x > 0$, $x < 0$, and $x = 0$. It is obvious that an integral curve starting above (in A_1) or below (in A_2) the t-axis, ends on this axis upon reaching it, since any extension of it would contradict the given direction field.

However, the actual process does continue in some way. So, for an adequate description of this process, the notion of solution of (4.5) should be revised. Note that an "instantaneous" switching of the force from $F_0 - F_1$ (for $x > 0$) to $F_0 + F_1$ (for $x < 0$) means that in reality the force changes from one value to another in a negligibly small time interval, on which it must take all intermediate values. Hence the process can be restricted by the DI

$$k\dot{x} \in \begin{cases} \{F_0 - (\mathrm{sgn}\, x)F_1\}, & x \neq 0, \\ [F_0 - F_1, F_0 + F_1], & x = 0. \end{cases} \tag{4.6}$$

The solutions of (4.6) can be regarded as generalized solutions of (4.5), and have arbitrary long prolongations. In particular, in Fig. 3.4.1, upon reaching B_i the integral curve of the DI starting at A_i $(i = 1, 2)$ continues along the t-axis.

So the evolution of the process under consideration consists of two stages. Initially the motion is described by (4.5), and after reaching the level $x = 0$ (i.e. the discontinuity point of the righthand side of the equation) it is described by the DI (4.6). Of course, (4.6) can also be used for the first stage.

For more complicated processes, described by ODEs with discontinuities on surfaces in phase space, the situation is very similar. Namely, such a process will consist of a number of stages, and is on certain stages represented by DEs, and on other stages (during which the representing point moves along the above mentioned discontinuity surfaces) by corres-

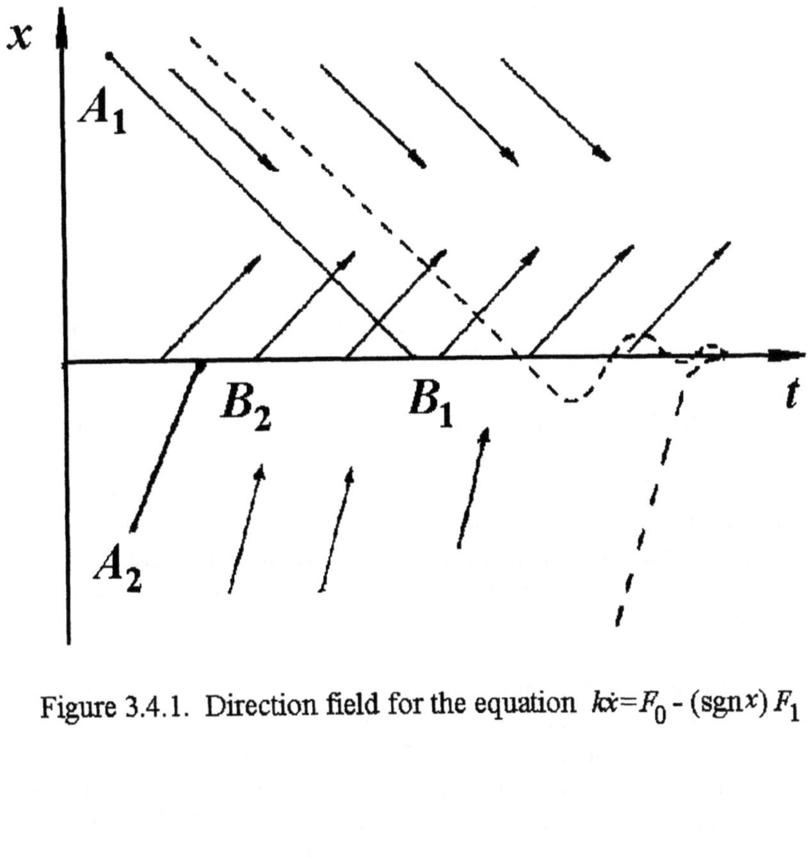

Figure 3.4.1. Direction field for the equation $k\dot{x}=F_0 - (\operatorname{sgn} x) F_1$

ponding DIs.

Stages of the second kind are called *sliding modes*, because it is as if the representative point in a phase space slides along the discontinuity surface, squeezed from both sides of this surface by the direction field.

Sliding modes are usually obtained as limits of a more detailed description of the process not using DIs. E.g., the solutions of (4.5) can be obtained as the limits as $m \to 0^+$ of the solutions of the equation

$$m\ddot{x} + k\dot{x} = F_0 - (\mathrm{sgn}x)F_1, \tag{4.7}$$

taking into account a small inertia, or as the limits as $a \to 0^+$ of the solutions of the equation

$$k\dot{x} = F_0 - \frac{2}{\pi}\left(\arctan\frac{x}{a}\right)F_1, \tag{4.8}$$

which models a fuzzy set of external force switching. Approximate graphs of solutions of these more detailed equations are indicated in Fig. 3.4.1 by dashed lines.

For RDIs we have a problem of precisely the type discussed above. Assume that in the problem (2.1), (2.2) the functional $(t, \psi) \mapsto F(t, \psi)$ is measurable in t for any fixed $\psi \in C[-h, 0]$, is bounded modulo by an integrable continuous almost everywhere function of t on every bounded subset of $[t_0, \infty) \times C[-h, 0]$, while at the same time there is no continuous dependence on ψ for almost all $t \in [t_0, \infty)$, as stipulated in Thm. 2.2. Then, even for $\phi \in C[-h, 0]$, a solution to the problem can be absent. (See (4.5) with $x(t_0) = 0$.)

To overcome this obstacle we use models of actual phenomena in the same way as for (4.5). Let, for $(t, \psi) \in [t_0, \infty) \times C[-h, 0], \Phi_F(t, \psi)$ be the convex hull of the limit sets of all convergent sequences $\{F(t_i, \psi_i)\}$, where $t_i \to t, \psi_i \to \psi$ as $i \to \infty$. (Recall that the *convex hull* of subset M of a linear space is the smallest convex set containing M.) Then the conditions of Thm. 4.1 generalized for the case of Caratheodory conditions for the functional Φ (see the end of Subs. 4.3) hold for the problem

$$\dot{x}(t) \in \Phi_F(t, x_t), \quad x_{t_0} = \phi. \tag{4.9}$$

It is natural to call solutions of (4.9) *generalized solutions* of (2.1), (2.2). So, these generalized solutions have all the properties stated in Thm. 4.1.

In certain cases the introduction of delays into an eqution eliminates the sliding modes. E.g., in (4.5) let us introduce a delay $h > 0$, and obtain

$$k\dot{x}(t) = F_0 - [\mathrm{sgn}x(t - h)]F_1. \tag{4.10}$$

It is readily verified that if the initial function $\phi \in C[-h, 0]$ and $\phi(\theta) \neq 0$, then the generalized solution can only have isolated zeros (and must be periodic from some t onwards, see Fig. 3.4.2). So the generalized solution is a solution in the sense of Subs. 2.4, although the conditions of Thm. 2.2 do not hold. However, there are "proper" generalized solutions of (4.10), e.g., $x(t) \equiv 0$. These solutions are in certain sense exceptional, because they may be transformed into ordinary solutions by an arbitrarily small variation of the initial function.

This example is typical. Thus, if the righthand side of the RDE (1.1.4) depends discontinuously on $x(t - h_i(t))$ for $h_i(t) > 0$ only, then generalized solutions as a rule do not have sliding modes and are ordinary solutions in the sense of Subs. 2.4. If, on the other hand, this dependence is discontinuous for $h_i(t) \equiv 0$, then sliding modes are an typical property of generalized solutions, and along them the corresponding RDI is satisfied usually but not (1.1.4).

We can similarly investigate RDIs (4.1) under the conditions of Thm. 4.1 except the requirement of convexity of the sets $\Phi(t, \psi)$. This case is a practical possibility. First of all we encounter the fact that, as for (4.5), the solution of (4.1), (4.4) need not exist. The simplest example of this kind is the ODI (without delay)

$$\dot{x}(t) \in \begin{cases} \{-\operatorname{sgn} x(t)\}, & x(t) \neq 0, \\ \{-1, 1\}, & x(t) = 0, \end{cases}$$

with $x(t_0) = 0$. Even the additional requirements that the set $\Phi(t, \psi)$ be connected and depend continuously on (t, ψ) hardly help, although the counterexamples seem rather artifical. (A map Φ from a metric space X into $K(\mathbf{R}^n)$ is said to be continuous if for all $x_0 \in X$,

$$\inf\{\varepsilon : \Phi(x) \subset U(\Phi(x_0), \varepsilon),\ \Phi(x_0) \subset U(\Phi(x), \varepsilon)\} \to 0 \quad \text{as} \quad x \to x_0.)$$

Moreover, if (4.1), (4.4) has solutions, then the solution set is, in general, nonclosed with respect to uniform convergence. This can be demonstrated using the following scalar DI (without delay) with constant righthand side:

$$\dot{x}(t) \in \{-1, 1\}.$$

A solution of this DI is an absolutely continuous function with $|\dot{x}(t)| = 1$ almost everywhere. However, an arbitrary function satisfying the Lipschitz condition with constant 1 can be uniformly approximated to arbitrary accuracy using such functions. Thus the solution set is nonclosed.

These obstacles can be overcome as before. Namely, suppose that all conditions of Thm. 4.1 except convexity of the sets $\Phi(t, \psi)$ hold for (4.1),

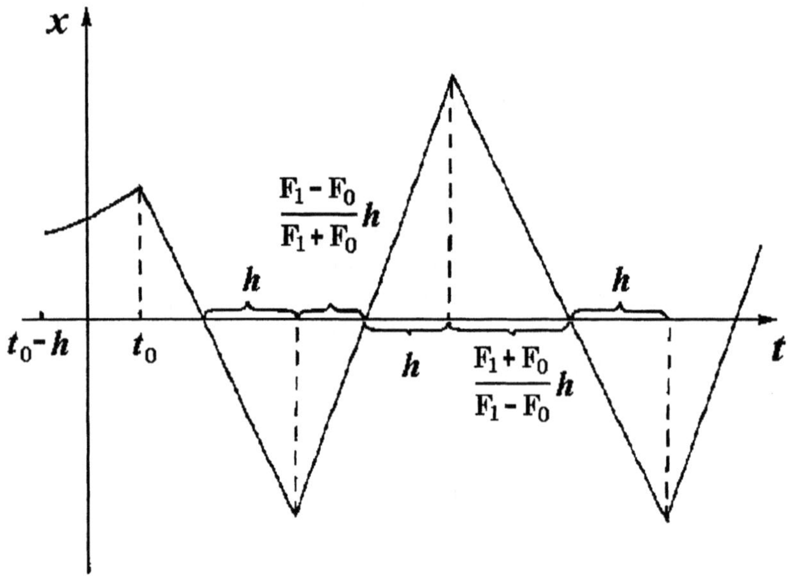

Figure 3.4.2. Trajectory of the equation $k\dot{x}(t)=F_0-[\operatorname{sgn} x(t-h)]F_1$

(4.4). Let, for any $(t,\psi) \in [t_0,\infty) \times C[-h,0]$, $\mathrm{co}\Phi(t,\psi)$ be the convex hull of the set $\Phi(t,\psi)$. Then all conditions of Thm. 4.1 hold for the problem

$$\dot{x}(t) \in \mathrm{co}\Phi(t,x_t), \quad x_{t_0} = \phi. \tag{4.11}$$

In the case under consideration, the solutions of (4.11) are naturally called *generalized solutions* of (4.1), (4.4). So these generalized solutions have all the properties described in Thm.4.1.

Note that a generalized solution of (4.1), (4.4) can be approximated by "proper" solutions of this problem to arbitrary accuracy by an arbitrarily small variation of the righthand side, e.g. by introducing an arbitrarily small additional delay. This fact, together with structural robustness of mathematical nodels of actual phenomena (i.e. stability of the description), give supplementary justification for using generalized solutions.

Similarly the notion of the generalized solution of Pr. (4.1), (4.4) is introduced for the most general case, even if *apriori* it is not assumed that $\Phi(t,\psi)$ be convex and compact and the map $\Phi(t,\cdot)$ be upper semicontinuous.

5. GENERAL LINEAR FDEs WITH AFTEREFFECT

5.1. *CAUCHY PROBLEM FOR LINEAR RDES*

In almost all applications linear RDEs can be represent in the following general form:

$$\dot{x}(t) = \sum_{j=1}^{m} A_j(t)x(t-h_j(t)) + \int_{-h(t)}^{0} A(t,\theta)x(t+\theta)d\theta + f(t), \qquad t_0 \le t < \infty. \tag{5.1}$$

Here $x(t) \in \mathbf{R}^n, n \ge 1, A_j, A$ are square n-order matrix functions, all $h_j(t) \ge 0, \quad h(t) \ge 0$. The Cauchy problem for (5.1) is to construct a solution satisfying

$$x(t) = \phi(t), \quad -\infty < t \le t_0, \tag{5.2}$$

where the inerval $(-\infty,t_0]$ is chosen for simplicity reasons only. In fact the function ϕ may be given only on an interval J with righthand end $t_0 \in J$ and containing all values $t-h(t), t-h_j(t) \le t_0 \quad (j=1,\dots,m; \quad t \ge t_0)$ Usually J is finite.

There are two theorems about the solvability of (5.1), (5.2) that are used often. One of them is considered with classical solutions, i.e. functions of class C^1 satisfying (5.1), (5.2) for all t for which they are defined and equal to $\phi(t_0)$ at $t=t_0$. The other theorem, which deals with absolutely continuous solutions satisfying (5.1) only almost everywhere, covers equations with discontinuous locally summable coefficients.

Theorem 5.1 Assume that all $A_j, h_j, h, f \in C[t_0, \infty), \phi \in C(-\infty, t_0]$, *while* A *is continuous on the set* $\{(t,\theta) : t_0 \le t < \infty, -h(t) \le \theta \le 0\}$. *Then there is a classical solution of (5.1), (5.2) on the interval* $[t_0, \infty)$. *On any interval* $[t_0, t_1] \subset [t_0, \infty)$ *this solution is unique and depends continuously on all prescribed functions.*

The last means that for any $t_1 \in (t_0, \infty), \varepsilon > 0$ there is a $\delta > 0$ such that when the functions A_j, h_j, A, h, f, ϕ in (5.1), (5.2) are replaced by continuous functions $\overline{A}_j, \overline{h}_j, \overline{A}, \overline{h}, \overline{f}, \overline{\phi}$ satisfying

$$\overline{h}_j(t) \ge 0, \quad \overline{h}(t) \ge 0,$$

$$\left.\begin{aligned} &\int_{t_0}^{t_1} |\overline{A}_j(t) - A_j(t)|dt < \delta, \quad |\overline{h}_j(t) - h_j(t)| < \delta, \quad |\overline{h}(t) - h(t)| < \delta, \\ &\qquad t_0 \le t \le t_1, \quad j = 1, \ldots, n, \\ &\int_{t_0}^{t_1} |\overline{f}(t) - f(t)|dt < \delta, \quad \int_{t_0}^{t_1} dt \int_{-\infty}^{0} |\tilde{\overline{A}}(t,\theta) - \tilde{A}(t,\theta)|d\theta < \delta, \end{aligned}\right\} \tag{5.3}$$

where

$$\tilde{A}(t,\theta) = \begin{cases} A(t,\theta), & -h(t) \le \theta \le 0, \\ 0, & -\infty < \theta < -h(t), \end{cases}$$

$$\tilde{\overline{A}}(t,\theta) = \begin{cases} \overline{A}(t,\theta), & -\overline{h}(t) \le \theta \le 0, \\ 0, & -\infty < \theta < -\overline{h}(\theta), \end{cases}$$

and $|\overline{\phi}(t) - \phi(t)| < \delta, \quad -\infty < t \le t_0$, then $|\overline{x}(t) - x(t)| < \varepsilon, \quad t_0 \le t \le t_1$.

Proof of Thm. 5.1. The assertion of Thm. 5.1 about existence and uniquiness of the solution follow from the assertions *a)* and *b)* of Thm. 2.1 and Subs. 2.2(.4), 2.2(.5). More exactly, Thm. 2.1 has a condition $h < \infty$; but existence of the solution is sufficiently to prove on every bounded interval $[t_0, t_1]$ where we can put $h = \max\{h_1(t), \ldots, h_m(t), h(t) : t_0 \le t \le t_1\}$. So, the condition $h < \infty$ holds.

Prove that the solution of Pr. (5.1), (5.2) depends continuously on the given functions. Take $t_1 \in (t_0, \infty)$ and put

$$y(t) := \sup\{|\overline{x}(\tau) - x(\tau)| : -\infty < \tau \le t\}, \quad t_0 \le t \le t_1, \tag{5.4}$$

where x is a solution of nonperturbed problem (5.1), (5.2) with fixed given functions, and $\overline{x}$ is a solution of perturbed problem, satisfying (5.3) with $\| \tilde{\phi} - \phi \| < \delta$ (initial conditions are taken into account in definition of $y(t)$).

The perturbed and unperturbed equations (5.1) imply that if $y(t) > \delta$ for some $t \in [t_0, t_1]$, then there is $\tau = \tau(t) \in (t_0, t]$ such that

$$y(t) = |\overline{x}(\tau) - x(\tau)| = |[\overline{\phi}(t_0) + \int_{t_0}^{\tau} \sum_{j=1}^{m} \overline{A}_j(\tau_1)\overline{x}(\tau_1 - \overline{h}_j(\tau_1))d\tau_1 +$$

$$+ \int_{t_0}^{\tau} d\tau_1 \int_{-\overline{h}(\tau_1)}^{0} \overline{A}(\tau_1, \theta)\overline{x}(\tau_1 + \theta)d\theta + \int_{t_0}^{\tau} \overline{f}(\tau_1)d\tau_1] -$$

$$-[\phi(t_0) + \int_{t_0}^{\tau} \sum_{j=1}^{m} A_j(\tau_1)x(\tau_1 - h_j(\tau_1))d\tau_1 +$$

$$+ \int_{t_0}^{\tau} d\tau_1 \int_{-h(\tau_1)}^{0} A(\tau_1, \theta)x(\tau_1 + \theta)d\theta + \int_{t_0}^{\tau} f(\tau_1)d\tau_1]| \leq$$

$$\leq |\overline{\phi}(t_0) - \phi(t_0)| + \sum_{j=1}^{m} \int_{t_0}^{\tau} |\overline{A}_j(\tau_1) - A_j(\tau_1)|\ |x(\tau_1 - \overline{h}_j(\tau_1))|d\tau_1 +$$

$$+ \sum_{j=1}^{m} \int_{t_0}^{\tau} |\overline{A}_j(\tau_1) - A_j(\tau_1)|\ |\overline{x}(\tau_1 - \overline{h}_j(\tau_1)) - x(\tau_1 - \overline{h}_j(\tau_1))|d\tau_1 +$$

$$+ \sum_{j=1}^{m} \int_{t_0}^{\tau} |A_j(\tau_1)|\ |\overline{x}(\tau_1 - \overline{h}_j(\tau_1)) - x(\tau_1 - \overline{h}_j(\tau_1))|d\tau_1 +$$

$$+ \sum_{j=1}^{m} \int_{t_0}^{\tau} |A_j(\tau_1)|\ |x(\tau_1 - \overline{h}_j(\tau_1)) - x(\tau_1 - h_j(\tau_1))|d\tau_1 +$$

$$+ \int_{t_0}^{\tau} d\tau_1 \int_{-\infty}^{0} |\overline{\widetilde{A}}(\tau_1, \theta) - \widetilde{A}(\tau_1, \theta)|\ |x(\tau_1 + \theta)|d\theta +$$

$$+ \int_{t_0}^{\tau} d\tau_1 \int_{-\infty}^{0} |\overline{\widetilde{A}}(\tau_1, \theta) - \widetilde{A}(\tau_1, \theta)|\ |\overline{x}(\tau_1 + \theta) - x(\tau_1 + \theta)|d\theta +$$

$$+\int_{t_0}^{T} d\tau_1 \int_{-\infty}^{0} |\widetilde{A}(\tau_1,\theta)|\,|\overline{x}(\tau_1+\theta)-x(\tau_1+\theta)|d\theta + \int_{t_0}^{T} |\overline{f}(\tau_1)-f(\tau_1)|d\tau_1 \leq$$

$$\leq \delta + m\delta\alpha + \sum_{j=1}^{m}\int_{t_0}^{t} |\overline{A}_j(\tau)-A_j(\tau)|y(\tau)d\tau + \sum_{j=1}^{m}\int_{t_0}^{t} |A_j(\tau)|y(\tau)d\tau +$$

$$+\sum_{j=1}^{m}\int_{t_0}^{t_1} |A_j(t)|\beta(\delta)dt + \delta\alpha + \int_{t_0}^{t}\Big(\int_{-\infty}^{0} |\widetilde{\overline{A}}(\tau,\theta)-\widetilde{A}(\tau,\theta)|d\theta\Big)y(\tau)d\tau + \tag{5.5}$$

$$+\int_{t_0}^{t}\Big(\int_{-h(\tau)}^{0} |A(\tau,\theta)|d\theta\Big)y(\tau)d\tau + \delta;$$

here $\alpha := \max|x(t)|$ on the minimal inerval I, which contains all the values $t - h_j(t)$ and $t - h(t)$ for $t_0 \leq t \leq t_1$, and

$$\beta(\delta) := \min\{|x(t) - x(t-\eta)| : t \in I,\ \ 0 \leq \eta \leq \delta\}.$$

For a given t_1 both α and $\beta(\delta)$ are characteristics of the nonperturbed solution with $\beta(\delta) \to 0$ as $\delta \to 0$. From (5.5) and the Bellman inequality we have for $t_0 \leq t \leq t_1$

$$y(t) \leq [\delta + m\delta\alpha + \beta(\delta)\sum_{j=1}^{m}\int_{t_0}^{t_1} |A_j(t)|dt + \delta\alpha + \delta]\times$$

$$\times \exp\int_{t_0}^{t}\{\sum_{j=1}^{m}[|\overline{A}_j(\tau)-A_j(\tau)| + |A_j(\tau)|]+$$

$$+\int_{\infty}^{0} |\widetilde{\overline{A}}(\tau,\theta)-\widetilde{A}(\tau,\theta)|d\theta + \int_{-h(\tau)}^{0} |A(\tau,\theta)|d\theta\}d\tau \leq$$

$$\leq \{[2+(m+1)\alpha]\delta + \beta(\delta)\sum_{j=1}^{m}\int_{t_0}^{t_1} |A_j(t)|dt\}\exp[(m+1)\delta+$$

$$+\sum_{j=1}^{m}\int_{t_0}^{t_1} |\overline{A}_j(\tau)-A_j(\tau)|d\tau + \int_{t_0}^{t_1} dt \int_{-h(t)}^{0} |A(t,\theta)|d\theta] \to 0 \quad \text{as} \quad \delta \to 0.$$

Thm. 5.1 is proven. □

Theorem 5.2 Assume that all $A_j, f \in L_1[t_0,\infty), h_j, h \in C[t_0,\infty)$*, the function* A *is defined and measurable on the set* $\{(t,\theta) : t_0 \le t < \infty, -h(t) \le \theta \le 0\}$ *with* $(t \mapsto \int_{-h(t)}^{0} |A(t,\theta)|d\theta) \in L_1[t_0,\infty)$*, the function* ϕ *has only first kind discontinuity points, whose set is denoted by* $S_\phi \subset (-\infty, t_0]$*, and the set* S_ϕ *and all sets* $\{t : t - h_j(t) \in S_\phi\}$ *do not have limit points. Then there is an absolutely continuous solution of (5.1), (5.2) on the interval* $[t_0,\infty)$*. On any interval* $[t_0, t_1] \subset [t_0,\infty)$ *this solution is unique and depends continuously on all prescribed functions.*

The last means that for any $t_1 \in (t_0,\infty), \varepsilon > 0, M > 0$ there is a $\delta > 0$ such that when the functions A_j, h_j, A, h, f, ϕ in (5.1), (5.2) are replaced by functions $\overline{A}_j, \ldots, \overline{\phi}$ satisfying the same requirements as do $A_j, \ldots, \phi$, for which the inequalities (5.3) holds, and also

$$\left.\begin{aligned} &|\overline{\phi}(t) - \phi(t)| < \delta,\ t \in \{t_0\} \bigcup (-\infty, t_0) \setminus U(S_\phi, \delta), \\ &|\overline{\phi}(t) - \phi(t)| < M,\ t \in U(S_\phi, \delta) \bigcap (-\infty, t_0) \end{aligned}\right\} \tag{5.6}$$

we obtain $|\overline{x}(t) - x(t)| < \varepsilon, \quad t_0 \le t \le t_1.$

Proof of Thm. 5.2. Existence and uniqueness of the absolute continuous solution follow from the generalization of Thm. 2.2 (see the end of Subs. 2.4). Continuous dependence of the solution on the given functions can be proven similar to the proof of Thm. 5.1 with the following modifications. First of all, in the definition (5.4) of $y(\cdot)$ the values $|\overline{\phi}(\tau) - \phi(\tau)|$ for $\tau \in U(S_\phi, \delta) \bigcap (-\infty, t_0]$ have to be ignored. This gives the following additional summands in the process of transition from the next to the last part of estimate (5.5) to the last one

$$\begin{aligned} M m\delta + M \sum_{j=1}^{m} \int_{V_{\delta,j}} |A_j(t)|dt + \sum_{j=1}^{m} \int_{V_{\delta,j}^1} |\phi(t - \overline{h}_j(t)) - \phi(t - h(t_j))|dt + \\ + M\delta + M \int_{t_0}^{t_1} dt \int_{U(S_\phi,\delta)-t} |\tilde{A}(t,\theta)|d\theta; \end{aligned} \tag{5.7}$$

here

$$V_{\delta,j} := \{t \in [t_0, t_1] : t - \overline{h}_j(t) \in U(S_\phi, \delta) \bigcap (-\infty, t_0],$$

$$V_{\delta,j}^1 := V_{\delta,j} \bigcap \{t \in [t_0, t_1] : t - h_j(t) \in U(S_\phi, \delta)\} \bigcap (-\infty, t_0],$$

$$U(S_\phi, \delta) - t := \{\tau - t : \tau \in U(S_\phi, \delta) \bigcap (-\infty, t_0]\}.$$

The set $S_\phi^j := \{t \in [t_0, t_1] : t - h_j(t) \in S_\phi\}$ is finite for each $j = 1, \ldots, m$. In addition, the sets $V_{\delta,j}$ and $V_{\delta,j}^1$ for sufficiently small $\delta > 0$ are contained in arbitrary small neighbourhood of the set S_ϕ^j, which is independent on the choice of $h(\cdot)$. Hence, the second and third summands in the righthand side of (5.7) tend to zero as $\delta \to 0$. Similar behaviour of the last summand in (5.7) takes place because for it the measure of the plane domain of integration is not greater than $2(t_1 - t_0)\delta$ card $(S_\phi \bigcap I)$, where I is the same interval as in the proof Thm. 5.1.

So, in spite of the additional summands (5.7), using the Bellman inequality we obtain that $y(t) \to 0$ as $\delta \to 0$ for every $t \in [t_0, t_1]$.

Theorem 5.2 is proven. □

5.2. *GENERALIZATION*

The linear RDE (5.1) admits a generalization which is widespread in mathematical papers and is founded on the use of the Stieltjes integral. We first recall some definitions.

The *variation $V_J r$ of a function $r : J \to \mathbf{R}^n$*, where J is an interval, is equal to

$$V_J r := \sup \sum_{i=1}^{k} |r(t_i) - r(t_{i-1})|.$$

Here the supremum is taken over all $k \geq 1$ and all $t_0 < t_1 < \ldots < t_k$ in J. If $V_J r < \infty$, then we say that r is *bounded variation* (sometimes *finite variation*) on J. The set of functions $r : J \to \mathbf{R}^n$ with $V_{j_1} r < \infty$ for all closed intervals $J_1 \subseteq J$ is denoted by BV$(J, \mathbf{R}^n)$(or BV(J), BV J). If J is a closed interval, BV J is a Banach space, with norm

$$\| r \|_{\mathrm{BV}(J)} := V_J r + |r(t_0)|,$$

where $t_0 \in J$ is some fixed point.

A function $r \in$BV$(J, \mathbf{R}^n)$ has a unique (up to constant terms) *Lebesgue decomposition*

$$r = r_a + r_d + r_s. \tag{5.8}$$

Here, r_a is absolutely continuous on every closed interval $J_1 \subseteq J$ and

$$V_J r_a = \int_J |\dot{r}_a(t)| dt.$$

The *jump function* r_d has the following structure:

$$r_d(t) = \begin{cases} -\sum\limits_{j:t_j\in(t,\bar{t}]} \alpha_j - \sum\limits_{j:t_j\in[t,\bar{t})} \beta_j + c, & t < \bar{t}, \\ c, & t = \bar{t}, \\ \sum\limits_{j:t_j\in(\bar{t},t]} \alpha_j + \sum\limits_{j:t_j\in[\bar{t},t)} \beta_j + c, & t > \bar{t}, \end{cases} \qquad (5.9)$$

where $\{t_1, t_2, \ldots\}$ is an at most countable set of points $t_j \in J, \bar{t} \in J, \alpha_j, \beta_j, c \in \mathbf{R}^n$, and the series are absolutely convergent. Moreover, $\alpha_j = 0$ ($\beta_j = 0$) if $t_j = \inf J$ ($t_j = \sup J$), and

$$V_J r_d = \sum_j (|\alpha_j| + |\beta_j|).$$

In particular, if the set $\{t_j\}$ does not have limit points, then r_d is piecewise constant. Finally, r_d is a *singular function*, i.e. $r_s \in C(J)$ and $\dot{r}_s(t) = 0$ for almost all $t \in J$. With the representation (5.8) we have the formula

$$V_J r = V_J r_a + V_J r_d + V_J r_s.$$

This implies that if the lefthand side is finite, then so are all terms at the righthand side.

In applications the decomposition (5.8) of the function $r \in$BV(J) usually contains only the first two terms.

The space BV$(J, \mathrm{R}^{n\times m})$, where $\mathrm{R}^{n\times m}$ is the space of real $(n \times m)$ – matrices, can be defined similarly to BV$(J, \mathbf{R}^n)$.

If $f \in C(J, \mathbf{R}^n)$, $\sup\limits_{t\in J} |f(t)| < \infty$, $r : J \to \mathrm{R}^{n\times m}, V_J r < \infty$, then the limit

$$\lim \sum_{i=1}^{k} [r(t_i) - r(t_{i-1})] f(\tau_i)$$

exists, where $t_0 < \tau_1 < t_1 < \tau_2 < \ldots < \tau_k < t_k$ are points in J and the limit is taken for $k \to \infty$, $t_0 \to \inf J$, $t_k \to \sup J$, $\max\limits_i (t_i - t_{i-1}) \to 0$. This limit is called the *Stiltjes integral of f with respect to r*, and is denoted by

$$\int_J [dr(t)] f(t).$$

The usual properties of the (Rieman) integral can be generalized to Stiltjes integrals in a natural way. We mention the estimate

$$\left| \int_J [dr(t)] f(t) \right| \leq \int_J |f(t)| \cdot |dr(t)|$$

where, by definition,

$$\int_J |f(t)| \cdot |dr(t)| := \liminf \sum_{i=1}^{k} |f(\tau_i)| \cdot |r(t_i) - r(t_{i-1})|.$$

If in (5.8) the function r_s is absent and r_d has the form (5.6), then

$$\int_J [dr(t)]f(t) = \int_J \dot{r}_a(t)f(t)dt + \sum_j (\alpha_j + \beta_j)f(t_j). \qquad (5.10)$$

The generalization of (5.1) mentioned at the beginning of Subs.5.2 takes the form

$$\dot{x}(t) = \int_{-\infty}^{0} [d_\theta R(t,\theta)]x(t+\theta) + f(t), \quad t_0 \le t < \infty, \qquad (5.11)$$

under the initial condition (5.2). Here d_θ means that integration is with respect to θ; $f \in L_1[t_0,\infty)$, and $\phi \in C(-\infty,t_0]$ with $\sup|\phi(t)| < \infty$. It is also assumed that the following *condition A* be fulfilled:

Condition A. The *kernel* $R : [t_0,\infty) \times (-\infty,0] \to \mathbf{R}^{n\times m}$ is measurable and

$$V_{(-\infty,0]}R(t,\cdot) \le \lambda(t), \quad t_0 \le t < \infty, \qquad (5.12)$$

where $\lambda \in L_1[t_0,\infty)$.

(Note that the requirements on ϕ may be relaxed when those on $R(t,\cdot)$ are made stronger, and vice versa, so as to ensure convergence of the integral in (5.11). E.g., the condition $\sup|\phi(t)| < \infty$ may be dropped if for any $t \in [t_0,\infty)$ the function $\theta \mapsto R(t,\theta)$ is constant for all $\theta \in (-\infty,-h(t)], h(t) \in [0,\infty)$. In this case the integral (5.11) has the form $\int_{-h(t)}^{0}$.)

Under the above assumptions Pr. (5.11), (5.2) has an absolutely continuous solution on the interval $[t_0,\infty)$, and on every interval $[t_0,t_1] \subset [t_0,\infty)$ this solution is unique and depends continuously on the kernel R and the functions f,ϕ (in a natural sense, not elaborated here). If, in addition, *condition B* below holds, then this solution is a classical solution:

Condition B. The functions $R(\cdot,0), \lambda$ (see (5.12)) and f belong to $C[t_0,\infty)$, the function ϕ is uniformly continuous, and

$$\forall t \in [t_0,\infty) \Rightarrow \int_{-\infty}^{0} |R(t,\theta) - R(t_1,\theta)|d\theta \to 0 \ \text{ as } t_1 \to t.$$

Using (5.10) the Eq. (5.1) follows from (5.11) if

$$R(t,\theta) = \int_{-\infty}^{0} \tilde{A}(t,\tau)d\tau + \sum_{j=1}^{m} H(\theta + h_j(t))A_j(t),$$

$$t_0 \leq t < \infty, \quad -\infty < \theta \leq 0,$$

where H is the Heaviside function and

$$\tilde{A}(t,\tau) = \begin{cases} A(t,\tau), & -h(t) \leq \tau \leq 0, \\ 0, & -\infty < \tau < -h(t). \end{cases}$$

5.3. *INTEGRAL REPRESENTATION FOR THE SOLUTION OF THE CAUCHY PROBLEM (VARIATION OF CONSTANTS FORMULA)*

Here we assume that all functions given in (5.1), (5.2) satisfy the conditions of Thm. 5.2. Then the solution of this problem can be represented as $x = x_f + x_\phi$, where $x_f(x_\phi)$ is the solution of Pr. (5.1), (5.2) with $\phi = 0$ (respectively, $f = 0$). We will consider integral representations for x_f and x_ϕ.

Suppose that for any $s \in [t_0, \infty)$ the sets $\{t : t - h_i(t) = s\}$ do not have limit points. For such an s we denote by $(t \mapsto C(t,s)) : [s, \infty) \to \mathbf{R}^{n\times m}$ the solution of the matrix equation

$$\frac{\partial C}{\partial t} = \sum_{j=1}^{m} A_j(t)C(t - h_j(t); s) + \int_{-h(t)}^{0} A(t,\theta)C(t+\theta; s)d\theta, \tag{5.13}$$

$$s \leq t < \infty,$$

under the initial condition

$$C(t;s) = \begin{cases} 0, & -\infty < t < s, \\ I, & t = s, \end{cases}$$

where I is the identity matrix. Writing (5.13) out in columns, we obtain homogeneous equations of the type (5.1). Hence, by Thm. 5.2 there is a unique solution of the problem stated, called the *Cauchy matrix* (or *fundamental matrix*) of Pr. (5.1), (5.2). It serves in the integral representation for x_f:

$$x_f(t) = \int_{t_0}^{t} C(t;s)f(s)ds, \quad t_0 \leq t < \infty. \tag{5.14}$$

Moreover, for any $t \in [t_0, \infty)$ the function $C(t;\cdot) : [t_0, t] \to \mathbf{R}^{n \times m}$ is bounded and measurable.

To obtain an integral representation for x_ϕ we assume in addition to the conditions of Thm. 5.2 that for any $s \in (-\infty, t_0]$ the sets $\{t : t - h_j(t) = s\}$ do not have limit points and that the function ϕ has *compact support* (i.e. vanishes outside a finite interval), $\phi \in$BV$(-\infty, t_0]$, and its Lebesque decomposition (5.8) does not include a singular term. For a fixed $s \in (-\infty, t_0]$ we let $(t \mapsto G(t;s)) : [t_0, \infty) \to \mathbf{R}^{n \times m}$ be the solution of (5.13) on the interval $t_0 \leq t < \infty$ satisfying the initial condition

$$G(t;s) = \begin{cases} 0, & -\infty < t < s, \\ I, & s \leq t \leq t_0. \end{cases}$$

Then

$$x_\phi(t) = G(t;t_0)[\phi(t_0) - \phi(t_0^-)] + \sum_i G(t;s_i)[\phi(s_i^+) - \phi(s_i^-)] + \\ + \int_{-\infty}^{t_0} G(t;s)\dot{\phi}_a(s)ds. \tag{5.15}$$

Here, summation is over all discontinuity points $s_i < t_0$ of the function ϕ, and ϕ_a is the absolutely continuous term in the Lebesque decomposition of ϕ.

Under certain additional restrictions (e.g., all $h_j \in C^1[t_0, \infty)$ and $\dot{h}_j(t) < 1$) formula (5.15) can be brought to the form

$$x_\phi(t) = G(t;t_0)\phi(t_0) - \int_{-\infty}^{t_0} \frac{\partial G(t;s)}{\partial s}\phi(s)ds, \quad t_0 \leq t < \infty, \tag{5.16}$$

with an integrable integrand. Formula (5.16) can be used without the assumption that ϕ has compact support if all $\inf_{t \geq t_0}\{t - h_j(t)\}$, $\inf_{t \geq t_0}\{t - h(t)\}$ are finite (because then $\partial G(t;s)/\partial s = 0$ for all $s < 0$ with sufficiently large modulus). Note that $G(t;t_0) = C(t;t_0)$.

For the particular case when all h_j, h in (5.1) are independent of t, the righthand side of (5.16) can be written in terms of the function C:

$$x_\phi(t) = C(t;t_0)\phi(t_0) + \sum_{j=1}^{m} \int_{t_0 - h_j}^{t_0} C(t;\alpha + h_j)A_j(\alpha + h_j)\phi(\alpha - t_0)d\alpha + \\ + \int_{t_0 - h}^{t_0} \left[\int_{t_0}^{\alpha + h} C(t;\tau)A(\tau, \alpha - \tau)d\tau\right]\phi(\alpha - t_0)d\alpha, \quad t_0 \leq t < \infty.$$

5.4. *ADJOINT EQUATION. PERIODIC SOLUTIONS*

Here we consider Eq. (5.1) with $f(t) \equiv 0$, where, for simplicity, all $h_j(\cdot)$ and $h(\cdot)$ are assumed to be constants:

$$\dot{x}(t) = \sum_{j=1}^{m} A_j x(t - h_j) + \int_{-h}^{0} A(t,\theta)x(t+\theta)d\theta, \quad t_0 \le t < \infty. \tag{5.17}$$

Suppose A_j and A satisfy the conditions of Thm. 5.2.

Following A. Halanay, we call the equation

$$\dot{y}(t) = -\sum_{j=1}^{m} A_j'(t+h_j)y(t+h_j) - \int_{-h}^{0} A'(t-\theta,\theta)y(t-\theta)d\theta \tag{5.18}$$

the (Lagrange) *adjoint* of (5.17). (Here, as before, prime denotes transposition.) Hence (5.18) is an ADE (see Subs. 1.1.3), which becomes an RDE when replacing t by $-t$. Consequently the Cauchy problem for (5.18) is well-posed for decreasing t direction.

A basic relation between solutions of both equations is as follows. Let x be a solution of (5.14) with some initial condition, and y a solution of (5.18) with some finite condition on an interval J. Then

$$y'(t)x(t) + \sum_{j=1}^{m} \int_{t}^{t+h_j} y'(\theta)A_j(\theta)x(\theta - h_j)d\theta +$$

$$+ \int_{-h}^{0} d\tau \int_{t}^{t-\tau} y'(\theta)A(\theta,\tau)x(\theta+\tau)d\theta \equiv \text{const}, \quad t \in J.$$

Adjoint equations are used for investigating boundary value problems for FDEs, and we will discuss this explicitly in Part IV of this book. Mention here as an example only the important problem of the existence of a periodic solution of (5.1) under the assumption that all h_j, h are independent of t, and that

$$A_j(t+T) \equiv A_j(t), \quad j = 1, \ldots, m,$$

$$A(t+T,\theta) \equiv A(t,\theta), \quad f(t+T) \equiv f(t), \quad f \in L_1[0,T]$$

for some $T > 0$ and all $t \in \mathbf{R}$. In this case the following modification of the Fredholm alternative holds.

Theorem 5.3 The linear spaces of T-periodic solutions of Eqs. (5.17) and (5.18) have the same finite dimension. If this dimension is equal to zero (i.e. these equations do not have nontrivial T-periodic solutions), then for any T-periodic function f the Eq. (5.1) has only one T-periodic solution. If, on the other hand, this dimension is equal to $d > 0$, then (5.1) has T-periodic solutions if and only if

$$\int_0^T y_i'(t)f(t)dt = 0, \quad i = 1, \ldots, d.$$

Here $y_1, \ldots, y_d$ is a basis in the space of T-periodic solutions of (5.18).

Various variants of this theorem were proven by many authors (see, e.g. [238] and references therein).

Thm. 5.3 holds also for the general equation (5.11) with $r(t+T, \theta) \equiv r(t, \theta)$ and correspondingly defined adjoint equation.

For the approximate construction of a periodic solution one can use a finite–dimensional approximation. E.g., the method of a harmonic balance in which all functions under consideration are represented as finite segments of T-periodic Fourier series. If (5.1) is an asymptotically stable equation, then it suffices to numerically construct a solution for any initial function, because for such equations any solution tends to a T-periodic solution with exponential rate as $T \to \infty$.

5.5. *LINEAR NDES*

As it was mentioned earlier, two general forms of NDEs, namely (3.1) and (3.13) are applied now. Moreover the latter is reduced to the former only for "sufficiently smooth" equations and generally speaking the former is not reduced to the latter even in the presence of this smoothness. But the consequent forms of linear equations, analogous (5.1), namely

$$\begin{aligned} \dot{x}(t) = \sum_{j=1}^{m} C_j(t)x(t-h_j(t)) + \sum_{j=1}^{m_1} D_j(t)\dot{x}(t-g_j(t)) + \\ + \int_{-h(t)}^{0} A(t,\theta)x(t+\theta)d\theta + f(t), \quad t_0 \le t < \infty \end{aligned} \tag{5.19}$$

(here an integral term with $\dot{x}$ is absent, because it may be eliminated by integration by parts), and

$$\left[x(t)-\sum_{j=1}^{m_1}B_j(t)x(t-g_j(t))\right]^{\bullet}=\sum_{j=1}^{m}A_j(t)x(t-h_j(t))+ \\ +\int_{-h(t)}^{0}A(t,\theta)x(t+\theta)d\theta+f(t),\quad t_0\le t<\infty \tag{5.20}$$

can be easy reduced to one another under assumptions natural in applications. But for discontinuous initial function ϕ the solution of Cauchy problem (5.19), (5.2) may depends on its interpretation that was discussed in Subs. 1.1.2. Therefore, (5.20) will be accepted as the main form of a linear NDE and the solutions of (5.19) are defined as the solutions of Eq. (5.20) which is formally equivalent to (5.19).

A function $r: J\to\mathbf{R}^n$ (where $J\subseteq\mathbf{R}$ is an interval) is called *piecewise absolutely continuous* (PAC) if $r\in$BV(J), there is no singular term in the expansion (5.8), and set of discontinuity points of the function r has no limit points on J. A function x defined on an interval J with left end $t_\Omega\in J$ is called as a *PAC-solution* of Pr. (5.20), (5.2) if x is a PAC function and the function

$$(t\mapsto x(t)-\sum_{j=1}^{m_1}B_j(t)x(t-g_j(t))):J\to\mathbf{R}^n \tag{5.21}$$

(where $x(t-g_j(t)):=\phi(t-g_j(t))$ if $t-g_j(t)<t_\Omega$) is absolutely continuous on J and satisfies Eq. (5.20) for almost all $t\in J$, and also $x(t_0)=\phi(t_0)$.

Theorem 5.4 Assume that all the functions B_j,ϕ are PAC ones; $g_j\in C^1[t_0,\infty)$ and also $g_j(t)>0$, $\dot{g}_j(t)<1$; functions A_j,h_j,A,h,f satisfy the conditions of Thm. 5.2 and also for any $s\in S_\phi\bigcup[t_0,\infty)$ all sets $\{t: t-h_j(t)=s\}$ have no limit points. Then there exists PAC-solution of Pr. (5.20), (5.2) on the interval $[t_0,\infty)$. On any interval $[t_0,t_1]\subset[t_0,\infty)$ this solution is unique and if $t_0-h_j(t_0)\notin S_\phi$ $(j=1,\dots,m_1)$ then it continuously depends on all prescribed functions (in the sense being given later).

For more detailed formulation designate by $S_B\subset[t_0,\infty)$ the set of all discontinuity points of the functions B_j and by $S\subset\mathbf{R}$ the least set containing S_ϕ,S_B and such that if $t-g_j(t)\in S$ for some $t\in(t_0,\infty)$, $j=1,\dots,k$, then $t\in S$. From continuity of the function (5.21) it follows that $S\bigcap(t_0,\infty)$ is a set of possible discontinuity points of the solution x.

Continuous dependence of the solution on all prescribed function means that for any $t_1\in(t_0,\infty)$, $\varepsilon>0$, $M>0$ there are $\delta>0$, $N>0$

(where N does not depend on δ) such that if in (5.20), (5.2) functions $B_j, g_j, A_j, h_j, A, h, f, \phi$ are replaced by $\overline{B}_j, \ldots, \overline{\phi}$ satisfying the same restrictions as $B_j, \ldots, \phi$, inequalities (5.3), (5.4) are valid and also

$$\left.\begin{aligned} &|\overline{B}_j(t) - B_j(t)| < \delta, \quad t \in [t_0, \infty) \setminus U(S_B, \delta), \\ &|\overline{B}_j(t) - B_j(t)| < M, \quad t \in [t_0, \infty) \bigcap U(S_B, \delta), \\ &|\overline{g}_j(t) - g_j(t)| < \delta, \quad t \in [t_0, \infty), \end{aligned}\right\} j = 1, \ldots, m_1,$$

$$\left.\begin{aligned} &|[\overline{x}(t) - \sum_{j=1}^{m_1} \overline{B}_j(t)\overline{x}(t - \overline{g}_j(t))] - [x(t) - \sum_{j=1}^{m_1} B_j(t)x(t - g_j(t))]| < \varepsilon, \\ &\qquad t_0 \le t \le t_1, \\ &|\overline{x}(t) - x(t)| < \varepsilon, \quad t \in [t_0, t_1] \setminus U(S, \varepsilon), \\ &|\overline{x}(t) - x(t)| < N, \quad (t \in [t_0, t_1] \bigcap U(S, \varepsilon). \end{aligned}\right\} \tag{5.22}$$

Proof of Thm. 5.4. Problem (5.20), (5.2) is equivalent to the equation

$$\begin{aligned} x(t) = \phi(t_0) + \sum_{j=1}^{m_1} B_j(t)x(t - g_j(t)) - B_j(t_0)\phi(t_0 - g_j(t_0))] + \\ + \sum_{j=1}^{m} \int_{t_0}^{t} A_j(\tau)x(\tau - h_j(\tau))d\tau + \\ + \int_{t_0}^{t} d\tau \int_{-h(\tau)}^{0} A(\tau, \theta)x(\tau + \theta)d\theta + \int_{t_0}^{t} f(\tau)d\tau, \\ t_0 \le t < \infty, \end{aligned} \tag{5.23}$$

where $x(s) := \phi(s)$, $s \le t_0$. Moreover, on each interval of absolute continuity of all functions B_j and $t \mapsto x(t - g_j(t))$ Eq. (5.20) is equivalent to the equation

$$\begin{aligned} \dot{x}(t) = \sum_{j=1}^{m_1} [B_j(t)x(t - g_j(t))]^{\bullet} + \sum_{j=1}^{m} A_j(t)x(t - h_j(t)) + \\ + \int_{-h(t)}^{0} A(t, \theta)x(t + \theta)d\theta + f(t). \end{aligned} \tag{5.24}$$

Denote by $S^0 \subset \mathbf{R}$ the set constructed similarly to the set S, where $S_\phi \bigcup S_B$ is replaced by $S_\phi \bigcup S_B \bigcup \{t_0\}$. It is clear that $S \subseteq S^0$ and S^0 does not have limit points. Denote by $t^0(= t_0) < t^1 < \dots$ the sequential points of the set $S^0 \bigcap [t_0, \infty)$. (Note that this sequence could be finite or even consists of a unique point t_0.) The proof of Thm. 5.4 will be obtained by analogy with the steps method.

If $t^0 < t < t^1$, then the first summand in the righthand side of (5.24) is a known integrable function which can be combined with f. Hence, Thm. 5.2 can be applied for $[t_0, t_1] \subset [t_0, t^1)$, from which for

$$x(t_0^+) = \phi(t_0) + \sum_{j=1}^{m_1} [B_j(t_0^+) - B_j(t_0)]\phi(t_0 - g_j(t_0))$$

it follows existence and uniqueness of the solution and its continuous dependence on the given functions.

For $t = t^1$ the solution x, generally, has a jump defined by (5.23)

$$\begin{aligned} x(t^{1+}) - x(t^{1-}) = \sum_{j=1}^{m_1} \{ \lim_{t \to t^1+} [B_j(t)x(t - g_j(t))] - \\ - \lim_{t \to t^1-} [B_j(t)x(t - g_j(t))] \}. \end{aligned} \tag{5.25}$$

Now if $t^1 < t < t^2$, then all $t - g_j(t) < t^1$, and we can combine the summands, as previously and apply Thm. 5.2 with $x(t^{1+})$ being obtained from (5.25). Note that for any sufficiently small $\varepsilon > 0$, $\overline{x}(t^1 + \varepsilon) - \overline{x}(t^1 - \varepsilon)$ is uniformly arbitrarily close to the righthand side (5.25) (for all perturbed solutions), if $\delta > 0$ is sufficiently small. The values $|\overline{x}(t) - x(t)|$ on $(t^1 - \varepsilon, t^1 + \varepsilon)$ are uniformly bounded, and on $[t^0, t^1 - \varepsilon]$ - are uniformly arbitrarily small. This gives the possibility to apply Thm. 5.2, with $t \mapsto x(t)$ $(-\infty < t \leq t^1)$ as nonperturbed initial function. In addition, the difference

$$z = x(t) - \sum_{j=1}^{m_1} B_j(t)x(t - g_j(t))$$

does not have a jump for $t = t^1$, and the difference

$$\overline{z} = \overline{x}(t) - \sum_{j=1}^{m_1} \overline{B}_j(t)\overline{x}(t - \overline{g}_j(t))$$

does not have a jump for $t = \overline{t}^1$ ($\overline{t}^1$ is a perturbed value of t^1). Hence, $|z - \overline{z}|$ is arbitrarily small near t^1 for sufficiently small δ.

Further let us calculate $x(t^{2+})$ and consider the solution x on the interval (t^2, t^3), etc.. If the sequence $\{t^k\}$ is infinite, then $t^k \to \infty$ as $k \to \infty$, and any $t_1 > t_0$ can be achieved in a finite number of steps. If the sequence $\{t^k\}$ is finite and finishes on t^{k_0}, then for $t > t^{k_0}$ we can apply Thm. 5.2 on arbitrarily large finite interval. Thm. 5.4 is proven. □

If in addition all functions B_j, ϕ are assumed to be continuous then the set S is empty and instead of (5.22) we get

$$|\overline{x}(t) - x(t)| < \varepsilon \quad (t_\Omega \le t \le t_1).$$

In this case PAC–solution becomes AC-solution, one can open the square brackets at the lefthand side of Eq. (5.20) and pass to the form (5.19). If also all functions A_j, A, f are continuous, $B_j, \phi \in C^1$ and the sewing condition holds (see Subs. 1.4), then the solution of Pr. (5.20), (5.2) becomes classical one.

Cauchy function C and the function G for Pr. (5.20), (5.2) can be introduced in much the same way as in Subs. 5.3 (with natural replacement of Eq. (5.13)). Integral representations (5.14), (5.15) remain valid for the considered problem.

5.6. *SIMPLEST NONAUTONOMOUS RDES OF THE FIRST AND SECOND ORDERS*

Given the complexity which can arise even from the simplest FDEs, it is hopeless to try to describe fully the behaviour of the solution which might arise from more complicated FDEs. So it is reasonable to consider first special FDEs, in the hope that through a better understanding of them we should be able to gain insight into more general FDEs. An extremely fruitful avenue of study is to consider some examples of investigations of linear, homogeneous, nonautonomous RDEs.

To facilitate exposition, the problems chosen for treatment are the simplest cases which capture some qualitative features of the general problem. This section is not intended to obtain the most general known results, but rather an attempt to outline some ways of the qualitative investigations.

Thus far, in this section all the methods discussed have been limited to linear RDEs of the form

$$\dot{x}(t) - a(t)x(t - h(t)) = 0, \tag{5.26}$$

$$\dot{x}(t) + a(t)x(t - h(t)) = 0, \tag{5.27}$$

$$\ddot{x}(t) + a(t)x(t - h(t)) = 0. \tag{5.28}$$

Assume that

$$a, h \in C([t_0, \infty), [0, \infty)), \quad A_0 := \sup a(t) \le \infty, \quad H_0 := \sup h(t) < \infty,$$

and an initial continuous function ϕ is given on $[t_0 - H_0, t_0]$. In what follows it is assumed that $H_0 > 0$ and $A_0 > 0$ because otherwise the assertions being given become trivial or empty. An interval $[\alpha, \beta]$ will be called a *positive* (resp. *negative*) *semicycle* for the solution $x : [t_0, \infty) \to \mathbf{R}$, if $x(\alpha) = x(\beta) = 0$ and $x(t) > 0$ (resp. $x(t) < 0$) on (α, β). The semicycle $[\alpha, \beta]$ is called a *large* one, if $t - h(t) > \alpha$ for all $t \geq \beta$; otherwise it will be called a *small semicycle.*

5.6.1 Equation (5.26).

Theorem 5.5 If $\phi(t) \geq 0$ (resp. $\phi(t) \leq 0$) $(t_0 - H_0 \leq t \leq t_0)$ then the solution x of Eq. (5.26) does not decrease (resp. does not increase) for $t_0 \leq t < \infty$.

Proof Let for the definiteness $\phi(t) \geq 0$ and assume firstly that $\phi(t_0) > 0$. Let us prove that $x(t) > 0$ $(t_0 \leq t < \infty)$. Suppose the opposite and introduce $t_1 = \min\{t > t_0 : x(t) \leq 0\}$. Then $x(t) > 0$ for $t_0 \leq t < t_1$. Hence by virtue of Eq. (5.26) the solution $x(t)$ does not decrease for $t_0 \leq t \leq t_1$. But this contradicts to the relations $x(t_0) > 0,\ x(t_1) = 0$.

If $\phi(t_0) = 0$ then we consider the solution x_ε of Eq. (5.26) defined by the initial function $t \mapsto \phi(t) + \varepsilon$ with arbitrary $\varepsilon > 0$. As it was proven above, the function x_t does not decrease. Moreover $x_\varepsilon(t) \to x(t)$ as $\varepsilon \to 0^+$ by virtue of continuous dependence of the solution on the initial function. Hence x_ε is a nondecreasing function. Thm. 5.5 is proven.□

Translating the initial point and disregarding too early values of the initial function we obtain

Corollary 5.1 Solutions x of Eq. (5.22) can be only one of the following three types:

a) eventually nonzero (i.e., $x(t) \neq 0$ for all sufficiently large t);

b) eventually zero;

c) oscillating; in this case x has at least one zero on each interval $[t - H_0, t](t \geq t_0)$ but $x(t) \not\equiv 0$ there.

Theorem 5.6 There exists a constant $\kappa > 0$ such that for $A_0 H_0 < \kappa$ any oscillating solution x of Eq. (5.22) tends to zero as $t \to \infty$.

Proof Let us prove Thm. 5.6 for $\kappa = 1$. Denote

$$m_k := \max_{[t_0 + kH_0, t_0 + (k+1)H_0]} |x(t)| \quad (k = 0, 1, \ldots).$$

Let t_k be a point of the maximum of the last expression, and $\alpha_k \in [t_k - H_0, t_k]$ is a zero of the function x nearest to t_k. Integrating both parts of Eq. (5.22) from α_k to t_k, we obtain for $k \geq 2$, that

$$m_k \leq A_0 H_0 \max\{m_{k-1}, m_{k-2}\} \quad (k = 2, 3, \ldots).$$

The assertion of Thm. 5.6 follows from here for $\kappa = 1$.□

Remark 5.1 In the proof of Thm. 5.5 the condition $a(t) \geq 0$ is not essential. It is only important that all semicycles are small (as it takes place for Eq. (5.26) by virtue of Thm. 5.5) and also we must put $A_0 := \sup|A(t)|$.

Thm. 5.6 is proven in [381.3] for $\kappa = 2$. J. Lillo[1] obtained an unimprovable value of $\kappa = \frac{11}{4} + \ln 2$ (≈ 3.443).

Theorem 5.7 (comparison of solutions). Let $a_1, h_1 \in C([t_0,\infty),[0,\infty))$, $a_1(t) \geq a(t)$, $0 \leq h_1(t) \leq h(t)$ ($t_0 \leq t < \infty$). Let x_1 be a finally nonzero solution of the equation

$$\dot{x}_1(t) - a_1(t)x_1(t - h_1(t)) = 0$$

and x be any solution of Eq. (5.26). Then the function $t \mapsto y(t) := x(t)/x_1(t)$ is bounded for all sufficiently large t.

Proof Let $x_1(t) \neq 0$ for $t_1 \leq t < \infty$. Then for $t \geq t_1$ we have

$$\begin{aligned}\dot{y}(t) &= [a(t)x(t-h(t))x_1(t) - x(t)a_1(t)x_1(t-h_1(t))]/x_1^2(t) = \\ &= [a(t)x_1(t-h(t))y(t-h(t)) - a_1(t)x_1(t-h_1(t))y(t)]/x_1(t).\end{aligned} \tag{5.29}$$

Prove that $\sup\{y(t) : t_1 \leq t < \infty\} < \infty$. Assume the opposite and let y_0,

$$y_0 > \max\{y(t) : t_1 \leq t \leq t_1 + 2H_0\}$$

be a nonstationary value of $y(t)$. (Remind that due to the Sard theorem[2] the set of stationary values of the function from C^1 has the zero measure.) Denote $t_2 = \min\{t > t_1 : y(t) = y_0\}$. Then

$$\dot{y}(t_2) > 0, \;\; y(t_2 - h(t_2)) \leq y(t_2),$$

$$a(t_2) \leq a_1(t_2),$$

$$|x_1(t_2 - h(t_2))| \leq |x_1(t_2 - h_1(t_2))|.$$

This contradicts to the equality (5.29). Similarly can be proven that $\inf\{y(t) : t_1 \leq t < \infty\} > -\infty$. Thm. 5.7 is proven.□

Corollary 5.2 Any two eventually nonzero solutions of Eq. (5.26) have the same order of value as $t \to \infty$.

It can be proved that if $A_0 < \infty$ then $x_2(t)/x_1(t) \to \text{const} \neq 0$ as $t \to \infty$ for any two eventually nonzero solution x_1, x_2 of Eq. (5.26).

[1] J. Lillo, Oscillatory solutions of equation $\dot{y}(x) = m(x)y(x - n(x))$, J. Diff. Eqs., 6 (1969), pp. 1-35

[2] A. Sard, Bull. Amer. Math. Soc., 48 (1942), 883-890

Corollary 5.3 If $A_0 < \infty$, then for any eventually nonzero solution x of Eq. (5.26) there are such constants c_1, c_2 that for all sufficiently large t

$$c_1 e^{p(a_0,H_0)t} < |x(t)| < c_2 e^{A_0 t}, \tag{5.30}$$

where $a_0 = \inf a(t)$ and $p(a_0, H_0)$ is a unique real root of the equation $p = a_0 \exp(-H_0 p)$. (Remark that the lefthand side of Eq. (5.30) is a solution of the equation $\dot{x}(t) = a_0 x(t - H_0)$.)

5.6.2 Equation (5.27). First of all we shall investigate the behaviour of Eq. (5.27) solutions eventually different from zero.

Theorem 5.8 Let $x(t)$ be a solution of Eq. (5.27), (5.2) such that $x(t) \neq 0$ $(t_1 \leq t < \infty,\ t_1 = \text{const} \geq t_0 - H_0)$. Then

$$|x(t)| \leq |x(t_1 + 2H_0)| \exp(-\int_{t_1+2H_0}^{t} a(\tau)d\tau)\ (t_1 + 2H_0 \leq t < \infty).$$

Proof Eq. (5.27) yields that if $x(t) \neq 0$ $(t_1 \leq t < \infty)$ then for $t_1 + H_0 \leq t < \infty$ the function $t \mapsto x(t)$ is nonincreasing. Hence, for $t_1 + 2H_0 \leq t < \infty$ we have $|x(t - h(t))| \geq |x(t)|$ and also

$$|x(t)|^{\bullet} = -a(t)|x(t - h(t))| \leq -a(t)|x(t)|.$$

From here it follows the assertion of Thm. 5.8.□

Corollary 5.4 If

$$\int^{\infty} a(s)ds = \infty$$

then all the solutions of Eq. (5.27) which finally are different from zero tend to zero as $t \to \infty$.

Lemma 5.1 Assume that $1 \leq A_0 H_0 < \infty$ and x is a solution of (5.27), (5.2) such that $x(t_1) = 0$, $x(t) > 0$ $(\forall t \in (t_1, t_2])$ for some $t_1 \geq t_0 + A_0^{-1}$, $t_2 > t_1$; $x_0 := -\min\{x(t) : t_1 - H_0 - A_0^{-1} \leq t \leq t_1\}$. Then

$$x(t_2) \leq \begin{cases} A_0(t_2 - t_1)x_0 & (\text{if } t_2 - t_1 \leq H_0 - A_0^{-1}), \\ [A_0 H_0 - \\ -\frac{1}{2}(1 + A_0^2(t_2 - t_1 - H_0)^2)] & (\text{if } H_0 - A_0^{-1} \leq t_2 - t_1 \leq H_0), \\ (A_0 H_0 - \frac{1}{2})x_0 & (\text{if } t_2 - t_1 \geq H_0). \end{cases} \tag{5.31}$$

Proof Assumptions of Lemma 5.1 and Eq. (5.27) mean that for $t_2 - t_1 \geq H_0$

$$\dot{x}(t) \leq A_0 x_0 \quad (t_1 - A_0^{-1} \leq t \leq t_2).$$

From here and $x(t_1) = 0$ it follows the estimate

$$x(t) \begin{cases} \geq \max\{A_0(t-t_1)x_0, -x_0\} & (t_1 - H_0 - A_0^{-1} \leq t \leq t_1), \\ \leq A_0(t-t_1)x_0 & (t_1 \leq t \leq t_2). \end{cases} \tag{5.32}$$

Second estimate (5.32) proves the first estimate (5.31). Consider now the case $t_2 - t_1 > H_0 - A_0^{-1}$. Then if $t > t_1 + H_0 - A_0^{-1}$ then $A_0(t - H_0 - t_1)x_0 > -x_0$. Hence from the first estimate (5.32) we obtain

$$\dot{x}(t) \leq -A_0^2(t - H_0 - t_1)x_0 \quad (t_1 + H_0 - A_0^{-1} \leq t \leq \min\{t_1 + H_0, t_2\}.$$

Therefore for these t we have the second estimate (5.31) because

$$x(t) \leq A_0(H_0 - A_0^{-1})x_0 - \int\limits_{t_1+H_0-A_0^{-1}}^{t} A_0^2(\tau - H_0 - t_1)x_0 d\tau =$$

$$= (A_0 H_0 - \frac{1}{2})x_0 - A_0^2(t - H_0 - t_1)^2 x_0/2.$$

At last, if $t_2 - t_1 > H_0$ then the function $t \mapsto x(t)$ for $t_1 + H_0 \leq t \leq t_2$ is nonincreasing and so

$$x(t) \leq x(t_1 + H_0) = (A_0 H_0 - 1/2)x_0$$

which is the third estimate (5.31). Lemma 5.1 is proven.□

Theorem 5.9 If $A_0 H_0 \leq 3/2$ then all the solutions of Eq. (5.27) are bounded; in addition the constant 3/2 *is unimprovable. If $A_0 H_0 < 3/2$ and $\int^{\infty} a(t)dt = \infty$ then all these solutions tend to zero as $t \to \infty$. Here the sign "$<$" can not be replaced by "$=$".*

Proof Assume that $A_0 H_0 \leq 3/2$ and $x(t)$ is a solution of (5.27), (5.2). Without loss of generality we can suppose that $A_0 H_0 \geq 1$. Then we can prove the estimate

$$|x(t)| \leq \max\{|x(\tau)| : t_0 - H_0 \leq \tau \leq t_0 + A_0^{-1} + H_0\} \quad (t_0 + A_0^{-1} + H_0 \leq t < \infty). \tag{5.33}$$

In fact, assume the opposite. Then there is $t_2 > t_0 + A_0^{-1} + H_0$ such that

$$|x(t_2)| > |x(t)| \; (\forall t \in [t_0 - H_0, t_2)). \tag{5.34}$$

For the definiteness consider the case $x(t_2) > 0$ (otherwise the function $t \mapsto -x(t)$ has to be considered). If on an interval $[t_2 - H_0 - \varepsilon, t_2]$ $(0 < \varepsilon \leq t_2 - t_0)$ the function x has no zeros then x is nonincreasing on $[t_2 - \varepsilon, t_2]$ that contradicts to (5.34). Let t_1 be the greatest from these zeros. Then $t_1 \in [t_2 - H_0, t_2)$. Hence $t_1 > t_0 + A_0^{-1}$ and $x(t) > 0$ for $t_1 < t \leq t_2$. At least let us introduce x_0 as in Lemma 5.1. Then all the conditions of Lemma 5.1 are fulfilled and $x_0 < x(t_2)$ by virtue of (5.34). But from (5.32) and inequality $A_0 H_0 \leq 3/2$ it follows that $x(t_2) \leq x_0$. This contradiction proves (5.33), i.e. the first assertion of Thm. 5.9.

Consider now the case $A_0 H_0 < 3/2$, $\int^\infty a(t)dt = \infty$. Let x be a solution of (5.27), (5.2). We assume as before that $A_0 H_0 \geq 1$ and prove that $x(t) \to 0$ as $t \to \infty$. If $x(t)$ is finally nonzero then this assertion is proven in Corollary 5.4. If finally $x(t) \equiv 0$ then the assertion is trivial. So we assume that both of these conditions are not valid. Then we can take a sequence $\tau_1 (\geq t_0 + A_0^{-1}) < \tau_2 < \dots$ such that $x(\tau_k) = 0$ and $\tau_{k+1} - \tau_k \geq H_0 + A_0^{-1}$. Let us denote

$$m_k := \max\{|x(t)| : \tau_k \leq t \leq \tau_{k+1}\}(> 0).$$

Suppose that this maximum is achieved in the point τ_k^*. Take any fixed $k \geq 2$, put $t_2 := \tau_k^*$, $t_1 := \max\{t < \tau_k^* : x(t) = 0\}$, and apply Lemma 5.1 to $x(t)$ if $x(\tau_k^*) > 0$ or to $-x(t)$ if $x(\tau_k^*) < 0$. As a result we get

$$m_k \leq (A_0 H_0 - 1/2) \max\{m_k, m_{k-1}\} \quad (k = 2, 3, \dots).$$

From here it follows that $m_k \to 0$ as $k \to \infty$, i.e. $x(t) \to 0$ as $t \to \infty$.

Let us prove that the constant 3/2 in the first assertion of Thm. 5.9 is unimprovable. For this we take a number $p > 3/2$ and define the function

$$t \mapsto y(t) = 1 - t \ (0 \leq t \leq p), = 1 - (p+1)t + (p^2 + t^2)/2 \quad (p \leq t \leq p+1).$$

This function is a solution of Eq. (5.27) on $[0, p+1]$ without previous history with $a(t) \equiv 1$ and

$$h = \tilde{h}(t) = t \ (0 \leq t \leq p), = p \ (p \leq t \leq p+1).$$

Moreover $y(p+1) = -(p-1)/2$).

Its being $(p+1)$-periodically extended on the whole axis, the function $\tilde{h}$ has jumps for $t = k(p+1) (k \in \mathbf{Z})$. By this reason we introduce for each $\varepsilon \in (0,1)$ the function

$$t \mapsto h(t) = \tilde{h}(t) \quad (0 \leq t \leq p+1-\varepsilon), \ = (p+1-\varepsilon)p/\varepsilon \quad (p+1-\varepsilon \leq t \leq p+1).$$

Let y_ε be a solution of Eq. (5.27) with $a(t) \equiv 1$ and initial condition $y_\varepsilon(0) = 1$. Then $y_\varepsilon(p+1)$ is arbitrarily close to $y(p+1)$ if $\varepsilon > 0$ is sufficiently small. Choose $\varepsilon > 0$ such that $y_\varepsilon(p+1) < -1$, and extend h from $[0, p+1]$ on $\mathbf{R}$ as a $(p+1)$-periodic function. Set

$$x(t) = (-y_\varepsilon(p+1))^k y_\varepsilon(t - k(p+1)) \quad (k(p+1) \leq t \leq (k+1)(p+1);\ k \in \mathbf{Z}).$$

The function $x : \mathbf{R} \to \mathbf{R}$ is a solution of (5.27) with $a(t) \equiv 1$ and previously constructed $h(t)$. This solution is unbounded as $t \to \infty$, but $A_0 H_0 = p$.

Impossibility to replace the sign "<" by "=" in the second assertion of Thm. 5.9 is proven similarly. Firstly, the functions y and $\tilde{h}$ for $p = 3/2$ are constructed as it was described previously; then all the values $|y(3k/2)| (k = 0, 1, \ldots)$ turn out to be equal with each other. In order to pass to the continuous function h we choose numbers $\alpha_k \in (0,1) (k \in N)$ such that

$$\prod_{k=1}^{\infty} \alpha_k > 0.$$

Next, the function $\tilde{h}$ must be corrected on every interval $J_k := [\frac{5}{2}(k-1), \frac{5}{2}k]$ near the point $\frac{5}{2}k$ with its own $\varepsilon = \varepsilon_k$, as it was described previously, in such order that for the corrected function h_{ε_k} and solution y_{ε_k} the following inequality $|y_{\varepsilon_k}(\frac{5}{2}k)| > \alpha_k |y_{\varepsilon_k}(\frac{5}{2}(k-1))|$ will be hold. (Remind that on each interval J_k RDE (5.27) is an equation without previous history). Let us put

$$h(t) = h_{\varepsilon_k}(t) \ (t \in J_k; k \in \mathbf{N}).$$

Then the solution x of Eq. (5.27) with $a(t) \equiv 1$, $t_0 = 0$ satisfies the relation

$$|x(\frac{5}{2}k)| = |x(0)| \prod_{i=1}^{k} [|x\left(\frac{5}{2}i\right)| / |x\left(\frac{5}{2}(i-1)\right)|] \geq$$

$$\geq |x(0)| \prod_{i=1}^{k} \alpha_i.$$

Hence if $x(0) \neq 0$, then $\limsup |x(t)| > 0$ as $t \to \infty$, but $A_0 H_0 = 3/2$.

Thm. 5.9 is proven. □

Denote $a_0 := \inf a(t)$, $h_0 := \inf h(t)$.

Theorem 5.10 If $a_0 h_0 > 1/e$, then any solution $x : [t_0, \infty) \to \mathbf{R}$ of (5.27) has an unbounded set of roots. Here the constant $1/e$ is unimprovable.

Proof The second assertion of Thm. 5.10 holds because of the function $t \mapsto e^{-et}$ is a solution of the equation $\dot{x}(t) + x(t - 1/e) = 0$.

Let us prove the first assertion. Assume that $a_0h_0 > 1/e$, and there is a solution $x : [t_0, \infty) \to \mathbf{R}$ of (5.27) such that $x(t) > 0$ on $[t_1, \infty) \subseteq [t_0, \infty)$. Choose two numbers $r \in (0, a_0)$ and $s \in (0, h_0)$ such that $rs > 1/e$ and among the roots of the equation $z + re^{-sz} = 0$ there are two roots $u \pm iv$ with $v \in (0, \pi/(2h_0))$. (The possibility of such choice follows from the continuous dependence of these roots on r and s, e.g., we can put $r = (1+\varepsilon)[a_0/(eh_0)]^{1/2}$, $s = (1+\varepsilon)[h_0/(ea_0)]^{1/2}$ for any sufficiently small $\varepsilon > 0$.) Then the function

$$t \mapsto y(t) = ce^{ut} \sin\{v[t - (t_1 + h_0)]\} \quad (c > 0;\ t \in \mathbf{R})$$

is a solution of the equation

$$\dot{y}(t) + ry(t - s) = 0,$$

and this solution has a positive semicycle $[t_1 + h_0, t_1 + h_0 + \pi/v]$, on which x by virtue of (5.27) is decreasing. Set

$$c := \min_{[t_1+h_0, t_1+h_0+\pi/v]} \left(x(t)\{e^{ut} \sin v[t - (t_1 + h_1)]\}^{-1}\right).$$

Assume that this minimum will be achieved at $t = \bar{t}$. Then $x(\bar{t}) = y(\bar{t})$, $\dot{x}(\bar{t}) = \dot{y}(\bar{t})$. This yields $\bar{t} > t_1 + h_0 + \pi/(2v) > t_1 + h_0 + s$, $a(\bar{t})x(\bar{t} - h(\bar{t})) = ry(\bar{t} - s)$. Hence $y(\bar{t} - s) > x(\bar{t} - h(\bar{t}))$. But $x(\bar{t} - h(\bar{t})) > x(\bar{t} - s) \geq y(\bar{t} - s)$. We obtain the contradiction. Thm. 5.10 is proven.□

Theorem 5.11 Assume that $a(t) > 0$, $h(t) > 0$, and there is a solution $x_1(t)([t_0 - H_0 \leq t \leq \infty))$ of constant signs. Then for any oscillating solution $x(t)$ of Eqs. (5.27) the function $y(t) := x(t)/x_1(t)$ is bounded.

Proof Denote $y_0 := \max\{|y(t)| : t_0 - H_0 \leq t \leq t_0\}$ and prove that $|y(t)| \leq y_0$ $(t_0 \leq t < \infty)$. Assume the opposite. Then there are $y_1 > y_0$ and the first moment $t_1 > t_0$ such that $|y(t_1)| = y_1$. Assume for the definiteness that $y(t_1) > 0$. But the function $y(t)$ is oscillating. Hence, $\dot{y}(t)$ has zeros for $t \geq t_1$, $y(t) \geq y_1$. Let t_2 be the least of these zeros. Then $y(t) < y(t_2)$ for all $t \in [t_0 - H_0, t_2)$. But $\dot{y}(t) = -a(t)[x_1(t - h(t))/x_1(t)][y(t - h(t)) - y(t)]$ and for $t = t_2$ we obtain a contradiction. Thm. 5.11 is proven.□

Theorem 5.12 Let $x(t)$ be a solution of (5.27), (5.2) and $x_1(t)$ is a solution of the problem

$$\dot{x}_1(t) + a_1(t)x_1(t - h_1(t)) = 0 \quad (t_0 \leq t < \infty), \quad x_1(t) = \phi_1(t)\ (-\infty < t \leq t_0)$$

where

$$a_1, h_1 \in C([t_0, \infty) \to [0, \infty)),\ \phi_1 \in C(-\infty, t_0],$$

$$a_1(t) \geq a(t),\ a_1(t) > 0,\ h_1(t) \geq h(t)\ (t_0 \leq t < \infty);$$

$$\phi_1(t) \geq \phi(t)\ (t_0 - H_0 \leq t \leq t_0),\ \phi_1(t_0) = \phi(t_0).$$

Assume also that x_1 (equal ϕ_1 for $t \le t_0$) is a positive, nonincreasing function on the whole axis. Then the function $t \mapsto x(t)/x_1(t)$ $(t_0 \le t < \infty)$ is nondecreasing.

Proof The function $y(t) := x(t)/x_1(t)$ satisfies the relation

$$\dot{y}(t) = [a_1(t)y(t)x_1(t-h_1(t)) - a(t)y(t-h(t))x_1(t-h(t))]/x_1(t) \qquad (5.35)$$
$$(t_0 < t < \infty).$$

Assume for a moment that $a_1(t) \ge ka(t)$, where $k > 1$. Note that $y(t) \le 1$ $(t_0 - H_0 \le t \le t_0)$, $y(t_0) = 1$, and let us prove that $\dot{y}(t) > 0$ $(t_0 < t < \infty)$. We have

$$\dot{y}(t_0^+) \ge a_1(t_0)[k\phi_1(t_0 - h_1(t_0)) - y(t_0 - h(t_0))\phi_1(t_0 - h(t_0))]/\phi_1(t_0) > 0.$$

Let our assertion be false and $t_1 = \min\{t > t_0 : \dot{y}(t) = 0\}$. Then

$$y(t_1) \ge y(t_1 - h(t_1)),\ x_1(t_1 - h_1(t_1)) \ge x_1(t_1 - h(t_1)),\ a_1(t_1) \ge ka(t_1)$$

and we obtain a contradiction to (5.35).

For the proof of Thm. 5.12 without assumption $a_1(t) \ge ka(t)$ let us replace Eq. (5.27) by the following one

$$\dot{x}(t) + \frac{a(t)}{k}x(t-h(t)) = 0$$

with a $k > 1$ and the same initial function. Applying the assertion proved above and passing on to the limit as $k \to 1^+$ we finish the proof of Thm. 5.12.□

Remind that two functions $t \mapsto f(t),\ g(t)$ have the same order for $t \to \infty$ if $f(t) = O(g(t))$ and $g(t) = O(f(t))$ as $t \to \infty$. The function $f(t)$ is infinitesimal in comparison with $g(t)$ for $t \to \infty$ if $f(t) = o(g(t))$ $(t \to \infty)$.

Theorem 5.13 *Let $a(t) > 0$, $h(t) > 0$. Then for any two eventually positive solutions of Eq. (5.27) there are two ways of their behaviour: or both of them have the same order for $t \to \infty$, or one of them is infinitesimal comparing with the other one for $t \to \infty$.*

Moreover there does not exist more than two eventually positive solutions of Eq. (5.27) of unequal order for $t \to \infty$.

Proof If the statement of Thm. 5.13 is invalid for the solutions x_1, x_2 then one of them, e.g. x_2, is unbounded comparing with the other one, but $y(t) := x_2(t)/x_1(t) \not\to \infty$ as $t \to \infty$. It means that the sets $\{t > t_0 : y(t) \le N\}$, $\{t \ge t_0 : y(t) \ge N\}$ are unbounded for any sufficiently large N. Hence the set $\{t \geq t_0 : y(t) \geq N,\ \dot{y}(t) = 0\}$ is also unbounded. Let $t_1 \geq t_0$ be such that $x_1(t) > 0$ and $x_2(t) > 0$ for $t_1 \leq t < \infty$. Take

$$N > \max\{y(t) : t_1 \leq t \leq t_1 + H_0\},$$

$$t_2 := \min\{t \geq t_1 + H_0 : y(t) \geq N,\ \dot{y}(t) = 0\}.$$

Then $y(t) < y(t_2)$ for all $t \in [t_1, t_2)$. Hence putting $t = t_2$ in (5.35) we obtain a contradiction.

Assume now that there are three solutions x_1, x_2, x_3 of Eq. (5.27), all $x_j(t) > 0$ for $t \geq t_1 (\geq t_0)$, and $x_3(t) = o(x_2(t))$, $x_2(t) = o(x_1(t))$ as $t \to \infty$. Let $k > 0$ be so large that $kx_3(t) > x_2(t)(t_1 \leq t \leq t_1 + H_0)$, and $t_2 > t_1 + H_0$ be such that $kx_3(t_2) < x_2(t_2)$. Then for sufficiently small $\varepsilon > 0$ we have $\varepsilon x_1(t_2) + kx_3(t_2) < x_2(t_2)$. Define t_3

$$t_3 := \min\{t \in [t_1 + H_0, t_2] : \varepsilon x_1(t) + kx_3(t) = x_2(t)\}.$$

Then $\varepsilon x_1 + kx_3(t) > x_2(t)$ for $t_3 - H_0 \leq t \leq t_3$. Let us transfer the initial point to t_3 and apply Thm. 5.12 to the solutions $t \mapsto \varepsilon x_1(t) + kx_3(t)$, $t \mapsto x_2(t)$ $(t_3 - H_0 \leq t < \infty)$, ignoring their values for $t < t_3 - H_0$. Then we obtain $x_2(t) \geq \varepsilon x_1(t) + kx_3(t)$ $(t_3 \leq t < \infty)$. But this is impossible because $x_2(t) = o(x_1(t))$ as $t \to \infty$. Theorem 5.13 is proven. □

The following interesting conclusion about structure of the set of Eq. (5.27) solutions was obtained by E. Kozakiewicz.

Theorem 5.14 Let

$$a_0 h_0 > 0, \quad A_0 H_0 \leq \frac{1}{e}. \tag{5.36}$$

Then there are two finally nonzero solutions x_1, x_2 of Eq. (5.27) such that $x_2(t) = o(x_1(t))(t \to \infty)$. In addition for any solution of Eq. (5.27) there exist two constants c_1, c_2 such that

$$x(t) = c_1 x_1(t) + c_2 x_2(t) + o(x_2(t)) \quad (t \to \infty). \tag{5.37}$$

Remark The sense of the constant $1/e$ in (5.36) follows from Thm. 5.9. Note that equation

$$\dot{x}(t) + \alpha x(t - \beta) = 0 \ \ (\alpha, \beta = \text{const} > 0) \tag{5.38}$$

has, in particular, the exponential solutions $t \mapsto x_j(t; \alpha, \beta) = \exp[-\alpha \nu_j(\alpha\beta)\, t]$ for $\alpha\beta < 1/e$, where $\nu_1(p) < \nu_2(p)$ are real roots of the equation $\nu = e^{p\nu}$. These roots depend continuously on $p \in (0, 1/e)$ and $\nu_1(p) \to e^-$, $\nu_2(p) \to e^+$ as $p \to (1/e)^-$. So we put $\nu_j(1/e) = e$. For $\alpha\beta = 1/e$ Eq. (5.38) has, in particular, the solutions $x_1(t) = te^{-\alpha e t}$ and $x_2(t) = e^{-\alpha e t}$. Using the Laplace transformation method one can conclude that any solution of Eq. (5.38) for $\alpha\beta \in (0, 1/e]$ is a linear combination of x_1 and x_2 with the precision to $o(x_2(t))(t \to \infty)$. Therefore, Thm. 5.14 claims that in this sense the structure of the solutions of Eqs. (5.27) and (5.38) under conditions (5.36) is the same.

Proof of Thm. 5.14. Let x^λ be a solution of Eq. (5.27) with the initial function $t \mapsto 1 - \lambda(t - t_0)$. Comparing x^λ with the solution $t \mapsto x_2(t - t_0; A_0, H_0)$ by the aid of Thm. 5.12 (in which the second solution is taken as x_1) we obtain for $0 \le \lambda \le A_0\nu_2(A_0H_0)$, $t_0 \le t < \infty$ that

$$x^\lambda(t) \ge \exp[-A_0\nu_2(A_0H_0)t]$$

and so $x^\lambda(t) > 0$. Let Λ be a set of those $\lambda \ge 0$ for which the last inequality holds for all $t \in [t_0, \infty)$. From the inequality $a_0h_0 > 0$ it follows that Λ is bounded. Denote $\overline{\lambda} := \sup \Lambda$ and check that $\overline{\lambda} \in \Lambda$. In fact, if this is not a case then the function $x^{\overline{\lambda}}$ has zeros. Define $\overline{t} = \min\{t > t_0 : x^{\overline{\lambda}}(t) = 0\}$. We have $x^{\overline{\lambda}}(\overline{t} + a_0) < 0$ because the function $x^{\overline{\lambda}}$ is decreasing for $t \in [\overline{t}, \overline{t} + a_0]$. Hence $x^\lambda(\overline{t} + a_0) < 0$ for all sufficiently small $|\lambda - \overline{\lambda}|$ (by virtue of continuous dependence of the solution on the initial function) that is impossible. Therefore, $\overline{\lambda} \in \Lambda$. Denote $x_2 := x^{\overline{\lambda}}$ and let x_1 be a solution of Eq. (5.27) with the initial function $t \mapsto t - t_0$ $(t_0 - H_0 \le t \le t_0)$. Then the relation $x_1(t) = o(x_2(t))(t \to \infty)$ is impossible. In fact, in the opposite case the solution $x_2 - \delta x_1$, defined by the initial function $t \mapsto 1 - (\overline{\lambda} + \delta)(t - t_0)$, does not have zeros for all sufficiently small $\delta > 0$. This contradicts to the definition of $\overline{\lambda}$. Further, by Thm. 5.10 the solution x_1 is finally nonzero, and by Thm. 5.13 $x_2(t) = o(x_1(t))(t \to \infty)$. Using (5.35) and proof by contradiction we obtain that $x_1(t)/x_2(t)$ increases for $t_0 - H_0 \le t < \infty$, and therefore $x_1(t) > 0$ $(t_0 \le t < \infty)$.

Let now x be an arbitrary solution of Eq. (5.27). Let us prove the existence of $c_1 = \text{const}$ such that

$$x(t) = c_1x_1(t) + \widetilde{x}(t), \quad \widetilde{x}(t) = O(x_2(t)) \quad (t \to \infty). \tag{5.39}$$

If the function x is oscillating then (5.39) follows from Thm. 5.10 (with $c_1 = 0$). If x is finally nonzero then $x(t) = O(x_1(t))(t \to \infty)$ by Thm. 5.13. If

$$\alpha := \liminf_{t \to \infty}[x(t)/x_1(t)] < \beta := \limsup_{t \to \infty}[x(t)/x_1(t)],$$

then taking any $\gamma \in (\alpha, \beta)$ we get that the function $t \mapsto x(t) - \gamma x_1(t)$ is oscillating. Thus by Thm. 5.10

$$x(t) - \gamma x_1(t) = O(x_2(t)) \;\; (t \to \infty) \Rightarrow [x(t)/x_1(t)] \to \gamma \;\, (t \to \infty).$$

But this is a contradiction. Therefore $\alpha = \beta =: c_1$. From here $\widetilde{x}(t) := x(t) - c_1x_1(t) = o(x_1(t))(t \to \infty)$. Then $\widetilde{x}(t) = O(x_2(t))(t \to \infty)$: if $\widetilde{x}$ is oscillating this relation follows from Thm. 5.10, and if $\widetilde{x}(t)$ is finally nonzero – from Thm. 5.13. So (5.39) is proven.

At last, let us show that any solution $\widetilde{x}$ of (5.27) such that $\widetilde{x}(t) = O(x_2(t))$ $(t \to \infty)$ can be represented in the form

$$\widetilde{x}(t) = c_2 x_2(t) + O(x_2(t)) \ \ (t \to \infty) \tag{5.40}$$

where $c_2 =$ const. Put

$$c_2 := \liminf_{t\to\infty}[\widetilde{x}(t)/x_2(t)], \quad x_3(t) := \widetilde{x}(t) - c_2 x_2(t).$$

Then

$$\liminf_{t\to\infty}[x_3(t)/x_2(t)] = 0.$$

Hence by virtue of Thm. 5.12 if the function x_3 is not finally zero then it is oscillating. In the latter case we apply Thm. 5.12, where δx_2 with an arbitrary small $\delta =$ const > 0, taken as a nonincreasing solution and compared with x_3 or $-x_3$, yields that $x_3(t)$ has values of both signs on each interval $[t, t + H_0](t_0 - H_0 \le t < \infty)$. Assume that (5.40) is not true. Then for any $\varepsilon \in (0, \limsup_{t\to\infty}[x_3(t)/x_2(t)]$ a set of the solutions of the equation $x_3(t) - \varepsilon x_2(t) = 0$ is unbounded. Comparing the lefthand side with the decreasing function $\varepsilon x_2(T)$ and using Thm. 5.12 we obtain that the inequality $x_3(\tau) \le \varepsilon x_2(\tau)(\forall \tau \in [t, t + H_0])$ is impossible for any $t \in [t_0 - H_0, \infty)$.

Denote for $k = 0, 1, \dots$

$$p_k := \max\{x_3(t) : t_0 + kH_0 \le t \le t_0 + (k+1)H_0\},$$

$$-q_k := \min\{x_3(t) : t_0 + kH_0 \le t \le t_0 + (k+1)H_0\},$$

$$r_k = \varepsilon x_2(t_0 + kH_0)(> 0).$$

The features of oscillations of x_3 indicated above yield that $p_k, q_k > 0, r_{k+1} < r_k, p_k > r_{k+1}$, and $q_k = o(r_k)(k \to \infty)$. Consider the relation between $p_k(k \ge 2)$ and $\{q_j\}(j \le k)$. Let $p_k = x_3(\tau_k), t_0 + kH_0 \le \tau_k \le t_0 + (k+1)H_0$. Then x_3 has zeros on $[t_0 - H_0, \tau_k]$. Denote by θ_k the greatest of these zeros. We have $\tau_k - \theta_k < H_0$ and

$$p_k = \int_{\theta_k}^{\tau_k} \dot{x}_3(t)dt = -\int_{\theta_k}^{\tau_k} a(t)x_3(t - h(t))dt < A_0 H_0 \max\{q_k, q_{k-1}, q_{k-2}\}.$$

From here and the estimates proved above it follows that for $k \ge 3$

$$r_k < p_{k-1} < A_0 H_0 \max\{q_{k-1}, q_{k-2}, q_{k-3}\} = o(r_{k-3}) \ (k \to \infty).$$

Therefore, $r_k \to 0$ as $k \to \infty$ faster than any geometric progression. But it is impossible because later on we shall obtain the estimate $x_2(t)$ from below by a positive exponent. This contradiction proves (5.40). Representation (5.37) follows from (5.39), (5.40). Thm. 5.14 is proven.□

The next theorem gives the exponential estimate of the functions x_1 and x_2 in Thm. 5.14 from below and above. In addition the notations used coincide with those of in Eq. (5.38).

Theorem 5.15 Under conditions of Thm. 5.14 there are positive constants $M_1, M_2 > 0$ *such that*

$$M_1 \exp[-A_0\nu_1(A_0H_0)t] < x_1(t) < M_2 \exp[-a_0\nu_1(a_0h_0)t],$$

$$M_1 \exp[-a_0\nu_2(a_0h_0)t] < x_2(t) < M_2 \exp[-A_0\nu_2(A_0H_0)t] \qquad (5.41)$$

for all sufficiently large t, where in the estimation of $x_1(t)$ *if* a_0h_0 *or* A_0H_0 *are equal to* $1/e$ *then the corresponding exponent must be multiplied on* $(t - t_0)$.

Proof Let us define

$$y_i(t) := \exp[-A_0\nu_i(A_0H_0)t], \quad z_i(t) := \exp[-a_0\nu_i(a_0h_0)t] \ (i = 1, 2)$$

where for $i = 1$ if a_0h_0 or A_0H_0 equal $1/e$ then the corresponding exponent must be multiplied on $(t - t_0)$. The function $t \mapsto y_i(t)(t \mapsto z_i(t))$ is a solution of Eq. (5.38) with $\alpha = A_0$, $\beta = H_0$ (respectively $\alpha = a_0$, $\beta = h_0$). The estimate $x_1(t)$ from above follows from Thm. 5.12 if we take $t \mapsto x_1(t)(t_0 + H_0 \le t < \infty)$ as a decreasing solution, compare it with $\tilde{z}(t) = k[z_1(t-T) - z_2(t-T)]$ $(z(t) = kz_1(t-T)$, if $a_0h_0 = 1/e)$ and transfer the initial point to the point $\min\{t > T : x_1(t) = \tilde{z}(t)\}$. Here $aT \ge t_0 + 2H_0$ and $k > 0$ is so large that

$$\{t > T : x_1(t) = \tilde{z}(t)\} \ne \emptyset.$$

The estimate $x_1(t)$ from below for $A_0H_0 < 1/e$ follows from Thm. 5.12 if we take y_1 as a decreasing solution and compare it with the solution of Eq. (5.27) under initial condition $x(t) \equiv y_1(t_0)$ $(t \le t_0)$, and then take into account Thm. 5.13. If $A_0H_0 = 1/e$, then y_1 has to be replaced by $y_1(t) + (H_0 + 1/A_0)y_2(t)$ which decreases for $t \ge t_0 - H_0$.

To estimate $x_2(t)$ from above we must assume that the second estimate in (5.41) is untrue for all $M_2 > 0$ and apply Thm. 5.12, where the decreasing solution y_2 must be compared with $\tilde{x} := \varepsilon_1 x_2 - \varepsilon_2 x_1$ and $t_1 := \min\{t \ge t_0 : \tilde{x}(t) = y_2(t)\}$ is a new initial point. Here $\varepsilon_1, \varepsilon_2 > 0$ are such that $\tilde{x}(t) < y_2(t)(t_0 \le t \le t_0 + H_0)$, $\tilde{x}(t_2) > y_2(t_2)$, and $t_2 > t_0 + H_0$, chosen from the condition $\varepsilon_1 x_2(t_2) > y_2(t_2)$, exists by the assumption. By Thm. 5.12

$$y_2(t) \leq \tilde{x}(t) = \varepsilon_1 x_2(t) - \varepsilon_2 x_1(t) \quad (t_1 \leq t < \infty).$$

This leads to the contradiction as $t \to \infty$, since $x_2(t) = o(x_1(t))(t \to \infty)$.

At last the estimate $x_2(t)$ from below can be obtained as in the previous item, where y_2, x_2, x_1 are replaced by x_2, z_2, z_1 respectively. Thm. 5.15 is proven.□

Corollary 5.5 (Exponential dichotomy of the solutions set of Eq. (5.27) for a sufficiently small delay).

Let $a_0 h_0 > 0$, $A_0 H_0 < 1/e$. Then any solution x of Eq. (5.27) satisfies for all sufficiently large t and a constant $c = c_x > 0$ one of the following estimates: $|x(t)| \geq c\exp[-A_0\nu_1(A_0H_0)t]$ or $|x(t)| \leq c\exp[-A_0\nu_2(A_0H_0)t]$. This "splitting" of the infinite dimensional set of all solutions on "slowly decreasing" and "fastly decreasing" solutions becomes the more evident, the less A_0H_0, since $\nu_1(p) \to p$, $\nu_2(p) \to \infty$ as $p \to 0$.

It can be proven that c_1 and c_2 in (5.37) depend continuously on the initial function. It means that a slow decreasing of the solution is a stable property which keeps under sufficiently small variation of the initial function. In contrast to this, the property of a fast decreasing is unstable with respect to the variation of the initial function.

5.6.3 Equation (5.28).

Theorem 5.16 Let

$$A_0^{1/2} H_0 < 2^{1/2} + \pi/2 \quad (\approx 2.985). \tag{5.42}$$

Assume that, $[\alpha, \beta]$ is a large semicycle of the solution $x : [t_0, \infty) \mapsto \mathbf{R}$ of Eqs. (5.24) and that x has at least one root on (β, ∞). Then β is the left end of another large semicycle of x.

Proof Assume for definiteness that $x(t) < 0$ $(\alpha < t < \beta)$. Then by (5.28) the graph of x is convex downwards (in nonstrict sense) for all $t \leq \beta$ sufficiently close to β. This yields $x(\beta) > 0$. Hence, if there are roots of x on (β, ∞) then there exists the least root γ. Let δ be a point of the maximum of the solution x on $[\beta, \gamma]$ and $\varepsilon \in (\beta, \delta]$ is the least root of the function $\dot{x}$ on $[\beta, \gamma]$. Then we have on $[\beta, \varepsilon]$

$$\ddot{x}(t) = -a(t)x(t - h(t) \geq -A_0 x(t).$$

Writing this inequality in the form

$$\ddot{x}(t) + A_0 x(t) =: f(t) \geq 0 \quad (\beta \leq t \leq \varepsilon),$$

and applying integral representation of Cauchy of the function $\dot{x}$ with $x(\beta) = 0$, $\dot{x}(\beta) =: \dot{x}_0 > 0$ we obtain that $\varepsilon - \beta \geq \pi/(2A_0^{1/2})$. But $\ddot{x} \geq -A_0 x(\delta)$ on $[\delta, \gamma]$. Hence

$$\dot{x}(t) \geq -A_0 x(\delta)(t-\delta), \quad x(t) \geq x(\delta) - A_0 x(\delta)(t-\delta)^2/2 \ (\delta \leq t \leq \gamma).$$

This means that $\gamma - \delta \geq (2/A_0)^{1/2}$. So, $\gamma - \beta \geq (2\sqrt{2} + \pi/2)A_0^{-1/2}$. Therefore, if (5.25) holds then $\gamma - \beta > H_0$. Thm. 5.11 proven. □

In [381.3] it was shown that the constant $\sqrt{2} + \pi/2$ in Thm. 5.16 is unimprovable.

Corollary 5.6 Under conditions (5.42) the solutions of Eq. (5.28) can be only one of the following four types:

a) the solutions have infinite sequence of the adjoining large semicycles, on which the signs of the solution are alternating;

b) eventually zero;

c) the solutions do not have large semicycles, but they have at least one zero on each interval $[t_0 - H_0, t](t \geq t_0)$. (In [381.3] it is proven that for $A_0^{1/2} H_0 < 2\sqrt{2}$ such solution and its derivative tend to zero as $t \to \infty$ nonslower than an exponent.)

d) eventually different from the zero (if $a_0 > 0$ *then such solutions are absent).*

Consider the influence of delay in Eq. (5.28) on stability of its solutions. For simplicity we investigate this question for equations with the constant coefficient $a(t)$:

$$\ddot{x}(t) + x(t - h(t)) = 0. \tag{5.43}$$

It can be shown shown that for $h(t) \equiv \text{const} > 0$ Eq. (5.43) is always unstable (see Subs. 4.3.4). In the book [381.3] there is an example of Eq. (5.43) with the variable $h(t)$, which has a solution in the form of the large semicycles tending to zero as $t \to \infty$ with the rate of exponent. Of course this does not mean the stability. Moreover, the existence of asymptotically stable equations of the form (5.43) seems to be doubtful, though in general this problem is not investigated.

Let us show that for an equation of the type (5.43) with an autonomous autoregulated delay $h(x(t), \dot{x}(t))$ it can happen that all the solutions tend to zero as an exponent (more precisely, the zero solution is globally exponentially stable). So let us construct an example of such equation of the form

$$\ddot{x}(t) + x(t - h(x(t), \dot{x}(t))) = 0. \tag{5.44}$$

Take any $p_1 \in (0,1), p_2 \in (\max\{2p_1 - 1, 0\}, 1)$ and define $h(t)$

$$\begin{aligned} h(t) &= 0 \ (0 \leq t \leq \pi/2), \ = t - t_1 \quad (\pi/2 < t \leq t_2), \\ &= 0 \quad (t_2 < t \leq T) \end{aligned} \tag{5.45}$$

where $t_1 := \arcsin p_1$, $t_2 := (\pi/2)+[2(1-p_2)/p_1]^{1/2}$, $T := t_2+\mathrm{arctg}\{p_2[2p_1\times (1-p_2)]^{-1/2}\}$. The function $h(t)$ has jumps for $t = \pi/2$ and $t = t_2$. Hence, the general properties of Eq. (5.43) solutions are given by Thm. 2.2 after passing to the system of equations of the first order. Eq. (5.43) thus obtained is an RDE without previous history. Therefore, the initial conditions can be taken as follows

$$x(0) = 0, \quad \dot{x}(0) = 1.$$

The corresponding solution x_0 is

$$x_0(t) = \sin t \ (0 \le t \le \pi/2), \ = 1 - p_1(t-\pi/2)^2/2 \ (\pi/2 \le t \le t_2),$$

$$= p_2 \cos(t-t_2) - \sqrt{2p_1(1-p_2)} \sin(t-t_2) \ (t_2 \le t \le T).$$

The functions $x_0, \dot{x}_0$ are continuous on $[0,T]$ and $x_0(T) = 0$, $x_0(t) > 0$ $(0 < t < T)$, $\dot{x}_0(T) = -(2p_1(1-p_2)+p_2^2)^{1/2} > -1$. Continue $h(t)$ periodically with a period T on the semiaxis $t \ge 0$. Then the solution x_0 will be obtain the additional multipliers $\dot{x}_0(T)$ on every sequential interval $[kT, (k+1)T]$ $(k \in \mathbf{N})$, i.e.

$$\max\{|x_0(t)|, |\dot{x}_0(t)|\} \le c\exp[-|\ln|\dot{x}_0(T)||t/T] \ (0 \le t < \infty).$$

To construct Eq. (5.44) with the same decaying particular solution x_0 let us note that the values $t - kT(kT < t < (k+1)T)$ correspond one–to–one the values $u := \dot{x}_0(t)/x_0(t)$, and this correspondence does not depend on k. In addition, the values $t = \pi/2$ and $t = t_2$ correspond to $u = 0$ and $u = -[2p_1(1-p_2)]^{1/2}/p_2$. The values t and u for $\pi/2 \le t \le t_2$ are connected by the equation

$$u = -p_1(t-\pi/2)[1-p_1(t-\pi/2)^2/2]^{-1}.$$

If we find from here $t - \pi/2$ as a function of u, we obtain by virtue of (5.45) the function $h(\cdot,\cdot)$ with which Eq. (5.45) is satisfied by x_0:

$$h(v,w) = \begin{cases} 0, \text{ if } v = 0, \text{or } v \ne 0 \text{ and } w/v \notin (-[2p_1(1-p_2)]^{1/2}/p_2, 0), \\ (\pi/2) - \arcsin p_1 - 2[p_1 + (p_1^2 + 2p_1 w^2/v^2)^{1/2}]^{-1}(w/v), \\ \qquad \text{if } v \ne 0 \text{ and } w/v \in (-[2p_1(1-p_2)]^{1/2}/p_2, 0) \end{cases} \tag{5.46}$$

This function has two lines of discontinuity. Hence Eq. (5.44) has to be understood in the generalized sense from Subs. 4.4 modified for equations of the second order. So, x and $\dot{x}$ are assumed to be continuous but $\ddot{x}$ may not exist if $\dot{x}(t) = 0$ and $\dot{x}(t) = -[2p_1(1-p_2)]^{1/2}p_2^{-1}x(t)$. Drawing the trajectories on the phase plane $x, \dot{x}$ we obtain that the sliding modes in this example are absent. So the singularities of Eq. (5.44) solutions are reduced to the first kind discontinuites of $\ddot{x}$.

From (5.46) it follows that if the solution x of Eq. (5.44) with the initial condition

$$x(t) = \phi(t) \ \ (t_0 - H_0 \leq t \leq t_0), \quad \dot{x}(t_0^+) = \psi_0 \tag{5.47}$$

has a zero $t = \tilde{t} \geq t_0$, then $x(t) \equiv \dot{x}(\tilde{t})x_0(t - \tilde{t}) \ \ (\tilde{t} \leq t < \infty)$. Here $H_0 = \pi/2 - \arcsin p_1 + [2(1 - p_2)/p_1]^{1/2}$ by (5.46).

Let us estimate $\tilde{t} := \min\{t \geq t_0 : x(t) = 0\}$ and $|\dot{x}(\tilde{t})|$, simultaneously proving the existence of zeros of the function x. Put

$$\Phi_0 := \max\{|\phi(t)| : t_0 - H_0 \leq t \leq t_0\}.$$

Assume for the definiteness that $\phi(t_0) \geq 0$. If $\tilde{t} \in [t_0, t_0 + H_0]$ then by (5.44) $|\dot{x}(\tilde{t})| \leq |\dot{x}(t_0^+)| + H_0$. If $x(t) > 0$ $(t_0 \leq t \leq t_0 + H_0)$ then on any interval $[t_0 + H_0, T]$, where $\min x(t) > 0$ we have $\sup \ddot{x}(t) < 0$. From here it follows that in this case there are $\tilde{t} > t_0 + H_0$ and $\bar{t} \in [t_0 + H_0, \tilde{t})$ such that x is increasing on $[t_0 + H_0, \bar{t}]$, x is decreasing on $[\bar{t}, \tilde{t}]$, $x(\tilde{t}) = 0$, and the graph of the function $x|_{[t_0 + H_0, \tilde{t}]}$ is upwards convex.

Prove the uniform boundedness of $\bar{t}$ with respect to all initial conditions. In fact, let $\bar{t} > \tau := [(t_0 + H_0) + \bar{t}]/2 + H_0$. From the convexity of $x|_{[t_0 + H_0, \tilde{t}]}$ it follows the inequality

$$\dot{x}(\tau) \leq [x(\tau) - x(t_0 + H_0)][\tau - (t_0 + H_0)]^{-1}.$$

Hence, for $\tau \leq t \leq \bar{t}$ we have

$$t - h(x(t), \dot{x}(t)) \geq [(t_0 + H_0) + \bar{t}]/2 \Rightarrow \ddot{x}(t) \leq -x(\bar{t})/2 \Rightarrow$$

$$\Rightarrow \dot{x}(t) \leq [x(\tau) - x(t_0 + H_0)][\tau - (t_0 + H_0)]^{-1} - x(\bar{t})(t - \tau)/2 \Rightarrow x(\bar{t}) \leq$$

$$\leq x(\tau) + [x(\tau) - x(t_0 + H_0)][\tau - (t_0 + H_0)]^{-1}(\bar{t} - \tau) - x(\bar{t})(\bar{t} - \tau)^2/4.$$

Since $x(\tau) < x(\bar{t})$, then transferring the lefthand side to the right and dividing by $x(\bar{t})(\bar{t} - \tau)/4$ we obtain

$$\bar{t} - \tau < 4[x(\tau) - x(t_0 + H_0)]/x(\bar{t})[\tau - (t_0 + H_0)] < 4/[\tau - (t_0 + H_0)].$$

From here substituting τ we have

$$(\bar{t} - t_0 - 3H_0)/2 < 8/(\bar{t} - t_0 + H_0) \Rightarrow \bar{t} - t_0 < H_0 + 2(H_0^2 + 4)^{1/2}.$$

This inequality means that $\bar{t}$ is uniformly bounded.

Prove now the uniform boundedness of $\widetilde{t}-\bar{t}$. If $\widetilde{t} > \overline{\tau} := \bar{t} + H_0$ then for $\overline{\tau} \le t \le \widetilde{t}$ by virtue of (5.44) $\ddot{x}(t) \le -x(t)$, i.e. $\ddot{x}(t) + x(t) \le 0$. This fact and the Chaplygin's theorem on differential inequalities give us that

$$x(t) \le x(\overline{\tau})\cos(t-\overline{\tau}) + \dot{x}(\overline{\tau})\sin(t-\overline{\tau}) \quad (\overline{\tau} \le t \le \min\{\overline{\tau}+\pi, \widetilde{t}\}).$$

Since $x(\overline{\tau}) > 0$, $\dot{x}(\overline{\tau}) < 0$, the righthand side is equal to zero for $t - \overline{\tau} \in (0, \pi/2)$, i.e. $\widetilde{t} - \overline{\tau} < \pi/2$. Q.E.D.

Thus for the solution of (5.44), (5.47) we have $\widetilde{t} - t_0 \le \theta$, where $\theta > 0$ does not depend on ϕ. But

$$|\ddot{x}(t)| \le \max\{|x(\tau)| : t_0 - H_0 \le \tau \le t\} \quad (t_0 \le t < \infty).$$

Hence,

$$|x(t)| \le \Phi_0 \mathrm{ch}(t-t_0) + |\dot{x}(t_0^+)|\mathrm{sh}(t-t_0),$$

$$|\dot{x}(t)| \le \Phi \mathrm{sh}(t-t_0) + |\dot{x}(t_0^+)|\mathrm{ch}(t-t_0) \quad (t_0 \le t \le \widetilde{t}).$$

Therefore,

$$|x(t)| \le (\Phi_0 + |\dot{x}(t_0^+)|)e^{\theta}, \quad |\dot{x}(t)| \le (\Phi_0 + |\dot{x}(t_0^+)|)e^{\theta} \quad (t_0 \le t \le \widetilde{t}).$$

But $x(t) \equiv \dot{x}(\widetilde{t})x_0(t-\tau)$ for $t \ge \widetilde{t}$, i.e.

$$\max\{|x(t)|, |\dot{x}(t)|\} \le (\Phi_0 + |\dot{x}(t_0^+)|)e^{\theta} c \exp[-|\ln|\dot{x}_0(T)||(t-\theta)/T]$$

$$(\widetilde{t} \le t < \infty).$$

Exponential stability of the zero solution of Eq. (5.44) follows from this estimate.

Note in conclusion that using smoothing of function (5.45), an analogous example can be constructed with the function $h(\cdot,\cdot)$ having the only discontinuity point at $(0,0)$.

6. LINEAR AUTONOMOUS FDEs

6.1. *EXPONENTIAL SOLUTIONS OF LINEAR AUTONOMOUS RDES*

According to Subs. 5.2, the general linear homogeneous autonomous RDE with finite delay has the form

$$\dot{x}(t) = \int_{-h}^{0} [dR(\theta)]x(t+\theta). \tag{6.1}$$

The corresponding inhomogeneous equation has the form

$$\dot{x}(t) = \int_{-h}^{0} [dR(\theta)]x(t+\theta) + f(t). \tag{6.2}$$

Here, $R : [-h, 0] \to \mathbf{R}^{n\times n}$ is a matrix–function of bounded variation, and $f \in L_1[t_0, \infty)$. For the Cauchy problem we must prescribe the initial condition:

$$x(t) = \phi(t) \quad (t_0 - h \leq t \leq t_0), \tag{6.3}$$

where it is assumed that $\phi \in C[t_0 - h, t_0]$. By Subs. 5.2 there is a unique solution of the Pr. (6.2), (6.3) on $[t_0, \infty)$.

The main property of (6.1) is the presence of a sufficiently representative set of exponential solutions

$$x_k(t) = e^{z_k t} c_k, \quad k = 1, 2, \ldots, \tag{6.4}$$

giving rise to a *normal motion* in real system modelled by (6.1). We now assume that $z_k \in \mathbf{C}$, $c_k \in \mathbf{C}^n$, $c_k \neq 0$. If $z_k \neq \mathbf{R}$, then, as for ODEs, the real and imaginary parts of the solutions (6.4) do play a main role.

Substitution of (6.4) into (6.1) shows that (6.1) has a solution of the form (6.4) for a given z_k if and only if $z = z_k$ is an eigenvalue of the matrix $\int_{-h}^{0} e^{z\theta} dR(\theta)$, i.e. z_k must be a root of the *characteristic equation*

$$\det \Psi(z) = 0, \quad \text{where } \Psi(z) := zI - \int_{-h}^{0} e^{z\theta} dR(\theta). \tag{6.5}$$

In this case c_k is the corresponding eigenvector. If the eigenvalue z_k has algebraic multiplicity $m_k > 1$, then next to the solutions (6.4) equation (6.1) can have solutions $e^{z_k t} c_k(t)$, where $c_k(t)$ is a vector polynomial in t of degree lower than m_k. This polinomial can be found by the method of indefinite coefficients. Moreover, the total number of linearity independent solutions of such a kind is, for a given z_k, equal to m_k.

We denote the $\det \Psi(z)$ in (6.5) by $g(z)$. The function $z \mapsto g(z)$ is an entire function (i.e. it is analytic and single–valued for all $z \in \mathbf{C}$). In case (6.1) involves only finitely many discrete delays, the expression $g(z)$ is a *quasipolynomial*, i.e. a linear combination of products $z^k e^{\alpha_k z}$ (with integer $k \geq 0$). Therefore, in general $g(z)$ may also be called a (generalized) *characteristic quasipolynomial*. It can be shown that, with the exception of the degenerate case (see below), for an RDE this quasipolynomial has infinitely many zeros $z_1, z_2, \ldots$, some of which may be repeated, i.e. be multiple. In addition,

$$\mathrm{Re} z_k \to -\infty \quad \text{as} \quad k \to \infty.$$

In the degenerate case $g(z)$ is a polynomial of degree n, and the solution manifold of (6.1), (6.3) becomes n – dimensional for arbitrary ϕ and sufficiently large t. For such t (6.1) reduces to an ODE. E.g., in example 2.2 we have $g(z) = z^n$ for (2.8). For $t \geq n-1$, (2.8) implies the ODE $x_n^{(n)}(t) = 0$. Having solved this equation, we obtain the remaining components from the formulas $x_i(t) = x_n^{(n-i)}(t+n-i)(i = 1, \ldots, n-1; i-1 \leq t < \infty)$.

Suppose that the kernel, and hence the quasipolynomial g, depends on parameters $p_1, \ldots, p_d$, where the function $(z; p_1, \ldots, p_d) \mapsto g(z; p_1, \ldots, p_d)$ is continuous. Then general theorems in analytic function theory imply that the roots z_k of (6.5) depend continuously on $p_1, \ldots, p_d$. If the parameters tend to some set $p_1^0, \ldots, p_d^0$, then some of these roots z_k may tend to infinity, and then $\mathrm{Re} z_k \to -\infty$. This implies that if $\inf\{\mathrm{Re} z : z \in G\} > -\infty$ for some open domain $G \subset \mathbf{C}$, then the number of roots of (6.5) belonging to G may vary under parameter variation only when at least one of these roots hits the boundary ∂G.

This property is the basis of the so-called *D-subdivision method*, in which the space of parameters on which the kernel R depends is partitioned into subsets on which the *instability degree* of (6.1) is constant. This is defined as the number of roots z_k of (6.5) with $\mathrm{Re} z_k > 0$. Now we briefly indicate how this is done, assuming (to be specific) that $d = 2$ and p_1, p_2 are arbitrary real numbers. We represent z and g as

$$z = \alpha + i\beta, \quad g(z; p_1, p_2) = g_1(\alpha, \beta; p_1, p_2) + i g_2(\alpha, \beta; p_1, p_2).$$

By virtue of the above mentioned property, a change in the instability degree $\kappa(p_1, p_2)$ of (6.1) in every neighbourhood of certain values of p_1, p_2 is possible only if the equations

$$g_j(0, \beta; p_1, p_2) = 0, \quad j = 1, 2, \tag{6.6}$$

have common solution β.

This condition defines in the parameter (p_1, p_2)-plane a collection of curves, dividing this plane into subdomains. This subdivision is called a D-subdivision. We will clarify how $\kappa(p_1, p_2)$ changes when (p_1, p_2) crosses a curve L dividing domains G_1 and G_2 in the D-subdivision. For this we must find the sign of the derivative $q^0 := (\partial g_1/\partial l)|_{(0,\beta^0;p_1^0,p_2^0)}$ with respect to some nontangential direction l to L (pointing from G_1 to G_2), with $(p_1^0, p_2^0) \in L$ chosen and β^0 as defined by (6.6) with $p_j = p_j^0$. If there is only one β^0 (which must equal zero in this case), then $\kappa(p_1, p_2)$ increases by 1 for $q^0 > 0$ and decreases by 1 for $q^0 < 0$ under transition through L from G_1 to G_2. If $q^0 = 0$, then higher–order derivatives must be used. If there are several β^0, then each of these should be verified. In particular, if there are two β^0's, then they differ only in sign and have the same q^0. Hence, for $q^0 \neq 0$ the values of $\kappa(p_1, p_2)$ on G_1 and G_2 differ by 2.

Now we consider (6.2) under assumption that $f(t)$ is a quasipolynomial. Then a solution of quasipolynomial type can be found by the method of indefinite coefficients, similarly as for ODEs. First of all, since the equation is linear, it suffices to study the case

$$f(t) = e^{zt} P(t),$$

where $z \in \mathbf{C}$ and P is a vector polynomial of degree $m \geq 0$. If there is no *resonance* (i.e. $g(z) \neq 0$), then (6.2) has exactly one particular solution of the form

$$x(t) = e^{zt} Q(t). \tag{6.7}$$

Here, the vector–polynomial Q also has degree m. In the presence of resonance the degree of Q exceeds or, seldom, is equal to m, but is not larger than $m + k$ (for a scalar equation (6.2) it is equal to $m + k$), where k is the algebraic multiplicity of z as a root of (6.5).

In particular, if $m = 0$ (i.e. $f(t) = e^{zt} c$) and $g(z) \neq 0$, then the solution of (6.2) has the form

$$x(t) = e^{zt} \Psi^{-1}(z) c.$$

From this we can derive the following assertion about periodic solutions of (6.2). Suppose f is periodic with period $T > 0$, and has Fourier expansion

$$f(t) = \sum_{k=-\infty}^{\infty} e^{(2k\pi i/T)t} c_k,$$

with

$$|c_k| = O(k^{-\varepsilon}) \ \ (k \to \infty)$$

for some $\varepsilon > 0$ (this condition can be weakend). Then in the nonresonance case (i.e. when $g(2k\pi i/T) \neq 0$ for all k) equation (6.2) has

$$x(t) = \sum_{k=-\infty}^{\infty} e^{(2k\pi i/T)t} \Psi^{-1} \left(\frac{2k\pi i}{T} \right) c_k \tag{6.8}$$

as unique T–periodic solution.

However, if $g(2k\pi i/T) = 0$ for some k, then for the existence of a T–periodic solution it is necessary and sufficient that $c_k' d = 0$ for this k and any eigenvector d of the corresponding matrix $(\psi(2k\pi i/T))'$, and prime is a transposition sign. (For the scalar case it follows from here that $c_k = 0$.)

If this condition holds, then in (6.8) for the resonance values of k we must take an arbitrary eigenvector of the matrix $\int_{-h}^{0} e^{(2k\pi i/T)\theta} dR(\theta)$ (or the zero-vector) instead of the factors $\Psi^{-1}(2k\pi i/T)c_k$.

6.2. *SOLUTION OF THE CAUCHY PROBLEM*

Using the Laplace integral transform it is easy to obtain an explicit formula for the solution of the Pr. (6.2), (6.3). Recall that for an arbitrary function $h \in AC[0,\infty)$ satisfying for some $\alpha_0 \in \mathbf{R}$ the estimate

$$\int_0^\infty |h(t)|e^{-\alpha t}dt < \infty \quad (\forall \alpha > \alpha_0),$$

its *Laplace transform* is the following function of the complex variable z:

$$z \mapsto \tilde{h}(z) = (Lh)(z) := \int_0^\infty e^{-zt}h(t)dt.$$

This function is analytic in the halfplane $\alpha(:= \mathrm{Re} z) > \alpha_0$. Conversely, a function h is determined by its Laplace transform $\tilde{h}$ by the formula

$$h(t) = (L^{-1}\tilde{h})(t) := \frac{1}{2\pi i} \int_{\alpha - i\infty}^{\alpha + i\infty} e^{zt}\tilde{h}(z)dz,$$

where the integration is over a straight line parallel to the β-axis, and $\alpha > \alpha_0$ is arbitrary.

Suppose that the solution f in (6.2) satisfies the estimate $|f(t)| = O(e^{\gamma t})$ $(t \to \infty)$. Then the solution of (6.2), (6.3) satisfies the exponential estimate $|x(t)| = O(e^{\max\{\gamma, v\}t})$, where $v = V_{[-h,0]}R$ and for $\gamma = v$ the righthand side of this estimate must be multiplied by the factor t. Replacing t by $t + t_0$ in (6.2) and applying Laplace transformation to both sides of the equation, we can find the Laplace transform of $x(\cdot + t_0)$. Applying the inversion formula, passing from $t + t_0$ to t, and using the initial condition

(6.3), we arrive at a formula for the solution of (6.2), (6.3):

$$x(t) = \frac{1}{2\pi i} \int\limits_{\alpha - i\infty}^{\alpha + i\infty} e^{zt} \Psi^{-1}(z) \times$$

$$\times \left\{ e^{-zt_0}\phi(t_0) + \int\limits_{-h}^{0} [dR(\theta)] \left[e^{z\theta} \int\limits_{t_0+\theta}^{t_0} e^{-z\tau}\phi(\tau)d\tau \right] + \int\limits_{t_0}^{\infty} e^{-z\tau} f(\tau)d\tau \right\} dz,$$

$$t_0 \le t < \infty, \tag{6.9}$$

where $\alpha > \max\{\gamma, v\}$.

Comparing (6.9) and (5.11), we obtain the Cauchy matrix for (6.2), (6.3):

$$C(t;s) = \frac{1}{2\pi i} \int\limits_{\alpha - i\infty}^{\alpha + i\infty} e^{z(t-s)} \Psi^{-1}(z) dz, \quad \alpha > v.$$

This equality holds for all $s, t \in \mathbf{R}, t \neq s$ (if $t = s$, then the righthand side equals to $I/2$). An explicit expression for the matrix $G(t;s)$ in Subs. 5.3 can be given in much the same way.

Formula (6.9) can be transformed by distinguishing at the righthand side the principal terms as $t \to \infty$. For simplicity we assume that $f(t) \equiv 0$, and put

$$\Phi(z) := \Psi^{-1}(z) \left\{ e^{-zt_0}\phi(t_0) + \int\limits_{-h}^{0} [dR(\theta)] \left[e^{z\theta} \int\limits_{t_0+\theta}^{t_0} e^{-z\tau}\phi(\tau)d\tau \right] \right\}. \tag{6.10}$$

This is an analytic function of z in the whole complex plane with the exception of the zeros $z_1, z_2, \ldots,$ of the quasipolynomial $g(z)$ (see Subs. 6.1), at which it has poles; also $\alpha_k (= \operatorname{Re} z_k) \le v$.

Having chosen an arbitrary $\tilde{\alpha} < \alpha, \tilde{\alpha} \notin \{\alpha_k\}$, and using standard methods of function theory, we obtain

$$x(t) = \frac{1}{2\pi i} \int\limits_{\alpha - i\infty}^{\alpha + i\infty} e^{zt} \Phi(z) dz =$$

$$= \sum_{k: \tilde{\alpha} < \alpha_k < \alpha} \operatorname{Res}_{z=z_k} [e^{zt}\Phi(z)] + \frac{1}{2\pi i} \int\limits_{\tilde{\alpha} - i\infty}^{\tilde{\alpha} + i\infty} e^{zt} \Phi(z) dz. \tag{6.11}$$

Here, $\mathrm{Res}_{z=z_k}[e^{zt}\Phi(z)]$ is the *residue* of the analytic function $z \mapsto e^{zt}\Phi(z)$ at its singularity z_k. If z_k is a pole of g of order $l \geq 1$, then this residue can be computed by the formula

$$\mathrm{Res}_{z=z_k}[e^{zt}\Phi(z)] = \frac{1}{(l-1)!}\lim_{z\to z_k}\frac{d^{l-1}}{dz^{l-1}}[(z-z_k)^l e^{zt}\Phi(z)]. \tag{6.12}$$

For the most important case $l = 1$ this formula becomes

$$\mathrm{Res}_{z=z_k}[e^{zt}\Phi(z)] = \lim_{z\to z_k}[(z-z_k)\Phi(z)]e^{z_k t}. \tag{6.13}$$

The last term in (6.11) satisfies the estimate

$$\left|\frac{1}{2\pi i}\int_{\tilde{\alpha}-i\infty}^{\tilde{\alpha}+i\infty} e^{zt}\Phi(z)dz\right| = O(e^{\tilde{\alpha}t}) \ \ (t\to\infty). \tag{6.14}$$

The formulas (6.10)–(6.14) show that for any of the above $\tilde{\alpha}$ the solution of the Pr. (6.1), (6.3) can be represented as

$$x(t) = \sum_{k:\alpha_k>\tilde{\alpha}} Q_k(t)e^{z_k t} + O(e^{\tilde{\alpha}t}) \ \ (t\to\infty). \tag{6.15}$$

Here, the vector–polynomial Q_k has degree less than l_k and can be given in explicit form, using (6.12), and l_k is the multiplicity of z_k as a root of (6.5). Under certain additional assumptions one can pass to the limit in (6.15) as $\tilde{\alpha} \to -\infty$, and obtain a representation of the solution as the sum of a series of exponential solutions of (6.1) and the quasipolynomial solutions adjoint to them. However, in applications this series is not commonly used, as the most intrinsic properties of the solution (estimates as $t \to \infty$, oscillation, etc.) are in fact determined by the (really existing) summands on the righthand side of (6.15) that have largest α_k. (The remark about the existence is made because the polynomials Q_k are determined by the initial function ϕ, and so some of them may vanish identically.)

We list all roots z_k of g according to nonincrease of real parts, taking into account algebraic multiplicity, i.e. $\alpha_1 \geq \alpha_2 \geq \alpha_3 \geq \ldots$. Then there are two 'rough' (structurally stable) cases relative to arbitrary sufficiently small variations of the kernel R. Namely, $\alpha_1 > \alpha_2$ (then $\beta_1 := \mathrm{Im} z_1 = 0$) and $\alpha_1 = \alpha_2 > \alpha_3$ (then $\beta_1 = -\beta_2 \neq 0$). In the first case all 'typical' real solutions have the form $ce^{\alpha_1 t} + o(e^{\alpha_1 t})(c \neq 0)$, and in the second $ce^{\alpha_1 t}\sin(\beta_1 t + \psi) + o(e^{\alpha_1 t})(c \neq 0)$ as $t \to \infty$. This implies oscillation properties of the solutions. In particular, it makes it possible to determine z_1 approximately by numerically solving the Cauchy problem (6.1), (6.3) for various initial functions ϕ on a time interval of sufficient length.

6.3. *EXAMPLE OF A SHOWERING PERSON*

Consider again equation (1.1.10), modeling the behaviour of a showering person. We set $x(t) = T_m(t) - T_d$, $\gamma = k\kappa > 0$. Then (1.1.10) takes the form

$$\dot{x}(t) = -\gamma x(t-h). \tag{6.16}$$

Suppose the water temperature before regulation is constant, i.e. the initial condition is given by the relation

$$x(t) = x_0 = \text{const} \neq 0 \quad (t_0 - h \leq t \leq t_0). \tag{6.17}$$

For small $t - t_0$ the solution can be obtained by the steps method (see Subs. 1.3). We will consider its behaviour for increasing t.

Equation (6.16) is obtained from (6.1) if

$$n = 1, \quad r(\theta) = \begin{cases} 0, & \theta = -h, \\ -\gamma, & -h < \theta \leq 0. \end{cases}$$

Hence in this case characteristic equation (6.5) takes form

$$-\gamma e^{-hz} - z = 0, \quad \text{or} \quad z = -\gamma e^{-hz}. \tag{6.18}$$

It is easy to show that (6.18) has a multiple (double) root if and only if $\gamma h = 1/e$. Depicting the graphs of the lefthand and righthand sides of (6.18) for $z \in \mathbf{R}$, we see that for $\gamma h < 1/e$ (6.18) has two real roots, z_1, z_2: $z_2 < z_1 < 0$. For a fixed $h > 0$ ($\gamma > 0$), these roots depend monotonically on $\gamma(h)$; moreover, $z_1 \to 0$ ($z_1 \to -\gamma$), $z_2 \to -\infty$ as $\gamma \to 0$ (as $h \to 0$), and $z_1 \to -h^{-1}, z_2 \to -h^{-1}$ as $\gamma \to [(eh)^{-1}]^-$ (as $h \to [(e\gamma)^{-1}]^-$). For $\gamma h = 1/e$ we obtain $z_1 = z_2 = -h^{-1}$; for $\gamma h > 1/e$ there are no real roots.

Put $z = \alpha + i\beta$. We verify that for $\gamma h \leq 1/e$ all roots except z_1, z_2 satisfy the inequality $\alpha < z_2$. In fact, (6.18) implies that for such roots

$$|\alpha| < |z| = \gamma e^{-h\alpha}, \quad \text{i.e.} \quad \alpha > -\gamma e^{-h\alpha}.$$

Examination of the above described graphs shows that if we would have $\alpha \geq z_2$ for some $\gamma \leq 1/(eh)$, then $\alpha > z_1$. Since the root under consideration depend continuously on γ, the last inequality remains valid as γ decreases to 0. But this is impossible, since for $\gamma = 0$ (6.18) has only the root $z_1 = 0$.

Hence, for $\gamma h \leq 1/e$ the listing of the roots z_1, z_2 corresponds to the rule given at the end of Subs. 6.2.

We show that in the representation

$$x(t) = ce^{z_1 t} + o(e^{z_1 t}), \quad t \to \infty,$$

for the solution of (6.16)–(6.17) $c \neq 0$ for $\gamma h < 1/e$. In fact, (6.10) implies

$$\Phi(z) = \frac{x_0}{z}(\gamma e^{-hz} + z)^{-1}(z + \gamma e^{-zt} - \gamma)e^{-t_0 z}.$$

This, (6.11), (6.13), and (6.18) imply

$$c = -\frac{x_0 \gamma}{z_1(1 + hz_1)} e^{-z_1 t_0} \neq 0. \text{ Q.E.D.}$$

Therefore, for $\gamma h < 1/e$, from some moment onwards the regulated temperature will monotonically tend to the desired value. It is interesting to note that as long as the last inequality is valid, z_1 is a decreasing function of both γ and h, i.e. with the increase in sensitivity or time delay, the stability of the system improves, contrary to the existing common opinion.

When γh increases and passes through the value $1/e$, the roots z_1, z_2 become imaginary conjugates. For a further investigation of the change of the roots we separate the real and imaginary parts in (6.18). We obtain

$$\alpha = -\gamma e^{-\alpha h} \cos \beta h, \quad \beta = \gamma e^{-\alpha h} \sin \beta h. \tag{6.19}$$

This implies that $\alpha_k = \alpha_l$ if and only if $\beta_k = \pm \beta_l$. Hence under continuous variation of γ and h the ordering of the α_k remains the same: $\alpha_1 = \alpha_2 > \alpha_3 = \alpha_4 > \dots$ and stability is determined by the roots $z_{1,2}$, which were initially real. Eqs. (6.19) lead to the conclusion that the stability of the system deteriorates when h passes through $1/(e\gamma)$. The main term of the solution of the Pr. (6.16), (6.17) now takes the form

$$x(t) = 2\mathrm{Re}\,\mathrm{Res}_{z=z_1}[e^{zt}\Phi(z)] + o(e^{\alpha_1 t}) =$$

$$= -2\gamma x_0[(\alpha_1 + h\alpha_1^2 - h\beta_1^2)^2 + (\beta_1 + 2h\alpha_1\beta_1)^2]^{-1} \times$$

$$\times [(\alpha_1 + h\alpha_1^2 - h\beta_1^2)\cos \beta_1(t - t_0) + (\beta_1 + 2h\alpha_1\beta_1)\sin \beta_1(t - t_0)] \times$$

$$\times e^{\alpha_1(t - t_0)} + o(e^{\alpha_1 t}) \quad (t \to \infty). \tag{6.20}$$

Thus, as long as $\alpha_1 < 0$, the regulated temperature, being a damped oscillation, asymptotically tends to the desired value.

It is well known that first–order autonomous ODEs cannot have oscillating solutions. Therefore the *appearance of oscillations for $h > (e\gamma)^{-1}$ is a pure delay effect.* Since a considerable delay of information of water temperature takes place at the mixer output, a person regulating this temperature will endlessly turn the mixer handle through the optimal position from one side to the other.

Eq. (6.19) also implies that the D–subdivision (see Subs. 6.1) of the parameter (γ, h)–plane (more exactly, of the first quadrant of this plane) is performed by the lines

$$\gamma h = \frac{\pi}{2}(4j-3) \quad (j = 1, 2, \ldots).$$

Moreover, the readily verified equality

$$\gamma(1 + h^2\beta^2)d\alpha|_{\alpha=0} = \beta^2 d(\gamma h)$$

implies that each root z_k lying on the imaginary axis for some γ, h moves to the right halfplane as γh increases, and this transition is always accomplished simultaneously for conjugate pairs of roots. Hence, for subsequent domains in the D–division,

$$0 < \gamma h < \frac{\pi}{2}, \quad \frac{\pi}{2} < \gamma h < 5\frac{\pi}{2}, \quad 5\frac{\pi}{2} < \gamma h < 9\frac{\pi}{2}, \ldots,$$

the instability degree of the system equals $0, 2, 4, \ldots$, respectively.

In particular, by fixing γ and increasing h we can see that the critical value h after which the system is unstable equals $\pi/(2\gamma)$. The information delay then becomes so large that a person regulating the temperature will twist the mixer handle further and further from one side to the other, and then jumps out from the shower. The same result appears if the delay is 'moderate' but the person has a very hot temperament.

It is interesting that if we keep γ constant and increase h further, the value of $\alpha_1(h)$ tends to zero, after achieving a possible maximum, i.e. instability relaxes in an exponential rate as $h \to \infty$. This is understandable, because for sufficiently large delays the rate of change of a process is influenced by its previous states when the oscillations amplitude is relatively small. However, the situation does not really improve, because for the same reason the amplitude of oscillation becomes very large even in the first oscillation cycle. (Increase of the coefficient at the slowly growing exponential also follows from (6.20).)

Note that from the analysis given here some important conclusions follow about the possibility of a principal influence of even small time delays on the solutions of the equations of mathematical physics. Consider the simplest initial boundary value problem for the heat equation

$$\frac{\partial u}{\partial t} = \eta \frac{\partial^2 u}{\partial x^2} \quad (0 < t < \infty, \ 0 \le x \le l; \ \ \eta = \text{const} > 0),$$

$$u|_{x=0,l} = 0 \ (0 \le t < \infty); \quad u|_{t=0} = u_0 \in C[0, l].$$

It is well known that the solution of this problem exists and is exponentially stable (with exponent $\pi^2\eta/l^2$) with respect to u_0. Assume now that we would like to take into consideration the possibility of delays anyhow, and to this end we have changed the equation as follows:

$$\frac{\partial u(t,x)}{\partial t} = \eta \frac{\partial^2 u(t-h,x)}{\partial x^2} \quad (h > 0).$$

As for ODEs, the continuous initial function u_0 will be prescribed on $[-h,0]\times[0,l]$. First of all we can conclude that the solution can be obtained by the method of steps:

$$u(t,x) = u_0(0,x) + \eta \int_0^t \frac{\partial^2 u_0(\tau - h, x)}{\partial x^2} d\tau \quad (0 \le t \le h),$$

etc., where the boundary condition must be ignored. Being continued with respect to t, the smoothness of the solution with respect to x becomes worse (the faster the smaller h). Therefore, for the existence of continuous solutions on $[0,\infty)\times[0,l]$ we must assume that u_0 has continuous derivatives with respect to x of arbitrary orders. To satisfy the boundary condition we must have

$$\left.\frac{\partial^{2k} u_0}{\partial x^{2k}}\right|_{x=0,l} = 0 \quad (-h \le t \le 0;\ k = 0,1,\ldots).$$

Assume that these conditions hold. The particular solutions of the form $u(x,t) = cv(t)\sin(n\pi x/l)(n = 1,2,\ldots)$ lead to the following equation for v:

$$\dot{v}(t) = -\eta \frac{n^2\pi^2}{l^2} v(t-h),$$

which is of the form (6.16). By investigating this equation we find that for $n > l/\sqrt{2\pi\eta h}$ it has an oscillating solution with exponentially increasing amplitude. But n is arbitrary. Hence the solution of the initial–boundary value problem is always unstable.

It is interesting that under the additional requirement $\sup_{x,t,k} |\partial^k u_0(x,t)/\partial x^k| < \infty$, the solution of the problem with delay h tends uniformly on any fixed finite time interval to the solution of the original initial–boundary value problem for the heat equation as $h \to 0$. Hence, for small fixed $h > 0$ at the initial stage of the process the solution decays, but in its spectral expansion when t increases there is redistribution of amplitudes on behalf of high frequency summands. Moreover, under unrestricted development of the process these summands become dominant, generating instability.

All these complications could not have appeared if the delay is taken into account in another way, e.g. if the heat equation is replaced by

$$\frac{\partial u(x,t)}{\partial t} = \eta \frac{\partial^2 u(t,x)}{\partial x^2} + \alpha \frac{\partial^2 u(t-h,x)}{\partial x^2},$$

where $|\alpha| < \eta$ (see Subs. 3.2.3, Example 1).

6.4. *LINEAR AUTONOMOUS NDES*

We do not aim at studying linear autonomous NDEs in the most common form, but rather restrict ourselves to the autonomous homogeneous equations of the form (5.17):

$$\left[x(t) - \sum_{j=1}^{m_1} B_j x(t-g_j)\right]^{\bullet} = \sum_{j=1}^{m} A_j x(t-h_j) + \int_{-h}^{0} A(\theta) x(t+\theta) d\theta, \quad (6.21)$$

$$\left[x(t) - \sum_{j=1}^{m_1} B_j x(t-g_j)\right]^{\bullet} = \sum_{j=1}^{m} A_j x(t-h_j) + \int_{-h}^{0} A(\theta) x(t+\theta) d\theta + f(t). \quad (6.22)$$

It is assumed that all $g_j > 0, h_j \geq 0, A(\cdot) \in L_1[-h,0], f \in L_1[t_0,\infty)$; without loss of generality we can also assume that all $g_j \leq h, h_1 \leq h$ (if this does not hold, increase h). The initial condition has the form (6.3), where ϕ is supposed to be a PAC function.

By Thm. 5.4 there is a unique PAC solution of the Pr. (6.22), (6.3) on $[t_0,\infty)$.

To find the exponential solutions (6.4) of (6.21) we have to consider the characteristic equation

$$\det\left(zI - z\sum_{j=1}^{m_1} e^{-g_j z} B_j - \sum_{j=1}^{m} e^{-h_j z} A_j - \int_{-h}^{0} e^{\theta z} A(\theta) d\theta\right) = 0.$$

In the nondegenerate case, this equation has, similarly to (6.5), an infinite number of roots $\{z_k\}$ without (finite) limit points. However, in contrast to (6.5), the set $\{\mathrm{Re} z_k\}$, which is bounded above, may have limit points. For example, such is the case when all g_j are different and $\det B_j \neq 0$ for at least one j.

Suppose we have an estimate $|f(t)| = O(e^{\gamma t})(t \to \infty)$. Then the solution of (6.22), (6.3) also satisfies an exponential estimate. Hence we can apply the Laplace transformation, and we obtain, similarly to Subs. 6.2,

$$x(t) = \frac{1}{2\pi i} \lim_{\beta\to\infty} \int_{\alpha-i\beta}^{\alpha+i\beta} e^{zt} \Phi(z) dz + \frac{1}{2\pi i} \int_{\alpha-i\infty}^{\alpha+i\infty} e^{zt} |\Psi(z)|^{-1} \left[\int_{t_0}^{\infty} e^{-z\tau} f(\tau) d\tau\right] dz \quad (t_0 < t < \infty), \quad (6.23)$$

where

$$\Psi(z) := zI - z\sum_{j=1}^{m_1} e^{-g_j z} B_j - \sum_{j=1}^{m} e^{-h_j z} A_j - \int_{-h}^{0} e^{z\theta} A(\theta) d\theta,$$

$$\Phi(z) := [\Psi(z)]^{-1} \left\{ e^{-t_0 z} \left[\phi(t_0) - \sum_{j=1}^{m_1} B_j \phi(t_0 - g_j) \right] + \right.$$

$$+ z\sum_{j=1}^{m_1} e^{-g_j z} B_j \int_{t_0 - g_j}^{t_0} e^{-z\tau} \phi(\tau) d\tau + \sum_{j=1}^{m} e^{-h_j z} A_j \int_{t_0 - h_j}^{t_0} e^{-z\tau} \phi(\tau) d\tau +$$

$$\left. + \int_{-h}^{0} e^{z\theta} A(\theta) \left[\int_{t_0+\theta}^{t_0} e^{-z\tau} \phi(\tau) d\tau \right] d\theta \right\}$$

and $\alpha > \max\{\gamma, \sup_k \alpha_k\}$. The limit sign on the righthand side of (6.23) is necessary because for a discontinuous function ϕ the solution x will also have discontinuity points. At these points the corresponding integral with infinite integration limits diverges, but its so–called *principal value*, given in (6.23), exists. For any $t > t_0$ the value of the righthand side of (6.23) is equal to $[x(t^-) + x(t^+)]/2$. This means that at continuity points of the solution it is equal to $x(t)$, while at discontinuity points, which can be of the first kind only, it equals the arithmetic mean of the left and right limit values of x at t. For $t = t_0$ the righthand side equals $[\phi(t_0^-) + x(t_0^+)]/2$.

For $f(t) \equiv 0$ there can be a transition similarly to (6.11), giving an estimate like (6.14), without 'comments' if $\tilde{\alpha} \notin \{\alpha_k\}$ and the strip $\tilde{\alpha} - \varepsilon <$ Re$z < \alpha (\varepsilon > 0)$ contains only finitely many roots z_k. In this case, if the set of such roots is nonempty, one can distinguish the principal exponential or quasipolynomial terms of the solution as $t \to \infty$ (except in the case when all coefficients turn out to be zero). If the given strip contains infinitely many roots z_k, then in the analog of (6.11) the sums become infinite series, and as a rule application of it is not very expedient. In particular, it may happen that only a subsequence of the partial sums of this series is convergent.

In contrast, to RDEs, for NDEs the estimate $|x(t)| = O(e^{\tilde{\alpha} t})$ $(t \to \infty)$ does not follow from the absence of roots z_k in the strip $\tilde{\alpha} \le$ Re$z < \alpha$. E.g., W. Brumley [66] has constructed an NDE for which all Re$z_k < 0$ but for which not all solutions are bounded. P. Gromova and A. Zverkin [215] have constructed an NDE for which all roots z_k are simple and Re$z_k = 0$, but for which not all solutions are bounded. Consequently, if it is impossible to distinguish finitely many principal terms of the solution, then assertions about its asymptotic behaviour must be given with great care.

7. HOPF BIFURCATION OF FDEs

7.1. *INTRODUCTION*

Here we consider the autonomous vector RDE

$$\dot{x}(t) = F(x_t; \alpha); \quad x_t := x(t+\theta), \quad -h \le \theta \le 0, \tag{7.1}$$

with sufficiently smooth righthand side F depending on the parameter $\alpha \in (\underline{\alpha}, \overline{\alpha})$. We assume that $F(0;\alpha) \equiv 0$, i.e. for all α the zero function $t \mapsto 0$ is a solution of (7.1). In less formal language one could say that the system under consideration, which depends on a parameter, has a fixed equilibrium position for all parameter values. Extracting the linear part from (7.1), we can rewrite it as

$$\dot{x}(t) = L(x_t; \alpha) + f(x_t; \alpha),$$

where $L(\cdot;\alpha)$ is a linear functional and $|f(u;\alpha)| = o(\| u \|) \ (\| u \| \to 0)$ for any fixed α.

Here we will use the basic notations connected with the stability of this equilibrium position. The respective definitions are quite the same as for ODEs, and will be discussed in detail in Chap. 3. In particular, an analog of Liapunov's stability theorem of first approximation is valid: if for some α all roots λ_k of the characteristic equation

$$\det[\lambda I - P(\lambda;\alpha)] = 0, \tag{7.2}$$

where

$$P(\lambda;\alpha) := e^{-\lambda t} L(t \mapsto e^{\lambda t}; \alpha),$$

are such that $\mathrm{Re}\lambda_k < 0$, then for this α the rest point is asymptotically stable; however, if $\mathrm{Re}\lambda_k > 0$ for at least one k, then it is unstable.

Now we assume that there is an $\alpha_0 \in (\underline{\alpha}, \overline{\alpha})$ such that for $\alpha < \alpha_0$ the first case holds, and for $\alpha > \alpha_0$ the second. Then the parameter value α_0 is called a *critical value*: if α increases and crosses the value α_0, then there is stability loss of the equilibrium position under consideration. This can happen in various ways.

E.g., it may happen that despite the stability loss for $\alpha > \alpha_0$ of the trivial solution, the solution of the Cauchy problem for (7.1), posed for $t_0 - h \le t \le t_0$, will be arbitrarily small on the whole interval $[t_0, \infty)$ if the initial function and the value of $\alpha - \alpha_0$ are sufficiently small. Loosely speaking, small initial disturbances cannot lead to a principal change of the picture; in particular, this cannot lead to a catastrophe under small transitions of the parameter through the critical value. This kind of stability loss is called *safe*. In contrast to this, in the case of a *dangerous stability loss* an arbitrary small transition through the critical value and arbitrarily small disturbances can lead to essential consequences.

Another classification of the types of stability loss reflects the behaviour of the roots of (7.2) when α passes through the critical value. In practice, the two following cases are most met:

A. If α increases and passes through α_0, one root of (7.2) turns from the negative to positive values while all other roots belong to the open left halfplane (i.e. for them $\text{Re}\lambda_k < 0$) for all sufficiently small $|\alpha - \alpha_0|$. In this case one says that the *principle of stability change* holds for the system with parameter under consideration. Arbitrarily small initial disturbances lead to the emergence of an accelerated aperiodic process, carrying the system away from its initial equilibrium position. (For a special set of initial disturbances, in theory, the process asymptotically tends to this position, but such processes are unstable and do not change the general picture.)

B. If α increases and passes through α_0, two conjugate nonreal roots of (7.2) turn from the open left halfplane into the open right halfplane, while the remaining roots remain in the open left halfplane. After such a transition, stability loss takes place in the following manner: arbitrarily small initial disturbances generate an oscillatory process with increasing amplitude and (circular) frequency near ω_0, where $\pm\omega_0 i$ are the values of the above-mentioned two roots for $\alpha = \alpha_0$.

The latter type of stability loss occurs under Hopf bifurcation, and we discuss it below. More complicated ways of stability loss are described in stability theory.

7.2. *EXAMPLE*

We will illustrate the main idea in the study of *Hopf bifurcation* by means of the following scalar ODEs:

$$\ddot{x} - \alpha\dot{x} + \dot{x}^3 + x + x^2 + x^3 = 0, \tag{7.3}$$

involving a parameter α. Eq. (7.3) is equivalent to the following system of first-order ODEs:

$$\dot{x}_1 = x_2, \quad \dot{x}_2 = -x_1 - x_1^2 - x_1^3 + \alpha x_2 - x_2^3. \tag{7.4}$$

One could say that this represents an oscillator with additional nonlinear dissipation and nonlinear restoring force which, as compared to the linear ones, are weaker for small velocities and deviations and stronger for high velocities and large deviations. There is only one rest point, $x = 0$, of the system, for any α.

It is clear that for small absolute values of $\alpha < 0$, i.e. when the linear dissipative term of the equation corresponds to small positive friction, the system has damped oscillation when taken off the rest point. This also follows from the expression for the roots $\lambda_{1,2}$ of the characteristic equation corresponding to (7.3), linearized about the origin (the nonlinear terms $\dot{x}^3$ and $x^2 + x^3$ are dropped in this linearization):

$$\lambda_{1,2} = \frac{1}{2}\alpha \pm i\sqrt{1 - \frac{1}{4}\alpha^2}. \tag{7.5}$$

It is clear that the damping decrement of oscillations tends to zero near the rest point, while the frequency tends to $\omega_0 = 1$ as $\alpha \to 0^-$.

When α passes through $\alpha_0 = 0$ and becomes positive, the friction becomes negative for small velocities, and the rest point $x = 0$ is unstable. This is also obvious from (7.5), because the roots $\lambda_{1,2}$ pass from the left to the right halfplane. Hence, for small initial disturbances there emerge oscillations with increasing amplitudes in the system. However, this increase cannot be unbounded, because for large velocities the nonlinear dissipative term annihilates the effect of the negative linear friction. Using results from the theory of oscillations of autonomous systems with one degree of freedom, this implies that the oscillations mentioned above tend to periodic oscillations as $t \to \infty$. An approximate phase portrait corresponding to (7.4) is depicted in Fig. 3.7.1. We see that a *limit cycle*, represented by the heavy line, appears in the phase plane. It contracts towards a point as $\alpha \to 0^+$. Thus, one can say that a limit cycle emerges from the rest point $(0, 0)$ in the phase plane when α passes through the value $\alpha_0 = 0$. In other words, it appears as if there is dichotomy (*bifurcation*) from the stable rest point to an unstable rest point and a limit cycle. (In this example it is possible to verify that there is only one limit cycle.)

For not too large values of α an asymptotic representation of the emerging cycle can be obtained by the method of the small parameter. However, the following feature must be taken into consideration. Let us look for a solution of (7.3) as the sum of a series:

$$x(t) = \alpha x_1(t) + \alpha^2 x_2(t) + \alpha^3 x_3(t) + \dots \tag{7.6}$$

(the term $x_0(t)$ is absent here, since for $\alpha = 0$ we must have $x(t) \equiv 0$). Substituting this expansion into (7.6) and equating coefficients at equal powers of α, we arrive at the system of equations

$$\ddot{x}_1 + x_1 = 0, \quad \ddot{x}_2 + x_2 = \dot{x}_1 - x_1^2, \quad \ddot{x}_3 + x_3 = \dot{x}_2 - \dot{x}_1^3 - 2x_1x_2 - x_1^3, \dots \tag{7.7}$$

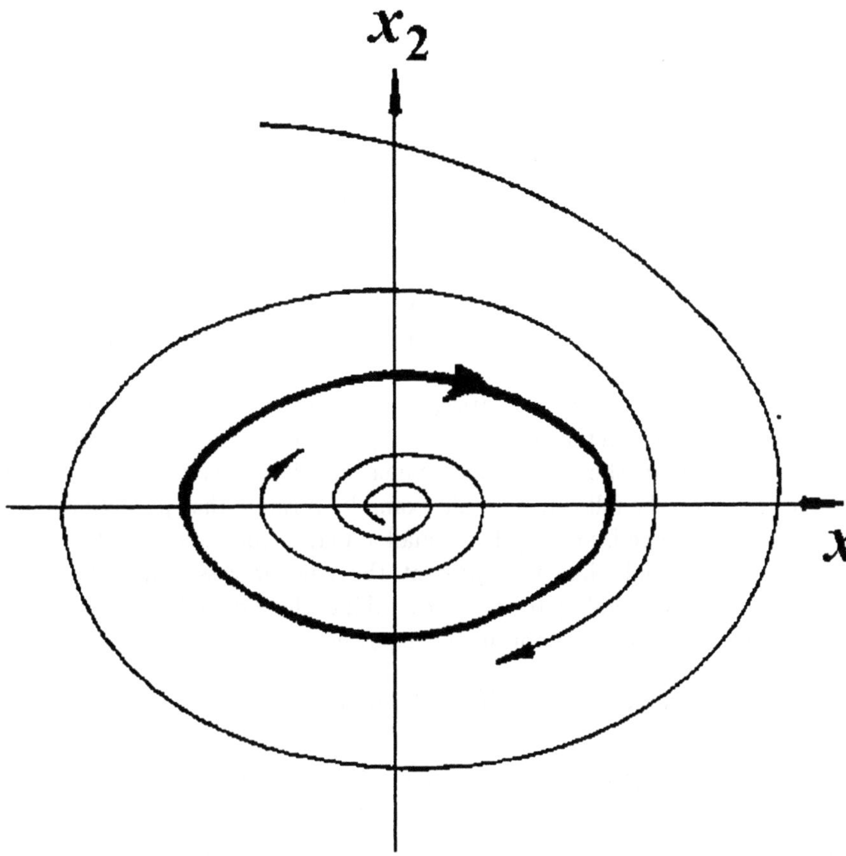

Figure 3.7.1. Approximate phase portrait of the system

$$\dot{x}_1 = x_2,\ \dot{x}_2 = -x_1 - x_1^2 - x_1^3 + \alpha x_2 - x_2^3$$

The first equation implies the quite natural formula $x_1 = C_1 \cos t + C_2 \sin t$. However, substitution of this formula in the next equation gives

$$x_2 = -\frac{1}{2}t(C_1 \cos t + C_2 \sin t) - \frac{1}{6}(C_1^2 + C_2^2) -$$

$$-\frac{1}{6}(C_1^2 - C_2^2) \cos 2t + \frac{1}{3}C_1 C_2 \sin 2t + C_3 \cos t + C_4 \sin t.$$

This expression is not acceptable as a representation of the cycle, because its righthand side is unbounded as $t \to \infty$. Of course, we can set $C_1 = C_2 = 0$, but then the same complication would arise for x_3, etc. As a result we would obtain the solution $x(t) \equiv 0$.

A similar difficulty always arises when we would like to use the method of the small parameter to describe a periodic process whose frequency is only known in the zeroth approximation. The reason for this can be understood by the following simple example. Assume that purely harmonic oscillations change frequency by a small amount, from ω to $\omega + \alpha (|\alpha| \ll \omega)$, and that we would like to represent the new process using expansions in power of the small parameter α, taking the original process as zero approximation. Then we obtain the expansion

$$M \sin(\omega + \alpha)t = M(\sin \omega t \cdot \cos \alpha t + \cos \omega t \cdot \sin \alpha t) =$$

$$= M\left(\sin \omega t + \alpha t \cos \omega t - \frac{1}{2}\alpha^2 t^2 \sin \omega t - \ldots\right).$$

The partial sums of this series expansion uniformly approximate the function $t \mapsto M \sin(\omega + \alpha)t$ on any given finite interval, but describe incorrectly the behaviour of this function on the infinite t-interval. The reason for this is that the partial sums of the Taylor series of the function $t \mapsto \sin t$ describe incorrectly its behaviour as $t \to \infty$.

To avoid this difficulty we make the linear change $t = c\tau$ of the variable t so that the unknown oscillation frequency at the 'new time' τ is independent of α. Then there appears an unknown coefficient c, but we have got rid of the expansions of periodic functions by Taylor series in their argument.

Because the oscillations frequency described by the cycle tends to 1 as $\alpha \to 0^+$, the coefficient c is assumed to be such that this oscillation frequency is equal to 1 at time τ for all $\alpha > 0$. This means that $c|_{\alpha=0} = 1$. Further, the first equation in (7.7) implies that we can accept as first approximation, up to phase shift (i.e. time shift), the function $t \mapsto \xi \cos \tau$, where ξ is related to α and $\xi|_{\alpha=0} = 0$.

It turns out to be more convenient to consider expansion with respect to ξ rather than with respect to α. The reason for this is as follows. The change $\xi \to -\xi$ is equivalent to phase shift of the oscillations by π, which preserves the invariance of the cycle and hence does not change the values of α and c. Hence α has the same order of value as ξ^2, and therefore ξ – as $\sqrt{\alpha}$. Thus, in the expansion in powers of α, not only integer but also half–integer powers are needed; in the expansion in powers of ξ, however, only integer powers are needed. (A similar situation takes place in overwhelming majority of examples in applications. However, it is theoretically possible that in the expansion in powers of α we must use *Puiseux series*, i.e. series of the form $\sum_{r=0}^{\infty} a_r \alpha^{r/m}$, where $m \in \mathbf{N}$ may be large than 2.)

Upon changing $t = c\tau$, Eq. (7.3) becomes

$$c\frac{d^2x}{d\tau^2} - \alpha c^2 \frac{dx}{d\tau} + \left(\frac{dx}{d\tau}\right)^3 + c^3(x + x^2 + x^3) = 0. \tag{7.8}$$

Keeping in mind the previous section we look for a solution and constants in the form

$$\left.\begin{aligned} x &= \xi \cos\tau + \xi^2 x_2(\tau) + \xi^3 x_3(\tau) + \dots, \\ c &= 1 + c_2\xi^2 + c_4\xi^4 + \dots, \quad \alpha = \alpha_2\xi^2 + \alpha_4\xi^4 + \dots \end{aligned}\right\} \tag{7.9}$$

(of course, here $x_2, x_3, \dots$ are different from those in (7.6)). Substituting these expansions into (7.8) and equating coefficient in front of equal powers of ξ, we arrive at the sequence of equations

$$\left.\begin{aligned} \frac{d^2x_2}{d\tau^2} + x_2 &= -\cos^2\tau, \\ \frac{d^2x_3}{d\tau^2} + x_3 &= -2c_2\cos\tau - \alpha_2\sin\tau + \sin^3\tau - 2x_2\cos\tau - \cos^3\tau, \dots \end{aligned}\right\} \tag{7.10}$$

All functions $x_2, x_3, \dots$ must be periodic with period 2π. It is possible to assume that the Fourier expansions of these functions do not contain terms with $\sin\tau$ and $\cos\tau$ (since we can change the parameter ξ and make an additional phase shift). Consequently, the first of the equations (7.10) gives

$$x_2(\tau) = -\frac{1}{2} + \frac{1}{6}\cos 2\tau.$$

Substituting this into the second of Eqs. (7.10) and writing the righthand side of this equation as a trigonometric polynomial, we obtain

$$\frac{d^2x_3}{d\tau^2} + x_3 =$$

$$= \left(-2c_2 + \frac{1}{12}\right)\cos\tau + \left(-\alpha_2 + \frac{3}{4}\right)\sin\tau - \frac{5}{12}\cos 3\tau - \frac{1}{4}\sin 3\tau.$$

The Fourier expansion of the righthand side should not contain terms with $\sin\tau$ and $\cos\tau$, because otherwise the solution would have nonperiodic resonance terms. This implies that

$$c_2 = \frac{1}{24},\ \alpha_2 = \frac{3}{4},\ x_3(\tau) = \frac{5}{96}\cos 3\tau + \frac{1}{32}\sin 3\tau.$$

This procedure allows us to continue the expansion (7.9) indefinitely, in principle, using further equations (7.10). Here $x_k(\tau)$ represents a linear combination of the functions $\cos j\tau$ and $\sin j\tau$ with $j \leq k$ and j and k of the same parity. In constructing $x_k(\tau)$ with k odd, the values of c_{k-1} and α_{k-1} are determined from the condition that resonance is absent. Similar assertions can be proved to hold in the construction of a generating cycle for general RDEs (see Subs. 7.3).

We restrict ourselves to the values obtained. Then the generating cycle can, for small $\alpha > 0$, be represented as

$$x = \xi\cos\tau + \xi^2\left(-\frac{1}{2} + \frac{1}{6}\cos 2\tau\right) +$$

$$+\xi^3\left(\frac{5}{96}\cos 3\tau + \frac{1}{32}\sin 3\tau\right) + O(\xi^4),$$

where

$$\tau = \left[1 + \frac{1}{24}\xi^2 + O(\xi^4)\right]t,\quad \alpha = \frac{3}{4}\xi^2 + O(\xi^4).$$

Elimination of ξ gives a representation of the cycle in terms of the basic parameter α:

$$x = \frac{2}{\sqrt{3}}\sqrt{\alpha}\cos\tau + \left(-\frac{2}{3} + \frac{2}{9}\cos 2\tau\right)\alpha + O(\alpha^{3/2}),$$

$$\tau = \left[1 + \frac{1}{18}\alpha + O(\alpha^2)\right]t.$$

In conclusion we note that there is a safe stability loss of the rest point in passage through the value $\alpha = 0$ (see Subs.7.1). This can be easily proved by contradiction.

7.3. *GENERAL CASE*

Here we describe a method for the construction of a limit cycle generating from a stable rest point under passage of the parameter of the system through a critical value, for the general scalar nth order autonomous RDE

$$\left.\begin{aligned} x^{(n)}(t) &= F(x_t, \dot{x}_t, \ldots, x_t^{(n-1)}; \alpha); \\ x_t^{(j)}(\theta) &:= x^{(j)}(t+\theta),\quad -h \leq \theta \leq 0, \end{aligned}\right\} \tag{7.11}$$

with sufficiently smooth righthand side. (This last requirement can be made more precise, but we do not need this precision; see details in the books [238, 281].) As above we assume that $F(0,\dots,0;\alpha)\equiv 0$, i.e. the zero function, which corresponds to the equilibrium point of the system under consideration, is a solution of (7.11) for all parameter values α. Then we can rewrite (7.11) as

$$x^{(n)}(t) = L(x_t,\dot{x}_t,\dots,x_t^{(n-1)};\alpha) + f(x_t,\dot{x}_t,\dots,x_t^{(n-1)};\alpha),$$

where L is a linear functional depending on the parameter α, and

$$f(u_1,\dots,u_n;\alpha) = o(\| u_1 \| + \dots + \| u_n \|) \quad (\| u \| + \dots + \| u_n \| \to 0).$$

The characteristic equation of the linear approximation to (7.11) has the form

$$\lambda^n = e^{-\lambda t}L(t\mapsto e^{\lambda t}, t\mapsto \lambda e^{\lambda t},\dots,t\mapsto \lambda^{n-1}e^{\lambda t};\alpha)$$

(it is obvious that t does not really participate at the righthand side). We assume that there is α_0 such that for $\alpha<\alpha_0$ all roots λ_k of the characteristic equation belong to the open left halfplane, whereas for $\alpha=\alpha_0$,

$$\left.\begin{aligned} &\lambda_{1,2}|_{\alpha=\alpha_0} = \pm i\omega_0; \quad \omega_0>0; \\ &\left.\frac{d\mathrm{Re}\lambda_{1,2}(\alpha)}{d\alpha}\right|_{\alpha=\alpha_0} > 0, \quad \mathrm{Re}\lambda_j|_{\alpha=\alpha_0} < 0 \quad (j>2). \end{aligned}\right\} \tag{7.12}$$

If, under these conditions, α increases and passes through the value α_0, then the zero solution, which was stable, becomes unstable, i.e. $\alpha=\alpha_0$ is a critical value. In addition, as a rule (exact assumptions will be given later), when α passes through α_0 in one of the directions, a periodic solution (the limit cycle in phase space) separates from zero. This solution is stable if it arises for $\alpha>\alpha_0$ and unstable in the opposite case. Theoretically it is possible that several cycles separate from zero.

To construct, as in Subs.7.2, a generating cycle we change the independent variable $t=c\tau$ in (7.11) in such a way that the oscillation period $2\pi/\omega_0$ corresponding to this cycle at the 'new time' τ is fixed. We also put $x(c\tau)=y(\tau)$. The a priori unknown function relationship between c and α must be defined such that $c|_{\alpha=\alpha_0}=1$. After this we obtain the equation

$$\left.\begin{aligned} &y^{(n)}(\tau) = c^nF_c(y_\tau,c^{-1}y_\tau^{(1)},\dots,c^{-n+1}y_\tau^{(n-1)};\alpha); \\ &y_\tau^{(j)}(\theta) := y^{(j)}(\tau+\theta) \quad (c^{-1}h\le\theta\le 0) \end{aligned}\right\} \tag{7.13}$$

where

$$F_c(u_1,\ldots,u_n;\alpha) := F(u_1^c,\ldots,u_n^c;\alpha);\quad u_j^c(\tau) := u_j(c\tau)\quad (c^{-1}h \le \tau \le 0).$$

As in Subs. 7.2 we look a periodic solution of (7.13) in the form (up to time shift)

$$y(\tau) = \xi\cos\omega_0\tau + \xi^2 y_2(\tau) + \xi^3 y_3(\tau) + \ldots. \tag{7.14}$$

The factor c of the oscillation period and the parameter α are looked for in the form

$$c = 1 + c_2\xi^2 + c_4\xi^4 + \ldots;\quad \alpha = \alpha_0 + \alpha_2\xi^2 + \alpha_4\xi^4 + \ldots. \tag{7.15}$$

Substituting these expressions into (7.13), expanding the righthand sides in powers of ξ, and equating coefficients at equal powers of ξ gives the sequence of equations of the form

$$\begin{gathered} y_{2j}^{(n)}(\tau) = L((y_{2j})_\tau, (y_{2j}^{(1)})_\tau,\ldots,(y_{2j}^{(n-1)})_\tau;\alpha_0)+ \\ +\phi_{2j}(y_1,y_2,\ldots,y_{2j-1};c_0,c_2,\ldots,c_{2j-2},\quad \alpha_0,\alpha_2,\ldots,\alpha_{2j-2}), \end{gathered} \tag{7.16}$$

$$\begin{gathered} y_{2j+1}^{(n)}(\tau) = L((y_{2j+1})_\tau, (y_{2j+1}^{(1)})_\tau,\ldots,(y_{2j+1}^{(n-1)})_\tau;\alpha_0)+ \\ +\phi_{2j+1}(y_1,y_2,\ldots,y_{2j};c_0,c_2,\ldots,c_{2j},\quad \alpha_0,\alpha_2,\ldots,\alpha_{2j}),\quad j=1,2,\ldots, \end{gathered} \tag{7.17}$$

where $y_1(\tau) := \cos\omega_0\tau$, $c_0 := 1$, and ϕ_{2j},ϕ_{2j+1} are nonlinear functional differential operators transforming the set of $(2\pi/\omega_0)$–periodic functions into itself and depending on the given parameters. (In concrete examples these operators can be given explicitly, as has been done in (7.10).)

The equations (7.16) and (7.17) allow us to successively determine the functions $y_2, y_3, \ldots$ similarly as in Subs. 7.2; using the absence of terms $\cos\omega_0\tau$ and $\sin\omega_0\tau$ in their Fourier expansions, these functions are determined uniquely. In addition, it turns out that the Fourier expansion of each next function ϕ_k includes only the terms $\cos m\omega_0\tau$ and $\sin m\omega_0\tau$ with $m \le k$ and m and k of the same parity. Hence the Fourier expansion of y_k has this property too. However, to construct a periodic solution of (7.17) there should be no resonance at all (there can be no resonance in (7.16)), i.e. the expansion of ϕ_{2j+1} should not contain $\cos\omega_0\tau$ and $\sin\omega_0\tau$. This restriction can be written as

$$\left.\begin{aligned}&\int_0^{2\pi/\omega_0} \phi_{2j+1}(y_1, y_2, \dots, y_{2j}; c_0, c_2, \dots, c_{2j}, \alpha_0, \alpha_2, \dots, \alpha_{2j})(\tau)\times \\ &\qquad\qquad \times \cos\omega_0\tau d\tau = 0, \\ &\int_0^{2\pi/\omega_0} \phi_{2j+1}(y_1, y_2, \dots, y_{2j}; c_0, c_2, \dots, c_{2j}, \alpha_0, \alpha_2, \dots, \alpha_{2j})(\tau)\times \\ &\qquad\qquad \times \sin\omega_0\tau d\tau = 0.\end{aligned}\right\} \quad (7.18)$$

Because all functions and constants on the lefthand side are assumed to be known from the previous steps, except for c_{2j} and α_{2j}, (7.18) is a system of equations in two unknowns. One can show that both equations are linear, and that the system has nonzero determinant. This makes it possible to determine c_{2j}, α_{2j} in this step, and to determine y_{2j+1}. Thus, the expansions (7.14) and (7.15) can be indefinitely continued, according to the scheme

$$(y_1, c_0, \alpha_0) \to y_2 \to (c_2, \alpha_2) \to y_3 \to y_4 \to (c_4, \alpha_4) \to \dots .$$

Theorem 7.1 Assume that all conditions (7.12) are met, and that $\alpha_2 \neq 0$. If $\alpha_2 > 0 (\alpha_2 < 0)$ and α increases (decreases) and passes through $\alpha = \alpha_0$, then exactly one periodic solution (up to time shift) of Eq. (7.11) arises near the solution $x(t) \equiv 0$ of the same equation. This solution is stable (unstable), and has asymptotic representation as $\alpha \to \alpha_0^+ (\alpha \to \alpha_0^-)$

$$x(t) = \left(\frac{\alpha - \alpha_0}{\alpha_2}\right)^{1/2} \cos\frac{\omega_0}{c} t + O(|\alpha - \alpha_0|),$$

$$c = 1 + \frac{c_2}{\alpha_2}(\alpha - \alpha_2) + O((\alpha - \alpha_0)^2)$$

(the similar assertion is true for $\dot{x}(t), \dots, x^{(n-1)}(t)$). In addition, any solution of (7.11) on the whole t-axis and sufficiently close to zero tends either asymptotically to zero or asymptotically to this periodic solution as $t \to \infty$ and $t \to -\infty$.

Here, closeness of solutions is understood in the sense of uniform closeness of the functions and their derivatives up to order $n - 1$, inclusively.

Proof of Thm. 7.1 can be found in [238].

7.4. *VARIANTS*

7.4.1 Equations with several parameters. Assume that the righthand side of (7.11) contains, besides α, parameters $\beta_1, \dots, \beta_k (k \geq 1)$, varying near values $\beta_{01}, \dots, \beta_{k0}$. Without loss of generality we may assume that all $\beta_{j0} = 0$, and that for any value $\beta = (\beta_1, \dots, \beta_k)$ the critical value $\alpha_0 = 0$.

If the functional F depends sufficiently smoothly on all its function arguments and parameters, and for $\beta = 0$ condition (7.12) holds, then for any β which is sufficiently small (in modulus) we can construct expansions (7.14) and (7.15), with coefficients depending sufficiently smoothly on β. In particular, the second expansion in (7.15) now takes the form

$$\alpha = \alpha_2(\beta)\xi^2 + \alpha_4(\beta)\xi^4 + \dots . \tag{7.19}$$

It turns out that for all sufficiently small $|\alpha|, |\beta|$ and for every sufficiently small $\xi > 0$ for which (7.19) is valid, there is (up to time shift) a unique solution of (7.11) of the form (7.14) with given α, β. In addition, distinct ξ give distinct solutions. This makes it possible to infer assertions about the number of periodic solutions bifurcating from the zero solution, by investigating (7.19) as an equation in α, β, ξ in a neighbourhood of the origin of the space $\mathbf{R}^{l+k+1}$ (more exactly, of the halfspace, since $\xi > 0$).

The most important case in applications is when $\alpha_2(0) \neq 0$. Direct use of the implicit function theorem shows that only one periodic solution emerges, and only if $\operatorname{sgn}\alpha = \operatorname{sgn}\alpha_2(0)$. However, if $\alpha_2(0) = 0$, then the picture is more complicated. For example, if the righthand side of (7.19) has the form $\beta_1\xi^2 - \xi^4$, then for $\alpha < 0$ only one periodic solution emerges from the zero solution; if $\alpha > 0$, the relation between α and β_1 plays a role: if $\alpha < \beta_1^2/4$ then there arise two such solutions, while for $\alpha > \beta_1^2/4$ none.

A detailed theoretical analysis of the general case, as well as some examples of applications, are given in [281].

7.4.2 Vector RDEs. The study of (7.1) under the condition (7.12) when $\{\lambda_k\}$ is the set of roots of (7.2) is done similarly as in Subs. 7.3. The expansion (7.14) must be replaced by

$$y(\tau) = \xi \operatorname{Re}(a_0 e^{i\omega_0\tau}) + \xi^2 y_2(\tau) + \xi^3 y_3(\tau) + \dots ,$$

where a_0 is an eigenvector of the matrix $P(i\omega_0, \alpha_0)$ (see (7.2)) corresponding to the eigenvalue $i\omega_0$, and $y_2, y_3, \dots$ are real-valued vector functions. The form of the expansions (7.15) does not change. Instead of (7.18), here we have the following condition for absence of resonance:

$$\int_0^{2\pi/\omega_0} \tilde{a}_0' e^{-i\omega_0\tau} \phi_{2j+1}(y_1, y_2, \dots, y_{2j}; c_0, c_2, \dots, c_{2j}, \alpha_0, \alpha_2, \dots, \alpha_{2j})(\tau) d\tau = 0. \tag{7.20}$$

Here $\tilde{\alpha}_0$ is an eigenvector of the matrix $[P(i\omega_0, \alpha_0)]'$ corresponding to the eigenvalue $i\omega_0$, and the prime denotes transposition, as before. Since the first two factors in the integrand are complex-valued functions, (7.20) is a system of two real equations for determining c_{2j} and α_{2j}.

7.4.3 Bifurcation for NDEs. In [281] it has been shown that all main assertions of Subs. 7.3 and 7.4 hold for vector-valued autonomous NDEs in Hale's form:

$$\left[x(t) - \sum_{j=1}^{m} A_j(\alpha)x(t-h_j) - f(x(t-h_1), \ldots, x(t-h_k); \alpha)\right]^{\bullet} = \tag{7.21}$$

$$= \sum_{j=1}^{m} B_j(\alpha)x(t-h_j) + g(x(t-h_1), \ldots, x(t-h_k); \alpha).$$

It is assumed that all matrix functions A_j, B_j and vector functions f, g are analytic; $0 < h_1 < \ldots < h_m; \alpha \in (\underline{\alpha}, \overline{\alpha})$;

$$|f(u_1, \ldots, u_m; \alpha)| + |g(u_1, \ldots, u_m; \alpha)| = o(|u_1| + \ldots + |u_m|)$$

$$(|u_1| + \ldots + |u_m| \to 0);$$

for some $\alpha_0 \in (\underline{\alpha}, \overline{\alpha})$ we have $\lambda_{1,2}|_{\alpha=\alpha_0} = \pm i\omega_0(\omega_0 > 0)$, $d\mathrm{Re}\lambda_{1,2}(\alpha)/d\alpha|_{\alpha=\alpha_0} > 0$; and $\sup\limits_{k>2, \alpha_0 \in (\underline{\alpha}, \overline{\alpha})} \lambda_k(\alpha) < 0$. Here $\{\lambda_1, \lambda_2, \ldots\}$ is the set of roots of the characteristic equation

$$\det\left(\lambda I - \sum_{j=1}^{m} e^{-h_j\lambda}[\lambda A_j(\alpha_0) + B_j(\alpha_0)]\right) = 0$$

of the linear approximation of (7.21).

7.5. *EXAMPLE OF AN RDE WITH CONSTANT DELAY: INTRASPECIFIC STRUGGLE FOR A COMMON FOOD*

In Subs.7.5–7.7 we will consider three substantive examples related to the emergence of periodic solutions, following [281].

The simplest model of intraspecific struggle for a common food when there are limited self-renewing food resources has the form (2.5.1):

$$\dot{N}(t) = \gamma[1 - K^{-1}N(t-h)]N(t).$$

Here $N(t)$ is the volume of the population at time t, γ is the Malthus coefficient, K is the equilibrium value of the population, h is the average time of development of food resources. Because the system under consideration is near its equilibrium position $N(t) \equiv K$, we introduce a new function x by $N = K(1+x)$, and to reduce the number of parameters we introduce dimensionless variables $t_l = t/h, \alpha = \gamma h$. We thus obtain

$$\dot{x}(t_l) = -\alpha x(t_l - 1) - \alpha x(t_l)x(t_l - 1). \qquad (7.22)$$

By Subs. 6.3, the critical value for the linearized equation is $\alpha = \alpha_0 := \pi/2$, and the function $x = C\cos\left(\frac{\pi}{2}t_l + \text{const}\right)$ is its periodic solution. Using the algorithm in Subs. 7.3 we obtain

$$x(t_l) = \xi\cos\frac{\pi}{2}\tau + \xi^2 x_2(\tau) + \xi^3 x_3(\tau) + \ldots,$$

$$t_l = (1 + c_2\xi^2 + c_4\xi^4 + \ldots)\tau,$$

$$x(t_l - 1) = \xi\cos\frac{\pi}{2}(\tau - (1 + c_2\xi^2 + \ldots)^{-1}) +$$

$$+\xi^2 x_2(\tau - (1 + c_2\xi^2 + \ldots)^{-1}) + \ldots,$$

$$\alpha = \frac{\pi}{2} + \alpha_2\xi^2 + \alpha_4\xi^4 + \ldots$$

(In the expression for $x(t_l - 1)$ there arises an additional expansion in powers of ξ, due to the transition to time τ.) Substituting these expressions into (7.22) and equating coefficients at equal powers of ξ gives the sequence of equations

$$\dot{x}_2(\tau) = -\frac{\pi}{2}x_2(\tau - 1) - \frac{\pi}{2}\cos\frac{\pi}{2}\tau \cdot \sin\frac{\pi}{2}\tau,$$

$$\dot{x}_3(\tau) = -\frac{\pi}{2}x_3(\tau - 1) - \frac{\pi}{2}[x_2(\tau - 1)\cos\frac{\pi}{2}\tau +$$

$$+x_2(\tau)\sin\frac{\pi}{2}\tau + c_2\sin\frac{\pi}{2}\tau] - \frac{\pi^2}{4}c_2\cos\frac{\pi}{2}\tau - \alpha_2\sin\frac{\pi}{2}\tau,$$

......

The first equation implies that

$$x_2(\tau) = 0.2\cos\pi\tau + 0.1\sin\pi\tau.$$

After substituting this expression into the second equation, the condition of absence of resonance (i.e. absence of terms $\cos\frac{\pi}{2}\tau$ and $\sin\frac{\pi}{2}\tau$ in the Fourier expansion of the righthand side) implies $c_2 = 0.1/\pi, \alpha_2 = 0.075\pi - 0.05$. So $\alpha_2 > 0$, and Thm. 7.1 asserts that there emerges a cycle for $\alpha > \alpha_0$, i.e.

$\gamma h > \pi/2$. This cycle is stable, and can be represented as (after elimination of ξ):

$$N = K\left[1 + (0.075\pi - 0.05)^{-1/2}\left(\gamma h - \frac{\pi}{2}\right)^{1/2}\cos\frac{\pi}{2}\tau + \right.$$

$$+(0.075\pi - 0.05)^{-1}\left(\gamma h - \frac{\pi}{2}\right)(0.2\cos\pi\tau + 0.1\sin\pi\tau) + \tag{7.23}$$

$$\left. + O\left(\left(\gamma h - \frac{\pi}{2}\right)^{3/2}\right)\right],$$

where

$$\tau = \frac{1}{h}\left[1 - \frac{0.1}{\pi}(0.075\pi - 0.05)^{-1}\left(\gamma h - \frac{\pi}{2}\right) + O\left(\left(\gamma h - \frac{\pi}{2}\right)^2\right)\right]t.$$

Therefore, if the coefficient γ of species reproduction passes through the critical value $\pi/2h$, then the equilibrium position fails to be stable, and periodic oscillations as described by (7.23) arise. Of course, this is a pure delay effect, because for $h = 0$ the solutions cannot oscillate. Such oscillations have been actually discovered in experiments, and the coincidence of theory and experiments proved to be sufficiently good even for large amplitude oscillations [281]. However, the parameter values of the problem had to be adjusted to the experimental results.

7.6. *EXAMPLE OF AN RDE WITH AUTOREGULATIVE DELAY: COMBUSTION IN THE CHAMBER OF A TURBOJET ENGINE*

Because we are only interested in the mathematical formalism, we do not give a derivation of the basic equation, nor the actual meaning of all the quantities involved (see, e.g., [281] for this information). The equation under consideration is:

$$\ddot{x}(t) + (\alpha + \beta p)\dot{x}(t) + \alpha\beta p x(t) + \gamma x(t - \Delta) = \delta_1 x(t)\dot{x}(t) + \delta_2 x^2(t). \tag{7.24}$$

Here, $x(t)$ is a small deviation in the velocity $v(t)$ of the liquid fuel in the line supply from its stationary value $v_0, p > 0$ is the stationary pressure in the combustion chamber (this is taken to be the basic parameter of the process), $\alpha, \beta, \gamma, \delta_1, \delta_2$ are fixed parameters, characterizing the fuel and the chamber, Δ is the time interval from the formation of a portion of the combustion product to the moment of outflow through the nozzle.

The essential distinguishing feature of this example is the dependence of Δ on the unknown solution x. It is possible to assume that

$$\Delta = \Delta(p) + h_1 p\Delta'(p)x(t) + h_2\Delta'(p)\dot{x}(t) +$$

$$+0.5[h_1^2 p^2 \Delta''(p) + h_3\Delta'(p)]x^2(t) + h_4 p\Delta''(p)x(t)\dot{x}(t) + 0.5h_2^2\Delta''(p)\dot{x}^2(t),$$

where h_j are given parameters, $p \mapsto \Delta(p)$ is an empirical function whose graph is similar to that of $\Delta = pe^{-p}$, and Δ' and Δ'' are its derivatives. The corresponding linearized equation is

$$\ddot{x}(t) + (\alpha + \beta p)\dot{x}(t) + \alpha\beta x(t) + \gamma x(t - \Delta(p)) = 0. \qquad (7.25)$$

Investigation of the roots of the characteristic equation shows that (7.25) is stable for $0 \leq p \leq p_l$ and $p_h \leq p < \infty$, where $p_l > 0$ and $p_h > p_l$ are obtained by solving a transcendental equation in p involving the quantity $\Delta(p)$. For $p = p_l$ ($p = p_h$) the characteristic equation has a pair of conjugate purely imaginary roots $\pm i\omega_l$ (resp., $\pm i\omega_h$).

The study of the complete equation (7.24), which could in principle be done as in Subs. 7.5, is more cumbersome, in particular because of the complicated structure of the delay. For p near p_l we set in (7.24):

$$x(t) = \xi \cos \omega_l \tau + \xi^2 x_2(\tau) + \xi^3 x_3(\tau) + \dots,$$

$$t = (1 + c_2\xi^2 + c_4\xi^4 + \dots)\tau,$$

$$p = p_l + p_2\xi^2 + p_4\xi^4 + \dots,$$

then perform the transition from $\dot{x}, \ddot{x}$ to derivatives with respect to τ, and make a corresponding change in $x(t - \Delta)$. Then we expand all terms in powers of ξ, and construct the periodic solutions of the derived equations. The rather cumbersome expressions for c_2, p_2, and hence of the periodic solutions of the first approximation, have been written out in [281].

We stress that the coefficient which are derived do not only contain the values of the empirical function $p \mapsto \Delta(p)$, which is known with little accuracy, but also of its derivatives Δ' and Δ''. Therefore the final result is likely to be unreliable, and certain conclusions can only be given on the basis of empirical evidence. It seems that for increasing p there appear first low frequency oscillations, these becoming high frequency oscillations until for p sufficiently large the combustion becomes stationary. These experimental conclusions are in agreement with the mathematical theory.

7.7. *EXAMPLE NDE: AUTO-OSCILLATIONS IN A LONG LINE WITH TUNNEL DIOD*

The model equations of this system have been described in Subs. 2.3.4, and take the form (see (2.3.4)):

$$\begin{aligned}\dot{x}(t_l) - K\dot{x}(t_l - h) + (1-\alpha)x(t_l) + K(1+\alpha)x(t_l - h) = \\ = [x(t_l) - Kx(t_l - h)]^2.\end{aligned} \tag{7.26}$$

In the sequel it is assumed that $K > 0, \alpha < 1$, since otherwise the stationary solution is certainly unstable. Study of the roots of characteristic equation corresponding to the linearization of (7.26) shows that stability loss in (7.26) occurs when K passes through

$$K_0 := [\omega_0^2 + (1-\alpha)^2]^{1/2}[\omega_0^2 + (1+\alpha)^2]^{-1/2}.$$

Here ω_0 is the smallest positive root of the equation with respect to ω:

$$\cos \omega h = [\omega^2 - 1 + \alpha^2][\omega^2 + (1-\alpha)^2]^{-1/2}[\omega^2 + (1+\alpha)^2]^{-1/2}.$$

For $K = K_0$ the characteristic equation has two purely imaginary roots, $\pm i\omega_0$, and the set of real parts of the remaining roots has negative least upper bound.

The study of the complete equation (7.26) for K near K_0 can be done along the same lines as before, using the expansions

$$x(t_l) = \xi \cos \omega_0 \tau + \xi^2 x_2(\tau) + \xi^3 x_3(\tau) + \dots,$$

$$t_l = (1 + c_2\xi^2 + c_4\xi^4 + \dots)\tau, \quad K = K_0 + K_2\xi^2 + K_4\xi^4 + \dots.$$

Substitution of these expansions into (7.26) allows us to obtain explicit expressions for $x_2(\tau), c_2$, and K_2 (here we do not give these, see [281]). In particular, the expression for K_2 implies $K_2 < 0$ for sufficiently small fixed $\alpha > 0$, i.e. for such α there emerges a cycle for $K < K_0$, and this cycle is unstable.

The total bifurcation picture for $\alpha \in (0,1)$ has not yet been obtained. It has only been shown that $K_2|_{\alpha=0} = K_4|_{\alpha=0} = 0$. It is conjectured in [281] that K_2 becomes positive when α passes through some value $\alpha_0 \in (0,1)$, i.e. that auto-oscillations arise in the system.

8. STOCHASTIC RETARDED DIFFERENTIAL EQUATIONS (SRDEs)

The functioning of many actual delay systems are subject to unknown disturbances, which may be regarded as stochastic. For the description of such systems we can use SRDEs.

An example of this kind is the model of a motion with viscous aftereffect given in Subs. 2.2.1. Another example is the average of a stock market, which depends on many random factors, as well as on the previous history. Furthermore, in many actual situations the value of the delay is imperfectly known, and may be regarded as being random.

In addition, the solution of a stochastic ODE can be approximated by those of SRDEs with small delays. E.g., in [40] the solution of the equation $dx(t) = g(x(t))d\xi(t)$ is approximated by $\{x^n(t)\}$, where $dx^n(t) = g(x^n(t - 1/n))d\xi(t)$. Of course, a similar procedure can be used for deterministic equations.

8.1. *INITIAL VALUE PROBLEM*

We will consider SRDEs of the form

$$\left.\begin{aligned} dx(t) &= a(t, x_t)dt + b(t, x_t)d\xi(t), \quad t \geq 0, \\ x_t(\theta) &:= x(t+\theta), \quad -\infty < \theta \leq 0. \end{aligned}\right\} \tag{8.1}$$

(Here we assume, without loss of generality, that $t_0 = 0$.) In this equation $a, b : [0, \infty) \times C(-\infty, 0] \to \mathbf{R}$ are continuous functionals (see Subs. 2.5) and $t \mapsto \xi(t) \in \mathbf{R}^l$ is the standard Wiener process, defined on a probability space $(\Omega, \mathcal{A}, \mathcal{P})$. We recall some basic definitions (see, e.g., [257]). Points $\omega \in \Omega$ are recalled *elementary events*. $\mathcal{A}$ is a sigma–algebra of subsets of Ω; these subsets are called *random events*. The measure $\mathcal{P}$ is defined on (the elements of) $\mathcal{A}$, such that $\mathcal{P}(\Omega) = 1$; it is called a *probability measure*. A real–valued function $\omega \mapsto \zeta(\omega)$ from Ω to $\mathbf{R}$ which is *measurable with respect to* $\mathcal{A}$ (i.e. $\{\omega \in \Omega : \zeta(\omega) \leq x\} \in \mathcal{A}$ for all $x \in \mathbf{R}$) is called a *random variable*. The *distribution function* F of ζ is defined by

$$F(x) = \mathcal{P}\{\omega \in \Omega : \zeta(\omega) \leq x\}, \quad x \in \mathbf{R}.$$

A function $\omega \mapsto \zeta(\omega) \in \mathbf{R}^n$ is called a *random vector* if every component $\zeta_i, i = 1, \ldots, n$, is a random variable. The distribution function F of a random vector $\omega \mapsto \zeta(\omega) \in \mathbf{R}^n$ is the *joint distribution function* of the random variables $\zeta_1, \ldots, \zeta_n$, defined as follows:

$$F(x_1, \ldots, x_n) = \mathcal{P}\{\omega \in \Omega : \zeta_1(\omega) \leq x_1, \ldots, \zeta_n \leq x_n\},$$

$$x_i \in \mathbf{R}, \quad i = 1, \ldots, n.$$

A family of random variables $\omega \mapsto \zeta(t, \omega) \in \mathbf{R}^n$ defined on some t–interval J and, in general, correlated with each other for distinct t, is called a *stochastic process* or a *random process*. This correlation gives rise to the probability space $(\overline{\Omega}, \overline{\mathcal{A}}, \overline{\mathcal{P}})$, where $\overline{\Omega}$ is the space of functions $J \to \Omega$ and $\overline{\mathcal{A}}, \overline{\mathcal{P}}$ are σ–algebra and measure on $\overline{\Omega}$. We do not consider the details of this construction. Thus, a random process is formally defined as a function $(\overline{\omega} \mapsto \zeta(\cdot, \overline{\omega})) : \overline{\Omega} \to \overline{\Omega}$, i.e. $((t, \overline{\omega}) \mapsto \zeta(t, \overline{\omega})) : J \times \overline{\Omega} \to \mathbf{R}^n$. The argument $\overline{\omega}$ is usually omitted, and one simply writes $t \mapsto \zeta(t)$. For a fixed $t \in J$ the quantity $\zeta(t)$ can be identified with stochastic quantity $\omega \mapsto \zeta(t, \omega)$, which is known as a *section of the random process* under consideration. As a function of t for fixed ω, the process $t \mapsto \zeta(t, \omega)$ is called a *trajectory of the random process.*

A process $t \mapsto \zeta(t)$ is said to be *random process with independent increments* if for any finite set $(t_1, \ldots, t_k)$, $k \geq 3$, such that $t_1 < t_2 < \ldots < t_k$, the random variables $\zeta(t_2) - \zeta(t_1), \ldots, \zeta(t_k) - \zeta(t_{k-1})$ are independent. The *standard Wiener process* $t \mapsto \xi(t), t \geq 0$, is the random process with independent stationary gaussian increments such that

$$\xi(0) = 0, \quad \mathcal{E}\xi(t) = 0,$$

$$\mathcal{E}[(\xi(t) - \xi(s))(\xi(t) - \xi(s))'] = (t - s)I, \quad t \geq s \geq 0.$$

Here $\mathcal{E}$ denotes mathematical expectation, I is the identity matrix, and the prime denotes transposition. The Wiener process $t \mapsto \xi(t)$ is completely determined by these conditions as a concrete random process. The trajectories of the Wiener process $t \mapsto \xi(t)$ are continuous functions (of t) that are, with probability 1, nowhere differentiable, satisfy the Hölder condition with index$< 1/2$, and have infinite variation on any bounded time interval. This process was introduced by N. Wiener to model Brownian motion of physical particles suspended in a liquid. (The term *Brownian motion* honours the botanist R. Brown (XIXs) who observed that the pollen suspended in water form a very complicated trajectory. Later (1905) A. Einstein showed that a molecular model of water predicts the existence of the Brownian motion.) The properties of the trajectories of the process $t \mapsto \xi(t)$ mean that in this model the particles do not have velocities. A natural way to overcome this deficiency is to take into consideration viscous aftereffect, which is always present in actual motions, i.e. to use FDEs (see Subs. 2.2.1).

Let $B_{t_1,t_2}(d\xi) \subset \mathcal{A}$ be the minimal σ–algebra with respect to which all random variables $\xi(s) - \xi(t)$ are measurable, for all $s, t \in [t_1, t_2]$.

We return to equation (8.1). As initial condition we take

$$x(\theta) = \phi(\theta) \quad (-\infty < \theta \leq 0), \tag{8.2}$$

where ϕ is a process with continuous trajectories. It is assumed that the process $\theta \mapsto \phi(\theta)$, $-\infty < \theta \leq 0$, is independent on the increments of the Wiener process $t \mapsto \xi(t)$, i.e. the σ–algebra $B_{-\infty 0}(\phi)$ is independent of

$B_{0\infty}(d\xi)$. We say that the process $t \mapsto x(t)$ is a solution of (8.1)–(8.2) if $B_{-\infty t}(x) \bigcup B_{0t}(d\xi)$ is independent of $B_{t\infty}(d\xi)$ and if with probability 1 the following equality holds for all $t \geq 0$:

$$x(t) = x(0) + \int_0^t a(s, x_s)ds + \int_0^t b(s, x_s)d\xi(s). \tag{8.3}$$

The second integral in (8.3) is called *Itô's stochastic integral*, and is defined by a special limiting procedure. The following two properties of this integral are used quite often ($T > 0$):

$$\mathcal{E} \int_0^T f_1(t)d\xi(t) = 0,$$

$$\mathcal{E}\left\{\int_0^T f_1(t)d\xi(t)\left[\int_0^T f_2(t)d\xi(t)\right]'\right\} = \int_0^T \mathcal{E}[f_1(t)f_2'(t)]dt.$$

Here, the continuous processes $f_i : [0,T] \to \mathbf{R}^{n\times l}$ should be measurable with respect to $B_{0t}(d\xi)$ and

$$\int_0^T \mathcal{E}|f_i(t)|^2 dt < \infty, \quad i = 1, 2.$$

8.2. *EXISTENCE AND UNIQUENESS OF SOLUTION*

Suppose the continuous functionals a and b in (8.1) satisfy for arbitrary continuous functions $\psi, \chi \in C(-\infty, 0]$ and all $t \in [0, \infty)$ the conditions

$$|a(t,\psi) - a(t,\chi)|^2 \leq \int_{-\infty}^0 |\psi(\theta) - \chi(\theta)|^2 dK_1(\theta),$$

$$|b(t,\psi) - b(t,\chi)|^2 \leq \int_{-\infty}^0 |\psi(\theta) - \chi(\theta)|^2 dK_2(\theta).$$

Here $K_{1,2}$ are nondecreasing bounded functions. Then [257] there is for any initial process ϕ with

$$\sup_{\theta \leq 0} \mathcal{E}|\phi(\theta)|^4 < \infty$$

a unique solution of the Pr. (8.1), (8.2) on $[0,\infty)$. This solution has bounded fourth moment on any finite time interval $[0,t]$ and is *measurable with respect to* $\phi(\theta), \theta \le 0$, and $\xi(s), 0 \le s \le t$, i.e. $B_{-\infty t}(x) \subset B_{-\infty 0}(\phi) \bigcup B_{0t}(d\xi)$.

Note that the functionals a and b defined on $[0,\infty) \times C(-\infty,0]$ may be regarded as Volterra type operators mapping the random function x to the random function $t \mapsto a_i(t,x_t)$ $(a_1 = a, a_2 = b)$, and the probability distributions of the latter are induced by the probability distribution of x.

8.3. *SOME CHARACTERISTICS OF SOLUTIONS OF LINEAR EQUATIONS*

Here we consider the linear SRDE with discrete delay of the form (8.1):

$$dx(t) = A(t)x(t-h)dt + \sigma(t)d\xi(t), \quad 0 \le t < \infty,$$

where $h > 0, A$ and σ are given continuous matrix functions. It is obvious that the function $t \mapsto m(t) := \mathcal{E}x(t)$ satisfies the relation

$$\dot{m}(t) = A(t)m(t-h), \quad 0 \le t < \infty; \quad m(\theta) = \mathcal{E}\phi(\theta), \quad \theta \le 0.$$

Given a time moment $T > 0$ and a matrix $H \in \mathbf{R}^{n \times n}$, the quantity $\delta := \mathcal{E}[x'(T)Hx(T)]$ is defined by the formula (Tr denotes the trace)

$$\delta = \mathcal{E}\left[\phi'(0)P\phi(0) + \phi'(0)\int_{-h}^{0} Q(\theta)\phi(\theta)d\theta + \right.$$

$$\left. + \int_{-h}^{0} \phi'(\theta)Q'(\theta)\phi(0)d\theta + \int_{-h}^{0} d\theta \int_{-h}^{0} \phi'(\theta)R(\theta,\tau)\phi(\tau)d\tau \right] +$$

$$+ \int_{0}^{T} \mathrm{Tr}[B'(t)HB(t)\sigma(t)\sigma'(t)]dt.$$

Here the matrix $B(t)$ is a solution of the problem

$$\dot{B}(t) = -B(t+h)A(t+h), \quad B(T) = I, \quad B(t) = 0, \quad t > T.$$

The matrices P, Q, R are given by:

$$P = B'(0)HB(0), \quad Q(t) = B'(0)H\dot{B}(t), \quad R(t,\tau) = \dot{B}'(t)H\dot{B}'(\tau).$$

Part III

STABILITY

CHAPTER 4

STABILITY OF RETARDED DIFFERENTIAL EQUATIONS

The following four chapters deal with stability problems for FDEs. Stability of a process (in particular, of a stationary state) is the ability of the process to resist a priory unknown, small influences (disturbances). A process is said to be stable if such disturbances do not essentially change it. This property turns out to be of utmost importance. We emphasize that an individual predictable process can be physically realized only if it is stable in the corresponding natural sense.

The great variety of types of processes and disturbances has led to numerous mathematical notions of stability, among them stability for hereditary processes [1-4, 6, 11–13, 17, 19, 20, 25, 26, 29, 31–33, 36–39, 41–43, 47–51, 58, 62, 64, 68, 69, 72–74, 76–78, 84, 108, 109, 111, 118, 120–125, 127, 128, 131–134, 144–146, 148, 149, 154, 155–160, 164, 167, 168, 170, 172, 175, 177, 179, 180,186–192, 194, 196, 197, 203–211, 213, 215, 219, 227, 233–238, 241, 241, 243, 246–249, 254, 256, 259, 261, 267, 268, 271, 272, 274–283, 287–294, 297–299, 301, 302, 304, 307–311, 323–325, 332, 341, 343, 345, 350, 355, 357, 369, 371, 372, 374, 378, 379, 381–383, 386, 389, 391, 402, 407, 408, 410, 417, 422–424, 428–430, 440, 447, 449–452, 455, 458, 463, 465, 470, 471, 475, 476, 481, 483–487, 501–506, 526, 529–532, 540–544].

In the sequel we will give our main attention to notions and assertions that are likely to be useful from the point of view applications. In particular, we give attention to linear autonomous equations and the corresponding quasipolynomials, Liapunov's direct method, and concrete hereditary systems, as well as to the notions of absolute, stochastic, etc. stability.

1. LIAPUNOV'S DIRECT METHOD

1.1. *STABILITY DEFINITIONS*

We begin with the basic stability definitions for RDEs with bounded delay. Consider the following Cauchy problem for $x : J_x \to \mathbf{R}^n$:

$$\left.\begin{aligned} \dot{x}(t) = f(t, x_t) \ \ (t \geq t_0), \ \ x_t(\theta) := x(t+\theta), \ \ -h \leq \theta \leq 0, \\ x_{t_0} = \phi. \end{aligned}\right\} \tag{1.1}$$

Here, $f : \mathbf{R} \times C[-h,0] \to \mathbf{R}^n$ is a continuous functional which is locally Lipschitz in the second argument; $\phi \in C[-h,0]$.

As usual, the study of the stability of any solution $x^0 : [t_0, \infty) \to \mathbf{R}^n$ of (1.1), corresponding to ϕ^0, can be reduced to the study of the stability of the trivial (zero) solution, by changing the unknown and initial functions: $x(t) = x^0(t) + y(t)$, $\phi(\theta) = \phi^0(\theta) + \psi(\theta)$. For y we obtain

$$\dot{y}(t) = f(t, x_t^0 + y_t) - f(t, x_t^0) =: \tilde{f}(t, y_t) \quad (\Rightarrow \tilde{f}(t,0) \equiv 0), \; y_{t_0} = \psi.$$

Hence in the sequel we may take $f(t,0) \equiv 0$ in (1.1). Thus, we have to study only the stability of the trivial solution of (1.1) corresponding to the zero initial function. Let $t \mapsto x(t; t_0, \phi)$ be the solution of (1.1), and let

$$Q_\delta := \{u \in C[-h,0] : \| u \| < \delta\}.$$

Definition 1.1 The trivial solution of (1.1) is called

A) *stable* for a given $t_0 \in \mathbf{R}$ if for any $\varepsilon > 0$ there is a $\delta = \delta(\varepsilon, t_0)$ such that $|x(t; t_0, \phi)| \le \varepsilon$ for any initial function $\phi \in Q_\delta$ and $t \in [t_0, \infty)$; in the opposite case the solution is called *unstable*;

B) *uniformly stable* if for any $\varepsilon > 0$ there is a $\delta(\varepsilon) > 0$ independent of t_0 such that $|x(t; t_0, \phi)| \le \varepsilon$ for any initial function $\phi \in Q_\delta, t_0 \in \mathbf{R}, t \in [t_0, \infty)$;

C) *asymptotically stable* if for a given $t_0 \in \mathbf{R}$ if it is stable and there is a $\Delta = \Delta(t_0) > 0$ such that $\lim_{t\to\infty} x(t; t_0, \phi) = 0$ for any initial function $\phi \in Q_\Delta$; the set of initial functions ϕ for which $\lim_{t\to\infty} x(t; t_0, \phi) = 0$ is called the *attraction domain* of the trivial solution for initial time t_0, and is denoted by $\Omega(t_0) \subseteq C[-h,0]$;

D) *uniformly asymptotically stable* if it is uniformly stable and there is a $H > 0$ such that for any $\gamma > 0$ there is a $T(\gamma) > 0$ such that $|x(t; t_0, \phi)| \le \gamma$ for any $t_0 \in \mathbf{R}$, $t \ge t_0 + T(\gamma)$ and $\phi \in Q_H$. Note that $Q_H \subseteq \Omega(t_0)$ for all $t_0 \in \mathbf{R}$.

Other, sometimes used, notions of stability, such as stability under persistently acting disturbances, global stability, exponential stability, stability under delay disturbances, stability for part of the variables, etc., can be similarly given. E.g., the trivial solution (1.1) is *uniformly exponentially stable* if there are constants $H > 0, C_1 > 0, C_2 > 0$ such that for any $t_0 \in \mathbf{R}$ and $\phi \in Q_H$ the solution $x(t; t_0, \phi)$ of this equation satisfies the inequality

$$|x(t; t_0, \phi)| \le C_1 \| \phi \| \exp[-C_2(t - t_0)], \quad t_0 \le t < \infty.$$

Remark 1.1 The requirement of uniform asymptotic stability is stronger than that of asymptotic stability. But for autonomous or periodic equations (1.1), asymptotic stability of the trivial solution implies uniform asymptotic stability of this solution. A similar relation holds for stability and uniform stability.

Show, e.g., that for autonomous ($f(t,x_t) = f(x_t)$) or ω–periodic $f(t+\omega, x_t) \equiv f(t,x_t)$) equation (1.1) from stability of the trivial solution for some $\bar{t}_0$ it follows its uniform stability. For any solution $x(t;t_0,\phi)$ of ω–periodic Eq. (1.1) and any integer $m > 0$ we have

$$x((t;t_1,x_{t_1}(\cdot;t_0,\phi)) = x(t;t_0,\phi), \quad t_0 \leq t_1 \leq t,$$

$$x(t \pm m\omega; t_0 \pm m\omega, \phi) = x(t;t_0,\phi).$$

Hence for the proof of the uniform stability it is sufficient to investigate only initial moments $t_0 \in [\bar{t}_0 - \omega, \bar{t}_0]$. Take any $\varepsilon > 0$ and define a $\eta(\bar{t}_0, \varepsilon) > 0$ such that $|x(t;\omega,\phi)| \leq \varepsilon$ for $t \geq \bar{t}_0$, $\| \phi \| \leq \eta(\bar{t}_0,\varepsilon)$. Existence of such η follows from the stability for initial moment $t_0 = \bar{t}_0$. Choose a $\delta(\varepsilon)$ (see Definition B)) such that

$$\max_{\bar{t}_0-\omega\leq t_0\leq \bar{t}_0} \max_{\|\phi\|\leq\delta(\varepsilon)} |x(\bar{t}_0;t_0,\phi)| \leq \eta(\bar{t}_0,\varepsilon).$$

The existence of this $\delta(\varepsilon) > 0$ follows from the continuous dependence of the solutions at the moment $t = t_0$ on the initial data. In fact, in the theorem on continuous dependence the value of t_0 was fixed; but using the proof by contradiction we can show that $\delta(\varepsilon)$ is independent on $t_0 \leq \bar{t}_0$ on any finite interval of variation of t_0. Hence, for $\bar{t}_0 \leq t < \infty$ we have

$$|x(t;t_0,\phi)| = |x(t;\bar{t}_0, x_{\bar{t}_0}(\cdot;t_0,\phi))| \leq \varepsilon$$

for $t_0 \in [\bar{t}_0 - \omega, \bar{t}_0]$ and $\| \phi \| \leq \delta(\varepsilon)$. □

The following examples show the differences between stability and uniform stability.

Example 1.1 The solution of the scalar Cauchy problem

$$\dot{x}(t) = e^{-t}x(t), \quad x(t_0) = x_0,$$

is

$$x(t) = x_0 \exp(e^{-t_0} - e^{-t}).$$

Therefore, for any $x_0 \neq 0$ we find unimprovable estimate

$$|x(t)| < |x_0| \exp e^{-t_0}, \quad t_0 \leq t < \infty,$$

and for any $\varepsilon > 0$ the unimprovable value of $\delta(\varepsilon,t_0)$ is $\varepsilon \exp(-e^{-t_0})$. We find that the trivial solution is stable, but not uniformly stable because $\delta(\varepsilon,t_0) \to 0$ as $t_0 \to -\infty$. □

Remark 1.2 This remark is connected with the dependence of stability on the initial moment t_0. By Thm. 2.2.1 the trivial solution, which can be indefinitely extended forward in time, depends continuously on the initial function, on every bounded interval. Hence, if the trivial solution is stable with respect to disturbances of the initial function for time t_0, then it must be stable with respect to disturbances for any previous initial moment $t_1 < t_0$. However, since not all solutions of RDEs can be indefinitely extended backwards, the trivial solution may become unstable at a future initial moment $t_1 > t_0$. Luckly, such asymmetry in the stability property is artifical and is unlikely to be met in practical applications.□

Example 1.2 Consider the scalar equation

$$\dot{x}(t) = a(t)x(t - h(t))$$

with $a, h \in C^1(\mathbf{R})$, $h(t) \geq 0, \dot{h}(t) < 1$, whose zero solution is asymptotically stable for arbitrary $t_0 \leq -1$ and unstable for any $t_0 \geq 0$. Take for h a $C^1(\mathbf{R})$–function such that

$$h(t) = 0, \quad -\infty < t \leq -1, \quad 0 \leq t < \infty,$$

$$h(t) = (t^2 \ln t^{-2})(1 + t \ln t^{-2})^{-1}, \quad -e^{-2} \leq t < 0.$$

Let the function $x_0 \in C^2(-\infty, 0]$ be defined by

$$x_0(t) > 0, \quad -\infty < t < 0; \quad x_0(t) \equiv 1, \quad -\infty < t \leq -1;$$

$$x_0(t) = \exp t^{-1}, \quad -0.5 \leq t < 0.$$

Finally we choose $a \in C^1(\mathbf{R})$ such that

$$a(t) = \dot{x}_0(t)[x_0(t - h(t))]^{-1}, \quad -\infty < t < 0,$$

$$a(t) = 1, \quad 1 \leq t < \infty.$$

It is possible to choose such an $a(t)$, since

$$a(t) \equiv -1, \quad -e^{-2} \leq t < 0.$$

If $t_0 \leq -1$, then the solution x of the Cauchy problem for $t_0 \leq t < 0$ is proportional to $x_0(t)$. Consequently, $x(0) = 0$, and also $x|_{[0,\infty)} \equiv 0$. But if $t_0 \geq 0$, then it is obvious that the trivial solution is unstable.□

Example 1.3 Consider the scalar equation

$$\dot{x}(t) = a(t)x(t - 3\pi/2), \quad a \in C(\mathbf{R}).$$

Here $a(t) = 0$ for $t \leq 3\pi/2$; $a(t) = -\cos t$ for $3\pi/2 \leq t \leq 3\pi; a(t) = 1$ for $t \geq 3\pi$. Take $t_0 = 0$ and any $\phi \in C[-3\pi/2, 0]$. It is easy to see that the solution $x(t; 0, \phi)$ equals $\phi(0)$ for $0 \leq t \leq 3\pi/2$ and $x(t; 0, \phi) = -\phi(0) \sin t$

for $t \geq 3\pi/2$. So $|x(t;0,\phi)| \leq |\phi(0)|$ for $t \geq 0$. Hence the trivial solution is stable for the initial moment $t_0 = 0$. Now consider another initial moment $t_1 \geq 3\pi$ and define initial function $\theta \mapsto \phi_1(\theta) := \delta\exp(\lambda\theta)$ for $-3\pi/2 \leq \theta \leq 0$. Here λ is a positive root of the equation $z = \exp(-3\pi z/2)$. We get that $x(t;t_1,\phi_1) = \delta\exp[\lambda(t-t_1)]$ for $t \geq t_1 - 3\pi/2$ and also $|x(t;t_1,\phi_1)| \to \infty$ as $t \to \infty$, $\delta \neq 0$. Thus the trivial solution is unstable for any initial moment $t_1 \geq 3\pi$. □

Example 1.4 It shows that in the definition of asymptotic stability the first assumption generally does not follow from the second one.

Consider the equation

$$\dot{x}(t) = t^{-1}x(t) - 27t(t+2)^{-3}x^3((t+2)/3) = t^{-1}x(t)-$$

$$-27t(t+2)^{-3}x^3(t-h(t)), \quad t \geq t_0 = 1, \quad h(t) = (2t-2)/3$$

with initial condition $x(1) = x_0$. The solution equals

$$x(t;1,x_0) = x_0 t\exp(-x_0^2(t-1)), \quad t \geq 1.$$

So any solution $x(t;1,x_0) \to 0$ as $t \to \infty$. But for $|x_0| = \delta$ and $t = 1 + \delta^{-2}$ will be

$$|x(1+\delta^{-2};1,x_0)| = \delta(1+\delta^{-2})\exp(-\delta^2(1+\delta^{-2}-1)) = (\delta+\delta^{-1})/e \geq 2/e.$$

Hence the zero solution is unstable.□

Stability with respect to the perturbations of delays leads us to the following problem. Assume that linear system is asymptotically stable for every constant delay from a given interval. Will this system be stable for variable delay from the same interval? Consider an example, showing that it is not true in general.

Example 1.5 Investigate the scalar equation

$$\dot{x}(t) = -x(t-h(t)), \quad x(0) = x_0, \quad t \geq 0.$$

Take a number $a \in (3/2, \pi/2)$. Let the delay $h(t)$ be constant: $h(t) = h =$const. Then for $0 \leq h \leq a$ the trivial solution is asymptotically stable (see Fig. 4.3.1). But in Subs. 3.5.6.1 there is an example of this type of equations with $h(t) \in [0,a]$ for which the trivial solution is unstable. □

The definitions of stability can by immediately generalized to RDEs with infinite delay. In this case, however, the space of initial disturbances ϕ (see Subs. 3.2.5) must be specified. In this connection we note that a solution which is stable under a certain class of disturbances is also stable under a narrower class of disturbances. On the other hand, a solution which is unstable for a certain class of disturbances may well be stable under given a narrower class of disturbances. We consider by a simple example the dependence of stability on the choice of the initial function space.

Example 1.6 Given the Cauchy problem

$$\dot{x}(t) = -\int_{-t}^{t} x(\tau)d\tau, \quad 0 \leq t < \infty;$$

$$x(t) = \phi(t) \ (-\infty < t \leq 0), \quad \phi \in C(-\infty, 0].$$

Its solution has the form

$$x(t) = -\int_{0}^{t} \sin(t-\tau)\phi(-\tau)d\tau + \phi(0)\cos t, \quad 0 \leq t < \infty.$$

This implies that the trivial solution is stable (unstable) if the Banach space of functions $\phi \in C(-\infty, 0]$ with finite norm $\| \phi \| = \sup_{-\infty<t\leq 0} (|\phi(t)|e^{-t})$ (respectively, $\sup_{-\infty<t\leq 0} |\phi(t)|$) is taken as the set of initial functions.□

1.2. *STABILITY THEOREMS FOR EQUATIONS WITH BOUNDED DELAY*

We will consider the stability problem for the following equation in $\mathbf{R}^n$ with bounded delay $h > 0$:

$$\left.\begin{aligned} &\dot{x}(t) = f(t, x_t), \quad t \geq t_0, \qquad f(t,0) \equiv 0; \\ &x_t(\theta) := x(t+\theta), \quad -h \leq \theta \leq 0; \quad x_{t_0} = \phi. \end{aligned}\right\} \tag{1.2}$$

We assume that $\phi \in C[-h, 0]$, and that for some $H > 0$ the functional $((t, \psi) \mapsto f(t, \psi)) : \mathbf{R} \times Q_H \to \mathbf{R}^n$ is continuous and bounded, and locally Lipschitz with respect to

$$\psi \in Q_H := \{\psi \in C[-h, 0] : \| \psi \| < H\}.$$

From the stability theory of ODEs the profound efficiency of Liapunov's direct method is well known, based as it is on deep energy analogies. N. Krasovskii was the first to generalize this method to RDEs. In addition, since to each solution of an RDE there is naturally related an integral curve in the space $\mathbf{R} \times C[-h, 0]$ (see Subs. 3.2.3), it is natural to use functionals in this space instead of Liapunov functions. Such functionals are often called *Liapunov functionals.*

Let $V : \mathbf{R} \times Q_H \to \mathbf{R}$ be a continuous functional such that $V(t, 0) \equiv 0$. In what follows $V(\cdot, \cdot)$ is always interpreted as a functional, possessing these properties. Substituting in the second argument of V the solution $(t \mapsto x_t) : J_x \to \mathbf{R}^n$ of (1.2), we obtain a composite function of $t, [t \mapsto V(t) := V(t, x_t)] : J_x \to \mathbf{R}$. By definition, $\dot{V}(t_0)$ is the right upper derivative:

$$\dot{V}(t_0) := \limsup_{\Delta t \to 0+} \frac{1}{\Delta t}[V(t_0 + \Delta t) - V(t_0)]; \tag{1.3}$$

it is called the *derivative of* $V(\cdot,\cdot)$ *along the solution* of (1.2).

It is obvious that $\dot{V}(t_0)$ depends on $\phi \in Q_H$. Since we can proceed similarly for arbitrary $t \in \mathbf{R}$, $\psi \in Q_H$, we obtain a generalized functional $V : \mathbf{R} \times Q_H \to [-\infty, \infty]$, with values

$$\dot{V}(t, \psi) := \limsup_{\Delta t \to 0+} \frac{1}{\Delta t}[V(t + \Delta t, x_{t+\Delta t}(\cdot; t, \psi)) - V(t, \psi)]. \tag{1.4}$$

The functional $\dot{V}$ is generalized because it can take infinite values.

We give some examples. Let $n = 1$ and $V(t, \psi) := q(\psi(\alpha))$, where $q \in C^1(\mathbf{R})$, $-h \leq \alpha < 0$. Then

$$\dot{V}(t, \psi) = \begin{cases} \dot{q}(\psi(\alpha))\dot{\psi}^{ur}(\alpha) & \text{if } \dot{q}(\psi(\alpha)) > 0, \\ \dot{q}(\psi(\alpha))\dot{\psi}^{lr}(\alpha) & \text{if } \dot{q}(\psi(\alpha)) < 0, \end{cases}$$

where ur means 'upper right' and lr 'lower right'; the case $\dot{q}(\psi(\alpha)) = 0$ is not considered here. We can see that in this example $\dot{V}(t, \psi)$ does not take finite values on all of $\mathbf{R} \times Q_H$, but only on a subspace of it, since $\dot{\psi}^{ur}$ and $\dot{\psi}^{lr}$ may become infinite.

As another example we take $V(t, \psi) := q(t, \psi(0))$ for any n, where $q \in C^1(\mathbf{R} \times Q_H)$. Then

$$\dot{V}(t, \psi) = q_t(t, \psi(0)) + [\text{grad}q(t, \psi(0))]' f(t, \psi),$$

where q_t denotes partial derivative with respect to t and the gradient is calculated for t fixed. As a result we obtain a functional that is finite and continuous on $\mathbf{R} \times Q_H$.

Finally, for $n = 1$ we consider the functional

$$V(t, \psi) := \int_{-h}^{0} K(t, \theta)\psi(\theta) d\theta, \tag{1.5}$$

where $K \in C^1(\mathbf{R} \times [-h, 0])$. Using (1.4) we find

$$\dot{V}(t, \psi) = \int_{-h}^{0} K_t(t, \theta)\psi(\theta)d\theta + \left[\frac{d}{d\tau}\int_{-h}^{0} K(t, \theta)x_\tau(\theta; t, \psi)d\theta\right]_{\tau=t} =$$

$$= \int_{-h}^{0} K_t(t, \theta)\psi(\theta)d\theta + \left[\frac{d}{d\tau}\int_{-h}^{0} K(t, \theta)x(\tau + \theta; t, \psi)d\theta\right]_{\tau=t}.$$

Here we cannot directly use Leibnitz's rule, since in general $\dot{x}(t + \theta; t, \psi)$ need not exist. Therefore we first integrate by parts:

$$\int_{-h}^{0} K(t, \theta)x(\tau + \theta; t, \psi)d\theta = K(t, 0)\int_{0}^{\tau} x(s; t, \psi)ds -$$

$$-K(t, -h)\int_{0}^{\tau-h} x(s; t, \psi)ds - \int_{-h}^{0} K_\theta(t, \theta)\left[\int_{0}^{\tau+\theta} x(s; t, \psi)ds\right]d\theta.$$

Differentiation with respect to τ and substituting $\tau = t$ leads to

$$\dot{V}(t, \psi) = K(t, 0)\psi(0) - K(t, -h)\psi(-h) + \\ + \int_{-h}^{0}[K_t(t, \theta) - K_\theta(t, \theta)]\psi(\theta)d\theta. \tag{1.6}$$

(The same result is obtained by formal using Leibnitz's rule and then integrating by parts; a justification of this can be given using distribution theory.) So, in this example the functional $\dot{V}(t, \psi)$ is finite on the whole space $\mathbf{R} \times Q_H$, and is independent of the righthand side of Eq. (1.2).

We can similarly prove the following general rule, which holds in many actual examples. Assume that for arbitrary n the functional $V : \mathbf{R} \times C[-h, 0] \to \mathbf{R}$ can be represented in the form $V(t, \psi) = W(t, \psi(0), \psi)$, where the functional $W : \mathbf{R} \times \mathbf{R}^n \times C[-h, 0]$ has continuous derivative in the first argument, continuous gradient in the second (vectorial) argument, while in the third (functional) argument the following linearization holds:

$$W(t, \psi^0, \psi + \delta\psi) - W(t, \psi^0, \psi) =$$

$$= \int_{-h}^{0} K'(t, \psi^0, \theta, \psi(\theta))\delta\psi(\theta)d\theta + o(\| \delta\psi \|) \quad \text{as} \quad \| \delta\psi \| \to 0.$$

Here, $K' = (K_1, \dots, K_n)$ is a continuous vector function on $\mathbf{R} \times \mathbf{R}^n \times [-h, 0] \times \mathbf{R}^n$. Assume that there is a scalar continuous function S, depending on the same arguments, which is of class C^1 in the last two arguments and such that $\partial S/\partial \psi_i \equiv K_i \quad (i = 1, \dots, n)$. Then

$$\dot{V}(t,\psi) = W_t(t,\psi(0),\psi) + [\mathrm{grad}_{\psi(0)} W(t,\psi(0),\psi)]' f(t,\psi) +$$

$$+ S(t,\psi(0),0,\psi(0)) - S(t,\psi(0),-h,\psi(-h)) - \int_{-h}^{0} S_\theta(t,\psi(0),\theta,\psi(\theta))d\theta.$$

For functionals V of this kind we can construct $\dot{V}$ as follows. Replace ψ by x_t in $V(t,\psi)$, pass to x, differentiate in t (so that $\dot{x}$ appears only for the current (not delayed) value of t), substitute $\dot{x}(t)$ from the RDE, pass from x to x_t, and replace x_t by ψ. We illustrate this procedure for the functional (1.5):

$$V(t,x_t) = \int_{-h}^{0} K(t,\theta)x_t(\theta)d\theta = \int_{-h}^{0} K(t,\theta)x(t+\theta)d\theta =$$

$$= \int_{t-h}^{t} K(t,s-t)x(s)ds$$

(the last transformation is done to avoid the appearance of $\dot{x}(t+\theta)$ after differentiation). This implies

$$\dot{V}(t,x_t) = K(t,0)x(t) - K(t,-h)x(t-h) +$$

$$+ \int_{t-h}^{t} [K_I(t,s-t) - K_{II}(t,s-t)]x(s)ds =$$

$$= K(t,0)x_t(0) - K(t,-h)x_t(-h) + \int_{-h}^{0} [K_t(t,\theta) - K_\theta(t,\theta)]x_t(\theta)d\theta,$$

where K_I and K_{II} are the partial derivatives of K in the first and second argument, respectively. As a result we obtain (1.6). Having in mind this procedure, the expressions for V and $\dot{V}$ are (in practice) often immediately written in terms of x, keeping in mind the transition from x to ψ. E.g., in this example, $V(t,x_t) = \int_{-h}^{0} K(t,\theta)x(t+\theta)d\theta, \dot{V}(t,x_t) = K(t,0)x(t) - \dots$.

Let Ω be the set of scalar increasing functions $\omega \in C[0,\infty)$ such that $\omega(r) > 0$ for $r > 0$ and $\omega(0) = 0$.

A functional $(t,u) \mapsto V(t,u)$ is called *positive define (negative define*, or is said to have *infinitesimal upper bound*) if there is a function $\omega \in \Omega$ such that $V(t,\psi) \geq \omega(|\psi(0)|)$ (respectively, $V(t,\psi) \leq -\omega(|\psi(0)|)$, $|V(t,\psi)| \leq \omega(\| \psi \|)$) for all $t \in \mathbf{R}$, $\psi \in Q_H$. Sometimes the functional V having infinitesimal upper limit is called "decrescent". Let us note that the functional V which is independent on t is always decrescent in some domain Q_H.

Theorem 1.1 Assume that for some $H > 0$ there is a positive define continuous functional $((t,\psi) \mapsto V(t,\psi)) : \mathbf{R} \times Q_H \to \mathbf{R}$ which has infinitesimal upper bound and whose derivative $\dot{V}$ is a negative define functional on $\mathbf{R} \times Q_H$. Then the trivial solution of (1.2) is uniformly asymptotically stable.

Proof Let $\omega_1(|\psi(0)|) \leq V(t,\psi) \leq \omega_2(\| \psi \|)$ for all $t \in \mathbf{R}$, $\psi \in Q_H$, where $\omega_1, \omega_2 \in \Omega$. Take an $\varepsilon \in (0,H)$, and choose $\delta > 0$ satisfying $\omega_2(\delta) < \omega_1(\varepsilon)$. Then $\delta < H$, and for an arbitrary function $\phi \in Q_\delta$ and corresponding solution of (1.2), we find that for $t \geq t_0$ and $x(\tau) \in Q_H$, $\forall \tau \in [t_0, t]$:

$$\omega_1(|x(t)|) \leq V(t,x_t) \leq V(t_0,\phi) \leq \omega_2(\| \phi \|) \leq \omega_2(\delta) < \omega_1(\varepsilon).$$

Hence $|x(t)| < \varepsilon$ for all $t \in [t_0,\infty)$, since ω_1 is monotone. This implies uniform stability of the trivial solution.

For the proof of the uniform asymptotic stability take a $\gamma \in (0,H)$ and denote x a solution of Pr. (1.2) for some $t_0 \in \mathbf{R}$, $\phi \in Q_H$. Choose $l > 0 : \omega_2(l) < \omega_1(\gamma)$. The arguments used in the previous item show that if $\| x_{\bar{t}} \| < l$ for some $\bar{t} \in [t_0,\infty)$, then $|x(t)| < \gamma$ for all $t \in [\bar{t},\infty)$. Note that if $V(\tilde{t}, x_{\tilde{t}}) < \omega_1(l)$ for some $\tilde{t} \in [t_0,\infty)$, then from lack of increase of the function $t \mapsto V(t,x_t)$ it follows that $|x(t)| < l$ for all $t \geq \tilde{t}$. Because of this we can put $\bar{t} = \tilde{t} + h$. Assume, however, that for some $t_1 \in [t_0,\infty)$ we have $V(t,x_t) \geq \omega_1(l)$ for all $t \in [t_0,t_1]$. Choose $m > 0 : \omega_2(m) < \omega_1(l)$. Then $\| x_t \| \geq m \ \ (t_0 \leq t \leq t_1)$. Denote

$$C := \sup\{|f(t,\psi)| : t \in \mathbf{R},\ \psi \in Q_H\}, \ \ \bar{h} := \min\{h, m/(2C)\}.$$

Let $t_1 \geq t_0 + 3\bar{h}$ and $\max\{|x(t)| : t_0 + \bar{h} \leq t \leq t + 2\bar{h}\}$ be achieved at $t = \tau$. Then for $|t - \tau| \leq \bar{h}$ we obtain $|x(t) - x(\tau)| \leq Cm/(2C)$. From here $|x(t)| \geq m/2$. Hence, $\dot{V}(t,x_t) \leq -\omega_3(m/2)$, where ω_3 is a function from the definition of negative definiteness of $\dot{V}$. This yield

$$V(t_0+3\overline{h}, x_{t_0+3\overline{h}}) - V(t_0,\phi) = \int_{t_0}^{t_0+3\overline{h}} \dot{V}(t,x_t)dt \leq$$

$$\leq \int_{\tau-\overline{h}}^{\tau+3\overline{h}} \dot{V}(t,x_t)dt \leq -2\overline{h}\omega_3(m/2).$$

Therefore

$$2\overline{h}\omega_3(m/2) \leq V_0 := \sup\{V(t,\psi) : t \in \mathbf{R},\ \psi \in Q_H\}.$$

Similar estimate on the intervals $[t_0+3\overline{h}, t_0+6\overline{h}]$, and so on shows that

$$2[(t_1-t_0)/3\overline{h}]\overline{h}\omega_3(m/2) \leq V_0$$

where $[\cdot]$ is an integral part. This implies that

$$t_1 < t_0 + 3\overline{h} + (3/2)V_0/\omega_3(m/2) =: \tilde{t}.$$

Thus by virtue of the previous item

$$|x(t)| < \gamma \ \ (\forall t \geq t_0 + T(\gamma),\ T(\gamma) := 3\overline{h} + (3/2)V_0/\omega_3(m/2) + h).$$

From here it follows uniform asymptotic stability, because of independence of $T(\gamma)$ on t_0 and ϕ.□

We can similarly state other stability conditions. Some of these are as follows.

Theorem 1.2 If there is a continuous positive define functional $V : \mathbf{R} \times Q_H \to \mathbf{R}$ *with* $\dot{V} \leq 0$, *then the trivial solution of (1.2) is stable.*

Proof Given any $\varepsilon \in (0,H)$. By virtue of $V(t_0,0) \equiv 0$ we can take $\delta(\varepsilon,t_0) \in (0,H)$ such that

$$\max_{\|\phi\| \leq \delta(\varepsilon,t_0)} V(t_0,\phi) \leq \omega_1(\varepsilon).$$

From condition $\dot{V} \leq 0$ it follows that V is non-increasing along the trajectories of (1.1). Hence

$$\omega_1(|x(t)|) \leq V(t,x_t) \leq V(t_0,x_{t_0}) = V(t_0,\phi) \leq \omega_1(\varepsilon), \quad t \geq t_0.$$

Monotonicity of ω_1 implies that $|x(t)| \leq \varepsilon, \quad t \geq t_0$. Thm. 1.2 is proven.□

Remark If in addition to the assumptions of Thm. 1.2 functional $V(t,\phi)$ is decrescent than the trivial solution is uniformly stable.

Theorem 1.3 A necessary and sufficient condition for the exponential stability of the trivial solution of (1.2) is that there exists a functional $V : \mathbf{R} \times Q_H \to \mathbf{R}$ *such that*

$$C_1 \parallel \psi \parallel \leq V(t,\psi) \leq C_2 \parallel \psi \parallel, \quad \dot{V}(t,\psi) \leq -C_3 \parallel \psi \parallel,$$

$$|V(t,\psi) - V(t,\xi)| \leq C_4 \parallel \psi - \xi \parallel .$$

Here, C_i *are some positive constants, and* ψ, ξ *are arbitrary functions from* Q_H.

Example 1.4 Consider the following scalar equation with continuous coefficients a and b:

$$\dot{x}(t) = -a(t)x(t) - b(t)x(t-h), \quad t \geq t_0, \quad h > 0. \tag{1.7}$$

To derive stability conditions for (1.7), we introduce a functional $V : \mathbf{R} \times C[-h,0] \to \mathbf{R}$ by

$$V(t,\psi) = \psi^2(0) + \int_{-h}^{0} |b(t+\theta+h)|\psi^2(\theta)d\theta.$$

To compute $\dot{V}$ we apply (1.4), putting $x(\tau) = x(\tau; t, \psi)$ $(\tau \geq t)$. Changing $\tau + \theta = s$ under the integral sign, we obtain

$$\dot{V}(t,\psi) = \left[\frac{d}{d\tau}V(\tau, x_\tau)\right]_{\tau=t} =$$

$$= \left[\frac{d}{d\tau}\left(x^2(\tau) + \int_{\tau-h}^{\tau} |b(s+h)|x^2(s)ds\right)\right]_{\tau=t} =$$

$$= x^2(t)[-2a(t) + |b(t+h)|] - 2b(t)x(t)x(t-h) - |b(t)|x^2(t-h) =$$

$$= \psi^2(0)[-2a(t) + |b(t+h)|] - 2b(t)\psi(0)\psi(-h) - |b(t)|\psi^2(-h) \leq$$

$$\leq -\gamma(t)\psi^2(0),$$

$$\gamma(t) := 2a(t) - |b(t)| - |b(t+h)|.$$

Therefore the trivial solution of (1.7) is stable if $\gamma(t) \geq 0$, by Thm. 1.2, and uniformly asymptotically stable if $\sup_{t \geq t_0} |b(t)| < \infty$, $\inf_{t \geq t_0} \gamma(t) > 0$, by Thm. 1.1. Note that these stability conditions are due to the negative instantaneous feedback $-a(t)x(t)$. The following sufficient condition for asymptotic stability was obtained in [294] for the case $a(t) \geq 0$, $b(t) \geq 0$:

$$h \sup_{t \geq t_0} [a(t) + b(t)] < 1.5. \quad \square$$

Example 1.5 Consider the problem of the stationary values V_a^0 and V_b^0 of arterial and venous pressure. By (2.6.1) and (2.6.2), these are equal to

$$V_a^0 = (Q_i + k_5)(k_4 k_a)^{-1} =: x_0,$$

$$V_b^0 = [Q_i(k_2 - k_3 x_0) + k_a x_0][k_b(1 + k_1(k_2 - k_3 x_0))]^{-1} =: y_0.$$

Sufficient stability conditions for the stationary solution have the form

$$\left.\begin{aligned} &-2(k_1 k_b + k_4 k_a) + \frac{1}{2}\alpha_0 k_b + 3\alpha_0 k_3 (k_a x_0 - k_b y_0)(k_2 - k_3 x_0) < 0, \\ &-k_1 k_b + \frac{1}{2}\alpha_0 k_a + \alpha_0 k_3 (k_a x_0 - k_b y_0)(k_2 - k_3 x_0)^{-1} < 0. \end{aligned}\right\} \quad (1.8)$$

Here $\alpha_0 := (k_2 - k_3\delta)^{-1}$, and $\delta \in (0, k_2/k_3)$ is a constant. To prove this it suffices to use Thm. 1.2, applied to the functional

$$V(x_t, y_t) = x_t^2(0) + y_t^0(0) + C_0 C_1^{-1}[x_t(0) + y_t(0)]^2 + 2\alpha_0 C_2 \int_{-\tau}^{0} x_t^2(\theta)d\theta,$$

where $x := V_a - x_0$, $y := V_b - y_0$, $C_0 := k_1 k_b$, $C_1 := k_4 k_a$, $C_2 := k_3(k_a x_0 - k_b y_0)(k_2 - k_3 x_0)^{-1}$.

Let us examine the Eqs. (1.8). We note that the stationary value V_a^0 increases as Q_i and k_5 increase while k_4 decreases. By (1.8) the rate with which any solution tends to a stationary value decreases if k_3, x_0 increase, while k_1, k_2, k_a, k_b decrease. These conclusions have been experimentally confirmed [206]. $\square$

1.3. *STABILITY OF EQUATIONS WITH UNBOUNDED DELAY*

Stability problems for RDEs with unbounded delay can be stated in much the same way as for RDEs with bounded delay.

Let $\mathcal{K}$ be a metric space of continuous functions ψ: $(-\infty, 0] \to \mathbf{R}^n$, with metric ρ, satisfying the conditions in Subs. 3.2.5. Consider the following Cauchy problem for $x : J_x \to \mathbf{R}^n$:

$$\left.\begin{aligned} &\dot{x}(t) = f(t, x_t), \quad t \geq t_0, \\ &x_t(\theta) := x(t + \theta), \quad -\infty < \theta \leq 0, \\ &x_{t_0} = \phi \in \mathcal{K}. \end{aligned}\right\} \quad (1.9)$$

Here, $f : \mathbf{R} \times \mathcal{K} \to \mathbf{R}^n$ is a continuous functional which is locally Lipschitz in the second argument and $f(t, 0) \equiv 0$.

The definitions of stability of the trivial solution of (1.9) are the same as in the case of Eq. (1.1) with bounded delay, with $C[-h, 0]$ replaced by $\mathcal{K}$ and $\| \psi \|$ replaced by $\rho(\psi, 0)$, so that

$$Q_H := \{\psi \in \mathcal{K} : \rho(\psi, 0) < H\}.$$

We state some conditions for the stability of the trivial solution of (1.9); these can be proved similar to Thm. 1.1.

Theorem 1.4 Assume that for some $H > 0$ and $\omega_i \in \Omega$ there is a functional $((t, \psi) \mapsto V(t, \psi)) : \mathbf{R} \times Q_H \to \mathbf{R}$ such that

$$\omega_1(|\psi(0)|) \le V(t, \psi) \le \omega_2(\rho(\psi, 0)),$$

$$\dot{V}(t, \psi) \le -\omega_3(|\psi(0)|).$$

Then the trivial solution of (1.9) is uniformly asymptotically stable.

Example 1.6 We will derive stability conditions for the equilibrium $\psi = 0, \quad \phi = 0 \quad (\psi, \phi : \mathbf{R} \to \mathbf{R}^n)$ of the system of equations (2.2.1), which describes the motion of a rigid body with PID regulator:

$$u = -A_1\phi - A_2\dot{\phi} - \int_0^t g(t-s)\phi(s)ds, \quad A_1, A_2 > 0.$$

Here $A_1, A_2 \in \mathbf{R}^{n \times m}$ are symmetric matrices, and the function $g : (-\infty, 0] \to \mathbf{R}$ is integrable. In the sequel we put $H = H(\phi(t)), \quad \dot{\phi} = \dot{\phi}(t), \quad \dot{\psi} = \dot{\psi}(t), \quad L = L(\phi, \psi) \ (H \to \mathbf{R}^{n \times m}, \ L \to \mathbf{R})$, and also $\phi(s) = 0, \quad \psi(s) = 0$ for $s < 0$. Then we can write (2.2.1) as

$$\left.\begin{aligned} \dot{\psi} &= -L_\phi - A_1\phi - A_2H^{-1}\psi - \int_0^\infty g(s)\phi(t-s)ds, \\ \dot{\phi} &= H^{-1}\psi, \quad t \ge 0, \\ 2L &= \psi' H^{-1}(\phi)\psi, \quad L_\phi = \frac{\partial L}{\partial \phi}. \end{aligned}\right\} \tag{1.10}$$

Put

$$\rho_0 = \int_0^\infty |g(s)|ds.$$

Since the operator on the righthand side of (1.10) satisfies the local Lipschitz condition with respect to ϕ, ψ there is a unique solution on a certain interval $[0, \delta)$, $\delta > 0$ for any $\phi(0)$, $\psi(0)$ (see the end of Subs. 3.2.4). Let $\mathcal{K}$ be the space $C(-\infty, 0]$ with $\| f \| := \sup_{-\infty < t \leq 0} |f(t)| < \infty$. This space does not satisfy all the conditions of Subs. 3.2.5, but this is not essential because of linearity of (1.10), the assertions in Subs. 3.5.2, and the condition $\rho_0 < \infty$.

To study the possibility of prolongation of a solution and the stability of the trivial solution, we use the functional (where ε is a positive constant)

$$\left.\begin{aligned} &V_1(v, w; \varepsilon) := V(v(0), w(0); \varepsilon) + \frac{1+\varepsilon}{2} \int_0^\infty ds \int_{-s}^0 |g(s)| \cdot |v(\theta)|^2 d\theta, \\ &V(v(0), w(0); \varepsilon) := L(v(0), w(0)) + \frac{1}{2} v'(0) A_1 w(0) + \varepsilon v'(0) w(0), \\ &v, w \in \mathcal{K}. \end{aligned}\right\} \tag{1.11}$$

Note that for sufficiently small $\varepsilon > 0$ the function $(\phi, \psi) \mapsto V(\phi, \psi; \varepsilon)$ is positive define in ϕ, ψ and has an infinitesimal upper bound.

First we consider the case $g(s) \equiv 0$. We compute the derivative $\dot{V} = \dot{V}(\phi, \psi; \varepsilon)$ of the function V along the trajectories of (1.10). We then find

$$\begin{aligned} \dot{V} &= \dot{\phi}' L_\phi + \dot{\psi}' L_\psi + \dot{\phi}' A_1 \phi + \varepsilon \dot{\psi}' \phi + \varepsilon \psi' \dot{\phi} = \\ &= \psi' H^{-1} L_\phi - (L_\phi + A_2 H^{-1} \psi + A_1 \phi)' H^{-1} \psi + \\ &+ \psi' L^{-1} A_1 \phi - \varepsilon (L_\phi + A_2 H^{-1} \psi + A_1 \phi)' \phi + \varepsilon \psi' H^{-1} \psi = \\ &= \psi' (H^{-1} A_2 H^{-1} - \varepsilon H^{-1}) \psi - \varepsilon \phi' L_\phi - \varepsilon \phi' A_1 \phi - \varepsilon \phi' A_2 H^{-1} \psi. \end{aligned}$$

Using the estimates

$$\varepsilon \phi' A_2 H^{-1} \psi \leq \frac{1}{2} \varepsilon^2 \phi' A_2 \phi + \frac{1}{2} \psi' H^{-1} A_2 H^{-1} \psi,$$

$$\varepsilon \phi' L_\phi = -\frac{1}{2} \varepsilon \psi' H^{-1} \left(\sum_{i=1}^n \phi_i \frac{\partial H}{\partial \phi_i} \right) H^{-1} \psi$$

we obtain

$$\dot{V} \le W,$$

$$W = W(\phi, \psi; \varepsilon) := -\varepsilon\phi'(A_1 - \frac{1}{2}\varepsilon A_2)\phi -$$

$$-\psi' H^{-1}\left[\frac{1}{2}A_2 - \varepsilon H - \frac{1}{2}\varepsilon \sum \phi_i \frac{\partial H}{\partial \phi_i}\right] H^{-1}\psi.$$

Together with (1.11) this implies that the function $(\phi, \psi) \mapsto V(\phi(0), \psi(0); \varepsilon)$ satisfies the conditions of the theorem on global asymptotic stability [298].

For an arbitrary scalar function $s \mapsto g(s)$ which is absolutely integrable on $[0, \infty)$ the derivative of the functional (1.11) along the trajectories of (1.10) satisfies the inequality

$$\dot{V}_1 \le W - \psi' H^{-1} \int_0^\infty g(s)\phi(t-s)ds - \varepsilon\phi' \int_0^\infty g(s)\phi(t-s)ds +$$

$$+\frac{1+\varepsilon}{2}\rho_0|\phi(t)|^2 - \frac{1+\varepsilon}{2}\int_0^\infty |g(s)| \cdot |\phi(t-s)|^2 ds.$$

Note that

$$\varepsilon\left|\phi'(t)\int_0^\infty g(s)\phi(t-s)ds\right| \le \frac{\varepsilon}{2}\rho_0|\phi(t)|^2 + \frac{\varepsilon}{2}\int_0^\infty |g(s)|\cdot|\phi(t-s)|^2 ds,$$

$$\left|\psi'(t)H^{-1}\int_0^\infty g(s)\phi(t-s)ds\right| \le$$

$$\le \frac{\rho_0}{2}\psi'(t)H^{-2}\psi(t) + \frac{1}{2}\int_0^\infty |g(s)|\cdot|\phi(t-s)|^2 ds.$$

This implies the inequality

$$\dot{V}_1 \le -\phi'\left[\varepsilon A_1 - \frac{1}{2}\varepsilon^2 A_2 - (\varepsilon + \frac{1}{2})\rho_0 I\right]\phi -$$

$$-\psi' H^{-1}\left[\frac{1}{2}A_2 - \varepsilon H - \frac{1}{2}\varepsilon\sum \phi_i \frac{\partial H}{\partial \phi_i} - \frac{1}{2}\rho_0 I\right] H^{-1}\psi, \tag{1.12}$$

where I is the identity matrix.

Let D_0 be a bounded domain in the (ϕ, ψ)–space. We show that we can choose the function g such that the equilibrium position of (1.10) will be asymptotically stable, with domain of attraction containing D_0. Let $M := \sup_{(\phi,\psi)\in D_0} V(\phi, \psi)$. There is a ball D outside which the inequality $V(\phi, \psi) > M$ is valid for all sufficiently small $\varepsilon > 0$. By the choice of g we can make ρ_0 arbitrarily small. Let ε and ρ_0 be so small that the matrices between square brackets in (1.12) are positive definite. Then $\dot{V}_1$ is negative definite in D. We verify that trajectories of (1.10) starting in D_0 do not leave D. In fact, for $t = 0$ we have $V_0 = V \leq M$. Further, the functional V_1 does not increase in D, and on ∂D the inequality $V_1 \geq V \geq M$ holds. This and (1.11) imply the asymptotic stability of the trivial solution, with domain of attraction containing D_0.□

Example 1.7 Suppose we are given the scalar equation

$$\dot{x}(t) = -a(t)x(t) + b(t)\int_0^\infty \alpha(s)x(t-s)ds, \quad t \geq 0, \tag{1.13}$$

with functions $a, b, \alpha \in L_1[0,\infty)$ and b, α integrable on $[0,\infty)$. Let $\mathcal{K}$ be the space of bounded continuous functions on $(-\infty, 0]$, with norm $\| \phi \|= \sup_{-\infty<t\leq 0} |\phi(t)|$. To obtain a stability condition we consider the functional

$$V(t,\psi) := \psi^2(0) + \int_0^\infty |\alpha(s)| \left[\int_{t-s}^{t} |b(\tau+s)|\psi^2(\tau-t)d\tau\right] ds.$$

Using the procedure of Subs. 1.2 (which can be used for unbounded delay as well), we find

$$V(t,x_t) = x^2(t) + \int_0^\infty |\alpha(s)| \left[\int_{t-s}^{t} |b(\tau+s)|x^2(\tau)d\tau\right] ds,$$

$$\dot{V}(t,x_t) = 2x(t)\dot{x}(t) + \int_0^\infty |\alpha(s)|[|b(t+s)|x^2(t) - |b(t)|x^2(t-s)]ds =$$

$$= 2x(t)\left[-a(t)x(t) + b(t)\int_0^\infty \alpha(s)x(t-s)ds\right] +$$

$$+\int_0^\infty |\alpha(s)|[|b(t+s)|x^2(t)-|b(t)|x^2(t-s)]ds \le$$

$$\le -2a(s)x_t^2(0)+\int_0^\infty |\alpha(s)|[|b(s)|+|b(t+s)|]ds\cdot x_t^2(0)-$$

$$-|b(t)|\int_0^\infty |\alpha(s)|[x_t(0)-x_t(-s)]^2 ds.$$

Replacing x_t by ψ we obtain an estimate for $\dot{V}(t,\psi)$. One can easily see that the trivial solution (1.13) is stable (hence all its solutions are stable), provided that

$$2a(t)\ge\int_0^\infty |\alpha(s)|[|b(s)|+|b(t+s)|]ds,\quad 0\le t<\infty.\square$$

Note that in this example the function $t\mapsto\dot{V}(t,x_t)$ is, for a given initial condition, only defined for almost all $t\in[0,\infty)$, in general. However, this does not impair the use of the stability theorem, since $t\mapsto V(t,x_t)$ is absolutely continuous.

1.4. *STABILITY OF LINEAR NONAUTONOMOUS RDES*

Here we will consider the Cauchy problem for the linear equation

$$\dot{x}(t)=\int_{-\infty}^{0}[d_\theta R(t,\theta)]x(t+\theta),\quad t\ge t_0,\quad x(t)\in\mathbf{R}^n, \tag{1.14}$$

$$x(t_0+\theta)=\phi(\theta),\quad -\infty<\theta\le 0. \tag{1.15}$$

We assume that the matrix–kernel R satisfies the condition of Subs. 3.5.2 related to (3.5.11). For the space $\mathcal{K}$ of initial functions ϕ we take $\{\phi\in C(-\infty,0]:\|\phi\|:=\sup|\phi(t)|<\infty\}$.

It is clear that if the solution of (1.14) is stable (asymptotically stable) for some initial function, then the same holds for any initial function, and the same is true for the inhomogeneous equation (3.5.11). Therefore, for linear equations it is possible to talk of the stability of not only an individual solution, but also of the *stability (asymptotic stability) of the equation* itself.

The following simple, but useful result is due to A. Zverkin.

Theorem 1.5 Eq. (1.14) is stable if and only if for every initial function $\phi \in \mathcal{K}$ the corresponding solution $x(t; t_0, \phi)$ is bounded.

Proof Necessity is obvious. To prove sufficiency we define for each $t \geq t_0$ a functional U_t on the Banach space $\mathcal{K}$ by $U_t\phi = x(t; t_0, \phi)$. Since the solution depends continuously on the initial data (cf. the end of Subs. 3.5.2), this U_t is continuous.

The assumptions of Thm. 1.5 imply that also $\sup_{t \geq t_0} |U_t\phi| < \infty$ for all $\phi \in \mathcal{K}$. Using the Banach – Steinhaus theorem we find that the norms of the functionals U_t are bounded in totality. This implies stability.□

Corollary 1.1 If $x(t; t_0, \phi) \to 0$, $t \to \infty$, for any initial function $\phi \in \mathcal{K}$, then (1.14) is asymptotically stable

1.5. *STABILITY OF LINEAR PERIODIC RDES*

Here we consider the Pr. (1.2), where $f(t, \psi)$ is, in addition, linear in ψ and periodic in t, with period $\omega > 0$. We introduce the *monodromy operator* (also called *shift operator on the period*) $U(t) : C[-h, 0] \to C[-h, 0]$ by $U(t)\phi = x_{t+\omega}(t + \omega; t, \phi)$. Here, $C[-h, 0]$ must be regarded as a space of complex–valued vector functions.

Theorem 1.6 The equation in (1.2) with functional f linear in x_t and periodic in t is:

1) asymptotically stable if and only if the spectrum of the operator $U(t)$ lies inside the unit circle;

2) stable if and only if the eigenvalues ρ_i of $U(t)$ are such that $|\rho_i| \leq 1$, while the eigenvalues ρ_j with $|\rho_j| = 1$ have simple elementary divisors.

Proof Let us remind at first some necessary facts from the theory of linear operators and linear algebra.

Let U be a linear operator which maps some complex Banach space M in itself, and moreover operator U^k for some natural k is completely continuous. Then the spectrum $\sigma(U)$ of the operator U consists of only on the eigenvalues. If the set $\sigma(U)$ is infinite then it forms a sequence $\{\lambda_j\}$, where $\lambda_j \to 0$ as $j \to \infty$. Moreover,

$$\lim_{m \to \infty} \| U^m \|^{1/m} = \max_j |\lambda_j| \tag{1.16}$$

(if $\sigma(U) = \emptyset$, this limit is equal to zero). In addition, if $\lambda_{\bar{j}} \neq 0$ is some of this eigenvalues, then the space M can be represented as a direct sum of its subspaces M' and M'' (i.e. every element $v \in M$ can be represented uniquely in the form $v_1 + v_2 : v_1 \in M'$, $v_2 \in M''$), where $UM' \subseteq M'$, $UM'' \subseteq$

M'', $\dim M' < \infty$, $\sigma(U|_{M'}) = \{\lambda_{\widetilde{j}}\}$, $\sigma(U|_{M''}) = \sigma(U)\setminus\{\lambda_{\widetilde{j}}\}$. It is clear that the operator $(U|_{M''})^k$ is completely continuous.

Let U be a linear map of the finite–dimensional complex linear space M in itself which has the only eigenvalue $\widetilde{\lambda}$. Then M can be represented as the direct sum of the subspaces $M_1, \ldots, M_r$ $(r \geq 1)$ such that $\forall UM_j \subseteq M_j$ and any map $U|_{M_j}$ has the only (up to the scalar factor) eigenvector u_j^1. Besides, if $d_j := \dim M_j > 1$, then there exist vectors $u_j^2, \ldots, u_j^{d_j}$ (they are called as vectors, adjointed to the eigenvector), for which the equalities

$$Uu_j^s = u_j^{s-1} + \widetilde{\lambda} u_j^s \quad (s = 2, \ldots, d_j) \tag{1.17}$$

hold. The numbers d_j for given eigenvalue $\widetilde{\lambda}$ are the exponents of the elementary divisors $(\lambda - \widetilde{\lambda})^{d_j}$ which are made up with definite rule. The vectors $u_j^1, u_j^2, \ldots, u_j^{d_j}$ form a basis in M_j.

Apply these general properties to the monodromy operator $U(t)$. First of all, it is evident that $(U(t))^k$ with $k = [h/\omega] + 1$ ($[\cdot]$ denote the integral part) translate every bounded subset of $C[-h, 0]$ in the bounded set of functions with uniformly bounded derivatives; therefore the last operator is completely continuous.

We show that the spectrum $\sigma(U(t))$ does not depend on t, up to the value $\lambda = 0$. First of all, it is obvious that the operators $U(t + k\omega)$ are identical, if $k \in \mathbf{Z}$. Therefore it is sufficient to prove that $\sigma(T(t_1)) \setminus \{0\} \subseteq \sigma(T(t_2))$, if $t_1 < t_2 < t_1 + \omega$. Let be $\lambda \in \sigma(U(t_1))\setminus\{0\}$ and $\phi \in C[-h, 0]\setminus\{0\}$ is such that $U(t_1)\phi = \lambda\phi$, i.e. $T(t_1 + \omega, t_1)\phi = \lambda\phi$ where T is the general shift operator (see Subs. 3.2.3). Then we have

$$T(t_2+\omega, t_1+\omega)T(t_1+\omega, t_1)\phi = \lambda T(t_2+\omega, t_1+\omega)\phi, \quad \text{i.e. } T(t_2+\omega, t_2)\psi = \lambda\psi, \tag{1.18}$$

where $\psi := T(t_2, t_1)\phi$. But $\psi \neq 0$ because otherwise we have

$$T(t_2, t_1)\phi = 0 \Rightarrow T(t_1 + \omega, t_2)T(t_2, t_1)\phi = 0 \Rightarrow$$

$$\Rightarrow T(t_1 + \omega, t_1)\phi = 0 \Rightarrow \lambda\phi = 0 \Rightarrow \phi = 0$$

what is impossible. Then we have from (1.18) that $\lambda \in \sigma(T(t_2))$, i.e. the independence $\sigma(T(t)) \setminus \{0\}$ on t is proved. Let $\sigma := \sigma(T(t)) \setminus \{0\}$.

If $\lambda \in \sigma$, then there exist linearly independent functions $\phi_1, \ldots, \phi_r \in C[-h, 0] \setminus \{0\}$ such that $\forall [x(\cdot; 0, \phi_j)]_\omega = \lambda\phi_j$. It follows the ω-periodicity of equation in (1.2) that the function $x(\cdot + \omega; 0, \phi_j)$ satisfies this equation; but this function coincide with $\lambda\phi_j$ on $[-h, 0]$, and therefore

$$x(t + \omega; 0, \phi_j) \equiv \lambda x(t; 0, \phi_j) \tag{1.19}$$

for $t \in [0, \infty)$, by virtue of uniqueness of the solution of Pr. (1.2). If we continue every such function x by means of (1.19) on the whole t–axis, we obtain the function which satisfies the equation of Pr. (1.2) for all $t \in \mathbf{R}$, by virtue of periodicity of this equation. The nonzero solution $x : \mathbf{R} \to \mathbf{R}^n$ of Eq. in (1.2) which have the property

$$x(t + \omega) \equiv \lambda x(t), \quad -\infty < t < \infty,$$

are called as *Floquet solutions,* and the corresponding numbers λ are called the *characteristic (Floquet) multipliers.* It is clear that this solution can be represented in the form

$$x(t) = e^{\mu t} y(t), \quad -\infty < t < \infty,$$

where $\mu := (\ln \lambda)/\omega$ ($\ln \lambda$ is an arbitrary value of logarithm) and $y(t+\omega) \equiv y(t)$.

If the operator $U(0)$ possesses, besides the eigenvector u_j^1, the adjoined to it vectors $u_j^2, \ldots, u_j^d$ for some eigenvalue λ, then it can be checked by (1.17) and induction on r that the equation in (1.2) has the solution

$$x_j^r(t) = e^{\lambda t} \sum_{s=0}^{r-1} \left[\prod_{q=0}^{s-1} (t - q\omega) \right] y_j^{r-s}(t)/s!(\lambda\omega)^s, \quad r = 1, \ldots, d$$

where $y_j^l(t + \omega) \equiv y_j^l(t) \not\equiv 0$ and $\prod_{q=0}^{-1} (t - q\omega) := 1$. Functions $x_j^2, \ldots, x_j^d$ are called as *Floquet adjoined functions.*

If $\sigma = \emptyset$, then it follows from (1.16) with $U = U(0)$ that $x(t) = o(e^{-\alpha t})$ as $t \to \infty$ for any solution x of Pr. (1.2) and any $\alpha \in \mathbf{R}$. If $\sigma \neq \emptyset$, we list all distinct eigenvalues λ_j of $U(0)$ in the order of nonincreasing of their modules, without taking into account their multiplicity:

$$|\lambda_1| \geq |\lambda_2| \geq \ldots .$$

The space $M := C[-h, 0]$ can be represented as a direct sum of subspaces M_1' and M_1'' where $U(0)M_1' \subseteq M_1'$, $U(0)M_1'' \subseteq M_1''$, $\dim M_1 < \infty, U(0)|_{M_1'}$ has only eigenvalue λ_1, the spectrum of $U(0)|_{M_1''}$ is $\sigma(U(0)) \setminus \{\lambda_j\}$. If $\phi \in M_1'$, then the solution of Pr. (1.2) is the linear combination of the Floquet solutions corresponding to λ_1 and the solutions adjoined to them.

If there are another eigenvalues in σ besides λ_1, we reperesent M_1'' as the direct sum of subspaces M_2' and M_2'' which corresponds to the eigenvalue λ_2 etc. There can be two cases. If $\text{card}\sigma = k < \infty$, then $U(0)|_{M_k''}$ has no nonzero eigenvalues, and therefore the solutions with $\phi \in M_k''$ tends to zero faster than any exponential as $t \to \infty$. As a result, we obtain the representation of the solution of Pr. (1.2) in the form of the linear combination of the Floquet solutions, adjoined solutions, corresponding to all eigenvalues $\lambda \in \sigma$, and the solution which tends to zero faster than any exponential as $r \to \infty$.

Consider now the second case, when card$\sigma = \infty$; then $\lambda_k \to 0$ as $k \to \infty$. Being given any $\varepsilon \in (0, |\lambda_1|)$, $\varepsilon \notin \{|\lambda_k| : \lambda_k \in \sigma\}$, we denote $k_\varepsilon :=$ card$\{\lambda_k : |\lambda_k| > \varepsilon\}$. Then, after k_ε steps described above, we obtain by virtue of (1.16) with $U = U(0)|_{M''_{k_\varepsilon}}$ the representation of the solution of Pr. (1.2) in the form

$$x(t) = \sum_{j=1}^{m_\varepsilon} a_j x^j(t) + o\left(e^{((\ln \varepsilon)/\omega)t}\right).$$

Here x^j are all Floquet solutions and adjoined to them which correspond to the eigenvalues $\lambda_k : |\lambda_k| > \varepsilon$; m_ε is the number of such solutions, and coefficients a_j depend on solution x. If ε decreases then new items are added to the sum but coefficients a_j in terms already written out are not changed. Going over to the limit as $\varepsilon \to 0^+$, we obtain asymptotical representation

$$x(t) \sim \sum_j a_j x^j(t)$$

in which the sum is extended on all Floquet solutions of equation in (1.2) and adjoined to them. The coefficients a_j are uniquelly defined by the solution x. If we take in the righthand side partial sum with sufficiently many terms, we obtain the representation of solution x with any preassigned accuracy in exponential scale, as $t \to \infty$.

The assertions of Thm. 1.6 are evident conclusions of this result, Thm. 1.5 and Cor. 1.1.□

Theorem 1.6 reduces the problem of RDEs stability to the investigation of spectrum $\sigma(U(t))$. Usually the problem of determination of $\sigma(U(t))$ is very complicated but for some particular classes of RDEs this problem is studied better. Consider the scalar equation with delays divisible by period.

$$\dot{x}(t) + b_0(t)x(t) + b_1(t)x(t - m_1\omega) +$$

$$+ b_2(t)x(t - m_2\omega) = 0, \quad \forall b_i(t + \omega) = b_i(t),$$

where m_1 and m_2 are positive integer numbers. Let us seek the Floquet solution of the form

$$x(t) = p(t)\exp(\lambda t), \quad p(t + \omega) = p(t).$$

For $p(t)$ we get the equation

$$\dot{p}(t) + [\lambda + g(t, \lambda)]p(t) = 0,$$

$$g(t, \lambda) = b_0(t) + b_1(t)\exp(-m_1\lambda\omega) + b_2(t)\exp(-m_2\lambda\omega).$$

Integrating this equation we conclude that the function

$$p(t) = c\exp[-\lambda t - \int_0^t g(s,\lambda)ds]$$

must be ω–periodic. Hence we have following equation for λ:

$$\lambda\omega + \int_0^\omega g(s,\lambda)ds = 0.$$

Lefthand side of this eqution is a quasipolynomial. Therefore it has at most countable number of roots and each of them has finite multiplicity (see Sec. 2). By Thm. 1.6 for asymptotic stability it is necessary and sufficient that all these roots λ_j satisfy the condition $\mathrm{Re}\lambda_j < 0$.

Consider as an example the scalar equation

$$\dot{x}(t) = 2\pi(\cos(2\pi t))x(t-1).$$

The only Floquet solution of this equation is

$$\psi_1(t) = \exp(\sin 2\pi t).$$

So the system of the Floquet solutions is incomplete, but for any solution $x(t)$ of this equation and arbitrary $\beta > 0$ we have $x(t) = c_1\psi_1(t) + o(\exp(-\beta t))$. Consider for example the solution x_1 defined by the initial condition $x_1(t) = 1$, $0 \le t \le 1$. By the method of steps (see Sec. 3.1) we get

$$x_1(t) = \sum_{j=0}^{[t]} \frac{1}{j!}(\sin 2\pi t)^j, \quad t \ge 0.$$

Expanding exponential in Taylor series with Lagrangian remainder term we obtain

$$x_1(t) - \psi_1(t) = -\frac{1}{[t+1]!}(\sin(2\pi t))^{[t+1]}\exp(\xi\sin(2\pi t)) =$$

$$= o\left(\frac{1}{\Gamma(t)}\right), \quad 0 < \xi < 1.$$

Here $\Gamma(t)$ is gamma–function growing quicker than any exponent. Therefore $\psi_1(t)$ is asymptotic representation for the solution $x_1(t)$. Remark that $x_1(t) - \psi_1(t)$ is decreasing quicker than any exponent. Equations with constant coefficients have no quickly decreasing solutions. Hence RDEs with periodic coefficients can not be reduced in general case to the stationary equations by the change of variants unlike to ODEs.

1.6. *APPLICATION OF COMPARISON THEOREMS*

In certain cases we can obtain stability conditions by comparing the solutions of the initial problem with solutions of auxiliary problems. We give an example of this using the linear equation

$$\dot{x}(t) = A(t)x(t) + \int_{-\infty}^{0} [d_\theta R(t,\theta)]x(t+\theta), \quad t \geq t_0. \tag{1.20}$$

Here, $A = (a_{ij})$ is a continuous matrix function with $a_{ij} \geq 0$ for $i \neq j$. The kernel $R(t,\theta)$ and initial function ϕ satisfy the conditions A and B of Subs. 3.5.2.

Along with (1.20) we consider the inequality

$$\dot{z}(t) \geq A(t)z(t) + \int_{-\infty}^{0} [d_\theta V(t,\theta)]z(t+\theta), \quad t \geq t_0 \quad z(t) \in \mathbf{R}^n \tag{1.21}$$

(we write $u \geq v$ for $u, v \in \mathbf{R}^n$ if $u_1 \geq v_1, \ldots, u_n \geq v_n$). Here, the entry $v_{ij}(t,\theta)$ of the matrix V is equal to the variation of the function $r_{ij}(t,\cdot)$ on $[\theta, 0]$.

Theorem 1.7 If there is a function $z \in C(\mathbf{R}, \mathbf{R}^n)$ *such that* $z|_{[t_0,\infty)} \in C^1[t_0, \infty)$, $z_i(t) \geq |\phi_i(t)|$ $(t \leq t_0)$, $z_i(t) > 0$ $(t \geq t_0)$, $i = 1, \ldots, n$, *and satisfying (1.21) for all* $t \geq t_0$, *then* $|x_i(t)| \leq z_i(t)$ $(t \geq t_0)$, $i = 1, \ldots, n$, *where* x *is the solution of (1.20), (1.15).*

Proof Let $\varepsilon > 0$ be arbitrary. Let I be the identity matrix, and x^ε the solution of the equation

$$\dot{x}^\varepsilon(t) = [A(t) - \varepsilon I]x^\varepsilon(t) + \int_{-\infty}^{0} [d_\theta R(t,\theta)]x^\varepsilon(t+\theta), \quad t \geq t_0,$$

with initial condition $x^\varepsilon(t_0 + \theta) = (1-\varepsilon)\phi(t_0+\theta)(-\infty < \theta \leq 0)$, where $\varepsilon \in (0,2)$. We check that $|x_j^\varepsilon(t)| < z_j(t)$, $t \geq t_0$, $\forall j = 1, \ldots, n$.

This assertion holds for $t = t_0$. Suppose, on the contrary, that it does not hold for all $t > t_0$, and let $\tau > t_0$ be the first moment such that $|x_i^\varepsilon(\tau) > 0$.

Then

$$\dot{x}_i^\varepsilon(\tau) - \dot{z}_i(\tau) \le$$

$$\le -\varepsilon x_i^\varepsilon(\tau) + \sum_{j=1}^{n} \left[\int_{-\infty}^{0} x_j^\varepsilon(\tau+\theta) d_\theta r_{ij}(\tau,\theta) - \int_{-\infty}^{0} z_j(\tau+\theta)|d_\theta r_{ij}(\tau,\theta)| \right] \le$$

$$\le -\varepsilon x_i^\varepsilon(\tau) < 0.$$

This contradicts the definition of the moment τ. Now Thm. 3.2.3 implies $x(t) = \lim x^\varepsilon(t)$, $\varepsilon \to 0^+$. Hence $|x_i(t)| \le z_i(t)$, $i = 1, \ldots, n$.□

Example A sufficient condition for stability of (1.20) with $A \in C([t_0,\infty), \mathbf{R}^{n\times m})$, $a_{ij}(t) \ge 0$ $(i \ne j)$ is that

$$a_{ii}(t) \le -\sum_{j \ne i} a_{ij}(t) - \sum_{j=1}^{n} V_{(-\infty,0]} r_{ij}(t,\cdot), \quad i = 1, \ldots, n. \tag{1.22}$$

(It suffices to note that under the conditions (1.22) the assumptions of Thm. 1.8 hold for the condition $z(t) = c > 0$.)□

1.7. *STABILITY IN THE FIRST APPROXIMATION*

Here we consider the equation in (1.1) with righthand side disturbed by a functional $G(t, x_t)$:

$$\dot{x}(t) = f(t, x_t) + G(t, x_t), \quad t \ge t_0. \tag{1.23}$$

Here f is the same as in (1.2), and $G \in C([t_0,\infty) \times C[-h,0],\ \mathbf{R}^n)$ satisfies the local Lipschitz condition in the second argument and

$$|G(t,\psi)| \le \beta(\psi) \parallel \psi \parallel. \tag{1.24}$$

Theorem 1.8 ([298]). Assume that $\beta(\psi) \to 0$ as $\parallel \psi \parallel \to 0$ in (1.24), and that the trivial solution of (1.2) with $\phi = 0$ is uniformly asymptotically stable. Then the trivial solution of (1.23) is asymptotically stable.

If f is linear in the second argument, then (1.2) is called as an *equation of the first approximation* with respect to Eq. (1.24).

Example 1.8 We investigate the stability of the nonzero stationary solution $x_0 \equiv K$ of the logistic equation (2.5.1):

$$\dot{x}(t) = \gamma[1 - K^{-1}x(t-h)]x(t), \quad \gamma > 0, \quad K > 0, \quad h \ge 0. \tag{1.25}$$

We define a new variable $y(t)$ by

$$x(t) = K[1 + y(t)].$$

For $y(t)$ we obtain

$$\dot{y}(t) = -\gamma y(t-h) - \gamma y(t)y(t-h).$$

We will study the first approximation $\dot{y}(t) = -\gamma y(t-h)$ in Subs. 3.4(.1) (see Fig. 4.2.1). There we will show that this equation is exponentially stable, if only $2\gamma h < \pi$. By Thm. 1.9 this implies that the stationary solution $x_0 \equiv K$ of (1.25) is asymptotically stable if $\gamma h < \pi/2$. A more refined analysis [261] has shown that the solution $x_0 \equiv K$ is globally asymptotically stable under all admissible disturbances of the initial position satisfying $x(0) > 0$, if $\gamma h < 37/24$. For $\gamma h > \pi/2$ there is a nonconstant periodic solution of (1.25).□

Remark 1.3 Liapunov's direct method is used not only in stability theory, but also in other areas, such as solvability theorems, existence of stationary and periodic measures induced by a solution, asymptotic properties of solutions, etc. [288]. In these cases the properties of Liapunov functionals and the manners of their application can essentially differ from those described above.

E.g., in [83, 348] a Liapunov functional with discrete values is introduced for studying the exponential decay rate of the solution x of the scalar RDE

$$\dot{x}(t) = f(t, x(t), x(t-1)), \quad t \geq 0, \quad f(t, 0, 0) \equiv 0.$$

The *exponential decay rate* of x is defined as

$$\alpha(x) = \inf\{\alpha \in \mathbf{R} : \lim_{t\to\infty} e^{-\alpha t} x(t) = 0\}.$$

The solution x is called e *superexponential solution* if $\alpha(x) = -\infty$ and $\| x|_{[t,\infty)} \| > 0$ for all $t \in [0, \infty)$.

In this case the Liapunov functional V is defined as the number of zeros of x in some unit time interval. It has been proved [83] that a solution x is superexponential if and only if V tends to infinity as $t \to \infty$.□

1.8. *CASE OF NON-POSITIVE DERIVATIVE*

Construction of Liapunov functionals with negative definite derivative is connected with some difficulties. Sometimes it is easier to construct the functionals with non–positive derivative. Let us formulate some correspon- ding statements.

1.8.1 Consider the autonomous RDEs

$$\dot{x}(t) = f(x_t), \ t \geq 0, \ x_t(\theta) = x(t+\theta), \ \ -h \leq \theta \leq 0, \ \ f(0) = 0 \qquad (1.26)$$

with a local Lipschitz functional $f : C[-h, 0] \to \mathbf{R}^n$. The set $Q \subseteq C[-h, 0]$ is called an *invariant set* (for Eq.(1.26)) if for any $\phi \in Q$ there exists a function $x \in C^1(-\infty,\infty) : x_t \in Q$ for all $t \in (-\infty,\infty)$, $x_0 = \phi$, such that x satisfies Eq. (1.26) for all $t \in \mathbf{R}$.

Theorem 1.9 [238] Let there exist the continuous functional $V(\phi) : C[-h,0] \to \mathbf{R}$ such that

$$\omega_1(|\phi(0)|) \le V(\phi) \le \omega_2(\| \phi(\theta) \|),$$

$\omega_1(r) \to \infty$, as $r \to \infty$, and also $\dot V(x_t) \le 0$. Let Z be a set of those elements from $C[-h,0]$ for which $\dot V = 0$ and Q is the greatest invariant set situated in Z. Then all solutions of (1.26) tend to Q as $t \to \infty$. In particular if the set Q has the only zero element then the trivial solution of (1.26) is asymptotically stable.

1.8.2 Other approach is connected with the using of vector Liapunov functionals. Let $V, W : [t_0,\infty) \times C[-h,0] \to \mathbf{R}$ be continuous functionals. The derivative $\dot W$ along the solutions of (1.26) is called *integrally unbounded* in a set $G \subseteq Q_H (:= \{\psi \in C[-h,0] : \| \psi \| \le H\})$ if for any $B > 0$ there exist a number $T(B) > 0$ and a function $\xi \in C([0,\infty),\mathbf{R})$ such that $\dot W(t,x_t) \le \xi(t)$, $\int_t^{t+T(B)} \xi(s)ds \le -B$, for $x_t \in G$, $t \ge t_0$. Denote $\rho(\psi, G)$ the distance between $\psi \in C[-h,0]$ and the set G.

Theorem 1.10 [289.3] It is necessary and sufficient for uniform asymptotic stability of the trivial solution of (1.26) that there exist the functionals V and W such that for $t \ge t_0$, $\psi \in Q_H$

1) *$\omega_1(|\psi(0)|) \le V(t,\psi) \le \omega_2(\| \psi \|)$, $t \ge t_0$, $\psi \in Q_H$,*
2) *$\dot V(t,x_t) \le \overline{\omega}_3(x_t) \le 0$ where $\overline{\omega}_3$ is a continuous functional defined on Q_H,*
3) *$\sup\{|W(t,\psi)| : t \ge t_0,\ \psi \in Q_H\} < \infty$,*
4) *for any $\mu \in (0,H)$ there exists a $\rho > 0$ such that the derivative $\dot W$ is integrally unbounded in the set*

$$E(\mu,\rho) := \{\psi \in Q_H : d(\psi, Q(\overline{\omega}_3 = 0)) \le \rho,\ \| \psi \| \ge \mu\},$$

where $Q(\overline{\omega}_3 = 0) := \{\psi \in Q_H : \overline{\omega}_3(\psi) = 0\}$. In addition the attraction domain of the trivial solution contains the sphere Q_K, where $K < H$ and $\omega_2(K) \le \omega_1(H)$.

Example Consider the system of two scalar equations

$$\left.\begin{aligned} \dot x(t) &= -ax(t) + b(t)x(t-h) + c(t)y(t), \\ \dot y(t) &= -\gamma c(t)x(t),\ t \ge 0,\quad h \ge 0, \end{aligned}\right\} \tag{1.27}$$

where $b, c \in C[0,\infty)$. Show that sufficient conditions of its uniform asymptotic stability are $-a+|b(t)| \leq c_1 < 0$, $c_3 \geq |c(t)| \geq c_2 > 0$, $\gamma > 0$. Introduce the functional V:

$$V(t, x_t, y_t) := \gamma x^2(t) + y^2(t) + \gamma a \int_{t-h}^{t} x^2(s)ds.$$

We have

$$\dot{V}(t, x_t, y_t) \leq \overline{\omega}_3(x_t) := \gamma c_1(x^2(t) + x^2(t-h)) \leq 0.$$

The set $Q(\overline{\omega}_3 = 0) \subset C[-h,0]$ is in our case the set of such functions $\psi \in C[-h,0]$ that $\psi(0) = \psi(-h) = 0$. The set $Q(\overline{\omega}_3 = 0)$ does not coincide with the origin. As a second functional we take

$$W(t, x_t, y_t) := -y(t)x(t)\mathrm{sgn}c(t).$$

This functional is bounded in any sphere Q_H and also

$$\dot{W}(t, x_t, y_t) = -x(t)(-\gamma c(t)x^2(t))\mathrm{sgn}c(t) - y(t)\mathrm{sgn}c(t)\times$$

$$\times[-ax(t) + b(t)x(t-h) + c(t)y(t)].$$

The set $E(\mu,\rho)$ is defined by inequalities $|x(t)| \leq \rho$, $|x(t-h)| \leq \rho$, $\mu^2 \leq y^2(t) \leq H^2$. In the set $E(\mu,\rho)$ we have

$$\dot{W}(t, x_t, y_t) \leq -c_2\mu^2 + a\rho H + |b(t)|\rho H + \gamma c_3\rho^2 \leq -c_2\mu^2/2$$

where ρ is such that

$$a\rho H + |b(t)|\rho H + \gamma c_3\rho^2 \leq 2a\rho H + \gamma c_3\rho^2 \leq c_3\mu^2/2.$$

Hence the derivative $\dot{W}$ is integrally unbounded in the $E(\mu,\rho)$. Asymptotic stability of the trivial solution (1.27) follows now from Thm. 1.10.

2. LINEAR AUTONOMOUS RDEs

2.1. *LAPLACE TRANSFORMATION*

In this section we continue consideration of Sect. 3.6 and derive stability conditions for equations

$$\left.\begin{aligned} \dot{x}(t) &= \int_{-\infty}^{0} [dR(\theta)]x(t+\theta), \quad x(t) \in \mathbf{R}^n, \quad t \geq 0, \\ x(\theta) &= \phi(\theta), \quad -\infty < \theta \leq 0. \end{aligned}\right\} \tag{2.1}$$

We assume that the matrix entries r_{ij} of R are functions of bounded variation, and

$$\int_{-\infty}^{0} |\theta| \cdot |dr_{ij}(\theta)| < \infty, \quad i,j = 1,\ldots,n.$$

The initial functions are chosen to belong to the linear space of continuous functions ϕ with finite norm $\| \phi \|_0$ given by

$$\| \phi \|_0 = |\phi(0)| + \left(\int_0^{\infty} |F(t)|^2 dt \right)^{1/2} . \tag{2.2}$$

Here $|\phi(0)|$ denotes the euclidean norm in $\mathbf{R}^n$ and the integral

$$F(t) := \int_{-\infty}^{-t} [dR(\theta)]\phi(t+\theta).$$

is assumed to be absolutely convergent. Under these assumptions there exists a unique solution $t \mapsto x(t)(= x(t,\phi))$ of Pr. (2.1) such that

$$|x(t)| + |\dot{x}(t)| \leq c_1 \| \phi \|_0 \exp(c_2 t)$$

where c_i are some positive constants. Hence there are the Laplace transformations of x, F and R. Let $\widetilde{x}(z), \widetilde{F}(z)$ and $\widetilde{R}(z)$ be the Laplace transforms of these functions for Re$z > c_2$:

$$\widetilde{x}(z) := \int_0^{\infty} e^{-zt} x(t)dt, \quad \widetilde{F}(z) := \int_0^{\infty} e^{-zt} F(t)dt, \quad \widetilde{R}(z) := \int_{-\infty}^{0} e^{-z\theta} dR(\theta).$$

By (2.1)

$$[zI - \widetilde{R}(z)]\widetilde{x}(z) = \phi(0) + \widetilde{F}(z), \tag{2.3}$$

with I the identity matrix.

In fact, from (2.1) it follows that

$$\int_0^{\infty} e^{-zt}\dot{x}(t)dt = \int_0^{\infty} e^{-zt}dt \int_{-\infty}^{0} [dR(\theta)]x(t+\theta).$$

Integrating by parts

$$\int_0^{\infty} e^{-zt}\dot{x}(t)dt = z\widetilde{x}(z) - x(0).$$

Using Fubini theorem

$$\int_0^\infty e^{-zt}dt \int_{-\infty}^0 [dR(\theta)]x(t+\theta) = \int_0^\infty F(t)e^{-zt}dt +$$

$$+ \int_{-\infty}^0 [dR(\theta)] \int_{-\theta}^\infty e^{-zt}x(t+\theta)dt = \widetilde{F}(z) + \widetilde{R}(z)\widetilde{x}(z).$$

From here it follows (2.3).

The expression

$$\Delta(z) = \det[Iz - \widetilde{R}(z)]$$

is called a *characteristic function,* corresponding to Equation (2.1).

Remark 2.1 Usually, for retarded systems the equation $\Delta(z) = 0$ has infinitely many roots. However, there do exist RDEs (2.1) (called *degenerate RDEs*) whose characteristic function has only finitely many roots (see Subs. 3.6.1). It can be proved that in this case the function Δ is a polinomial of degree n. E.g., consider the equation $\dot{x}(t) = Ax(t) + Bx(t-h)$, $n = 3$. If

$$A = \begin{pmatrix} 0 & 1 & 0 \\ -0.5 & 0 & 1 \\ 0 & -0.5 & 0 \end{pmatrix}, \quad B = \begin{pmatrix} 0 & 1 & 0 \\ 0 & 0 & -1 \\ 0 & 0 & 0 \end{pmatrix},$$

then $\Delta(z) = z(z^2+1)$. If

$$A = \begin{pmatrix} 0 & 1 & 0 \\ 0 & 0 & 0 \\ 0 & 0 & 1 \end{pmatrix}, \quad B = \begin{pmatrix} -e & -e & -e^2 \\ 0 & e & e^2 \\ 1 & 0 & 0 \end{pmatrix},$$

then $\Delta(z) = z(z^2-1)$.

If the RDE is degenerate, then for sufficiently large t its solution space is finite dimensional, and we can give an equivalent ODE for each component of a solution. E.g., for the first example above such equations have the form

$$\ddot{x}_1 + x_1 = 0, \quad \ddot{x}_2 + x_2 = 0, \quad \dddot{x}_3 + \dot{x}_3 = 0. \quad \square$$

2.2. *STABILITY CONDITIONS*

The definitions of stability for (2.1) coincide with the definitions A–D in Subs. 1.1 and 1.3, with $\rho(\phi, 0)$ replaced by $\| \phi \|_0$. Note first of all that if the kernel $R(\theta)$ is a constant for $\theta \leq -h < 0$ then some stability conditions would follow immediately from Subs. 3.6.2, e.g., the following theorem holds.

Theorem 2.1 Suppose that $R(\theta) \equiv \text{const}$ *for* $\theta \leq -h < 0$ *and the characteristic function* $\Delta(z)$ *is zero-free in the halfplane* $Rez \geq 0$*. Then (2.1) is asymptotically stable. Moreover,* $|x(t)| \leq c_\phi \exp(-\alpha t)$*,* $\alpha > 0$*, if the equation* $\Delta(z) = 0$ *has no roots in the halfplane* $Rez \geq -\alpha$*.*

Let us consider some stability conditions for the general case.

Definition 2.1 Eq. (2.1) is called L_2*– stable* if

$$\int_0^\infty |x(t)|^2 dt < \infty$$

for any initial function $\phi, \| \phi \|_0 < \infty$.

Theorem 2.2 Let the characteristic function $\Delta(z)$ *have no zeros in the halfplane* $Rez \geq 0$*, and* $\phi(\theta)$ *and* $R(\theta)$ *satisfy conditions of Subs. 2.1. Then the trivial solution of Eq. (2.1) is* L_2*–stable.*

Proof By virtue of inversion theorem [316] we have for some $\gamma > 0$

$$x(t) = \frac{1}{2\pi i} \int_{\gamma - i\infty}^{\gamma + i\infty} \widetilde{x}(z) e^{zt} dt.$$

Under the assumptions of Thm. 2.2 the function $\widetilde{x}(z)$ is analytic for Re$z > 0$ and continuous for Re$z \geq 0$. Actually

$$\widetilde{x}(z) = [Iz - \widetilde{R}(z)]^{-1}[\phi(0) + \widetilde{F}(z)]. \tag{2.4}$$

Further the matrix $[Iz - \widetilde{R}(z)]^{-1}$ is analytic for Re$z > 0$ and continuous for Re$z \geq 0$ since function $\Delta(z)$ does not vanish for Re$z \geq 0$ and $\widetilde{R}(z)$ is analytic for Re$z > 0$ and continuous for Re$z \geq 0$. Hence using Cauchy theorem ([316], p.44)

$$x(t) = \frac{1}{2\pi} \int_{-\infty}^{\infty} \widetilde{x}(i\omega) e^{i\omega t} d\omega \tag{2.5}$$

By virtue of Plancherel theorem it is sufficient for L_2–stability that

$$\int_{-\infty}^{\infty} |\widetilde{x}(i\omega)|^2 d\omega < \infty \quad \text{if} \quad \| \phi \|_0 < \infty. \tag{2.6}$$

Show (2.6). Since the function $\Delta(z)$ has no zeros on imaginary axis and the matrix $\widetilde{R}(z)$ is bounded on it

$$\left. \begin{aligned} &\| [i\omega I - \widetilde{R}(i\omega)]^{-1} \| \leq c|\omega|^{-1}, \quad |\omega| \to \infty, \\ &\| [i\omega I - \widetilde{R}(i\omega)]^{-1} \| \leq c_1, \quad -\infty < \omega < \infty. \end{aligned} \right\} \tag{2.7}$$

From (2.7)

$$\int_{-\infty}^{\infty} \|[Ii\omega - \widetilde{R}(i\omega)]^{-1}\phi(0)|^2 d\omega < \infty. \tag{2.8}$$

From (2.7) and the Plancherel theorem

$$\int_{-\infty}^{\infty} \|[i\omega I - \widetilde{R}(i\omega)]^{-1}\widetilde{F}(i\omega)|^2 d\omega \le c_1^2 \int_{-\infty}^{\infty} |\widetilde{F}(i\omega)|^2 d\omega =$$

$$= c_1^2 2\pi \int_0^{\infty} |F(t)|^2 dt \le c_1^2 2\pi \parallel \phi \parallel_0^2 . \tag{2.9}$$

Using (2.4), (2.8) and (2.9) we obtain (2.6). Thm. 2.2 is proven. □

Theorem 2.3 Under assumptions of Thm. 2.2 the trivial solution of (2.1) is asymptotically stable.

Proof Show that for $\parallel \phi \parallel_0 < \infty$

$$x(t) \to 0, \quad t \to \infty. \tag{2.10}$$

From (2.6) it follows the validity of representation (2.5) where the integral is understood in the sense of the principal value. This and one theorem on Fourier integrals ([181], p.529) give us that relation (2.10) holds if each component $\widetilde{x}_j(i\beta)$ can be represented as a sum of two functions, the first of which is absolutely integrable on $(-\infty, \infty)$ and the second one is bounded, monotone for all sufficiently large $|\omega|$ and tends to zero for $|\omega| \to \infty$.

On the basis of Plancherel theorem, and (2.7),

$$\int_{-\infty}^{\infty} |(i\omega I - \widetilde{R}(i\omega))^{-1}\widetilde{F}(i\omega)| d\omega \le$$

$$\le (\int_{-\infty}^{\infty} |\widetilde{F}(i\omega)|^2 d\omega)^{1/2} (\int_{-\infty}^{\infty} \parallel i\omega I - \widetilde{R}(i\omega) \parallel^{-2} d\omega)^{1/2} \le$$

$$\le 2\pi (\int_0^{\infty} |F(t)|^2 dt)^{1/2} (\int_{-\infty}^{\infty} \parallel i\omega I - \widetilde{R}(i\omega) \parallel^{-2} d\omega)^{1/2} < \infty,$$

$$\parallel R \parallel := \sum_{i,j=1}^{n} |R_{ij}|.$$

This yields

$$\int_{-\infty}^{\infty} |i\omega I - \widetilde{R}(i\omega))^{-1}\widetilde{F}(i\omega)| d\omega < \infty.$$

Consider the function

$$q_j(\omega) := [(i\omega I - \widetilde{R}(i\omega))^{-1}x(0)]_j = \sum_{l=0}^{n}[A_{lj}(i\omega)/\Delta(i\omega)]x_l(0) =$$
$$= \sum_{l=0}^{n} a_{lj}(i\omega)^{n-1}x_l(0)/\Delta(i\omega) + O(|\omega|^{-2}) \quad (|\omega| \to \infty). \tag{2.11}$$

Here $A_{lj}(i\omega) = a_{lj}(i\omega)^{n-1} + O(|\omega|^{n-2})$ $(j, l = 1, \dots, n)$ are cofactors of the matrix $(i\omega I - \widetilde{R}(i\omega))$. If $\sum_{l=0}^{n} a_{lj}x_l(0) = 0$ then because of (2.11)

$$|q_j(\omega)| \le c_4|\omega|^{-2}, \quad \int_{-\infty}^{\infty} |q_j(\omega)|^2 d\omega < \infty.$$

If $\sum_{l=0}^{n} a_{lj}x_l(0) \neq 0$, then from (2.11) it follows that $q_j(\omega)$ is bounded and for large $|\omega|$

$$q_j(\omega) = [(i\omega)^{n-1}\sum_{l=0}^{n} a_{lj}x_l(0)]/[(i\omega)^n + O(|\omega|^{n-1})] + O(|\omega|^{-2}) =$$
$$= [\sum_{l=0}^{n} a_{lj}x_l(0)]/(i\omega) + O(|\omega|^{-2}).$$

Hence $|q_j(\omega)|$ is expanded in the sum of two functions: one of them for large $|\omega|$ is equal to $\omega^{-1}|\sum_{l=0}^{n} a_{lj}x_l(0)|$ and another is absolutely integrable. So (2.10) follows from [181].

Now asymptotic stability follows from (2.10) similar to Cor. 1.1 from Thm. 1.5. Thm. 2.3 is proven. □

Theorem 2.4 Let $\phi(\theta)$ and $R(\theta)$ satisfy conditions of Subs. 2.1 and for some $\gamma > 0$

$$\int_{-\infty}^{0} e^{-\gamma\theta}|dr_{ij}(\theta)| < \infty, \quad i.j = 1, \dots, n.$$

Assume also that $\Delta(z)$ has no zeros in the halfplane $Re z \ge -\gamma$. Then $|x(t, \phi)| \le C_\phi \exp(-\gamma t)$, $0 \le t < \infty$.

Proof Consider the function $t \mapsto y(t) := x(t)\exp(\gamma t)$. Substituting $y(t)$ in (2.1) we obtain equation of the same type with the characteristic function

$$\Delta_\gamma(z) = \det[(z-\gamma)I - \widetilde{R}(z-\gamma)].$$

This and Thm. 2.4 assumptions imply that $\Delta_\gamma(z)$ has no zeros in the halfplane $\mathrm{Re}z \geq 0$. So by virtue of Theorem 2.3 we get $\lim_{t\to\infty} y(t) = 0$. Thm. 2.4 is proven. □

3. STABILITY INVESTIGATION METHODS FOR LINEAR AUTONOMOUS RDEs

3.1. *INTRODUCTION*

Thms. 2.1–2.4 show that the stability depends essentially on the absence of the roots in the halfplane $\mathrm{Re}z \geq 0$ of the function $\Delta(z)$:

$$\Delta(z) =: \det[zI - \widetilde{R}(z)]. \tag{3.1}$$

Note that the roots of (3.1) are such that each halfplane $\mathrm{Re}z \geq$const contains only finitely many of them.

Many stability criteria for linear autonomous equations can be derived by methods of complex function theory (e.g., from Rouche's theorem, the argument principle, the residue theorem, etc.). Below we consider some of them.

Let C_R be a contour consisting of the interval $[-iR, iR]$ and the halfcircle $z = R\exp(i\psi)$, $-\pi/2 \leq \psi \leq \pi/2$, $R > 0$, $i^2 = -1$. Let L be the limit of the images of the contours C_R as $R \to \infty$ under the map $z \mapsto \Delta(z)$. The argument principle implies:

Theorem 3.1 Suppose L does not encircle the point $z = 0$. Then (2.1) is asymptotically stable.

3.2. *MIKHAILOV CRITERIUM*

Put $z = i\omega$, and represent $\Delta(z)$ as

$$\Delta(i\omega) = u(\omega) + iv(\omega). \tag{3.2}$$

The graph of the function (3.2) in the complex plane is usually called the *Mikhailov hodograph.*

Theorem 3.2 Equation (2.1) is asymptotically stable if and only if $|\Delta(i\omega)|^2$ $(= u^2(\omega) + v^2(\omega)) > 0$ $(\omega \in \mathbf{R})$, *while the variation of* $\arg\Delta(i\omega)$ *when* ω

varies from 0 to ∞ is equal to $n\pi/2$ ($n = \dim(2.1)$). The last condition is equivalent to the equality

$$\int_{\omega=0}^{\infty} K(\omega) = \frac{n\pi}{2}; \quad K(\omega) := \frac{u(\omega)dv(\omega) - v(\omega)du(\omega)}{u^2(\omega) + v^2(\omega)}. \tag{3.3}$$

Proof Proof of Thm. 3.2 follows also from the argument principle.
Remark 3.1 Choose a number s such that

$$\left| \int_{s}^{\infty} K(\omega) \right| < \pi/2$$

where $R(\omega)$ is defined by (3.3). Then instead of condition (3.3) it is sufficient to verify the inequality

$$\int_{0}^{s} K(\omega) > (n-1)\pi/2.$$

3.3. *SCALAR N - TH ORDER EQUATIONS*

Here we consider the problem of the stability of a scalar linear equation

$$\left. \begin{aligned} x^{(n)}(t) &= \sum_{j=0}^{n-1} \int_{-\infty}^{0} x^{(j)}(t+\theta)dr_j(\theta), \quad t \geq 0, \\ x^{(j)}(t) &:= \frac{d^j x}{dt^j}, \quad x(t) \in \mathbf{R}. \end{aligned} \right\} \tag{3.4}$$

Put

$$\alpha_{lj} := \int_{-\infty}^{0} |\theta|^l |dr_j(\theta)| (\leq \infty), \quad \beta_{lj} := \int_{-\infty}^{0} |\theta|^l dr_j(\theta) \quad (\text{if} \quad \alpha_{lj} < \infty).$$

Since (3.4) can be reduced to a system of the form (2.1), here we will only consider the problem of when the assumptions of Thm. 2.3 are valid for this system. Characteristic function Δ_n of (3.4) has the form

$$\Delta_n(z) = z^n - \sum_{j=0}^{n-1} z^j \int_{-\infty}^{0} e^{z\theta} dr_j(\theta).$$

3.3.1 Case $n = 1$. (a) We prove that if $\beta_{00} < 0,\ \alpha_{10} < 1$, then $\Delta_1(z)$ has no zeros with $\mathrm{Re}z \geq 0$. Let $z = \alpha + i\beta$, and write $\Delta_1(z) = 0$ as

$$\mathrm{Im}\Delta_1(z) = \beta - \int_{-\infty}^{0} e^{\alpha\theta} \sin \beta\theta dr_0(\theta) = 0, \tag{3.5}$$

$$\mathrm{Re}\Delta_1(z) = \alpha - \int_{-\infty}^{0} e^{\alpha\theta} \cos \beta\theta dr_0(\theta) = 0. \tag{3.6}$$

If $\beta \neq 0$, then (3.5) implies

$$1 - \int_{-\infty}^{0} e^{\alpha\theta} \frac{\sin \beta\theta}{\beta\theta} \theta dr_0(\theta) = 0.$$

This equation is root–free for $\alpha \geq 0$, since its lefthand side is positive for $\alpha \geq 0$. However, $\beta = 0,\ \alpha \geq 0$ Eq. (3.6) is root–free because the function

$$\Phi(\alpha) := \alpha - \int_{-\infty}^{0} e^{\alpha\theta} dr_0(\theta) = \mathrm{Re}\Delta_1(z)|_{\beta=0}$$

satisfies the estimates

$$\Phi(0) = -\beta_{00} > 0, \quad \frac{d\Phi(\alpha)}{d\alpha} = 1 - \int_{-\infty}^{0} \theta e^{\alpha\theta} dr_0(\theta) \geq 1 - \alpha_{10} > 0.$$

Hence, under the conditions $\beta_{00} < 0,\ \alpha_{10} < 1$ the Eqs. (3.5) and (3.6) cannot be simultaneously satisfied in the domain $\mathrm{Re}z \geq 0$. □

(b) Suppose r_0 is nonincreasing, $r_0(\theta) \equiv$const for $\theta < -h \leq 0$ and $\beta_{00} < 0,\ h\alpha_{00} < \pi/2$. Then the function $\Delta_1(z)$ has n_0 zeros with $\mathrm{Re}z \geq 0$.

Proof Put $z = \alpha + i\beta$, and write the equation $\Delta_1(z) = 0$ in the equivalent form (3.5), (3.6). We show that these equations cannot be simultaneously satisfied for $\alpha \geq 0$. If $|\beta| < \pi/(2h),\ \alpha \geq 0$, then (3.6) has no roots, since

$$\alpha - \int_{-\infty}^{0} e^{\alpha\theta} \cos \beta\theta dr_0(\theta) \geq -\int_{-h}^{0} e^{\alpha\theta} \cos \beta\theta dr_0(\theta) \geq -\beta_{00} \cos \beta h > 0.$$

If $|\beta| \geq \pi/(2h)$, then (3.5) has no roots, since

$$\left| \int_{-\infty}^{0} e^{\alpha\theta} \sin \beta\theta dr_0(\theta) \right| \leq \left| \int_{-h}^{0} dr_0(\theta) \right| < \frac{\pi}{2h}.$$

Hence (3.5) and (3.6) are incompatible for $\alpha \geq 0$. Note that the bound $h\alpha_{00} < \pi/2$ is sharp (see Subs. 2.6.3).□

3.3.2 Case $n = 2$. (a) If the function r_1 has a jump $a_0 < 0$ at the point $\theta = 0$, and $a_{11} < 2$, $\beta_{00} < 0$,

$$\gamma := a_0 + \alpha_{10} + \int_{-\infty}^{0^-} |dr_1(\theta)| < 0,$$

then $\Delta_2(z)$ has no zeros with $\mathrm{Re}z \geq 0$.

Proof Put $z = \alpha + i\beta$. If $\beta = 0$, $\alpha \geq 0$, then

$$\mathrm{Re}\Delta_2(z)(= \Delta_2(z)) = \alpha^2 - \alpha \int_{-\infty}^{0} e^{\alpha\theta} dr_1(\theta) - \int_{-\infty}^{0} e^{\alpha\theta} dr_0(\theta) =$$

$$= \alpha^2 - \alpha a_0 - \alpha \int_{-\infty}^{0^-} e^{\alpha\theta} dr_1(\theta) - \int_{-\infty}^{0} (e^{\alpha\theta} - 1) dr_0(\theta) - \int_{-\infty}^{0} dr_0(\theta) \geq$$

$$\geq \alpha^2 - \alpha \left(a_0 + \int_{-\infty}^{0^-} |dr_1(\theta)| + \alpha_{10} \right) - \beta_{00} > 0.$$

On the other hand, for $\beta \neq 0$, $\alpha \geq 0$,

$$\frac{\mathrm{Im}\Delta_2(z)}{\beta} =$$

$$= 2\alpha - \alpha \int_{-\infty}^{0} e^{\alpha\theta} \frac{\sin \beta\theta}{\beta} dr_1(\theta) - \int_{-\infty}^{0} (\cos \beta\theta) e^{\alpha\theta} dr_1(\theta) - \int_{-\infty}^{0} e^{\alpha\theta} \frac{\sin \beta\theta}{\beta} dr_0(\theta) \geq$$

$$\geq \alpha \left(2 - \int_{-\infty}^{0} |\theta| \cdot |dr_1(\theta)| \right) - \gamma > 0.$$

So the equations $\mathrm{Re}\Delta_2(z) = 0$ and $\mathrm{Im}\Delta_2(z) = 0$ are incompatible for $\alpha \geq 0$.□

(b) Assume that

$$\beta_{00} < 0, \quad 2\alpha_{11} + \alpha_{20} < 2, \quad \beta_{10} + \beta_{01} < 0. \tag{3.7}$$

Then the function $\Delta_2(z)$ has no zeros with $\mathrm{Re}z \geq 0$.

Proof Put $\Delta_2(z) = \psi_1(z) + \psi_2(z)$, where

$$\psi_1(z) := z^2 - z(\beta_{01} + \beta_{10}) - \beta_{00},$$

$$\psi_2(z) := -z \int_{-\infty}^{0} (e^{z\theta} - 1)dr_1(\theta) - \int_{-\infty}^{0} (e^{z\theta} - 1 - z\theta)dr_0(\theta).$$

Under conditions (2.7) the functions ψ_1 has no zeros with Re$z \geq 0$,

$$|\psi_2(z)| \leq |z|^2 \left(\alpha_{11} + \frac{1}{2}\alpha_{20}\right).$$

Consequently, on the boundary of the domain Re$z \geq 0$ the inequality $|\psi_1(z)| > |\psi_2(z)|$ holds. Rouche's theorem now implies that the conditions (3.7) are sufficient for Δ_2 to be zero-free in the halfplane Re$z \geq 0$. □

3.4. *EQUATIONS WITH DISCRETE DELAYS*

Here we will consider Eq. (2.1) with a piecewise constant kernel $R(\theta)$ which may have jumps at finitely many points. In this case the characteristic function (2.3), now also called a quasipolynomial, can be represented as

$$\Delta(z) = z^n + \sum_{l=0}^{n-1} \sum_{j=1}^{m(l)} a_{lj} z^l e^{-zh_{lj}}, \quad h_{lj} \geq 0. \tag{3.8}$$

Numerous studies have been devoted to the investigation of the zeros of the quasipolynomial [42, 207, 238, 287.1, 289.1–289.3, 471].

As has been pointed out in Subs. 3.6.1, all zeros of the function (3.8) lie in the halfplane Re$z \leq \alpha$, for a certain α. In general, only finitely many zeros, located near the origin, can be found either approximately or numerically (these are called *nonasymptotic zeros*). The *asymptotic zeros* of the function (3.8) are such that Re$z_k \to -\infty$ as $r \to \infty$, and are distributed along a finite number of chains. Any sector containing the imaginary axis also contains all zeros of (3.8), except finitely many.

Not that the distance between two asymptotic roots of the equation $\Delta(z) = 0$ with a single delay is always larger than a positive number. In general, however, the distance between two zeros of (3.8) may become arbitrarily small.

We will describe some methods for determining conditions under which (3.8) has no roots in the halfplane Re$z \geq 0$.

3.4.1 The method of D-subdivision. This method has already been described in Subs. 3.6.1. Recall that it requires one to consider in the parameter (a_{lj}, h_{lj})-space of the function (3.8) all hypersurfaces whose points correspond to at least one pure imaginary root of this function. This procedure was called *D–subdivision.* Now assume that the parameters a_{lj}, h_{lj} vary. Since the zeros of (3.8) depend continuously on the parameters, the number of zeros with $\mathrm{Re}\, z \geq 0$ can only change when certain zeros cross the imaginary axis. Hence, each domain in the D–subdivision corresponds to quasipolynomials with the same number of such zeros, counted with their multiplicity. In particular, the domain without zeros with $\mathrm{Re}\, z \geq 0$ is the stability region.

We will consider some concrete examples.

1) Let us find the stability domain for the scalar equation

$$\dot{x}(t) + ax(t) + bx(t-h) = 0. \tag{3.9}$$

The characteristic quasipolynomial of this equation is

$$\Delta(z) = z + a + be^{-hz}.$$

The equality $\Delta(i\omega) = 0$ implies that for fixed $h > 0$ the boundaries of the domains in a D–subdivision of the (a, b)–plane are formed by the straight line $a + b = 0$ and the parametric curve $a = -\omega \coth h\omega$, $b = \omega / \sin(h\omega)$, where $\omega \in \mathbf{R}$, which has infinitely many branches (see Fig. 4.3.1, in which one branch is depicted and the domain of asymptotic stability is shaded).

For given coefficients a and b, (3.9) is unstable if $a + b < 0$ and stable if $a + b > 0$, $a > b$. Suppose $a + b > 0$, $b > a$, i.e. $b > |a|$. Put

$$h_0 := q^{-1} \arccos\left(-\frac{a}{b}\right), \quad q := (b^2 - a^2)^{1/2},$$

where $\arccos(-a/b) \in (0, \pi)$. Then, for $h < h_0$ all roots of the equation $\Delta(z) = 0$ have negative real parts; for $h = h_0$ this equation has two purely imaginary roots, $\pm iq$; and the equation acquires two additional roots with positive real parts if h passes through the values $h_0 + 2\pi j$, $j = 0, 1, \ldots$, in increasing direction.

Note that by replacing t by ht we can take $h = 1$ in (3.9).

2) The stability domain of the equation

$$\ddot{x}(t) + a\dot{x}(t-1) + bx(t-1) = 0$$

is depicted in Fig. 4.3.2, in which the upper bound (the curve q) is given by the relations $a = \omega \sin \omega$, $b = \omega^2 \cos \omega$, $0 \leq \omega \leq \pi/2$. Note that the slope of the curve q (i.e. db/da) is equal to 1 for $a = 0$ and to $-(\pi/2)^2$ for $a = \pi/2$.

3) The stability domain of the equation

$$\ddot{x}(t) + a\dot{x}(t-1) + bx(t) = 0$$

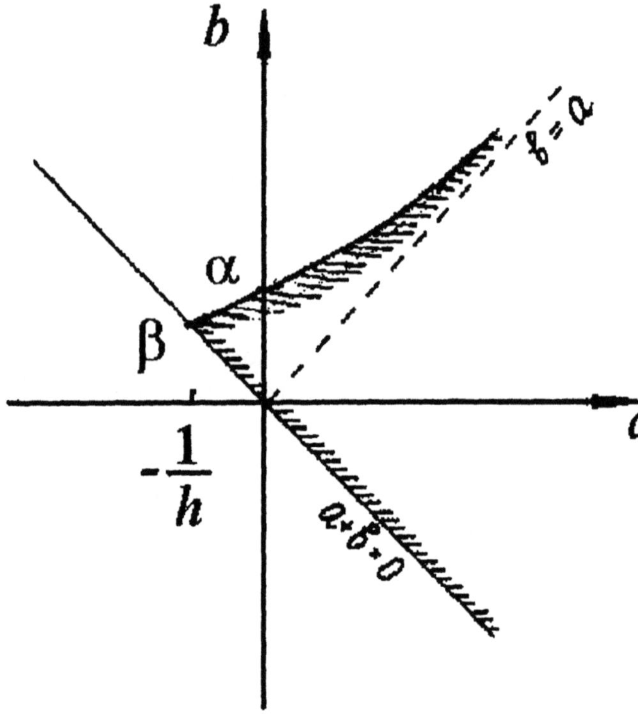

Figure 4.3.1. $\beta=(-\frac{1}{h},\frac{1}{h})$, $\alpha=(0,\frac{\pi}{2h})$. Stability domain for the equation $\dot{x}(t)+ax(t)+bx(t-h)=0$

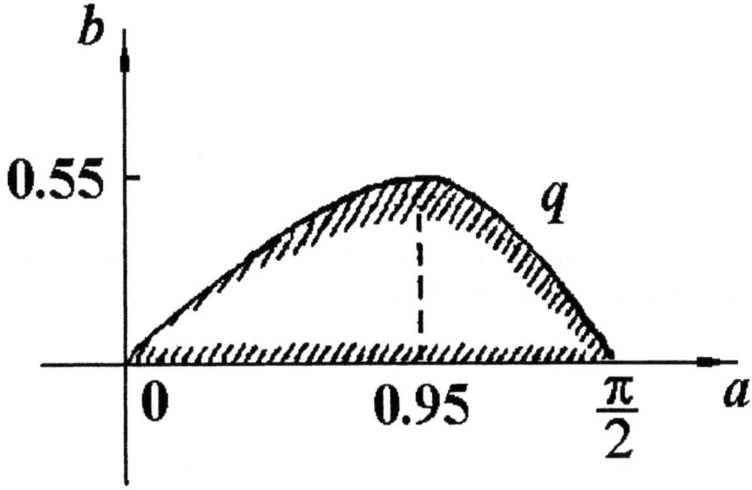

Figure 4.3.2. Stability domain for the equation
$\ddot{x}(t)+a\dot{x}(t-1)+bx(t-1)=0$

is depicted in Fig. 4.3.3. It consist of a sequence of triangles, formed by parts of the a– and the b–axis and the line segments

$$b = (-1)^{l+1}\frac{2l+1}{2}\pi a + \left(\frac{2l+1}{2}\pi\right)^2, \quad l = 0, 1, \dots$$

The points β_l have coordinates

$$\beta_l = \left((-1)^{l+1}\pi, \frac{(2l+1)(2l+3)}{4}\pi^2\right), \quad l = 0, 1, \dots$$

4) The stability domain of the equation

$$\ddot{x}(t) + a\dot{x}(t-1) + bx(t-1) = 0$$

is depicted in Fig. 4.3.4. The boundary curve q is given by the parametric equations

$$a = \omega \tan\omega, \quad b = \frac{\omega^2}{\cos\omega}, \quad 0 \le \omega < \infty.$$

The curve q is downwards convex, has slope $db/da = 1$ for $a = 0$, and tends to the straight line with slope $\pi/2$ as $\omega \to \pi/2$ (the dashed line in Fig. 4.3.4).

5) The equation

$$\ddot{x}(t) + 2\dot{x}(t-1) + x(t-2) + \alpha x(t) = 0$$

is stable for $-1 < \alpha < \alpha_0$, $\alpha_{2l-1} < \alpha < \alpha_{2l}$, $l = 1, 2, \dots$, where

$$\alpha_j = \left(\frac{\pi}{2} + j\pi - (-1)^j\right)^2, \quad j = 0, 1, \dots.$$

3.4.2 Generalization of the Routh–Hurwitz conditions. Chebotarev's theorem. We write the quasipolynomial (3.8) as the product of a quasipolynomial $\Delta_1(z)$ and the function $z \mapsto \exp(-zh_{pq})$, where $h_{pq} = \max_{l,j} h_{lj}$, i.e.

$$\Delta(z) = e^{-zh_{pq}}\Delta_1(z).$$

By expanding all exponents in Taylor series we obtain

$$\Delta_1(z) = \sum_{j=0}^{\infty} a_j z^j.$$

It is obvious that the zeros of $\Delta(z)$ coincide with those of $\Delta_1(z)$. We further consider functions $y \mapsto u(y)$ and $y \mapsto v(y)$, $y \in \mathbf{R}$, such that $\Delta_1(iy) = u(y) + iv(y)$,

$$u(y) = \sum_{j=0}^{\infty}(-1)^j a_{2j} y^{2j}, \quad v(y) = \sum_{j=0}^{\infty}(-1)^j a_{2j+1} y^{2j+1},$$

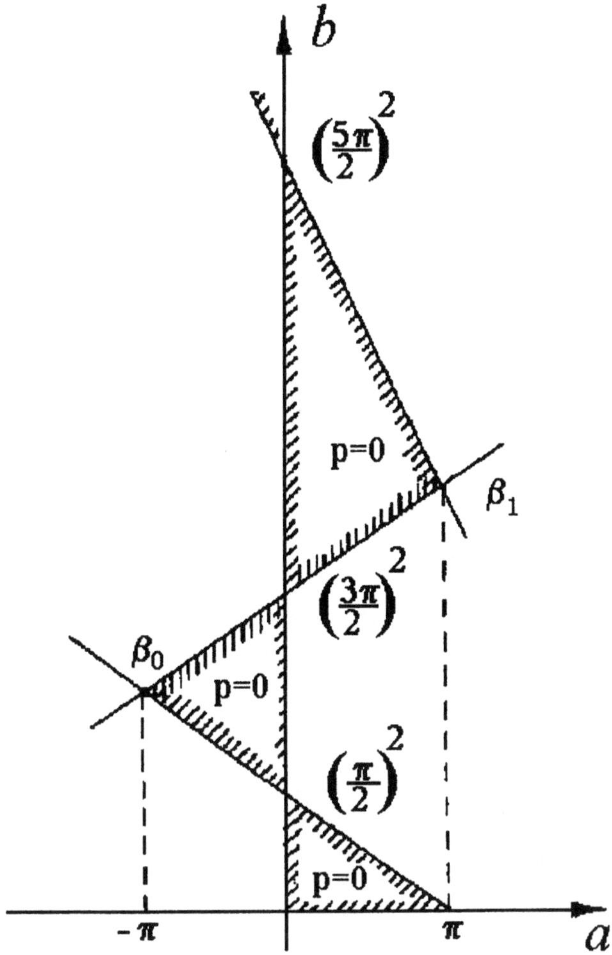

Figure 4.3.3. Stability domain for the equation
$\ddot{x}(t)+a\dot{x}(t-1)+bx(t)=0$

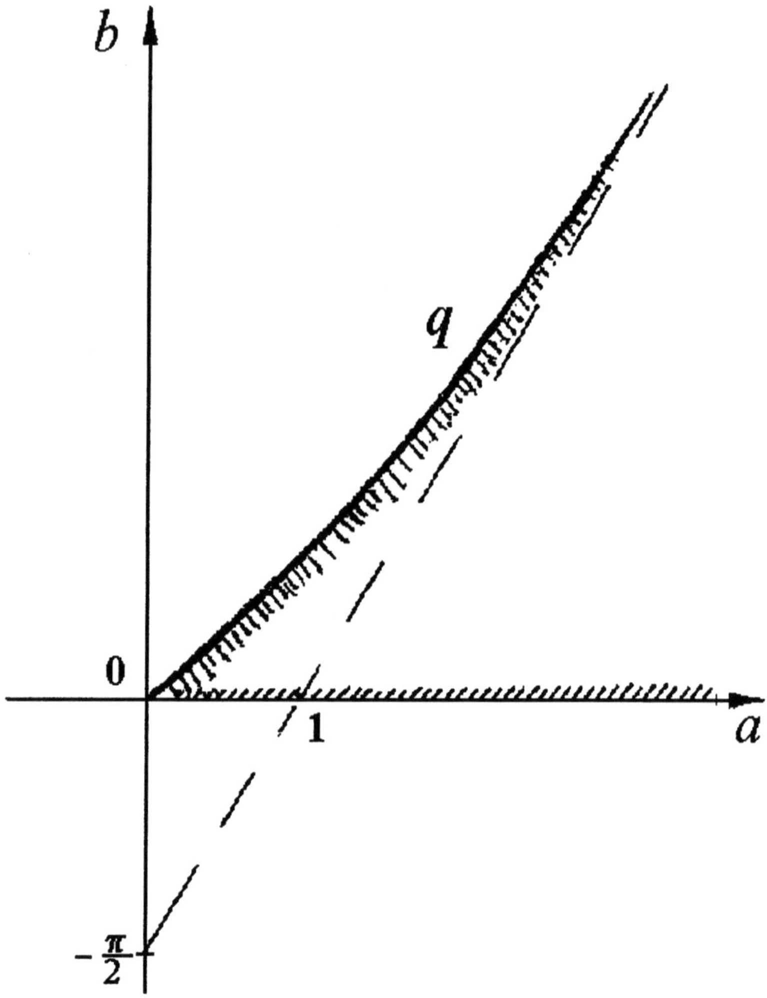

Figure 4.3.4. Stability domain for the equation

$$\ddot{x}(t)+a\dot{x}(t)+bx(t-1)=0$$

and define the matrices $Q_j := (a_{2k-i})_{i,k=1}^{j}$, $j = 1, 2, \ldots$, and put $a_{2k-i} := 0$ for $2k - i < 0$. E.g.,

$$Q_1 = (a_1), \quad Q_2 = \begin{pmatrix} a_1 & a_3 \\ a_0 & a_2 \end{pmatrix}, \quad Q_3 = \begin{pmatrix} a_1 & a_3 & a_5 \\ a_0 & a_2 & a_4 \\ 0 & a_1 & a_3 \end{pmatrix}.$$

Theorem 3.3 [99] Suppose the functions v and u do not have common zeros. Then the function (3.8) has no roots in the halfplane $\mathrm{Re}z \geq 0$ *if and only if* $\det Q_j > 0$ *for all* $j = 1, 2, \ldots$.

Consequently every inequality $Q_j > 0$ is a necessary condition for the asymptotic stability of an RDE with characteristic quasipolynomial (2.8).

Remark 2.2 We stress that for arbitrary quasipolynomials Thm. 3.3 is not true, in general. The following example illustrates this. Consider equation (3.9) with $h = 1$, $a+b > 0$ and stability domain as depicted in Fig. 4.2.1. The quasipolynomial corresponding to (2.9) is $\Delta(z) = z + a + be^{-z}$. Expanding e^{-z} in a Taylor series, we find

$$\Delta(z) = (a+b) + (1-b)z + \frac{b}{2!}z^2 - \frac{b}{3!}z^3 + \ldots$$

Thm. 2.4 implies

$$1 - b > 0, \quad \begin{vmatrix} 1-b & -b/6 \\ a+b & b/2 \end{vmatrix} = \frac{b}{6}(3 - 2b + a) > 0.$$

If $a > -1$ this yields that a necessary condition for asymptotic stability would be $0 < b < 1$. However, this is not true, since for $a = 1$, $b = -0.1$ all roots of the equation $\Delta(z) = 0$ belong to the left halfplane. The reason for this mistake is that we have not passed from $\Delta(z)$ to $\Delta_1(z)$ (see above).□

3.4.3 Quasipolynomial with commensurate delays. If the delays h_{lj} in (3.8) are commensurate, then by a linear change of t in (2.1) the quasipolynomial (3.8) can be written in the form

$$\Delta(z) = z^n + \sum_{l=0}^{n-1} \sum_{j=1}^{m(l)} a_{lj} z^l e^{-jz}.$$

Again we introduce functions u and v by $\Delta_1(i\omega) = u(\omega) + iv(\omega)$, where $\omega \in \mathbf{R}$. The following theorem is a generalization of the Hermite – Biler theorem, and has been proved in [99].

Theorem 3.4 The function $\Delta(z)$ *has no roots in the halfplane* $\mathrm{Re}z \geq 0$ *if one of the following conditions holds:*

1) all zeros of the functions u or v, analytically continued to the whole complex plane, are real and simple, and for each zero ω the relation

$$\dot{u}(\omega)v(\omega) - \dot{v}(\omega)u(\omega) < 0, \quad \dot{u} := \frac{du(\omega)}{d\omega} \tag{3.10}$$

holds;

2) all zeros of the functions u and v are real, simple, alternate, and for at least one ω the relation (3.10) holds;

3) all zeros of the function u are real, and for each zero ω the inequality

$$\dot{u}(\omega)v(\omega) < 0$$

holds;

4) all zeros of the function v are real, and for each zero ω the inequality

$$\dot{v}(\omega)u(\omega) > 0$$

holds.

Example 3.1 Consider, as example, the equation

$$\dot{x}(t) = -bx(t-h), \quad b > 0. \tag{3.11}$$

The quasipolynomial $\Delta_1(a) = z\exp(hz) + b = 0$ corresponds to Eq. (3.11). By setting $z = i\omega$ we get

$$\Delta_1(i\omega) = b - \omega\sin\omega h + i\omega\cos\omega h = u(\omega) + iv(\omega),$$

$$u(\omega) = b - \omega\sin\omega h, \quad v(\omega) = \omega\cos\omega h.$$

For further investigation let us use variant *4)* of Thm. 3.4. Zeros of the function v equal $\tilde{\omega}_0 = 0, \quad \omega_{k+1} = (k\pi + \pi/2)h; \quad k = 0, \pm 1, \ldots$. So all the zeros of $v(\omega)$ are real. Then using the relation

$$\dot{v}(\omega)u(\omega) = (\cos\omega h - h\omega\sin\omega h)(b - \omega\sin\omega h)$$

we obtain

$$\dot{v}(\tilde{\omega}_0)u(\tilde{\omega}_0) = b > 0$$

$$\dot{v}(\omega_k)u(\omega_k) = -(2k+1)\frac{\pi}{2}(-1)^k[b - (2k+1)\frac{\pi}{2h}(-1)^k].$$

The last expression is positive if

$$b(2k+1)(-1)^k < (2k+1)^2\frac{\pi}{2h} \quad (k = 0, \pm 1, \ldots).$$

Necessary and sufficient condition for this is $b < \pi/(2h)$, i.e. $bh < \pi/2$, which coincides with that of in Fig. 4.3.1. So the trivial solution of Equation (3.11) is asymptotically stable for $2bh < \pi$.

Similarly we can prove that equation $\dot{x}(t) = ax(t)+bx(t-h)$ is asymptotically stable if $a < h^{-1}$, $a < -b < (a^2 + p^2h^{-2})^{1/2}$, where p is the root of equation $p = ah^{-1}\tan p$ such that $0 < p < \pi$. If $a = 0$ we set $p = \pi/2$.

3.4.4 Stability for arbitrary delays. Few papers have been devoted to conditions under which there is stability for arbitrary delays [11, 33, 56, 192, 279, 280, 290, 289.1, 289.3]. We will consider such conditions for a scalar equation

$$x^{(n)}(t) + \sum_{l=1}^{n} \left[a_l x^{(n-l)}(t) + \sum_{j=1}^{m} b_{lj} x^{(n-l)}(t-h_j) \right] = 0, \tag{3.12}$$

where a_l, b_{lj} and $h_j \geq 0$ are certain constants. The quasipolynomial $\Delta(z)$ corresponding to this equation is

$$\left.\begin{aligned} \Delta(z) &= P(z) + \sum_{j=1}^{m} Q_j(z) e^{-h_j z}, \\ P(z) &= z^n + \sum_{l=1}^{n} a_l z^{n-l}, \quad Q_j(z) = \sum_{l=1}^{n} b_{lj} z^{n-l}. \end{aligned}\right\} \tag{3.13}$$

Theorem 3.5 ([289.3]) Assume that $|b_{n1}| + |b_{n2}| + \ldots + |b_{nm}| < |a_n|$. *Then the Eq. (3.12) is asymptotically stable for all delays* $h_j \geq 0$ *if and only if the polynomial* $P(z)$ *is asymptotically stable (i.e.* Re$z < 0$ *for all the roots of the polynomial* $P(z)$*) and*

$$\sum_{j=1}^{m} |Q_j(i\omega)| < |P(i\omega)|, \quad i^2 = -1, \quad \omega > 0.$$

Example 3.2 Consider the equation

$$\ddot{x}(t) + a_1\dot{x}(t) + a_2 x(t) = b\dot{x}(t-h).$$

Here $P(z) = z^2 + a_1 z + a_2$, $Q(z) = bz$. By Thm. 3.5, this equation is stable for arbitrary delays $h \geq 0$ if and only if $a_1 > |b|$, $a_2 > 0$.□

The problems of stability for arbitrary delays of the system

$$\dot{x}(t) = Ax(t) + Bx(t-\tau), \quad x(t) \in \mathbf{R}^n$$

was investigated in [330]. It was proven that if the matrix A is stable and all the roots λ_j of equation

$$\begin{vmatrix} \lambda A - B & -\lambda\mu E \\ -\lambda\mu E & \lambda A - B \end{vmatrix} = 0$$

lie inside the unit circle for all real μ then the system under consideration is stable for all delays $\tau \geq 0$.

3.4.5 Robust stability. In [11,33,279] the following *robust stability problems* for a class of *uncertain delay equations* were considered. It was assumed that the coefficients of the equations are known only within certain bounds. In general, stability of uncertain FDEs is not implied by checking their stability for extremal values of coefficients. However, some boundary theorems are valid for Eqs. (3.12) with fixed constant delays and

$$\underline{a}_l \leq a_l \leq \overline{a}_l, \quad \underline{b}_{lj} \leq b_{lj} \leq \overline{b}_{lj}, \tag{3.14}$$

where $\underline{a}_l, \overline{a}_l, \underline{b}_{lj}$, and $\overline{b}_{lj}$ are given constants.

We introduce four polynomials $P_k(z), k = 1,2,3,4$, associated with the polynomial $P(z)$ in (3.13). The coefficients of $P_1(z)$ are:

$$a_{n-2l} = \begin{cases} \overline{a}_{n-2l} & \text{if } l \text{ is even,} \\ \underline{a}_{n-2l} & \text{if } l \text{ is odd,} \end{cases} \qquad a_{n-2l-1} = \begin{cases} \overline{a}_{n-2l-1} & \text{if } l \text{ is even,} \\ \underline{a}_{n-2l-1} & \text{if } l \text{ is odd.} \end{cases}$$

The coefficients of $P_2(z)$ are:

$$a_{n-2l} = \begin{cases} \underline{a}_{n-2l} & \text{if } l \text{ is even,} \\ \overline{a}_{n-2l} & \text{if } l \text{ is odd,} \end{cases} \qquad a_{n-2l-1} = \begin{cases} \underline{a}_{n-2l-1} & \text{if } l \text{ is even,} \\ \overline{a}_{n-2l-1} & \text{if } l \text{ is odd.} \end{cases}$$

The coefficients of $P_3(z)$ are:

$$a_{n-2l} = \begin{cases} \underline{a}_{n-2l} & \text{if } l \text{ is even,} \\ \overline{a}_{n-2l} & \text{if } l \text{ is odd,} \end{cases} \qquad a_{n-2l-1} = \begin{cases} \overline{a}_{n-2l-1} & \text{if } l \text{ is even,} \\ \underline{a}_{n-2l-1} & \text{if } l \text{ is odd.} \end{cases}$$

The coefficients of $P_4(z)$ are:

$$a_{n-2l} = \begin{cases} \overline{a}_{n-2l} & \text{if } l \text{ is even,} \\ \underline{a}_{n-2l} & \text{if } l \text{ is odd,} \end{cases} \qquad a_{n-2l-1} = \begin{cases} \underline{a}_{n-2l-1} & \text{if } l \text{ is even,} \\ \overline{a}_{n-2l-1} & \text{if } l \text{ is odd.} \end{cases}$$

We similarly define four polynomials $Q_{jk}(z)$ for each $Q_j(z)$.

Theorem 3.6 ([289.3]) Assume that

$$\sum_{j=1}^{m} \max\{|\underline{b}_{nj}|, |\bar{b}_{nj}|\} < \min |\underline{a}_n|, |\bar{a}_n|\}.$$

Then the Eqs. (3.12), (3.14) are asymptotically stable if and only if each of the 4^{m+1} *"vertex" quasipolynomials*

$$P_k(z) + \sum_{j=1}^{m} Q_{jl_j}(z) e^{-h_j z}, \quad k, l_j = 1, 2, 3, 4,$$

is asymptotically stable.

Other robust stability conditions were obtained for Eq. (3.12) with known delays in which b_{lj} are replaced on $\gamma\lambda_l b_{lj}$. It is assumed that γ is a constant and

Theorem 3.7 [417] *Modified Eq. (3.12) with condition* $\max_{l=1,\dots,m} |\lambda_l| \le 1$ *is asymptotically stable if and only if the polynomial* $P(z)$ *is asymptotically stable and for all* $\omega \ge 0$

$$\max_{1\le j\le m} \frac{\left|\operatorname{Im} \dfrac{P(i\omega)}{Q_j(i\omega)}\right|}{\sum_{k=1}^{m} \left|\operatorname{Im} \dfrac{Q_k(i\omega)}{Q_j(i\omega)}\right|} > \gamma, \quad i^2 = -1.$$

4. RAZUMIKHIN'S METHOD

4.1. *INTRODUCTION*

B.S. Razumikhin proposed in 1956 a method to investigate stability of solutions of systems with delays. This method is based on application of Liapunov functions in combination with the general concept of "impossibility of the first breakdown" (see Subs. 4.6). This method was developed later both by Razumikhin [424] and by other authors (see, in particular, [238]). Sometimes this method is opposed to N.N. Krasovskii's method of functionals that seems to be unjustified (see Subs. 4.8).

Razumikhin's method essentially extends stability theorems in Liapunov's sense. In this case, it is necessary to pass from "guiding Liapunov functions" to arbitrary guiding functions.

These and related questions will be considered here. Since only basic concepts will be discussed, the statements will be presented far from a general form known now.

4.2. *GUIDING FUNCTIONS FOR SYSTEMS WITHOUT DELAY*

Let the state of some autonomous system be defined by the point $x = (x_1, \dots, x_n)$ of n–dimensional domain D, whereas its evolution is governed by the differential law, i.e. by the vector ODE

$$\dot{x} = f(x) \tag{4.1}$$

with smooth (i.e. continuously differentiable) righthand side. The solution $t \mapsto x(t)$ of Eq. (4.1) are interpreted as trajectories in the phase space D.

Let $V : D \to \mathbf{R}$ be a given smooth function, i.e. $V(x)$ is a scalar characteristic of a considered physical system at state x. Then the change rate of this characteristic at state x is a derivative of the composite function and is equal to:

$$dV(x(t))/dt = \sum_{i=1}^{n} V_{x_i}(x)\dot{x}_i = (\text{grad}V(x), f(x)),$$

where grad $V(x) := (V_{x_1}(x), \dots, V_{x_n}(x))$, and $x = x(t)$ is any solution of Eq. (4.1). This derivative is called the *derivative of the function V taken by virtue of Eq. (4.1)*; we shall denote it by

$$\dot{V}(x) := (\text{grad}V(x), f(x)).$$

Thus, if the value $x(t)$ of the solution is known for some t, then it is possible to get the value $dV(x(t))/dt$ for this t directly, without solving Eq. (4.1). This simple property is crucial for application of guiding functions and, in particular, guiding Liapunov functions.

We shall call a smooth function $V : D \to \mathbf{R}$ a *guiding function* for Eq. (4.1) if $\dot{V}(x) \leq 0$ in D; in other words, if it is a decreasing function in the weak sense (i.e. it can have intervals of constancy) along any trajectory of this system. The guiding function for a physical system is like a wind which drives it from the states with higher levels (of values of $V(x)$) to the states with lower levels; in any case, this level can remain constant, but cannot increase. It is natural that the knowledge of the guiding function for some system allows us to make essential conclusions about the character of its evolution.

It was A.M. Liapunov who for the first time began to apply systematically guiding functions in his famous studies of stability theory and it is usual to call such functions as *Liapunov functions.* We are going to use this term in a wider sense: namely, we call a *Liapunov function* in $\mathbf{R}^n$ any scalar smooth function V, defined in some neighbourhood G of the origin $0 \in \mathbf{R}^n$, such that $V(0) = 0, V(x) > 0 \ \ (\forall x \in G \setminus \{0\})$.

Let us recall Liapunov's theorems on stability and asymptotic stability for Eq. (4.1). (Liapunov proved them for the general case of non–autonomous systems of ODEs.)

Theorem 4.1 Let $0 \in G, f(0) = 0$ *and assume a Liapunov function* $V : G \to \mathbf{R}$ *exists, for which* $\dot{V}(x) \leq 0 \ \ (\forall x \in G)$ *(i.e. which is guiding for Eq. (4.1)). Then the solution* $x(t) \equiv 0$ *of Eq. (4.1) is stable.*

Theorem 4.2 Assume, in addition to the conditions of Thm. 4.1 that $\dot{V}(x) < 0 \ (\forall x \in G \setminus \{0\})$. *Then the solution* $x(t) \equiv 0$ *of Eq. (4.1) is asymptotically stable.*

These statements have a simple interpretation. The level surfaces of Liapunov function V (i.e. surfaces in $\mathbf{R}^n$ defined by the equation $V(x) =$ const.) are similar to the family of concentric spheres with center 0 and any sufficiently small radii. It follows from the negativeness of $\dot{V}$ that all those surfaces intersect trajectories of Eq. (4.1) towards the interior. (If $\dot{V} \leq 0$ then the trajectories can slide along the level surfaces.) Hence, the validity of the theorems mentioned becomes intuitively clear.

M.A. Krasnosel'skii was the first in 1958 to apply systematically guiding functions for studying various problems of the qualitative theory of ODEs. Let us give one example of such an application for Eq. (4.1).

Theorem 4.3 Let a function f *be defined on* $\mathbf{R}^n$ *and let the smooth function* $V : G \to \mathbf{R}$ *exist in some neighbourhood* G *of infinity (i.e.* G *contains all points* x *with sufficiently large* $|x|$*) such that* $V(x) \to \infty$ *as* $x \to \infty$ *and* $\dot{V}(x) < 0 \ \ (\forall x \in G)$. *Then Eq. (4.1) is dissipative, i.e. there exists an* $r > 0$ *such that any solution of Eq. (4.1) belongs to the sphere* $|x| < r$, *starting at some instant dependent on initial conditions.*

The intuitive interpretation of this theorem is similar to that of Thms. 4.1, 4.2 mentioned above, but level surfaces are similar now to the family of spheres of any sufficiently large radius.

4.3. *GUIDING FUNCTIONALS FOR RDES*

Let us investigate an autonomous RDE of the 1st order with single discrete constant delay:

$$\dot{x}(t) = f(x(t), x(t-h)), \quad h = \text{const} > 0. \tag{4.2}$$

Here, $x(t)$ takes values in $\mathbf{R}^n$, whereas the smooth function $(x, y) \mapsto f(x, y)$ with $2n$ scalar arguments is defined for all $x, y \in \mathbf{R}^n$. However, all considerations can be extended *mutatis mutandis* to other forms of RDEs.

As it was said in Subs. 3.1.1, it is natural to take the space of all continuous functions on some closed interval of length h valued in $\mathbf{R}^n$ with a norm of the uniform deviation as a phase space corresponding to Eq. (4.2). One usually takes $[-h, 0]$ as such interval. Then, the corresponding space is $C[-h, 0]$.

The trajectory in the space $K := C[-h, 0]$ corresponds to each solution $t \mapsto x(t)$ $(t_0 - h \le t < T)$ of Eq. (4.2) by the following rule: each $t \in [t_0, T)$ corresponds to the function $\theta \mapsto x_t(\theta) := x(t+\theta)$ $(-h \le \theta \le 0)$ as an element (point) of K. Exactly one trajectory starts at each point $\phi \in K$. It correspond to the solution of system (4.2) with initial function

$$x(t) = \phi(t - t_0) \quad (t_0 - h \le t \le t_0), \tag{4.3}$$

and the same trajectory corresponds to different values of t_0 but passing with translation in time. Thus the phase space K and the trajectories in it for Eq. (4.2) are similar in many aspects to the phase space $\mathbf{R}^n$ and the trajectories for ODE (4.1); but important distinctions still exist: the space K is infinite–dimensional, and trajectories can be continued only forward in time. Let $P : K \to \mathbf{R}$ be a scalar functional in the space K, i.e. the numer $P(\phi) \in \mathbf{R}$ corresponds to any $\phi \in K$. Then it is easy to verify that if the function $x : [t_0 - h, T) \to \mathbf{R}^n (t_0 < T)$ is continuous, then the function $t \mapsto P(x_t)$ $(t_0 \le t < T)$ is also continuous.

The derivative of the functional P taken by virtue of Eq. (4.2) is defined analogously to that of ODE (4.1). Namely, if $\phi \in K$ and x^ϕ is a solution of Eq. (4.2) under the initial condition (4.3), then

$$\dot{P}(\phi) := \limsup_{\Delta t \to 0+} \frac{1}{\Delta t}[P(x^\phi_{t+\Delta t}) - P(\phi)]. \tag{4.4}$$

Note that $\dot{P}$ is generalized functional $K \to \mathbf{R} \bigcup \{-\infty, \infty\}$ in the sense that it can assume $\pm\infty$ along with finite values.

From formula (4.4) and the autonomy of Eq. (4.2) it follows that the solution $x(\cdot)$ of Pr. (4.2), (4.3) satisfies the equation

$$\limsup_{\Delta t \to 0+} \frac{1}{\Delta t}[P(x_{t_0+\Delta t}) - P(x_t)] = \dot{P}(x_t) \quad (t_0 \le t < T).$$

The limit on the left–hand side is called *upper right derivative* of the function $t \mapsto P(x_t)$. The simple known lemma below makes use of this derivative.

Lemma 4.1 If the function $g : [\alpha, \beta] \to \mathbf{R}$ $(\alpha < \beta)$ is continuous and its upper right derivative $\dot{g}^{ur}(t) \le 0$ (resp. ≥ 0) for $\alpha \le t < \beta$, then the function g is nonincreasing (resp. nondecreasing) on $[\alpha, \beta]$. If in addition the set $\{t : \dot{g}^{ur}(t) = 0\}$ does not contain intervals, then the function g is decreasing (resp. increasing) on $[\alpha, \beta]$.

Analogously to ODEs, the functional P is called a *guiding functional* in some domain $H \subseteq K$ for RDE (4.2) if $\dot{P}(\phi) \le 0 (\forall \phi \in H)$. By virtue of Lemma 4.1 P is then a nonincreasing function of t along any arc of the trajectory of Eq. (4.2) entirely contained in H.

Guiding "Liapunov functionals" are an important example of guiding functionals; such functionals were introduced by N.N.Krasovskii and were successfully used by him and other authors in the theory of RDEs (see, e.g., Sec. 1). Let us underline that the guiding functionals can be used not only in stability theory. The following theorem can be obtained, for example, as an analog of Theorem 4.3.

Theorem 4.4 Let the functional $\phi \mapsto P(\phi)$ be defined for all $\phi \in K$ with sufficiently large $|\phi(0)|$, $P(\phi) \to \infty$ as $|\phi(0)| \to \infty$, and $\sup\{\dot{P}(\phi) : P(\phi) > M\} < 0$ for any sufficiently large M. Then there exists $r > 0$ such that the solution of Pr. (4.2), (4.3) will be contained in the sphere $|x| < r$ for any initial function, beginning at some t.

4.4. *DIRECT APPLICATION OF LIAPUNOV FUNCTIONS TO RDES*

Prior to N.N. Krasovskii's papers on Liapunov functionals, L.E. El'sgol'ts considered the stability problem of the solution $x(t) \equiv 0$ of RDEs by proving that the function $t \mapsto V(x(t))$ is decreasing. Here V is some Liapunov function and $x(\cdot)$ is any perturbed solution. He showed that it is only possible in some comparatively rare special cases.

We shall show this on the example of Eq. (4.2) with $f(0,0) = 0$ for $n = 1$, i.e. when it returns into a scalar equation. Let us assume that $V(x) = x^2$ which is typical Liapunov function for $n = 1$. Then we have for any solution $x(\cdot)$ of Eq. (4.2) that

$$[V(x(t))]^\bullet = 2x(t)\dot{x}(t) = 2x(t)f(x(t), x(t-h)).$$

For the validity of inequality $[V(x(t))]^\bullet \leq 0$, we need to require that $xf(x,y)$ be nonpositive for all sufficiently small $|x|$ and $|y|$. It follows then that $f(0,y) \equiv 0$ which essentially restricts the class of equations considered.

It should be noted that the function $t \mapsto V(x(t))$, with a given V, can be written in the form $t \mapsto P(x_t)$, if we define the functional P by $P(\phi) := V(\phi(0))$, because $P(x_t) = V(x_t(0)) = V(x(t))$. We get for this functional $\dot{P}(\phi) = (\text{grad}V(\phi(0)), f(\phi(0), \phi(-h)))$.

4.5. *THE MAIN IDEA OF B.S. RAZUMIKHIN*

The method proposed by B.S. Razumikhin allowed to "rehabilitate" an application of Liapunov functions to RDEs to a considerable extent. This application was found in some cases simpler and more visual than an application of general functionals.

The example from Subs. 4.4 helps understand easily the idea of Razumikhin's method. We need the inequality $[V(x(t))]^{\bullet} \leq 0$ in order to prove that a solution starting in a small neighbourhood of the zero function remains in it in the future. However, it does not require for the later that the value of $V(x(t))$ be decreasing (or at least nonincreasing) all the time; it is sufficient that the trajectory, which corresponds to the solution, will not cross the boundary of this neighbourhood. Hence, if a neighbourhood has the form $V(x) < \delta$ $(\delta = \text{const} > 0)$, it would be sufficient to require that $[V(x(t))]^{\bullet} < 0$ when $V(x(t)) = \delta$ and $V(x(t-h)) < \delta$, because the last inequality holds when the trajectory reaches the boundary for the first time. We show later that the inequality $\dot{V} < 0$ can be replaced here by $\dot{V} \leq 0$. Thus, for stability of the zero solution, the inequality $xf(x,y) \leq 0$ need not be valid for all x and y with sufficiently small modulo, but only for those with $|y| < |x|$. This (weaker) restriction extends considerably the number of cases for which stability of the zero solution of Eq. (4.2) can be proved by Liapunov functions.

Let us consider, as a particular case, the scalar RDE

$$x(t) = ax(t) + bx(t-h) \quad (a, b = \text{const}). \tag{4.5}$$

Stability of the solution $x(t) \equiv 0$ is guaranteed by approach in Subs. 4, with $V(x) = x^2$, only in the trivial case $b = 0$ and $a \leq 0$. Using Razumikhin's method, we conclude that the stability is ensured if $x(ax + bx) \leq 0$ for $|y| < |x|$, i.e. always when $|b| \leq -a$. (Of course, as for any linear equation RDEs, from stability of some solution of Eq. (4.5) it follows the stability of any other solution, i.e. here it is natural to consider the stability of this equations; the asymptotic stability of linear equations is analogously considered.)

4.6. *"IMPOSSIBILITY OF THE FIRST BREAKDOWN"*

The idea of Razumikhin's method can also be explained in more general situations. Let us consider a continuous evolution of some system, starting from some instant, and we wish to prove that it possesses some property S all the time. Let this property be stable, i.e. if the system possesses it in some state, then it would do so for all sufficiently close states.

Two ways can be used to prove, by contradiction, the presence of property S. First, it can be assumed simply that the property S is absent at some instant, and try to arrive at a contradiction. Second, it can be proved that S takes place at the initial instant, and then deduce a contradiction considering the *first* instant of breakdown of this property. It is clear that if we are able to prove that the first breakdown is impossible, then it would always be impossible.

The second approach is in some sense analogous to the mathematical induction method for continuous processes. It is effective for the cases when the assumption about the validity of the property S at all instants preceding the considered one, gives an information which facilitates to get a contradiction. Of course, this approach was used by many authors in different branches of mathematics. It is crucial for Razumikhin's method along with the idea of a guiding function.

4.7. *PRECISE FORMULATION*

To apply Razumikhin's method, one can use the following lemma, which is a refinement of a particular case of J. Kato's theorem [274.1, 274.2], but can easily be proved independently.

Lemma 4.2 Let the function $g : [\alpha - h, \beta) \to \mathbf{R}$ *be continuous for some* $h > 0, \quad \alpha < \beta \leq \infty$, *and* $\dot{g}^{ur}(t) \leq 0$ *for all values* $t \in [\alpha, \beta)$ *for which* $g(s) < g(t)$ $(\forall s \in [\alpha - h, t))$. *Then* $g(t) \leq \max_{s \in [\alpha - h, \alpha]} g(s) (\alpha \leq t < \beta)$.

Applied to Eq. (4.2) with $f(0,0) = 0$, this lemma leads to the theorem:

Theorem 4.5 Assume the Liapunov function $V : G \to \mathbf{R}$ *exists with* $x, y \in G$, $V(y) < V(x)$ *imply*

$$(\text{grad}V(x), f(x,y))(= (\text{grad}V(x))' f(x,y)) \leq 0. \tag{4.6}$$

Then the solution $x(t) \equiv 0$ *of Eq. (4.2) is stable.*

Proof Set $g(t) = V(x(t))$, where $x(\cdot)$ is any solution of Eq. (4.2) with initial function ϕ sufficiently close to zero, and apply Lemma 4.2. We see that even a stronger statement for such solutions holds: $V(x(t)) \leq \max_{[t_0 - h, t_0]} V(\phi)$. However, in general, the monotonicity of the function $t \mapsto V(x(t))$ does not hold unlike the situation arising under conditions of Thm. 4.1. Thus the function V turns out to be not a guiding function in the restricted sense, but a "barrier" function. □

It should be noted that the function $t \mapsto g(t)$, defined in the proof of Thm. 4.5, is smooth for $t \geq \alpha := t_0$. Lemma 4.2 follows easily under this additional assumption from the following particular case of A. Sard's theorem: *a set of stationary values of a smooth function of a single argument has the everywhere dense complement.*

Thm. 4.5 directly extended with natural modifications to any autonomous RDEs of 1st order, both with discrete and continuous delays, and to the non–autonomous case; see e.g. [238] and the references therein. Besides, if a non–autonomy of the system appears only in the dependence of the delay on t or even on the unknown solution, then no changes in the formulation and the proof of the theorem are required. Generalization for stochastic RDEs was done in [350].

4.8. *CONNECTION BETWEEN RAZUMIKHIN'S METHOD AND LIAPUNOV FUNCTIONALS*

Let a smooth function $V : \mathbf{R}^n \to \mathbf{R}$ be given. Define the functional $P : K \to \mathbf{R}$ as follows

$$P(\phi) := \max_{-h\leq\theta\leq 0} V(\phi(\theta)) \quad (\forall \phi \in K).$$

It is not difficult to verify that

$$\dot{P}(\phi) \begin{cases} \leq 0, & \text{if } V(\phi(0)) < P(\phi), \\ = \max\{(\text{grad}V(\phi(0)), f(\phi(0), \phi(-h))), 0\}, & \text{if } V(\phi(0)) = P(\phi). \end{cases}$$

Thus, $\dot{P}(\phi) > 0$ if and only if

$$V(\phi(0)) = \max_{-h\leq\theta\leq 0} V(\phi(\theta)) \quad \text{and} \quad (\text{grad}V(\phi(0)), f(\phi(0), \phi(-h))) > 0. \quad (4.7)$$

The function V can be defined on some domain $G \subset \mathbf{R}^n$. The functional P is then defined for $\phi \in K$ with values in G. Its properties, pointed out here, remain valid.

If the function V is such as in Thm. 4.5, then relations (4.7) are inconsistent. Indeed, if they are valid for some function $\phi \in K$, then we could obtain the inequality $V(\phi(-h)) < V(\phi(0))$ making ϕ arbitrarily small. At the same time, the second relation in (4.7) remains valid. It contradicts the condition of Thm. 4.5. Hence $\dot{P}(\phi) \leq 0$ for all ϕ. So, if $f(0,0) = 0$, then all the conditions of Thm. 3.1.2 are satisfied and therefore, the solution $x(t) \equiv 0$ of Eq. (4.2) is stable.

Thus, Razumikhin's method can be considered as a particular case of the method of Liapunov functionals. But this particular case possesses a feature which makes its independent study quite natural. This feature is based, first of all, on the possibility to apply the guiding functions in $\mathbf{R}^n$ (with $n = 1, 2$ etc. in many examples) which is well elaborated for ODEs.

4.9. *ASYMPTOTIC STABILITY*

Asymptotic stability of Eq. (4.2) zero solution can be proved under a stronger assumption than that of Thm. 4.5. Consider an example of such statement.

Theorem 4.6 Assume, in addition to the conditions of Thm. 4.5 that for any sufficiently small $p > 0$, there exists $q > p$ such that, if $V(x) = p$, $V(y) < q$, then $(\text{grad}V(x))'f(x,y) < 0$. Then the solution $x(t) \equiv 0$ of Eq. (4.2) is asymptotically stable.

Proof If the solution $t \mapsto x(t)$ of Eq. (4.2) is modulo sufficiently small and does not tend to zero as $t \to \infty$, then we denote

$$p := \limsup_{t \to \infty} V(x(t)),$$

and apply the additional condition of Theorem (4.6) which leads to the contradiction.□

In particular, we obtain for $n = 1$, taking $V(x) = x^2$, that the condition $|b| < -a$ is sufficient for the asymptotic stability of Eq. (4.5).

Remark 4.1 If the solution V is defined on the whole $\mathbf{R}^n$ in Thm. 4.6 and $V(x) \to \infty$ as $|x| \to \infty$, and $p > 0$ can be arbitrary, then the solution of Eq. (4.2) tends to zero as $t \to \infty$ for arbitrary initial function.

4.10. *REFINEMENT OF ESTIMATIONS*

Razumikhin also notes that many parameters estimates of RDEs, which guarantee stability, can be made much more accurate by constricting a set of points where the validity of inequalities (4.6) is required. Let us formulate a relevant theorem for Eq. (4.2) with $f(0,0) = 0$.

Theorem 4.7 Assume that the Liapunov function $V : G \to \mathbf{R}$ and a number $N > 0$ exist such that the inequality

$$(\text{grad} V(x(0)))' f(x(0), x(-h)) \leq 0, \tag{4.8}$$

holds for any continuous function $x(\cdot) : [-(N+1)h, 0] \to G$ for which $x(t) \in G$, $V(x(t)) < V(x(0))$ $(-(N+1)h \leq t < 0)$, and which satisfies (4.2) for $-Nh \leq t \leq 0$. Then the solution $x(t) \equiv 0$ of Eq. (4.2) is stable.

Proof Let $\varepsilon > 0$ be given with $\{x : |x| < \varepsilon\} \subset G$ and $\phi \in K$, $V(\phi(t)) < \delta$ $(-h \leq t \leq 0)$, where $\delta > 0$ is sufficiently small. Then the inequality $V(x(t)) < \varepsilon$ $(t_0 \leq t \leq t_0 + Nh)$ for the solution $x(\cdot)$ of Pr. (4.2), (4.3) follows from continuous dependence of the solution on the initial function. The impossibility of the breakdown of this inequality for $t > t_0 + Nh$ can be justified as in Thm. 4.5. □

4.11. *EXAMPLE*

Let us show how Thm. 4.7 can be applied to the scalar equation

$$\dot{x}(t) = -bx(t-h) \ (b = \text{const} > 0), \tag{4.9}$$

with $V(x) = x^2$. We are following Razumikhin's presentation with some refinements.

The application of Thm. 4.5 gives nothing in this example that can be seen from the consideration of equation (4.5) in Subs. 4.6. Let us apply Thm. 4.7. Observe that inequality (4.8) takes form

$$[2x(0)][-bx(-h)] \leq 0, \text{ i.e. } x(0)x(-h) \geq 0. \tag{4.10}$$

Let us take at first $N = 1$; then inequality (4.10) must be proven for any continuous function $x(\cdot) : [-2h, 0] \to \mathbf{R}$, for which $|x(t)| < |x(0)|$ $(-2h \leq t < 0)$, and Eq. (4.9) is satisfied for $-h \leq t \leq 0$.

If $x(0) = 0$ then the validity of (4.10) is obvious. Therefore, let $x(0) > 0$ (if $x(0) < 0$, we can multiply $x(\cdot)$ by -1, which does not affect (4.10).

It follows by virtue of (4.10) from inequality $|x(t)| < x(0)$ $(-2h \leq t \leq -h)$ that

$$\dot{x}(t) < bx(0) \quad (-h \leq t \leq 0).$$

Now integrating it from $-h$ to 0, we obtain $x(0) - x(-h) < bx(0)h$. Hence, $x(-h) > (1 - bh)x(0)$.

We see that if $bh \leq 1$, then inequality (4.10) holds true, and therefore, the condition $bh \leq 1$ is sufficient for the stability of Eq. (4.9).

This restriction can be essentially weakened, if we put $N = 2$. Inequality (4.10) must be verified now for any continuous function $x(\cdot) : [-3h, 0] \to \mathbf{R}$, for which $|x(t)| < |x(0)|$ $(-3h \leq t < 0)$, and Eq. (4.9) is satisfied for $-2h \leq t \leq 0$.

We can assume again $x(0) > 0$. Let us assume that (4.10) is not fulfilled, i.e. $x(-h) < 0$; then $bh > 1$ by virtue of considerations carried out for $N = 1$. We obtain by integrating (4.9) from any $t \in [-2h, -h]$ to $-h$ that

$$|x(-h) - x(t)| \leq bx(0)(-h - t) \; (-2h \leq t \leq -h).$$

It follows from here that

$$x(t) \geq \max\{-x(0), x(-h) - bx(0)(-h - t)\} \quad (-2h \leq t \leq -h). \tag{4.11}$$

Let $t = g$ be the value for which both quantities under max equal each other, i.e.

$$g := -h - \frac{1}{b} - \frac{x(-h)}{bx(0)}. \tag{4.12}$$

Since $bh > 1$, then $-2h < g < -h$ and therefore inequality (4.11) can be rewritten as follows:

$$x(t) \geq \begin{cases} -x(0) & (-2h \leq t \leq g), \\ x(-h) + bx(0)(h + t) & (g \leq t \leq -h). \end{cases}$$

Using this estimation we obtain by integrating both sides of Eq. (4.9) from $-h$ to 0 that

$$x(0) - x(-h) = -b\left[\int_{-2h}^{g} x(t)dt + \int_{g}^{-h} x(t)dt\right] \leq$$

$$\leq -b\{-x(0)(g+2h) + x(-h)(-h-g) + bx(0)[-\frac{1}{2}(h+g)^2]\}.$$

Substituting here the righthand side of (4.12) for g and grouping similar terms, we get:

$$x(0) - x(-h) \leq -bhx(0) - \frac{1}{2}x(0) - x(-h) - \frac{1}{2x(0)}[x(-h)]^2,$$

from which

$$(3/2 - bh)x(0) \leq -\frac{1}{2x(0)}[x(-h)]^2 < 0.$$

Hence, the described situation can occur only if $bh > 3/2$, and for $bh \leq 3/2$ condition (4.10) is satisfied, i.e. Eq. (4.9) is stable.

Thus, the transition from $N = 1$ to $N = 2$ widens to one and a half the interval of values b for which the stability of equation (4.9) is guaranteed with a given h. It should be noted that the precise estimate, for which stability takes place, is $bh \leq \pi/2$, which can be obtained by considering the corresponding characteristic quasipolynomial (see Subs. 3.4). Thus Razumikhin's method gives an estimate in this example which differs from the precise one by less than 5% .

It could be shown that the transition to $N = 3, 4$ etc. leads to further refinements of the estimate. However, this transition would require an essential complication of arguments and apparently noone did it until now.

The point is that the foregoing discussion of the cases $N = 1$ and $N = 2$ is valid for the RDE

$$\dot{x}(t) = -bx(t - h_1(t)) \tag{4.13}$$

with $0 \leq h_1(t) \leq h$ as well. (And it is a strong aspect of Razumikhin's method, because the method of characteristic quasipolynomials is no longer applicable to Eq. (4.13).) However, it appears (see Thm. 3.5.9) that the continuous function $h_1(\cdot)$ could be constructed for any $b > 3/2h$ for which Eq. (4.13) is unstable, i.e. the estimate $bh \leq 3/2$, obtained by Razumikhin's method and guaranteeing stability, is *best possible* for this equation. A refinement of the estimate $bh \leq 3/2$ requires arguments which use essentially the constancy of the delay or at least the impossibility of its considerable decrease in comparison with the estimating value.

4.12. *TRANSFORMATION OF RDES*

We can see from Subs. 4.11 that, although an application of Thm. 4.7 even for $N = 1$ leads to more precise estimates than Thm. 4.5, it requires essentially more complicated arguments and therefore is poorly algorithmized. For this reason, Razumikhin applied a method of preliminary transformations of RDEs, based on Lagrange formula, after which an application of Thm. 4.5 leads to refined estimates. Here we apply the same method of transformation of RDE.

Put $V(x) = x^2$. Then sufficient stability condition obtained in Subs. 4.5 for Eq. (4.5) can be generalized for scalar RDEs

$$\dot{x}(t) = ax(t) + f(t, x_t) \quad (t \geq t_0; a = \text{const} < 0; f(t,0) \equiv 0)$$

where $f : [t_0, \infty) \times C[-h, 0] \to \mathbf{R}$ is a continuous functional. This condition has a form

$$(\forall t \in [t_0, \infty), \quad \psi \in C[-h,0] : \| \psi \| = |\psi(0)| > 0) \Rightarrow a + \frac{|f(t,\psi)|}{\| \psi \|} \leq 0. \tag{4.14}$$

Let us transform Eq. (4.5) for $t \geq t_0 + h$ as follows

$$\dot{x}(t) = (a+b)x(t) - b\int_{t-h}^{t} \dot{x}(s)ds = (a+b)x(t) - \\ -b\int_{t-h}^{t} [ax(\tau) + bx(\tau - h)]ds. \tag{4.15}$$

Eq. (4.15) has delay $2h$ and initial point $t = t_0 + h$. Stability of Eq. (4.5) follows from stability of Eq. (4.15). By virtue of (4.14) Eq. (4.15) as well as Eq. (4.5) are stable under condition

$$a + b + h|ab| + hb^2 \leq 0 \tag{4.16}$$

Inequality (4.16) determines the domain on a, b – plane, which is bounded by the line $b = -1/h$ and the hyperbola $a = b(bh+1)/(bh-1)$; this domain is shaded steeply in Fig. 4.4.1. Together with domain of stability, known from direct application of Thm. 4.5 (it is shaded slatingly in Fig. 4.4.1), we obtain an essential extension of the latter domain. The exact domain of stability is shown on Fig. 4.4.1 for comparison; it is bounded by dashed lines – straight line $a+b = 0$ from above and line $a + b\cos(h\sqrt{b^2 - a^2})$ $= 0$ from below (see Subs. 3.4).

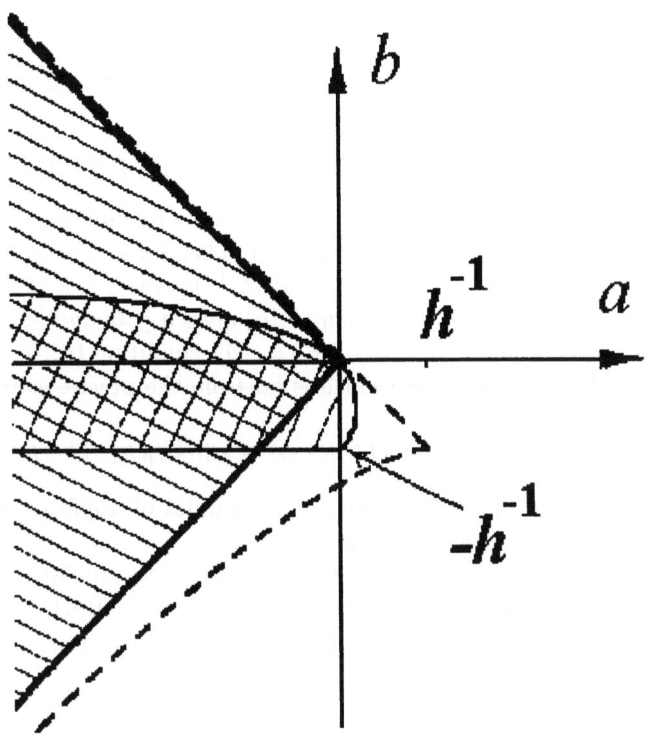

Figure 4.4.1

Eq. (4.15) can be transformed in the same way. As a result we obtain for $t \geq t_0 + 2h$ the equation

$$\dot{x}(t) = (a+b)x(t) - b\int_{t-h}^{t}\{ax(t) - a\int_{\tau}^{t}[ax(\tau_1) + bx(\tau_1 - h)]d\tau_1 + bx(t) -$$

$$-b\int_{\tau-h}^{t}[ax(\tau_1) + bx(\tau_1 - h)]d\tau_1\}d\tau = (a + b - hab - hb^2)x(t) +$$

$$+ab\int_{t-h}^{t}d\tau\int_{\tau}^{t}[ax(\tau_1) + bx(\tau_1 - h)]d\tau_1 + b^2\int_{t-h}^{t}d\tau\int_{\tau-h}^{t}[ax(\tau_1) + bx(\tau_1 - h)]d\tau_1.$$

By virtue of (4.14) sufficient condition of stability of Eq. (4.5) is given by the inequality

$$a + b - hab - hb^2 + h^2(a^2|b| + 4|a|b^2 + 3|b|^3)/2 \leq 2. \tag{4.17}$$

Sometimes (4.17) can give a refinement of condition (4.16) especially for small values of h. Take, e.g., $h = 0.1$. Then for $b = -5$ condition (4.16) gives stability condition $a \leq 1.667$, but (4.17) gives $a \leq 2.720$ (exact estimate is $a \leq 5$); for $b = -10$ condition (4.16) gives $a \leq 0$, and (4.17) gives $a \leq 2.361$ (exact estimate is $a \leq 10$); for $b = -11$ condition (4.16) gives nothing, and (4.17) gives $a \leq 0.688$ (which is far from the precise estimate $a \leq 8.056$).

4.13. *OTHER APPLICATIONS OF RAZUMIKHIN'S METHOD*

In this section we consider some examples of application of Razumikhin's method for other problems.

The following theorem concerns a property similar to exponential stability of the solution $x(t) \equiv 0$ of Eq. (4.2) with $f(0,0) = 0$.

Theorem 4.8 Assume that the Liapunov function $V : G \to \mathbf{R}$ and a number $p > 0$ exist such that

$$[V(y) < e^{ph}V(x), (x,y) \in G] \Rightarrow [(\text{grad}V(x))'f(x,y) + pV(x) \leq 0].$$

Then for each $\varepsilon > 0$ there is a $\delta > 0$ such that

$$[|x(t)| < \delta(t_0 - h \leq t \leq t_0)] \Rightarrow [V(x(t)) < \varepsilon e^{-pt}, \quad t_0 \leq t < \infty].$$

The proof is based on the formulas

$$[V(x(t))e^{pt}]^{\bullet} = [(\text{grad}V(x(t)))'f(x(t), x(t-h)) + pV(x(t))]e^{pt},$$

$$[V(x(t-h))e^{p(t-h)} < V(x(t))e^{pt}] \Leftrightarrow [V(x(t-h)) < V(x(t))e^{pt}],$$

and on application of Lemma 4.2 to the function $t \mapsto V(x(t))e^{pt}$.

It follows from Thm. 4.8, for instance, that Eq. (4.5) is exponentially stable if $|b| < -a$.

Similar theorems can also be proven about boundedness of solutions and dissipativity of Eq. (4.2).

Theorem 4.9 Let the smooth function $V : \mathbf{R}^n \to \mathbf{R}$ *exist such that* $V(x) \to \infty$ *for* $|x| \to \infty$ *and* $(\text{grad}V(x))'f(x,y) \leq 0$ *for* $V(y) < V(x)$ *if* $|x|$ *is sufficiently large. Then* $(\forall M > 0)(\exists N > 0)$: *if* $\phi \in K, |\phi(\theta)| < M$ $(-h \leq \theta \leq 0)$, *then we have* $|x(t)| < N$, $t_0 \leq t < \infty$ *for the solution of Pr. (4.2), (4.3).*

Theorem 4.10 Let all the conditions of Thm. 4.9 be valid and for any sufficiently large $p > 0$, *there exists* $q > p$, *such that* $(\text{grad}V(x))'f(x,y) < 0$, *if* $V(x) = p$, $V(y) < q$. *Then Eq. (4.2) is dissipative.*

Remark 4.2 A similar idea can be applied to prove that for any solution $x(\cdot)$ of 1st order RDE, the point $x(t)$ remains, for $t_0 < t < \infty$, in a given part of the space if values of the initial function belong to it. Consider e.g. Eq. (4.2). Let the function $V : \mathbf{R}^n \to \mathbf{R}$ be smooth, $V(x) < 0$ in some (connected open) domain G and $V(x) = 0$ on its boundary ∂G. Let $\phi(\theta) \in \overline{G} := G \bigcup \partial G$ $(-h \leq \theta \leq 0)$ and

$$(x \in \partial G, y \in \overline{G}) \Rightarrow (\text{grad}V(x))'f(x,y) \leq 0. \tag{4.18}$$

Then, from Lemma 4.2, it follows that $x(t) \in \overline{G}$ on the whole interval of existence of the solution $x(\cdot)$ of Pr. (4.2), (4.3). If $\phi(\theta) \in G(-h \leq \theta \leq 0)$ and condition (4.18) is satisfied, then $x(t) \in G$ on this whole interval.

For example we obtain for the scalar RDE (4.2), taking $V(x) := x$, that if $\phi(\theta) \leq 0$ (resp.< 0) for all $\theta \in [-h, 0]$ and $f(0, y) \leq 0$ for all $y \leq 0$, then $x(t) \leq 0$ (resp.< 0) for $t > t_0$ on the whole existence interval of the solution $x(\cdot)$ of problem (4.2), (4.3).

This remark means in essence that any guiding function for an ODE can be used as a guiding function for an RDE under appropriate assumptions about righthand sides of both equations. E.g., if applied to Eqs. (4.1) and (4.2), the condition $(\text{grad}V(x))'f(x,y) \leq 0$ must be replaced by

$$[V(y) < V(x)] \Rightarrow [(\text{grad}V(x))'f(x,y) \leq 0].$$

Of course, refinements are possible here in the fashion of Thm. 4.7.

Remark 4.3 Razumikhin's method adapted to difference equations (DcEs) turns out into a variant of the usual method of mathematical induction. Let us illustrate this on the example of DcE

$$x_{m+2} = f(x_m, x_{m+1}) \quad (m = 0, 1, \ldots),$$

where $f : \mathbf{R}^n \times \mathbf{R}^n \to \mathbf{R}^n$.

Let the function $V : \mathbf{R}^n \to \mathbf{R}$ be given. We want to prove that $V(x_m) \leq 0$ for all $m \geq 0$. Then, taking into account discreteness of m, the natural analog of Razumikhin's method is the following: we must check that $V(x_m) \leq 0$ $(m = 0, 1)$, and in addition that

$$[V(x) \leq 0, V(y) \leq 0] \Rightarrow [V(f(x, y)) \leq 0] \quad (\forall x, y \in \mathbf{R}^n).$$

But, if we substitute $x = x_m, y = x_{m+1}$ here, we obtain the method of induction.

The refinement of estimates in Subs. 4.11 can be useful for DcEs as well. Let us consider, for example, the scalar equation

$$x_{m+2} = ax_m + bx_{m+1} \; (m = 0, 1, \ldots), \tag{4.19}$$

for which the stability domain is

$$|a| \leq 1, \; |b| \leq 1 - a. \tag{4.20}$$

The direct approach described in the previous paragraph, with $V(x) = |x|$ or with $V(x) = x^2$, gives smaller domain:

$$|a| + |b| \leq 1. \tag{4.21}$$

Let us iterate Eq. (4.19) to widen this domain. It gives

$$x_{m+3} = (a + b^2)x_{m+1} + abx_m.$$

We obtain from here a condition on a, b which is sufficient for stability of Eq. (4.19)

$$|a + b^2| + |ab| \leq 1.$$

It is not difficult to verify that the (a, b) – domain, defined by this inequality, is wider than that of (4.21), but nevertheless narrower than the exact domain (4.20).

CHAPTER 5

STABILITY OF RDEs WITH AUTONOMOUS LINEAR PART [1]

General methods for stability analysis, described previously, can be essentially modified for particular clases of FDEs. One of these classes is considered in this chapter. This class consist of nonlinear RDEs with autonomous linear parts. Explicit stability conditions are proposed. They are formulated in terms of the roots of the characteristics polynomials and based on an estimate for the matrix resolvent.

1. NOTATIONS

Let A be $n \times n$ – complex matrix, and let $\lambda_k(A)(k = 1, \dots, n)$ denote the eigenvalues of A including their multiplicities. *The following quantity plays an essential role* hereafter

$$g(A) := (N^2(A) - \sum_{k=1}^{n} |\lambda_k(A)|^2)^{1/2}, \tag{1.1}$$

where $N(A)$ is the *Hilbert – Schmidt (Frobenius) norm* of A, i.e. $N^2(A) := \mathrm{Tr}(AA^*)$. Here $\mathrm{Tr}A$ is a trace of A, and A^* is an adjoint matrix. The following properties of $g(A)$ are proved in [203]: $g^2(A) \le N^2(A) - |\mathrm{Tr}A^2|$; if $A = (a_{jk})_{j,k=1}^{n}$ is a triangular matrix: $a_{jk} = 0$ $(1 \le k < j \le n)$, then

$$g^2(A) = \sum_{1 \le j < k \le n} |a_{jk}|^2. \tag{1.2}$$

If A is normal: $A^*A = AA^*$, then $g(A) = 0$, and also the relations

$$g(A) \le \sqrt{1/2}N(A^* - A) \quad \text{and} \quad g(Ae^{i\omega} + zI) = g(A)$$
$$\text{for every } \omega \in (-\infty, \infty), \; z \in \mathbf{C} \tag{1.3}$$

are true. Here I is the unit matrix and z is a complex number. Our results below are based on the following estimate for the Euclidean norm $|\cdot|$ of the

[1]This chapter is written jointly with Prof. M.I. Gil' (Ben Gurion University).

resolvent of an $n \times n$ matrix A:

$$|(A-\lambda I)^{-1}| \leq \sum_{k=0}^{n-1} \frac{g^k(A)}{\sqrt{k!}\rho^{k+1}(A,\lambda)} \quad \text{for all regular } \lambda, \tag{1.4}$$

where $\rho(A,\lambda)$ is the distance between the spectrum of A and a complex point λ (see [203.1]).

Let us consider in the Euclidean space $\mathbf{R}^n$ a homogeneous linear equation

$$\dot{x}_0(t) = \int_{-h}^{0} [dR(\theta)]x_0(t+\theta) \quad (0 < h = \text{const} < \infty; \quad t \geq 0) \tag{1.5}$$

where $R(\cdot)$ is a matrix–valued function of bounded variation

$$VR = \int_{-h}^{0} |dR(s)|.$$

Further, $K(z)$ be defined as follows

$$K(z) := \widetilde{R}(z) - zI, \text{ where } \widetilde{R}(z) := \int_{-h}^{0} e^{z\theta} dR(\theta) \quad (z \in \mathbf{C}).$$

In addition, denote

$$\Gamma(K(z)) := \sum_{k=0}^{n-1} \frac{g^k(\widetilde{R}(z))}{\sqrt{k!}d^{k+1}(K(z))} \quad (z \in \mathbf{C}),$$
$$\text{and } \Gamma_0(K) := \sup_{\omega} \Gamma(K(i\omega)) \quad (-\infty < \omega < \infty) \tag{1.6}$$

where $d(K(z)) = \rho(K(z),0)$ is the smallest modulus of eigenvalues of $K(z)$:

$$d(K(z)) := \min_{k=1,\dots,n} |\lambda_k(K(z))|.$$

Everywhere below in this chapter it is assumed that all the zeros of $\det K(z)$ lie in the open left half–plane.

Let us give a few estimates for $\Gamma_0(K)$. Put

$$v(\widetilde{R}) := \sup_{\omega \in \mathbf{R}} g(\widetilde{R}(i\omega)) \text{ and } d_0(K) := \inf_{-2V(R) \leq \omega \leq 2V(R)} d(K(i\omega)).$$

Lemma 8.2.3 from [203.1], yields

$$\inf_{-2V(R) \leq \omega \leq 2V(R)} d(K(i\omega)) = \inf_{-\infty \leq \omega \leq \infty} d(K(i\omega)).$$

Thus,

$$\Gamma_0(K) \leq \overline{\Gamma}_0(K) := \sum_{k=0}^{n-1} \frac{v^k(R)}{\sqrt{k!}d_0^{k+1}(K)}.$$

By the definition of $g(A)$, it can be written

$$g(B(i\omega)) \leq N(B(i\omega)) \leq \sqrt{n}|B(i\omega)| \leq \sqrt{n} \int_{-h}^{0} |dR(s)| = \sqrt{n}VR.$$

Let us give a more precise estimate for $g(B(i\omega))$ in the case

$$K(z) = \sum_{k=0}^{m} A_k \exp(-zh_k) - zI, \tag{1.7}$$

where A_k are constant matrices; $h_0 = 0$ and $h_k > 0 \;\; (k = 1, \ldots, m)$ are constants. Due to (1.3)

$$g(\widetilde{R}(i\omega)) \leq \sqrt{1/2} \sum_{k=0}^{m} N(A_k \exp(-i\omega h_k) - A_k^* \exp(i\omega h_k)).$$

Hence,

$$v(\widetilde{R}) \leq \sqrt{1/2} N(A_0 - A_0^*) + \sqrt{2} \sum_{k=1}^{m} N(A_k).$$

Denote in this Chapter $L_2(J, \mathbf{R}^n)$ the space of n–vector–valued functions which are defined on an interval J, and have finite norm

$$\| w \|_{L_2(J,\mathbf{R}^n)} = \left[\int_J |w(t)|^2 dt \right]^{1/2} \quad (w \in L_2(J, \mathbf{R}^n)).$$

For the sake of simplicity denote the norms in $L_2([0,\infty), \mathbf{R}^n)$ and $L_2([-h, 0), \mathbf{R}^n)$ by $\| \cdot \|_L$ and $\zeta(\cdot)$, respectively:

$$\| w \|_L := \| w \|_{L_2([0,\infty),\mathbf{R}^n)} = \left[\int_0^\infty |w(t)|^2 dt \right]^{1/2}$$

and

$$\zeta(\phi) := \| \phi \|_{L_2([-h,0],\mathbf{R}^n)} = \left[\int_{-h}^{0} |\phi(s)|^2 ds \right]^{1/2}$$

with $w \in L_2([0,\infty),\mathbf{R}^n)$ and $\phi \in L_2([-h,0],\mathbf{R}^n)$. In addition, $C([a,b],\mathbf{R}^n)$ is the space of continuous n-vector-valued functions defined on a segment $[a,b]$ and equipped with norm

$$\| w \|_{C([a,b],\mathbf{R}^n)} := \max_{t\in[a,b]} |w(t)| \quad (w \in C([a,b],\mathbf{R}^n)).$$

2. L^2-STABILITY

Let us consider in $\mathbf{R}^n$ the equation

$$\dot{x}(t) = \int_{-h}^{0} [dR(\theta)]x(t+\theta) + f(t,x_t) \quad (t \geq 0, x_t(\theta) := x(t+\theta), \quad -h \leq \theta \leq 0), \tag{2.1}$$

where f continuously maps $[0,\infty) \times C([-h,0],\mathbf{R}^n)$ into $\mathbf{R}^n$ and $VR < \infty$. Take the initial condition

$$x(\theta) = \phi(\theta) \quad \text{for} \quad -h \leq \theta \leq 0. \tag{2.2}$$

Here and below $\phi(\cdot)$ is a given continuous vector-valued function.

A solution of (2.1) is an absolutely continuous function which satisfies this equation on $[0,\infty)$ almost everywhere with a given initial condition. In addition, it is assumed that for all $u \in L_2([-\eta,\infty],\mathbf{R}^n)$, the inequality

$$\| f(\cdot,u.) \|_L \left(= [\int_0^\infty |f(t,u_t)|^2 dt]^{1/2} \right) \leq q \| u \|_L + M(u) \tag{2.3}$$

holds, where $M(\cdot)$ is a continuous (generally nonlinear) functional defined on the space $L_2([-\eta,0],\mathbf{R}^n)$.

For instance, let there be constants $b_j > 0$ and $q_j \geq 0$ $(j = 1,\ldots,m < \infty)$ such that the relation

$$|f(t,u_t)| \leq \sum_{k=1}^{m} q_j |u(t-h_j(t))| \quad \text{for all} \quad u \in L_2([-h,\infty),\mathbf{R}^n), \quad t \in [0,\infty) \tag{2.4}$$

holds, where $h_j(\cdot)$ $(j = 1,\ldots,m)$ are differentiable scalar-valued functions with the properties

$$1 - \dot{h}_j(t) \geq b_j > 0, \quad \text{and} \quad 0 \leq h_j(t) \leq h \quad (t \geq 0, \ j = 1,\ldots,m). \tag{2.5}$$

Theorem 2.1 Let relations (2.4) and (2.5) be fulfilled. Then condition (2.3) holds with

$$q := \sum_{j=1}^{m} q_j b_j^{-1/2}, \tag{2.6}$$

and

$$M(u) := q\left[\int_{-h}^{0}|u(t)|^2dt\right]^{1/2} = q\zeta(u) \; (u \in L_2([-\eta,0],\mathbf{R}^n)).$$

Proof Relation (2.5) gives us the inequality

$$b_j \parallel u(\cdot - h_j(\cdot)) \parallel_L^2 \leq \int_0^\infty |u(t-h_j(t))|^2(1-\dot{h}_j(t))dt.$$

Thus,

$$b_j \parallel u(\cdot - h_j(\cdot)) \parallel_L^2 \leq \int_{-h_j(0)}^{\infty} |u(s)|^2 ds =$$

$$= \parallel u \parallel_L^2 + \int_{-h_j(0)}^{\infty} |u(s)|^2 ds \leq \parallel u \parallel_L^2 + \zeta^2(u)$$

for $j = 1, \ldots, m$. Therefore by condition (2.4),

$$\parallel f(\cdot, u_\cdot) \parallel_L \leq \sum_{k=1}^{m} q_j \parallel u(\cdot - h_j(\cdot)) \parallel_L \leq \sum_{k=1}^{m} q_j (b_j)^{-1/2}(\parallel u \parallel_L + \zeta(u)).$$

As claimed.□

Now we can formulate

Theorem 2.2 Let the conditions (2.3) and

$$q\Gamma_0(K) < 1 \tag{2.7}$$

hold. Then any solution $x(\cdot)$ of Pr. (2.1), (2.2) belongs to $L_2([0,\infty),\mathbf{R}^n)$. Moreover, the bound

$$\parallel x \parallel_L \leq (1-\Gamma_0(K)q)^{-1}(\Gamma_0(K)M(\phi) + \parallel x_0 \parallel_L) \tag{2.8}$$

is valid, where $x_0(\cdot)$ is the solution of Eq. (1.5) with the initial function ϕ.

It is simple to check that x_0 belongs to $L_2([0,\infty),\mathbf{R}^n)$ (see also Lemma 4.2 below). We have divided the proof of the theorem into a series of lemmas.

Lemma 2.1 Any solution $u(\cdot)$ of the equation

$$\dot{u}(t) = \int_{-h}^{0}[dR(\theta)]u(t+\theta) + f(t) \tag{2.9}$$

with $f \in L_2([0,\infty),\mathbf{R}^n)$, and the initial condition (2.2) satisfies the inequality $\| u \|_L \leq \tilde{b}_0 \| f \|_L + \zeta(\phi)$, where

$$\tilde{b}_0 := \sup_{\omega \in \mathbf{R}} |K^{-1}(i\omega)|.$$

Proof Solution of (2.9) admits the representation

$$u(t) = -\frac{1}{2\pi}\int_{-\infty}^{\infty}[K(i\omega)]^{-1}\exp(i\omega t)\tilde{f}(i\omega)d\omega + x_0(t), \tag{2.10}$$

where $\tilde{f}$ is the Laplace transform of f. Due to the Parseval equality $\tilde{f}(i\omega) \in L_2((-\infty,\infty),\mathbf{R}^n)$, and the equality

$$\frac{1}{2\pi}\int_{-\infty}^{\infty} |[K(i\omega)]^{-1}\tilde{f}(i\omega)|^2 d\omega = \| Q \|_L^2$$

is valid. Here

$$Q(t) := \frac{1}{2\pi}\int_{-\infty}^{\infty}[K(i\omega)]^{-1}\tilde{f}(i\omega)\exp(i\omega t)d\omega.$$

Due to (2.10) we have $y(t) = Q(t) + x_0(t)$. Once more exploiting Parseval's equality, we have

$$\| Q \|_L^2 \leq \tilde{b}_0^2 \frac{1}{2\pi}\int_{-\infty}^{\infty} |\tilde{f}(i\omega)|^2 d\omega = \tilde{b}_0^2 \| f \|_L^2 .$$

This proves the result.□

Lemma 2.2 Let $\Gamma_0(K) < \infty$. Then the inequality $\tilde{b}_0 \leq \Gamma_0(K)$ is true.

Proof For any invertible $n \times n$ matrix A, inequality (1.4) yields

$$|A^{-1}| \leq \sum_{k=0}^{n-1}\frac{g^k(A)}{\sqrt{k!}d^{k+1}(A)}$$

where $d(A)$ is the smallest modulus of eigenvalues of A:

$$d(A) := \min_{k=1,\dots,n} |\lambda_k(A)|.$$

Thus,

$$|[K(z)]^{-1}| \leq \sum_{k=0}^{n-1}\frac{g^k(K(z))}{\sqrt{k!}d^{k+1}(K(z))} \quad (z \in \mathbf{C}).$$

But from (1.3) it follows that $g(K(z)) = g(\widetilde{R}(z) - zI) = g(\widetilde{R}(z))$ $(z \in \mathbf{C})$. Hence according to (1.6),

$$|[K(i\omega)]^{-1}| \leq \Gamma(K(i\omega)) \leq \Gamma_0(K) \ \ (-\infty < \omega < \infty).$$

As claimed.□

Lemmas 2.1 and 2.2 imply

Corollary 2.1 Any solution $u(\cdot)$ of Eq. (2.9) with initial condition (2.2) satisfies the inequality $\| u \|_L \leq \Gamma_0 \| f \|_L + \| x_0 \|_L$.

Lemma 2.3 Under the hypothesis of Theorem 2.2, let the inequality

$$q\tilde{b}_0 < 1 \tag{2.11}$$

hold. Then the estimate

$$\| x \|_L < (1 - \tilde{b}_0 q)^{-1}(\tilde{b}_0 M(\phi) + \| x_0 \|_L)$$

is valid for any solution $x(\cdot)$ of Eq. (2.1) with initial condition (2.2).

Proof For a positive $a < \infty$ let us introduce the scalar–valued function by the relations $\chi_a(t) = 1$ $(0 \leq t \leq a)$ and $\chi_a(t) = 0$ $(t > a)$. Consider the equation

$$\dot{u}(t) = \int_{-h}^{0} [dR(\theta)]Iu(t+\theta) + \chi_a(t) f(t, u_t) \ (t \geq 0) \tag{2.12}$$

with the initial function ϕ. Clearly, any solution $w_a(\cdot)$ of (2.12) for $t \geq a$ coincides with a solution x_0 of the homogeneous equation (1.5) considered on $[a, \infty)$ with the corresponding initial condition. Therefore, $w_a(\cdot) \in L_2([0, \infty), \mathbf{R}^n)$. Rewrite Eq. (2.12) in the form (2.9) with $f(t) = \chi_a(t) f(t, u_t)$. Lemma 2.1 and restriction (2.3) imply

$$\| w_a \|_L \leq \tilde{b}_0 \| f(.,(w_a).) \|_L + \| x_0 \|_L \leq \tilde{b}_0(q \| w_a \|_L + M(w_a)) + \| x_0 \|_L .$$

But by (2.2) $M(w_a) = M(\phi)$. Now condition (2.11) ensures the inequality

$$\| w_a \|_L \leq (1 - \Gamma_0(K)q)^{-1}(\Gamma_0(K)M(\phi) + \| x_0 \|_L).$$

Passing at the left part of that inequality to the limit as $a \to \infty$, and taking into account that $w_a(t)$ for $t \leq a$ coincides with a solution $x(t)$ of (2.1), we arrive to the result. □

Proof of Thm. 2.2 The assertion of Thm. 2.2 immediately follows from Lemmas 2.2 and 2.3.□

3. ESTIMATES FOR THE GREEN FUNCTION

Let $G(\cdot)$ be the Green function of Eq. (1.5). By the Laplace transformation, it can be written

$$G(t) = -\frac{1}{2\pi}\int_{-\infty}^{\infty}[K(i\omega)]^{-1}\exp(i\omega t)d\omega \tag{3.1}$$

(see Subs. 2.6.2).

Theorem 3.1 The Green function $G(\cdot)$ of Eq. (1.5) satisfies the inequality

$$\| G \|_L \leq W_0(K) := \sqrt{2\Gamma_0(K)[VR\Gamma_0(K)+1]}.$$

Lemma 3.1 Let $f \in L_2([0,\infty),\mathbf{R}^n)$, and let the solution $u(t)$ of Eq. (2.9) with the zero initial condition belong to $L_2([0,\infty),\mathbf{R}^n)$. Then the inequality $\| \dot{u} \|_L \leq VR \| u \|_L + \| f \|_L$ is true.

Proof From (2.9) it follows that $\| \dot{u} \|_L \leq I_1 + \| f \|_L$, where

$$I_1 = \left\{\int_0^{\infty} \| \int_{-h}^{0}[dR(\theta)]u(t+\theta) \|^2 dt\right\}^{1/2}.$$

We have

$$I_1 \leq \int_{-h}^{0}|dR(\theta)|[\max_{-h\leq\theta\leq 0}\int_0^{\infty} \| u(t+\theta) \|^2 dt]^{1/2}.$$

Taking into account that u has the zero initial condition, we get

$$\int_0^{\infty} \| u(t+\theta) \|^2 dt = \| u \|_L^2 .$$

Consequently, $I_1 \leq VR \| u \|_L$. This gives the assertion of the lemma.□

Proof of Theorem 3.1 Set $\psi(z) := (z+b)^{-1}I$ ($z \in \mathbf{C}$) with a positive constant b. Therefore, due to (3.1)

$$G(t) = -\frac{1}{2\pi}\int_{-\infty}^{\infty}[K(i\omega)]^{-1}\exp(i\omega t)(i\omega+b)\psi(i\omega)d\omega.$$

Hence

$$G(t) = \dot{w}(t) + bw(t), \tag{3.2}$$

where

$$w(t) := -\frac{1}{2\pi}\int_{-\infty}^{\infty} [K(i\omega)]^{-1} \exp(i\omega t)\psi(i\omega)d\omega.$$

Since Ie^{-bt} is the Laplace original for $\psi(z)$, the function w is the solution of the equation

$$\dot{w}(t) = \int_{-h}^{0} [dR(\theta)]w(t+\theta) + Ie^{-bt}$$

with the zero initial condition $w(t) = 0$ for $t \leq 0$ and $w(0^+) = 0$. By Corollary 2.1 the inequality

$$\| w \|_L \leq \Gamma_0(K) \| e^{-b\cdot} \|_L = \Gamma_0(K)(2b)^{-1/2}$$

holds. This inequality and Lemma 3.1 yield

$$\| \dot{w} \|_L \leq V(R) \| w \|_L + \| e^{-b\cdot} \|_L \leq \frac{VR\Gamma_0(K)+1}{\sqrt{2b}}.$$

Now (3.2) gives

$$\| G \|_L \leq \| \dot{w} \|_L + b \| w \|_L \leq \frac{VR\Gamma_0(K)+1+b\Gamma_0(K)}{\sqrt{2b}}.$$

Taking

$$b = \Gamma_0^{-1}(K)(V(R)\Gamma_0(K)+1),$$

we obtain the required inequality.□

Theorem 3.2 The derivative of the Green function of Eq. (1.5) satisfies the inequality

$$\| \dot{G} \|_L \leq W_0(K)VR.$$

Proof Take into account that $G(t)$ is the solution of (1.5) with the initial condition $G(0) = I$, $G(t) = 0$ $(t \leq 0)$. Then

$$\| \dot{G} \|_L^2 = \| \int_{-h}^{0} [dR(\theta)]G(t+\theta) \|_L^2 = \int_0^{\infty} | \int_{-h}^{0} [dR(\theta)]G(t+\theta)|^2 dt.$$

But $G(t+\theta) = 0$ $(t+\theta < 0)$. Hence

$$\| \dot{G} \|_L^2 \leq \left(\int_{-h}^{0} |dR(s)| \right)^2 \sup_{-h\leq\theta\leq 0} \int_0^{\infty} |G(t+\theta)|^2 dt = (VR)^2 \| G \|_L^2 .$$

Now the previous lemma yields the result.□

To prove the estimate for the sup–norm of the Green function we will apply the following

Lemma 3.2 Let locally absolutely continuous vector–valued function α be defined on $[0,\infty)$, and its derivative $\dot{\alpha}$ belong to space $L_2([0,\infty),\mathbf{R}^n)$. Then

$$|\alpha(t)|^2 \le 2\left[\int_t^\infty |\alpha(s)|^2 ds \int_t^\infty |\dot{\alpha}(s)|^2 ds\right]^{1/2}.$$

Proof Since α is a continuous function and $\| \alpha \|_L < \infty$, due to the mean value theorem it can be written

$$\int_t^{t+1} |\alpha(s)|^2 ds = |\alpha(\theta)|^2 \to 0 \text{ as } t \to \infty \ (t \le \theta \le t+1).$$

Consequently, $\alpha(\infty) = 0$. Obviously,

$$|\alpha(t)|^2 = -\int_t^\infty \frac{d|\alpha(s)|^2}{ds} ds = -2\int_t^\infty |\alpha(s)|\frac{d|\alpha(s)|}{ds} ds.$$

Taking into account that

$$\left|\frac{d}{ds}|\alpha(s)|\right| \le |\dot{\alpha}(s)|$$

we get the result due to Schwartz's inequality.□

By the latter lemma,

$$\sup_{t\ge 0} |G(t)| \le \sqrt{2 \| G \|_L \| \dot{G} \|_L}. \tag{3.3}$$

Thms. 3.1 and 3.2 imply

Corollary 3.1 The Green function of Eq. (3.1) satisfies the inequality

$$\sup_{t\ge 0} |G(t)| \le a_0(K) := 2\sqrt{VR\Gamma_0(K)[VR\Gamma_0(K)+1]}.$$

Now we are going to establish an estimate for the Liapunov exponent of the Green function.

Let all the zeros of $\det K(z)$ lie in the half–plane

$$\mathbf{C}_{\alpha_0} := \{z \in \mathbf{C} : \text{Re} z \le -\alpha_0\}$$

with a positive α_0. Further, for a non–negative number ν, put

$$\Gamma_\nu(K) := \sup_\omega \Gamma(K(-\nu+\omega i)),\ V_\nu(R) := \nu + \int_{-h}^0 e^{-\nu s}|dR(s)|,$$

and

$$\alpha_\nu(K) := 2\sqrt{V_\nu(R)\Gamma_\nu(K)[V_\nu(R)\Gamma_\nu(K)+1]}.$$

That is $V_0(R) = VR$. Assume that

$$0 < \nu < \alpha_0, \tag{3.4}$$

and substitute the equality

$$G_\nu(t) := e^{\nu t}G(t) \tag{3.5}$$

in Eq. (1.5). Take into account that $G(t)$ is a solution of (1.5), then

$$\dot{G}_\nu(t) = \nu G_\nu(t) + \int_{-h}^{0} e^{-\nu s}dR(s)G_\nu(t+s) \quad (t \geq 0). \tag{3.6}$$

Since $G_\nu(t) = G(t) = 0$ $(t < 0)$ and $G_\nu(0) = G(0) = I$, function $G_\nu(\cdot)$ is the Green function of Eq. (3.6). Put

$$K_\nu(z) := B_\nu(z) - zI \quad \text{with} \quad B_\nu(z) := \nu I + \int_{-h}^{0} e^{-(\nu-z)s}dR(s).$$

Or

$$B_\nu(z) = \int_{-h}^{0} e^{zs}dR_\nu(s) \quad \text{with} \quad R_\nu(t) := \nu\chi(t)I + \int_{0}^{t} e^{-\nu s}dR(s),$$

where $\chi(t) := 0$ if $t < 0$, and $\chi(t) := 1$ if $t \geq 0$. We can write

$$\int_{-h}^{0} |dR_\nu(s)| \leq \nu + \int_{-h}^{0} e^{-\nu s}|dR(s)| = V_\nu(R).$$

Clearly, $K_\nu(i\omega) = K(-\nu + i\omega)$. According Cor. 3.5 to Eq. (3.6), and taking into account the second relation from (1.3), we get the estimate $|G_\nu(t)| \leq a_\nu(K)$ $(t \geq 0)$. Hence (3.5) yields the following

Lemma 3.3 Let all the zeros of $\det K(z)$ *lie in half-plane* $\mathbf{C}_{\alpha_0}$. *Then under condition (3.4), the Green function (3.1) is*

$$|G(t)| \leq a_\nu(K)e^{-\nu t} \; (t \geq 0).$$

The latter inequality gives an estimate for the Liapunov exponent of the Green function.

4. A BOUND FOR A REGION OF ATTRACTION

Let us consider in $\mathbf{R}^n$ the equation

$$\dot{x}(t) = \int_{-h}^{0} [dR(\theta)]x(t+\theta) + f(t, x(t-h_1(t)), \ldots, x(t-h_m(t))) \; (t \geq 0), \quad (4.1)$$

where $h_j(t)$ $(j = 1, \ldots, m)$ are functions satisfying (2.4), (2.5), and f continuously maps $[0, \infty) \times \mathbf{R}^{nm}$ into $\mathbf{R}^n$. Take the initial condition (2.2) and denote

$$\Omega(r) := \{\phi \in C([-h, 0], \mathbf{R}^n) : \| \phi \| \leq r\}$$

for a positive number $r < \infty$; $\Omega(\infty) := \mathbf{R}^n$. It is assumed that there are positive constants $q_j (j = 1, \ldots, m)$ such that

$$|f(t, y_1, \ldots, y_m)| \leq \sum_{j=1}^{m} q_j |y_j| \text{ for all } y_j \in \Omega(r) \; (j = 1, \ldots, m), \; t \geq 0. \quad (4.2)$$

Let the zero solution of (4.1) be asymptotically stable. We shall say that the ball $\{\phi \in C([-\eta, 0], \mathbf{R}^n) : \| \phi \|_{C([-\eta,0],\mathbf{R}^n)} \leq r_0\}$ belongs to *a region of attraction* of the zero solution if the condition

$$\max_{-h \leq \theta \leq 0} |\phi(t)| \leq r_0$$

for a given continuous initial function ϕ implies the relation $|x(t)| \to 0$ as $t \to \infty$ for the corresponding solution x.

Theorem 4.1 Let q be defined by (2.6), and conditions (4.2) and (2.7) hold. Then the zero solution of (4.1) is asymptotically stable. Moreover, any initial function ϕ satisfying the inequality

$$\theta(K, \phi, q) := \frac{1 + VR\Gamma_0(K)}{1 - q\Gamma_0(K)} [2|\phi(0)| + \zeta(\phi)\sqrt{2(VR + q)}] \leq r, \quad (4.3)$$

belongs to the region of attraction of the zero solution, and the bound

$$|x(t)| \leq \theta(K, \phi, q) \; (t \geq 0) \quad (4.4)$$

is true for the corresponding solution of Eq. (4.1).

We divide the proof of Thm. 4.1 into a series of lemmas.

Lemma 4.1 Let $x_0(t)$ be a solution of the homogeneous equation (1.5) with initial condition (2.2). Then

$$\| x_0(t) \|_L \leq W_0(K)|\phi(0)| + \Gamma_0(K)\zeta(\phi)VR.$$

Proof Let $\tilde{x}_0(z)$ be the Laplace transform of $x_0(t)$. We use the representation

$$\tilde{x}_0(z) = -K^{-1}(z)[\phi(0) + \int_{-h}^{0} dR(s) \int_{s}^{0} e^{-z(t-s)}\phi(\tau)d\tau]$$

(see Subs. 2.6.2). The inverse Laplace transformation yields

$$\phi(t) = G(t)\Phi(0) - \frac{1}{2\pi}\int_{-\infty}^{\infty} [K(i\omega)]^{-1}\exp(i\omega t)\int_{-h}^{0} dR(s)\int_{s}^{0} e^{-z(t-s)}\phi(\tau)d\tau d\omega,$$

since $-K(z)$ is the Laplace transform of $G(t)$. Hence

$$x_0(t) = G(t)\phi(0) + \int_{-h}^{0} dR(s)\int_{s}^{0} G(t-\tau+s)\phi(\tau)d\tau. \tag{4.5}$$

Denote

$$J_s(t) := \int_{s}^{0} G(t-\tau+s)\phi(\tau)d\tau.$$

Clearly,

$$J_s(t) = \int_{0}^{-s} G(t-\tau)\Phi(\tau+s)d\tau = \int_{0}^{\infty} G(t-\tau)\Phi_s(\tau)d\tau$$

with the notation $\Phi_s(\tau) := \phi(\tau+s)$ for $0 \leq \tau \leq -s$, and $\Phi_s(\tau) := 0$ for $\tau \geq -s$. Since $G(t) = 0$ for $t < 0$, we get

$$J_s(t) = \int_{0}^{t} G(t-\tau)\Phi_s(\tau)d\tau.$$

That is, $J_s(t)$ is a solution of the equation

$$\dot{x}(t) = \int_{-h}^{0} [dR(\theta)]x(t+\theta) + \Phi_s(t)$$

with the zero initial condition. By Cor. 2.1 we get $\| J_s \|_L \leq \Gamma_0(K) \times$

$\| \Phi_s(\cdot) \|_L$. But

$$\| \Phi_s(\cdot) \|_L^2 = \int_0^{-s} |\phi(\tau - s)|^2 d\tau \leq \zeta^2(\phi).$$

Thus $\| J_s \|_L \leq \Gamma_0(K)\zeta(\phi)$, and by (4.5)

$$\| x_0 \|_L \leq \| G \|_L |\phi(0)| + \| \int_{-h}^{0} dR(s) J_s(t) \|_L \leq \| G \|_L |\phi(0)| +$$

$$+ \int_{-h}^{0} |dR(s)| \max_{-h \leq s \leq 0} \| J_s(\cdot) \|_L \leq \| G \|_L |\phi(0)| + VR\Gamma_0(K)\zeta(\phi).$$

Now Theorem 3.1 yields the result.□

Lemma 4.2 Assume that inequality (4.2) holds with $\Omega(r) = \mathbf{R}^n$. Then under condition (2.7), estimate (4.4) is valid for any solution $x(t)$ of Eq. (4.1).

Proof For the brevity, in the sequel we set $\Gamma_0(K) := \Gamma_0$, and

$$\| y \|_M = \| y(\cdot) \|_{C([0,\infty),\mathbf{R}^n)}$$

for every continuous vector–valued function y defined on $[0, \infty)$. Further, by Thm. 2.1,

$$\| f(\cdot, u) \|_L = \left[\int_0^\infty |f(t, u(t - h_1(t)), \ldots, u(t - h_m(t)))|^2 dt \right]^{1/2} \leq \tag{4.6}$$

$$\leq q(\| u \|_L + \zeta(u))$$

with an arbitrary $u \in L_2([-\eta, 0], \mathbf{R}^n)$. By Thm. 2.2 we get

$$\| x \|_L \leq (1 - \Gamma_0 q)^{-1}(\Gamma_0 q \zeta(\phi) + \| x_0 \|_L). \tag{4.7}$$

Furthermore, take into account that

$$\| \int_{-h}^{0} [dR(s)] x(t + s) \|_L^2 = \int_0^\infty | \int_{-h}^{0} [dR(s)] x(t + s)|^2 dt \leq$$

$$\leq (\int_{-h}^{0} |dR(s)|)^2 \sup_{-h \leq s \leq 0} \int_0^\infty |x(t + s)|^2 dt = (VR)^2(\| x \|_L^2 + \zeta^2(\phi)).$$

According to (4.1) and (4.6), this relation yields

$$\| \dot{x}(\cdot) \|_L \leq \| \int_{-h}^{0} [dR(s)]x(\cdot - s) \|_L + \| f(\cdot, x_\cdot) \|_L \leq (VR + q)(\| x \|_L + \zeta(\phi)).$$

Further, by Lemma 3.2 $\| x \|_M \leq \sqrt{2 \| x \|_L \| \dot{x} \|_L}$. Consequently,

$$\| x \|_M \leq \sqrt{2(VR + q)(\| x \|_L + \zeta(\phi)) \| x \|_L} \leq$$

$$\leq \sqrt{2(VR + q)}(\| x \|_L + \zeta(\phi)).$$

Now (4.7) gives

$$\| x \|_M \leq \sqrt{2(VR + q)}[(1 - q\Gamma_0)^{-1}(\| x_0 \|_L + q\Gamma_0\zeta(\phi)) + \zeta(\phi)] =$$

$$= \sqrt{2(VR + q)}(1 - q\Gamma_0)^{-1}[\| x_0 \|_L + \zeta(\phi)].$$

Lemma 4.1 implies

$$\| x \|_M \leq \frac{\sqrt{2(VR + q)}}{1 - q\Gamma_0}[W_0(K)|\phi(0)| + (\Gamma_0 VR + 1)\zeta(\phi)]. \tag{4.8}$$

But according to (2.7),

$$\sqrt{2(VR + q)}W_0(K) = 2\sqrt{(VR + q)\Gamma_0(VR\Gamma_0 + 1)} \leq 2(VR\Gamma_0 + 1).$$

Thus, by (4.8) $\| x \|_M \leq \theta(K, \phi, q)$. As claimed.□

Lemma 4.3 Let f continuously map $[0, \infty) \times \mathbf{R}^{nm}$ into $\mathbf{R}^n$ and satisfy (4.2) with a $r < \infty$. Then there is a continuous function

$$f_1 : [0, \infty) \times \mathbf{R}^{nm} \to \mathbf{R}^n$$

with the properties

$$|f_1(t, y_1, \ldots, y_m)| \leq \sum_{j=1}^{m} q_j |y_j| \quad for\ all \quad y_j \in \mathbf{R}^n \quad (j = 1, \ldots, m; t \geq 0), \tag{4.9}$$

and

$$f_1(t, y_1, \ldots, y_m) = f(t, y_1, \ldots, y_m) \quad for \quad y_j \in \Omega(r)\ (j = 1, \ldots, m, t \geq 0). \tag{4.10}$$

Proof It is sufficiently to put

$$f_1(t, y_1, \ldots, y_m) := f(t, \min\{|y_1|, r\}(\mathrm{sgn} y_1), \ldots, \min\{|y_m|, r\}(\mathrm{sgn} y_m)).\Box$$

Proof of Thm. 4.1. In the case $r = \infty$ the theorem is valid due to Lemma 4.2. Consider the equation

$$\dot{x}(t) = \int_{-h}^{0} dR(s)x(t+s) + f_1(t, x(t-h_1(t)), \ldots, x(t-h_m(t))) \ (t \geq 0), \quad (4.11)$$

where the vector-valued function f_1 defined on $[0, \infty) \times \mathbf{R}^{n \times m}$ satisfies conditions (4.9) and (4.10). The function f_1 can be constructed due to Lemma 4.3. Thus, for a solution of (4.11) Lemma 4.2 gives estimate (4.4). Condition (4.3) and estimate (4.4) guarantee that the solution of (4.11) lies in $\Omega(r)$. But according to (4.10) it is the solution of (4.1), simultaneously. This proves Thm. 4.1.$\Box$

CHAPTER 6

LIAPUNOV FUNCTIONALS FOR CONCRETE FDEs

In this chapter the procedure to construct Liapunov functionals for some FDEs is proposed and stability conditions are obtained.

1. STATEMENT OF THE PROBLEM

Let $\mathbf{R}^n$ be a real n–dimensional linear vector space with a norm $|\cdot|$. Consider the Cauchy problem for RDE

$$\dot{x}(t) = f(t, x_t),\ t > t_0,\ x_t(\theta) := x(t+\theta)\ (-\infty < \theta \leq 0), \tag{1.1}$$

$$x_{t_0} = \phi : (-\infty, 0] \to \mathbf{R}^n. \tag{1.2}$$

Here $x(t) \in \mathbf{R}^n$, continuous functional $f : [t_0, \infty) \times C(-\infty, 0] \to \mathbf{R}^n$, $f(t, 0) \equiv 0$, and f satisfies local Lipschitz condition with respect to the second argument. Space $C(-\infty, 0]$ in this chapter is a linear space of continuous functions $\phi : (-\infty, 0] \to \mathbf{R}^n$ equipped with some metric d satisfying the conditions of Subs. 3.2.5.

Function $x(\cdot; t_0, \phi) \in C^1[t_0, T)$ is a solution of the problem (1.1), (1.2) if it satisfies Eq. (1.1) at all points of $[t_0, T)$, with $x(t+\theta) = \phi(t+\theta-t_0)$ for $t + \theta < t_0$, and $x(t_0) = \phi(0)$.

Because of the assumption $f(t, 0) \equiv 0$ there exists the zero solution of Eq. (1.1). Definitions of stability of the zero solution are given in Subs. 4.1.1.

The most powerful method for studying the stability properties of FDEs is the Liapunov's direct method. It consists in the use of an auxiliary functional defined on the solutions of Eq. (1.1) (see Sect. 4.1). Moreover some general results using Liapunov's approach were obtained for the problems of existence of solutions defined in the future, existence of periodic solutions, periodic and stationary measures induced by solutions at the phase space, etc.

Along with these general theorems some concrete equations were investigated by construction of the appropriate functionals V, allowing to formulate effective stability conditions immediately in terms of the characteristics of equations under consideration.

But the process of construction of a functional V is often connected with so many difficulties that some authors compared it with an art (see, e.g. [72], p.166).

Nevertheless, the natural wish is to determine some formal procedure or algorithm leading to the construction of functional V. Positive answer over the whole class of FDEs cannot be expected. Therefore one should try to investigate the indicated problem for special classes of FDEs.

Analysis of the existing Liapunov functionals allows us to perceive some common features in their structure and formulate appropriate formal procedure.

2. FORMAL DESCRIPTION OF THE PROCEDURE

The proposed procedure to construct a Liapunov functional V consists in the following steps:

1) the right hand side of Eq. (1.1) (i.e. functional f) must be transformed in such a way that the transformed equation, equivalent to (1.1), is written in the form

$$\dot{z}(t,x_t) = f_1(t,x(t)) + f_2(t,x_t), \quad (\dot{z}(t,x_t) := (z(t,x_t))^{\bullet})$$

where $f_1 : [t_0,\infty) \times \mathbf{R}^n \to \mathbf{R}^n$ is a function depending only on t and on the current values of the solution $x(t), z(t,0) \equiv 0, f_1(t,0) \equiv 0, f_2(t,0) \equiv 0$, and $z, f_2 : [t_0,\infty) \times C(-\infty,0]$ are some continuous functionals.

2) Assume that the zero solution of the auxiliary ODE $\dot{y}(t) = f_1(t,y(t))$ is uniformly asymptotically stable. Then, due to the inversion theorem, there exists a Liapunov function $(t,y) \mapsto v(t,y)$ which is positive definite, decrescent and for which $\dot{v}(t,y)$ is negative definite. The aim of the second step is to construct the function v.

3) Changing the argument y of the function $(t,y) \mapsto v(t,y)$ to the value of the functional $z(t,x_t)$, obtained at the first step, one obtains a functional V_1 which is the basic component of the desired functional V. Fitting V_2 in such a way that $V_1 + V_2$ satisfies the assumptions of some stability theorem, one obtains the functional V in the form $V = V_1 + V_2$.

Remark that many previously invented functionals V could be obtained by using basically the above–described procedure. It is also worth noting that the form of transformation and Liapunov function v can be (generally speaking) picked out by different ways and each of them could lead to different stability conditions. Let us illustrate the last assertion by some examples.

Example 2.1 Consider a scalar RDE with constant coefficients a, b and constant delay $h \geq 0$

$$\dot{x}(t) = -ax(t) - bx(t-h), \quad t \geq 0.$$

This equation is already represented in the desired form

$$\dot{z}(t,x_t) = f_1(t,x(t)) + f_2(t,x_t)$$

with

$$f_1(t,x(t)) := -ax(t),\ f_2(t,x(t)) := -bx(t-h),\ z(t,x_t) := x(t).$$

So, we can pass to the second step of the procedure. Auxiliary ODE $\dot{y}(t) = -ay(t)$ is asymptotically stable if $a > 0$. Liapunov function for this ODE can be taken in the form $v(t,y) := y^2$. Hence according to the procedure we must put $V_1(t,x_t) := x^2(t)$. The functional V_1 is positive definite, decrescent. Therefore, the last step is to choose V_2 such that $\dot{V}$ will be negative definite. But

$$\dot{V}_1(t,x_t) = -2ax^2(t) - 2bx(t)x(t-h) \leq (-2a + |b|)x^2(t) + |b|x^2(t-h).$$

This inequality means that the functional V_2 has to be chosen in the form

$$V_2(t,x_t) := |b| \int_{t-h}^{t} x^2(s)ds.$$

As a result we obtain that $\dot{V}(t,x_t) \leq -2(a - |b|)x^2(t)$. Thus application of this functional V gives us asymptotic stability conditions $a > |b|$ which is independent on the value of delay h.

In order to obtain other stability conditions we can use another types of transformation. Let us take, e.g., the following transformed equation (which is an equation of a neutral type)

$$\dot{z}(t,x_t) = f_1(t,x(t)),\ z(t,x_t) := x(t) - b\int_{t-h}^{t} x(s)ds,$$

$$f_1(t,x_t) := -(a+b)x(t),\ f_2(t,x_t) :\equiv 0.$$

Then the auxiliary ODE has the similar form as previously $\dot{y}(t) = -(a+b)y(t)$ for which the Liapunov function is $v(t,y) := y^2$. Hence $V_1(t,x_t) := z^2(t,x_t)$ and for $a+b \geq 0$ we get

$$\dot{V}_1(t,x_t) = -2z(t,x_t)(a+b)x(t) \leq -(a+b)(2-|b|h)x^2(t)+$$

$$+(a+b)|b| \int_{t-h}^{t} x^2(s)ds.$$

Hence in order to guarantee negative–definiteness of $\dot{V}$ we can take

$$V_2(t,x_t) := (a+b)|b| \int_{t-h}^{t} ds \int_{s}^{t} x^2(\tau)d\tau.$$

We have for $V := V_1 + V_2$ that

$$\dot{V}(t,x_t) \le -2(a+b)(1-|b|h)x^2(t).$$

So $\dot{V}(t,x_t)$ is negative if $a+b>0,\ |b|h<1$.

Let us pay attention that the functional V is only nonnegative (not positive definite). Because of this some additional arguments must be involved to prove stability. For NDEs it is sufficient to assume that the trivial solution of the functional inequality $|z(t,x_t)| \le f(t)$ is asymptotically stable (see Thm. 8.1.1). The latter assumption will be valid if $h|b|<1$. As a result we obtain, using the second form of the functional V, that the sufficient asymptotic stability conditions are $a+b>0,\ |b|h<1$. This gives us additional part of stability domain comparing with that of defined by the inequality $a>|b|$.

Note that using explicit form of the functional V we can obtain the same asymptotic stability conditions without reference to Thm. 8.1.1. In fact, from inequality $V(t,x_t) \le V(0,\phi)$ it follows that $(1-|b|h)|x(t))| \le V(0,\phi)^{1/2}$. This implies stability and also boundedness of any solution and its derivative. From the estimate of $\dot{V}(t,x_t)$ we have $\int_0^\infty x^2(t)dt < \infty$. Hence $x(t) \to 0$ as $t \to \infty$ for any solution.

Consider briefly two more examples in which the way of application of the procedure is the same as in example 2.1.

Example 2.2 Consider the one–dimensional equation

$$\dot{x}(t) = \int_0^h f(x(t-s))dK_0(s). \quad t \ge 0,\ h>0. \tag{2.1}$$

Here the continuously differentiable function $f : \mathbf{R} \to \mathbf{R}$ satisfies the conditions

$$|f_x(x)| \le \gamma_1, \quad xf(x) \le -\gamma_2 x^2, \quad x \in \mathbf{R}, \quad f(0)=0,$$

where γ_i are some nonnegative constants, $K_0 : [0,h] \to \mathbf{R}$ is a function with bounded variation on $[0,h]$, and $f_x(x) := df(x)/dx$.

We set

$$\alpha_{j,0} := \int_0^h s^j\,|dK_0(s)|, \quad \beta_{j,0} := \int_0^h s^j\,dK_0(s).$$

In order to obtain some stability conditions for (2.1), let us apply the proposed procedure. Take the transformed equation as follows

$$\dot{z}(t) = \beta_{0,0} f(x(t)), \quad z(t) := \int_0^h dK_0(s) \int_{t-s}^t f(x(\tau))d\tau + x(t). \tag{2.2}$$

Then the auxiliary ODE is

$$\dot{y}(t) = \beta_{0,0} f(y(t)). \tag{2.3}$$

As a Liapunov function v for Eq. (2.3) we can take $v(t,y) := y^2$. The component V_1 of the functional V is always obtained as a result replacing y in the function v by the functional z defined in (2.2). Hence we get

$$V_1(x_t) := \left[x(t) + \int_0^h dK_0(s) \int_{t-s}^t f(x(\tau))d\tau \right]^2 .$$

Now in order that $\dot{V} = \dot{V}_1 + \dot{V}_2$ would be negative definite we have to choose

$$V_2(x_t) := \gamma_1^2 \beta_{0,0} \int_0^h |dK_0(s)| \int_{t-s}^t (\tau - t + s) x^2(\tau) d\tau.$$

From the properties of the functional $V := V_1 + V_2$ it follows that the asymptotic stability conditions obtained from the transformed Eq. (2.2) are

$$\beta_{0,0} > 1, \quad \gamma_2 > \gamma_1^2 \alpha_{1,0}, \quad \gamma_1 \alpha_{1,0} < 1.$$

The transformation we have just described is not the only one available. Take another transformed equation, which is valid for $t \geq h$

$$\dot{x}(t) = \beta_{0,0} F(x(t)) - \int_0^h dK_0(s) \int_{t-s}^t f_x(x(\tau))d\tau \int_0^h f(x(\tau - s_1))dK_0(s_1). \tag{2.4}$$

Then the auxiliary ODE keeps the previous form (2.3). Hence $V_1(x_t) := x^2(t)$. Because of this

$$V_2(x_t) := \gamma_1^2 \int_0^h |dK_0(s)| [\int_{t-s}^t d\tau_0 \int_{\tau_0}^t d\tau \int_0^h x^2(\tau - s_1) dK_0(s_1) +$$

$$+ \alpha_{1,0} \int_{t-s}^t x^2(\tau) d\tau].$$

So using the transformed equation (2.4) we get asymptotic stability condition

$$\gamma_2\beta_{0,0} > \gamma_1^2\alpha_{0,0}\alpha_{1,0}.$$

Example 2.2 Consider the second order equation which is equivalent to the system

$$\dot{x}_1(t) = x_2(t), \quad \dot{x}_2(t) = -ax_2(t) + \int_0^h x_2(t-s)dK_0(s) - f(x_1(t)), \; t \geq 0. \quad (2.5)$$

Here a is a constant, the kernel K_0 is the same as in equation (2.1) and the continuous function $f : \mathbf{R} \to \mathbf{R}$ is such that $xf(x) > 0$ for all $x \neq 0$.

From (2.5) it follows the auxiliary ODEs system

$$\dot{y}_1(t) = y_2(t), \quad \dot{y}_2(t) = -ay_2(t) - f(y_1(t)). \quad (2.6)$$

A Liapunov function v for (2.6) can be taken as

$$v(y_1, y_2) := (ay_1 + y_2)^2 + 4\int_0^{y_1} f(s)ds + y_2^2.$$

Hence in view of (2.5), (2.6)

$$V_1(x_{1t}, x_{2t}) := [x_2(t) + ax_1(t)]^2 + 4\int_0^{x_1(t)} f(s)ds + x_2^2(t).$$

Because of this

$$V_2(x_{1t}, x_{2t}) := (2+a)\int_0^h |dK(s)| \int_{t-s}^t x_2^2(\tau)d\tau.$$

Using the functional $V = V_1 + V_2$ we obtain asymptotic stability conditions

$$a > \left(2 + \frac{a}{2}\right)\alpha_{0,0} \geq 0, \quad 2xf(x) \geq (\varepsilon + \alpha_{0,0})x^2$$

where ε is a positive number.

3. DISSIPATIVE SYSTEMS

3.1. *STABILITY*

We want to apply the proposed procedure to FDEs (1.1) for which the auxiliary ODEs are dissipative. For simplicity of presentation consider (1.1) with one discrete delay

$$\dot{x}(t) = f(t, x(t-h)), \quad t > t_0, \quad x(t) \in \mathbf{R}^n, \quad f(t,0) \equiv 0, \quad (3.1)$$

Here $f : [t_0, \infty) \times \mathbf{R} \to \mathbf{R}$ is a continuous function with continuous derivative $\partial f(t,x)/\partial x =: F(t,x)$ which is called as a Jacobian matrix. The matrix $F(t,x)$ is called ***strictly dissipative*** with respect to some vector norm $\| \cdot \|$ if there exist some positive constant ε and a neighbourhood $D \subset \mathbf{R}^n$ of the origin, such that

$$\| \exp((\tau - t_0)F(t,x)) \|_1 \leq \exp(-\varepsilon(\tau - t_0)), \quad \forall t \geq t_0, \ \tau \geq t_0, \ \ x \in D. \quad (3.2)$$

Here $\| \cdot \|_1$ is the matrix norm induced by the vector norm $\| \cdot \|$:

$$\| F(t,x) \|_1 := \sup_{y \neq 0} \frac{\| F(t,x)y \|}{\| y \|} = \sup_{y : \|y\| = 1} \| F(t,x)y \| .$$

As a constant ε in (3.2) any positive number can be chosen which is less or equal $-\gamma(F(t,x))$ for all $t \geq t_0$, $x \in D$, where $\gamma(F(t,x))$ is the *logarithmic norm* of the matrix $F(t,x)$. Remind (see. e.g. [67,81]) that

$$\gamma(f(t,x)) := \lim_{\Delta \to 0^+} \frac{1}{\Delta}(\| I + \Delta f(t,x) \|_1 - 1),$$

here I is the $n \times n$ identity matrix.

So dissipativity means that

$$a := - \sup_{t \geq t_0, \ x \in D} \gamma(F(t,x)) > 0. \quad (3.3)$$

In the following table there are examples of the logarithmic norms [67] (see p. 286). Here α_i are any positive constants, $\rho(A)$ is the spectral norm of the matrix A, $\lambda_{\max}(A'A)$ is the maximum eigenvalue of the matrix $A'A$, and B is a matrix with the entries $b_{ij} = \alpha_j a_{ij} (i,j = 1, \ldots, n)$. First norm in the table 3.1 is called *octahedral*, the second *cubic*, the third *Euclidean* or *spheric*, the fourth *romboidal*, the fifth *rectangular*, and the sixth *elliptic*. Denote

$$L := \sup_{t \geq t_0, \ x \in D} \| F(t,x) \|_1 .$$

Then the following estimate

$$\| f(t,x) - f(t,y) \| \leq L \| x - y \|, \quad \forall t \geq t_0, \ \ x, y \in D \quad (3.4)$$

holds.

TABLE 3.1

	Norm of the vector $\| x \|$	Operator norm of the matrix $\| A \|_1$	Logarithmic norm $\gamma(A)$
1.	$\sum_{i=1}^{n} \|x_i\|$	$\max_{1\le j\le n} \sum_{i=1}^{n} \|a_{ij}\|$	$\max_{1\le j\le n} (a_{jj} + \sum_{i\ne j} \|a_{ij}\|)$
2.	$\max_{1\le i\le n} \|x_i\|$	$\max_{1\le i\le n} \sum_{j=1}^{n} \|a_{ij}\|$	$\max_{1\le i\le n} (a_{ii} + \sum_{j\ne i} \|a_{ij}\|)$
3.	$(x'x)^{1/2} = \left(\sum_{i=1}^{n} x_i^2\right)^{1/2}$	$\rho(A) = (\lambda_{\max}(A'A))^{1/2}$	$\frac{1}{2}\lambda_{\max}(A+A') = \lambda_{\max}\left(\frac{A+A'}{2}\right)$
4.	$\sum_{i=1}^{n} \frac{\|x_i\|}{\alpha_i}$	$\max_{1\le j\le n} \sum_{i=1}^{n} \frac{\alpha_i}{\alpha_j} \|a_{ij}\|$	$\max_{1\le j\le n} (\alpha_j a_{jj} + \sum_{i\ne j} \alpha_i \|a_{ij}\|)$
5.	$\max_{1\le i\le n} \frac{\|x_i\|}{\alpha_i}$	$\max_{1\le i\le n} \sum_{j=1}^{n} \frac{\alpha_j}{\alpha_i} \|a_{ij}\|$	$\max_{1\le i\le n} (\alpha_i a_{ii} + \sum_{j\ne i} \alpha_j \|a_{ij}\|)$
6.	$\left(\sum_{i=1}^{n} \frac{x_i^2}{\alpha_i}\right)^{1/2}$	$(\lambda_{\max}(B'B))^{1/2}$	$\frac{1}{2}\lambda_{\max}(B+B') = \lambda_{\max}\left(\frac{B+B'}{2}\right)$

Theorem 3.1 Assume that the function $f : [t_0,\infty)\times\mathbf{R}^n \to \mathbf{R}^n$ is continuous, continuously differentiable with respect to the components of the second argument and

$$a > hL^2. \tag{3.5}$$

Then the zero solution of Eq. (3.1) is uniformly asymptotically stable.

Proof Using the proposed procedure, let us construct appropriate Liapunov functional V. The simplest way to obtain an auxiliary ODE is to put $h = 0$. It means that in the transformed equation $f_1(t,x) := f(t,x)$ and $f_2(t,x_t) := f(t,x(t-h)) - f(t,x(t))$. So we get

$$\dot{y}(t) = f(t,y(t)), \quad t \ge t_0, \quad y \in \mathbf{R}^n. \tag{3.6}$$

As a Liapunov function v for Eq. (3.6) take $v(y) :=\| y \|$. Hence $V_1(x_t) :=$

$\| x(t) \|$. The right upper derivative d^+V_1/dt for an arbitrary norm in the space $\mathbf{R}^n$ is equal to

$$\frac{d^+V_1}{dt} = \frac{d^+ \| x(t) \|}{dt} = Q[x(t), \dot{x}(t)]$$

(see [81]), where the function $Q[x(t), \dot{x}(t)]$ is defined by the relation

$$Q(x,y) := \lim_{\Delta \to 0+} \frac{1}{\Delta}[\| x + \Delta y \| - \| x \|] \quad (\forall x, y \in \mathbf{R}^n). \tag{3.7}$$

(It is easy to prove that the limit in (3.7) does exist for any norm $\| \cdot \|$). Replacing the derivative $\dot{x}(t)$ in the right side of (3.7) according to Eq. (3.1), we get

$$\frac{d^+V_1(x_t)}{dt} \le \lim_{\Delta \to 0+} \frac{1}{\Delta}[\| x(t) + \Delta f(t, x(t)) \| - \| x(t) \|] +$$

$$+ \| f(t, x(t)) - f(t, x(t-h)) \| =$$

$$= Q[x(t), f(t, x(t))] + \| f(t, x(t)) - f(t, x(t-h)) \| .$$

Now verify the validity of the upper bound

$$Q[x(t), f(t, x(t))] \le -a \| x(t) \| . \tag{3.8}$$

From the definition (3.7) of the function $t \mapsto Q[x(t), f(t, x(t))]$ it follows the possibility to interpret this function as the right upper derivative $d^+ \| y \| /dt$ of the norm $\| y \|$ along the solutions of the auxiliary ODE (3.6) at the point (t, y), where $y = x(t)$. So

$$\frac{d^+ \| y \|}{dt} = Q[y, f(t, y)], \quad y = x(t). \tag{3.9}$$

Furthermore

$$f(t, y) = \left(\int_0^1 F(t, sy) ds \right) y. \tag{3.10}$$

In addition, by virtue of [81], for any $t \ge t_0$, $y \in \mathbf{R}^n$

$$\gamma(F(t, y)) = \sup_z \{ \| z \|^{-1} Q[z, F(t, y)z] \}, \quad \| z \| \ne 0, \; z \in \mathbf{R}^n.$$

Taking into account the relations (3.3), (3.9), (3.10) and the convexity of the logarithmic norm, one can conclude that

$$\frac{d^+ \| y \|}{dt} = Q\left[y, \left(\int_0^1 F(t,sy)ds\right) y\right] \le \gamma \left[\int_0^1 F(t,sy)ds\right] \| y \| \le$$

$$\le \int_0^1 \gamma[F(t,sy)]ds \| y \| \le -a \| y \|, \quad y = x(t).$$

Estimate (3.8) is then proved.

Relations (3.8), (3.4), (3.1) mean that for $t \ge h$

$$\frac{d^+ V_1(x_t)}{dt} \le -a \| x(t) \| + L \| x(t) - x(t-h) \| =$$

$$= -a \| x(t) \| + L \| \int_{t-h}^{t} \dot{x}(s)ds \| =$$

$$= -a \| x(t) \| + L \| \int_{t-h}^{t} f(s, x(s-h))ds \| \le$$

$$\le -a \| x(t) \| + L^2 \int_{t-2h}^{t-h} \| x(s) \| ds.$$

So it is sufficient to take the functional $V := V_1 + V_2$, where

$$V_2(x_t) := L^2 \left(\int_{t-2h}^{t-h} ds \int_{s}^{t-h} \| x(\tau) \| d\tau + h \int_{t-h}^{t} \| x(s) \| ds \right),$$

because

$$\frac{d^+ V(x_t)}{dt} \le -(a - hL^2) \| x(t) \| .$$

By virtue of Thm. 4.1.1 Thm. 3.1 is proven.□

Remark 3.1 Note that the dissipativity of the Jacobian matrix $F(t,x)$ must take place with respect to some particular matrix norm $\| \cdot \|_1$ induced by some vector norm $\| \cdot \|$. For example, if $y' f(t,x) y \le -cy'y$ for any $t \ge t_0, x \in D, \ y \in \mathbf{R}^n$ and some $c > 0$, where prime denotes transposition, then $F(t,x)$ must be strictly dissipative with respect to the Euclidean norm and $\gamma(F(t,x)) \le -c$.

Remark 3.2 The same arguments as in proof of Thm. 3.1 can be used to obtain stability conditions for another classes of FDEs (e.g., equations with many discrete delays, with distributed or unbounded delays etc.). Consider, for example, the equation

$$\dot{x}(t) = \int_0^h K(t,s)f(t,x(t-s))ds, \quad t \geq t_0, \quad x \in \mathbf{R}^n, \quad f(t,0) \equiv 0.$$

Here f is the same function as in Eq. (3.1) and the elements of the matrix $K : [t_0, \infty) \times [0,h] \to \mathbf{R}^{nm}$ are continuous functions.

From the transformed equation

$$\dot{x}(t) = \int_0^h K(t,s)ds f(t,x(t)) + \int_0^h K(t,s)[f(t,x(t-s)) - f(t,x(t))]ds$$

it follows that the stability condition keeps the previous form $a > hL^2$, where the constants a and L are defined by formulas (3.3), (3.4), in which $F(t,x)$ must be changed into

$$\left(\int_0^h K(t,s)ds\right) F(t,x).$$

Another example is connected with the equation

$$\dot{x}(t) = f_1(t,x_t)x(t) + f_2(t,x_t)f(t,x(t-h)), \quad x \in \mathbf{R}^n.$$

Here f_i are continuous bounded $n \times n$ matrices and f is as in the previous example. Assume also that for some positive constants a_1 and a_2

$$\gamma(f_1(t,x_t)) \leq -a_1 < 0, \quad \| f_2(t,x_t)f(t,x(t-h)) \| \leq a_2 \| x(t-h) \| .$$

Then, if $a_1 > a_2$, the zero solution is uniformly asymptotically stable. To prove this stability condition it is sufficient to use the functional

$$V(x_t) := \| x(t) \| + a_2 \int_{t-h}^{t} \| x(s) \| ds.$$

For illustration consider the two-dimensional equation $\dot{x}(t) = f_1(t,x_t) \times x(t) + f_2(t,x_t)x(t-h)$,

$$f_1(t,x_t) := \begin{pmatrix} -a_1 & g(t,x_t) \\ -g(t,x_t) & -a_1 \end{pmatrix},$$

$$f_2(t,x_t) := \begin{pmatrix} -a_2 \sin x_2(t) & 0 \\ 0 & a_2 \sin x_1(t) \end{pmatrix},$$

where g is a continuous bounded functional. Take as the norm $\| \cdot \|$ the Euclidean norm in $\mathbf{R}^2$. Then the logarithmic norm of the matrix f_1 is equal to $\gamma(f_1) = -a_1$ and the matrix norm $\| f_2 \|_1 \leq a_2$. Hence sufficient asymptotic stability condition is $a_1 > a_2$.

Remark 3.3 Denote $\| \cdot \|_i$ different norms in $\mathbf{R}^n$, c_i some positive constants, and $\gamma_i(A)$ logarithmic norm of the matrix A corresponding to the vector norm $\| \cdot \|_i$. Consider equation

$$\dot{x}(t) = f_1(t,x_t)x(t), \quad x \in \mathbf{R}^n, \quad t \geq t_0 \tag{3.11}$$

where F_1 is the same as in Remark 3.2.

The resonable choice of the Liapunov functional V for this equation is the product

$$V(x_t) := \| x(t) \|_1^{c_1} \ldots \| x(t) \|_m^{c_m} .$$

Function V is positive definite, decrescent and using (3.6)-(3.10) we get

$$\dot{V}(\psi) \leq [c_1\gamma_1(f_1(t,\psi)) + \ldots + c_m\gamma_m(f_1(t,\psi))]V(\psi).$$

Hence, we obtain

Theorem 3.2 Assume that there exist at least one set of the logarithmic norms $\gamma_1(A),\ldots,\gamma_m(A)$ and positive constants $c_1,\ldots,c_m$ such that for all $t \geq t_0, \phi \in C[-h,0]$

$$c_1\gamma_1(f_1(t,\phi)) + \ldots + c_m\gamma_m(f_1(t,\phi)) \leq -c.$$

Then the zero solution of Eq. (3.11) is exponentially stable if $c < 0$, and uniformly stable if $c = 0$.

Example 3.1 Let f_1 be an arbitrary skew–symmetric matrix, and $\| \cdot \|_1$ be the Euclidean norm. Then the logarithmic norm

$$\gamma(f_1) = \frac{1}{2}\lambda_{\max}(f_1' + f_1) = 0.$$

Hence the zero solution of Eq. (3.11) is uniformly stable.

Example 3.2 Consider Eq. (3.11) with

$$f_1(q) := \begin{pmatrix} -4 & 0 & 4\sin q \\ 0 & -1 & 0 \\ 0 & \cos q & -6 \end{pmatrix}, \quad q := x(t-h), \quad h \geq 0.$$

Let $\gamma_1(f_1)$ be the octahedral logarithmic norm and $\gamma_2(f_1)$ – the cubic one (see table 3.1). Then

$$\gamma_1(f_1(q)) = \max\{-4, -1 + |\cos q|, -6 + 4|\sin q|\} = -1 + |\cos q|,$$

$$\gamma_2(f_1(q)) = \max\{-4 + 4|\sin q|, -1, -6 + |\cos q|\} =$$

$$= \max\{-4 + 4|\sin q|, -1\}.$$

Hence

$$2\gamma_1(f_1(q)) + \gamma_2(f_1(q)) = \max\{2|\cos q| + 4|\sin q| - 6, -3 + |\cos q|\} < -1.$$

So by Thm. 3.2 the zero solution of Eq. (3.11) in this case is exponentially stable.

3.2. *EXPONENTIAL CONTRACTIVITY*

Definition 3.1 A norm $\| \cdot \|$ in $\mathbf{R}^n$ is called *exponentially contractive* for Eq. (3.1) if there exist a constant $\omega > 0$ such that for any two solutions x_1, x_2 of this equation the following inequality is valid

$$\| x_2(t) - x_1(t) \| \leq \sup_{\theta \leq 0} \| x_2(t_0 + \theta) - x_1(t_0 + \theta) \| \exp(-\omega(t - t_0)). \quad (3.12)$$

Denote

$$a_1 = - \sup_{t \geq t_0, x \in \mathbf{R}^n} \gamma(F(t,x)), \quad L_1 = \sup_{t \geq t_0, x \in \mathbf{R}^n} \| F(t,x) \|_1 .$$

Theorem 3.3 If $f : [t_0, \infty) \times \mathbf{R}^n \to \mathbf{R}^n$ is continuous and continuously differentiable with respect to the second argument then the norm $\| \cdot \|$ is exponentially contractive for Eq. (3.1), provided that

$$a_1 > hL_1^2. \quad (3.13)$$

Moreover, the constant ω in the inequality (3.12) can be taken as a unique positive root of the equation

$$\omega = a_1 - hL_1^2 \exp(2\omega h).$$

Proof Let $x_1(t)$ and $x_2(t)$ be two arbitrary solutions of Eq. (3.1). Defining $y(t) = x_2(t) - x_1(t)$, we get

$$\left.\begin{aligned} \dot{y}(t) &= A(t)y(t-h), \\ A(t) &:= \int_0^1 F[t, s(x_2(t-h) - x_1(t-h)) + x_1(t-h)]ds. \end{aligned}\right\} \quad (3.14)$$

Take the functional V

$$V(y_t) :=\| y(t) \| + L_1^2 \left(\int\limits_{t-2h}^{t-h} ds \int\limits_{s}^{t-h} \| y(\tau) \| d\tau + h \int\limits_{t-h}^{t} \| y(s) \| ds \right). \tag{3.15}$$

Then

$$\frac{d^+ V(y_t)}{dt} \leq -(a_1 - hL_1^2) \| y(t) \| . \tag{3.16}$$

From (3.15) it follows also

$$\frac{d^+ V(y_t)}{dt} = \frac{d^+ \| y(t) \|}{dt} + hL_1^2 \int\limits_{t-2h}^{t-h} \| y(\tau) \| d\tau. \tag{3.17}$$

Then, comparing (3.16) and (3.17), we find

$$\begin{aligned} \frac{d^+ \| y(t) \|}{dt} &\leq -a_1 \| y(t) \| + hL_1^2 \max_{t-2h \leq \tau \leq t-h} \| y(\tau) \| \leq \\ &\leq -a_1 \| y(t) \| + hL_1^2 \max_{t-2h \leq \tau \leq t} \| y(\tau) \| . \end{aligned}$$

Applying to this inequality Lemma 3.1 from [289.3],p.23, we obtain the assertion of Thm. 3.3.□

Remark 3.4 Using the same technique one should be able to obtain contractivity conditions for other classes of equations. Consider, for example, the two–dimensional system of equations

$$\left.\begin{aligned} \dot{x}_1(t) &= g_{11}(x_1(t)) + g_{12}(x_2(t)) + f_1(t, x(t-h)), \\ \dot{x}_2(t) &= g_{21}(x_1(t)) + g_{22}(x_2(t)) + f_2(t, x(t-h)), \quad t \geq t_0. \end{aligned}\right\} \tag{3.18}$$

Here $g_{ij}(0) = 0$, $f(t,0) \equiv 0$, $f = (f_1, f_2)$, $x = (x_1, x_2)$. Functions g_{ij} are continuously differentiable, f is continuous and $\| f(t,y) - f(t,x) \| \leq L \| y - x \|$, $\forall x, y \in \mathbf{R}^2$.

From the proof of Thm. 3.2 it follows that the norm $\| \cdot \|$ in $\mathbf{R}^2$ is exponentially contractive uniformly with respect to the delay $h \geq 0$ if

$$-\sup_{x \in \mathbf{R}^2} \gamma(G(x)) > L; \quad G(x) := \begin{pmatrix} \dfrac{\partial g_{11}(x_1)}{\partial x_1} & \dfrac{\partial g_{12}(x_2)}{\partial x_2} \\[2ex] \dfrac{\partial g_{21}(x_1)}{\partial x_1} & \dfrac{\partial g_{22}(x_2)}{\partial x_2} \end{pmatrix}. \tag{3.19}$$

Different contractivity conditions can be derived from (3.19) by concrete choice of the norm:

1. for octahedral norm $\| x \|= |x_1| + |x_2|$ we get

$$\gamma(G(x)) = \max\{G_{11}(x) + |G_{21}(x)|, G_{22}(x) + |G_{12}(x)|\},$$

2. for cubic norm $\| x \|= \max\{|x_1|, |x_2|\}$

$$\gamma(G(x)) = \max\{G_{11}(x) + |G_{12}(x)|, G_{22}(x) + |G_{21}(x)|]\},$$

3. for the Euclidean norm $\| x \|= \sqrt{x_1^2 + x_2^2}$ the logarithmic norm $\gamma(G(x))$ is equal to the maximal eigenvalue of the symmetrized matrix $G(x)$, i.e.

$$\gamma(G(x)) = \max \zeta \left(\frac{G(x) + G'(x)}{2} \right). \tag{3.20}$$

Let c_i be some positive constants and

$$\begin{aligned} &G_{11}(x_1) \leq -c_1, \quad G_{22}(x_2) \leq -c_2, \\ &|G_{12}(x_2)| \leq c_3 c_4, \quad |G_{21}(x_1)| \leq c_3 c_4 \end{aligned} \tag{3.21}$$

where

$$c_3 \in [0, 1), \quad c_4 := \left[\frac{1}{2}(c_1 + c_2) \min\{c_1, c_2\} \right]^{1/2}.$$

Under conditions (3.21) the matrix $G(x) + G'(x)$ is uniformly negative definite [141]. Hence in view of (3.20)

$$\gamma(G(x)) \leq - \min\{c_1, c_2\} + c_3 c_4.$$

From here and (3.19) it follows the exponential contractivity of Euclidean norm for Eq. (3.18) if

$$\min\{c_1, c_2\} - c_3 c_4 > L.$$

4. STABILITY IN THE FIRST APPROXIMATION

Consider the stability conditions for Eq. (1.1) with linear stable part

$$\dot{x}(t) = A(t)x(t) + f(t, x_t), \quad t > t_0. \tag{4.1}$$

Here $x(t) \in \mathbf{R}^n$, the matrix function $A : [t_0, \infty) \to \mathbf{R}^{n \times n}$ is continuous, the functional f is the same as in Eq. (1.1) and satisfies for arbitrary $\phi, \psi \in C(-\infty, 0]$ the inequality

$$\| f(t, \psi) - f(t, \phi) \| \leq \int_0^\infty \| \psi(-s) - \phi(-s) \| \, dk(s) \tag{4.2}$$

where k is a bounded nondecreasing function, for which

$$k(0) = 0, \quad \int_0^\infty s dk(s) < \infty.$$

4.1. *EXPONENTIALLY STABLE LINEAR PART*

Theorem 4.1 Assume that the logarithmic norm $\gamma(A(t))$ of the matrix A is such that

$$a := -\sup_{t \geq t_0} \gamma(A(t)) > k(\infty).$$

Then the zero solution of Eq. (4.1) is uniformly asymptotically stable in the whole.

Proof The method of stability investigation used is the same as that employed in previous paragraphs; for this reason the details of the proof will be discussed briefly. From (4.1) it follows that auxiliary ODE is

$$\dot{y}(t) = A(t)y(t) \tag{4.3}$$

for which $v(y) = \| y \|$. Hence $V_1(x_t) := \| x(t) \|$ and functional V according to the procedure is equal to

$$V(x_t) := \| x(t) \| + \int_0^\infty dk(s) \int_{t-s}^{t} \| x(\tau) \| d\tau. \tag{4.4}$$

Using the same arguments as in the proof of Thm. 3.1 we get

$$\frac{d^+V(x_t)}{dt} = \frac{d^+ \| x(t) \|}{dt} + k(\infty) \| x(t) \| - \int_0^\infty \| x(t-s) \| dk(s) \leq$$
$$\leq -(a - k(\infty)) \| x(t) \| . \tag{4.5}$$

By virtue of (4.4), (4.5) and Thm. 1.1 we obtain Thm. 4.1.□

4.2. *SMOOTH COEFFICIENTS*

In this section we make the additional assumption that $\mathbf{R}^n$ is an Euclidean space with the scalar product $x'y, x, y \in \mathbf{R}^n$, where prime is a transposition and $\| x \| = \sqrt{x'x}$. The functional f satisfies (4.2). The matrix $A(t)$ is continuously differentiable, and

$$\| A(t) \|_1 \leq c_1; \ \forall \mathrm{Re}\lambda_j(A(t)) \leq -c_2 \ \text{ for all } \ t \geq t_0. \tag{4.6}$$

Here and below c_i are some positive constants, $\lambda_j(A)$ are the eigenvalues of the matrix A and $\mathrm{Re}\lambda_j(A)$ the real parts of eigenvalues. Note that $\mathrm{Re}\lambda_j(A(t)) \le \gamma(A(t))$. Under these assumptions for every $t \ge t_0$ there exists a unique symmetric positive definite solution $p(t)$ of the matrix equation

$$A'(t)p(t) + p(t)A(t) = -I.$$

The matrix $p(t)$ satisfies the conditions [262]

$$\| \dot{p}(t) \|_1 \le c_3 \| \dot{A}(t) \|_1, \quad \| p(t) \| \le c_4. \tag{4.7}$$

Now let a Liapunov function $(t, y) \mapsto v(t, y)$ for auxiliary ODE (4.3) be the quadratic form

$$v(t, y) := y'p(t)y.$$

Then $d^+v(t, y)/dt$ along the solutions of (4.3) is

$$\frac{d^+v(y,t)}{dt} = - \| y \|^2 + y'\dot{p}(t)y.$$

According to the form of transformation $V_1(t, x_t) := x'(t)p(t)x(t)$. Hence V_2 can be taken

$$V_2(x_t) := c_4 k(\infty) \int_0^\infty dk(s) \int_{t-s}^{t} \| x(\tau) \|^2 \, d\tau.$$

For the functional $V := V_1 + V_2$ we get

$$\frac{d^+V(t, x_t)}{dt} \le -[1 - \| \dot{p} \|_1 - c_4 k^2(\infty) \| x(t) \|^2 .$$

So the following theorem is proven:

Theorem 4.2 Under assumptions (4.2), (4.6) the zero solution of Eq. (4.1) is uniformly asymptotically stable in the whole if

$$1 > c_4 k^2(\infty) + \sup_{t \ge t_0} \| \dot{p}(t) \|_1 . \tag{4.8}$$

Note that due to (4.7) it is sufficient to assume instead of (4.8) that

$$1 > c_4 k^2(\infty) + c_3 \sup_{t \ge t_0} \| \dot{A}(t) \|_1 .$$

5. SCALAR RDEs

5.1. *SCALAR EQUATIONS OF N-TH ORDER*

Consider stability of the scalar equation

$$x^{(n)}(t) := \sum_{j=1}^{n} \int_0^\infty x^{(j-1)}(t-s)dK_j(s), \quad t \geq 0, \tag{5.1}$$

where $x^{(j)}(t) := d^j x(t)/dt^j$, $n \geq 1$. Initial conditions for Eq. (5.1) have a form

$$x^{(j)}(\theta) := \phi^{(j)}(\theta) \quad \theta \leq 0 \tag{5.2}$$

where ϕ is a given $(n-1)$ times continuously differentiable function with a norm

$$\| \phi \| := \sup_{\theta \leq 0} (|\phi(\theta)| + \ldots + |\phi^{(n-1)}(\theta)|).$$

Kernels $K_j(s)$ are such that

$$\int_0^\infty s^i |dK_j(s)| < \infty \ (0 \leq i \leq n, \ 1 \leq j \leq n).$$

Put $x_j(t) := x^{(j-1)}(t)$ and rewrite Eq. (5.1) as a system

$$\dot{x}_i(t) = x_{i+1}(t) \ \ (i = 1, \ldots, n-1), \quad \dot{x}_n(t) = \sum_{j=1}^{n} \int_0^\infty x_j(t-s)dK_j(s). \tag{5.3}$$

Denote for the same i, j

$$\beta_{i,j} := \int_0^\infty s^i dK_j(s), \quad \alpha_{i,j} := \int_0^\infty s^i |dK_j(s)|.$$

Remark that for $0 \leq i \leq n-1, \ \ 1 \leq m \leq n$

$$\begin{aligned} \int_0^\infty x_{n-i}(t-s)dK_m(s) &= \sum_{j=1}^{i+1} (-1)^{j-1} x_{n-i+j-1}(t)\beta_{j-1,m} \frac{1}{(j-1)!} + \\ &+ (-1)^{i+1} \frac{d}{dt} \left[\int_0^\infty dK_m(s) \int_{t-s}^t ds_0 \int_{s_0}^t ds_1 \ldots \int_{s_{i-2}}^t ds_{i-1} \int_{s_{i-1}}^t x_n(s_i)ds_i \right]. \end{aligned} \tag{5.4}$$

Moreover

$$\int_0^\infty dK_m(s) \int_{t-s}^t ds_0 \int_{s_0}^t ds_1 \dots \int_{s_{i-1}}^t x_n(s_i)ds_i =$$
$$= \int_0^\infty dK_m(s) \int_{t-s}^t x_n(\tau)\frac{(\tau - t + s)^i}{i!}d\tau. \tag{5.5}$$

By virtue of (5.4), (5.5) Eq. (5.3) takes the form

$$\dot{x}_i(t) = x_{i+1}(t), \quad \frac{d}{dt}[x_n(t) - z(t)] = \sum_{l=0}^{n-1} x_{n-l}(t)r_l. \tag{5.6}$$

Here

$$z(t) := \sum_{l=0}^{n-1}(-1)^{l+1}\int_0^\infty dK_{n-l}(s)\int_{t-s}^t x_n(\tau)\frac{(\tau-t+s)^l}{l!}d\tau,$$
$$r_l := \sum_{i=1}^{n-1}(-1)^{l+i}\beta_{i-1,n-i}\frac{1}{(i-l)!}, \quad l = 0,1,\dots,n-1. \tag{5.7}$$

Using (5.6) we obtain auxiliary system of ODE

$$\dot{y}_i(t) = y_{i+1}(t) \quad (i = 1,\dots,n-1), \quad \dot{y}_n(t) = \sum_{l=0}^{n-1} y_{n-l}(t)r_l. \tag{5.8}$$

Let y be a vector with the components $(y_1,\dots,y_n)$, and A be a matrix $(n \times n)$ such that

$$A = \begin{pmatrix} 0 & 1 & 0 & \dots & 0 \\ 0 & 0 & 1 & \dots & 0 \\ \dots & \dots & \dots & \dots & \dots \\ 0 & 0 & 0 & \dots & 1 \\ r_{n-1} & r_{n-2} & r_{n-3} & \dots & r_0 \end{pmatrix}.$$

Note that

$$\det(\lambda I - A) = \lambda^n - \sum_{j=1}^n r_{j-1}\lambda^{n-j}.$$

By (5.8) we have $\dot{y} = Ay$.

Now we must assume that A is the Hurvitz matrix. Then there exists a unique positive definite symmetric matrix B satisfying the Liapunov matrix equation

$$A'B + BA = -I \tag{5.9}$$

where I is the identity matrix and prime is the transposition sign. Because of (5.9) we have

$$\dot{v}(y) = -|y|^2 \quad \text{where} \quad v(y) := y'By.$$

Now according to the procedure we must put $V := V_1 + V_2$, where

$$V_1(x_t) := (x_1(t), \dots, x_n(t) - z(t))'B(x_1(t), \dots, x_n(t) - z(t)). \tag{5.10}$$

From (5.10) it follows that $\dot{V}_1$ with respect to Eq. (5.6) equals

$$\dot{V}_1(x_t) = -|x(t)|^2 - 2\sum_{l=1}^{n} x_l(t)(BA)_{nl}z(t), \tag{5.11}$$

where $(BA)_{nl}$ is nl–th entry of the matrix BA. But

$$\begin{aligned} 2|x_l(t)z(t)| &= 2\left|x_l(t)\sum_{l=0}^{n-1}\int_0^\infty dK_{n-i}(s)\int_{t-s}^{t} x_n(\tau)\frac{(\tau-t+s)^i}{i!}d\tau\right| \le \\ &\le \sum_{l=0}^{n-1}\left[x_l^2(t)\frac{\alpha_{i+1,n-i}}{(i+1)!} + \int_0^\infty |dK_{n-i}(s)|\int_{t-s}^{t} x_n^2(\tau)\frac{(\tau-t+s)^i}{i!}d\tau\right]. \end{aligned} \tag{5.12}$$

Hence V_2 has to be chosen as follows

$$V_2(x_t) := \sum_{l=1}^{n}|(BA)_{nl}|\sum_{i=0}^{n-1}\int_0^\infty |dK_{n-i}(s)|\int_{t-s}^{t} x_n^2(\tau)\frac{(\tau-t+s)^i}{i!}d\tau. \tag{5.13}$$

In fact, relations (5.10)–(5.13) give us

$$\dot{V}(x_t) \le -|x(t)|^2 + \sum_{l=1}^{n}|(BA)_{nl}|\sum_{i=0}^{n-1}\frac{\alpha_{i+1,n-i}}{(i+1)!}[x_l^2(t) + x_n^2(t)]. \tag{5.14}$$

From here it follows that $\dot{V}$ is negative definite if the absolute $\alpha_{i,j}$ of the kernels K_j are sufficiently small. At last we must pay attention that the functional V is positive definite with respect to all variables x_i but x_n. For the proof that $x_n(t) \to 0$ as $t \to \infty$ let us assume that $\dot{V}$ is negative definite and

$$\sum_{i=0}^{n-1}\frac{\alpha_{i+1,n-i}}{(i+1)!} < 1. \tag{5.15}$$

From (5.14) it follows that

$$\int_0^\infty x_n^2(t)dt < \infty.$$

In addition due to (5.14), (5.15)

$$|x_n(t)| \leq c \parallel \phi \parallel .$$

Hence because of the boundedness of $\dot{x}_n(t)$ we get

$$\lim_{t\to\infty} x_n(t) = 0.$$

So we have proven

Theorem 5.1 System (5.1) is asymptotically stable if:
1) ODE (5.8) is asymptotically stable;
2) quadratic form in the righthand side of (5.14) is negative definite;
3) inequality (5.15) holds.

Remark Stability conditions in Thm. 5.1 use the form of transform (5.4) where integrals in the righthand side depends only on x_n for all i. Following the same procedure we can obtain other stability conditions using one of the following representations for $m = n - i + l,\ 1 \leq l \leq i \leq n-1$

$$J = (-1)^{l+1}\frac{d}{dt}\left[\int_0^\infty dK_{n-i}(s)\int_{t-s}^t x_m(\tau)\frac{(\tau - t + s)^l}{l!}d\tau\right] \quad (m \leq n),$$

$$J = (-1)^{l+1}\int_0^\infty dK_{n-i}(s)\int_{t-s}^t x_{m+1}(\tau)\frac{(\tau - t + s)^l}{l!}d\tau \ (m \leq n-1). \tag{5.16}$$

Here

$$J := \int_0^\infty x_{n-i}(t-s)dK_{n-i}(s) - \sum_{j=1}^{l+1}(-1)^j x_{m-l+j-1}(t)\beta_{j-1,n-i}\frac{1}{(j-1)!}.$$

5.2. *SCALAR EQUATIONS OF SECOND ORDER*

Consider in details the procedure for scalar second order equations which can be represented in the form of the system

$$\left.\begin{aligned} &\dot{x}_1(t) = x_2(t),\\ &\dot{x}_2(t) = -\int_0^\infty x_2(t-s)dK_2(s) - \int_0^\infty x_1(t-s)dK_1(s),\ t \geq 0, \end{aligned}\right\} \tag{5.17}$$

where K_i are functions of bounded variation on $[0,\infty)$ with finite first absolute moments.

We have

$$\int_0^\infty x_2(t-s)dK_2(s) = -\frac{d}{dt}\int_0^\infty dK_2(s)\int_{t-s}^t x_2(\tau)d\tau + \beta_{0,2}x_2(t) =$$

$$= \frac{d}{dt}\int_0^\infty x_1(t-s)dK_2(s), \tag{5.18}$$

$$\int_0^\infty x_1(t-s)dK_1(s) = \frac{d}{dt}\int_0^\infty dK_1(s)\int_{t-s}^t (\tau - t + s)x_2(\tau)d\tau -$$

$$-\beta_{1,1}x_2(t) + \beta_{0,1}x_1(t) = -\frac{d}{dt}\int_0^\infty dK_1(s)\int_{t-s}^t x_1(\tau)d\tau + \beta_{0,1}x_1(t) =$$

$$= -\int_0^\infty dK_1(s)\int_{t-s}^t x_2(\tau)d\tau + \beta_{0,1}x_1(t). \tag{5.19}$$

Taking various combinations of the integrals in (5.18) we obtain different transformed systems.

Consider one of them:

$$\dot{x}_1(t) = x_2(t), \quad \dot{z}(t) = -ax_2(t) - bx_1(t), \tag{5.20}$$

here

$$a := \beta_{0,2} - \beta_{1,1}, \quad b := \beta_{0,1},$$

$$z(t) := x_2(t) - \int_0^\infty dK_2(s)\int_{t-s}^t x_2(\tau)d\tau + \int_0^\infty dK_1(s)\int_{t-s}^t (\tau - t + s)x_2(\tau)d\tau.$$

Auxiliary system of ODE obtained from (5.20) is the following

$$\dot{y}_1(t) = y_2(t), \quad \dot{y}_2(t) = -ay_2(t) - by_1(t). \tag{5.21}$$

Necessary and sufficient conditions for system (5.21) to be asymptotically stable are

$$a > 0, \; b > 0. \tag{5.22}$$

The solution of Eq. (5.9) corresponding (5.21) gives us the following Liapunov function v

$$v(y) := \frac{1}{2ab}[y_1^2(a^2+b^2+b) + 2ay_1y_2 + (b+1)y_2^2]. \tag{5.23}$$

Now according to the procedure we must replace in (5.23) the function y_2 on $z(t)$. As a result we obtain the component V_1 of the functional $V := V_1 + V_2$

$$V_1(x_t) := \frac{1}{2ab}[x_1^2(a^2+b^2+b) + 2ax_1(t)z(t) + (b+1)z^2(t)]. \tag{5.24}$$

Hence

$$\dot V_1(x_t) = -x_1^2(t) - x_2^2(t) - x_2(t)\,(z(t) - x_2(t)) - \frac{(b+1)}{a}x_1(t)\,(z(t)-x_2(t)).$$

Note that due to (5.12)

$$2|x_i(t)\,(z(t)-x_2(t))| \le qx_i^2(t) + \int_0^\infty |dK_2(s)| \int_{t-s}^t x_2^2(\tau)d\tau +$$

$$+\int_0^\infty |dK_1(s)| \int_{t-s}^t x_2^2(\tau)(\tau - t + s)d\tau, \quad q := \alpha_{1,2} + \frac{\alpha_{2,1}}{2}, \quad i = 1,2.$$

Hence

$$V_2(x_t) := \frac{1}{2}\left(\frac{b+1}{a}+1\right)\left[\int_0^\infty |dK_1(s)| \int_{t-s}^t x_2^2(\tau)\frac{(\tau-t+s)^2}{2}d\tau + \right.$$

$$\left. + \int_0^\infty |dK_2(s)| \int_{t-s}^t x_2^2(\tau)(\tau-t+s)d\tau\right].$$

So

$$\dot V(x_t) \le -x_1^2(t) - x_2^2(t) + \frac{(b+1)}{2a}x_1^2(t)q + x_2^2(t)q\left(\frac{b+1}{2a}+1\right).$$

Also inequality (5.15) leads us to the condition $q < 1$.

As a result we obtain that system (5.17) is asymptotically stable under conditions

$$q := \alpha_{1,2} + \frac{1}{2}\alpha_{2,1} < 1, \quad \beta_{0,1} > 0, \quad 2(\beta_{0,2} - \beta_{1,1})(1-q) > (1+\beta_{0,1})q. \tag{5.25}$$

Consider now other type of transform of system (5.17) when the transformed system has the same form (5.20) where

$$a := \beta_{1,1}, \quad b := \beta_{0,1},$$

$$z(t) := x_2(t) - \int_0^\infty x_1(t-s)dK_2(s) + \int_0^\infty dK_1(s) \int_{t-s}^t (\tau - t + s)x_2(\tau)d\tau. \tag{5.26}$$

Hence the auxiliary system keeps the same form (5.21) with a and b from (5.26). It follows that V_1 is given by (5.24) with a, b, z from (5.26). Therefore

$$V_2(x_t) := \frac{b+1+a}{2a}[\int_0^\infty |dK_2(s)| \int_{t-s}^t x_1^2(\tau)d\tau + \int_0^\infty |dK_1(s)| \times$$

$$\times \int_{t-s}^t x_2^2(\tau)\frac{(\tau - t + s)^2}{2}d\tau].$$

Then $\dot{V} = \dot{V}_1 + \dot{V}_2$ satisfies the inequality

$$\dot{V}(x_t) \leq -x_1^2(t) - x_2^2(t) + \frac{1}{2}x_2^2(t)\left(\alpha_{0,2} + \alpha_{2,1} + \frac{(b+1)}{2a}\alpha_{2,1}\right) +$$

$$+\frac{1}{2}x_1^2(t)\left(\alpha_{0,2} + \frac{b+1}{a}\left(2\alpha_{0,2} + \frac{\alpha_{2,1}}{2}\right)\right).$$

Hence stability conditions of Eqs. (5.17) are given by the relations

$$-\beta_{1,1} > 0, \ \beta_{0,1} > 0, \ -2\beta_{1,1} > \max\{-\beta_{1,1}(\alpha_{0,2} + \alpha_{2,1}) +$$

$$+(1 + \beta_{0,1})\frac{\alpha_{2,1}}{2} - \beta_{1,1}\alpha_{0,2} + (1 + \beta_{0,1})\left(2\alpha_{0,2} + \frac{\alpha_{2,1}}{2}\right)\}. \tag{5.27}$$

Consider some other possibilities. Take the transformed system (5.20) with

$$a := \beta_{0,2}, \quad b := \beta_{0,1},$$

$$z(t) := x_2(t) - \int_0^\infty dK_2(s) \int_{t-s}^t x_2(\tau)d\tau - \int_0^\infty dK_1(s) \int_{t-s}^t x_2(\tau)d\tau. \tag{5.28}$$

Then V_1 is given by (5.24) with a, b, z from (5.28). Hence

$$V_2(x_t) := \frac{b+1+a}{2a}\left[\int_0^\infty |dK_2(s)| \int_{t-s}^t x_2^2(\tau)(\tau - t + s)d\tau + \right.$$

$$\left. + \int_0^\infty |dK_1(s)| \int_{t-s}^t x_1^2(\tau)(\tau - t + s)d\tau \right].$$

So we obtain the following stability conditions of Eq. (5.17)

$$\beta_{0,1} > 0, \quad \alpha_{1,2} < 1, \quad -2\beta_{0,2} > \max\{(\beta_{0,1}+1)(\alpha_{1,1}+2\alpha_{1,1}), (\beta_{0,1}+1)\alpha_{1,2}+$$

$$+\beta_{0,2}(\alpha_{1,1}+2\alpha_{1,2})\}. \tag{5.29}$$

In Subs. 5.2 it was noted that the choice of the Liapunov function can be done nonuniquely. This source of nonuniqueness must also be used.

Take, e.g., the transformed system (5.20) and auxiliary system of ODE (5.21). As a Liapunov function for (5.21) take

$$y \mapsto v(y) := 2by_1^2 + y_2^2 + (y_2 + ay_1)^2.$$

For $\dot{v}$ along the solution of (5.21) we get

$$\dot{v}(y) = -2a(by_1^2 + y_2^2).$$

Hence, according to the procedure, $V = V_1 + V_2$, where

$$V_1(x_t) := 2bx_1^2(t) + z^2(t) + (z(t) + ax_1(t))^2,$$

$$V_2(x_t) := (a+2b)\left[\int_0^\infty |dK_2(s)| \int_{t-s}^t x_2^2(\tau)(\tau - t + s)d\tau + \right.$$

$$\left. + \frac{1}{2}\int_0^\infty |dK_1(s)| \int_{t-s}^t x_2^2(\tau)(\tau - t + s)^2 d\tau \right].$$

Using these functionals we obtain the following stability conditions of Eq. (5.17)

$$q := \alpha_{1,2} + \tfrac{1}{2}\alpha_{2,1} < 1, \quad \beta_{0,1} > 0,$$

$$\beta_{0,2} - \beta_{1,1} > \max\{q, (1-q)^{-1}\beta_{0,1}q\}. \tag{5.30}$$

5.3. *STABILITY OF CHEMOSTAT*

Equations of the model are [282.2]

$$\dot{S}(t) = D(C - S(t)) - \gamma U(S(t))N(t) + bD_1 \int_0^\infty F(\tau)N(t-\tau)d\tau,$$

$$\dot{N}(t) = N(t)\left(-D - D_1 + m\int_0^\infty G(\tau)U(S(t-\tau))d\tau\right). \tag{5.31}$$

Here $U : [0,\infty) \to [0,1)$ is a smooth function, $U(0) = 0$, $\dot{U}(s) := dU(s)/ds > 0$, $U(s) \to 1$ as $s \to \infty$, all constants C, D, D_1, γ, m are positive, $b \in [0,1]$, functions G, and F are continuous and also

$$\int_0^\infty G(\tau)d\tau = \int_0^\infty F(\tau)d\tau = 1, \; F, G \geq 0, \; \int_0^\infty (F^2(\tau)d\tau + G^2(\tau))d\tau < \infty.$$

Stationary positive solution S^0, N^0 of Eq. (5.31) existing under conditions

$$q_0 := (D + D_1)m^{-1} < 1, \; \gamma_1 := \gamma q_0 - bD > 0, \; C > U^{-1}(q_0),$$

$$(\text{or } q_0 < 1, \gamma_1 < 0, \;\; C < U^{-1}(q_0))$$

($u^{-1} : [0,1) \to [0,\infty)$ is a function inverse to U) equals

$$S^0 = U^{-1}(q_0), \;\; N^0 = D(C - S^0)\gamma_1^{-1}.$$

Let us linearize Eqs. (5.31) in the neighbourhood of (S^0, N^0). Then we obtain that asymptotic stability of (S^0, N^0) with respect to the perturbations of the initial conditions is consequence of the asymptotic stability of the linear equations

$$\dot{x}_1(t) = x_2(t), \; \dot{x}_2(t) = -a\dot{x}_2(t) - \int_0^\infty x_1(t-s)K_1(s)ds. \tag{5.32}$$

Here

$$a := D + \gamma N^0 \dot{U}(S^0), \;\; K_1(s) := -\int_0^s g(\tau)f(s-\tau)d\tau + a_1 g(s),$$

$$a_1 := \gamma U(S^0) = \gamma q_0, \;\; f(s) := bD_1 F(s), \;\; g(s) := mN^0 \dot{U}(S^0)G(s).$$

System (5.32) is a particular case of (5.17) if in (5.17) the kernel $K_2(s)$ is a constant for $s > 0$, and has a jump a at the point $s = 0$. Hence, stability conditions of (5.32) are given by one of the relations (5.25), (5.27), (5.29), (5.30). This means that the chemostat is stable if $\beta_{0,1} > 0$, and one of the following conditions holds.

1)$a > \alpha_{1,1}$;

2)$\alpha_{2,1} < 2, \quad 2(a - \beta_{1,1})(1 - \frac{1}{2}\alpha_{2,1}) > \frac{1}{2}\alpha_{2,1}(1 + \beta_{0,1})$;

3)$\beta_{1,1} < 0, \quad -2\beta_{1,1} > \max\{-\beta_{1,1}(a + \alpha_{2,1}) + \frac{1}{2}(1 + \beta_{0,1})\alpha_{2,1} - \beta_{1,1}a+$ $+(1 + \beta_{0,1})(2a + \frac{1}{2}\alpha_{2,1})\}$;

4)$\alpha_{2,1} < 2, \quad 4(a - \beta_{1,1}) > \max\{\alpha_{2,1}, (1 - \frac{1}{2}\alpha_{2,1})^{-1}\beta_{0,1}\alpha_{2,1}\}$.

Remark Sometimes it can happen that the different transformed systems (5.6) (i.e. systems (5.6) with different z) have the same auxiliary system of equations (5.8). If so, then V_1 can be obtained from the Liapunov function v by replacing y_n onto $x_n(t) - z(t)$ with various z.

E.g., the transformed system, corresponding (5.32), can be taken in the form

$$\dot{x}_1(t) = x_2(t), \quad \dot{z}_1(t) = -ax_2(t) - \beta_{0,1}x_1(t) + \int_0^\infty K_1(s)ds \int_{t-s}^t x_2(\tau)d\tau,$$

$$z_1(t) = x_2(t),$$

or in the form

$$\dot{x}_1(t) = x_2(t), \quad \dot{z}_2(t) = -ax_2(t) - \beta_{0,1}x_1(t),$$

$$z_2(t) = x_2(t) - \int_0^\infty K_1(s)ds \int_{t-s}^t x_1(\tau)d\tau.$$

In both cases the auxiliary system is the same:

$$\dot{y}_1 = y_2, \quad \dot{y}_2 = -ay_2 - \beta_{0,1}y_1.$$

Liapunov function v for this system is

$$v(y) := 2\beta_{0,1}y_1^2 + y_2^2 + (y_2 + ay_1)^2. \tag{5.33}$$

According to the procedure for obtaining V_1 we must replace y_1 by $x_1(t)$ and y_2 by $z_1(t)$ or by $z_2(t)$. But it is also possible to replace y_2 by $z_1(t)$ in the second term in (5.33), and y_2 by $z_2(t)$ in the third term. As a result we obtain

$$V_1(x_t) := 2\beta_{0,1}x_1^2(t) + x_2^2(t) + \left[x_2(t) - \int_0^\infty K_1(s)ds \int_{t-s}^t x_1(\tau)d\tau + ax_1(t)\right]^2.$$

Hence

$$V_2(x_t) := \int_0^\infty |K_1(s)|ds \int_{t-s}^t d\tau_1 \int_{\tau_1}^t (x_2^2(\tau_2) + \beta_{0,1} x_1^2(\tau_2))d\tau_2.$$

Applying the functional $V := V_1 + V_2$ we obtain that the chemostat is stable under conditions $a > \alpha_{1,1}$; $\beta_{0,1} > 0$.

CHAPTER 7

RICCATI TYPE STABILITY CONDITIONS OF SOME LINEAR SYSTEMS WITH DELAY[1]

1. INTRODUCTION

Direct Liapunov method represents a powerful tool for determining systems stability. Along with the general theorems it allows for linear equations to formulate stability conditions in terms of existence of the positive definite solutions of some auxiliary matrix equations.

Consider a familiar example, the linear ODE

$$\dot{x}(t) = Ax(t), \tag{1.1}$$

where A is a constant $(n \times n)$ matrix, $x(t) \in \mathbf{R}^n$ and $\mathbf{R}^n$ is the n dimensional real vector space. A reasonable of a Liapunov function V for Eq. (1.1) is a quadratic form,

$$V(x) := x'Px, \tag{1.2}$$

where prime is a transposition sign and P is a symmetric matrix that yields

$$A'P + PA = -Q. \tag{1.3}$$

It is well known that necessary and sufficient conditions for asymptotic stability of Eq. (1.1) can be formulated in terms of existence of a positive definite solution P to Eq. (1.3) for any positive definite matrix Q (see, e.g. [4]). But for equations with delay the situation is different. Indeed, consider the simplest case,

$$\dot{x}(t) = Ax(t) + Bx(t-h), \tag{1.4}$$

where $x(t) \in \mathbf{R}^n$, $h \geq 0$ is a constant delay and A, B are constant $(n \times n)$ matrices. By analogy to (1.2), consider a Liapunov functional V for Eq.

[1]This chapter is written jointly with Prof. J.–P. Richard (Ecole Centrale de Lille).

(1.4) as a generalized quadratic form,

$$V(x_t) := x'(t)Px(t) + x'(t)\int_{-h}^{0} P_1(s)x(t+s)ds + \int_{-h}^{0} x'(t+s)P_1'(s)dsx(t) + \int_{-h}^{0}\int_{-h}^{0} x'(t+s)P_2(s,\tau)x(t+\tau)dsd\tau. \tag{1.5}$$

Let $\dot{V}$ denote the full derivative of V along the solution of Eq. (1.4). Then, $\dot{V} < 0$ implies that the matrices P, P_1, P_2 have satisfy a complicated system of algebraic, ordinary and partial differential equations with appropriate boundary conditions (see Ch. 14). Because the analysis of this system and construction of P, P_1, P_2 are rather difficult, some authors prefer to invent concrete functionals V for special classes of delayed equations (see, for instance, [289.3]) and obtain stability conditions formulated directly in terms of their coefficients.

In the paper [290.2], using a particular Liapunov functional V, delay–independent stability condition for Eq. (1.4) were formulated in terms of existence of a positive definite solution to an auxiliary algebraic Riccati matrix equation. We emphasize that delay–independent stability conditions can be valid only for systems whose right hand side contains instantaneous negative stabilizing term. For example, Eq. (1.4) must contain the term $Ax(t)$, where A is a hurwitzian matrix.

Delay–independent stability conditions for linear Eq. (1.4) with hurwitzian matrix A and finite number of discrete delays are contained in the book [345], p. 87. They are formulated in terms of existence of two matrices satisfying two auxiliary algebraic equations. This result is also obtained by a particular choice of the functional V. It can be reformulated, after appropriate change of variables, in terms of existence of a positive definite solution to a Riccati equation.

For linear neutral type equations similar approach and delay–independent stability conditions were proposed in [41]. Some authors (see, e.g. [238]) have formulated delay–independent stability conditions in terms of existence of positive definite solution of linear matrix inequalities (LMIs) which can be transformed into Riccati equations. This LMI approach show that the form of Riccati equation depends, essentially, on a particular choice of Liapunov functional V. Moreover, if for the same equation we have various functionals V, then we should be able to obtain different Riccati equations and, as a result, to enlarge the estimate of the stability domain in the space of parameters. A formal procedure to construct Liapunov functionals V for concrete equations with delay was described in Ch. 6.

A similar procedure is used in this chapter to obtain both delay–dependent and delay–independent stability conditions formulated in terms of existence of positive definite solutions of Riccati equations for the linear equations of the form

$$\dot{x}(t) = \sum_{i=1}^{m} B_i x(t - h_i) + \sum_{j=1}^{r} C_j \int_{t-\tau_j}^{t} x(s)ds, \quad t \geq 0. \tag{1.6}$$

Here, $x(t) \in \mathbf{R}^n$, constant matrices B_i and C_j have dimension $(n \times n)$, integers m and r are given, delays h_i and τ_j are some nonnegative constants. The initial conditions for Eq. (1.6) must be prescribed on the interval $[-h_0, 0]$, where $h_0 = \max_{i,j}\{h_i, \tau_j\}$, and have a form

$$x(\theta) = \phi(\theta), \quad -h_0 \leq \theta \leq 0, \tag{1.7}$$

ϕ being a given continuous function.

2. SPECIAL CASE

Consider special case of Eq. (1.6) in the form

$$\dot{x}(t) = B_1 x(t - h_1), \quad t \geq 0, \quad x \in \mathbf{R}^n. \tag{2.1}$$

We now apply the procedure and obtain an appropriate Riccati equation for the Eq. (2.1).

2.1. *THE STABILITY CONDITION FOR THIS CASE*

Rewrite Eq. (2.1) for $t \geq h_1$ as

$$\dot{x}(t) = B_1 x(t) - B_1 \int_{t-h_1}^{t} \dot{x}(s)ds = B_1 x(t) - B_1^2 \int_{t-h_1}^{t} x(s - h_1)ds. \tag{2.2}$$

Then, the functional V for Eq. (2.2) can be represented as a sum $V := V_1 + V_2$, where

$$V_1(x_t) := x'(t)Px(t), \quad V_2(x_t) := \int_{h_1}^{2h_1} ds \int_{t-s}^{t} x'(\tau)Rx(\tau)d\tau \tag{2.3}$$

Here P and R are constant symmetric positive definite matrices. Simple calculations show that, by virtue of (2.2), (2.3),

$$\dot{V}_1(x_t) = x'(t)(B_1'P + PB_1)x(t) - \int_{t-2h_1}^{t-h_1} x'(s)B_1^{\prime 2}Px(t)ds - x'(t)PB_1^2 \int_{t-2h_1}^{t-h_1} x(s)ds, \tag{2.4}$$

and, similarly,

$$\dot V_2(x_t) = h_1 x'(t)Rx(t) - \int\limits_{t-2h_1}^{t-h_1} x'(s)Rx(s)ds. \tag{2.5}$$

Relations (2.4), (2.5) mean that $\dot V = \dot V_1 + \dot V_2$, and

$$\begin{aligned} \dot V(x_t) &= x'(t)(B_1'P + PB_1 + h_1R)x(t) - \\ &- \int\limits_{t-2h_1}^{t-h_1} [x'(s){B_1'}^2Px(t) + x'(t)PB_1^2x(s) + x'(s)Rx(s)]ds. \end{aligned} \tag{2.6}$$

Adding and subtracting the term $h_1x'(t)PB_1^2R^{-1}{B_1'}^2Px(t)$ to the right hand side of (2.6), we get

$$\begin{aligned} \dot V(x_t) &= x'(t)[B_1'P + PB_1 + h_1R + h_1PB_1^2R^{-1}{B_1'}^2P]x(t) - \\ &- \int\limits_{t-2h_1}^{t-h_1} [Rx(s) + {B_1'}^2Px(t)]'R^{-1}[Rx(s) + {B_1'}^2Px(t)]ds. \end{aligned} \tag{2.7}$$

Taking into account (2.3), (2.7) and Thm. 4.1.1, we obtain the following theorem.

Theorem 2.1 Assume that for some symmetric matrices $Q > 0, R > 0$, there exists a symmetric solution, $P > 0$, of the matrix Riccati equation

$$B_1'P + PB_1 + h_1PB_1^2R^{-1}{B_1'}^2P + h_1R = -Q. \tag{2.8}$$

Then, Eq. (2.1) is asymptotically stable.

2.2. *AN APPLICATION OF A FORM OF NDE*

Consider now a different approach to the construction of the functional V leading to a different Riccati equation. Represent the RDE (2.1) as a NDE

$$\frac{d}{dt}\left(x(t) + B_1\int\limits_{t-h_1}^{t} x(s)ds\right) = B_1x(t). \tag{2.9}$$

Put $V := V_1 + V_2$, where

$$\left.\begin{aligned} V_1(x_t) &:= \left(x(t) + B_1 \int_{t-h_1}^{t} x(s)ds\right)' P \left(x(t) + B_1 \int_{t-h_1}^{t} x(s)ds\right), \\ V_2(x_t) &:= \int_0^{h_1} ds \int_{t-s}^{t} x'(\tau)Rx(\tau)d\tau. \end{aligned}\right\} \tag{2.10}$$

Then, by virtue of (2.9), (2.10),

$$\left.\begin{aligned} \dot V_1(x_t) &= x'(t)(B_1'P + PB_1)x(t) + x'(t)B_1'PB_1 \int_{t-h_1}^{t} x(s)ds + \\ &+ \int_{t-h_1}^{t} x'(s)dsB_1'PB_1x(t), \\ \dot V_2(x_t) &= x'(t)Rx(t)h_1 - \int_{t-h_1}^{t} x'(s)Rx(s)ds. \end{aligned}\right\} \tag{2.11}$$

Hence, adding and subtracting the term $h_1x'(t) + B_1'PB_1R^{-1}B_1'PB_1x(t)$ to the righthand side of (2.11), we obtain

$$\begin{aligned} \dot V(x_t) &= x'(t)[B_1'P + PB_1 + h_1B_1'PB_1R^{-1}B_1'PB_1 + h_1R]x(t) - \\ &- \int_{t-h_1}^{t} [(B_1'PB_1x(t) + Rx(s))'R^{-1}(B_1'PB_1x(t) + Rx(s))]ds. \end{aligned} \tag{2.12}$$

In view of (2.10), the functional V belongs to the class of the so-called degenerated functionals (i.e. nonpositive definite).

Thm. 4.1.1 together with (2.10), (2.12), shows the validity of the following theorem.

Theorem 2.2 Assume that the equation

$$y(t) + B_1 \int_{t-h_1}^{t} y(s)ds = 0, \quad t \geq 0, \tag{2.13}$$

is asymptotically stable, and that for some symmetric matrices $Q > 0, R > 0$, there exists a symmetric solution, $P > 0$, of the Riccati equation:

$$B_1' P + P B_1 + h_1 B_1' P B_1 R^{-1} B_1' P B_1 + h_1 R = -Q. \tag{2.14}$$

Then Eq. (2.1) is asymptotically stable.

Note that stability definitions of Eq. (2.13) with respect to the perturbations of the initial conditions are quite similar to those of from Subs. 4.1.1.

Remark Eq. (2.13) is asymptotically stable if for at least one matrix norm $\| \cdot \|$, $h_1 \| B_1 \| < 1$.

2.3. *ONE MORE STABILITY CONDITION*

Yet, another Riccati equation can be obtained if we use the following representation of Eq. (2.1) for $t \geq h_1$:

$$\dot{x}(t) = B_1 x(t) - B_1 \int_{t-h_1}^{t} \dot{x}(s) ds. \tag{2.15}$$

Put $V := V_1 + V_2$, where

$$V_1(x_t) := x'(t) P x(t),$$

$$V_2(x_t) := \int_0^{h_1} ds \int_{t-s}^{t} \dot{x}'(\tau) R \dot{x}(\tau) d\tau + h_1 \int_{t-h_1}^{t} x'(s) B_1' R B_1 x(s) ds. \tag{2.16}$$

Take into account that

$$\dot{x}'(t) R \dot{x}(t) = x'(t-h) B_1' R B_1 x(t-h).$$

Hence,

$$\dot{V}(x_t) = x'(t)[B_1' P + P B_1 + h_1 B_1' R B_1] x(t) -$$

$$- \int_{t-h_1}^{t} [\dot{x}'(s) B_1' P x(t) + x'(t) P B_1 \dot{x}(s) + \dot{x}'(s) R \dot{x}(s)] ds =$$

$$= x'(t)[B_1' P + P B_1 + h_1 P B_1 R^{-1} B_1' P + h_1 B_1' R B_1] x(t) -$$

$$- \int_{t-h_1}^{t} \left(B_1' P x(t) + R \dot{x}(s)\right)' R^{-1} \left(B_1' P x(t) + R \dot{x}(s)\right) ds.$$

From Thm. 4.1.1, (2.15), and (2.16) we obtain the following result.

Theorem 2.3 Eq. (2.1) is asymptotically stable if for some symmetric matrices $R > 0$ and $Q > 0$ there exists a symmetric solution, $P > 0$, of the Riccati equation

$$B_1'P + PB_1 + h_1PB_1R^{-1}B_1'P + h_1B_1'RB_1 = -Q. \tag{2.17}$$

Remark If $h_1 = 0$, then all Riccati equations (2.8), (2.14), and (1.17) coincide with the linear Liapunov equation (1.3) for the linear RDE (1.1).

Example 2.1 Consider a scalar equation of the form (2.1),

$$\dot{x}(t) = b_1x(t - h_1). \tag{2.18}$$

Then, Eq. (2.8) takes the form

$$2b_1p + h_1b_1^4p^2r^{-1} + h_1r = -q, \tag{2.19}$$

where $b_1, h_1 \geq 0,\ r > 0, q > 0$ are scalars.

A solution, $p > 0$, of Eq. (2.19) exists if and only if

$$b_1 < 0,\ |b_1|h_1 < 1. \tag{2.20}$$

By virtue of Theorem 2.1, inequalities (2.20) are sufficient conditions for asymptotic stability of Eq. (2.18).

Now, apply Theorem 2.2. In this case, Eq. (2.14) is the same as (2.19). Furthermore, due to the remark after Thm. 2.2, equation (2.13) is asymptotically stable if $|b_1|h_1 < 1$. Hence Thm. 2.2 gives the same stability condition (2.20) for Eq. (2.18). In view of (2.17), this condition (2.20) results also from Thm. 2.3.

3. DISCRETE DELAY–INDEPENDENT STABILITY CONDITIONS

We already noted that delay–independent stability conditions may be valid only for systems with instantaneous negative feedback. Let us derive such conditions following the general framework of this Chapter for Eq. (1.6) with the instantaneous feedback $Ax(t)$:

$$\dot{x}(t) = Ax(t) + \sum_{i=1}^{m} B_ix(t - h_i) + \sum_{j=1}^{r} C_j \int_{t-\tau_j}^{t} x(s)ds. \tag{3.1}$$

Put $V := V_1 + V_2$, where for symmetric matrices $P > 0, R_i > 0, S_j > 0$,

$$\left.\begin{aligned} &V_1(x_t) := x'(t)Px(t), \\ &V_2(x_t) := \sum_{i=1}^{m} \int_{t-h_i}^{t} x'(s)R_ix(s)ds + \sum_{j=1}^{r} \int_{0}^{\tau_j} \int_{t-s}^{t} x'(\tau)S_jx(\tau)d\tau. \end{aligned}\right\} \tag{3.2}$$

It is clear that

$$\dot{V}(x_t) = x'(t)\left(A'P + PA + \sum_{i=1}^{m} R_i + \sum_{j=1}^{r} \tau_j S_j\right) x(t) -$$

$$- \sum_{i=1}^{m} x'(t-h_i) R_i x(t-h_i) - \sum_{j=1}^{r} \int_{t-\tau_j}^{t} x'(s) S_j x(s) ds + \alpha + \alpha', \tag{3.3}$$

where

$$\alpha := \left[\sum_{i=1}^{m} B_i x(t-h_i) + \sum_{j=1}^{r} C_j \int_{t-\tau_j}^{t} x(s) ds\right]' Px(t). \tag{3.4}$$

But,

$$-x'(t-h_i)R_i x(t-h_i) + x'(t-h_i)B_i'Px(t) + x'(t)PB_i x(t-h_i) =$$

$$= -(B_i'Px(t) - R_i x(t-h_i))'R^{-1}(B_i'Px(t) - R_i x(t-h_i)) + \tag{3.5}$$

$$+ x'(t)PB_i R_i^{-1} B_i' Px(t),$$

and, similarly,

$$-x'(s)S_j x(s) + x'(s)C_j'Px(t) + x'(t)PC_j x(s) =$$

$$= -(C_j'Px(t) - S_j x(s))'S_j^{-1}(C_j'Px(t) - S_j x(s)) + x'(t)PC_j S_j^{-1} C_j' Px(t). \tag{3.6}$$

Relations (3.3)–(3.6) imply that

$$\dot{V}(x_t) \leq x'(t) \times$$

$$\times \left[A'P + PA + \sum_{i=1}^{m}(R_i + PB_i R_i^{-1} B_i' P) + \sum_{j=1}^{r} \tau_j (S_j + PC_j S_j^{-1} C_j' P)\right] x(t).$$

Hence, taking into account Thm. 4.1.1, we obtain the following theorem.

Theorem 3.1 Assume that for some symmetric $R_i > 0, S_j > 0$ and $Q > 0$ there exists a symmetric solution, $P > 0$, of the equation

$$A'P + PA + \sum_{i=1}^{m}(R_i + PB_i R_i^{-1} B_i' P) + \sum_{j=1}^{r} \tau_j (S_j + PC_j S_j^{-1} C_j' P) = -Q. \tag{3.7}$$

Then Eq. (3.1) is asymptotically stable.

Example 3.1 Consider a two–dimensional equation,

$$\dot{x}(t) = Ax(t) + Bx(t-h),\ A = \begin{pmatrix} -a_1 & 0 \\ 0 & -a_2 \end{pmatrix},\ B = \begin{pmatrix} b_1 & b_2 \\ -b_2 & b_1 \end{pmatrix}, \quad (3.8)$$

with constant coefficients a_i and b_i $(i = 1, 2)$.

Let $Q = qI, R_1 = rI$ $(q, r > 0)$, where I is the (2×2) identity matrix. The Riccati equation (3.7) corresponding to (3.8) takes the form

$$A'P + PA + rI + r^{-1}PBB'P = -qI. \quad (3.9)$$

Entries p_{ij} of the matrix P satisfying (3.9) are $p_{12} = p_{21} = 0$, and p_{ij} $(i = 1, 2)$ is a positive root of the equation

$$p_{ii}^2[r^{-1}(b_1^2 + b_2^2)] - a_i p_{ii} + r = -q < 0. \quad (3.10)$$

Positive roots p_{ii} of (3.10) exist under condition

$$a_i > (b_1^2 + b_2^2)^{\frac{1}{2}},\quad i = 1, 2. \quad (3.11)$$

Hence Eq. (3.8) is asymptotically stable if both inequalities (3.11) hold.

Let us compare this result with previous ones based on M–matrices [374] that give the delay–independent stability conditions:

$$(a_1 - |b_1|)(a_2 - |b_2|) > b_2^2,$$
$$a_1 > |b_1|, \quad (3.12)$$
$$a_2 > |b_2|.$$

If $a_1 = a_2 = 1.5$, $b_1 = b_2 = 1$, then (3.12) does not hold, but (3.11) does. If $a_1 = 2$, $a_2 = 15$, $b_1 = 1$, $b_2 = 3$, then (3.12) holds, but (3.11) does not. Therefore, the two approaches are complementary.

Moreover, if we now consider the equation

$$\dot{x}(t) = Ax(t) + Bx(t-h) + C\int_{t-\tau}^{t} x(s)ds,\ C = \begin{pmatrix} c_1 & c_2 \\ -c_2 & c_1 \end{pmatrix} \quad (3.13)$$

with A, B previously defined in (3.8), Riccati equation (3.7) with $Q = qI, R_1 = S_1 = I$ leads to the sufficient condition of discrete delay–independent stability for (3.13):

$$a_i > (1+\tau)^{\frac{1}{2}}[(b_1^2 + b_2^2) + \tau(c_1^2 + c_2^2)]^{\frac{1}{2}},\quad \forall i = 1, 2.$$

Example 3.2 Consider the two–dimensional equation

$$\dot{x}(t) = Ax(t) + Bx(t-h), \quad A := \begin{pmatrix} -a_1 & 0 \\ 0 & -a_2 \end{pmatrix}, \quad B := \begin{pmatrix} b_1 & b_3 \\ -b_4 & b_1 \end{pmatrix} \tag{3.14}$$

with constant $a_i (i = 1, 2)$ and $b_i (i = 1, 3, 4)$ and

$$b_3 b_4 > 0. \tag{3.15}$$

This example is similar to the previous one (3.8), but with more general off diagonal entries in the matrix B.

Note that (3.14), (3.15) can be reduced to (3.8) with $b_2 = \sqrt{b_3 b_4}$ using a change of basis,

$$x(t) = Tx(t), \quad T := \begin{pmatrix} 1 & 0 \\ 0 & \sqrt{b_2 b_4^{-1}} \end{pmatrix}, \tag{3.16}$$

and that for any nonsingular matrix T, the Riccati equation (3.7) applied to equation

$$\dot{x}(t) = TAT^{-1}x(t) + TBT^{-1}x(t-h), \tag{3.17}$$

leads to

$$A'\widetilde{P} + \widetilde{P}A + \widetilde{R} + \widetilde{P}B\widetilde{R}^{-1}B'\widetilde{P} = -\widetilde{Q}, \tag{3.18}$$

with $\widetilde{P} := T'PT, \widetilde{R} := T'RT, \widetilde{Q} := T'QT$.

Then, using previous condition (3.11) of Example 3.1, we obtain the following, independent of delay stability conditions for (3.14):

$$a_i > (b_1^2 + b_3 b_4)^{\frac{1}{2}}, \quad \forall i = 1, 2.$$

Of course, this condition could be directly obtained by using the following matrices:

$$P := \begin{pmatrix} p_{11} & 0 \\ 0 & p_{22} b_3 b_4^{-1} \end{pmatrix}, \quad Q := q \begin{pmatrix} 1 & 0 \\ 0 & b_3 b_4^{-1} \end{pmatrix}, \quad R := r \begin{pmatrix} 1 & 0 \\ 0 & b_3 b_4^{-1} \end{pmatrix},$$

for solving the Riccati equation (3.7) corresponding to equation (3.14).

4. DELAY–DEPENDENT STABILITY CONDITIONS FOR EQUATIONS WITH DISCRETE DELAYS

In this section, we obtain stability conditions for equation (1.6) with purely discrete delays,

$$\dot{x}(t) = \sum_{i=1}^{m} B_i x(t - h_i). \tag{4.1}$$

The general way to obtain Riccati equations for (4.1) is the same as previously, but here the number of possibilities is much larger, because every term in the righthand side of (4.1) can be transformed independently. We shall describe only the cases when all the terms in (4.1) are transformed simultaneously and in a unified manner, similar to Sec. 2. Moreover, we give sketches of the proofs paying attention only to differences with those from Sec. 2.

4.1. *THE STABILITY CONDITION*

We represent Eq. (4.1) in the form

$$\dot{x}(t) = Ax(t) - \sum_{i,j=1}^{m} B_{ij} \int_{t-h_{ij}}^{t-h_j} x(s)ds.$$

Here and below in this section we are using notations:

$$A := \sum_{i=1}^{m} B_i, \; B_{ij} := B_i B_j, \; h_{ij} := h_i + h_j, \; h := \sum_{i=1}^{m} h_i. \qquad (4.2)$$

Take as a functional $V := V_1 + V_2$, with

$$V_1(x_t) := x'(t)Px(t),$$

$$V_2(x_t) := \sum_{i,j=1}^{m} \int_{h_j}^{h_{ij}} ds \int_{t-s}^{t} x'(\tau)Rx(\tau)d\tau.$$

Then,

$$\dot{V}(x_t) = x'(t)\left[A'P + PA + mRh + P\sum_{i,j=1}^{m} h_i B_{ij} B_{ij}' P\right] x(t) -$$

$$- \sum_{i,j=1}^{m} \int_{t-h_{ij}}^{t-h_j} [Rx(s) + B_{ij}' Px(t)]' R^{-1} [Rx(s) + B_{ij}' Px(t)] ds.$$

Hence, we obtain the following theorem.

Theorem 4.1 Assume that for some symmetric matrices $R > 0$ and $Q > 0$ there exists a symmetric solution, $P > 0$, of the equation

$$\left.\begin{aligned} A'P + PA + mRh + P\sum_{i,j=1}^{m} h_i B_i B_j R^{-1} B_j' B_i' P = -Q, \\ A := \sum_{i=1}^{m} B_i, \; h := \sum_{i=1}^{m} h_i. \end{aligned}\right\} \tag{4.3}$$

Then Eq. (4.1) is asymptotically stable.

4.2. *AN APPLICATION OF A FORM OF NDE*

We now obtain a generalization of Thm. 2.2 for Eq. (4.1) using its representation as an NDE,

$$\frac{d}{dt}\left[x(t) + \sum_{i=1}^{m} B_i \int_{t-h_i}^{t} x(s)ds\right] = Ax(t).$$

Then, put $V := V_1 + V_2$, where

$$V_1(x_t) := \left(x(t) + \sum_{i=1}^{m} B_i \int_{t-h_i}^{t} x(s)ds\right)' P \left(x(t) + \sum_{i=1}^{m} B_i \int_{t-h_i}^{t} x(s)ds\right),$$

$$V_2(x_t) := \sum_{i=1}^{m} \int_{0}^{h_i} ds \int_{t-s}^{t} x'(\tau) R_i x(\tau) d\tau.$$

For $\dot{V}$, we get

$$\dot{V}(x_t) = x'(t)\left[A'P + PA + \sum_{i=1}^{m} R_i h_i + \sum_{i=1}^{m} A'PB_i R_i^{-1} B_i' PAh_i\right] x(t) -$$

$$-\sum_{i=1}^{m} \int_{t-h_i}^{t} [B_i'PAx(t) - R_i x(s)]' R_i^{-1} [B_i'PAx(t) - R_i x(s)].$$

Thus we obtain the following result.

Theorem 4.2 Assume that for some symmetric matrices $R_i > 0$ and $Q > 0$ there exists a symmetric solution, $P > 0$, of the equation

$$\begin{aligned} &A'P + PA + \sum_{i=1}^{m} R_i h_i + \sum_{i=1}^{m} A'PB_iR^{-1}B_i'PAh_i = -Q, \\ &A = \sum_{i=1}^{m} B_i, \; h = \sum_{i=1}^{m} h_i. \end{aligned} \tag{4.4}$$

Then Eq. (4.1) is asymptotically stable if the equation

$$x(t) + \sum_{i=1}^{m} B_i \int_{t-h_i}^{t} x(s)ds = 0$$

is asymptotically stable.

4.3. *ONE MORE STABILITY CONDITION*

Finally, we consider modification of Thm. 2.3 for Eq. (4.1), represented as follows with notation (4.2):

$$\dot{x}(t) = Ax(t) - \sum_{i=1}^{m} B_i \int_{t-h_i}^{t} \dot{x}(s)ds.$$

Let $V := V_1 + V_2 + V_3$ where

$$V_1(x_t) := x'(t)Px(t), \quad V_2(x_t) := \sum_{i=1}^{m} \int_0^{h_i} ds \int_{t-s}^{t} \dot{x}'(\tau)R\dot{x}(\tau)d\tau,$$

$$V_3(x_t) := mh \sum_{i=1}^{m} \int_{t-h_i}^{t} x'(s)B_i'RB_ix(s)ds.$$

Note that $\dot{x}(t)$ in V_2 must be substituted from (4.1). We have,

$$\dot{V}_1(x_t) = x'(t)(A'P + PA)x(t) - \sum_{i=1}^{m} \int_{t-h_i}^{t} [\dot{x}'(s)B_i'Px(t) + x'(t)PB_i\dot{x}(s)]ds,$$

$$\dot{V}_2(x_t) = h\dot{x}'(t)R\dot{x}(t) - \sum_{i=1}^{m} \int_{t-h_i}^{t} \dot{x}'(s)R\dot{x}(s)ds.$$

Furthermore,

$$-\dot{x}'(s)B_iPx(t) - x'(t)PB_i\dot{x}(s) - \dot{x}'(s)R\dot{x}(s) =$$

$$= -[B_i'Px(t) + R\dot{x}(s)]'R^{-1}[B_i'Px(t) + R\dot{x}(s)] + x'(t)PB_iR^{-1}B_i'Px(t).$$

But,

$$h\dot{x}'(t)R\dot{x}(t) + \dot{V}_3(x_t) = h\sum_{i=1}^{m} x'(t-h_i)B_i'RB_jx(t-h_j)+$$

$$+mhx'(t)\sum_{i=1}^{m} B_i'RB_ix(t) - mh\sum_{i=1}^{m} x'(t-h_i)B_i'RB_ix(t-h_i).$$

Note that

$$\sum_{i,j=1}^{m} x'(t-h_i)B_i'RB_jx(t-h_j) - m\sum_{i=1}^{m} x'(t-h_i)B_i'RB_ix(t-h_i) =$$

$$= -(m-1)\sum_{i=1}^{m} x'(t-h_i)B_i'RB_ix(t-h_i)+$$

$$+ \sum_{i,j=1,\ i\neq j}^{m} x'(t-h_i)B_i'RB_jx(t-h_j) =$$

$$= - \sum_{i=1,j=i+1}^{m} (B_ix(t-h_i) - B_jx(t-h_j))'R(B_ix(t-h_i) - B_jx(t-h_j)).$$

Here for $i = m$ the summand equals zero. As a result, we obtain

$$\dot{V}(x_t) = x'(t)[A'P + PA + \sum_{i=1}^{m}(h_iPB_iR^{-1}B_i'P + mhB_i'RB_i)]x(t)-$$

$$-\sum_{i=1}^{m}\int_{t-h_i}^{t} [B_i'Px(t) + R\dot{x}(s)]'R^{-1}[B_i'Px(t) + R\dot{x}(s)]ds-$$

$$-h\sum_{i=1,j=i+1}^{m} (B_ix(t-h_i) - B_jx(t-h_j))'R(B_ix(t-h_i) - B_jx(t-h_j)).$$

So the generalization of Theorem 2.3 for Eq. (4.1) results.

Theorem 4.3 Assume that for some symmetric matrices $R > 0, Q > 0$ there exists a symmetric solution, $P > 0$, of the equation

$$\left.\begin{aligned} A'P + PA + \sum_{i=1}^{m}(h_i PBR^{-1}B_i' + mhB_i'RB_i) = -Q, \\ A := \sum_{i=1}^{m} B_i, \ h := \sum_{i=1}^{m} h_i. \end{aligned}\right\} \tag{4.5}$$

Then Eq. (4.1) is asymptotically stable.

Remark For the case $m = 1$ Eqs. (4.3)–(4.5) coincide with those of (2.8), (2.14), (2.17).

Example 4.1 Consider the two–dimensional equation

$$\dot{x}(t) = B\sum_{i=1}^{m} x(t - h_i), \quad B = \begin{pmatrix} -\alpha & \beta \\ -\beta & -\alpha \end{pmatrix}, \tag{4.6}$$

where α, β are some constants and $h_i \geq 0, \ \sum\limits_{i=1}^{m} h_i =: h$.

For stability investigation of Eq. (4.6) let us apply Thm. 4.3. The Eq. (4.5) with $R := rI, Q := qI, P := pI$, takes the form

$$-2\alpha mp + hp^2(\alpha^2 + \beta^2)r^{-1} + m^2h(\alpha^2 + \beta^2)r = -q, \tag{4.7}$$

which has a positive solution, p, if and only if

$$h < \frac{\alpha}{\alpha^2 + \beta^2}. \tag{4.8}$$

This condition is sufficient for the asymptotic stability of (4.6) independently of the number of delays, m. Note that it reduces to the necessary and sufficient one as $h \to 0$.

5. DISTRIBUTED DELAY

The transition from the case $m = 1$ to the case of arbitrary m, fulfilled in Sec. 4, shows that it is accompanied, as a rule, only by some additional cumbersome calculations. While, the basic features of the approach remain the same. Thus, for the case of distributed delay, we shall consider, in detail, only the Eq. (1.6) of the form

$$\dot{x}(t) = C_1 \int_{t-\tau_1}^{t} x(s)ds, \quad x(t) \in \mathbf{R}^n, \ t \geq 0. \tag{5.1}$$

The types of transformation of Eq. (5.1) correspond to those of Sec. 2.

5.1. *THE STABILITY CONDITION*

Represent Eq. (5.1) for $t \geq \tau_1$ as follows (compare with (2.2)):

$$\dot{x}(t) = \tau_1 C_1 x(t) - C_1^2 \int_{t-\tau_1}^{t} (s - t + \tau_1) ds \int_{s-\tau_1}^{s} x(s_1) ds_1. \tag{5.2}$$

Let $V := V_1 + V_2 + V_3$, where for some symmetric $P > 0$, $R > 0$,

$$V_1(x_t) := x'(t)Px(t),$$

$$V_2(x_t) := \int_0^{\tau_1} ds \int_{t-s}^{t} (s_1 - t + s) ds_1 \int_{s_1-\tau_1}^{s_1} x'(s_2) R x(s_2) ds_2,$$

$$V_3(x_t) := \frac{1}{2}\tau_1^2 \int_{t-\tau_1}^{t} (s - t + \tau_1) x'(s) R x(s) ds.$$

Then, by virtue of (5.2),

$$\dot{V}_1(x_t) = \tau_1 x'(t)(C_1' P + PC_1)x(t) - x'(t)PC_1^2 \int_{t-\tau_1}^{t} (s - t + \tau_1) ds \times$$

$$\times \int_{s-\tau_1}^{s} x(s_1) ds_1 - \int_{t-\tau_1}^{t} (s - t + \tau_1) ds \int_{s-\tau_1}^{s} x'(s_1) ds_1 C_1'^2 Px(t),$$

and, also,

$$\dot{V}_2(x_t) = -\int_0^{\tau_1} ds \int_{t-s}^{t} ds_1 \int_{s_1-\tau_1}^{s_1} x'(s_2) R x(s_2) ds_2 + \frac{1}{2}\tau_1^2 \int_{t-\tau_1}^{t} x'(s) R x(s) ds.$$

Hence,

$$\dot{V}_2(x_t) + \dot{V}_3(x_t) = \frac{1}{2}\tau_1^3 x'(t) R x(t) - \int_{t-\tau_1}^{t} (s - t + \tau_1) ds \int_{s-\tau_1}^{s} x'(s_1) R x(s_1) ds_1.$$

But,

$$x'(t)PC_1^2 x(s) + x'(s)PC_1'^2 Px(t) + x'(s)Rx(s) =$$

$$= [C_1'^2 Px(t) + Rx(s)]' R^{-1} [C_1'^2 Px(t) + Rx(s)] - x'(t)PC_1^2 R^{-1} C_1'^2 Px(t).$$

As a result, we obtain

$$\dot{V}(x_t) = \tau_1 x'(t)[C_1'P + PC_1 + \frac{1}{2}\tau_1^2(R + PC_1^2R^{-1}C_1'P)]x(t) -$$

$$-\int_{t-\tau_1}^{t}(s-t+\tau_1)ds\int_{s-\tau_1}^{s}[C_1'^2Px(t)+Rx(s_1)]'R^{-1}[C_1'^2Px(t)+Rx(s_1)]ds_1.$$

From the last relations and Thm. 4.1.1 the following result is obtained.

Theorem 5.1 Assume that $\tau_1 \geq 0$ and for some symmetric $R > 0, Q > 0$ there exists a symmetric solution, $P > 0$, of the Riccati equation

$$C_1'P + PC_1 + \frac{1}{2}\tau_1^2(R + PC_1^2R^{-1}C_1'^2P) = -Q. \tag{5.3}$$

Then Eq. (5.1) is asymptotically stable.

5.2. *AN APPLICATION OF A FORM OF NDE*

Rewrite Eq. (5.1) as a NDE,

$$\frac{d}{dt}\left[x(t) + C_1\int_{t-\tau_1}^{t}(s-t+\tau_1)x(s)ds\right] = \tau_1 C_1 x(t).$$

Put $V := V_1 + V_2$, where

$$V_1(x_t) := \left(x(t) + C_1\int_{t-\tau_1}^{t}(s-t+\tau_1)x(s)ds\right)' \times$$

$$\times P\left(x(t) + C_1\int_{t-\tau_1}^{t}(s-t+\tau_1)x(s)ds\right),$$

$$V_2(x_t) := \tau_1\int_{0}^{\tau_1}ds\int_{t-s}^{t}(s_1-t+s)x'(s_1)Rx(s_1)ds_1.$$

Then,

$$\dot{V}_1(x_t) = \tau_1\left[x'(t)(C_1'P + PC_1)x(t) + x'(t)C_1'PC_1\int_{t-\tau_1}^{t}(s-t+\tau_1)x(s)ds + \right.$$

$$\left. + \int_{t-\tau_1}^{t}(s-t+\tau_1)x'(s)dsC_1'PC_1x(t)\right],$$

Furthermore,

$$\dot{V}_2(x_t) = \tau_1 \left[\frac{1}{2}\tau_1^2 x'(t)Rx(t) - \int\limits_{t-\tau_1}^{t} (s - t + \tau_1)x'(s)Rx(s)ds \right].$$

But,

$$x'(t)C_1'PC_1x(s) + x'(s)C_1'PC_1x(t) - x'(s)Rx(s) =$$

$$= -[C_1'PC_1x(t) - Rx(s)]'R^{-1}[C_1'PC_1x(t) - Rx(s)] +$$

$$+ x'(t)C_1'PC_1R^{-1}C_1'PC_1x(t).$$

Hence,

$$\dot{V}(x_t) = \tau_1 x'(t)[C_1'P + PC_1 + \frac{1}{2}\tau_1^2(R + C_1'PC_1R^{-1}C_1'PC_1)] -$$

$$- \int\limits_{t-\tau_1}^{t} (s - t + \tau_1)[C_1'PC_1x(t) - Rx(s)]'R^{-1}[C_1'PC_1x(t) - Rx(s)]ds.$$

Thus the following theorem is proven.

Theorem 5.2 Assume that $\tau_1 > 0$ *and for some symmetric* $R > 0, Q > 0$ *there exists a symmetric solution,* $P > 0$, *of the Riccati equation*

$$C_1'P + PC_1 + \frac{1}{2}\tau_1^2(R + C_1'PC_1R^{-1}C_1'PC_1) = -Q. \tag{5.4}$$

Then Eq. (5.1) is asymptotically stable if the equation

$$x(t) + C_1 \int\limits_{t-\tau_1}^{t} (s - t + \tau_1)x(s)ds = 0$$

is asymptotically stable.

5.3. *ONE MORE STABILITY CONDITION*

As we pointed out earlier, using other types of transformations we should be able to obtain other stability conditions. We illustrate this for Eq. (5.1), represented for $t \geq \tau_1$ as

$$\dot{x}(t) = \tau_1 C_1 x(t) - C_1 \int\limits_{t-\tau_1}^{t} (s - t + \tau_1)\dot{x}(s)ds.$$

Put $V_1(x_t) := x'(t)Px(t)$. Then,

$$\dot{V}_1(x_t) = \tau_1 x'(t)(C_1'P + PC_1)x(t) -$$

$$-\int_{t-\tau_1}^{t} (s - t + \tau_1)(\dot{x}'(s)C_1'Px(t) + x'(t)PC_1\dot{x}(s))ds.$$

Hence it is natural to choose as V_2 the functional

$$V_2(x_t) := \int_0^{\tau_1} ds \int_{t-s}^{t} (s_1 - t + s)\dot{x}'(s_1)R\dot{x}(s_1)ds_1.$$

We get,

$$\dot{V}_2(x_t) = \frac{1}{2}\tau_1^2\dot{x}'(t)R\dot{x}(t) - \int_{t-\tau_1}^{t} (s - t + \tau_1)\dot{x}'(s)R\dot{x}(s)ds.$$

Note that

$$\dot{x}'(s)C_1'Px(t) + x'(t)PC_1\dot{x}(s) + \dot{x}'(s)R\dot{x}(s) =$$

$$= [C_1'Px(t) + R\dot{x}(s)]' R^{-1}[C_1'Px(t) + R\dot{x}(s)] - x'(t)PC_1R^{-1}C_1'Px(t).$$

Now, taking into account (5.1), it turns out that

$$J := \frac{1}{2}\tau_1^2\dot{x}'(t)R\dot{x}(t) = \frac{1}{2}\left[\int_{t-\tau_1}^{t} x'(s)ds\right] C_1'RC_1 \left[\int_{t-\tau_1}^{t} x(s_1)ds_1\right]\tau_1^2.$$

Because of the last relation we must introduce the functional

$$V_3(x_t) := \frac{\tau_1^2}{2}\int_{t-\tau_1}^{t}\left[\int_s^t x'(s_1)ds_1\right] C_1'RC_1 \left[\int_s^t x(s_2)ds_2\right] ds,$$

for which,

$$\dot{V}_3(x_t) = -J + \left[x'(t)C_1'RC_1\int_{t-\tau_1}^{t}(s - t + \tau_1)x(s)ds + \right.$$

$$\left. + \int_{t-\tau_1}^{t}(s - t + \tau_1)x'(s)dsC_1'RC_1x(t)\right]\frac{\tau_1^2}{2}.$$

The final step of our constructions is the functional V_4, which is defined as

$$V_4(x_t) := \int_0^{\tau_1} ds \int_{t-s}^{t} (s_1 - t + s)x'(s_1)Rx(s_1)ds_1 \frac{\tau_1^2}{2},$$

for which,

$$\dot{V}_4(x_t) = -\frac{\tau_1^2}{2} \int_{t-\tau_1}^{t} (s - t + \tau_1)x'(s)Rx(s)ds + \frac{1}{4}\tau_1^4 x'(t)Rx(t).$$

Furthermore, similar to (2.7), we have,

$$x'(t)C_1'RC_1x(s) + x'(s)C_1'RC_1x(t) - x'(s)Rx(s) =$$

$$= -[C_1'RC_1x(t) - Rx(s)]'R^{-1}[C_1'RC_1x(t) - Rx(s)] +$$

$$+x'(t)C_1'RC_1R^{-1}C_1'RC_1x(t).$$

As a result, we obtain that the full derivative $\dot{V}$ of the functional $V := V_1 + V_2 + V_3 + V_4$ satisfies the inequality

$$\dot{V}(x_t) \leq \tau_1 x'(t)\Big[C_1'P + PC_1 + \frac{\tau_1}{2}PC_1R^{-1}C_1'P +$$

$$+ \frac{\tau_1^3}{4}(R + C_1'RC_1R^{-1}C_1'RC_1)\Big] x(t).$$

Consequently, the following theorem is proven.

Theorem 5.3 Assume that $\tau_1 > 0$ *and for some symmetric* $R > 0, Q > 0$, *there exists a symmetric solution,* $P > 0$, *of the Riccati equation*

$$C_1'P + PC_1 + \frac{\tau_1}{2}PC_1R^{-1}C_1'P + \frac{\tau_1^3}{4}(R + C_1'RC_1R^{-1}C_1'RC_1) = -Q.$$

Then Eq. (5.1) is asymptotically stable.

Example 5.1 Consider a scalar equation of the form (5.1),

$$\dot{x}(t) = c_1 \int_{t-\tau_1}^{t} x(s)ds. \tag{5.5}$$

Eq. (5.3) (Thm. 5.1), corresponding to (5.5), takes the form

$$2c_1p + \frac{1}{2}\tau_1^2 r + \frac{1}{2}r^{-1}\tau_1^2 p^2 c_1^4 + q = 0, \tag{5.6}$$

where r and q are some positive constants. A positive solution of Eq. (5.6) exists if

$$c_1 < 0, \quad |c_1|\tau_1^2 < 2.$$

Hence, Eq. (5.5) is asymptotically stable under condition

$$-2 < c_1\tau_1^2 < 0. \tag{5.7}$$

Note that application of Thm. 5.2 results in the same stability condition (5.7).

Remark In some sense, the Riccati equations obtained are similar to those encountered in the theory of optimal control for linear quadratic problem, but with the opposite sign. Comparing with control theory where determination of the solution of Riccati equation is absolutely necessary, here it is essential only to know conditions under which this positive definite symmetric solution does exist.

CHAPTER 8

STABILITY OF NEUTRAL TYPE FUNCTIONAL DIFFERENTIAL EQUATIONS

In this chapter we give some methods for investigating the stability of various classes of NDEs.

1. DIRECT LIAPUNOV'S METHOD

1.1. *DEGENERATE LIAPUNOV FUNCTIONALS*

$$[x(t) - g(t, x_t)]^{\bullet} = f(t, x_t), \quad t \geq t_0, \tag{1.1}$$

$$x_{t_0} = \phi \in C[-h, 0], \quad x_t(\theta) = x(t + \theta), \quad -h \leq \theta \leq 0. \tag{1.2}$$

Here, $f, g : [t_0, \infty) \times Q_H \to \mathbf{R}^n$ are continuous maps, $Q_H = \{u \in C[-h, 0] : \| u \| < H\}$, $0 < H < \infty$, satisfying all the general conditions in Subs. 3.3.5, as well as

$$f(t, 0) \equiv g(t, 0) \equiv 0, \quad |f(t, \psi)| \leq C, \quad |g(t, \psi)| \leq C, \quad t \geq t_0, \quad \psi \in Q_H. \tag{1.3}$$

The definitions of stability for NDEs and the notation are the same as for RDEs, given in Sec. 4.1. In particular, we have:

Definition 1.1 The zero solution of Pr. (1.1)-(1.3) is called: (a) *stable*, if for any $\varepsilon > 0$ there exists $\delta(\varepsilon, t_0) > 0$ such that $|x(t; \phi)| \leq \varepsilon$ for $t \geq t_0$ and $\| \phi \| \leq \delta(\varepsilon, t_0)$; (b) *asymptotically stable*, if it is stable and besides $\lim_{t \to \infty} x(t; \phi) = 0$, for any initial function ϕ with sufficiently small $\| \phi \|$; (c) *uniformly stable* with respect to the initial moment t_0 if $\delta(\varepsilon, t_0) \equiv \delta(\varepsilon)$.

NDEs have the properties that are similar in some aspects to the properties of difference equations and in the other aspects are similar to the properties of differential equations. This duality of NDE is essential for a study of asymptotic behaviour of their solutions. Below some results of both types are stated. The method of investigation that we use is based essentially on degenerate Liapunov functionals. It is possible with the aid of degenerate functionals to separate the cases when the asymptotic behaviour of the solutions of NDEs is similar to that of difference equations from the cases when this is not valid.

Let us underline that some theorems of the direct Liapunov's method could be obtained by formal generalization of the corresponding assertions for RDEs. However, the results thus obtained have limited applications because of the difficulties in the construction of the positive definite Liapunov functionals for NDEs. It is the main reason to investigate stability of NDEs by using positive semidifinite functionals. So, the method given below for studying stability is based on the use of such functionals and the study of the stability of the following type of functional inequalities:

$$|z(t,y_t)| \leq \overline{f}(t), \quad t \geq t_0; \quad y_{t_0} = \phi, \tag{1.4}$$

where $z(t,y_t) := y(t) - g(t,y_t)$, functional g is the same as in Eq. (1.1) and $\overline{f} : [t_0,\infty) \to [0,\infty)$ is a continuous function.

Definition 1.2 The trivial solution of inequality (1.4) is said to be:

1) *stable* if for any $\varepsilon > 0$ there is a $\delta(\varepsilon, t_0) > 0$ such that for any solution y: $[t_0, T) \to \mathbf{R}^n (t_0 < T \leq \infty)$ of (1.4) we have $|y(t)| \leq \varepsilon$, $t_0 \leq t < T$, for all ϕ and $\overline{f}$, $\| \phi \| \leq \delta(\varepsilon, t_0)$, $\| \overline{f} \| (:= \sup_{t_0 \leq t < T} |\overline{f}(t)|) \leq \delta(\varepsilon, t_0)$;

2) *asymptotically stable* if it is stable, $T = \infty$ and $\lim_{t\to\infty} y(t) = 0$ for all ϕ, f, $\| \phi \| < \delta_0$, $\| f \| < \delta_0$ (δ_0 fixed), where $\overline{f}(t) \to 0$ as $t \to \infty$.

In Theorems 1.1, 1.2, 1.4, 1.6 all $\omega_i \in \Omega$ (see Subs. 4.1.2).

Theorem 1.1 Suppose that the trivial solution of (1.4) is stable and that there exists a continuous functional $V : [t_0,\infty) \times Q_H \times \mathbf{R}^n \to \mathbf{R}$ *such that*

$$\omega_1(|z(t,\psi)|) \leq V(t,\psi,z(t,\psi)) \leq \omega_2(\| \psi \|), \tag{1.5}$$

$$\dot{V} := \frac{d}{dt} V(t, x_t, z(t, x_t)) \leq 0 \tag{1.6}$$

for all solutions x of Eq. (1.1), where ω_i and d/dt have the same meaning as in Subs. 3.1.2. Then the trivial solution of (1.1) is stable. If the trivial solution of (1.4) is asymptotically stable, V satisfies (1.5), and

$$\dot{V}(t,x_t,z(t,x_t)) \leq -\omega_3(|z(t,x_t)|),$$

then the trivial solution of (1.1) is asymptotically stable.

Proof Take $\varepsilon \in (0,H)$ arbitrary, and choose $\delta_1 = \delta_1(\varepsilon) > 0$ such that $|y(t)| \leq \varepsilon$, $t \geq t_0$, if $\| \overline{f} \| \leq \delta_1$, $\| \phi \| \leq \delta_1$. Define $\delta_2 \in (0,\delta_1)$ by $\omega_2(\delta_2) \leq \omega_1(\delta_1)$. Consider an arbitrary initial function ϕ with $\| \phi \| \leq \delta_2$, and suppose the solution x of (1.1), (1.2) is defined and satisfies the inequality

$|x(t)| < H$ on an interval $[t_0, T)$. Then, taking into account (1.5), (1.6), we find that for $t_0 \le t < T$,

$$\omega_1(|z(t, x_t)|) \le V(t, x_t, z(t, x_t)) \le V(t_0, \phi, z(t_0, \phi)) \le \omega_2(\delta_2) \le \omega_1(\delta_1).$$

Putting $\overline{f}(t) := |z(t, x_t)|$ and using the definition of δ_1 we obtain $|x(t)| \le \varepsilon,\ t_0 \le t \le T$. Hence this inequality holds for $t_0 \le t < \infty$ and arbitrary ϕ with $\| \phi \| \le \delta_2$.

This proves stability.

To prove asymptotic stability we take $\delta > 0$ such that $|x(t; t_0, \phi)| \le H,\ t \ge t_0$, for any $\phi \in Q_\delta$. It suffices to verify that

$$z(t, x_t(t; t_0, \phi)) \to 0, \quad \phi \in Q_\delta, \quad t \to \infty. \tag{1.7}$$

Suppose (1.7) is not true. Then we can find a $\gamma \in (0, H)$ and a sequence $t_i \to \infty$ such that $|z(t_i, x_{t_i})| \ge \gamma$. The conditions (1.1)–(1.3) give

$$|\dot{z}(t, x_t)| = |f(t, x_t)| \le C, \quad \phi \in Q_\delta.$$

Using now the same arguments as in the proof of Thm. 4.1.1, we obtain

$$V(t, x_t, z(t, x_t)) - V(t_0, \phi, z(t_0, \phi)) \to -\infty, \quad t \to \infty.$$

This relation contradicts (1.5), hence (1.7) holds. Relation (1.7) and the asymptotic stability of (1.4) imply the asymptotic stability of the trivial solution of (1.1).□

We can similarly prove

Theorem 1.2 Suppose the conditions of Thm. 1.1 hold with $C[-h, 0]$ *instead of* Q_H, $\sup\limits_{t_0 \le t < \infty, \|\psi\| < H} |f(t, \psi)| < \infty$ *for all* $H > 0, \omega_1(r) \to \infty$ *as* $r \to \infty$, *and any solution of (1.4) is bounded on any bounded interval of its existence. Then* $x(t; t_0, \phi) \to 0,\ t \to \infty$, *for any initial function* $\phi \in C[-h, 0]$.

We give some stability conditions for inequality (1.4) and Eq. (1.1).

Theorem 1.3 Suppose that the continuous functional $g : [t_0, \infty) \times Q_H \to \mathbf{R}^n$ *satisfies the condition*

$$|g(t, \psi)| \le \gamma \| \psi \|, \quad \gamma < 1,$$

if $\| \psi \| \le \delta_0$, *for some* $\delta_0 \in (0, H)$. *Then the trivial solution of (1.4) is asymptotically stable.*

Proof Let $y : [t_0, T) \to \mathbf{R}^n$ be a solution of (1.4). Denote $\alpha(t) := \sup\limits_{t_0 \le s \le t} |y(s)|$

$(t_0 \le t < T)$ and take a $\delta \in (0, \delta_0)$. Let us assume that $\| \phi \| \le \delta, \| \bar{f} \| \le \delta$, and $\alpha \le \delta_0$ for some $t \in (t_0, T)$. Then for any $\eta > 0$ and some $t^\eta \in [t_0, t]$ we have

$$\alpha(t) \le |y(t^\eta)| + \eta \le |y(t^\eta) - g(t^\eta, y_{t^\eta})| + |g(t^\eta, y_{t^\eta})| + \eta \le \delta + \gamma \| y_{t^\eta} \| +$$

$$+\eta \le \delta + \gamma \max\{\alpha(t), \| \phi \|\} + \eta \le \delta + \gamma\alpha(t) + \gamma\delta + \eta.$$

Hence, by virtue of arbitrariness of η

$$(1 - \gamma)\alpha(t) \le (1 + \gamma)\delta.$$

From here it follows that if $(1-\gamma)^{-1}(1+\gamma)\delta < \delta_0$ then $\alpha(t) \le (1-\gamma)^{-1}(1+\gamma)\delta$ on $[t_0, T)$. This means stability of the trivial solution of (1.4).

In addition, let us suppose that $T = \infty$, $\bar{f}(t) \to 0$ as $t \to \infty$, and $\| \psi \|$ and $\| \bar{f} \|$ are so small that $|y(t)| \le \delta_0 (t_0 \le t < \infty)$. If $y(t) \not\to 0$ as $t \to \infty$, then $\limsup_{t\to\infty} |y(t)| =: p > 0$. Let $t^i \to \infty$, $|y(t^i)| \to p$. Then

$$|y(t^i)| \le |y(t^i) - g(t^i, y_{t^i})| + |g(t^i, y_{t^i})| \le \bar{f}(t^i) + \gamma \| y_{t^i} \| .$$

From here it follows that for any $\eta > 0$ and all sufficiently large i

$$p - \eta \le |y(t^i)| \le \bar{f}(t^i) + \gamma(p + \eta) \Rightarrow \quad p \le (1 - \gamma)^{-1}[f(t^i) + (1 + \gamma)\eta].$$

We now let i go to infinity and next $\eta \to 0$ to obtain a contradiction. Hence we conclude that $y(t) \to 0$ as $t \to \infty$. □

With natural modifications, Thms. 1.1 – 3 can be immediately generalized to systems with infinite delay (in particular, Thm. 1.3 then only concerns stability).

Theorem 1.4 Suppose the functionals $f, g : [t_0, \infty) \times C[-h, 0] \to \mathbf{R}^n$ are continuous, and

$$|g(t + s, \psi) - g(t, \chi)| \le \omega_1(s) + \gamma \| \psi - \chi \|,$$

$$0 < \gamma < 1, \quad \forall \psi, \chi \in Q_H, \quad t \ge t_0, \quad s \ge 0.$$

Assume that there is a continuous functional V satisfying condition (1.5) and

$$\dot{V}(t, x_t) \le -\omega_2(|x(t)|).$$

Then the trivial solution of the Pr. (1.1), (1.2) is asymptotically stable.

Proof By virtue of Thms. 1.1, 1.3 the trivial solution of (1.1) is stable. Hence $x(t; \phi) \in Q_H$ for $\phi \in Q_\delta, \delta > 0$. Show that $x(t; \phi) \to 0$ as $t \to \infty$.

Suppose the contrary. Then there exist a number $\nu > 0$ and a sequence $\{t_i\}$, $t_i \to \infty$ such that all $|x(t_i)| > \nu$. From (1.1) it follows

$$x(t) = g(t, x_t) + \int_0^t f(s, x_s)ds.$$

Here $x(s) = \phi(s)$ for $s \leq 0$. Then taking into account (1.1) we have for $\Delta \geq 0$

$$|x(t+\Delta) - x(t)| = |g(t+\Delta, x_{t+\Delta}) - g(t, x_t) +$$

$$+ \int_t^{t+\Delta} f(s, x_s)ds \leq \alpha \parallel x_{\tau+\Delta} - x_\tau \parallel + \omega_3(\Delta) + C\Delta.$$

Designate

$$\rho(\eta) := \sup_{t\geq 0} \sup_{\Delta \leq \eta} |x(t+\Delta) - x(t)|.$$

Then

$$\rho(\eta) \leq (1-\alpha)^{-1}[\alpha \max_{\eta \geq \Delta} \max_{-h \leq \theta \leq 0} |x(\theta+\Delta) - \phi(\theta)| + \omega_3(\eta) + C\eta].$$

So, due to the uniform continuity of the function $x(t)$ on the closed interval $[-h, \eta]$, the righthand side of this inequality tends to zero as $\eta \to 0$. Hence, a $\overline{\eta} > 0$ exists such that $\rho(\overline{\eta}) \leq \nu/2$. In this case for $\tau \in [t_i - \overline{\eta}, t_i + \overline{\eta}]$ will be $x(\tau) > \nu/2$ uniformly for all i. From this it follows a contradiction: $V(t, x_t) \to -\infty$ for $t \to \infty$.□

Example 1.1 Consider the system of scalar equations

$$[x_1(t) + \sum_{j=1}^m \alpha_j x_1(t-h_j)]^\bullet = x_2(t), \quad \dot{x}_2(t) = -g(x_1(t) + \sum_{j=1}^m \alpha_j x_1(t-h_j)). \tag{1.8}$$

Here $0 < h_1 < \ldots < h_m$, $g \in C(\mathbf{R})$, $yg(y) > 0$ for $y \neq 0$,

$$\sum_{j=1}^m |\alpha_j| < 1.$$

We will prove that under these conditions the trivial solution of (1.8) is stable. Define a functional V for arbitrary $x_1, x_2 \in C[-h, 0]$ by :

$$V(x_{1t}, x_{2t}) := \frac{1}{2}[x_2(t)]^2 + \int_0^{z(t)} g(s)ds, \quad z(t) := x_1(t) + \sum_{j=1}^m \alpha_j x_1(t-h_j).$$

Then $\dot{V} = x_2\dot{x}_2 + g(z)\dot{z} = 0$. Hence, by Thms. 1.1, 1.3 system (1.8) is stable. □

Example 1.2 Consider the scalar NDE

$$\dot{x}(t) = -b(t)x(t-h) + c\dot{x}(t-h), \quad t \geq 0.$$

The function $b(t) \geq 0$ is assumed to be continuous. Stability conditions of this system are given by the inequalities

$$\sup_{t\geq 0} \left\{ |c| + \int_t^{t+h} b(s)ds \right\} < 1,$$

$$\sup_{t\geq 0}[-2b(t+h) + |c|(b(t+h) + b(t+2h)) +$$

$$+b(t+h)\int_{t-h}^{t} (b(s+h) + b(s+2h))ds] < 0.$$

For the proof it is sufficient to use the functional

$$V(x_t) := \left[x(t) - cx(t-h) - \int_{t-h}^{t} b(s+h)x(s)ds \right]^2 +$$

$$+|c|\int_{t-h}^{t} b(s+2h)x^2(s)ds + \int_{t-h}^{t} b(t_1+2h)dt_1 \int_{t_1}^{t} b(t_2+h)x^2(t_2)dt_2.$$

Corollary 1.1 Consider a special class of NDEs

$$\dot{z}(t,x_t) = f(t, z(t,x_t)), \quad t \geq t_0, \quad x_{t_0} = \phi, \quad z(t,x_t) := x(t) - g(t,x_t).$$

Here $f : [t_0, \infty) \times \mathbf{R}^n \to \mathbf{R}^n$ *is a continuous function satisfying the local Lipschitz condition in the second argument,* $f(t,0) \equiv 0$, *and* g *satisfies the assumptions of Thm. 1.4. The trivial solution of this equation is asymptotically stable if the trivial solution of the ordinary differential equation*

$$\dot{y}(t) = f(t, y(t)), \quad t \geq t_0, \quad y(t_0) = y_0,$$

is uniformly asymptotically stable with respect to the initial moment t_0.

Proof By virtue of the inversion theorem for ODEs (see, e.g. [298]) there exists a function $(t,y) \mapsto v(t,y)$, which is positive definite, decrescent, and $\dot{v}(t,y)$ is negative definite. Hence, the functional $(t,x_t) \mapsto V(t,x_t) := v(t,z(t,x_t))$ satisfies the assumptions of Thm. 1.4.□

Example 1.3 Consider the NDE

$$\ddot{z}(t,x_t) + g(z(t,x_t))\dot{z}(t,x_t) + f(z(t,x_t)) = 0, \quad f(0) = 0$$

where $g, f : \mathbf{R} \to \mathbf{R}$, and

$$z(t,x_t) := x(t) + \frac{1}{2}[\exp(\sin t)]x(t-1).$$

The zero solution of this equation is asymptotically stable under conditions

$$xf(x) > 0, \quad x \neq 0, \quad g(x) > 0$$

because under these conditions the zero solution of ODE

$$\ddot{x}(t) + g(x(t))\dot{x}(t) + f(x(t)) = 0,$$

is uniformly asymptotically stable.

1.2. *STABILITY IN A FIRST APPROXIMATION*

Here we consider an equation of the type (1.1) with finite delay:

$$[x(t) - g(t,x_t) - g_1(t,x_t)]^\bullet = f(t,x_t) + f_1(t,x_t), \quad t \geq t_0, \tag{1.9}$$

$$x_{t_0} = \phi \in C[-h,0].$$

Along with the general assumptions of Subs. 1.1, the functionals g_1 and f_1 are assumed to satisfy the following condition uniformly for $t \in [t_0, \infty)$:

$$|f_1(t,\psi)| + |g_1(t,\psi)| = o(\|\psi\|) \quad (\|\psi\| \to 0).$$

The functionals $(t,\psi) \mapsto g(t,\psi)$ and $(t,\psi) \mapsto f(t,\psi)$ are assumed to be linear in ψ, and the linear system

$$[x(t) - g(t,x_t)]^\bullet = f(t,x_t)$$

(called the *system of first approximation*) is assumed to be uniformly exponentially stable.

Theorem 1.5 ([289.3]) Under the above assumptions the trivial solution of (1.9) is asymptotically stable.

Example 1.4 Consider the NDE (a variant of logistic equation)

$$\dot{x}(t) = \gamma x(t)(1 - K^{-1}x(t-h)) + \gamma c\dot{x}(t-h)$$

$$(\gamma, K, h, c = \text{const} > 0).$$

This equation has steady state equilibrium $x(t) \equiv K$; we study its stability. We write $x(t) = K(y(t)+1)$. Then the equation for $y(t)$ takes the form

$$\dot{y}(t) = -\gamma(y(t)+1)y(t-h) + \gamma c\dot{y}(t-h).$$

So the equation of first approximation is

$$\dot{y}(t) = -\gamma y(t-h) + \gamma c\dot{y}(t-h).$$

Sufficient conditions for the exponential stability of this equation are [289.3]:

$$0 < \gamma c < 1, \quad 1 < \gamma h < \beta(1-\gamma^2 c^2),$$

where $\beta \in (\pi/2, \pi)$ is the root of the equation $\beta\tan\beta = -h/c$. By Thm.1.5, these conditions give us those for asymptotic stability of the initial logistic equation.□

1.3. *THE USE OF FUNCTIONALS DEPENDING ON DERIVATIVES*

We consider the stability problem for NDEs of the form

$$\dot{x}(t) = f(t, x_t, \dot{x}_t), \quad x : [t_0 - h, \infty) \to \mathbf{R}^n, \quad f(t,0,0) \equiv 0, \qquad t \geq t_0, \quad (1.10)$$

$$x_{t_0} = \phi.$$

Let W be the Banach space of absolutely continuous functions $[-h, 0] \to \mathbf{R}^n$ with square–integrable derivative, with norm

$$\| \psi \|_W = \left[|\psi(0)|^2 + \int_{-h}^{0} |\dot{\psi}(\theta)|^2 d\theta \right]^{1/2}.$$

Let $Q_H \subset W$ be a ball, $Q_H := \{\psi : \| \psi \|_W < H\}$, $H > 0$. Let $f : [t_0, \infty) \times Q_H \times L_2[-h, 0] \to \mathbf{R}^n$ be a continuous functional which satisfies a Lipschitz condition in the second and third arguments, with Lipschitz constant for the third argument less than 1. The proof of the following theorem is similar to that of Thm. 1.1.

Theorem 1.6 If there is a continuous functional $V : [t_0, \infty) \times Q_H \times L_2[-h, 0] \to \mathbf{R}$ *satisfying the inequalities*

$$\omega_1(|\psi(0)|) \leq V(t, \psi, \chi) \leq \omega_2(\| \psi \|_W), \quad \dot{V} \leq 0,$$

then the solution $x(t) \equiv 0$ *is stable. If, moreover, we have* $\dot{V} \leq -\omega_3(|x(t)|)$, *then the trivial solution of (1.10) is asymptotically stable.*

Example 1.5 Consider the scalar equation

$$\dot{x}(t) = -a(t)x(t) + b(t)\dot{x}(t-h), \quad t \geq 0, \quad h > 0, \tag{1.11}$$

where a, b are bounded continuous functions, and $a(t) \geq c > 0$. Consider also the functional

$$V(t, x_t, \dot{x}_t) := x^2(t) + \int_{-h}^{0} \frac{\dot{x}^2(t+\theta)}{a(t+\theta)} d\theta.$$

We obtain

$$\dot{V}(t, x_t, \dot{x}_t) = 2x(t)\dot{x}(t) + \frac{\dot{x}^2(t)}{a(t)} - \frac{\dot{x}^2(t-h)}{a(t-h)}.$$

Eliminating $\dot{x}(t)$ according to (1.11) we find

$$\dot{V}(t, x_t, \dot{x}_t) = -a(t)x^2(t) - \left[\frac{1}{a(t-h)} - \frac{b^2(t)}{a(t)}\right] \dot{x}^2(t-h).$$

Hence, by Thm. 1.6 the trivial solution of (1.1) is asymptotically stable if

$$a(t) - b^2(t)a(t-h) \geq 0.$$

□

1.4. *INSTABILITY OF NDES*

Consider the Cauchy problem (1.1) – (1.3), in which without loss of generality $t_0 = 0$. Remind that in this chapter $z(t, \psi) := \psi(0) - g(t, \psi)$, $\forall (t, \psi) \in [0, \infty) \times Q_H$.

Definition 1.3 The zero solution of Pr. (1.1) – (1.3) is called *unstable* if it is not stable, i.e. if there is $\varepsilon > 0$ such that for any $\delta > 0$ there exists an initial function ϕ and an instant $t_1 = t_1(\delta, \phi) > 0$ such that $\| \phi \| < \delta$, but $|x(t_1; \phi)| \geq \varepsilon$.

Let us underline that stability (or instability) of the system depends essentially on the choice of the class of perturbations. It can happen that the system is stable with respect to one class of perturbations and unstable with respect to the other one.

Consider, e.g., the following model which was used to describe the behaviour of the central nervous system in the process of learning [452]

$$\dot{x}(t) = k[x(t) - x(t-1)][N - x(t)], \quad t > 1, \tag{1.12}$$

$$\left.\begin{aligned} \dot{x}(t) &= kx(t)[N - x(t)], \quad 0 \le t \le 1, \\ x(0) &= x_0, \quad 0 < x_0 < N, \quad k > 0. \end{aligned}\right\} \tag{1.13}$$

It was shown [289.3] that the solution $x(t) \equiv N$ of Eq. (1.12) is stable with respect to the perturbations coinciding with the set of the solutions of Pr. (1.13), and unstable if allowed perturbations coincide with the space $C[0,1]$.

1.4.1 Instability conditions. Consider a continuous functional $(t,\psi) \mapsto V(t,\psi)$ such that $V : [0,\infty) \times Q_H \to \mathbf{R}$. Let us denote

$$W(t;\phi) := V(t, x(\cdot;\phi)_t)$$

$$\dot{W}(t;\psi) := \liminf_{\Delta t \to 0+} \frac{1}{\Delta t}[W(t+\Delta t;\psi) - W(t;\psi)].$$

Theorem 1.7 Let there exist a continuous functional $(t,\psi) \mapsto V(t,\psi)$ and a set $\{V > 0\}(= \{t,\psi) : V(t,\psi) > 0\}) \subset [0,\infty) \times Q_H$ such that

1) $\psi = 0$ belongs to the boundary of the set $\{\psi : V(0,\psi) > 0\}$;

2) the functional V is bounded from above

$$V(t,\psi) \le q = \text{const}, \quad (t,\psi) \in \{V > 0\}; \tag{1.14}$$

3) if

$$W(t;\psi) \ge \alpha > 0 \tag{1.15}$$

then

$$\dot{W}(t;\psi) \ge \beta(\alpha) > 0 \tag{1.16}$$

where $\beta(\cdot)$ is a given function. Then the trivial solution of (1.1) is unstable.

Proof Let $\delta > 0$ be arbitrary small. There exists an initial function ϕ such that

$$\| \phi \| \le \delta, \quad V(0,\phi) = \alpha > 0. \tag{1.17}$$

Show that the solution $x(t;\phi)$ of (1.1) with such initial function reaches the boundary of the sphere Q_H as t increases. The point $(t, x(\cdot;\phi)_t)$ cannot leave the domain $\{V > 0\}$ by passing through the part of the boundary where $V = 0$. In fact, if $t_1 > 0$ is such that

$$W(t_1;\phi) = 0, \quad W(t;\phi) > 0 \text{ for } 0 \leq t < t_1 \tag{1.18}$$

then from (1.15), (1.16) it follows that $\dot{W}(t;\phi) \geq 0 \quad (0 \leq t < t_1)$. So, $W(t;\phi) \geq W(0;\phi) = \alpha > 0$, and by continuity $W(t_1;\phi) \geq W(0;\phi) = \alpha > 0$ that contradicts (1.18). Hence the point $(t, x(\cdot;\phi)_t)$ cannot leave the domain $\{V > 0\}$ across the part of the boundary on which $V = 0$. But the point $(t, x(\cdot;\phi)_t)$ cannot remain always in the domain $\{V > 0\}$. Assuming this we would successively obtain

$$\dot{W}(t;\phi) \geq 0, \quad W(t;\phi) \geq W(0;\phi) = \alpha > 0, \quad \dot{W}(t,\phi) \geq \beta(\alpha) > 0. \tag{1.19}$$

Integrating both parts of (1.19) we obtain that

$$W(t;\phi) \geq W(0;\phi) + t\beta(\alpha). \tag{1.20}$$

If $t > (q-\alpha)/\beta(\alpha)$ then (1.20) contradicts (1.14). So the solution must leave the domain $\{V > 0\}$ through the part of the boundary where $\| x_t \| = H$. This complete the proof of Thm. 1.7.□

Remark 1.1 Suppose that we can select a connected component of the domain $\{V > 0\}$ such that all the conditions of Thm. 1.7 are fulfilled in D. Then the assertion of Thm. 1.7 is valid.

Example 1.6 Consider equation

$$\dot{x}(t) - c\dot{x}(t-\tau) = a(t)x(t) + b(t)x(t-\tau), \tag{1.21}$$

$$c \geq 0, \quad a(t) \geq 0, \quad b(t) \geq 0, \quad a(t) + b(t) \geq A > 0, \quad \tau > 0. \tag{1.22}$$

Introduce the functional $V(t, x_t) := z(t, x_t) := x(t) - cx(t-\tau)$. Choose the initial function ϕ such that $\phi(\theta) > 0$, $\| \phi \| \leq \delta$, $z(0,\phi) = \phi(0) - c\phi(-\tau) = \alpha > 0$. Prove that $x(t,\phi) > 0$ for $t > 0$. Assume the contrary and let $t_1 > 0$ be the first instant such that $x(t_1,\phi) = 0$. For $0 < t < t_1$ we have

$$\dot{z}(t, x_t) = \dot{x}(t) - c\dot{x}(t-\tau) = a(t)x(t) + b(t)x(t-\tau) > 0.$$

Hence $z(t, x_t) > z(0,\phi) = \alpha > 0$. By going over to the limit as $t \to t_1 - 0$, we obtain $z(t_1, x_{t_1}) > \alpha > 0$. Therefore $x(t_1) = z(t_1, x_{t_1}) + cx(t_1 - \tau) > \alpha > 0$. The latter inequality contradicts the above assumption. Thus for any $t > 0$ we get $\dot{W}(t;\phi) = a(t)x(t) + b(t)x(t-\tau) \geq A\alpha > 0$. Using the same arguments as in the proof of Thm. 1.7 we obtain that the trivial solution of (1.21) is unstable under conditions (1.22).

In the same way it is possible to prove that the trivial solution of the equation

$$\dot{x}(t) - c\dot{x}(t-\tau) = g(x_t), \quad g(0) = 0, \quad c > 0$$

is unstable if the continuous functional $g(\phi) \geq \omega(\min \phi)$ $(\omega \in \Omega,\ \forall \phi :$ $\min \phi \geq 0)$.

1.4.2 Connection between instability of NDEs and ordinary differential equations. Similarly to Cor. 1.1 let us investigate the connection between the instability of ODE

$$\dot{x}(t) = f(t, x(t)),\quad t \geq 0,\quad x(t_0) = x_0,\quad t_0 \geq 0,\quad f(t, 0) \equiv 0 \tag{1.23}$$

and NDE of the form

$$\begin{aligned} \dot{z}(t, x_t) &= f(t, z(t, x_t)),\quad t \geq 0,\quad x_0 = \phi \in C[-h, 0], \\ z(t, x_t) &:= x(t) - g(t, x_t),\quad g(t, 0) \equiv 0. \end{aligned} \tag{1.24}$$

Theorem 1.8 If the trivial solution of ODE (1.23) is unstable then the trivial solution of NDE (1.24) is also unstable; f and g are the same as in Eq. (1.1).

Proof By the inversion theorem [298] there exists for (1.23) a function $(t, x) \mapsto v(t, x)$ which satisfies the conditions of Chetaev's theorem. In particular, by virtue of (1.23) the function

$$(t, x_t) \mapsto \dot{v}(t, x) = \frac{\partial v}{\partial t} + \sum_{i=1}^{n} \frac{\partial v}{\partial x_i} f_i = \Phi(t, x) \tag{1.25}$$

is positive definite in the domain $v > 0$ (see [298]). Now consider the functional $(t, x_t) \mapsto v(t, z(t, x_t))(= V(t, x_t))$. Using (1.24) and (1.25) we find $\dot{v}(t, z(t, x_t)) = \Phi(t, z(t, x_t))$. The validity of the other assumptions of Thm. 1.7 about the functional $v(t, z(t, x_t))$ is obvious. Hence, the functional $v(t, z(t, x_t))$ satisfies all the conditions of Thm. 1.7. So the trivial solution of (1.24) is unstable. □

Example 1.7 Consider the scalar equation of n–th order

$$\frac{d^n}{dt^n}[x(t) + cx(t - \tau)] = f(x(t) + cx(t - \tau)), \tag{1.26}$$

where f is a continuous function. Let c be a constant, $n \geq 3$, and $y = 0$ be an isolated root of the equation $f(y) = 0$. Then the trivial solution of the equation $y^{(n)}(t) = f(y)$ is unstable. ([432], p. 148). According to Thm. 1.8 the trivial solution of (1.26) is also unstable.

1.4.3 Another instability conditions. Remind that we denote $\omega_i : [0, \infty) \to$ **R** continuous nondecreasing functions of the argument $u \in [0, \infty)$ such that $\omega_i(0) = 0$ and $\omega_i(u) > 0$ for $u > 0$.

Theorem 1.9 Let all the conditions of Thm. 1.7 hold except the conditions (1.14) and (1.16) which are replaced in the domain $\{W > 0\}$ *by the following*

$$W(t;\phi) \le \omega_2(|z(t, x(\cdot;\phi)_t)|), \tag{1.27}$$

$$\dot{W}(t;\phi) \ge \omega_3(|z(t, x(\cdot;\phi)_t)|). \tag{1.28}$$

Then the trivial solution of (1.1) is unstable.

Proof Choose any initial function ϕ satisfying the conditions (1.17). Similarly to the proof of Thm. 1.7 we obtain that for this initial function the point $(t, x(\cdot;\phi)_t)$ cannot leave the region $\{V > 0\}$ across that part of the boundary at which $V = 0$. However this point cannot remain all the time in the region $\{V > 0\}$. Otherwise $\dot{W}(t,\phi) \ge 0$, $W(t;\phi) \ge W(0;\phi) = \alpha > 0$. Take $\beta > 0$ such that $\omega_2(\beta) = \alpha$. Then $\omega_2(\beta) = \alpha = W(0;\phi) \le W(t;\phi) \le \omega_2(|z(t,x_t)|)$. Hence $\beta \le |z(t,x_t)|$. Consequently

$$\dot{W}(t;\phi) \ge \omega_3(\beta) > 0. \tag{1.29}$$

Integrating (1.29) we obtain a contradiction with the boundedness of functional V in the sphere $Q_H : V(t,x_t) \le \omega_2(|z(t,x_t)|) \le \omega_2(H + c)$. Thm. 1.9 is proved. □

Example 1.8 Given the scalar equation

$$\frac{d}{dt}[x(t) - g(t, x(t-h))] = ax(t), \quad a > 0, \quad h > 0, \tag{1.30}$$

Here $g : \mathbf{R} \times \mathbf{R} \to \mathbf{R}$ is a continuous function, $g(t, 0) \equiv 0$. Let us introduce the functional

$$V(t,x_t) := z^2(t,x_t) - a\int_{t-h}^{t} x^2(s)ds,$$
$$z(t,x_t) := x(t) - g(t, x(t-h)). \tag{1.31}$$

Functional (1.31) satisfies condition (1.27). Further assume that $g^2(t, x(t-h)) \le x^2(t-h)$. Then

$$\dot{W}(t;\phi) = 2ax(t)z(t,x_t) - ax^2(t) + ax^2(t-h) = ax^2(t)-$$
$$-2ax(t)g(t,x(t-h)) + ag^2(t,x(t-h))-$$
$$-ag^2(t,x(t-h)) + ax^2(t-h) \ge az^2(t,x_t).$$

Hence by virtue of Thm. 1.9 the trivial solution of (1.30) is unstable.□

Theorem 1.10 Assume that there is a continuous functional $V : [0,\infty) \times Q_H \to \mathbf{R}$, and the domain $\{V > 0\}$ satisfies the conditions of Thm. 1.7. Let in the domain $\{V > 0\}$ the following inequalities hold

$$V(t,\phi) \leq \omega_1(\| \phi \|), \tag{1.32}$$

$$|g(t+s,\psi) - g(t,\chi)| \leq \omega_2(s) + \nu \| \psi - \chi \|, \tag{1.33}$$

$$\dot{W}(t;\phi) \geq \omega_3(|x(t;\phi)|), \tag{1.34}$$

$$0 < \nu < 1, \quad s \geq 0, \quad \forall \psi, \chi \in Q_H, \quad t \geq 0.$$

Then the trivial solution of (1.1) is unstable.

Proof The point $(t, x(\cdot;\phi)_t)$ cannot leave the region $\{V > 0\}$ through the part of the boundary $V = 0$. Show that this point cannot remain in the region $\{V > 0\}$. In fact, let the initial function satisfy (1.17) and $\mu > 0$ is such that $\omega_1(\mu) = \alpha$. Then $\omega_1(\mu) = \alpha \leq W(t; x(\cdot;\phi)_t) \leq \omega_1(\| x_t \|)$. Consequently, there exists a sequence $\{t_i\}$ such that $t_i \to \infty$ and $|x(t_i)| \geq \mu$. Show that the function $x(t) = x(t;\phi)$ is uniformly continuous in $[-h,\infty)$. Using (1.1), (1.2) and (1.33) for any $\Delta > 0$ we get

$$|x(t+\Delta) - x(t)| = |g(t+\Delta, x_{t+\Delta}) - g(t,x_t) + \int_t^{t+\Delta} f(s,x_s)ds|. \tag{1.35}$$

Denote $\rho(\eta)$ the function

$$\rho(\eta) := \sup_{t\geq 0} \max_{\Delta \leq \eta} |x(t+\Delta) - x(t)|.$$

From (1.35) and (1.33) it follows that

$$(1-\nu)\rho(\eta) \leq \max_{\Delta\leq\eta} \max_{-h\leq\theta\leq 0} |x(\theta+\Delta) - \phi(\theta)| + \eta c + \omega_2(\eta).$$

The righthand side of this inequality tends to zero as $\eta \to 0$ due to the uniform continuity of $x(t;\phi)$, $t \in [-h,\eta]$. Hence the uniform continuity of $x(t;\phi)$ with respect to $t \in [-h,\infty)$ is proven. Therefore there exists a η such that $|x(s)| \geq \mu/2$ for $t_i - \eta \leq s \leq t_i + \eta$ and any i. Going over to the subsequence of $\{t_i\}$ we can assume that all $t_{i+1} - t_i \geq 2\eta$. Then integrating (1.34) we obtain for $t \to \infty$

$$W(t;\phi) \geq \alpha + \int_0^t \omega_3(|x(s)|)ds \geq \alpha + 2N(t)\omega_3(\mu/2)\eta \to \infty \quad \text{as} \quad t \to \infty$$

where $N(t)$ is a number of points t_i such that $t_i \in [0,t]$. Thm. 1.10 is proven.□

Example 1.9 Obtain the instability conditions for a scalar NDE

$$\dot{x}(t) = ax(t) + \int_0^h \dot{x}(t-s)\lambda(s)ds$$

where $a > 0$ a constant and $\lambda : [0,h] \to \mathbf{R}$ is a continuous function. Consider the functional

$$W(t;\phi) = [x(t) - \int_0^h x(t-s)\lambda(s)ds]^2 + a\int_0^h |\lambda(s)|ds \int_{t-s}^t x^2(s_1)ds_1.$$

Then

$$\dot{W}(t;\phi) \geq 2ax^2(t)\left(1 - \int_0^h |\lambda(s)|ds\right).$$

Hence by Thm. 1.10 the instability sufficient conditions have a form

$$a > 0, \quad \int_0^h |\lambda(s)|ds < 1.$$

Remark Conditions of instability are sometimes equivalent to the conditions of the self–excitation of oscillations. In this way the self–excitation conditions of a distributed lossless line from Ch. 2 were obtained in [289.3].

2. STABILITY OF LINEAR NDEs

2.1. *LINEAR AUTONOMOUS NDES*

Suppose we are given the initial problem for the NDE

$$\dot{x}(t) = \int_{-\infty}^0 [dR_0(\theta)]x(t+\theta) + \int_{-\infty}^0 [dR_1(\theta)]\dot{x}(t+\theta), \quad t \geq 0,$$
$$x(\theta) = \phi(\theta), \quad -\infty < \theta \leq 0, \quad x : \mathbf{R} \to \mathbf{R}^n, \quad \phi \in C^1(-\infty, 0]. \tag{2.1}$$

Here the entries of the matrix function R_0 are functions of bounded variation and with finite first moments. Also, by definition,

$$\int_{-\infty}^0 [dR_1(\theta)]\dot{x}(t+\theta) = \sum_{j=1}^{\infty} L_j\dot{x}(t-h_j) + \int_{-\infty}^0 \lambda(\theta)\dot{x}(t+\theta)d\theta,$$

where the constants $h_j \geq 0$, the matrices L_j, and the absolutely integrable matrix λ with entries λ_{ij} satisfy

$$\sum_{j=1}^{\infty} \| L_j \| + \int_{-\infty}^{0} \| \lambda(\theta) \| \, d\theta < 1. \tag{2.2}$$

(Here $\| \cdot \|$ is the euclidean matrix norm.) Assume that the initial function ϕ satisfies

$$\| \phi \| < \infty, \quad \| \dot{\phi} \| < \infty,$$

$$\| \phi \|_0 := |\phi(0)| + \left[\int_0^{\infty} |F_0(t)|^2 dt \right]^{1/2} + \left[\int_0^{\infty} |F_1(t)|^2 dt \right]^{1/2} < \infty$$

where

$$F_j(t) := \int_{-\infty}^{-t} [dR_j(\theta)] \phi^{(j)}(t+\theta), \quad j = 0,1, \quad 0 \leq t < \infty.$$

The definitions of stability of the trivial solution of (2.1) coincide with those in Subs. 4.1.1, with $\| \phi \|_0$ instead of $\| \phi \|$.

Put

$$\widetilde{R}_j(z) := \int_{-\infty}^{0} e^{z\theta} dR_j(\theta), \quad \widetilde{F}_j(z) := \int_0^{\infty} e^{-zt} F_j(t) dt, \quad \widetilde{x}(z) := \int_0^{\infty} e^{-zt} x(t) dt.$$

Remark that under assumptions made above all these integrals are absolutely convergent for sufficiently large Rez, and Pr. (2.1) has a unique AC–solution $x(t;\phi)$ such that

$$|x(t;\phi)| + |\dot{x}(t;\phi)| \leq c_1 \| \phi \|_0 \exp(c_2 t), \quad 0 \leq t < \infty$$

where $c_1, c_2 > 0$ are some constants. Hence there exists the Laplace transform $x_t(z)$ of the solution $x(t;\phi)$.

Then

$$[Iz - \widetilde{R}_1(z)z - \widetilde{R}_0(z)]\widetilde{x}(z) = [I - \widetilde{R}_1(z)]\phi(0) + \widetilde{F}_0(z) + \widetilde{F}_1(z).$$

(Here I is the identity matrix.) Let

$$\Delta(z) = \det[Iz - \widetilde{R}_1(z)z - \widetilde{R}_0(z)]$$

be the so-called *characteristic function* of Eq. (2.1).

Theorem 2.1 Suppose the above stated conditions on the kernels R_0, R_1 *hold, and suppose also the function* $z \mapsto \Delta(z)$ *is zero-free in the domain* $\mathrm{Re}z \geq 0$. *Then (2.1) is asymptotically stable.*

Proof The proof of Thm. 2.1 is similar to those of Thms. 4.2.3. So we shall give only sketch of it. Assumption (2.2) means that for $\mathrm{Re}z \geq 0$

$$\| \widetilde{R}_1(z) \| = \| \int_0^\infty \exp(-zs) dR_1(s) \| \leq 1 - \gamma \quad (\gamma = \mathrm{const} > 0).$$

From here it follows the existence of the bounded matrix function $[I - \widetilde{R}_1]^{-1}$ for which

$$\| (I - \widetilde{R}_1(z))^{-1} \| \leq \gamma^{-1}, \quad \mathrm{Re}z \geq 0.$$

Hence

$$[Iz - (I - \widetilde{R}_1(z))^{-1} \widetilde{R}_0(z)] \widetilde{x}(z) = \phi(0) + (I - \widetilde{R}_1(z))^{-1} (\widetilde{F}_0(z) + \widetilde{F}_1(z)).$$

Let us introduce the matrix $\widetilde{R}_2(z) := (I - \widetilde{R}_1(z))^{-1} \widetilde{R}_0(z)$ and the function $z \mapsto \widetilde{R}_2(z) := (I - \widetilde{R}_1(z))^{-1} (\widetilde{F}_0(z) + \widetilde{F}_1(z))$ which are bounded for $\mathrm{Re}z \geq 0$. Then we have

$$(Iz - \widetilde{R}_2(z)) \widetilde{x}(z) = \phi(0) + \widetilde{F}_2(z).$$

This equation is just similar Eq. (4.2.4). So by virtue of Plancherel theorem and boundedness of the matrix $(I - \widetilde{R}_1(z))^{-1}$ will be

$$\int_{-\infty}^{\infty} |\widetilde{F}_2(i\beta)|^2 d\beta \leq 2\pi \gamma^{-2}.$$

As in Thm. 4.2.2 we can prove that

$$\int_{-\infty}^{\infty} |\widetilde{x}(i\beta)|^2 d\beta < \infty, \quad \text{if} \quad \| \phi \|_0 < \infty.$$

L_2-stability is proved. The asymptotic stability proof is just the same as in Thm. 2.1 is proven. □

Remark 2.1 From the proof of Thm. 2.1 it follows that assumption (2.2) can be replaced by

$$\sum_{j=1}^{\infty} \| L_j \| < 1, \quad \int_{-\infty}^{0} \| \lambda(s) \| ds < \infty.$$

But entirely assumption (2.2) can not be dropped out. Without (2.2) the trivial solution may be unstable even if all zeros of all function $\Delta(z)$ lie from the left of imaginary axis. E.g., the trivial solution of equation

$$L(L(L(x))) = 0, \quad L(x) := \dot{x}(t) - \dot{x}(t-1) + \pi^2 x(t)$$

is Liapunov unstable. But at the same time all the zeros of the quasipolynomial corresponding to this equation lie from the left of imaginary axis [463].

2.2. *SCALAR NDES*

Here we consider the Cauchy problem for a scalar equation of type (2.1):

$$x^{(n)}(t) = \sum_{j=0}^{n} \int_{-\infty}^{0} x^{(j)}(t+\theta)dr_j(\theta), \quad t \geq 0,$$

$$x^{(j)}(\theta) = \phi^{(j)}(\theta), \quad \theta \leq 0, \quad j = 0, \ldots, n.$$

Put

$$\alpha_{lj} := \int_{-\infty}^{0} |\theta|^l |dr_j(\theta)|, \quad \beta_{lj} := \int_{-\infty}^{0} |\theta|^l dr_j(\theta) \text{ if } \alpha_{lj} < \infty,$$

$$\Delta_n(z) := z^n[1 - \widetilde{r}_n(z)] - \sum_{j=0}^{n-1} z^j \widetilde{r}_j(z).$$

Here we assume that all functions r_j are of bounded variation on $(-\infty, 0]$, $\alpha_{0n} < 1$, and r_n does not contain a singular component (see Subs. 3.5.2). The initial function $\phi \in C^n(-\infty, 0]$, and

$$\sum_{j=0}^{n-1} |\psi^j(0)| + \sum_{j=0}^{n} \left(\int_{0}^{\infty} |F_j(t)|^2 dt \right)^{1/2} < \infty$$

where

$$F_j(t) := \int_{-\infty}^{-t} \phi^{(j)}(t+\theta)dr_j(\theta),$$

and last integrals are absolutely convergent for all $j = 0, \ldots, n$ and $t \geq 0$.

Thm. 2.1 implies that if the function $z \mapsto \Delta_n(z)$ is zero–free in the half–plane $\mathrm{Re}z \geq 0$, then the trivial solution of the equation under consideration is asymptotically stable. We will consider some concrete cases.

2.2.1 Case $n = 1$. For $n = 1$ we obtain

$$\Delta_1(z) = z\left[1 - \int_{-\infty}^{0} e^{z\theta}dr_1(\theta)\right] - \int_{-\infty}^{0} e^{z\theta}dr_0(\theta).$$

A. Assume that

$$\beta_{00} < 0, \quad \alpha_{10} + \alpha_{01} < 1.$$

Then the function Δ_1 is zero–free in $\mathrm{Re}z \geq 0$.

For the proof we choose functions $z \mapsto \psi_1(z)$ and $z \mapsto \psi_2(z)$ as follows

$$\psi_1(z) := z - \beta_{00}, \quad \psi_2(z) := \Delta_1(z) - \psi_1(z).$$

The function ψ_1 is zero–free in $\mathrm{Re}z \geq 0$, and $|\psi_2(z)| \leq (\alpha_{10}+\alpha_{01})|z|$, $|\psi_1(z)| > |z|$ for $\mathrm{Re}z \geq 0$. Hence $|\psi_1(z)| > |\psi_2(z)|$ for $\mathrm{Re}z \geq 0$. It means that $\psi_1(z) + \psi_2(z) \neq 0$ for $\mathrm{Re}z \geq 0$.

B. Assume that the kernel $r_0(\theta)$ has a jump $a_0 < 0$ at the point $\theta = 0$, and that

$$\gamma := a_0(1 - \alpha_{01}) + (1 + 2\alpha_{01})(\alpha_{00} + a_0) < 0. \tag{2.3}$$

Then the function Δ_1 is zero–free in $\mathrm{Re}z \geq 0$.

Proof Assume the contrary, i.e. the equation

$$\Delta_1(z) = 0 \tag{2.4}$$

has a root $z_0 = \alpha + i\beta$ with real part $\alpha \geq 0$. Then

$$\mathrm{Re}z_0 \left|1 - \int_{-\infty}^{0} e^{z_0\theta}dr_1(\theta)\right|^2 \geq 0. \tag{2.5}$$

We prove that (2.5) is impossible. Eq. (2.4) implies

$$z_0\left[1 - \int_{-\infty}^{0} e^{z_0\theta}dr_1(\theta)\right] = a_0 + \int_{-\infty}^{0^-} e^{z_0\theta}dr_0(\theta). \tag{2.6}$$

Multiplying (2.6) by the complex conjugate of $1 - \int_{-\infty}^{0} e^{z_0\theta} dr_1(\theta)$, we find by (2.3)

$$\mathrm{Re} z_0 \left| 1 - \int_{-\infty}^{0} e^{z\theta} dr_1(\theta) \right|^2 =$$

$$= \left[a_0 + \int_{-\infty}^{0^-} e^{\alpha\theta} \cos \beta\theta dr_0(\theta) \right] \cdot \left[1 - \int_{-\infty}^{0} e^{\alpha\theta} \cos \beta\theta dr_1(\theta) \right] -$$

$$- \int_{-\infty}^{0^-} e^{\alpha\theta} \sin \beta\theta dr_0(\theta) \cdot \int_{-\infty}^{0} e^{\alpha\theta} \sin \beta\theta dr_1(\theta) \le \gamma < 0,$$

contradicting (2.5). □

C. Assume that the kernels $r_0(\theta), r_1(\theta)$ are constant for $\theta \le -h < 0$, nonincreasing, and

$$\beta_{00} < 0, \quad \alpha_{00} < \frac{\pi}{2h} \left[1 - \left(1 + \frac{1}{e} \right) \alpha_{01} \right]. \tag{2.7}$$

Then equation (2.4) does not have roots in $\mathrm{Re} z \ge 0$.

Proof For $z = \alpha + i\beta$, (2.4) implies the two equations

$$q_1(\alpha, \beta) := \alpha \left[1 - \int_{-h}^{0} e^{\alpha\theta} \cos \beta\theta dr_1(\theta) \right] +$$

$$+\beta \int_{-h}^{0} e^{\alpha\theta} \sin \beta\theta dr_1(\theta) - \int_{-h}^{0} e^{\alpha\theta} \cos \beta\theta dr_0(\theta) = 0, \tag{2.8}$$

$$q_2(\alpha, \beta) := \beta \left[1 - \int_{-h}^{0} e^{\alpha\theta} \cos \beta\theta dr_1(\theta) \right] +$$

$$- \alpha \int_{-h}^{0} e^{\alpha\theta} \sin \beta\theta dr_1(\theta) - \int_{-h}^{0} e^{\alpha\theta} \sin \beta\theta dr_0(\theta) = 0. \tag{2.9}$$

Eq. (2.8) does not have a solution in the domain $2h|\beta| < \pi$, $\alpha \ge 0$, since under the conditions (2.7) we have $q_1(\alpha, \beta) > 0$ in this domain. We verify that for $2h|\beta| \geq \pi$, $\alpha \geq 0$ the Eq. (2.9) also does not have a solution. In fact, we obtain

$$\left|\alpha \int_{-h}^{0} e^{\alpha\theta} \sin \beta\theta dr_1(\theta)\right| = \left|\beta \int_{-h}^{0} \alpha\theta e^{-\alpha\theta} \frac{\sin \beta\theta}{\beta\theta} dr_1(\theta)\right| \leq \frac{|\beta|}{e}\alpha_{01}.$$

This and (2.9) imply that for $2h\beta \geq \pi, \alpha \geq 0$ we have $q_2(\alpha, \beta) \geq \beta[1 - \alpha_{01}(1 + 1/e)] - \alpha_{00} > 0$, while if $2h\beta \leq -\pi$, $\alpha \geq 0$ we have $q_2(\alpha, \beta) < 0$. Hence Eqs. (2.8) and (2.9) are not compatible for $\alpha \geq 0$. □

2.2.2 Case $n = 2$. The function Δ_2 is equal to

$$\Delta_2(z) = z^2 \left[1 - \int_{-\infty}^{0} e^{z\theta} dr_2(\theta)\right] - z \int_{-\infty}^{0} e^{z\theta} dr_1(\theta) - \int_{-\infty}^{0} e^{z\theta} dr_0(\theta).$$

Assume that

$$\alpha_{02} + \alpha_{11} + \frac{\alpha_{20}}{2} < 1, \quad \beta_{10} - \beta_{01} > 0, \quad \beta_{00} < 0.$$

Then the function Δ_2 is zero–free in the halfplane $\mathrm{Re}z \geq 0$.

To prove this we write Δ_2 as a sum $\Delta_2(z) = \psi_1(z) + \psi_2(z)$, where

$$\psi_1(z) := z^2 - (\beta_{01} + \beta_{10})z - \beta_{00},$$

$$\psi_2(z) := -z^2 \int_{-\infty}^{0} e^{z\theta} dr_2(\theta) - \int_{-\infty}^{0} (e^{z\theta} - 1) dr_1(\theta) - \int_{-\infty}^{0} (e^{z\theta} - 1 - z\theta) dr_0(\theta).$$

Note that under the conditions imposed the function ψ_1 is zero–free in the domain $\mathrm{Re}z \geq 0$, and also $|\psi_1(z)| > |\psi_2(z)|$ on the boundary of this domain. Hence, by Rouche's theorem the function Δ_2 is zero–free in this halfplane as well.

2.3. *STABILITY OF NDES WITH DISCRETE DELAYS*

Consider the following scalar equation with discrete delays $h_j \geq 0$ and constant coefficients a_l and b_{lj}:

$$x^{(n)}(t) + \sum_{l=1}^{n} a_l x^{(n-l)}(t) + \sum_{l=0}^{n} \sum_{j=1}^{m} b_{lj} x^{(n-l)}(t - h_j) = 0. \tag{2.10}$$

Assume that

$$\sum_{j=1}^{m} |b_{0j}| < 1, \quad \sum_{j=1}^{m} |b_{nj}| < |a_n|.$$

Then for (2.10) to be asymptotically stable it is necessary and sufficient that the polynomial $P(z)$ be asymptotically stable and that

$$\sum_{j=1}^{m} |Q_j(i\omega)| < |P(i\omega)|, \quad i^2 = -1, \quad \omega > 0,$$

where

$$P(z) := z^n + \sum_{l=1}^{n} a_l z^{n-l}, \quad Q_j(z) := \sum_{l=0}^{n} b_{lj} z^{n-l}. \tag{2.11}$$

Robust stability of NDEs. Consider (2.10) with coefficients a_l and b_l unknown but satisfying the conditions (3.2.13). Define the four polynomials $P_k(z)$, $k = 1, 2, 3, 4$, using the relation (2.11), by (3.2.14) – (3.2.17). Define for each Q_j the $Q_{jk}(z)$ in a similar manner.

Theorem 2.2 ([279]) Assume that

$$\sum_{j=1}^{m} \beta_{0j} < 1, \quad \sum_{j=1}^{m} \beta_{nj} < \alpha_n.$$

Then all Eqs. (2.10), (3.2.13) are asymptotically stable if and only if each of the 4^{m+1} *"vertex" quasipolynomials*

$$P_k(z) = \sum_{j=1}^{m} Q_{jl}(z) e^{-h_j z}$$

is asymptotically stable.

Some remarks about the roots of quasipolynomials of NDEs with discrete delays.

Numerous investigations have been devoted to the zeros of the characteristic function [42]

$$\Delta(z) = \sum_{l=0}^{m} \sum_{j=1}^{r(l)} a_{lj} z^l e^{-h_{lj} z}.$$

It has been proved that if the delays h_{nj} are commensurate, then the asymptotic roots z_k of the equation $\Delta(z) = 0$ belong to certain chains, and $\lim \mathrm{Im} z_k = \pm\infty$, as $k \to \infty$. The real parts of the roots z_k on each chain are either uniformly bounded, or tend to $-\infty$ as $k \to \infty$.

If the delays h_{nj} are incommensurate, then there are, in addition to these chains, finitely many strips parallel to the imaginary axis containing certain zeros of the function Δ. The real parts of these zeros densely fill the intervals on the real axis corresponding to these strips. The imaginary parts of these zeros tend to an almost asymptotically periodic point set. The maximal multiplicity of a zeros of Δ is at most $n + (n+1) \max_{0 \le l \le n} r(l)$.

Suppose we have a single delay $h > 0$, i.e. $m = 1$ in (2.10), and $b_{01} > 0$. By the solution $x(t) = b_{01}^{t/h} y(t)$ we get

$$\Delta(z) = P(z) + Q(z)e^{-hz},$$

$$P(z) = z^n + \sum_{l=1}^{m} a_l z^{n-l}, \quad Q(z) = z^n + \sum_{l=1}^{m} b_l z^{n-l}.$$

Assume that the equation $\Delta(z) = 0$ has a root z_0 of maximal possible multiplicity $2n+1$. Then all roots z_k of this equation are such that $\mathrm{Re} z_k = \mathrm{Re} z_0$ [408]. From this we can derive that certain NDEs have infinitely many fundamental frequencies (defined as the positive roots of $\Delta(i\omega) = 0$). A necessary and sufficient condition for this is that $|P(i\omega)| = |Q(i\omega)|$ for all $\omega \in \mathbf{R}$. If, in addition, $P(z) \not\equiv Q(z)$, then the roots ω_k of the equation $\Delta(i\omega) = 0$ satisfy

$$\omega_k = -\frac{2k+1}{h}\pi - \frac{(b_1 + a_1)h}{(2k+1)\pi} + O\left(\frac{1}{k^2}\right), \quad k \to \pm\infty.$$

On the other hand, if $P(z) \equiv Q(z)$, then

$$\omega_k = -\pi \frac{2k+1}{h} i, \quad k = 0, \pm 1, \ldots$$

Note that the case when the number of fundamental frequencies is infinite is exceptional. In general, the number of distinct fundamental frequencies ω does not exceed n, since ω is a root of the algebraic equation $|P(i\omega)|^2 = |Q(i\omega)|^2$, which is of order $2n$ and involves only even powers of ω.

2.4. *THE INFLUENCE OF SMALL DELAYS ON STABILITY*

Many actual phenomena involve small delays which are often neglected in the process of mathematical modeling. Sometimes this leads to false conclusions. As an example we will consider a scalar NDE which is asymptotically stable for $h = 0$ but unstable for arbitrary $h > 0$:

$$\ddot{x}(t) + \dot{x}(t) + x(t) = a[\ddot{x}(t-h) + \dot{x}(t-h) + x(t-h)], \tag{2.12}$$

where $a > 1$. If $h = 0$, Eq. (2.12) is asymptotically stable. The characteristic quasipolynomial of (2.12) is $\Delta(z) = (z^2 + z + 1)(1 - ae^{-hz})$, and has the following zeros with positive real part if $h > 0$:

$$z_k = \frac{1}{h}(\ln a + 2k\pi i), \quad i^2 = -1, \quad k = 0, \pm 1, \ldots.$$

So, for any $h > 0$ the trivial solution of (2.12) is unstable.

2.5. *LINEAR INHOMOGENEOUS NDES*

Let us discuss briefly connection between boundedness and stability for linear NDE

$$\begin{gathered}\dot{x}(t) = L(t, x_t, \dot{x}_t) + f(t), \quad t > 0, \\ x_0(\theta) = \phi(\theta), \quad \dot{x}_0(\theta) = \dot{\phi}(\theta), \quad -h \leq \theta \leq 0.\end{gathered} \tag{2.13}$$

Here the functional $(t, x_t, \dot{x}_t) \mapsto L(t, x_t, \dot{x}_t) = A_1(t, \dot{x}_t) + A_2(t, x_t)$ is linear with respect to the second and third arguments, satisfies the assumptions of Thm. 3.3.2, and $\phi \in C^1[-h, 0]$. The proof of the following theorem coincides with that of Thm. 3.1.6.

Theorem 2.3 If any solution of (2.13) with $f(t) = 0$ is bounded for $t \geq 0$, then its trivial solution is stable.

Note that in general the boundedness of all the solutions of nonlinear equation does not imply their stability.

Let the functional $(t, \phi, \chi) \mapsto L(t, \phi, \chi)$ be ω–periodic in t, i.e. $L(t + \omega, \phi, \chi) \equiv L(t, \phi, \chi)$ $(\forall t, \phi, \chi)$. Consider the functional equation

$$z(t) = A_1(t, z_t) + f(t), \quad t > 0, \quad z(\theta) = 0, \quad -h \leq \theta \leq 0. \tag{2.14}$$

Pr. (2.14) is said to satisfy the *Perron condition* if its solution defined by any functions $f \in L_\infty[0, \infty)$ with $\| f \|_\infty < \infty$ and $\phi = 0$ is bounded in $[0, \infty)$.

Theorem 2.4 [544] Let the functional L in Eq. (2.13) satisfies the assumptions formulated above. Assume that Pr. (2.14) satisfies the Perron condition and Eq. (2.13) has a bounded solution for any function $f \in L_\infty[0, \infty), \| f \|_\infty$ and $\phi = 0$. Then the trivial solution of (2.13) with $f = 0$ is asymptotically stable.

2.6. *BOUNDEDNESS OF DERIVATIVES FOR LINEAR NDES*

Consider ordinary differential equation with bounded continuous coefficients

$$\left.\begin{gathered}x^{(m)}(t) + A_1(t)x^{(m-1)}(t) + \ldots + A_m(t)x(t) = f(t), \\ x^{(j)} := d^j x / dt^j.\end{gathered}\right\} \tag{2.15}$$

Theorem of Esclangon [94] If there exist a bounded solution $t \mapsto x(t)$, $t_0 \leq t < \infty$ of (2.15) for some bounded function f, then the derivatives $\dot{x}, \ldots, x^{(m)}$,

$t_0 \leq t < \infty$, *are bounded.*

In general, this theorem is not valid for NDEs. Consider, e.g., the scalar NDE $(t \geq t_0 \geq \sqrt{3\pi/2})$

$$\dot{x}(t) - \dot{x}\left(\sqrt{t^2 - 2\pi}\right) = 2\left(t - \sqrt{t^2 - 2\pi}\right) x\left(\sqrt{t^2 - 3\pi/2}\right).$$

Coefficient of this equation is bounded for $t \geq t_0$ and $f(t) = 0$. But this equation has the bounded solution $x(t) = \sin(t^2)$, $t \geq t_0$, with unbounded derivative.

Let us formulate the Esclangon–type theorem for NDE

$$\begin{gathered} x^{(m)}(t) + A_1(t, x_t^{(m)}) + A_2(t, x_t, \ldots, x_t^{(m-1)}) = f(t), \quad t > 0 \\ x(\theta) = \dot{x}(\theta) = \ldots = x^{(m)}(\theta) = 0, \quad -h \leq \theta \leq 0. \end{gathered} \tag{2.16}$$

Here continuous functionals $A_1 : [0, \infty) \times L_\infty[-h, 0] \to \mathbf{R}^n$, $A_2 : [0, \infty) \times C[-h, 0] \times \ldots \times C[-h, 0] \to \mathbf{R}^n$ are linear in all argument beginning from the second. In addition $(t \mapsto A_1(t, \phi_t)) \in L_\infty[0, T]$ for any $\phi \in L_\infty[-h, T]$, $T > 0$, and $(t \mapsto A_2(t, \psi_t^{(1)}, \ldots, \psi_t^{(m)})) \in L_\infty[0, T]$ for any $\psi \in C^m[-h, T]$. Assume also that

$$\begin{gathered} \operatorname*{vrai\,sup}_{0 \leq t \leq T} |A_2(t, \psi_t^{(1)}, \ldots, \psi_t^{(m)})| \leq \\ \leq a(\| \psi^{(1)} \| + \ldots + \| \psi^{(m)} \|) \end{gathered} \tag{2.17}$$

where $a > 0$ does not depend on $T > 0$ and $\psi \in C^m[-h, T]$. Condition (2.17) replaces the assumption about boundedness of Eq. (2.15) coefficients. Consider the functional equation

$$\begin{gathered} z(t) + A_1(t, z_t) = R(t), \quad t > 0; \quad z(\theta) = 0, \quad -h \leq \theta \leq 0, \\ R(t) \in L_\infty[0, \infty]. \end{gathered} \tag{2.18}$$

Theorem 2.5 [289.3] Let the functionals A_1 and A_2 satisfy to the formulated requirements and Perron condition be met for Eq. (2.18). If there exists a bounded solution $x(t)$ of equation (2.16) for a bounded function $f \in L_\infty[0, \infty)$ then $\operatorname*{vrai\,sup}_{0 \leq t < \infty} [|\dot{x}(t)| + \ldots + |x^{(m)}(t)|] < \infty$.

2.7. *BOUNDEDNESS OF DERIVATIVES FOR NONLINEAR NDES*

Consider nonlinear NDE

$$A_1(t, x_t^{(m)}) + A_2(t, x_t, \ldots, x_t^{(m-1)}) = 0 \quad (0 \leq t < \infty), \tag{2.19}$$

where $m \geq 2$, $A_1 : [0,\infty) \times L_\infty[-h,0] \to \mathbf{R}^n$ is a linear functional, $A_2 : [0,\infty) \times (C[-h,0])^m \to \mathbf{R}^n$ is in general nonlinear functional. Function $x : [-h,\infty) \to \mathbf{R}^n$ is a solution of Eq. (2.19) if $x^{(m-1)} \in \mathrm{Lip}[-h,\infty)$ and x satisfies Eq. (2.19) for almost all $t \in [0,\infty)$. Consider also auxiliary inequality

$$|A_1(t,z_t)| \leq f(t) \quad (0 \leq t < \infty), \quad z_0 = \phi \tag{2.20}$$

where it is assumed that $z \in L_\infty[-h,\infty)$.

Theorem 2.6 Assume that the zero solution of inequality (2.20) is stable and for any $\alpha > 0$ and $\psi_1,\dots,\psi_m \in C[-h,0]$ with $\| \psi_1 \| \leq \alpha$ the values of $A_2(t,\psi_1,\dots,\psi_m)$ are bounded for bounded $\| \psi_2 \|,\dots,\| \psi_m \|$, $t \in [0,\infty)$ and also uniformly with respect to t and ψ_1 the following asymptotic estimate

$$|A_2(t,\psi_1,\dots,\psi_m)| = o\left(\sum_{i=2}^{m} \psi_i^{\frac{m}{i-1}}\right)$$

as

$$\sum_{i=2}^{m} \| \psi_i \| \to \infty$$

holds. Then any solution of Eq. (2.19) bounded on $[0,\infty)$ will have bounded derivatives of the order less or equal m.

Proof For the case $A_1(t,\psi) = \psi(0)$, i.e. for RDEs, the proof is given in the paper.[1] Let us use the Kolmogorov – Gorny inequality (see, e.g., G.H Hardy, J.E. Lettlewood, G. Polya, Inequalities, 1934):

$$\| \psi^{(i)} \| \leq a_{mi} \| \psi \|^{(m-i)/m} [\max\{\| \psi^{(m)} \|_\infty, m!(b-a)^{-m} \| \psi \|\}]^{i/m},$$
$$i = 1,\dots,m-1. \tag{2.21}$$

Here ψ is any function $\psi : [a,b] \to \mathbf{R}^n$, $\psi^{(m-1)} \in \mathrm{Lip}[a,b]$ and a_{mi} are absolute constants. (In the book mentioned above estimate (2.21) is proven for the scalar case. But it can be easily generalized for the vector case with the modified constants $a_{mi} (= a_{min})$.)

From linearity of the functional A_1 and stability of the trivial solution of inequality (2.20) it follows that for almost all $t \in [0,T)$

$$|z(t)| \leq k[\sup_{0 \leq t < T} |A_1(t,z_t)| + \| z_0 \|_\infty] \tag{2.22}$$

where z is any function $z \in L_\infty[0,T)$ and k is a constant which is independent on $T \in (0,\infty)$ and z. Hence, it is sufficient to prove the boundedness of all

[1] A. Ja. Lepin, A. D. Myshkis, Boundedness of derivatives of the bounded solutions of ordinary differential equations. Diff. Uravn. (Diff. Eqns.), 1965, 1,9, 1260–1263.

derivatives of the order less or equal $m-1$. Let us take any bounded solution $x : [-h, \infty) \to \mathbf{R}^n$ and a constant $\alpha > 0$. Then the following estimates are valid for almost all $t \in [0, \infty)$. First of all by virtue of the assumptions of Thm. 2.6

$$|A_1(t, x_t^{(m)})| \leq \alpha \sum_{i=1}^{m-1} \| x_t^{(i)} \|^{m/i} + c_\alpha$$

for some $c_\alpha > 0$. From here and (2.22) we obtain

$$|x^{(m)}(t)| \leq k\alpha \max_{0 \leq \tau \leq t} \sum_{i=1}^{m-1} \| x_\tau^{(1)} \|^{m/i} + kc_\alpha + \| x_0^{(m)} \|_\infty .$$

Righthand side of this inequality can be estimated on the successive intervals of the length h by the aid of (2.21). Hence

$$|x^{(m)}(t)| \leq k\alpha \sum_{i=1}^{m-1} [a_{mi} \| x \|^{(m-i)/m} \times$$

$$\times (\max\{\| x^{(m)}|_{[-h,t]} \|_\infty, m! h^{-m} \| x \|\})^{i/m}]^{m/i} + kc_\alpha + k \| x_0^{(m)} \|_\infty .$$

Denote

$$y(t) := \| x^{(m)}|_{[-h,t]} \|_\infty \quad (0 \leq t < \infty).$$

Then

$$y(t) \leq k\alpha \sum_{i=1}^{m-1} a_{mi}^{m/i} \| x \|^{(m-i)/i} \max\{y(t), m! h^{-m} \| x \|\} + kc_\alpha + k \| x_0^{(m)} \|_\infty .$$

If α is chosen such that

$$k\alpha \sum_{i=1}^{m-1} a_{mi}^{m/i} \| x \|^{(m-i)/i} < 1,$$

then the function y will be bounded in essential and the same is true with respect to $x^{(m)}$. From here and (2.21) it follows the boundedness of all derivatives $x^{(i)}$, $1 \leq i \leq m-1$. Thm. 2.6 is proven.□

2.8. *LINEAR PERIODIC NDES*

Consider the linear NDE

$$\frac{d}{dt}[x(t) - G(t, x_t)] = L(t, x_t). \tag{2.23}$$

Here $G : \mathbf{R} \times C[-h, 0] \to \mathbf{R}^n$, $L : \mathbf{R} \times C[-h, 0] \to \mathbf{R}^n$ are continuous in both arguments functionals, linear in the second argument and ω-periodic in t. Remind that $G(t,\phi)$ is called a stable operator if the trivial solution of the homogeneous FE

$$x(t) - G(t, x_t) = 0, \quad t \geq 0$$

is uniformly (with respect to the initial data) asymptotically stable. Denote by $U(t) : C[-h,0] \to C[-h,0]$ a *shift operator on the period* ω (also called a *monodromy operator*) along the solutions of Eq. (2.23)

$$U(t)\phi = x_{t+\omega}, \quad x_t = \phi.$$

Eigenvalues ρ of $U(t)$ do not depend on t. They are called *characteristic* (or *Floquet*) *multipliers* of Eq. (2.23). The solutions $x(t)$ of (2.23) such that

$$x_{t+\omega} = \rho x_t, \quad x_t \neq 0,$$

are called the *Floquet solutions*. The *Floquet exponent* λ equals $\lambda = \frac{1}{\omega} \ln \rho$. If ρ is an eigenvalue then the corresponding Floquet solution can be represented in the form

$$x(t) = p(t)\exp(\lambda t), \quad p(t+\omega) = p(t).$$

Theorem 2.7 [238] Let the above formulated assumption be met and the operator G be stable. Then there exists a number $a \in (0,1)$ such that Eq. (2.23) has a finite number of the Floquet multipliers ρ_j with $|\rho_j| \geq a$. The trivial solution of (2.23) is stable if and only if all multipliers ρ_j satisfy the condition $|\rho_j| \leq 1$ and for those of with $|\rho_j| = 1$ the elementary divisors must be simple. The trivial solution of (2.23) is asymptotically stable if and only if $|\rho_j| < 1$ for all ρ_j.

Usually the monodromy operator for NDEs is not completely continuous. This results in some phenomena which do not take place for RDEs. Let us illustrate some of them by examples (compare with Subs. 4.1.5).

Example 2.1 Floquet multiplier of NDE can have infinite multiplicity. For equation $\dot{x}(t) - \dot{x}(t-1) = (\cos 2\pi t)[x(t) - x(t-1)]$ Floquet multiplier $\rho = 1$ has infinite multiplicity since any continuously differentiable 1-periodic function $x(t)$ is its solution.

Example 2.2 Consider the scalar equation which has no Floquet solutions

$$\dot{x}(t) - \dot{x}(t-2\pi) = (\cos t)x(t-2\pi). \tag{2.24}$$

If $x(t) = \exp(\lambda t)y(t)$ is a Floquet solution of (2.24) then $y(t)$ must be 2π-periodic solution of equation

$$\begin{aligned} \dot{y}(t)[1 - \exp(-2\pi\lambda)] = [(\cos t)\exp(-2\pi\lambda) - \\ -\lambda(1 - \exp(-2\pi\lambda))]y(t). \end{aligned} \tag{2.25}$$

But for any complex number λ equation (2.25) has no nontrivial periodic solutions. Indeed, let be $\lambda = mi$, $m = 0, \pm 1, \pm 2, \ldots$. Then a unique solution of (2.25) having the form $0 = (\cos t)y(t)$ is $y(t) = 0$. If $\lambda \neq mi$, $m = 0, \pm 1, \pm 2, \ldots$ then the general solution of (2.25) is

$$y(t) = c \exp\left\{ \int_0^t \left[\frac{\exp(-2\pi\lambda)}{1 - \exp(-2\pi\lambda)} \cos s - \lambda \right] ds \right\}, \quad c = \text{const.}$$

This solution is 2π–periodic if and only if

$$\int_0^{2\pi} \left[\frac{\exp(-2\pi\lambda)}{1 - \exp(-2\pi\lambda)} \cos s - \lambda \right] ds = -2\pi\lambda = 0.$$

But it is impossible for $\lambda \neq mi$, $m = 0, \pm 1, \pm 2, \ldots$. Hence there are no nontrivial periodic solutions of (2.25) and hence the Floquet solutions of (2.24).

Example 2.3 Consider equation whose Floquet solutions does not give asymptotic representation for the other solutions:

$$2\dot{x}(t) - \dot{x}(t - 2\pi) = (\cos t)x(t - 2\pi). \tag{2.26}$$

Determination of the Floquet solutions of (2.26) is equivalent to the determination of such a λ that equation

$$\begin{aligned} \dot{y}(t)[2 - \exp(-2\pi\lambda)] = [(\cos t)\exp(-2\pi\lambda) - \\ -\lambda(2 - \exp(-2\pi\lambda))]y(t) \end{aligned} \tag{2.27}$$

has nontrivial periodic solution. If $2 - \exp(-2\pi\lambda) = 0$ then Eq. (2.27) has no nontrivial periodic solutions. If $2 - \exp(-2\pi\lambda) \neq 0$ then the periodic solution of (2.27) exists only for $\lambda = 0$. Then this periodic solution $y(t)$ is simultaneously the Floquet solution $x_0(t)$ of Eq. (2.27), $x_0(t) = y(t) = \exp(\sin t)$. Now find the solution $x_1(t)$ of (2.27) with initial data $x_1(t) = \sin t$, $\dot{x}_1(t) = \cos t$, $0 \le t \le 2\pi$. By the steps method we have

$$x_1(t) = \sum_{k=1}^{m+1} A_m^k (\sin t)^k, \; 2\pi m \le t \le 2\pi(m+1), \; m = 0, 1, \ldots. \tag{2.28}$$

Here A_m^k are defined by the formulae

$$A_m^k = A_{m-1}^k/2 + A_{m-1}^{k-1}/2k, \;\; A_0^1 = 1, \; A_m^0 = A_m^{m+2} = 0. \tag{2.29}$$

From (2.28), (2.29) it follows that $A_m^k > 0,\ m = 1,2,\ldots;\ 1 \le k \le m+1$ and also

$$\left.\begin{aligned} &\max_{2\pi m \le t \le 2\pi(m+1)} |x_1(t)| = x_1(t_m) = \sum_{k=1}^{m+1} A_m^k, \quad t_m = \frac{\pi}{2} + 2\pi m, \\ &\frac{1}{2}\sum_{k=1}^{m} A_{m-1}^k \le \sum_{k=1}^{m+1} A_m^k \le \frac{3}{4}\sum_{k=1}^{m} A_{m-1}^k. \end{aligned}\right\} \tag{2.30}$$

So

$$\frac{1}{2}x_1(t_{m-1}) < x_1(t_m) \le \frac{3}{4}x_1(t_{m-1}). \tag{2.31}$$

By (2.30), (2.31) we get $|x_1(t)| \le (3/4)^{(t/2\pi)-1}$. Hence $\lim x_1(t) = 0,\ t \to \infty$. But $\inf_t x_0(t) > 0$. Consequently function $cx_0(t)$ is the best approximation of the solution $x_1(t)$ if and only if $c = 0$. But Floquet representation obtained is not an asymptotic for $x_1(t)$. Indeed, from (2.31) it follows that $x_1(t) \ge 2^{-m}$. Therefore the difference between the solution $x_1(t)$ and its Floquet representation $cx_0(t) = 0$ decreases not quicker than $\exp(-\ln 2(t/2\pi - 1/4))$. So, the Floquet solution $x_0(t)$ does not give asymptotic representation for $x_1(t)$ (see Subs. 4.1.5).

CHAPTER 9

APPLICATIONS OF THE DIRECT LIAPUNOV METHOD

1. DESCRIPTION OF THE PROCEDURE

Applications of the general theorems of the direct Liapunov's method is connected with necessity to construct a suitable functional and, in general, this is a difficult problem.

Remind that the general theorems of Liapunov's second method are formulated for systems of neutral type in terms of the existence of either degenerate functionals that depend on both x_t and $\dot{x}_t$.

Functionals of these kind have actually been constructed for some specific systems, and stability conditions have been established with their help that depend directly on the characteristics of the equations themselves. In addition, the form of the functionals enables one to detect a certain link between these functionals and Liapunov functions for suitably chosen ordinary equations. This link allows us to perceive the existence of a formal procedure to construct Liapunov functionals for some classes of FDEs. In Chapter 6 this procedure was considered for RDEs and here for NDEs.

We propose to construct Liapunov functionals V in the form $V = V_1 + V_2$ for NDEs

$$\dot{x}(t) = f(t, x_t, \dot{x}_t), \quad t \geq t_0, \tag{1.1}$$

$$x_{t_0} = \phi, \quad \dot{x}_{t_0} = \dot{\phi} \tag{1.2}$$

where $x(t) \in \mathbf{R}^n$, $x_t = x(t+\theta)$, $\theta \in (-\infty, 0]$, and the initial function $\phi : (-\infty, 0] \to \mathbf{R}^n$ is absolutely continuous. Righthand side of Eq. (1.1) is a functional $(t, \psi, \chi) \mapsto f(t, \psi, \chi)$, defined on $[t_0, \infty) \times C(-\infty, 0] \times L_\infty(-\infty, 0]$ with $\| \psi \| < \infty$, $\| \chi \| < \infty$, and such that there exists a unique solution of the Cauchy problem (1.1), (1.2) (see, e.g., Thm. 3.3.2). Also as usual in stability theory, we assume without loss of generality that

$$f(t, 0, 0) \equiv 0, \quad t \geq t_0. \tag{1.3}$$

Definition 1.1 The trivial solution of Pr. (1.1)–(1.3) is said to be

A) *stable* if, for any $\varepsilon > 0$, a $\delta(\varepsilon) > 0$ exists such that $|x(t)| < \varepsilon$, $t \geq t_0$ for any initial conditions that satisfy the inequality

$$\| \psi \| + \| \dot{\psi} \|_\infty < \delta(\varepsilon);$$

B) *asymptotically stable* if it is stable and $\lim x(t) = 0$ as $t \to \infty$ for any initial conditions in some neighbourhood of $0 \in C(-\infty, 0] \times L_\infty(-\infty, 0]$.

The proposed procedure is as follows:

1) transform the righthand side F of Eq. (1.1) so that it is expressed as the sum of two terms, of which the first depends only on the present state of the system, i.e.

$$\dot{x}(t) = f(t, x(t), \dot{x}(t)), \quad f(t, x(t), \dot{x}(t)) := f_1(t, x(t)) + f_2(t, x_t, \dot{x}_t), \\ f_1(t, 0) \equiv 0. \quad f_2(t, 0, 0) \equiv 0; \tag{1.4}$$

2) delete f_2 from the transformed equation to obtain an auxiliary ordinary equation $\dot{y}(t) = f_1(t, y(t))$, and construct a Liapunov function $v(t, y)$ for that equation;

3) replace the second argument of $v(t, y)$ by a certain function, depending on the transformation used in the first step. Specifically: if

$$f_2(t, x_t, \dot{x}_t) = f_3(t, x_t, \dot{x}_t) + \frac{d}{dt} f_4(t, x_t) \tag{1.5}$$

then $V_1(t, x_t) := v(t, z(t, x_t))$, where $z(t, x_t) := x(t) - f_4(t, x_t)$. But if the component f_4 in representation (1.5) vanishes, then $V_1(t, x_t) = v(t, x(t))$. Now, adding to V_1 the component $V_2 = V_2(t, x_t, \dot{x}_t)$ in such a way as to satisfy the requirements of some stability theorem for system (1.1), we obtain a functional $V = V_1 + V_2$. The use of the functional $v(t, z(t))$ may necessitate a stability analysis for the trivial solution of the functional equation $z(t, x_t) = 0$.

We stress that the individual steps of the procedure may be implemented in more than one way – a feature that should be used to expand the stability domains obtained. Moreover, the procedure not only provides a unified formal approach to the construction of a series of already known functionals for equations of neutral type, but also enables one to investigate the stability of specific systems.

The individual steps of the procedure will be clarified by an example. Let $\dot{V}$ denote the right upper derivative of the functional V along trajectories of Eq. (1.1).

Example 1.1 Consider the scalar equation

$$\dot{x}(t) = -ax(t) + b\dot{x}(t-h), \quad t \geq 0 \tag{1.6}$$

where $a, b, h \geq 0$ are given constants, $|b| < 1$.

The righthand side of Eq. (1.6) is already expressed in the form of (1.3) when $f_1(x_t) := -ax(t)$, $f_2(x_t) := bx(t-h)$. We may therefore proceed to step 2. The auxiliary ordinary equation is $\dot{y}(t) = -ay(t)$, for which the

Liapunov function v may be taken as equal to $v(y) = y^2$. The realization of step 3 depends on the representation (1.5).

Suppose that in (1.5) $f_3(x_t) := 0$ and $f_4(x_t) := bx(t-h)$. Then $V_1(x_t) := (x(t) - bx(t-h))^2$. To construct the component V_2 of the functional $V := V_1 + V_2$, we calculate $\dot{V}_1$

$$\dot{V}_1(x_t) = -2ax(t)(x(t) - bx(t-h)) = -2ax^2(t) + 2abx(t)x(t-h) \leq$$

$$\leq (-2a + |ab|)x^2(t) + |ab|x^2(t-h).$$

Thus, if one puts

$$V_2(x_t) := |ab| \int_{t-h}^{t} x^2(\tau)d\tau$$

then $\dot{V}(x_t) \leq 2(-a + |ab|)x^2(t)$. Hence $\dot{V}$ is negative definite if

$$a > 0, \quad |b| < 1. \tag{1.7}$$

Since the functional V as constructed is only non-negative, we shall require in addition that the zero solution of the difference equation $x(t) - bx(t-h) = 0$ must be asymptotically stable with respect to the perturbations of the initial conditions. A necessary and sufficient condition for this is that $|b| < 1$. Thus, by virtue of Thm. 8.1.1 the functional V yields (1.7) as asymptotic stability conditions for Eq. (1.6).

We will now consider another implementation of the procedure. Suppose that the auxiliary equation $\dot{y}(t) = -ay(t)$, the Liapunov function for which may be taken as $v(y) = |y|$, and in (1.5) we have $f_4 := 0$, $f_3(x_t) := b\dot{x}(t-h)$. Then

$$V_1(x_t) := |x(t)|, \quad \dot{V}_1(x_t) \leq -a|x(t)| + |b\dot{x}(t-h)|. \tag{1.8}$$

Therefore

$$V_2(x_t) := c \int_{t-h}^{t} |\dot{x}(s)|ds, \quad c := \frac{|b|}{1-|b|}. \tag{1.9}$$

Consequently, taking (1.8) and (1.9) into account and putting $V = V_1 + V_2$, we have

$$\dot{V}(x_t) \leq -a|x(t)| + c|\dot{x}(t)| + |b\dot{x}(t-h)| - c|\dot{x}(t-h)| =$$

$$= -a|x(t)| + c|\dot{x}(t)| - bc|\dot{x}(t-h)|.$$

Replacing $\dot{x}(t)$ by the righthand side of Eq. (1.6), we obtain

$$\dot{V}(x_t) \leq (-a + |a|c)|x(t)|.$$

Hence, in view of (1.8) and (1.9), it is obvious that when $a > 0$, $|b| < 1/2$ the functional $V = V_1 + V_2$ is positive definite and its derivative is negative definite. Thus, when the functionals (1.8) and (1.9) are used, the asymptotic stability conditions (1.7) for Eq. (1.6) remain unchanged become worse: $a > 0$, $|b| < 1/2$.

We shall now consider some classes of Eqs. (1.1), confining our attention, for simplicity, to a few characteristic representatives. Note that our aim is not only to obtain stability conditions but also to illustrate possible ways of implementing the individual steps of the procedure and the functionals V thus constructed.

2. SCALAR NDEs OF n–TH ORDER

Consider the linear equation of the form (8.1.1)

$$\frac{d}{dt}[x^{(n-1)}(t) + \int_0^\infty x^{(n-1)}(t-s)dK_{n+1}(s)] = \sum_{j=1}^{n} \int_0^\infty x^{(j-1)}(t-s)dK_j(s), \quad (2.1)$$

where $x^{(j)}(t)$ denotes the jth derivative of a scalar function $x(t)$ and the kernels $K_j(s)$, $j = 1, \ldots, n$, have bounded variation on $[0, \infty)$ and finite absolute moments up to the order n. Given a $(n-1)$ times continuously differentiable function $\phi(t)$, the initial conditions for the Eq. (2.1) are defined by the relations

$$x^{(j)}(t) = \phi^{(j)}(t), \quad t \leq 0, \quad j = 0, 1, \ldots, n-1. \quad (2.2)$$

The norm of the initial function ϕ is given by

$$\| \phi \| = \sup_{t \leq 0}(|\phi(t)| + |\phi^{(1)}(t)| + \ldots + |\phi^{(n-1)}(t)|).$$

Under the further assumptions that $\| \phi \| < \infty$ and that the variation of the kernel K_{n+1} on $[0, \infty)$ is less than one, there exists a unique solution of Eq. (2.1) with initial conditions (2.2) (see, for example, Thm. 8.3.5).

According to the definition 8.1.1 Eq. (2.1) is called *stable* if for any positive ε there exists $\delta(\varepsilon) > 0$ such that if $\| \phi \| < \delta(\varepsilon)$, then $|x(t)| < \varepsilon$ for each $t \geq 0$. If in addition $x(t) \to 0$ as $t \to \infty$ then Eq. (2.1) is called *asymptotically stable.*

The problem is to obtain stability conditions for the Eq. (2.1), in terms of the characteristics of the kernels $K_j(s)$. To this aim, we use the proposed procedure:

(1) *Transformation of Eq. (2.1).* Define $x_j(t) := x^{(j-1)}(t)$, for $j = 1, \ldots, n$, and rewrite Eq. (2.1) as the system of n equations

$$\left.\begin{aligned} &\dot{x}_j(t) = x_{j+1}(t), \quad j = 1, \ldots, n-1, \\ &\frac{d}{dt}[x_n(t) + \int_0^\infty x_n(t-s)dK_{n+1}(s)] = \sum_{j=1}^{n} \int_0^\infty x_j(t-s)dK_j(s). \end{aligned}\right\} \tag{2.3}$$

Now we want to transform the last equation of system (2.3) in order to separate terms with instantaneous and non–instantaneous feedback. This can be done, for example, as follows: for $j = 1, \ldots, n+1, \ \ i \geq 0$, define the moments

$$\alpha_{i,j} := \int_0^\infty s^i |dK_j(s)| \quad \text{and} \quad \beta_{i,j} := \int_0^\infty s^i dK_j(s) \ \ (\text{if} \ \ \alpha_{i,j} < \infty),$$

and expanded every $x_j(t-s)$ in the Taylor series in s so as to have

$$\begin{aligned} \int_0^\infty x_j(t-s)dK_j(s) = \sum_{k=j}^{n} x_k(t)\frac{(-1)^{k-j}}{(k-j)!}\beta_{k-j,j} - \\ -\frac{(-1)^{n-j}}{(n-j)!}\frac{d}{dt}\left[\int_0^\infty \int_{t-s}^{t} (u-t+s)^{n-j} x_n(u)du dK_j(s)\right]. \end{aligned} \tag{2.4}$$

With this transformation, system (2.3) can be represented as

$$\left.\begin{aligned} &\dot{x}_j(t) = x_{j+1}(t), \quad j = 1, \ldots, n-1, \\ &\frac{d}{dt}[x_n(t) - z(x_t)] = \sum_{k=1}^{n} r_k x_k(t), \end{aligned}\right\} \tag{2.5}$$

where

$$\begin{aligned} z(x_t) := \sum_{k=j}^{n} \frac{(-1)^{n-j}}{(n-j)!}\frac{d}{dt}\left[\int_0^\infty \int_{t-s}^{t} (u-t+s)^{n-j} x_n(u)du\, dK_j(s)\right] - \\ -\left[\int_0^\infty x_n(t-s)dK_{n+1}(s)\right] \end{aligned} \tag{2.6}$$

and

$$r_k := \sum_{j=1}^{k} \frac{(-1)^{k-j}}{(k-j)!}\beta_{k-j,j}, \quad k = 1, \ldots, n. \tag{2.7}$$

The transformation (2.4) is not the unique one suitable to separate delayed and nondelayed terms. Consider for example a lower order expansion of $x_j(t-s)$ up to some order $n' < n$. In this case the delayed part of (2.4) will not involve the maximum order derivative x_n. It also possible to use for some $j = 2, \dots, n$ the relation

$$\int_0^\infty x_j(t-s)dK_j(s) = \frac{d}{dt}\int_0^\infty x_{j-1}(t-s)dK_j(s).$$

It is worth remarking again that different transformations give rise to different expression of $z(x_t)$ and hence of terms (2.7) in the system (2.5).

(2) *Auxiliary system without delay.* An auxiliary system without delay obtained from (2.5) by disregarding the term $z(x_t)$. This auxiliary system is

$$\left.\begin{aligned} \dot{y}_j(t) &= y_{j+1}(t), \quad j = 1, \dots, n-1, \\ \dot{y}_n(t) &= \sum_{k=1}^{n} r_k y_k(t). \end{aligned}\right\} \tag{2.8}$$

By introducing the $n \times n$ companion matrix

$$A := \begin{pmatrix} 0 & 1 & 0 & \dots & 0 \\ 0 & 0 & 1 & \dots & 0 \\ \dots & \dots & \dots & \dots & \dots \\ 0 & 0 & 0 & \dots & 1 \\ r_1 & r_2 & r_3 & \dots & r_n \end{pmatrix},$$

the system (2.8) can be written in the form

$$\dot{y}(t) = Ay(t),$$

where $y(t)$ is a column vector with components y_j, $j = 1, \dots, n$.

Assume that the system (2.8) is asymptotically stable, i.e., that all the eigenvalues of the matrix A have a negative real part. In this case we say that matrix A is *Hurwitzian.* Then there exists a unique symmetric positive definite matrix B satisfying the Liapunov equation

$$A'B + BA = -C, \tag{2.9}$$

where C is a positive definite matrix and $'$ is the transposition operator. Remark that $B = \int_0^\infty e^{A't}Ce^{At}dt$. Therefore the full derivative $\dot{v}(y)$ of the function $v(y) := y'By$ along the trajectories of the system (2.8), by virtue of (2.9), is equal to $-y'Cy$. In particular we choose C equal to the identity matrix I.

(3) *Construction of a Liapunov functional.* Define the functional V_1 from the function $y \mapsto v(y)$ defined above, where the y_i are changed with x_i for $i = 1, \dots, n-1$ and y_n is changed on $x_n - z$. As a result we obtain

$$V_1(t, x_t) := (x_1(t), \dots, x_{n-1}(t), x_n(t) - z(x_t))B\times \\ \times(x_1(t), \dots, x_{n-1}(t), x_n(t) - z(x_t))'. \tag{2.10}$$

By computing the full derivative $\dot{V}_1$ of the functional (2.10) along the trajectories of the system (2.5) and by (2.9), we get

$$\dot{V}_1(x_t) = -|x(t)|^2 - 2\sum_{k=1}^{n} x_k(t)(BA)_{n,k} z(x_t) \leq \\ \leq -|x(t)|^2 + 2\sum_{k=1}^{n} |x_k(t)||(BA)_{n,k}||z(x_t)|, \tag{2.11}$$

where $(BA)_{n,k}$, $k = 1, \dots, n$, are the elements of the last row of the matrix BA. By (2.6) and by the Cauchy inequality, the last addenda in (2.11) can be bounded as

$$2|x_k(t)z(t)| \leq 2|x_k(t)\sum_{j=1}^{n}\int_0^\infty\int_{t-s}^{t}(u-t+s)^{n-j}x_n(u)du\,dK_j(s)| + \\ +2|x_k(t)\int_0^\infty x_n(t-s)dK_{n+1}(s)| \leq x_k^2(t)\sum_{j=1}^{n+1}\frac{\alpha_{n+1-j,j}}{(n+1-j)!} + \\ +\sum_{j=1}^{n}\frac{1}{(n-j)!}\int_0^\infty\int_{t-s}^{t}(u-t+s)^{n-j}x_n^2(u)du\,|dK_j(s)| + \\ +\int_0^\infty x_n^2(t-s)|dK_{n+1}(s)|. \tag{2.12}$$

Thus we define the functional

$$V_2(x_t) := \left(\sum_{k=1}^{m}|(BA)_{n,k}|\right)\sum_{j=1}^{n+1}\int_0^\infty\int_{t-s}^{t}\frac{(u-t+s)^{n-j+1}}{(n-j+1)!}x_n^2(u)du\,|dK_j(s)| \tag{2.13}$$

by which we get the following bound for the derivative of $V := V_1 + V_2$

$$\dot{V}(x_t) \leq -|x(t)|^2 + \alpha \sum_{k=1}^{m} |(BA)_{n,k}|(x_k^2(t) + x_n^2(t)), \tag{2.14}$$

where

$$\alpha := \sum_{j=1}^{n+1} \frac{\alpha_{n+1-j,j}}{(n+1-j)!}. \tag{2.15}$$

Now we are in a position to prove the following Theorem 2.1:

Theorem 2.1 Consider Eq. (2.1) where the kernels $K_j(s)$, $j = 1, \ldots, n$, have bounded variation on $[0, +\infty)$, finite absolute moments up to the order n, and $K_{n+1}(s)$ has a variation less than one. If the auxiliary system (2.8) is asymptotically stable and inequality

$$\alpha \left(\sum_{k=1}^{n-1} |(BA)_{n,k}| + 1 \right) < 1 \tag{2.16}$$

is fulfilled, then Eq. (2.1) is asymptotically stable.

Proof We want to give only a sketch of the proof. Since $(BA)_{n,n} = -1/2$, (2.16) ensures that the righthand term of (2.14) is a quadratic form with a negative definite matrix, hence $\dot{V}$ is negative definite. Observe that, by (2.10) and by the positive definiteness of the matrix B, V is positive definite with respect to x_i, $i = 1, \ldots, n-1$, but in general not with respect to x_n.

In fact from (2.14), (2.16), and from the definition of V_1, V_2, we get

$$V_1(x_t) \leq V_1(x_t) + V_2(x_t) = V(x_t) \leq V(x_0), \text{ for every } t \geq 0.$$

So

$$\lambda_{\min} \left(\sum_{i=1}^{n-1} x_i^2(t) + (x_n(t) - z(t))^2 \right) \leq V(x_0),$$

where $\lambda_{\min}$ is the minimum eigenvalue of matrix B. From this inequality it follows that there exists a positive constant c such that

$$x_i^2(t) \leq cV(x_0) \text{ for } i = 1, \ldots, n-1 \Rightarrow |x_i(t)| < \infty, \; i = 1, \ldots, n-1;$$

$$(x_n(t) - z(x_t))^2 \leq cV(x_0) \Rightarrow |x_n| \leq |z| + V(x_0).$$

Now, if z does not depend on x_n, then the component x_n is also bounded, otherwise we rewrite $z(t)$ so that

$$|x_n| \leq \alpha |x_n| + z_1((x_1)_t, \ldots, (x_{n-1})_t) + V(x_0),$$

where α is defined in (2.15) and $z_1((x_1)_t, \ldots, (x_{n-1})_t)$ is the part of z depending on $x_1, \ldots, x_{n-1}$. In this case, if $\alpha < 1$, also x_n is bounded. Remark that condition $\alpha < 1$ is already guaranteed by (2.16), thus, by the boundedness of the $|x_i|$ and by standard argument, we get the boundedness of $\int_0^\infty x_i^2(s)ds$ and of $|\dot{x}_i|$, $i = 1, \ldots, n$, and the asymptotic stability of (2.1). □

Remark 2.2 Once more, we observe that the essential part of the described method to obtain stability is the construction of the auxiliary system without delay of the form (2.8). As remarked before, this construction is not unique and depends on the form of the Taylor expansion. This leads to different expressions of $z(x_t)$ and hence of r_k, $V(t, x_t)$, and α and to different formulations of inequality (2.14) for $\dot{V}$. Another source of non–uniqueness in the definition of the Liapunov functional is the choice of the matrix C in (2.9). Each of these choices gives rise to different conditions on the parameters and hence to different parts of the stability domain.

3. LINEAR NDEs

Using the procedure just described, let us establish stability conditions for the linear equation

$$\dot{x}(t) = A_0 x(t) + A_1 x(t - h_1) + A_2 \dot{x}(t - h_2), \quad t \geq 0, \quad x(t) \in \mathbf{R}^n \tag{3.1}$$

where A_i are constant $n \times n$ matrices and $h_i \geq 0$ are constants. Let $\|\cdot\|$ denote the matrix norm induced by some vector norm $|\cdot|$ in $\mathbf{R}^n$, i.e. $\| A_i \| = \max_x |A_i x|$, where the maximum is evaluated over all $x \in \mathbf{R}^n$ such that $|x| = 1$. Set $\alpha_i := \| A_i \|$. We will now consider possible versions of representations (1.4) and (1.5) and the resulting stability conditions and functionals $V = V_1 + V_2$, assuming that $\alpha_2 < 1$.

3.1. *THE STABILITY CONDITION*

Define $f_1(x) := A_0 x$, $f_2(x_t) := A_1 x(t - h_1) + A_2 \dot{x}(t - h_2)$, $f_4 := 0$. Then the auxiliary equation is $\dot{y}(t) = A_0 y(t)$, the Liapunov function v for which may be taken as $v(y) := |y|$. This implies $V_1(x_t) := |x(t)|$. For any norm $|\cdot|$ in $\mathbf{R}^n$ we have

$$d|x(t)|/dt = Q[x(t), \dot{x}(t)] \tag{3.2}$$

where the scalar function $Q[x, y]$ of two independent variables $x, y \in \mathbf{R}^n$ is defined by

$$Q[x, y] := \lim_{h \to 0+} \frac{1}{h}[|x + hy| - |x|]. \tag{3.3}$$

We recall, moreover, that for any square matrix A there exists

$$\gamma(A) := \sup_x |x|^{-1} Q[x, Ax], \quad |x| \neq 0, \quad x \in \mathbf{R}^n \tag{3.4}$$

where $\gamma(A)$ is the logarithmic norm of A. It follows from (3.1) – (3.3)

$$\dot{V}_1(x_t, \dot{x}_t) \leq \lim_{h \to 0+} \frac{1}{h}[|x + hy| - |x|] \leq \lim_{h \to 0+} \frac{1}{h}[|x + hA_0 y| - |x|] +$$

$$+|f_2(t, x_t, \dot{x}_t)| = Q[x, A_0 x] + |f_2(t, x_t, \dot{x}_t)|.$$

Hence, by (3.4), we obtain the inequality

$$\dot{V}_1(x_t, \dot{x}_t) \leq \gamma(A_0)|x(t)| + \alpha_1 |x(t-h_1)| + \alpha_2 |\dot{x}(t-h_2)|. \tag{3.5}$$

To ensure that $\dot{V}$ is negative definite, we define

$$V_2(x_t, \dot{x}_t) := (1-\alpha_2)^{-1} \left[\alpha_1 \int_{t-h_1}^{t} |x(s)| ds + \alpha_2 \int_{t-h_2}^{t} |\dot{x}(s)| ds \right]. \tag{3.6}$$

In view of (3.5) and (3.6)

$$\dot{V}(x_t, \dot{x}_t) \leq [\gamma(A_0) + \alpha_1(1-\alpha_2)^{-1}]|x(t)| + \alpha_2(1-\alpha_2)^{-1}|\dot{x}(t)| -$$

$$-\alpha_1\alpha_2(1-\alpha_2)^{-1}|x(t-h_1)| - \alpha_2^2(1-\alpha_2)^{-1}|\dot{x}(t-h_2)|.$$

By this inequality and the estimate implied by (3.1), we have

$$|\dot{x}(t)| \leq \alpha_0|x(t)| + \alpha_1|x(t-h_1)| + \alpha_2|\dot{x}(t-h_2)|.$$

We finally conclude that $\dot{V}(x_t, \dot{x}_t) \leq \alpha_3 |x_3(t)|$, where

$$\alpha_3 := \gamma(A_0) + (1-\alpha_2)^{-1}(\alpha_1 + \alpha_0\alpha_2).$$

We have thus shown that system (3.1) is asymptotically stable if

$$\alpha_3 < 0, \quad \alpha_2 < 1. \tag{3.7}$$

3.2. *ANOTHER STABILITY CONDITION*

Let

$$f_1(x_t) := (A_0 + A_1)x, \quad f_4 := 0,$$

$$f_2(x_t) := f_3(x_t) := -A_1 \int_{t-h_1}^{t} \dot{x}(s) ds + A_2 \dot{x}(t-h_2). \tag{3.8}$$

Then the auxiliary equation is $\dot{y}(t) = (A_0 + A_1)y(t)$ and the Liapunov function $v(y) := |y|$, i.e. in accordance with the procedure and formulae (3.8) we must define $V_1(x_t) := |x(t)|$. Then, proceeding as for (3.5), we obtain

$$\dot{V}_1(x_t, \dot{x}_t) \leq \gamma(A_0 + A_1)|x(t)| + \alpha_1 \int_{t-h_1}^{t} |\dot{x}(s)|ds + \alpha_2|\dot{x}(t - h_2)|. \tag{3.9}$$

Hence, as in the case of (3.6), we obtain an expression for V_2

$$V_2(x_t, \dot{x}_t) := (1-h_1\alpha_1-\alpha_2)^{-1} \left[\alpha_1 \int_{t-h_1}^{t} (s - t + h_1)|\dot{x}(s)|ds + \alpha_2 \int_{t-h_2}^{t} |\dot{x}(s)|ds\right].$$

Consequentially

$$\dot{V}_2(x_t, \dot{x}_t) = (1 - h_1\alpha_1 - \alpha_2)^{-1}\times$$

$$\times \left[(h_1\alpha_1 + \alpha_2)|\dot{x}(t)| - \alpha_1 \int_{t-h_1}^{t} |\dot{x}(s)|ds - \alpha_2|\dot{x}(t - h_2)|\right].$$

Replacing $|\dot{x}(t)|$ here by the righthand side of the estimate

$$|\dot{x}(t)| \leq \| A_0 + A_1 \| \, |\dot{x}(t)| + \alpha_1 \int_{t-h_1}^{t} |\dot{x}(s)|ds + \alpha_2|\dot{x}(t - h_2)|,$$

and taking (3.9) into consideration, we conclude that $\dot{V}(x_t, \dot{x}_t) \leq \alpha_4|x(t)|$, where

$$\alpha_4 := \gamma(A_0 + A_1) + \| A_0 + A_1 \| \, (1 - h_1\alpha_1 - \alpha_2)^{-1}(h_1\alpha_1 + \alpha_2).$$

Thus, Eq. (3.1) is asymptotically stable if

$$1 > h_1\alpha_1 + \alpha_2, \quad \alpha_4 < 0. \tag{3.10}$$

Conditions (3.7) and (3.10) are not identical.

Thus, for example, conditions (3.7) mean that A_0 is a Hurwitz matrix; conditions (3.10), however, may be satisfied even when all the eigenvalues of A_0 have positive real parts. It should also be noted that, under conditions (3.7), Eq. (3.1) is stable for any delays $h_1, h_2 \geq 0$, while inequalities (3.10) impose certain restrictions on h_1.

Using other representations of the righthand side of (3.1), one can deduce other stability conditions. To illustrate, here is on of them.

3.3. *THE SUMMARIZING RESULT*

Put

$$f_1(x) := (A_0 + A_1)x, \quad f_3 := 0, \quad f_2(x_t, \dot{x}_t) := \frac{d}{dt} f_4(x_t),$$

$$f_4(x_t) := -A_1 \int\limits_{t-h_1}^{t} |x(s)| ds + A_2 x(t - h_2). \tag{3.11}$$

The auxiliary system is $\dot{y}(t) = (A_0 + A_1)y(t)$, the Liapunov function $v(y)$ for which may be taken as the quadratic form $v(y) := y'By$, where the prime denotes transposition and B is a symmetric matrix. We set

$$(A_0 + A_1)'B + B(A_0 + A_1) =: -C. \tag{3.12}$$

Let us assume that $A_0 + A_1$ is a Hurwitz matrix and the constant matrix C is positive definite. Then Eq. (3.12) has a unique solution $B > 0$ for any given symmetric matrix $C > 0$. In accordance with the procedure, we must put $V_1(z_t) := v(z(t))$, $z(t) = x(t) - f_4(x_t)$. Choose $C = I$, where I is the identity matrix; we will work with the Euclidean norm $|x| := (x'x)^{1/2}$. Then, taking (3.11) and (3.12) into account, we have

$$\dot{V}_1(x_t) = x'(t)(A_1 + A_0)'Bz(t) + z'(t)B(A_0 + A_1)x(t) =$$

$$= -|x(t)|^2 - 2x'(t)(A_0 + A_1)'Bf_4(x_t).$$

Thus V_2 should be chosen in the form

$$V_2(\dot{x}_t) := \| (A_0 + A_1)B \| \left[\alpha_1 \int\limits_{t-h_1}^{t} (s - t + h_1)|\dot{x}(s)|^2 ds + \alpha_2 \int\limits_{t-h_2}^{t} |\dot{x}(s)|^2 ds \right].$$

The sum $V := V_1 + V_2$ then satisfies an estimate $\dot{V}(x_t, \dot{x}_t) \leq \alpha_5 |x(t)|^2$, where $\alpha_5 := -[1 - 2 \| (A_0 + A_1)B \| (h_1\alpha_1 + \alpha_2)]$.

Since the functional V is only sign–definite, we must still require the trivial solution of the functional equation $x(t) - F_4(t, x_t) = 0$ to be asymptotically stable. A sufficient condition to that end is $h_1\alpha_1 + \alpha_2 < 1$. Consequently, Eq. (2.1) will be asymptotically stable if

$$\forall \mathrm{Re}\lambda(A_0 + A_1) < 0, \quad h_1\alpha_1 + \alpha_2 < 1, \quad \alpha_5 < 0, \tag{3.13}$$

where $\lambda(A_0 + A_1)$ are the eigenvalues of the matrix $A_0 + A_1$ and $\mathrm{Re}\lambda$ are their real parts.

We have thus proved the following theorem.

Theorem 3.1 Eq. (3.1) is asymptotically stable if any of the conditions (3.7), (3.10) or (3.13) is satisfied.

4. NONLINEAR NDEs

We will now describe the application of the procedure to non–linear equations, establishing stability conditions for them, confining ourselves for simplicity to Eq. (1.1) of the form

$$\dot{x}(t) = A_1(t, x(t-h_1)) + A_2(t, \dot{x}(t-h_2)), \quad t \geq t_0, \quad x(t) \in \mathbf{R}^n \tag{4.1}$$

with initial conditions (1.2).

Here $h_i \geq 0$ are constants and $A_i : [t_0, \infty) \times \mathbf{R}^n \to \mathbf{R}^n$ $(i = 1, 2)$ are continuous functions, $A_i(t, 0) \equiv 0$. Let D denote the set of continuous functions $\psi : (-\infty, 0] \to \mathbf{R}^n$ such that $|\psi| < r$ for some $r > 0$ and $\dot{\psi}$ can have discontinuity only if the first kind. Here $h = \max\{h_1, h_2\}$. In addition $\dot{\psi}(0) := \dot{\psi}(0^-)$, $\dot{\psi}(-h_2) := \dot{\psi}((-h_2)^+)$. If initial function $\psi \in D$ then the solution of NDEs of the form (4.1) also belongs $D : x_t \in D$. Let us assume that the function $(t, x) \mapsto A_1(t, x)$ is continuously differentiable with respect to x, and for some constants $\alpha_i > 0$ and any $\phi, \psi \in D$ we have

$$|A_1(t, x_1) - A_1(t, x_2)| \leq \alpha_1 |x_1 - x_2|, \quad |A_2(t, x)| \leq \alpha_2 |x|, \quad \alpha_2 < 1, \tag{4.2}$$

where $|\cdot|$ is some vector norm in $\mathbf{R}^n$.

Let $F : [t_0, \infty) \times \mathbf{R}^n \to \mathbf{R}^{n \times n}$ denote the Jacobian, $F(t, x) = \partial A_1(t, x)/\partial x$, and $\gamma(F)$ the logarithmic norm of the matrix F. Let $\gamma_1 := \sup_{t,x} \gamma(F(t, x))$, $t \geq t_0$, $|x| \leq r$.

Theorem 4.1 Suppose that inequalities (4.2) hold and moreover

$$\gamma_2 < 0, \quad \alpha_2 + \alpha_1 h_1 < 1, \quad \gamma_2 := \gamma_1 + \alpha_1(\alpha_1 h_1 + \alpha_2)(1 - \alpha_2 - \alpha_1 h_1)^{-1}. \tag{4.3}$$

Then the trivial solution of Eq. (4.1) is asymptotically stable.

Proof Let $f_1(t, x) := A_1(t, x)$. This means that the auxiliary system has the form $\dot{y}(t) = A_1(t, y(t))$. Let $v(y) := |y|$. Then in formula (1.4) we must put $f_3(t, x_t, \dot{x}_t) := A_1(t, x(t-h_1)) - A_1(t, x(t)) + A_2(t, \dot{x}(t-h_2))$, $f_4 := 0$. Consequently, $V_1(x_t) := |x(t)|$. Therefore, by (4.2) and (4.3)

$$\dot{V}_1(t, x_t, \dot{x}_t) \leq \lim_{h \to 0+} \frac{1}{h}[|x(t) + hA_1(t, x(t))| - |x(t)|] + |f_3(t, x_t, \dot{x}_t)|. \tag{4.4}$$

Note that

$$A_1(t, x) = \left[\int_0^1 F(t, sx) ds \right] x.$$

Hence, from (4.4), we obtain the inequality

$$\dot{V}_1(t,x_t,\dot{x}_t) \le \lim_{h\to 0+}\frac{1}{h}\left[\left|x(t)+h\left(\int_0^1 f(t,sx)ds\right)x(t)\right|-|x(t)|\right]+ \tag{4.5}$$

$$+|f_3(t,x_t,\dot{x}_t)|.$$

In view of (3.3), (3.4) and (4.5), we have

$$\dot{V}_1(t,x_t,\dot{x}_t) \le Q\left[x(t),\left(\int_0^1 F(t,sx(t))ds\right)x(t)\right]+|f_3(t,x_t,\dot{x}_t)| \le$$

$$\le \gamma\left(\int_0^1 F(t,sx(t))ds\right)|x(t)|+|f_3(t,x_t,\dot{x}_t)| \le \tag{4.6}$$

$$\le \int_0^1 \gamma(F(t,sx(t)))ds|x(t)|+|f_3(t,x_t,\dot{x}_t)|.$$

Now consider the functional $V := V_1 + V_2$ in D, where

$$V_2(\psi) := (1-\alpha_2-\alpha_1 h_1)^{-1}\left[\alpha_1\int_{-h_1}^0 (\tau+h_1)|\dot{\psi}(\tau)|d\tau+\right.$$

$$\left.+\alpha_2\int_{-h_2}^0 |\dot{\psi}(\tau)|d\tau+(\alpha_1 h_1+\alpha_2)\alpha_1\int_{-h_1}^0 |\dot{\psi}(\tau)|d\tau\right]. \tag{4.7}$$

On the basis of (4.2) and (4.6)

$$\dot{V}_1(t,\psi,\dot{\psi}) \le \gamma_1|\psi(0)|+\alpha_2|\dot{\psi}(-h_2)|+\alpha_1|\psi(-h_1)-\psi(0)| \le$$

$$\le \gamma_1|\psi(0)|+\alpha_2|\dot{\psi}(-h_2)|+\alpha_1\int_{-h_1}^0 |\dot{\psi}(\tau)|d\tau. \tag{4.8}$$

We now calculate $\dot{V}_2$:

$$\dot{V}_2(\psi,\dot{\psi}) = (1-\alpha_2-\alpha_1 h_1)^{-1}[(\alpha_1 h_2+\alpha_2)|\dot{\psi}(0)|-\alpha_1\int_{-h_1}^0 |\dot{\psi}(\tau)|d\tau-$$

$$-\alpha_2|\dot{\psi}(-h_2)|+(\alpha_1 h_1+\alpha_2)\alpha_1(|\dot{\psi}(0)|-|\dot{\psi}(-h_1)|)]. \tag{4.9}$$

Note that, by (4.1) and (4.2)

$$|\dot{\psi}(0)| \leq \alpha_1|\psi(-h_1)| + \alpha_2|\dot{\psi}(-h_2)|. \tag{4.10}$$

Replacing $|\dot{\psi}(0)|$ in (4.9) by the righthand side of (4.10) and taking note of (4.8), we obtain $\dot{V}(t,\psi,\dot{\psi}) \leq \gamma_2|\psi(0)|$. Thus, V is positive definite in D, admits of an infinitely small upper limit, and its total derivative is negative definite when conditions (4.2) and (4.3) hold. Consequently, the trivial solution of Eq. (4.1) is asymptotically stable. This assertion can be proven similar to the proof of Thm. 8.1.1. □

Remark 4.1 In applications of the procedure to equations of neutral type it is sometimes useful to iterate the derivative on the right once or more times. For example, define $V_1(x_t) := |x(t)|$ for $t \geq t_0 + h_1$. Then, as in (4.8) and (4.10), integrating once, we conclude that

$$\dot{V}_1(t,x_t,\dot{x}_t) \leq \gamma_1|x(t)| + \alpha_2|\dot{x}(t-h_2)| + \alpha_1\int_{t-h_1}^{t}|\dot{x}(\tau)|d\tau \leq$$

$$\leq \gamma_1|x(t)| + \alpha_2|\dot{x}(t-h_2)| + \alpha_1^2\int_{t-2h_1}^{t-h_1}|x(\tau)|d\tau + \alpha_1\alpha_2\int_{t-h_1-h_2}^{t-h_2}|\dot{x}(\tau)|d\tau.$$

Thus

$$V_2(x_t,\dot{x}_t) = \alpha_1^2\left[\int_{t-2h_1}^{t-h_1}ds\int_{s_1}^{t-h_1}|x(s_1)|ds_1 + h_1\int_{t-h_1}^{t}|x(\tau)|d\tau\right] +$$

$$+\alpha_2(1-\alpha_2(1+\alpha_1h_1))^{-1}\left[\int_{t-h_2}^{t}|\dot{x}(\tau)|d\tau + \int_{t-h_1}^{t}|x(\tau)|d\tau\alpha_1\right] +$$

$$+\alpha_1\alpha_2\int_{t-h_1-h_2}^{t-h_2}ds\int_{s}^{t-h_2}|\dot{x}(s_1)|ds_1.$$

Setting $V := V_1 + V_2$, we obtain $\dot{V}(t,x_t,\dot{x}_t) \leq \gamma_3|x(t)|$, where

$$\gamma_3 := \gamma_1 + \alpha_1^2h_1 + \alpha_2(1-\alpha_2(1+\alpha_1h_1))^{-1}\alpha_1.$$

We have thus proved the following theorem.

Theorem 4.2 The trivial solution of Eq. (4.1) is asymptotically stable if inequalities (4.2) hold and $(\alpha_1h_1+1)\alpha_2 < 1$, $\gamma_3 < 0$.

Remark 4.2 For some equations of type (1.1) one has $V = V_1$. For example, suppose that the transformations (1.4) and (1.5) reduce Eq. (1.1) to the form $\dot{z}(t) = F_5(t, z(t))$, where $z(t) = x(t) - F_4(t, x_t)$, $F_4(t,0) = 0$, $F_5(t,0) = 0$. Then if the auxiliary equation which has the form $\dot{y}(t) = F_5(t, y(t))$, is uniformly asymptotically stable and the trivial solution of the equation $x(t) = F_4(t, x_t)$ is asymptotically stable, then the trivial solution of Eq. (1.1) is also asymptotically stable. This was pointed out in [289.3].

Example 4.1 Consider a shunted power transmission line, described by the equation [238]

$$\dot{x}(t) = -g(x(t)) + b\dot{x}(t-h), \quad t \geq 0, \quad x(t) \in \mathbf{R}, \quad |b| < 1, \quad g(0) = 0 \quad (4.11)$$

where $g : \mathbf{R} \to \mathbf{R}$ is a continuous function, $b, h \geq 0$ are given constants, and

$$xg(x) > 0, \quad x \neq 0. \quad (4.12)$$

The stability of the trivial solution of (4.11) was studied in [238], where it was established that sufficient conditions for stability are provided by inequalities (4.12) and $|b| < 1/2$. If $g(x) = ax$, these conditions reduce to $a > 0$, $|b| < 1/2$, i.e. they are not the same as (1.7), which were applied to the linear equation (1.6) in Example 1.1. Using the procedure proposed here, we will establish stability conditions for the trivial solution of Eq. (4.11) that reduce to (4.7) in the linear case. The auxiliary ordinary equation for (4.11) has the form $\dot{y}(t) = -g(y(t))$, a Liapunov function for which may be taken as

$$v(y) := \int_0^y g(s)ds. \quad (4.13)$$

By (4.12), $v(y)$ is positive definite and has an infinitesimal upper limit. On the basis of the procedure, $V_1(x_t) := v(x(t))$. Hence, via (4.11), it follows that $V = V_1 + V_2$ where

$$V_2(\dot{x}_t) := \frac{1}{2} \int_{t-h}^{t} \dot{x}^2(\tau)d\tau. \quad (4.14)$$

One finds then that the derivative $\dot{V}$ is given by

$$\dot{V}(x_t, \dot{x}_t) = g(x(t))[-g(x(t)) + b\dot{x}(t-h)] + \frac{1}{2}(\dot{x}^2(t) - \dot{x}^2(t-h)).$$

Replacing $\dot{x}(t)$ here by the righthand side of Eq. (4.11), we obtain

$$2\dot{V}(x_t, \dot{x}_t) = -g^2(x(t)) - (1 - b^2)\dot{x}^2(t-h).$$

Thus, the asymptotic stability conditions are the inequalities (4.12) and $|b| < 1$, which become (1.6) in the linear case $g(x) = ax$.

Eq. (4.11) with a non-linearity $g(x)$ satisfying (4.12) is an equation of gradient type, for which application of the above procedure also yields stability conditions for $n > 1$.

For example, consider the equation

$$\dot{x}(t) = -\nabla G(x(t)) + B\dot{x}(t-h), \quad t \geq t_0, \quad x(t) \in \mathbf{R}^n, \quad G(0) = 0 \qquad (4.15)$$

where $G : \mathbf{R}^n \to \mathbf{R}$ is a continuously differentiable function, B is a given constant matrix, $h \geq 0$, ∇G denotes the gradient of G and

$$G(x) > 0, \quad x \neq 0. \qquad (4.16)$$

As in (4.13) and (4.14), putting

$$V(x_t, \dot{x}_t) = G(x(t)) + \frac{1}{2}\int_{t-h}^{t} \dot{x}'(s)\dot{x}(s)ds$$

we conclude that the trivial solution of Eq. (4.15) is asymptotically stable if inequality (4.16) holds and $\| B \| < 1$.

Remark 4.3 The procedure is also applicable to certain equations with the variable coefficients. For example, consider the scalar equation

$$\dot{x}(t) = -a(t)x(t) + b\dot{x}(t-h), \quad t \geq t_0 \qquad (4.17)$$

where a is a non-negative continuous function, $b \in \mathbf{R}, h \geq 0$ are given constants, using the functional

$$V(x_t) := [x(t) + bx(t-h)]^2 + |b|\int_{t-h}^{t} a(s+h)x^2(s)ds$$

we arrive at the following sufficient asymptotic stability conditions

$$\inf_{t \geq t_0}[2a(t) - |b|(a(t) + a(t+h))] > 0, \quad |b| < 1.$$

These conditions reduce to the stability conditions (1.7) for Eq. (1.6) with constant coefficients.

5. STABILITY OF THE SECOND ORDER NDEs

Let us illustrate the described method for the second order scalar equation

$$\frac{d}{dt}\left[\dot{x}(t) + \int_0^\infty \dot{x}(t-s)dK_3(s)\right] = \int_0^\infty x(t-s)dK_1(s) + \int_0^\infty \dot{x}(t-s)dK_2(s), \quad (5.1)$$

where K_i, $i = 1,2,3$, are functions with bounded variation on $[0,\infty)$ and bounded moments up to the second order. Initial conditions for Eq. (5.1) satisfy the conditions from Sect. 2.

According to (2.3) we get the system

$$\left.\begin{aligned} &\dot{x}_1(t) = x_2(t),\\ &\frac{d}{dt}[x_2(t) + \int_0^\infty x_2(t-s)dK_3(s)] = \int_0^\infty x_1(t-s)dK_1(s) +\\ &+\int_0^\infty x_2(t-s)dK_2(s). \end{aligned}\right\} \tag{5.2}$$

The righthand side of the last equation can be transformed by (2.4) or by using Taylor expansions of $x_j(t-s)$ of different orders as

$$\left.\begin{aligned} &\int_0^\infty x_1(t-s)dK_1(s) =\\ &= \beta_{01}x_1(t) - \beta_{11}x_2(t) + \frac{d}{dt}\int_0^\infty\int_{t-s}^t x_2(u)(u-t+s)du\, dK_1(s);\\ &= \beta_{01}x_1(t) - \frac{d}{dt}\int_0^\infty\int_{t-s}^t x_1(u)du\, dK_1(s);\\ &= \beta_{01}x_1(t) - \int_0^\infty\int_{t-s}^t x_2(u)du\, dK_1(s); \end{aligned}\right\} \tag{5.3}$$

$$\left.\begin{aligned} \int_0^\infty x_2(t-s)dK_2(s) &= \beta_{02}x_2(t) - \frac{d}{dt}\int_0^\infty\int_{t-s}^t x_2(u)du\, dK_2(s);\\ &= \frac{d}{dt}\int_0^\infty x_1(t-s)dK_2(s). \end{aligned}\right\} \tag{5.4}$$

If we consider different combinations of these expansions, we get different transformations of (5.2) and then different conditions on the parameters for the asymptotic stability of (5.1).

For example, consider the transformed system

$$\dot{x}_1(t) = x_2(t), \quad \dot{x}_2(t) - \dot{z}(x_t) = r_1 x_1(t) + r_2 x_2(t). \tag{5.5}$$

Here

$$r_1 := \beta_{01}, \quad r_2 := \beta_{02} - \beta_{11},$$

$$z(x_t) := -\int_0^\infty x_2(t-s)dK_3(s) - \int_0^\infty \int_{t-s}^t x_2(u)du\, dK_2(s) + \tag{5.6}$$

$$+ \int_0^\infty \int_{t-s}^t x_2(u)(u-t+s)du\, dK_1(s).$$

The auxiliary linear system may be is obtained from (5.5) by rejecting the term $z(x_t)$ and has the form

$$\dot{y}_1(t) = y_2(t), \quad \dot{y}_2(t) = r_1 y_1(t) + r_2 y_2(t). \tag{5.7}$$

Necessary and sufficient conditions for asymptotic stability of the system (5.7) are given by the inequalities

$$r_1 < 0, \quad r_2 < 0. \tag{5.8}$$

Let us form and solve Eq. (2.9) with $C = I$ for the system (5.7). We obtain

$$B = \frac{1}{2r_1 r_2} \begin{pmatrix} r_1^2 + r_2^2 - r_1 & -r_2 \\ -r_2 & 1 - r_1 \end{pmatrix}.$$

Define $v(t, y) = y' B y$ and hence

$$v(t, y) = \frac{1}{2r_1 r_2}[(r_1^2 + r_2^2 - r_1)y_1^2 - 2r_2 y_1 y_2 + (1 - r_1)y_2^2]. \tag{5.9}$$

Now, according to the algorithm given in Sec. 1, we must replace in (5.9) the function $y_2(t)$ by $x_2(t) - z(x_t)$. As a result we obtain the component V_1 of the functional V in the form

$$V_1(x_t) := \frac{1}{2r_1 r_2}[(r_1^2 + r_2^2 - r_1)x_1^2(t) - 2r_2 x_1(t)(x_2(t) - z(x_t)) +$$

$$+ (1 - r_1)(x_2(t) - z(x_t))^2].$$

For the full derivative $\dot{V}_1$ by virtue of the system (5.5) we get

$$\dot{V}_1(x_t) = -x_1^2(t) - x_2^2(t) + x_2(t)z(x_t) + \frac{r_1 - 1}{r_2} x_1(t) z(x_t),$$

and, using the estimate (2.12), we can see that

$$2|x_k(t)z(x_t)| \leq$$

$$\leq \alpha x_k^2(t) + \int_0^\infty x_2^2(t-s)|dK_3(s)| + \int_0^\infty \int_{t-s}^t x_2^2(u)du\,|dK_2(s)| +$$

$$+ \int_0^\infty \int_{t-s}^t (u-t+s)x_2^2(u)du\,|dK_1(s)|,$$

where now $\alpha := \alpha_{03} + \alpha_{12} + \alpha_{21}/2$.

Therefore V_2 must be chosen as

$$V_2(x_t) := \frac{r_1 + r_2 - 1}{r_2} \int_0^\infty \int_{t-s}^t x_2^2(u)du\,|dK_2(s)| +$$

$$+ \int_0^\infty \int_{t-s}^t (u-t+s)x_2^2(u)du\,|dK_2(s)| + \int_0^\infty \int_{t-s}^t \frac{(u-t+s)^2}{2} x_2^2(u)du\,|dK_1(s)|.$$

Finally, for $V := V_1 + V_2$, we obtain that

$$\dot{V}(x_t) \leq -x_1^2(t) - x_2^2(t) + \frac{\alpha}{2}x_1^2(t) + \frac{r_1 - 1}{r_2}\frac{\alpha}{2}x_2^2(t) + \frac{r_1 + r_2 - 1}{r_2}\alpha x_2^2(t).$$

So we can conclude that conditions for the asymptotic stability of (5.1) (by using transformation (5.5)–(5.6)) have the form

$$\left. \begin{array}{c} \beta_{01} < 0, \quad \beta_{02} - \beta_{11} < 0, \quad \alpha := \alpha_{12} + \alpha_{03} + \alpha_{21}/2 < 1, \\ 2(\alpha - 1)(\beta_{02} - \beta_{11}) + 3\alpha(\beta_{01} - 1) > 0. \end{array} \right\} \tag{5.10}$$

If we consider another transformed system (5.5) with

$$r_1 := \beta_{0,1}, \quad r_2 := -\beta_{1,1},$$

$$z(x_t) := -\int_0^\infty x_2(t-s)dK_3(s) + \int_0^\infty x_1(t-s)dK_2(s) -$$

$$-\int_0^\infty \int_{t-s}^t (u-t+s)x_2(u)du\,dK_1(s),$$

then, by using the same procedure as above, the stability conditions have the form

$$\left.\begin{aligned} &\beta_{0,1} < 0, \quad \beta_{1,1} > 0, \\ &2\beta_{11} > \max\{\alpha(1-\beta_{01}) + \alpha_{02}(1+\beta_{11}-\beta_{01}), \\ &\beta_{11}(2\alpha_{03}+\alpha_{21}+\alpha_{02}) + (1-\beta_{01})(\alpha_{03}+\alpha_{21}/2)\}, \end{aligned}\right\} \tag{5.11}$$

where $\alpha := \alpha_{0,3} + \alpha_{0,2} + \alpha_{2,1}/2$.

A further possibility is given by

$$r_1 := \beta_{0,1}, \quad r_2 := \beta_{0,2},$$

$$z(x_t) := -\int_0^\infty x_2(t-s)dK_3(s) - \int_0^\infty \int_{t-s}^t x_2(u)du\,dK_2(s) -$$

$$-\int_0^\infty \int_{t-s}^t x_1(u)du\,dK_1(s),$$

for which stability conditions take the form

$$\left.\begin{aligned} &\beta_{0,1} < 0, \quad \beta_{0,2} < 0, \\ &-2\beta_{0,2} > \max\{\alpha_{1,1}(1-\beta_{0,1}-\beta_{0,2}) + \alpha(1-\beta_{0,1}), \\ &(\alpha_{1,2}+2\alpha_{0,3})(1-\beta_{0,1}-\beta_{0,2}) - \alpha\beta_{0,2}\}, \end{aligned}\right\} \tag{5.12}$$

where $\alpha := \alpha_{1,1} + \alpha_{1,2} + \alpha_{0,3}$.

As remarked earlier, another situation arises if the right side of Eq. (2.9) is varied. Consider for instance the transformed system (5.5)–(5.6); from the auxiliary system (5.7) we obtain the conditions (5.8) but, if we solve (2.9) with

$$C := \begin{pmatrix} 2r_1r_2 & 0 \\ 0 & -2r_2 \end{pmatrix},$$

we get

$$B = \begin{pmatrix} -2r_1 + r_2^2 & -r_2 \\ -r_2 & 2 \end{pmatrix},$$

and so we can define a new function

$$v(y) := y'By = -2r_1y_1^2 + y_2^2 + (y_2 - r_2y_1)^2.$$

By using the same procedure as above, we get the following new stability conditions

$$\left.\begin{array}{l}\beta_{01} < 0, \quad \beta_{02} - \beta_{11} < 0, \quad \alpha := \alpha_{03} + \alpha_{12} + \alpha_{21}/2 < 1, \\ -\beta_{02} + \beta_{11} > \max\{\alpha, (\alpha - 1)^{-1}\alpha\beta_{01}\}.\end{array}\right\}$$

6. AN ILLUSTRATIVE EXAMPLE FOR DIMENSION n = 3

Before presenting the example, let us make some general considerations for dimension $n = 3$.

The Hurwitz matrix associated with the characteristic polynomial of the companion matrix A in system (2.8) is

$$M = \begin{pmatrix} -r_3 & -r_1 & 0 \\ 1 & -r_2 & 0 \\ 0 & -r_3 & -r_1 \end{pmatrix},$$

and hence A is stable, i.e., it is Hurwitzian, if and only if

$$r_3 < 0, \quad r_2 r_3 + r_1 > 0, \quad r_1 < 0. \tag{6.1}$$

Observe that (6.1) implies $r_2 < 0$.

Consider now Eq. (2.9) with $C = I$. Since $B = B'$, we get $(BA)' + BA = -I$, and hence $(BA)_{ij} = -(BA)_{ji}$ for $i \neq j$, $(BA)_{ii} = -1/2$ for each i. So

$$BA = \begin{pmatrix} -1/2 & (BA)_{12} & (BA)_{13} \\ -(BA)_{12} & -1/2 & (BA)_{23} \\ -(BA)_{13} & -(BA)_{23} & -1/2 \end{pmatrix}. \tag{6.2}$$

By adding equation $A'(BA)A^{-1})' + (BA) = -I$ and its transpose $(A'(BA) \times A^{-1})' + (BA)' = -I$, where

$$A^{-1} = \frac{1}{r_1} \begin{pmatrix} -r_2 & -r_3 & 1 \\ r_1 & 0 & 0 \\ 0 & r_1 & 0 \end{pmatrix},$$

we get that also $A'(BA)A^{-1}$ is such that $(A'(BA)A^{-1})_{ij} = -(A'(BA)A^{-1})_{ji}$ for $i \neq j$, $(A'(BA)A^{-1})_{ij} = -1/2$ for each i, and this result is independent of the dimension of the system.

Now, to fulfil inequality (2.16) in Thm. 2.1, we need only the last row of matrix BA and, in the particular case of $n = 3$, we already know that $(BA)_{33} = -1/2$ and we can easily get the elements $(BA)_{13}$ and $(BA)_{23}$ from the equations $(A'(BA)A^{-1})_{11} = -1/2$, $(A'(BA)A^{-1})_{22} = -1/2$. In fact we have

$$\left.\begin{aligned} &r_2(BA)_{13} - r_1(BA)_{23} = -1/2, \\ &\frac{1}{r_1}(r_3/2 - r_1 r_2/2 + (r_1 + r_2 r_3)(BA)_{13}) = -1/2 \end{aligned}\right\}$$

and hence

$$(BA)_{13} = \frac{r_1 r_2 - r_1 - r_3}{2(r_1 + r_2 r_3)}, \quad (BA)_{23} = \frac{r_2^2 - r_2 + 1}{2(r_1 + r_2 r_3)}, \quad (BA)_{33} = -1/2. \tag{6.3}$$

Observe that, from (6.1), $(BA)_{13}$ and $(BA)_{23}$ are positive.

Consider the rectilinear motion of a unitary mass under a viscosity effect. Assuming that the system is controlled by two proportional integral regulators, one of them depending on the past, we get the equation

$$\ddot{x}(t) = c\dot{x}(t) + a\int_0^{t-\tau} x(s)ds + b\int_0^{t} x(s)ds$$

and hence

$$\frac{d^3}{dt^3}x(t) = ax(t-\tau) + bx(t) + c\frac{d^2}{dt^2}x(t) \tag{6.4}$$

which is type (2.1) with $K_1(s) = b\delta(s) + a\delta(s-\tau)$, $K_2(s) = 0$, $K_3(s) = c\delta(s)$, where the Heaviside function $\delta(s)$ is defined by

$$\delta(s) := \begin{cases} 0, & s \leq 0, \\ 1, & s > 0. \end{cases}$$

Note that (6.4) is not asymptotically stable if $\tau = 0$.

By means of the transformation

$$x(t-\tau) = x(t) - \tau\dot{x}(\tau) + \frac{d}{dt}\left[\int_t^{t-\tau}(t-\tau-u)\dot{x}(u)du\right], \tag{6.5}$$

(6.4) becomes

$$\frac{d^3}{dt^3}x(t) = (a+b)x(t) - a\tau\frac{d}{dt}x(t) + c\frac{d^2}{dt^2}x(t) + \frac{d}{dt}z(\dot{x}_t),$$

where

$$z(\dot{x}_t) := a \int_{t-\tau}^{t} (u - t + \tau)\dot{x}(u)du$$

and also we get the transformed system

$$\left.\begin{aligned} &\dot{x}_1(t) = x_2(t), \quad \dot{x}_2(t) = x_3(t), \\ &\frac{d}{dt}(x_3(t) - z(\dot{x}_t)) = (a+b)x_1(t) - a\tau x_2(t) + cx_3(t). \end{aligned}\right\}$$

By using the notation of (2.8), the companion matrix A for the auxiliary system without delay is such that $r_1 = a + b$, $r_2 = -a\tau$, $r_3 = c$. By (6.1), A is Hurwitzian if and only if

$$c < 0, \quad -ac\tau + a + b > 0, \quad a + b < 0. \tag{6.6}$$

Observe that the coefficient a must be positive.

Now, by using (2.10) we get

$$\dot{V}_1(x_t) = -|x(t)|^2 - 2\sum_{k=1}^{3}(BA)_{3,k}\left(x_k^2(t)\alpha + a\int_{t-\tau}^{t}(u - t + \tau)x_2^2(u)du\right),$$

where $\alpha := \int_{t-\tau}^{t} a(u - t + \tau)du = a\tau^2/2$, B is the solution of (2.9) with $C := I$, and the elements $(BA)_{3,k}$ for $k = 1, 2, 3$ are given in (6.3). Note that this transformation does not require the condition $\alpha < 1$ (see proof of Thm. 2.1). By defining

$$V_2(x_t) := \sum_{k=1}^{3}|(BA)|_{3,k}\int_{t-\tau}^{t} a\frac{(u - t + \tau)^2}{2}x_2^2(u)du$$

we easily obtain the following inequality for the full derivative of $V := V_1 + V_2$:

$$\dot{V}(x_t) \leq -|x(t)|^2 + \alpha\sum_{k=1}^{3}|(BA)_{3,k}|(x_k^2 + x_2^2).$$

By the Sylvester criterium, $\dot{V}$ is negative definite if and only if

$$(BA)_{13} + 2(BA)_{23} + (BA)_{33} < 1/\alpha$$

and hence, by (6.3), we get the further condition

$$b > \frac{a\tau^2(2a^2\tau^2 + 2a\tau + 2 - a^2\tau - c - ac\tau) - 4a + 4ac\tau}{4 + a^2\tau^3}. \tag{6.7}$$

We can conclude that, by Thm. 2.1, if (6.6) and (6.7) hold, then (6.4) is asymptotically stable.

On the other hand, by considering another transformation for the delayed term $x(t-\tau)$ in (6.4),

$$x(t-\tau) = x(t) - \tau\dot{x}(t) + \frac{\tau^2}{2}\ddot{x}(t) + \frac{d}{dt}\int\limits_t^{t-\tau} \frac{(t-\tau-u)^2}{2}\ddot{x}(u)du,$$

we get

$$\frac{d^3}{dt^3}x(t) = (a+b)x(t) - a\tau\frac{d}{dt}x(t) + \left(c + \frac{a\tau^2}{2}\right)\frac{d^2}{dt^2}x(t) + \frac{d}{dt}z(\dot{x}_t),$$

where

$$z(\dot{x}_t) := a\int\limits_{t-\tau}^{t} \frac{(u-t+\tau)^2}{2}\dot{x}(u)du.$$

Since $r_1 = a+b$, $r_2 = a\tau$, $r_3 = c + a\tau^2/2$, A is Hurwitzian if and only if

$$c + \frac{a\tau^2}{2} < 0, \quad -a\tau\left(c + \frac{a\tau^2}{2}\right) + a + b > 0, \quad a+b < 0. \tag{6.8}$$

Observe that the coefficient a is still positive. In this case $\alpha := a\int\limits_{t-\tau}^{t}((u-t+\tau)^2/2)du$, and hence we must require

$$\frac{a\tau^3}{6} < 1. \tag{6.9}$$

The condition for the negative definiteness of $\dot{V}$ is

$$(BA)_{13} + (BA)_{23} + 2(BA)_{33} < 1/\alpha$$

and so we get

$$\begin{aligned} b > a\tau^3(11 - 12c\tau + 13/2a\tau^2 - a\tau - a - a^2\tau^2 + \\ + a^2\tau + c + 2ac\tau + a^2\tau^3)(a\tau^3 - a^2\tau^4 - 12)^{-1}. \end{aligned} \tag{6.10}$$

Thus (6.4) is asymptotically stable if (6.8), (6.9), and (6.10) hold.

In Figs. 9.6.1 and 9.6.2, using MATLAB, we have fixed some values for the parameters of viscosity c and the delay τ and we have represented the portions of the asymptotic stability region for coefficients a and b, obtained respectively by (6.6)–(6.7) (see horizontal lines) and (6.8)–(6.10)

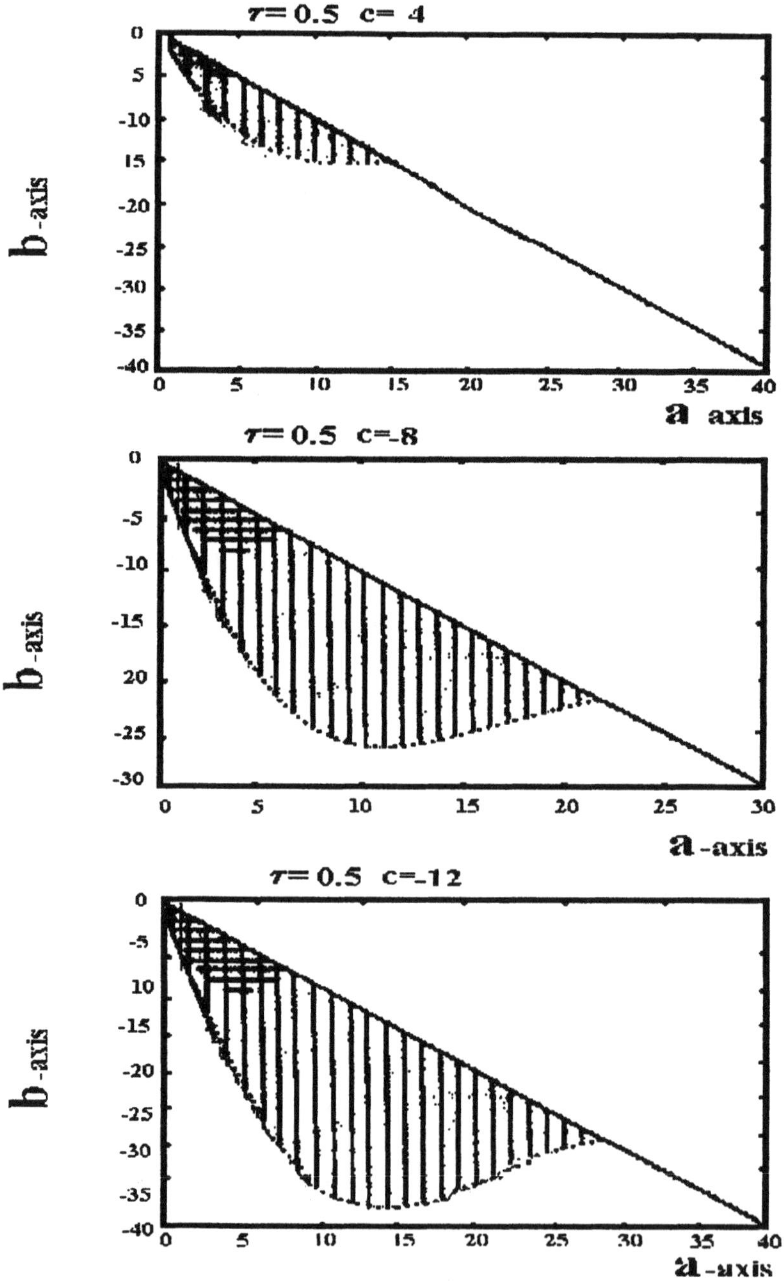

Figure 9.6.1. Stability regions determined by conditions (6.6), (6.7) (horizontal lines), and (6.8), (6.9), (6.10) (vertical lines).

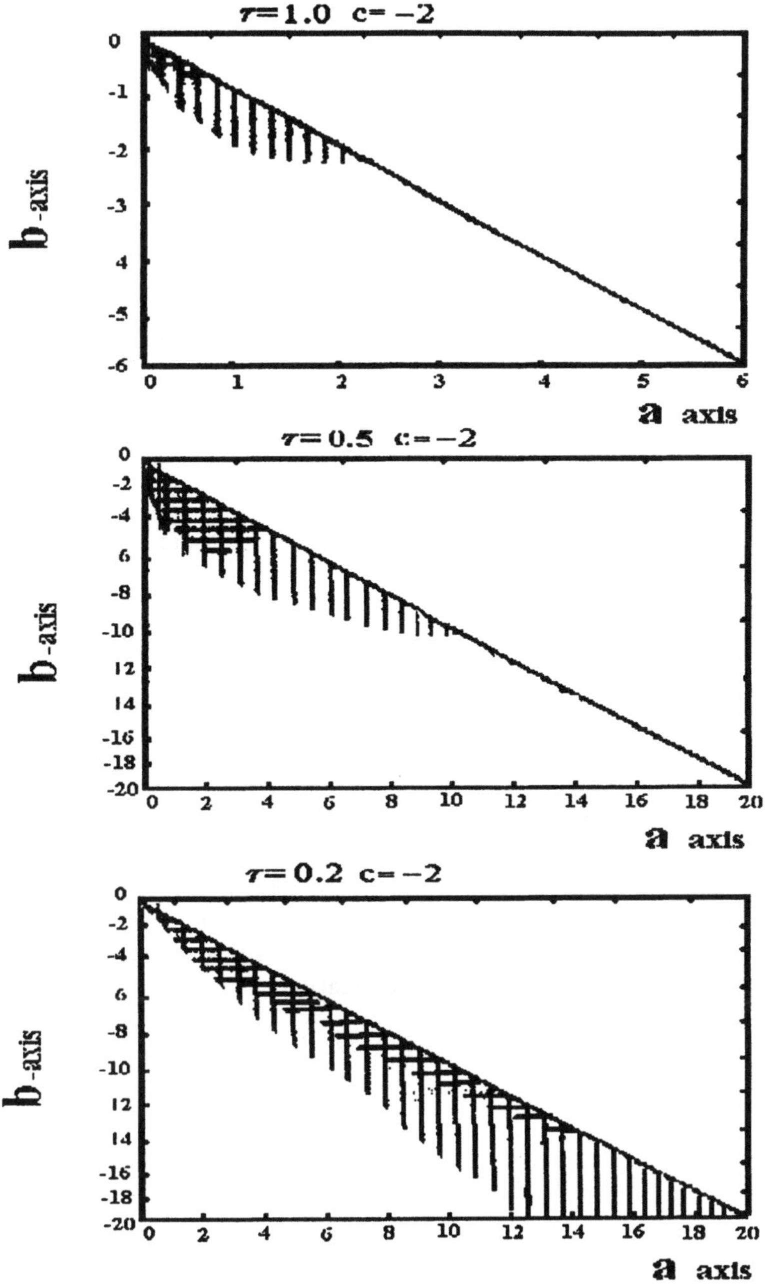

Figure 9.6.2. Stability regions determined by conditions (6.6), (6.7) (horizontal lines), and (6.8), (6.9), (6.10) (vertical lines).

(see vertical lines). Observe that in general by different transformations we get different conditions for the parameters involved, and thus we can enlarge the stability regions.

7. MATRIX RICCATI EQUATIONS IN STABILITY OF NDEs

For linear NDEs with constant coefficients we can formulate stability conditions in terms of the positive definite solutions of some auxiliary matrix Riccati equations. This approach to stability of RDEs was described in Ch. 6. Similar to Ch. 6 various Riccati equations are generated by the different Liapunov functions obtained by using the procedure from Sec. 1.

Let us illustrate this approach for NDEs of the form

$$\frac{d}{dt}[x(t) + B\int_{t-h}^{t} x(s)ds] = Ax(t), \quad x \in \mathbf{R}^n. \tag{7.1}$$

Here A, B are constant matrices, $h \geq 0$. Consider the functional $V := V_1 + V_2$:

$$V_1(x_t) := [x(t) + B\int_{t-h}^{t} x(s)ds]'P[x(t) + B\int_{t-h}^{t} x(s)ds],$$

$$V_2(x_t) := \int_{t-h}^{t} (s - t + h)x'(s)Rx(s)ds.$$

Using this functional and Thm. 1.1 we obtain that Eq. (7.1) is asymptotically stable if $h \parallel B \parallel < 1$ for at least one matrix norm $\parallel \cdot \parallel$ and for some symmetric matrices $Q > 0, R > 0$ there is a symmetric positive definite solution P of the Riccati matrix equation

$$A'P + PA + hR + hA'PBR^{-1}B'PA = -Q.$$

CHAPTER 10

STABILITY OF STOCHASTIC FUNCTIONAL DIFFERENTIAL EQUATIONS

1. STATEMENT OF THE PROBLEM

1.1. *DEFINITIONS OF STABILITY*

Here we will consider the Itô type SRDE

$$\left.\begin{aligned} dx(t) &= a_1(t, x_t)dt + a_2(t, x_t)d\xi(t), \quad t \geq t_0, \\ x_t(\theta) &= x(t+\theta), \quad -h \leq \theta \leq 0, \quad x : J_x \to \mathbf{R}^n. \end{aligned}\right\} \tag{1.1}$$

Here $\xi : [t_0, \infty) \to \mathbf{R}^l$ is the standard Wiener process, and the continuous functionals a_1, a_2 are defined on $[t_0, \infty) \times C[-h, 0]$. The initial condition for (1.1) is

$$x_{t_0} = \phi. \tag{1.2}$$

The initial function ϕ is a stochastic process with continuous trajectories, independent of the σ–algebra $B_{t_0\infty}(d\xi)$, where $B_{t_1t_2}(d\xi)$ is the minimal σ–algebra generated by the random variables $\xi(s) - \xi(s)$ for arbitrary s, t : $t_1 \leq t \leq s \leq t_2$. A solution of the problem (1.1)–(1.2) is a process $t \mapsto x(t)$ such that $B_{t\infty}(d\xi)$ is independent of $B_{-\infty t}(x) \bigcup B_{t_0 t}(d\xi), t \geq t_0$, and such that with probability 1,

$$x(t) = \phi(t_0) + \int_{t_0}^{t} a_1(s, x_s)ds + \int_{t_0}^{t} a_2(s, x_s)d\xi(s).$$

Here, the last term is an Itô's stochastic integral.

Similarly, the inequality

$$dx(t) \leq a_1(t, x_t)dt + a_2(t, x_t)d\xi(t)$$

means that for arbitrary $t, t_1 : t \geq t_1 \geq t_0$, we have with probability 1:

$$x(t) \leq x(t_1) + \int_{t_1}^{t} a_1(s, x_s)ds + \int_{t_1}^{t} a_2(s, x_s)d\xi(s),$$

where this inequality is interpreted componentwise.

As usual in stability theory one may assume that $a_{1,2}(t,0) \equiv 0$. Sometimes we will write $t \mapsto x(t;\phi)$ for the solution of (1.1)–(1.2).

Definition 1.1 The trivial solution of (1.1)–(1.2) is said to be

1) *p–stable* $(0 < p < \infty)$ if for each $\varepsilon > 0$ there is a $\delta > 0$ such that $x(t;\phi)$ exists for $t_0 \le t < \infty$ and $\mathcal{E}|x(t;\phi)|^p < \varepsilon (t_0 \le t < \infty)$, if only

$$\sup_{-h\le\theta\le 0} \mathcal{E}|\phi(\theta)|^p < \delta(\varepsilon), \qquad \sup_{-h\le\theta\le 0} \mathcal{E}|\phi(\theta)|^{2p} < \infty; \tag{1.3}$$

2) *asymptotically p–stable* if it is p–stable and there is an $\varepsilon > 0$ such that for all initial functions satisfying (1.3) we have

$$\lim_{t\to\infty} \mathcal{E}|x(t;\phi)|^p = 0 \tag{1.4}$$

(the set of all ϕ for which the last condition holds is called the domain of attraction of the trivial solution);

3) *exponentially p–stable* if there are positive constants C_1, C_2 such that

$$\mathcal{E}|x(t;\phi)|^p \le C_1 \sup_{-h\le\theta\le 0} \mathcal{E}|\phi(\theta)|^p e^{-C_2 t}, \quad t_0 \le t < \infty,$$

for any ϕ satisfying the second inequality in (1.3).

For $p = 2$ we obtain the definitions of *mean–square stability, asymptotical mean–square stability* and *exponential mean–square stability;*

4) *stable in probability* if for arbitrary $\varepsilon_1 > 0, \varepsilon_2 > 0$ there is a $\delta(\varepsilon_1, \varepsilon_2) > 0$ such that

$$\mathcal{P}\left\{ \sup_{t_0\le t<\infty} |x(t,\phi)| \le \varepsilon_1 \right\} \ge 1 - \varepsilon_2$$

for any initial function ϕ satisfying with probability 1 inequality

$$\sup_{-h\le\theta\le 0} |\phi(\theta)| \le \delta(\varepsilon_1, \varepsilon_2);$$

5) *asymptotically stable in probability* if it is stable in probability and for any positive λ_1, λ_2, and arbitrary initial function $\phi : \| \phi \| \le H$ there is $T(H, \lambda_1, \lambda_2)$ such that

$$\mathcal{P}\left\{ \sup_{t\ge T(H,\lambda_1,\lambda_2)} |x(t,\phi)| \le \lambda_1 \right\} \ge 1 - \lambda_2.$$

Without especially saying so, we assume in this chapter that the functionals a_i satisfy the following conditions:

$$|a_i(t,\psi) - a_i(t,\chi)|^2 \le \int_{-h}^{0} |\psi(\theta) - \chi(\theta)|^2 dr_i(\theta),$$

where the $r_i(\theta)$ are nondecreasing bounded functions.

All definitions above can be immediately generalized to equations with unbounded delay.

1.2. *ITÔ'S FORMULA*

In many studies related to stability and control problems for SRDEs it is useful to have the following Itô's formula [257], which gives an expression for the stochastic differential of the process $t \mapsto \eta(t) := u(t, x(t))$. Here, $t \mapsto x(t)$ is a solution of (1.1), (1.2) and the function $u : [t_0, \infty) \times \mathbf{R}^n \to \mathbf{R}$ has continuous partial derivatives u_t, u_x, u_{xx}. Then

$$d\eta(t) = \Big[u_t(t, x(t)) + u'_x(t, x(t))a_1(t, x_t) + \\ + \frac{1}{2}\mathrm{Tr}(a_2(t, x_t)a'_2(t, x_t)u_{xx}(t, x(t)))\Big] dt + u'_x(t, x(t))a_2(t, x_t)d\xi(t), \tag{1.5}$$

where Tr denotes the matrix trace and the prime denotes transposition.

2. LIAPUNOV'S DIRECT METHOD

2.1. *ASYMPTOTIC STABILITY*

Certain stability conditions for SRDEs can be stated in terms of Liapunov functionals, similar to the appropriate theorems for deterministic FDEs. In the sequel C_i are positive numbers, not always the same on each occurrence.

Theorem 2.1 Suppose there is a continuous functional $V : [t_0, \infty) \times C[-h, 0] \to \mathbf{R}$ *such that for any solution of (1.1), (1.2) the following inequalities hold:*

$$V(t, x_t) \geq C_1|x(t)|^2, \tag{2.1}$$

$$\mathcal{E}V(t, x_t) \leq C_2 \sup_{-h \leq \theta \leq 0} \mathcal{E}|x(t+\theta)|^2, \tag{2.2}$$

while for arbitrary $t \geq t_0, s \geq t$,

$$\mathcal{E}[V(s, x_s) - V(t, x_t)] \leq -C_3 \int_t^s \mathcal{E}|x(\tau)|^2 d\tau. \tag{2.3}$$

Then the trivial solution of (1.1) is asymptotically mean–square stable.

Proof By (2.3) we have

$$\mathcal{E}V(t, x_t) \leq \mathcal{E}V(t_0, \phi) - C_3 \int_{t_0}^t \mathcal{E}|x(s)|^2 ds.$$

It follows from (2.1), (2.2) that

$$C_1\mathcal{E}|x(t)|^2 \leq C_2 \sup_{-h\leq\theta\leq 0} \mathcal{E}|\phi(\theta)|^2.$$

This proves stability.

We verify (1.4). Note that by the previous two inequalities,

$$\int_{t_0}^{\infty} \mathcal{E}|x(t)|^2 dt < \infty, \quad \sup_{t\geq t_0} \mathcal{E}|x(t)|^2 < \infty. \tag{2.4}$$

Applying Itô's formula to the function $|x(t)|^2$ we find

$$d|x(t)|^2 = 2x'(t)dx(t) + \mathrm{Tr}[a_2(t,x_t)a_2'(t,x_t)]dt.$$

This implies that for arbitrary points $t_1, t_2 : t_0 \leq t_1 \leq t_2$,

$$\mathcal{E}[|x(t)|^2 - |x(t_1)|^2] = 2\mathcal{E}\int_{t_1}^{t_2} [x'(t)a_1(t,x_t) + \mathrm{Tr}(a_2(t,x_t)a_2'(t,x_t))]dt.$$

By (2.4) there is a constant C_4 such that

$$2\mathcal{E}|x'(t)a_1(t,x_t)| \leq \mathcal{E}[|x(t)|^2 + |a_1(t,x_t)|^2] \leq$$

$$\leq \mathcal{E}\left[|x(t)|^2 + \int_{-h}^{0} |x(t+\theta)|^2 dr_1(\theta)\right] \leq C_4,$$

$$\mathcal{E}|\mathrm{Tr}[a_2(t,x_t)a_2'(t,x_t)]| \leq \mathcal{E}\int_{-h}^{0} |x(t+\theta)|^2 dr_2(\theta) \leq C_4.$$

Hence,

$$|\mathcal{E}|x(t_2)|^2 - \mathcal{E}|x(t_1)|^2| \leq C_4(t_2 - t_1),$$

i.e. the function $\mathcal{E}|x(t)|^2$ satisfies the Lipschitz condition. This and (2.4) imply (1.4) with $p = 2$. □

2.2. *EXAMPLES*

Example 2.1 Consider the stochastic scalar equation

$$dx(t) = \left[-a(t)x(t) + \int_{-h}^{0} x(t+\theta)dr(\theta)\right] dt + b(t)x(t)d\xi(t), \tag{2.5}$$

where $\theta \mapsto r(\theta)$ is a function of bounded variation on $[-h, 0]$ and a, b are continuous functions. Consider the functional

$$\psi \mapsto V(\psi) = \psi^2(0) + \int_{-h}^{0} \left[\int_{\theta}^{0} \psi^2(\tau) d\tau \right] |dr(\theta)|.$$

For the stochastic differential of the functional V we find

$$dV(x_t) \leq -2x^2(t)\alpha(t)dt + 2b(t)x^2(t)d\xi(t),$$

where

$$\alpha(t) := a(t) - \int_{-h}^{0} |dr(\theta)| - \frac{1}{2}b^2(t).$$

So, by Thm. 2.1 the trivial solution of (2.5) is asymptotically mean–square stable if $\inf_{t \geq t_0} \alpha(t) > 0$. □

In this example the stability is due to the instantaneous negative feedback $-a(t)x(t)$. However, in certain cases one can use negative feedback with delay to stabilize a system, even if the system has only positive instantaneous feedback.

Example 2.2 Consider the stochastic scalar equation

$$dx(t) = -bx(t-h)dt + cx(t)d\xi(t), \quad t > t_0,$$

with constants $b, c, h > 0$.

Sufficient conditions for asymptotic mean–square stability of this equation are:

$$0 < bh < 1, \quad b(1 - bh) > \frac{c^2}{2}.$$

To prove this one can use the functional

$$V(\psi) = \left[\psi(0) - b \int_{-h}^{0} \psi(\theta) d\theta \right]^2 + b^2 \int_{-h}^{0} ds \int_{s}^{0} \psi^2(\theta) d\theta.$$

Using Itô's formula we find

$$dV(x_t) = 2\left[x(t) - b\int_{t-h}^{t} x(\theta)d\theta\right][dx(t) - bx(t)dt + bx(t-h)dt] +$$

$$+\left[c^2x^2(t) + b^2hx^2(t) - b^2\int_{t-h}^{t} x^2(\theta)d\theta\right]dt =$$

$$= 2\left[x(t) - b\int_{t-h}^{t} x(\theta)d\theta\right][cx(t)d\xi(t) - bx(t)dt] +$$

$$+\left[c^2x^2(t) + b^2hx^2(t) - b^2\int_{t-h}^{t} x^2(\theta)d\theta\right]dt.$$

However,

$$2b^2x(t)\int_{t-h}^{t} x(\theta)d\theta \le b^2\left[hx^2(t) + \int_{t-h}^{t} x^2(\theta)d\theta\right].$$

Therefore,

$$dV(x_t) \le 2c\left[x(t) - b\int_{t+h}^{t} x(\theta)d\theta\right]x(t)d\xi(t) - [2b(1-bh) - c^2]x^2(t)dt.$$

Integration of both parts of this inequality from $s \in [t_0, t]$ to t and taking the expectation yields

$$\mathcal{E}[V(x_t) - V(x_s)] \le -[2b(1-bh) - c^2]\int_{s}^{t} \mathcal{E}x^2(\tau)d\tau.$$

This inequality means that

$$\mathcal{E}V(x_t) \le \mathcal{E}V(x_{t_0}), \quad t \ge t_0.$$

Therefore

$$\mathcal{E}\left[x(t) - b\int_{t-h}^{t} x(\theta)d\theta\right]^2 \le \mathcal{E}V(x_{t_0}), \quad \int_{t_0}^{\infty} \mathcal{E}x^2(s)ds < \infty.$$

This and the condition $bh < 1$ implies mean–square stability, since

$$\sup_{t \geq t_0} \mathcal{E}x^2(t) \leq C_1 \sup_{-h \leq \theta \leq 0} \mathcal{E}\phi^2(\theta).$$

Note also that the function $t \mapsto \mathcal{E}x^2(t)$ satisfies the Lipschitz condition (see the proof of Thm. 2.1). Together with the relation

$$\int_{t_0}^{\infty} \mathcal{E}x^2(s)ds < \infty$$

this implies asymptotic mean–square stability, since

$$\lim_{t \to \infty} \mathcal{E}x^2(t) = 0. \quad \square$$

Example 2.3 We will derive mean–square stability conditions for the equation

$$\left.\begin{aligned} dx(t) = \left[\sum_{j=0}^{r} K_j x(t-h_j) + a(t, x_t)\right] dt + b(t, x_t)d\xi(t), \\ t \geq t_0, \quad x : J_x \to \mathbf{R}^n. \end{aligned}\right\} \tag{2.6}$$

Here the h_j are nonnegative constants, $h_0 = 0; a(t, x_t) = a(t, x(t), x(t - \tau_1(t)), \ldots, x(t - \tau_m(t))); b(t, x_t) = b(t, x(t), x(t - \theta_1(t)), \ldots, x(t - \theta_l(t)))$. The matrices K_j are assumed to be constant, and the nonnegative delays $\tau_i(t), \theta_j(t)$ are continuously differentiable and such that

$$0 \leq \dot{\tau}_i(t) \leq \alpha_i < 1, \quad i = 1, \ldots, m,$$

$$0 \leq \dot{\theta}_j(t) \leq \beta_j < 1, \quad j = 1, \ldots, l.$$

The functions a, b are assumed to be continuous and to satisfy the inequalities

$$|a(t, z_0, \ldots z_m)|^2 \leq \sum_{i=0}^{m} a_i^2 |z_i|^2, \quad |b(t, z_0, \ldots z_l)|^2 \leq \sum_{j=0}^{l} b_j^2 |z_j|^2,$$

where a_i and b_i are constants and z_i are arbitrary vectors in $\mathbf{R}^n$.

We introduce the following notations:

$$k_j := |K_j|, \quad K := \sum_{j=1}^{r} K_j, \quad (kh) := \sum_{j=1}^{r} k_j h_j, \quad k := \sum_{j=1}^{r} k_j,$$

$$\gamma_j := |(K_0 + K)' K_j|, \quad (\gamma h) := \sum_{j=1}^{r} \gamma_j h_j,$$

where $|\cdot|$ is the euclidean matrix norm. The nonnegative constants $\overline{a}$ and $\overline{b}$ are defined by the relations

$$\overline{a}^2 := a_0^2 + \sum_{i=1}^{m} a_i^2(1-\alpha_i)^{-1}, \quad \overline{b}^2 := b_0^2 + \sum_{j=1}^{l} b_j^2(1-\beta_j)^{-1}.$$

Theorem 2.2 The trivial solution of (2.6) is asymptotically mean–square stable if there is an $\varepsilon \geq 0$ *such that the matrix*

$$L := K_0 + K + (1+\varepsilon)^{-1}[kI - K + (\gamma h)\varepsilon I] + \frac{1}{2}I\overline{b}^2 +$$

$$+\overline{a}[1 + (1+\varepsilon)^{-1}(kh)\varepsilon]I$$

is negative define. Here I is the identity matrix.

To prove this it suffices to consider the functional

$$V(t,\psi) := |\psi(0)|^2 + \varepsilon\left|\psi(0) + \sum_{j=1}^{r}\int_{-h_j}^{0} K_j\psi(\theta)d\theta\right|^2 +$$

$$+\varepsilon\sum_{j=1}^{r}(\gamma_j + \overline{a}k_j)\int_{-h_j}^{0}(\theta - h_j - t)|\psi(\theta)|^2 d\theta + \sum_{j=1}^{r} k_j \int_{-h_j}^{0}|\psi(\theta)|^2 d\theta +$$

$$+\overline{a}^{-1}\left[1 + \varepsilon(1+(kh))\sum_{i=1}^{m} a_i^2(1-\alpha_i)^{-1}\int_{-\tau_i(t)}^{0}|\psi(\theta)|^2 d\theta\right] +$$

$$+(1+\varepsilon)\sum_{j=1}^{l} b_j^2(1-\beta_j)^{-1}\int_{-\theta_j(t)}^{0}|\psi(\theta)|^2 d\theta.$$

By computation we find for any $t_1 \in [t_0,\infty), t_2 \geq t_1$,

$$\mathcal{E}[V(t_2,x_{t_2}) - V(t_1,x_{t_1})] \leq 2(1+\varepsilon)\int_{t_1}^{t_2} x'(s)Lx(s)ds.$$

Hence, by Thm. 2.1, if the conditions of Thm. 2.2 hold, the trivial solution of (2.6) is asymptotically mean–square stable. □

2.3. *EXPONENTIAL STABILITY*

Similar to Thm. 2.1 we can state sufficient conditions for other types of stability, e.g., exponential p–stability, stability in probability, etc. Here we will consider sufficient conditions for the trivial solution of (1.1) to be exponentially mean–square stable. Our general assumptions are the same as in Subs. 1.1.

Theorem 2.3 Suppose there is a continuous functional $V : [t_0, \infty) \times C[-h, 0] \to \mathbf{R}$ *such that*

$$\left.\begin{aligned} C_1 \parallel \psi \parallel^2 \leq V(t,\psi) \leq C_2 \parallel \psi \parallel^2, \quad t_0 \leq t < \infty, \\ \lim\sup_{\Delta t \to 0+} \frac{1}{\Delta t}[V(t+\Delta t, y_{t+\Delta t}) - V(t,\psi)] \leq -C_3 \parallel \psi \parallel^2, \end{aligned}\right\} \tag{2.7}$$

where y *is a solution of Eq. (1.1) with initial condition* $y_t = \psi \in C[-h, 0]$. *Then the trivial solution of (1.1) is exponentially mean–square stable.*

Note that for an Eq. (1.1) with stationary coefficients (i.e. $a_i(t,\psi) \equiv a(\psi)$), the conditions (2.7) in Thm. 2.3 are not only sufficient, but also necessary for exponential stability [484].

2.4. *STABILITY IN THE FIRST APPROXIMATION*

The method of first approximation for investigating stability is based on the linearized initial equation. Under certain additional assumptions, stability of the initial equation follows from that of the linearized initial equation. To illustrate this method we consider the following equation of the form (1.1):

$$dx(t) = Ax(t)dt + a_2(t, x_t)d\xi(t), \quad t \geq t_0, \quad x : J_x \to \mathbf{R}. \tag{2.8}$$

Here, A is a constant matrix, and a_2 is a continuous matrix functional satisfying

$$|a_2(t,\psi)|^2 \leq \int_{-h}^{0} |\psi(\theta)|^2 dr(\theta), \tag{2.9}$$

where r is a nondecreasing function. As first approximation we take the deterministic equation

$$\dot{x}(t) = Ax(t). \tag{2.10}$$

We assume that (2.10) is uniformly asymptotically stable, and show that the trivial solution of (2.8) is asymptotically mean–square stable if $r_0 := r(0) - r(-h)$ is sufficiently small.

Our assumption implies the existence of two positive define symmetric matrices P and Q such that

$$A'P + PA = -Q. \tag{2.11}$$

Now we take the following functional:

$$V(t,\psi) := \psi'(0)P\psi(0) + |P| \int_{-h}^{0} dr(\theta) \int_{\theta}^{0} |\psi(\tau)|^2 d\tau,$$

where $|\cdot|$ is the euclidean matrix norm. The conditions (2.1)–(2.2) in Thm. 2.1 clearly hold. Using Itô's formula and (2.11) we find

$$dV(t,x_t) = [dx'(t)]Px(t) + x'(t)Px(t)dt +$$

$$+ \left[\mathrm{Tr}(Pa_2(t,x_t)a_2'(t,x_t)) - |P| \int_{-h}^{0} |x(t+\theta)|^2 dr(\theta) + r_0|P||x(t)|^2 \right] dt \leq$$

$$\leq d\xi'(t)a_2'(t,x_t)Px(t) + x'(t)Pa_2(t,x_t)d\xi(t) +$$

$$+[r_0|P||x(t)|^2 - x'(t)Qx(t)]dt. \tag{2.12}$$

Since all moments of the solution of (2.8) are bounded on any finite time interval, (2.12) implies that (2.3) holds. Hence, by Thm. 2.1 the system (2.8) is asymptotically mean–square stable.

This assertion about stability in the first approximation can be generalized to equation with infinite delay, provided that

$$\int_{-\infty}^{0} |\theta| dr(\theta) < \infty.$$

2.5. *STABILITY UNDER PERSISTENT DISTURBANCES*

The idea of this method is similar to the method of first approximation. Namely, a certain part of the equation is interpreted as being undisturbed, and the other part as the disturbance. Assuming that the undisturbed equation is stable and its righthand side majorizes, in a certain sense, the disturbances, this method makes it possible to obtain stability conditions.

Using this approach we will derive conditions for the mean–square stability of a scalar second–order system with constant coefficients a_0, a_1:

$$d\dot{x}(t) = -[a_0\dot{x}(t) + a_1x(t)]dt + a_2(t,\dot{x}_t)d\xi(t). \tag{2.13}$$

In this case the unperturbed system is

$$\ddot{x}(t) = -[a_0\dot{x}(t) + a_1x(t)],$$

and necessary and sufficient conditions for its asymptotic stability are $a_0 > 0, a_1 > 0$. The intensity $a_2(t, \dot{x}_t)$ of the disturbances is supposed to satisfy a condition like (2.9):

$$[a_2(t,\psi)]^2 \le \int_{-h}^{0} [\psi(\theta)]^2 dr(\theta), \tag{2.14}$$

where r is a nondecreasing function. Sufficient conditions for stability of the disturbed system take the form

$$a_0 > r_0 := r(0) - r(-h), \quad a_1 > 0. \tag{2.15}$$

To prove this one can use Thm. 2.1, applied to the functional $V : C^1[-h,0] \to \mathbf{R}$:

$$V(\psi) := 2a_1\psi^2(0) + \dot{\psi}^2(0) + 2\int_{-h}^{0} dr(\theta) \int_{\theta}^{0} \dot{\psi}^2(\tau)d\tau + [\dot{\psi}(0) + a_0\psi(0)]^2. \tag{2.16}$$

By Itô's formula, (2.13) and (2.14) we obtain

$$dV(x_t) = 2\dot{x}(t)d\dot{x}(t) +$$

$$+ \left[4a_1 x(t)\dot{x}(t) + 2a_2^2(t,x_t) + 2r_0\dot{x}^2(t) - 2\int_{-h}^{0} \dot{x}^2(t+\theta)dr(\theta)\right] dt +$$

$$+2[\dot{x}(t) + a_0 x(t)][-a_1 x(t)dt + a_2(t,x_t)d\xi(t)] \le$$

$$\le -2[(a_0 - r_0)\dot{x}^2(t) + a_1 a_0 x^2(t)]dt + 2[2\dot{x}(t) + a_0 x(t)]a_2(t,x_t)d\xi(t).$$

This, (2.15) and (2.16) imply that the conditions of Thm. 2.1 hold. Hence the trivial solution of (2.13) is asymptotically mean–square stable.

3. BOUNDEDNESS OF MOMENTS OF SOLUTIONS

3.1. *GENERAL CONDITIONS FOR BOUNDEDNESS OF MOMENTS*

By slightly modifying the proof of Thm. 2.1 we can justify the following conditions for boundedness of moments of solutions of (1.1) ([289.3]).

Theorem 3.1 Suppose there is a continuous functional $V : [t_0, \infty) \times C[-h, 0] \to \mathbf{R}^n$ *such that for any solution* x *of (1.1), (1.2) the following inequalities hold:*

$$V(t,x_t) \ge C_1|x(t)|^p, \quad \mathcal{E}V(t,x_t) \le C_2[1 + \sup_{-h\le\theta\le 0} \mathcal{E}|x(t+\theta)|^p], \tag{3.1}$$

and for any $t \geq t_0, s \geq t$,

$$\mathcal{E}[V(s,x_s) - V(t,x_t)] \leq -C_3 \int\limits_t^s \mathcal{E}|x(\tau)|^p d\tau + C_4(s-t). \tag{3.2}$$

(Here $p \geq 2$, and all $C_i > 0$.) Then

$$\mathcal{E}|x(t,\phi)|^p \leq \left[1 + C_4 C_3^{-1} + \sup_{-h\leq\theta\leq 0} \mathcal{E}|\phi(\theta)|^p\right] C_2 C_1^{-1}.$$

We will consider some applications of Thm. 3.1 to concrete systems.

3.2. *SCALAR SRDE*

We will obtain a condition for the boundedness of the second moment of solutions of the scalar equation

$$dx(t) = [-a_0(t)x(t) + a_1(t,x_t)]dt + a_2(t,x_t)d\xi(t). \tag{3.3}$$

Theorem 3.2 *Suppose that a_0 is a bounded measurable function, and*

$$[a_i(t,\psi)]^2 \leq q_i + \int\limits_{-h}^0 \psi^2(\theta) dr_i(\theta), \quad i = 1,2, \tag{3.4}$$

where q_i are constants and r_i nondecreasing functions. If the inequality

$$\inf_{t_0\leq t<\infty} \left[a_0(t) - \sqrt{\alpha_{01}} - \frac{1}{2}\alpha_{02}\right] =: q_3 > 0, \quad \alpha_{ij} := \int\limits_{-h}^0 |\theta|^i |dr_j(\theta)|, \tag{3.5}$$

holds, then the solution x of (3.3), (1.2) satisfies

$$\mathcal{E}x^2(t) \leq \overline{C}_1 \left[1 + \sup_{-h\leq\theta\leq 0} \mathcal{E}\phi^2(\theta)\right], \quad t_0 \leq t < \infty, \tag{3.6}$$

where the constant $\overline{C}_1$ is independent of ϕ.

Proof Consider the functional

$$V(\psi) = \psi^2(0) + \int\limits_{-h}^0 dr_3(\theta) \int\limits_\theta^0 \psi^2(\tau) d\tau, \quad r_3 := r_2 + \alpha_{01}^{-1/2} r_1.$$

Itô's formula and (3.3) imply that the stochastic differential dV satisfies

$$dV(x_t) = \Big[a_2^2(t,x_t) + x^2(t)(\sqrt{\alpha_{01}} + \alpha_{02} - 2a_0(t)) - \\ - \int_{-h}^{0} x^2(t+\theta)dr_3(\theta) + 2x(t)a_1(t,x_t)\Big] dt + 2x(t)a_2(t,x_t)d\xi(t). \tag{3.7}$$

We estimate the individual summands in (3.7). By (3.4) we have

$$|2x(t)a_1(t,x_t)| \le \sqrt{\alpha_{01}}x^2(t) + \alpha_{01}^{-1/2}a_1^2(t,x_t) \le$$

$$\le \sqrt{\alpha_{01}}x^2(t) + \alpha_{01}^{-1/2}\left[q_1 + \int_{-h}^{0} x^2(t+\theta)dr_1(\theta)\right].$$

Using for a_1 the estimate (3.4), we find

$$dV(x_t) \le [-2q_3x^2(t) + q_4]dt + 2x(t)a_2(t,x_t)d\xi(t), \\ q_4 := q_2 + q_1\alpha_{01}^{-1/2}. \tag{3.8}$$

Integrating both sides of this inequality from $t \ge t_0$ to $s \ge t$ and computing the expectation, by (3.5) we find that (3.2) is valid with $p = 2$. The inequalities (2.1) and (3.1) with $p = 2$ also hold. Hence, by Thm. 3.1, the estimate (3.6) holds. □

Now we will obtain conditions for boundedness of moments without the assumption of an instantaneous negative feedback. Consider the scalar equation

$$dx(t) = \left[-\int_{-h}^{0} x(t+\theta)dr_0(\theta) + a_1(t,x_t)\right] dt + a_2(t,x_t)d\xi(t).$$

The functionals a_i in this expression are the same as in (3.3), and r_0 is a function of bounded variation. In the sequel we put

$$\beta_{ij} = \int_{-h}^{0} |\theta|^i dr_j(\theta).$$

Theorem 3.3 The estimate (3.6) holds if

$$C_1 := (1-\alpha_{10})\beta_{00} - (1+\alpha_{10})\sqrt{\alpha_{01}} - \frac{1}{2}\alpha_{02} > 0, \quad \alpha_{10} < \frac{1}{2}. \tag{3.9}$$

Proof Consider the functional $V : C[-h,0] \to \mathbf{R}$:

$$V(\psi) := \left[\psi(0) - \int_{-h}^{0} dr_0(\theta) \int_{\theta}^{0} \psi(\tau)d\tau\right]^2 +$$

$$+(\beta_{00} + \sqrt{\alpha_{01}}) \int_{-h}^{0} |dr_0(\theta)| \int_{\theta}^{0} d\tau \int_{\tau}^{0} \psi^2(\tau_1)d\tau_1 +$$

$$+\int_{-h}^{0} [dr_2(\theta) + (1+\alpha_{10})\alpha_{10}^{-1/2} dr_1(\theta)] \int_{\theta}^{0} \psi^2(\tau)d\tau.$$

Using Itô's formula we obtain, similarly to (3.8),

$$dV(x_t) \leq [-C_1 x^2(t) + C_2]dt+$$

$$+2\left[x(t) - \int_{-h}^{0} dr_0(\theta) \int_{t+\theta}^{t} x(\tau)d\tau\right] a_2(t, x_t)d\xi(t), \tag{3.10}$$

$$C_2 := q_2 + q_1(\alpha_{01}^{-1/2} + \alpha_{10}).$$

Relations (3.9), (3.10) show that for some constant C we get

$$\mathcal{E}V(x_t) \leq C\left(1 + \sup_{-h\leq\theta\leq 0} [\phi(\theta)]^2\right) =: C_3.$$

Hence

$$\mathcal{E}\left[x(t) - \int_{-h}^{0} dr_0(\theta) \int_{t+\theta}^{t} x(\tau)d\tau\right]^2 \leq C_3.$$

This implies that

$$(1-\alpha_{10})\mathcal{E}x^2(t) \leq C_3 + \int_{-h}^{0} |dr_0(\theta)| \int_{t+\theta}^{t} \mathcal{E}x^2(\tau)d\tau.$$

Putting $\rho(t) = \max_{t_0\leq\tau\leq t} \mathcal{E}x^2(\tau)$ we obtain

$$(1-2\alpha_{10})\rho(t) \leq C_3 + \alpha_{10} \sup_{-h\leq\theta\leq 0} \mathcal{E}[\phi(\theta)]^2.$$

This proves Thm. 3.3. □

3.3. *SECOND ORDER SRDE*

Consider the system of equations

$$\dot{x}(t) = y(t),$$

$$dy(t) = \left[-a_0(t)y(t) - \int_{-h}^{0} y(t+\theta)dr_0(\theta) - \int_{-h}^{0} x(t+\theta)dr_1(\theta) \right] dt +$$

$$+ a_2(t, y_t)d\xi(t).$$

Here, $a_0(t)$ is a bounded measurable function, r_i are functions of bounded variation, and the functional a_2 is the same as in (3.3).

Theorem 3.4 Suppose that

$$\beta_{01} > 0, \quad \beta := \inf_{t \geq t_0} [a_0(t) - \alpha_{00} - \alpha_{11} - \alpha_{02}] > 0.$$

Then there is a constant C such that

$$\mathcal{E}[x^2(t) + y^2(t)] \leq C \left[1 + \sup_{-h \leq \theta \leq 0} \mathcal{E}(\phi^2(\theta) + \dot{\phi}^2(\theta)) \right]. \tag{3.11}$$

Proof By Thm. 3.1 it suffices to give a functional V satisfying (2.1), (3.1) and (3.2). Take

$$V(\psi, \chi) := 2\beta_{01}\psi^2(0) + \chi^2(0) + [V_0(\psi, \chi)]^2 +$$

$$+ 2\int_{-h}^{0} dr_2(\theta) \int_{\theta}^{0} \chi^2(\tau)d\tau + \int_{-h}^{0} |dr_0(\theta)| \int_{\theta}^{0} [\beta_{01}\psi^2(\tau) + \chi^2(\tau)]d\tau +$$

$$+ \int_{-h}^{0} |dr_1(\theta)| \int_{\theta}^{0} d\tau_1 \int_{\tau_1}^{0} [\beta_{01}\psi^2(\tau) + \chi^2(\tau)]d\tau,$$

where

$$V_0(\psi, \chi) := \chi(0) + \int_{-h}^{0} \psi(\theta)dr_0(\theta) + \int_{-h}^{0} dr_1(\theta) \int_{\theta}^{0} \psi(\tau)d\tau.$$

Using Itô's formula, the same calculation as in the proof of Thm. 3.2 gives

$$dV(x_t, y_t) \le -2\beta_{01}x^2(t)[a_0(t) - \alpha_{00} - \alpha_{11}]dt +$$

$$+2[q_2 - \beta y^2(t)] + 2a_2(t, y_t)[V_0(x_t, y_t) + y(t)]d\xi(t).$$

Hence, by Thm. 3.1 the estimate (3.11) holds. □

Remark 3.1 The methods for constructing the above functionals can also be used in other situations. E.g., (3.11) holds under the conditions [292.6]

$$a_0(t) \ge 0, \quad q_0 := \alpha_{10} + \frac{1}{2}\alpha_{21} < 1, \quad \beta_{01} > 0,$$

$$\beta_{00} - \beta_{11} > \max\{q_0, (\beta_{01}q_0 + \alpha_{02})(1 - q_0)^{-1}\}.$$

□

We also give certain conditions for the boundedness of the fourth moments of solutions of (3.3). Assume that

$$\inf_{t \ge t_0} a_0(t) > (\alpha_{01})^{1/4} + \frac{\sqrt{3}}{2}(\alpha_{02})^{1/2}.$$

Then for some constant C we have

$$\mathcal{E}x^4(t) \le C\left[1 + \sup_{-h \le \theta \le 0} \mathcal{E}\phi^4(\theta)\right].$$

Remark 3.4 Stability properties of SFDEs were investigated by many authors and we here mention [307, 350, 372, 430.1, 430.2]. Detailed investigation of exponential stability of SRDEs was fulfilled in [350]. Properties of SFDEs solutions with semimartingales were considered in [430.2].

4. CONSTRUCTION OF LIAPUNOV FUNCTIONALS FOR SNDEs

Similar to the deterministic case, the most difficult part of application of the direct Liapunov's method is to construct an appropriate Liapunov functional. Here we describe the procedure which allows us to obtain such functionals in some cases. It is interesting to note that the formal description of the procedure is approximately the same as for deterministic FDEs (see chapters 6 and 9).

4.1. *STATEMENT OF THE PROBLEM*

Consider stochastic equation

$$d(x(t) - G(t, x_t)) = F(t, x_t)dt + \sigma(t, x)d\xi(t), \quad t \ge 0, \tag{4.1}$$

$$x_0 = \phi \in C(-\infty, 0], \quad x_t(\theta) = x(t+\theta), \quad \theta \le 0. \tag{4.2}$$

Here $F, G : [0,\infty)\times C(-\infty,0] \to \mathbf{R}^n$ and $\sigma : [0,\infty)\times C(-\infty,0] \to \mathbf{R}^{n\times m}$ are continuous maps, $\xi : [0,\infty) \to \mathbf{R}^m$ is the standard Wiener process, $G(t,0) \equiv F(t,0) \equiv 0, \sigma(t,0) \equiv 0$. Here $C(-\infty,0]$ is a space of continuous functions equipped with the uniform norm. Also general conditions sufficient for the existence and uniqueness of the solution, hold. Initial process $\phi(\theta)$ belongs to the set H_0 of processes with continuous trajectories, which are independent of $\xi(t), t \ge 0$, as well as

$$\sup_{\theta\le 0} \mathcal{E}|\phi(\theta)|^4 < \infty.$$

Stability definitions of the zero solution of Pr. (4.1), (4.2) follow from those of Subs. 1.1. For the definiteness we shall describe the procedure with respect to the mean square stability. The zero solution of (4.1), (4.2) will be called:

1) *mean square stable* if for any $\varepsilon > 0$ there is a $\delta(\varepsilon) > 0$ such that

$$\mathcal{E}|x(t)|^2 < \varepsilon, \quad t \ge 0 \tag{4.3}$$

if

$$\rho(\phi) := \sup_{\theta\le 0} \mathcal{E}|\phi(\theta)|^2 < \delta(\varepsilon).$$

2) *asymptotically mean square stable* if it is mean–square stable and for all initial functions $\phi \in H_0$ we have

$$\lim_{t\to\infty} \mathcal{E}|x(t)|^2 = 0.$$

Consider also the inequality

$$\mathcal{E}|x(t) - G(t,x_t)|^2 \le f(t), \quad t \ge 0 \tag{4.4}$$

where $f(t) \ge 0$ is a scalar continuous bounded function. Initial conditions for (4.4) have a form (4.2). *Mean square stability* of the zero solution of (4.4) means that for any $\varepsilon > 0$ there is a $\delta(\varepsilon) > 0$ such that inequality (4.3) is valid for any solution of Pr. (4.2), (4.4) under conditions $\sup_{t\ge 0} f(t) < \delta$, $\rho(\phi) < \delta$. If, in addition, $\lim_{t\to\infty} \mathcal{E}|x(t)|^2 = 0$ for any $\phi \in H_0$, and $f(t) \to 0$ as $t \to \infty$, then the zero solution of (4.4) is called *mean square asymptotically stable.*

Similarly to Thm. 2.1 can be proven

Theorem 4.1 Assume that the zero solution of (1.4) is mean square asymptotically stable and there exists a continuous functional $W : [0,\infty)\times C(-\infty,$

0] $\to [0,\infty)$, *satisfying the condition* $\mathcal{E}W(t,\phi) \le C_1\rho(\phi)$, *and for the functional*

$$V(t,\phi) := W(t,\phi) + |\phi(0) - G(t,\phi)|^2 \tag{4.5}$$

the following estimates are valid

$$\mathcal{E}V(t,\phi) \le C_2\rho(\phi), \tag{4.6}$$

$$\mathcal{E}V(t,x_t) - \mathcal{E}V(\tau,x_\tau) \le -C_3 \int_\tau^t \mathcal{E}x^2(s)ds, \quad \tau \le t \tag{4.7}$$

where C_i are some positive constants. Then the zero solution of (4.1) is mean square asymptotically stable.

Remark 4.1 Condition (4.7) can be replaced by

$$LV(t,x_t) \le -C_3\mathcal{E}|x(t)|^2.$$

Here L is a *generator* of Eq. (4.1), defined by

$$LV(t,x_t) := \limsup_{\Delta t \to 0+} \frac{1}{\Delta t}\mathcal{E}[V(t+\Delta t, x_{t+\Delta t}) - V(t,x_t)]$$

4.2. *DESCRIPTION OF THE PROCEDURE*

Let us represent V as a sum $V = V_1 + V_2$. The following steps to construct V_1 and V_2 must be done:

1) Transform Eq. (4.1) so to represent it in the form

$$dz(t,x_t) = (b_1(t,x(t)) + a_1(t,x_t))dt + (b_2(t,x(t)) + a_2(t,x_t))d\xi(t). \tag{4.8}$$

Here $z, a_1 : [0,\infty) \times C(-\infty,0] \to \mathbf{R}^n$, and $a_2 : [0,\infty) \times C(-\infty,0] \to \mathbf{R}^{n\times m}$ are continuous functionals, $b_1 : [0,\infty) \times \mathbf{R}^n \to \mathbf{R}^n$ and $b_2 : [0,\infty) \times \mathbf{R}^n \to \mathbf{R}^{n\times m}$ are continuous functions, and $a_i(t,0) \equiv 0, b_i(t,0) \equiv 0, i = 1,2$.

2) Consider the Itô's stochastic equation without delay

$$dy(t) = b_1(t,y(t))dt + b_2(t,y(t))d\xi(t). \tag{4.9}$$

Assume that the zero solution of Eq. (4.9) is uniformly (with respect to the initial moment) mean square asymptotically stable. Then there exists the Liapunov function $v : [0,\infty) \times \mathbf{R}^n \to [0,\infty)$ such that

$$C_1|y|^2 \le v(t,y) \le C_2|y|^2, \quad L_0v(t,y) \le -C_3|y|^2. \tag{4.10}$$

Here L_0 is a generator of Eq. (4.9) and C_i are some positive constants.

3) Replace the second argument of $v(t, y)$ by the functional $z(t, x_t)$. As a result we obtain the main component V_1 of the functional V. Now, adding to V_1 the functional V_2 in such a way that the requirements of some stability theorem for (4.1) will be satisfied, we obtain the desired functional V.

Like the deterministic case, the various steps of this procedure can be implemented nonuniquely. This nonuniqueness must be used to extend the stability domain

4.3. *SCALAR SNDE*

Let us investigate asymptotic mean square stability of the scalar equation with constant coefficients

$$d(x(t) + cx(t-h)) + (ax(t) + bx(t-h))dt + \sigma x(t-\tau)d\xi(t) = 0. \quad (4.11)$$

According to the procedure let us rewrite (4.11) in the form

$$dz(x_t) = (-ax(t) - bx(t-h))dt - \sigma x(t-\tau)d\xi(t). \quad (4.12)$$

Here

$$z(x_t) := x(t) + cx(t-h).$$

Hence auxiliary equation (4.9) takes the form

$$\dot{y}(t) = -ay(t). \quad (4.13)$$

If $a > 0$ then the Liapunov function $v(t, y) := y^2$ satisfies (4.10). Because of this we have

$$V_1(x_t) := z^2(x_t) = (x(t) + cx(t-h))^2.$$

Further

$$LV_1(x_t) = 2(x(t) + cx(t-h))(-ax(t) - bx(t-h)) + \sigma^2 x^2(t-\tau) \leq$$

$$\leq (|ac+b| - 2a)x^2(t) + (|ac+b| - 2bc)x^2(t-h) + \sigma^2 x^2(t-\tau).$$

In order to satisfy (4.8) we put

$$V_2(x_t) := \rho \int_{t-h}^{t} x^2(s)ds + \sigma^2 \int_{t-\tau}^{t} x^2(s)ds.$$

Here $\rho := \max\{|ac+b| - 2bc, 0\}$. In addition $V := V_1 + V_2$,

$$LV(x_t) \leq -2x^2(t)(a - p - (|ac+b| + \rho)/2), \quad p := \sigma^2/2.$$

Note that inequality (4.4) having a form

$$\mathcal{E}(x(t)+cx(t-h))^2 \le f(t), \quad t \ge 0$$

is asymptotically mean square stable under condition $|c| < 1$. So equation (4.13) is mean square asymptotically stable if

$$a > p + (|ac+b| + \rho)/2, \quad |c| < 1. \tag{4.14}$$

Consider now another way of application of the procedure using the following representation of Eq. (4.11)

$$dz(x_t) = -(a+b)x(t)dt - \sigma x(t-\tau)d\xi(t).$$

Here

$$z(x_t) := x(t) + cx(t-h) - b\int_{t-h}^{t} x(s)ds,$$

$$b_1(x(t)) := -(a+b)x(t), \quad c_1(x_t) := 0,$$

$$b_2(x(t)) := 0, \qquad c_2(x_t) := -\sigma x(t-\tau).$$

Eq. (4.9) in this case is

$$\dot{y}(t) = -(a+b)y(t). \tag{4.15}$$

If $a+b>0$, then the function $v(t,y) := y^2$ satisfies (4.10). Therefore

$$V_1(x_t) := \left(x(t) + cx(t-h) - b\int_{t-h}^{t} x(s)ds\right)^2.$$

So

$$LV_1(x_t) = -2(a+b)\left(x(t) + cx(t-h) - b\int_{t-h}^{t} x(s)ds\right)x(t) + \sigma^2 x^2(t-\tau) \le$$

$$\le [-2(a+b) + (a+b)(|c| + |b|h)]x^2(t) + \sigma^2 x^2(t-\tau) +$$

$$+(a+b)|c|x^2(t-h) + (a+b)|b|\int_{t-h}^{t} x^2(s)ds.$$

From it follows that V_2 equals

$$V_2(x_t) := (a+b)|b| \int_{t-h}^{t} (s-t+h)x^2(s)ds+$$

$$+(a+b)|c| \int_{t-h}^{t} x^2(s)ds + \sigma^2 \int_{t-\tau}^{t} x^2(s)ds.$$

In addition $V := V_1 + V_2$,

$$LV(x_t) \leq -2x^2(t)((a+b)(1-|c|-|b|h)-p), \quad p := \sigma^2/2.$$

Under condition $|c| + |b|h < 1$ inequality

$$\mathcal{E}\left(x(t) + cx(t-h) - b\int_{t-h}^{t} x(s)ds\right)^2 \leq f(t), \quad t \geq 0$$

is mean square asymptotically stable. Hence Eq. (4.11) is asymptotically mean square if

$$a > p/(1-|c|-|b|h) - b, \quad |c| + |b|h < 1. \tag{4.16}$$

Stability domains given by relations (4.14) and (4.16) are shown in Figs. 10.4.1 – 10.4.7. All of them are located to the right from the depicted curves. In Fig. 10.4.1 stability domains were obtained for $p = 0, c = 0.5, h = 0.4$, and in Fig. 10.4.2 for $p = 0.2, c = 0.5, h = 0.4$ where curve 1) corresponds (4.14), and curve 2) – (4.16). Stability domains using both relations (4.14), (4.16) are shown in Figs. 10.4.3 – 10.4.7:

– in Fig. 10.4.3 for $c = 0.5, h = 0.4$ and various values of p: for curve 1) $p = 0$; 2)$p = 0.15$; 3) $p = 0.3$; 4)$p = 0.45$; 5) $p = 0.6$; 6) $p = 0.75$; 7) $p = 0.9$;

– in Fig. 10.4.4 for $c = -0.5$ and the same values of other parameters as in Fig. 10.4.3;

– in Fig. 10.4.5 for $h = 0.2, p = 0.8$: for curve 1) $c = -0.5$; 2) $c = 0.5$;

– in Fig. 10.4.6 for $p = 0.8, c = 0$: for curve 1) $h = 0$; 2) $h = 0.3$; 3) $h = 0.6$; 4)$h = 0.9$; 5) $h = 1.2$;

– in Fig. 10.4.7 for $p = 0.8, c = 0.4$: for curve 1) $h = 0$; 2) $h = 0.3$; 3) $h = 0.6$.

4.4. *NONLINEAR EXAMPLE*

Consider scalar equation

$$d[x(t) - cx(t-h)] = (b - ax(t))(x(t) - cx(t-h))(dt + \sigma d\xi(t)), \tag{4.17}$$

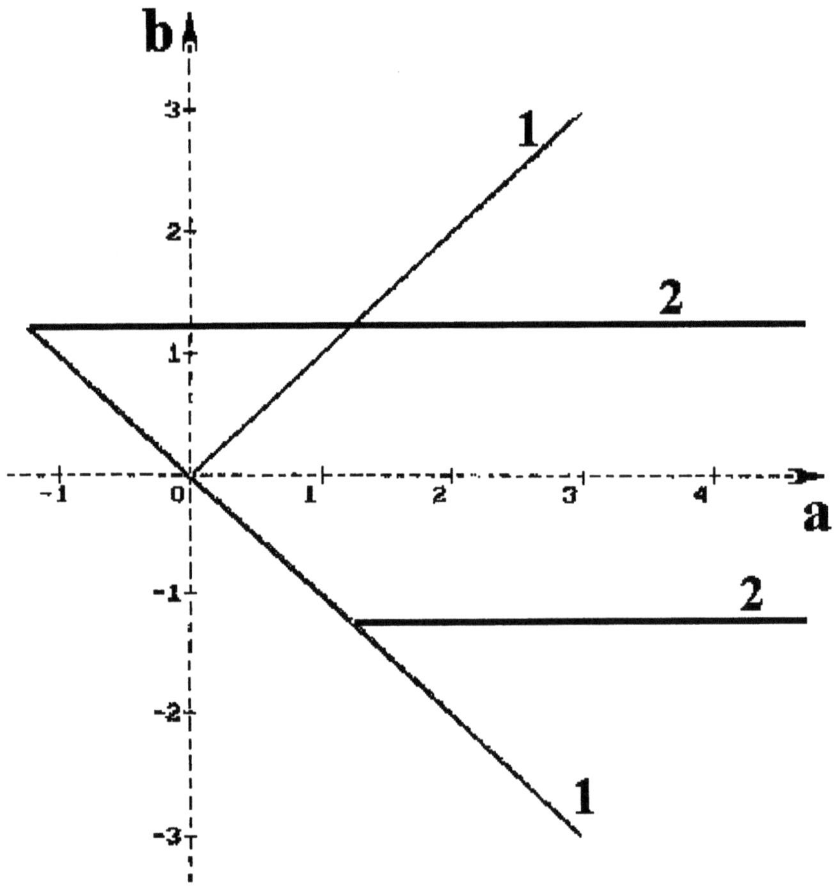

Figure 10.4.1

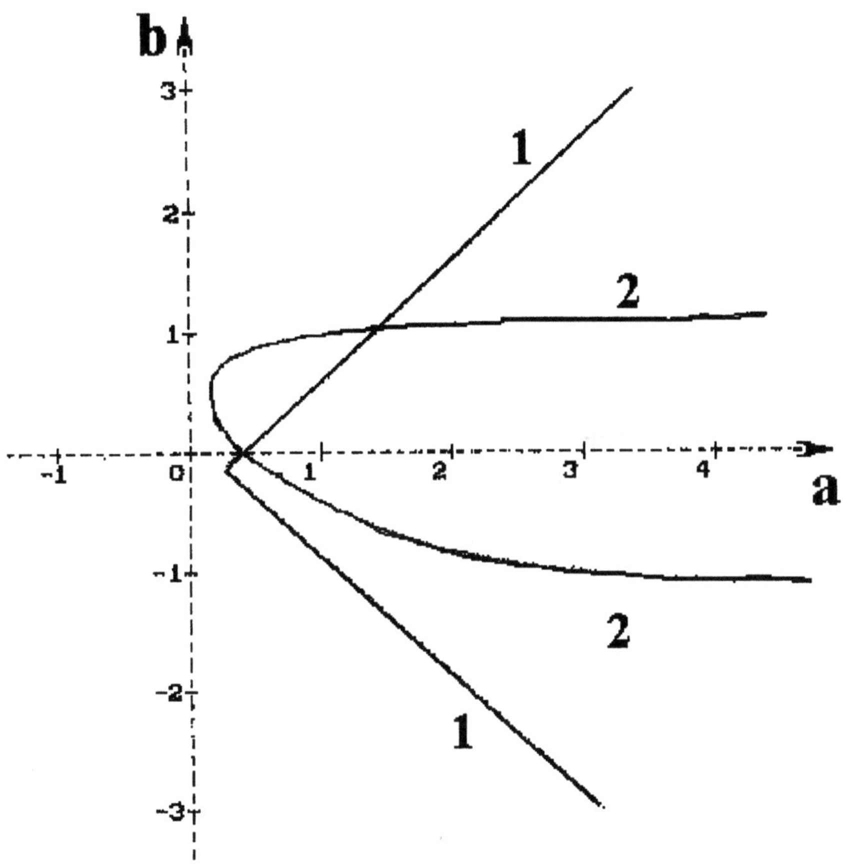

Figure 10.4.2

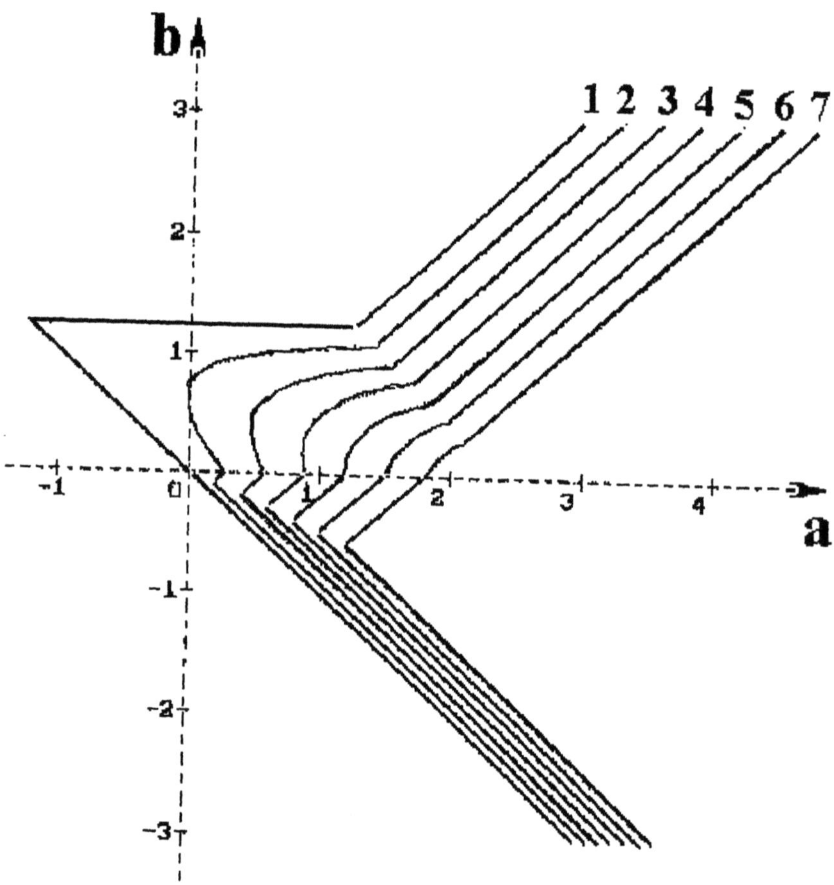

Figure 10.4.3

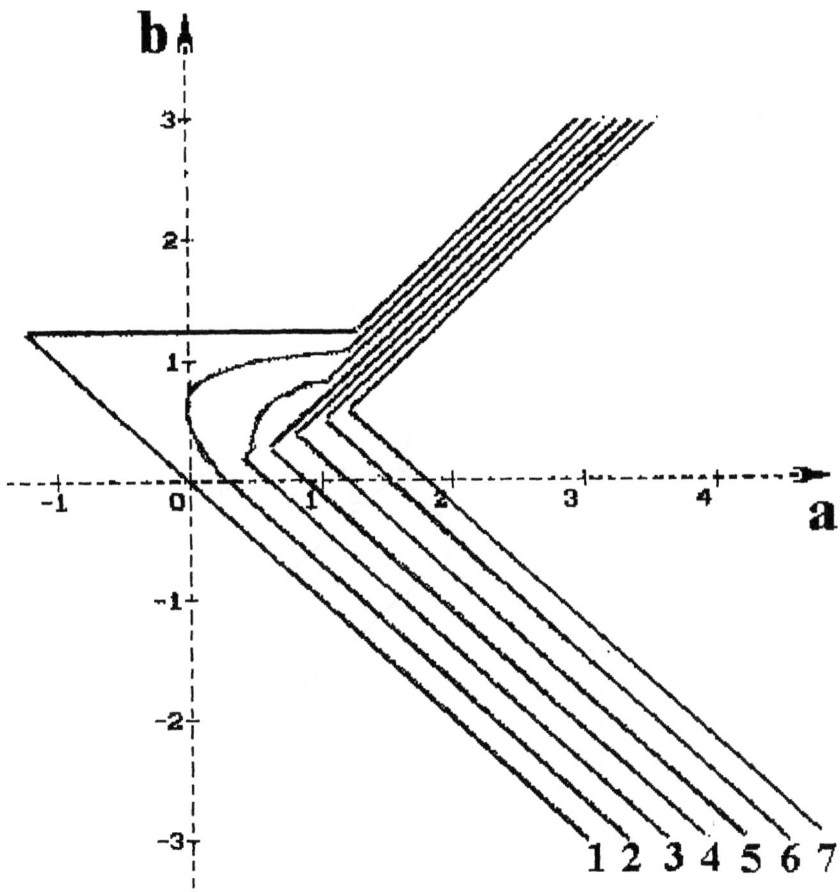

Figure 10.4.4

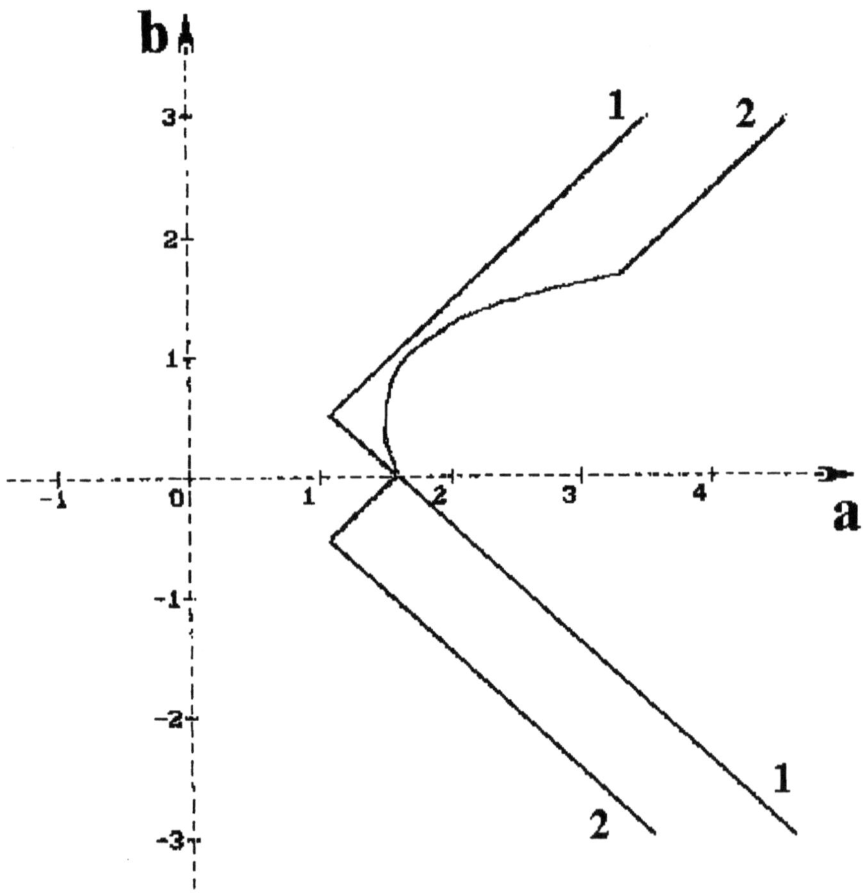

Figure 10.4.5

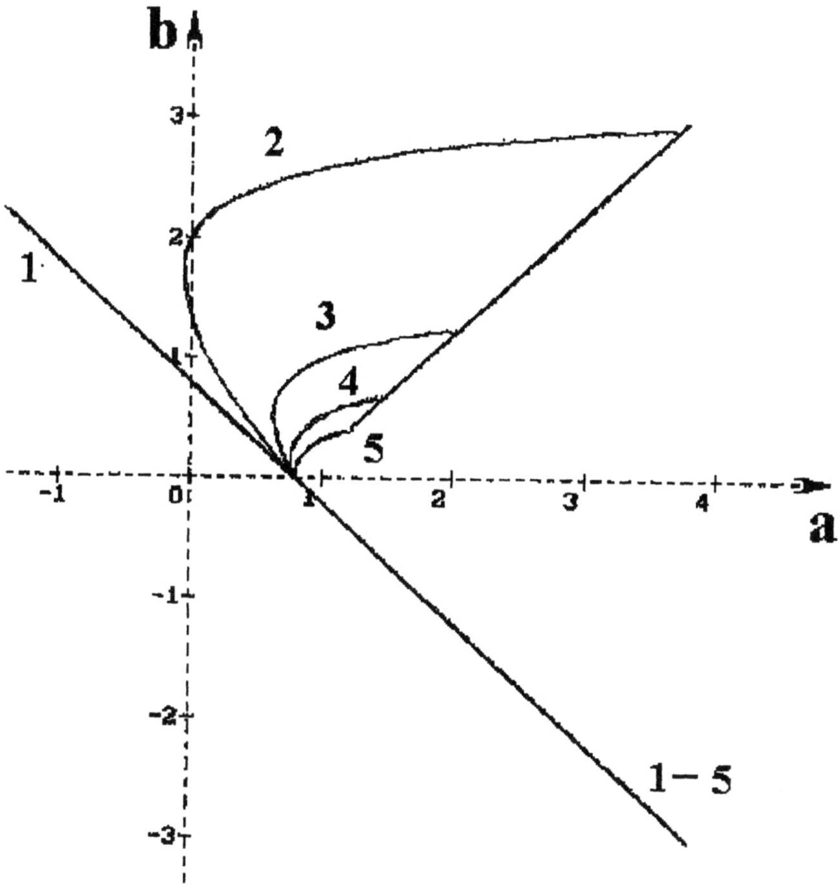

Figure 10.4.6

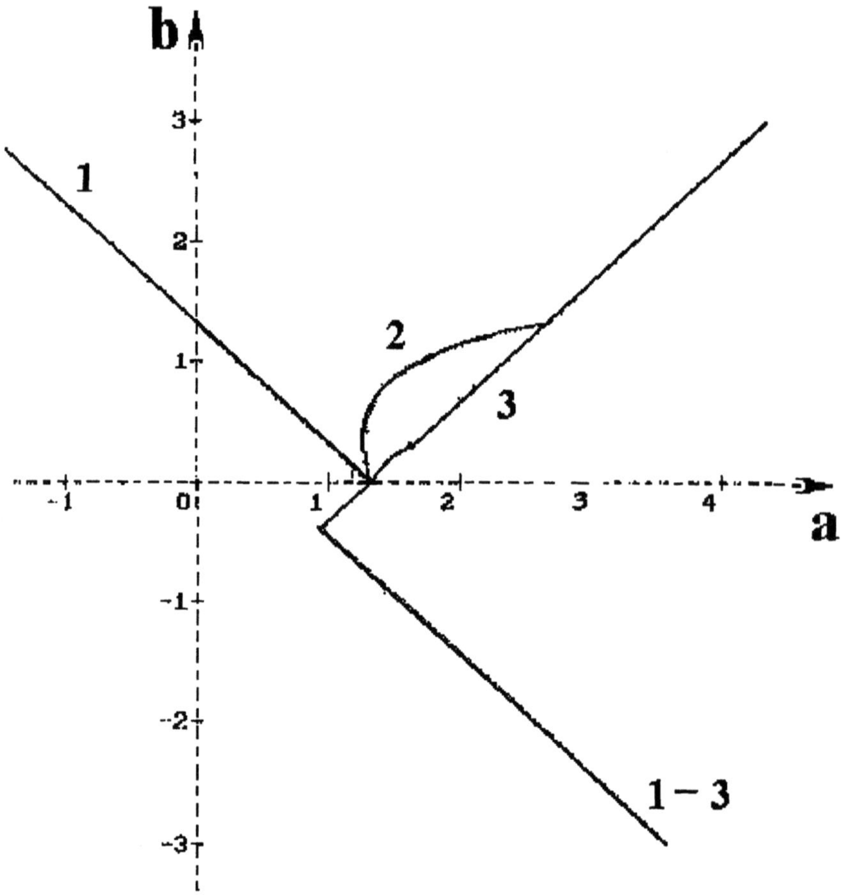

Figure 10.4.7

where $a > 0, b > 0, h \geq 0, c \in (-1,1), \sigma$ are some constants. Let us investigate mean–square stability of the solution $x_0(t) \equiv x_0 := b/a$ with respect to perturbations of the initial function ϕ, satisfying the following inequality

$$\phi(0) - c\phi(-h) > 0. \tag{4.18}$$

The application of the procedure leads us to the functional

$$V(x_t) := y(t) - y_0 - y_0 \ln(y(t)/y_0) + \frac{a|c|}{2} \int_{t-h}^{t} (x(s) - x_0)^2 ds$$

where $y(t) := x(t) - cx(t-h), y_0 := (1-c)x_0$ and

$$dy(t) = a(x_0 - x(t))y(t)(dt + \sigma d\xi(t)), \;\; y(0) = \phi(0) - c\phi(-h) > 0, \;\; y(t) > 0.$$

Hence

$$LV(x_t) = -a(x(t) - x_0)(y(t) - y_0) + a^2 y_0 p(x(t) - x_0)^2 +$$

$$+\frac{a|c|}{2}[(x(t) - x_0)^2 - (x(t-h) - x_0)^2] \leq$$

$$\leq -a(x(t) - x_0)^2 + a|c(x(t) - x_0)(x(t-h) - x_0)| +$$

$$+abp(1-c)(x(t) - x_0)^2 + \frac{a|c|}{2}[x(t) - x_0)^2 - (x(t-h) - x_0)^2] \leq$$

$$\leq a[-1 + |c| + bp(1-c)](x(t) - x_0)^2, \;\; p := \sigma^2/2.$$

From here it follows that the solution $x_0(t) \equiv b/a$ of Eq. (4.17) is mean square asymptotically stable under condition (4.18) if $a > 0, b > 0, |c| < 1, \; |c| + bp(1-c) < 1$.

5. RICCATI MATRIX EQUATIONS IN STABILITY OF LINEAR SRDEs

Liapunov's direct method, described previously, can be used for stability studying of various types of stochastic differential equations (SDEs). For linear SDEs it allows us to characterize asymptotic mean square stability in terms of existence of positive define solutions of Riccati matrix equations. Moreover, using the procedure from Subs. 4.2 we can obtain different Riccati equations for the some original SDEs. If SDE does not contain the delay then all these Riccati equations coincide with a unique linear matrix equation. The latter can be obtained by using a quadratic form as a Liapunov function.

Consider, e.g., the stochastic linear differential equation

$$dx(t) = Ax(t)dt + Cx(t)d\xi(t). \tag{5.1}$$

Here A and C are constant $n \times n$ – matrices, $x(t) \in \mathbf{R}^n$, $\xi(t)$ is a scalar standard Wiener process.

Denote $P > 0$ any symmetric positive definite matrix. Then an approximate Liapunov function V for the equation (1.1) is a quadratic form $V(x) = x'Px$, where a matrix P is a positive solution of the linear matrix equation (prime is a transposition sign)

$$A'P + PA + CPC = -Q. \tag{5.2}$$

The necessary and sufficient conditions of asymptotic mean square of the Eq. (5.1) can be formulated in terms of existence of a positive definite solution P of the matrix equation (5.2) for any positive definite matrix Q.

But for stochastic linear differential equations with delays, for example,

$$\begin{gathered} dx(t) = (Ax(t) + Bx(t-h))dt + Cx(t-\tau)d\xi(t), \quad t \geq 0 \\ x(\theta) = \phi(\theta), \quad \theta \leq 0 \end{gathered} \tag{5.3}$$

the situation is more complicated. In fact, formal application of a quadratic functional as a Liapunov functional for (5.3) leads us to a system of matrix, ordinary and partial differential equations. At the same time the approach, used in this chapter, gives Liapunov functionals defined by the solution of nonlinear (Riccati type) matrix equations.

Denote L a generator of Eq. (5.3) and consider the functional $V := V_1 + V_2$, where $V_1(x) := x'Px$. Calculating LV_1, we get

$$\begin{gathered} LV_1(x_t) = (Ax(t) + Bx(t-h))'Px(t) + x'(t)P(Ax(t) + Bx(t-h)) + \\ + x'(t-\tau)C'PCx(t-\tau) = \\ = x'(t)(A'P + PA)x(t) + x'(t-\tau)C'PCx(t-\tau) + \\ + x'(t-h)B'Px(t) + x'(t)PB(t-h). \end{gathered}$$

Remark that for any $R > 0$

$$\begin{gathered} x'(t-h)B'Px(t) + x'(t)PB(t-h) = \\ = x'(t)PBR^{-1}B'Px(t) + x'(t-h)Rx(t-h) - \\ -(Rx(t-h) - B'Px(t))'R^{-1}(Rx(t-h) - B'Px(t)) \leq \\ \leq x'(t)PBR^{-1}B'Px(t) + x'(t-h)Rx(t-h). \end{gathered} \tag{5.4}$$

Then

$$LV_1 \leq x'(t)(A'P + PA + PBR^{-1}B'P)x(t)+$$

$$+x'(t-h)Rx(t-h) + x'(t-\tau)C'PCx(t-\tau).$$

Choosing the functional V_2 in the form

$$V_2(x_t) := \int_{t-h}^{t} x'(s)Rx(s)ds + \int_{t-\tau}^{t} x'(s)C'PCx(s)ds,$$

we obtain

$$LV(x_t) \leq x'(t)(A'P + PA + C'PC + R + PBR^{-1}B'P)x(t).$$

Thus, it is proved

Theorem 5.1 Let for some symmetric matrices $Q > 0$ and $R > 0$ there exists a positive definite solution P of the Riccati matrix equation

$$A'P + PA + C'PC + R + PBR^{-1}B'P = -Q. \tag{5.5}$$

Then the zero solution of Eq. (5.3) is asymptotically mean square stable.

Remark 5.1 By analogy with inequality (5.4) we can use inequality

$$x'(t-h)B'Px(t) + x'(t)PB(t-h) =$$

$$= x'(t-h)B'PR^{-1}PBx(t-h) + x'(t)Rx(t)-$$

$$-(Rx(t) - PBx(t-h))'R^{-1}(Rx(t) - PBx(t-h)) \leq$$

$$\leq x'(t-h)B'PR^{-1}PBx(t-h) + x'(t)Rx(t).$$

In this case choosing the functional V_2 in the form

$$V_2(x_t) := \int_{t-h}^{t} x'(s)B'PR^{-1}PBx(s)ds + \int_{t-\tau}^{t} x'(s)C'PCx(s)ds,$$

we obtain

$$LV(x_t) \leq x'(t)(A'P + PA + C'PC + R + B'PR^{-1}PB)x(t).$$

Thus, in Theorem 5.1 instead of equation (5.5) can be used the equation

$$A'P + PA + C'PC + R + B'PR^{-1}PB = -Q. \tag{5.6}$$

Example 5.1 In scalar case a positive solution of the Eq. (5.5) (or (5.6)) exists if and only if

$$A + |B| + \frac{1}{2}C^2 < 0.$$

Consider now another way of construction of the Liapunov functional V leading to another Riccati equation.

Let us represent Eq. (5.3) as follows

$$\frac{d}{dt}\left(x(t) + B\int_{t-h}^{t} x(s)ds\right) = (A+B)x(t) + Cx(t-\tau)\dot{\xi}(t).$$

We construct the Liapunov functional V again in the form $V := V_1 + V_2$, but now the functional V_1 we choose in the form

$$V_1(x_t) := (x(t) + B\int_{t-h}^{t} x(s)ds)'P(x(t) + B\int_{t-h}^{t} x(s)ds). \tag{5.7}$$

Calculating LV_1, we get:

$$LV_1(x_t) = (x(t) + B\int_{t-h}^{t} x(s)ds)'P(A+B)x(t)+$$

$$+x'(t)(A+B)'P(x(t) + B\int_{t-h}^{t} x(s)ds) + x'(t-\tau)C'PCx(t-\tau) =$$

$$= x'(t)[P(A+B) + (A+B)'P]x(t) + x'(t-\tau)C'PCx(t-\tau)+$$

$$+\int_{t-h}^{t}[x'(t)(A+B)'PBx(s) + x'(s)B'P(A+B)x(t)]ds.$$

Remark that for any $R > 0$

$$\int_{t-h}^{t}[x'(t)(A+B)'PBx(s) + x'(s)B'P(A+B)x(t)]ds =$$

$$= hx'(t)(A+B)'PBR^{-1}B'P(A+B)x(t) + \int_{t-h}^{t} x'(s)Rx(s)ds-$$

$$-\int_{t-h}^{t}(B'P(A+B)x(t)-Rx(s))'R^{-1}(B'P(A+B)x(t)-Rx(s))ds \leq$$

$$\leq hx'(t)(A+B)'PBR^{-1}B'P(A+B)x(t)+\int_{t-h}^{t}x'(s)Rx(s)ds. \tag{5.8}$$

Then

$$LV_1(x_t) \leq x'(t)[(A+B)'P+P(A+B)+h(A+B)'PBR^{-1}B'P(A+$$

$$+B)]x(t)+\int_{t-h}^{t}x'(s)Rx(s)ds+x'(t-\tau)C'PCx(t-\tau).$$

Choosing the functional V_2 in the form

$$V_2(x_t) := \int_{t-h}^{t}(s-t+h)x'(s)Rx(s)ds+\int_{t-\tau}^{t}x'(s)C'PCx(s)ds,$$

we obtain for $V := V_1 + V_2$

$$LV(x_t) \leq x'(t)[(A+B)'P+P(A+B)+C'PC+hR+$$

$$+h(A+B)'PBR^{-1}B'P(A+B)]x(t).$$

So, using Thm. 4.1 we obtain

Theorem 5.2 Let for at least one matrix norm $\| B \|$ the inequality

$$h \| B \| < 1$$

hold and for some symmetric matrices $Q > 0$ and $R > 0$ there exists a positive definite solution P of the Riccati matrix equation

$$(A+B)'P+P(A+B)+C'PC+hR+h(A+B)'PBR^{-1}B'P(A+B) = -Q. \tag{5.9}$$

Then the zero solution of the Eq. (5.3) is asymptotically mean square stable.

Remark 5.2 By analogy with the inequality (5.8) we can use the inequality

$$\int_{t-h}^{t} [x'(t)(A+B)'PBx(s) + x'(s)B'P(A+B)x(t)]ds =$$

$$= hx'(t)Rx(t) + \int_{t-h}^{t} x'(s)B'P(A+B)R^{-1}(A+B)'PBx(s)ds -$$

$$- \int_{t-h}^{t} ((A+B)'PBx(s) - Rx(t))'R^{-1}((A+B)'PBx(s) - Rx(t))ds \leq$$

$$\leq hx'(t)Rx(t) + \int_{t-h}^{t} x'(s)B'P(A+B)R^{-1}(A+B)'PBx(s)ds.$$

In this case choosing the functional V_2 in the form

$$V_2(x_t) := \int_{t-h}^{t} (s-t+h)x'(s)B'P(A+B)R^{-1}(A+B)'PBx(s)ds +$$

$$+ \int_{t-\tau}^{t} x'(s)C'PCx(s)ds,$$

for $V := V_1 + V_2$ we obtain

$$LV(x_t) \leq x'(t)[(A+B)'P + P(A+B) + C'PC + hR +$$

$$+ hB'P(A+B)R^{-1}(A+B)'PB]x(t).$$

Thus, in Thm. 5.2 in place of the Eq. (5.9) can be used the equation

$$(A+B)'P + P(A+B) + C'PC + hR + hB'P(A+B)R^{-1}(A+B)'PB = -Q. \tag{5.10}$$

Example 5.2 In the scalar case a positive solution of equation (5.9) (or (5.10)) exists if and only if

$$(A+B)(1-h|B|) + \frac{1}{2}C^2 < 0, \quad h|B| < 1.$$

Example 5.3 Consider two-dimensional equation (5.3), where

$$A = \begin{pmatrix} -a_1 & 0 \\ 0 & -a_2 \end{pmatrix}, \quad B = \begin{pmatrix} b_1 & b_2 \\ -b_2 & b_1 \end{pmatrix}, \quad C = \begin{pmatrix} c_1 & 0 \\ 0 & c_2 \end{pmatrix}.$$

Let be $Q := qI, R := rI, q, r > 0, I$ is (2×2)–identity matrix. The positive solution $(P_{12} = 0, P_{11} > 0, P_{22} > 0)$ of the Riccati equation (5.5) exists if and only if

$$a_i > \sqrt{b_1^2 + b_2^2} + \frac{1}{2}c_i^2, \quad i = 1, 2.$$

Example 5.4 Consider Example 5.3: $b_1 = 0, b_2 = 1, 2a_1 = 6 + c_1^2, 2a_2 = \frac{3}{2} + c_2^2$. In this case the condition

$$a_i > 1 + \frac{1}{2}c_i^2, \quad i = 1, 2$$

is fulfilled. Therefore the Riccati equation (5.5) does not have a positive solution. But the Riccati equation (5.3) has positive solution $P_{11} = 1, P_{22} = 2, P_{12} = 0$ by $q = r = 1$.

Consider now more general linear SRDE

$$\dot{x}(t) = Ax(t) + \sum_{i=1}^{m} B_i x(t - h_i) + Cx(t - \tau)\dot{\xi}(t). \tag{5.11}$$

For the construction of Liapunov functionals for the Eq. (5.11) we use now both previous representations of initial equation. Namely, represent Eq. (5.11) in the form

$$\frac{d}{dt}\left(x(t) + \sum_{j=1}^{m_0} B_j \int_{t-h_j}^{t} x(s)ds\right) =$$

$$= (A + B)x(t) + \sum_{i=m_0+1}^{m} B_i x(t - h_i) + Cx(t - \tau)\dot{\xi}(t).$$

Here

$$B := \sum_{j=1}^{m_0} B_j, \quad 0 \le m_0 \le m.$$

We construct the Liapunov functional V in the form $V := V_1 + V_2$, where

$$V_1(x_t) := \left(x(t) + \sum_{j=1}^{m_0} B_j \int_{t-h_j}^{t} x(s)ds\right)' P \left(x(t) + \sum_{j=1}^{m_0} B_j \int_{t-h_j}^{t} x(s)ds\right).$$

In this case

$$LV_1(x_t) = ((A+B)x(t) + \sum_{i=m_0+1}^{m} B_i x(t-h_i))' P \, (x(t) +$$

$$+\sum_{j=1}^{m_0} B_j \int\limits_{t-h_j}^{t} x(s)ds \Bigg) + \left(x(t) + \sum_{j=1}^{m_0} B_j \int\limits_{t-h_j}^{t} x(s)ds \right)' P((A+B)x(t)+$$

$$+ \sum_{i=m_0+1}^{m} B_i x(t-h_i)) + x'(t-\tau)C'PCx(t-\tau) =$$

$$= x'(t)[(A+B)'P + P(A+B)]x(t) + x'(t-\tau)C'PCx(t-\tau)+$$

$$+ \sum_{i=m_0+1}^{m} [x'(t-h_i)B_i'Px(t) + x'(t)PB_i x(t-h_i)]+$$

$$+\sum_{j=1}^{m_0} \int\limits_{t-h_j}^{t} [x'(t)(A+B)'PB_j x(s) + x'(s)B_j'P(A+B)x(t)]ds+$$

$$+ \sum_{i=m_0+1}^{m} \sum_{j=1}^{m_0} \int\limits_{t-h_j}^{t} [x'(t-h_j)B_i'PB_j x(s) + x'(s)B_j'PB_i x(t-h_i)]ds.$$

By analogy with inequalities (5.4), (5.8) we can obtain the following inequalities. For any $R_i > 0, i = m_0+1, \ldots, m$,

$$\begin{aligned} &x'(t-h_i)B_i'Px(t) + x'(t)PB_i x(t-h_i) \leq \\ &\leq x'(t)PB_i R_i^{-1} B_i' P x(t) + x'(t-h_i)R_i x(t-h_i). \end{aligned} \tag{5.12}$$

For any $G_j > 0, j = 1, \ldots, m_0$,

$$\begin{aligned} &x'(t)(A+B)'PB_j x(s) + x'(s)B_j'P(A+B)x(t) \leq \\ &\leq x'(t)(A+B)'PB_j G_j^{-1} B_j' P(A+B)x(t) + x'(s)G_j x(s). \end{aligned} \tag{5.13}$$

For any $S_{ij} > 0, i = m_0+1, \ldots, m, j = 1, \ldots, m_0$,

$$\begin{aligned} &x'(t-h_i)B_i'PB_j x(s) + x'(s)B_j'PB_i x(t-h_i) \leq \\ &\leq x'(t-h_i)B_i'PB_j S_{ij}^{-1} B_j' PB_i' x(t-h_i) + x'(s)S_{ij}x(s). \end{aligned} \tag{5.14}$$

Using the inequalities (5.12)–(5.14) we obtain

$$LV_1(x_t) \le x'(t)[(A+B)'P + P(A+B) + \sum_{i=m_0+1}^{m} PB_iR_i^{-1}B_i'P +$$

$$+\sum_{j=1}^{m} h_j(A+B)'PB_jG_j^{-1}B_j'P(A+B)]x(t) + x'(t-\tau)C'PCx(t-\tau)+$$

$$+\sum_{i=m_0+1}^{m} x'(t-h_i)R_ix(t-h_i) + \sum_{j=1}^{m_0} \int_{t-h_j}^{t} x'(s)G_jx(s)ds+$$

$$+\sum_{i=m_0+1}^{m}\sum_{j=1}^{m_0}[h_jx'(t-h_i)B_i'PB_jS_{ij}^{-1}B_j'PB_ix(t-h_i) + \int_{t-h_j}^{t} x'(s)S_{ij}x(s)ds].$$

Choosing the functional V_2 in the form

$$V_2(x_t) := \int_{t-\tau}^{t} x'(s)C'PCx(s)ds+$$

$$+\sum_{i=m_0+1}^{m} \int_{t-h_i}^{t} x'(s)R_ix(s)ds + \sum_{j=1}^{m_0} \int_{t-h_j}^{t} (s-t+h_j)x'(s)G_jx(s)ds+$$

$$+\sum_{i=m_0+1}^{m}\sum_{j=1}^{m}[h_j \int_{t-h_i}^{t} x'(s)B_i'PB_jS_{ij}^{-1}B_j'PB_ix(s)ds+$$

$$+\int_{t-h_j}^{t} (s-t+h_j)x'(s)S_{ij}x(s)ds],$$

for $V := V_1 + V_2$ we obtain

$$LV(x_t) \le x'(t)[(A+B)'P + P(A+B) + C'PC+$$

$$+\sum_{i=m_0+1}^{m} (R_i + PB_iR_i^{-1}B_i'P) + \sum_{j=1}^{m_0} h_j(G_j + (A+B)'PB_jG_j^{-1}B_j'P(A+B))+$$

$$+\sum_{i=m_0+1}^{m}\sum_{j=1}^{m_0} h_j(S_{ij} + B_i'PB_jS_{ij}^{-1}B_j'PB_i)]x(t).$$

Thus, it is proved

Theorem 5.3 Let for some $m_0 \in [0, m]$ and for at least one matrix norm $\| B_j \|$ the inequality

$$\sum_{j=1}^{m_0} h_j \| B_j \| < 1$$

hold and for some symmetric matrices $R_i > 0, G_j > 0, S_{ij} > 0$ and $Q > 0$ there exists a positive definite solution P of the Riccati matrix equation

$$\begin{aligned}(A+B)'P + P(A+B) + C'PC + \sum_{i=m_0+1}^{m} (R_i + PB_iR_i^{-1}B_i'P) + \\ + \sum_{j=1}^{m_0} h_j(G_j + (A+B)'PB_jG_j^{-1}B_j'P(A+B)) + \\ + \sum_{i=m_0+1}^{m} \sum_{j=1}^{m_0} h_j(S_{ij} + B_i'PB_jS_{ij}^{-1}B_j'PB_i) = -Q,\end{aligned} \tag{5.15}$$

where

$$B := \sum_{j=1}^{m_0} B_j, \quad 0 \le m_0 \le m.$$

Then the zero solution of the Eq. (5.11) is asymptotically mean square stable.

Remark 5.3 By analogy with inequalities (5.12)– (5.14) we can use the inequalities:

$$\begin{aligned}x'(t-h_i)B_i'Px(t) + x'(t)PB_ix(t-h_i) \le \\ \le x'(t-h_i)B_i'PR_i^{-1}PB_ix(t-h_i) + x'(t)R_ix(t),\end{aligned} \tag{5.16}$$

$$\begin{aligned}x'(t)(A+B)'PB_jx(s) + x'(s)B_j'P(A+B)x(t) \le \\ \le x'(s)B_j'P(A+B)G_j^{-1}(A+B)'PB_jx(s) + x'(t)G_jx(t),\end{aligned} \tag{5.17}$$

$$\begin{aligned}x'(t-h_i)B_i'PB_jx(s) + x'(s)B_j'PB_ix(t-h_i) \le \\ \le x'(s)B_j'PB_iS_{ij}^{-1}B_i'PB_jx(s) + x'(t-h_i)S_{ij}x(t-h_i).\end{aligned} \tag{5.18}$$

Using different combinations of inequalities (5.12)–(5.14), (5.16)–(5.18) and choosing a corresponding form of the functional V_2 in place of the Eq. (5. 15) in Thm. 5.3 we can use one from the eight different Riccati matrix equations. For example, using in the place of all inequalities (5.12)–(5.14) inequalities (5.16)–(5.18) and choosing the functional V_2 in the form

$$V_2(x_t) := \int_{t-\tau}^{t} x'(s)C'PCx(s)ds + \sum_{i=m_0+1}^{m} \int_{t-h_i}^{t} x'(s)B_i'PR_i^{-1}PB_ix(s)ds+$$

$$+\sum_{j=1}^{m_0} \int_{t-h_j}^{t} (s-t+h_j)x'(s)B_j'P(A+B)G_j^{-1}(A+B)'PB_jx(s)ds+$$

$$+ \sum_{i=m_0+1}^{m} \sum_{j=1}^{m_0}[h_j \int_{t-h_i}^{t} x'(s)S_{ij}x(s)ds+$$

$$+ \int_{t-h_j}^{t} (s-t+h_j)x'(s)B_j'PB_iS_{ij}^{-1}B_i'PB_jx(s)ds],$$

we obtain the equation

$$(A+B)'P + P(A+B) + C'PC + \sum_{i=m_0+1}^{m} (R_i + B_i'PR_i^{-1}PB_i)+$$

$$+\sum_{j=1}^{m_0} h_j(G_j + B_j'P(A+B)G_j^{-1}(A+B)'PB_j)+$$

$$+ \sum_{i=m_0+1}^{m} \sum_{j=1}^{m_0} h_j(S_{ij} + B_j'PB_iS_{ij}^{-1}B_i'PB_j) = -Q.$$

Remark 5.4 It is easy to see that representing Eq. (5.11) in the form

$$\frac{d}{dt}\left(x(t) + \sum_{j=m_0+1}^{m} B_j \int_{t-h_j}^{t} x(s)ds\right) =$$

$$= (A+B)x(t) + \sum_{i=1}^{m_0} B_ix(t-h_i) + Cx(t-\tau)\dot{\xi}(t),$$

where

$$B := \sum_{j=m_0+1}^{m} B_j, \quad 0 \le m_0 \le m,$$

we obtain a new modification of Thm. 5.3 after replacement the all sums $\sum_{j=1}^{m_0}$ on $\sum_{j=m_0+1}^{m}$ and the all sums $\sum_{i=m_0+1}^{m}$ on $\sum_{i=1}^{m_0}$. Using other combinations of the summands for representations of the Eq. (5.11), we will obtain other modifications of Thm. 5.3.

Example 5.5 Consider the scalar case of Eq. (5.11) with $m = 2$. Using Thm. 5.3 and Remark 5.4 for different values of $m_0 = 0, 1, 2$, we obtain four different sufficient conditions of asymptotic mean square stability of the scalar equation (5.11) zero solution:

$$A + |B_1| + |B_2| + \frac{1}{2}C^2 < 0,$$

$$(A + B_1)(1 - h_1|B_1|) + |B_2|(1 + h_1|B_1|) + \frac{1}{2}C^2 < 0, \quad h_1|B_1| < 1,$$

$$(A + B_2)(1 - h_2|B_2|) + |B_1|(1 + h_2|B_2|) + \frac{1}{2}C^2 < 0, \quad h_2|B_2| < 1,$$

$$(A + B_1 + B_2)(1 - h_1|B_1| - h_2|B_2|) + \frac{1}{2}C^2 < 0, \quad h_1|B_1| + h_2|B_2| < 1.$$

Consider now the stochastic linear differential equation with distributed delay

$$\dot{x}(t) = Ax(t) + \sum_{i=1}^{m} B_i \int_{t-h_i}^{t} x(s)ds + C \int_{t-\tau}^{t} x(s)ds\dot{\xi}(t). \tag{5.19}$$

Represent it in the form

$$\frac{d}{dt}\left(x(t) + \sum_{j=1}^{m_0} B_j \int_{t-h_j}^{t} (s - t + h_j)x(s)ds \right) =$$

$$= (A + (B, h))x(t) + \sum_{i=m_0+1}^{m} B_i \int_{t-h_i}^{t} x(s)ds + C \int_{t-\tau}^{t} x(s)ds\dot{\xi}(t).$$

Here

$$(B, h) := \sum_{j=1}^{m_0} h_j B_j, \quad 0 \le m_0 \le m.$$

We will construct now a Liapunov functional V in the form $V := V_1 + V_2$, where

$$V_1(x_t) := \left(x(t) + \sum_{j=1}^{m_0} B_j \int_{t-h_j}^{t} (s-t+h_j)x(s)ds \right)' P\times$$

$$\times \left(x(t) + \sum_{j=1}^{m_0} B_j \int_{t-h_j}^{t} (s-t+h_j)x(s)ds \right).$$

In this case

$$LV_1(x_t) = \left((A+(B,h))x(t) + \sum_{i=m_0+1}^{m} B_i \int_{t-h_i}^{t} x(s)ds \right)' \times$$

$$\times P \left(x(t) + \sum_{j=1}^{m_0} B_j \int_{t-h_j}^{t} (s-t+h_j)x(s)ds \right) +$$

$$+ \left(x(t) + \sum_{i=1}^{m_0} B_j \int_{t-h_i}^{t} (s-t+h_j)x(s)ds \right)' \times$$

$$\times P \left((A+(B,h))x(t) + \sum_{i=m_0+1}^{m} B_i \int_{t-h_i}^{t} x(s)ds \right) +$$

$$+ \int_{t-\tau}^{t} x'(s)dsC'PC \int_{t-\tau}^{t} x(\theta)d\theta =$$

$$= x'(t)[(A+(B,h))'P + P(A+(B,h))]x(t) + \int_{t-\tau}^{t} x'(s)dsC'PC \int_{t-\tau}^{t} x(\theta)d\theta +$$

$$+ \sum_{i=m_0+1}^{m} \int_{t-h_i}^{t} [x'(s)B_i'Px(t) + x'(t)PB_ix(s)]ds +$$

$$+ \sum_{j=1}^{m_0} \int_{t-h_j}^{t} (s-t+h_j)[x'(t)(A+(B,h))'PB_jx(s) + x'(s)B_j'P(A+(B,h))x(t)]ds \cdot$$

$$+\sum_{i=m_0+1}^{m}\sum_{j=1}^{m_0}\int_{t-h_i}^{t}\int_{t-h_j}^{t}(s-t+h_j)[x'(\theta)B_i'PB_jx(s)+x'(s)B_j'PB_ix(\theta)]dsd\theta.$$

Remark, that for any $H>0$

$$\int_{t-\tau}^{t}x'(s)dsC'PC\int_{t-\tau}^{t}x(\theta)d\theta=$$

$$=\frac{1}{2}\int_{t-\tau}^{t}\int_{t-\tau}^{t}[x'(\theta)C'PCx(s)+x'(s)C'PCx(\theta)]dsd\theta=$$

$$=\frac{1}{2}\int_{t-\tau}^{t}\int_{t-\tau}^{t}[x'(\theta)C'PCH^{-1}C'PCx(\theta)+x'(s)Hx(s)-$$

$$-(Hx(s)-C'PCx(\theta))'H^{-1}(H(x(s)-C'PCx(\theta))]dsd\theta\le$$

$$\le\frac{\tau}{2}\int_{t-\tau}^{t}x'(\theta)C'PCH^{-1}C'PCx(\theta)d\theta+\frac{\tau}{2}\int_{t-\tau}^{t}x'(s)Hx(s)ds.$$

Using the inequalities similar to (5.12)–(5.14) we obtain

$$LV_1(x_t)\le x'(t)[(A+(B,h))'P+P(A+(B,h))+\sum_{i=m_0+1}^{m}h_iPB_iR_i^{-1}B_i'P+$$

$$+\frac{1}{2}\sum_{j=1}^{m_0}h_j^2(A+(B,h))'PB_jG_j^{-1}B_j'P(A+(B,h))]x(t)+$$

$$+\frac{\tau}{2}\int_{t-\tau}^{t}x'(s)(H+C'PCH^{-1}C'PC)x(s)ds+$$

$$+\sum_{i=m_0+1}^{m}\int_{t-h_i}^{t}x'(s)R_ix(s)ds+\sum_{j=1}^{m_0}\int_{t-h_j}^{t}(s-t+h_j)x'(s)G_jx(s)ds+$$

$$+\sum_{i=m_0+1}^{m}\sum_{j=1}^{m_0}[\frac{1}{2}h_j^2\int_{t-h_i}^{t}x'(s)B_i'PB_jS_{ij}^{-1}B_j'PB_ix(s)ds+$$

$$+\int_{t-h_j}^{t}(s-t+h_j)x'(s)S_{ij}x(s)ds].$$

Choosing the functional V_2 in the form

$$V_2(x_t) := \frac{\tau}{2}\int_{t-\tau}^{t}(s-t+\tau)x'(s)(H+C'PCH^{-1}C'PC)x(s)ds+$$

$$+\sum_{i=m_0+1}^{m}\int_{t-h_i}^{t}(s-t+h_i)x'(s)R_ix(s)ds+$$

$$+\frac{1}{2}\sum_{j=1}^{m_0}\int_{t-h_j}^{t}(s-t+h_j)^2x'(s)G_jx(s)ds+$$

$$+\sum_{i=m_0+1}^{m}\sum_{j=1}^{m_0}[\frac{1}{2}h_j^2\int_{t-h_i}^{t}(s-t+h_i)x'(s)B_i'PB_jS_{ij}^{-1}B_j'PB_ix(s)ds+$$

$$+\frac{1}{2}h_i\int_{t-h_j}^{t}(s-t+h_j)^2x'(s)S_{ij}x(s)ds],$$

for $V := V_1 + V_2$ we obtain

$$LV(x_t) \le x'(t)[(A+(B,h))'P+P(A+(B,h))+$$

$$+\frac{\tau^2}{2}(H+C'PCH^{-1}C'PC)+\sum_{i=m_0+1}^{m}h_i(R_i+PB_iR_i^{-1}B_i'P)+$$

$$+\frac{1}{2}\sum_{j=1}^{m_0}h_j^2(G_j+(A+(B,h))'PB_jG_j^{-1}B_j'P(A+(B,h)))+$$

$$+\frac{1}{2}\sum_{i=m_0+1}^{m}h_i\sum_{j=1}^{m_0}h_j^2(S_{ij}+B_i'PB_jS_{ij}^{-1}B_j'PB_i)]x(t).$$

Thus, it is proved

Theorem 5.4 Let for some $m_0 \in [0,m]$ and for at least one matrix norm $\| B_j \|$ the inequality

$$\sum_{j=1}^{m_0}h_j^2 \| B_j \| < 2$$

hold and for some symmetric matrices $R_i > 0, G_j > 0, S_{ij} > 0$ and $Q > 0$ there exists a positive definite solution P of the Riccati matrix equation

$$(A + (B,h))'P + P(A + (B,h)) + \frac{\tau^2}{2}(H + C'PCH^{-1}C'PC) +$$

$$+ \sum_{i=m_0+1}^{m} h_i(R_i + PB_iR_i^{-1}B_i'P) +$$

$$+\frac{1}{2}\sum_{j=1}^{m_0} h_j^2(G_j + (A + (B,h))'PB_jG_j^{-1}B_j'P(A + (B,h))) +$$

$$+\frac{1}{2}\sum_{i=m_0+1}^{m} h_i \sum_{j=1}^{m_0} h_j^2(S_{ij} + B_i'PB_jS_{ij}^{-1}B_j'PB_i) = -Q, \tag{5.20}$$

where

$$(B,h) := \sum_{j=1}^{m_0} h_jB_j, \quad 0 \le m_0 \le m.$$

Then the zero solution of the Eq. (5.19) is asymptotically mean square stable.

Remark 5.6 Using arguments from Remark 5.5, we can prove that in place of the Eq. (5.20) in Thm. 4.1 one from the eight different Riccati matrix equations can be used. For example,

$$(A + (B,h))'P + P(A + (B,h)) + \frac{\tau^2}{2}(H + C'PCH^{-1}C'PC) +$$

$$+ \sum_{i=m_0+1}^{m} h_i(R_i + B_i'PR_i^{-1}PB_i) +$$

$$+\frac{1}{2}\sum_{j=1}^{m_0} h_j^2(G_j + B_j'P(A + (B,h))G_j^{-1}(A + (B,h))'PB_j) +$$

$$+\frac{1}{2}\sum_{i=m_0+1}^{m} h_i \sum_{j=1}^{m_0} h_j^2(S_{ij} + B_j'PB_iS_{ij}^{-1}B_i'PB_j) = -Q.$$

Remark 5.7 Using representation of Eq. (5.19) in the form

$$\frac{d}{dt}\left(x(t) + \sum_{i=m_0+1}^{m} B_j \int_{t-h_j}^{t} (s-t+h_j)x(s)ds \right) =$$

$$= (A+(B,h))x(t) + \sum_{i=1}^{m_0} B_i \int_{t-h_i}^{t} x(s)ds + C\int_{t-\tau}^{t} x(s)ds\dot{\xi}(t),$$

where

$$(B,h) := \sum_{j=m_0+1}^{m} h_j B_j, \quad 0 \le m_0 \le m,$$

we obtain the new modification of Thm. 5.4 after replacement all the sums $\sum_{j=1}^{m_0}$ an $\sum_{j=m_0+1}^{m}$ and all the sums $\sum_{i=m_0+1}^{m}$ on $\sum_{i=1}^{m_0}$. Using other combinations of the summands for representations of the Eq. (5.19), we will obtain other modifications of Thm. 5.4.

Example 5.6 Consider the scalar case of Eq. (5.19) for $m = 2$. Using Thm. 5.4 and Remark 5.7 for different values of $m_0 = 0, 1, 2$, we obtain four different sufficient conditions of asymptotically mean square stability of the scalar equation (5.19) zero solution:

$$A + h_1|B_1| + h_2|B_2| + \frac{\tau^2}{2}C^2 < 0,$$

$$(A+h_1B_1)(1-\frac{1}{2}h_1^2|B_1|) + h_2|B_2|(1+\frac{1}{2}h_1^2|B_1|) + \frac{\tau^2}{2}C^2 < 0, \quad h_1^2|B_1| < 2,$$

$$(A+h_2B_2)(1-\frac{1}{2}h_2^2|B_2|) + h_1|B_1|(1+\frac{1}{2}h_2^2|B_2|) + \frac{\tau^2}{2}C^2 < 0, \quad h_2^2|B_2| < 2,$$

$$(A+h_1B_1+h_2B_2)(1-\frac{1}{2}h_1^2|B_1| - \frac{1}{2}h_2^2|B_2|) + \frac{\tau^2}{2}C^2 < 0,$$

$$h_1^2|B_1| + h_2^2|B_2| < 2.$$

Using the same approach, similar stability conditions can be obtained for systems with different combinations of discrete and distributed delays.

For example, consider the linear SRDE

$$\dot{x}(t) = Ax(t) + B_1x(t-h_1) + B_2\int_{t-h_2}^{t} x(s)ds+ \\ + \left(C_1x(t-\tau_1) + C_2\int_{t-\tau_2}^{t} x(s)ds\right)\dot{\xi}(t). \tag{5.21}$$

Using four different representations of this equation, we will construct four different Liapunov functionals for it.

At first consider Eq. (5.21). Using the functional $V := V_1 + V_2$, where $V_1(x_t) := x'(t)Px(t)$ and for any $R_i > 0, G_i > 0, i = 1,2$

$$V_2(x_t) := \int_{t-h_1}^{t} x'(s)R_1x(s)ds + \int_{t-h_2}^{t} (s-t+h_2)x'(s)R_2x(s)ds+$$

$$+\int_{t-\tau_1}^{t} x'(s)(C_1'PC_1 + \tau_2C_1'PC_2G_1^{-1}C_2'PC_1)x(s)ds+$$

$$+\int_{t-\tau_2}^{t} (s-t+\tau_2)x'(s)(G_1 + \frac{\tau_2}{2}(G_2 + C_2'PC_2G_2^{-1}C_2'PC_2))x(s)ds$$

we obtain

Theorem 5.5 Let for some symmetric matrices $R_i > 0, G_i > 0$ and $Q > 0$ there exists a positive definite solution P of the Riccati matrix equation

$$A'P + PA + C_1'PC_1 + R_1 + PB_1R_1^{-1}B_1'P+ \\ +h_2(R_2 + PB_2R_2^{-1}B_2'P) + \tau_2(G_1 + C_1'PC_2G_1^{-1}C_1'PC_1)+ \\ +\frac{\tau_2^2}{2}(G_2 + C_2'PC_2G_2^{-1}C_2'PC_2) = -Q. \tag{5.22}$$

Then the zero solution of Eq. (5.21) is asymptotically mean square stable.

Remark 5.7 Using arguments from Remark 5.5, we can prove that in place of Eq. (5.22) in Thm. 5.5 one from the eight different Riccati matrix equations can be used. For example

$$A'P + PA + C_1'PC_1 + R_1 + B_1'PR_1^{-1}PB_1 + h_2(R_2 + B_2'PR_2^{-1}PB_2)+$$

$$+\tau_2(G_1 + C_2'PC_1G_1^{-1}C_1'PC_2) + \frac{\tau_2^2}{2}(G_2 + C_2'PC_2G_2^{-1}C_2'PC_2) = -Q.$$

Represent Eq. (5.21) in the form

$$\frac{d}{dt}\left(x(t)+B_1\int_{t-h_1}^{t}x(s)ds\right)=(A+B_1)x(t)+B_2\int_{t-h_2}^{t}x(s)ds+$$

$$+\left(C_1x(t-\tau_1)+C_2\int_{t-\tau_2}^{t}x(s)ds\right)\dot{\xi}(t).$$

Using the functional $V := V_1 + V_2$ where

$$V_1(x_t) := \left(x(t)+\int_{t-h_1}^{t}B_1x(s)ds\right)' P\left(x(t)+\int_{t-h_1}^{t}B_1x(s)ds\right),$$

and for any $R_i > 0, i = 1,2,3, G_j > 0, j = 1,2$

$$V_2(x_t) := \int_{t-h_1}^{t}(s-t+h_1)x'(s)(R_1+h_2B_1'PB_2R_3^{-1}B_2'PB_1)x(s)ds+$$

$$+\int_{t-h_2}^{t}(s-t+h_2)x'(s)(R_2+h_1R_3)x(s)ds+$$

$$+\int_{t-\tau_1}^{t}x'(s)(C_1'PC_1+\tau_2C_1'PC_2G_1^{-1}C_2'PC_1)x(s)ds+$$

$$+\int_{t-\tau_2}^{t}(s-t+\tau_2)x'(s)(G_1+\frac{\tau_2}{2}(G_2+C_2'PC_2G_2^{-1}C_2'PC_2))x(s)ds,$$

we obtain

Theorem 5.6 Let for at least one matrix norm $\| B_1 \|$ the inequality

$$h_1 \| B_1 \| < 1$$

hold and for some symmetric matrices $R_i > 0, G_i > 0$, and $Q > 0$ there exists a positive definite solution P of the Riccati matrix equation

$$(A+B_1)'P+P(A+B_1)+C_1'PC_1+$$

$$+h_1(R_1+(A+B_1)'PB_1R_1^{-1}B_1'P(A+B_1))+$$

$$+h_2(R_2+PB_2R_2^{-1}B_2'P)+h_1h_2(R_3+B_1'PB_2R_3^{-1}B_2'PB_1)+$$

$$+\tau_2(G_1+C_1'PC_2G_1^{-1}C_2'PC_1)+\frac{\tau_2^2}{2}(G_2+C_2'PC_2G_2^{-1}C_2'PC_2)=-Q. \tag{5.23}$$

Then the zero solution of Eq. (5.21) is asymptotically mean square stable.

Remark 5.8 Using arguments from Remark 5.5, we can prove that in place of Eq. (5.23) in Thm. 5.6 one from the sixteen different Riccati matrix equations can be used. For example.

$$(A+B_1)'P+P(A+B_1)+C_1'PC_1+$$

$$+h_1(R_1+B_1'P(A+B_1)R_1^{-1}(A+B_1)'PB_1)+$$

$$+h_2(R_2+B_2'PR_2^{-1}PB_2)+h_1h_2(R_3+B_2'PB_1R_3^{-1}B_1'PB_2)+$$

$$+\tau_2(G_1+C_2'PC_1G_1^{-1}C_1'PC_2)+\frac{\tau_2^2}{2}(G_2+C_2'PC_2G_2^{-1}C_2'PC_2)=-Q.$$

Represent Eq. (5.21) in the form

$$\frac{d}{dt}\left(x(t)+B_2\int_{t-h_2}^{t}(s-t+h_2)x(s)ds\right)=(A+h_2B_2)x(t)+B_1x(t-h_1)+$$

$$+\left(C_1x(t-\tau_1)+C_2\int_{t-\tau_2}^{t}x(s)ds\right)\dot{\xi}(t).$$

Using the functional $V:=V_1+V_2$, where

$$V_1(x_t):=\left(x(t)+B_2\int_{t-h_2}^{t}(s-t+h_2)x(s)ds\right)'\times$$

$$\times P\left(x(t)+B_2\int_{t-h_2}^{t}(s-t+h_2)x(s)ds\right),$$

and for any $R_i > 0, i = 1,2,3, G_j > 0, j = 1,2$

$$V_2(x_t) := \int_{t-h_1}^{t} x'(s)(R_1 + \frac{1}{2}h_2^2 B_1' P B_2 R_3^{-1} B_2' P B_1) x(s) ds +$$

$$+\frac{1}{2} \int_{t-h_2}^{t} (s - t + h_2)^2 x'(s)(R_2 + R_3) x(s) ds +$$

$$+ \int_{t-\tau_1}^{t} x'(s)(C_1' P C_1 + \tau_2 C_1' P C_2 G_1^{-1} C_2' P C_1) x(s) ds +$$

$$+ \int_{t-\tau_2}^{t} (s - t + \tau_2) x'(s)(G_1 + \frac{\tau_2}{2}(G_2 + C_2' P C_2 G_2^{-1} C_2' P C_2)) x(s) ds,$$

we obtain

Theorem 5.7 Let for at least one matrix norm $\| B_2 \|$ *the inequality*

$$h_2^2 \| B_2 \| < 2$$

hold and for some symmetric matrices $R_i > 0, G_i > 0,$ *and* $Q > 0$ *there exists a positive definite solution* P *of the Riccati matrix equation*

$$(A + h_2 B_2)' P + P(A + h_2 B_2) + C_1' P C_1 + R_1 + P B_1 R_1^{-1} B_1' P +$$

$$+\frac{1}{2} h_2^2 (R_2 + R_3 + (A + h_2 B_2)' P B_2 R_2^{-1} B_2' P (A + h_2 B_2) +$$

$$+ B_1' P B_2 R_3^{-1} B_2' P B_1) +$$

$$+\tau_2 (G_1 + C_1' P C_2 G_1^{-1} C_2' P C_1) + \frac{\tau_2^2}{2}(G_2 + C_2' P C_2 G_2^{-1} C_2' P C_2) = -Q. \tag{5.24}$$

Then the zero solution of the Eq. (5.21) is asymptotically mean square stable.

Remark 5.9 Using arguments from Remark 5.5, we can prove that in place of the Eq. (5.24) in Thm. 5.7 one from the sixteen different Riccati matrix equations can be used. For example,

$$(A + h_2 B_2)' P + P(A + h_2 B_2) + C_1' P C_1 + R_1 + B_1' P R_1^{-1} P B_1 +$$

$$+\frac{1}{2}h_2^2(R_2+R_3+B_2'P(A+h_2B_2)R_2^{-1}(A+h_2B_2)'PB_2+$$

$$+B_2'PB_1R_3^{-1}B_1'PB_2)+$$

$$+\tau_2(G_1+C_2'PC_1G_1^{-1}C_1'PC_2)+\frac{\tau_2^2}{2}(G_2+C_2'PC_2G_2^{-1}C_2'PC_2)=-Q.$$

Represent Eq. (5.21) in the form

$$\frac{d}{dt}\left(x(t)+\int_{t-h_1}^{t}B_1x(s)ds+\int_{t-h_2}^{t}(s-t+h_2)B_2x(s)ds\right)=$$

$$=A_1x(t)+\left(C_1x(t-\tau_1)+C_2\int_{t-\tau_2}^{t}x(s)ds\right)\dot{\xi}(t),$$

where $A_1:=A+B_1+h_2B_2$. Using the functional $V:=V_1+V_2$, where

$$V_1(x_t):=\left(x(t)+\int_{t-h_1}^{t}B_1x(s)ds+\int_{t-h_2}^{t}(s-t+h_2)B_2x(s)ds\right)'\times$$

$$\times P\left(x(t)+\int_{t-h_1}^{t}B_1x(s)ds+\int_{t-h_2}^{t}(s-t+h_2)B_2x(s)ds\right)$$

and for any $R_i>0, G_i>0, i=1,2,$

$$V_2(x_t):=\int_{t-h_1}^{t}(s-t+h_1)x'(s)R_1x(s)ds+\frac{1}{2}\int_{t-h_2}^{t}(s-t+$$

$$+h_2)^2x'(s)R_2x(s)ds+$$

$$+\int_{t-\tau_1}^{t}x'(s)(C_1'PC_1+\tau_2C_1'PC_2G_1^{-1}C_2'PC_1)x(s)ds+$$

$$+\int_{t-\tau_2}^{t}(s-t+\tau_2)x'(s)(G_1+\frac{\tau_2}{2}(G_2+C_2'PC_2G_2^{-1}C_2'PC_2))x(s)ds,$$

we obtain

Theorem 5.8 Let for at least one matrix norm $\| B \|$ *the inequality*

$$h_1 \| B_1 \| + \frac{1}{2} h_2^2 \| B_2 \| < 1$$

hold and for some symmetric matrices $R_i > 0, G_i > 0$*, and* $Q > 0$ *there exists a positive definite solution* P *of the Riccati equation*

$$A_1' P + PA_1 + C_1' PC_1 + h_1(R_1 + A_1' PB_1 R_1^{-1} B_1' PA_1) +$$

$$+ \frac{1}{2} h_2^2 (R_2 + A_1' PB_2 R_2^{-1} B_2' PA_1) +$$

$$+ \tau_2 (G_1 + C_1' PC_2 G_1^{-1} C_2' PC_1) + \frac{\tau_2^2}{2} (G_2 + C_2' PC_2 G_2^{-1} C_2' PC_2) = -Q. \tag{5.25}$$

Then the zero solution of Eq. (5.21) is asymptotically mean square stable.

Remark 5.10 Using arguments from Remark 5.5, we can prove that in place of the Eq. (5.25) in Thm. 5.8 one from the eight different Riccati matrix equations can be used. For example,

$$A_1' P + PA_1 + C_1' PC_1 + h_1(R_1 + B_1' PA_1 R_1^{-1} A_1' PB_1) +$$

$$+ \frac{1}{2} h_2^2 (R_2 + B_2' PA_1 R_2^{-1} A_1' PB_2) +$$

$$+ \tau_2 (G_1 + C_2' PC_1 G_1^{-1} C_1' PC_2) + \frac{\tau_2^2}{2} (G_2 + C_2' PC_2 G_2^{-1} C_2' PC_2) = -Q.$$

Example 5.7 Consider the scalar case of Eq. (5.21). Using Thms. 5.5–5.8, we obtain four different sufficient conditions of asymptotically mean square stability of the scalar equation (5.1):

$$A + |B_1| + h_2 |B_2| + \frac{1}{2} C^2 < 0,$$

$$(A + B_1)(1 - h_1 |B_1|) + h_2 |B_2| (1 + h_1 |B_1|) + \frac{1}{2} C^2 < 0, \quad h_1 |B_1| < 1,$$

$$(A + h_2 B_2)(1 - \frac{1}{2} h_2^2 |B_2|) + |B_1| (1 + \frac{1}{2} h_2^2 |B_2|) + \frac{1}{2} C^2 < 0, \quad h_2^2 |B_2| < 2,$$

$$(A + B_1 + h_2 B_2)(1 - h_1 |B_1| - \frac{1}{2} h_2^2 |B_2|) + \frac{1}{2} C^2 < 0, \quad h_1 |B_1| + \frac{1}{2} h_2^2 |B_2| < 1.$$

Here

$$C := |C_1| + \tau_2 |C_2|.$$

Part IV

BOUNDARY VALUE PROBLEMS AND PERIODIC SOLUTIONS OF FUNCTIONAL DIFFERENTIAL EQUATIONS

Boundary value problems and problems of investigation of periodic solutions of FDEs are among the most important problems in the theory and applications of FDEs. Various methods were proposed to use in these problems, e.g., theory of the Fredholm operators, methods of small parameters, topological methods, asymptotic methods, theory of integral manifolds, numerical methods etc.

Detailed survey of these methods and corresponding results is, e.g., in [236, 238, 287.1, 289.1, 289.3].

CHAPTER 11

BOUNDARY VALUE PROBLEMS FOR FUNCTIONAL DIFFERENTIAL EQUATIONS

1. BOUNDARY VALUE PROBLEMS FOR FDEs OF EVOLUTIONARY TYPE

1.1. *INTRODUCTION*

Boundary value problems for FDEs can be roughly subdivided into problems for equations of evolutionary and nonevolutionary types. In the first case the independent variables t is interpreted as a time and the problem keeps features of evolution of some process. In the second case considered in the next section such direct connection is absent. However this subdivision is not precise because there are problems which can be related to the both types.

For evolutionary FDEs it is naturally to assume that the equation is RDE or NDE. Moreover here we shall consider equations of the first order with values of an unknown function in $\mathbf{R}^n$.

Consider general RDE

$$\dot{x}(t) = f(t, x_t) \quad (t_0 \leq t \leq T;\ x_t(\theta) := x(t+\theta), \quad -h \leq \theta \leq 0) \tag{1.1}$$

under assumptions of Subs. 3.2.1.

1.2. *PROBLEMS WITH A FINITE DEFECT*

Let us assume that the initial function $\phi : [-h, 0] \to \mathbf{R}^n$ includes $k \geq 1$ unknown scalar parameters $p_1, \ldots, p_k$. It means that the initial condition has a form

$$x_{t_0} = \phi(\cdot; p_1, \ldots, p_k). \tag{1.2}$$

Then it is quite reasonable to require that the solution should satisfy to k additional scalar conditions of the type of equalities. Thus, if the solution is constructed on the interval $[t_0, T]$ these conditions may have a form

$$Q_1(x_T) = q_1, \ldots, \quad Q_k(x_T) = q_k \tag{1.3}$$

where $Q_1, \dots, Q_k$ are scalar functionals, and $q_1, \dots, q_k$ are some prescribed values. Number k is called as a *defect* of this problem because k is a dimension missing to the completely defined Cauchy problem. (If the parameters p_i or the functionals Q_i are vectors then defect is defined regard to their dimensions.)

The problem with a finite number of unknown parameters in the equation can be also reduced to the problem described above. Let us illustrate this conjecture for the scalar problem

$$\dot{x}(t) = f(t, x_t; p_1) \ \ (t_0 \le t \le T), \ \ x(t) = \phi(t) \ \ (t_0 - h \le t \le T), \ \ x(T) = q_1.$$

Designate $y(t) := (x(t), p(t))'$. Then for y we obtain 2–dimensional boundary value problem with defect 1

$$\dot{y}(t) = (f(t, x_t; p(t)), 0)' \quad (t_0 \le t \le T),$$

$$y(t) = (\phi(t), p_1)' \quad (t_0 - h \le t \le t_0), \ \ Q(y_T) = q_1,$$

where

$$Q(v) := v_1(0), \quad \forall v = (v_1, v_2)' \in C([-h, 0], \mathbf{R}^2).$$

The case of any number of parameters in the equation can be considered in the same way.

Choosing the functions Q_j we can involve not only boundary value problems with conditions for $t \in [t_0 - h, t]$ and $t = T$ but also the de la Vallée – Poussin's problem, the problems with the conditions of the integral type etc.. For $h \ge T - t_0$ it is evident. If $h < T - t_0$ for Eq. (1.1) then we let formally assume that h is arbitrarily large.

Let us assume for simplicity that all the functionals $Q_i : C([-h, 0], \mathbf{R}^n) \to \mathbf{R}$. (Some of these functionals or even all of them can be defined on more narrow spaces if it is *apriori* known or it is required that the solution must belong them. For example under conditions of the smoothness increasing of the solution with the increasing of t (see 3.1.3) these functionals can be defined on the spaces of the functions which are differentiable some times.)

In addition it is supposed that all the functions under consideration are continuous, and a unique solution $t \mapsto x(t; \phi)$ of the Cauchy problem for Eq. (1.1) exists on the whole interval $[t_0, T]$ for any initial function ϕ. Then the problem (1.1)–(1.3) is reduced to the solution of the system of equation

$$Q_j(x_T(\cdot; \theta \mapsto \phi(\theta; p_1, \dots, p_k))) = q_j \quad (j = 1, \dots, k) \qquad (1.4)$$

where the number of unknown parameters is equal to the number of equations.

Let be additionally given that (1.1)–(1.3) is a linear problem, i.e. it has a form

$$\dot{x}(t) = L(t, x_t) + f(t) \quad (t_0 \le t \le T), \qquad (1.5)$$

$$x_{t_0} = \sum_{j=1}^{k} p_j \phi_j + \phi, \tag{1.6}$$

$$Q_i(x_T) = q_i \quad (i = 1, \dots, k). \tag{1.7}$$

Here $L(t, \cdot) : C([-h, 0], \mathbf{R}^n) \to \mathbf{R}^n$ is a linear functional which depends continuously on $t \in [t_0, T]$; function $f \in C[t_0, T]$; all functions $\phi_j, \phi \in C[-h, 0]$; linear functionals $Q_i : C([-h, 0], \mathbf{R}^n) \to \mathbf{R}$; at last q_j are arbitrary constants. Then the solution of Eq. (1.5) with initial condition $x_{t_0} = \phi$ has a form

$$x(t) = (\mathcal{L}\phi)(t) + (\mathcal{M}f)(t) \quad (t_0 \le t \le T),$$

where

$$\mathcal{L} : C[-h, 0] \to C[t_0, T], \quad \mathcal{M} : C[t_0, T] \to C[t_0, T]$$

are linear operators. In addition, the condition of existence and uniqueness of the solution on the interval $[t_0, T]$ are met here. Hence the solution of Pr. (1.5), (1.6) has a form

$$x(t) = \sum_{j=1}^{k} p_j (\mathcal{L}\phi)(t) + (\mathcal{L}\phi)(t) + (\mathcal{M}f)(t).$$

In order to determine the parameters p_j we obtain the system of linear algebraic equations

$$\sum_{j=1}^{k} Q_i((\mathcal{L}\phi_j)_T) p_j = q_i - Q_i((\mathcal{L}\phi)_T) - Q_i((\mathcal{M}f)_T) \quad (i = 1, \dots, k). \tag{1.8}$$

We can see, in particular, that for the existence and uniqueness of the solution of Eq. (1.8) (and, hence, Pr. (1.5)–(1.7)) with any functions f, ϕ and constants q_j it is necessary and sufficient that $\det(Q_i((\mathcal{L}\phi_j)_T)) \ne 0$, or, equivalently, that Pr. (1.5)–(1.7) has the only trivial solution for $f = 0, \phi = 0, \forall q_i = 0$.

The proof of theorems on solvability of nonlinear problem (1.1)–(1.3) can use some methods of finite–dimensional nonlinear analysis. Let us consider one example of such theorem the proof of which is founded on the notion of a rotation of the vector field.

Theorem 1.1 Let be given Pr. (1.1)–(1.3) of the form

$$\begin{gathered} \dot{x}(t) = L(t, x_t) + \tilde{f}(t, x_t) + f(t) \quad (t_0 \le t \le T), \\ x_t = \sum_{j=1}^{k} p_j \phi_j + \tilde{\phi}(\cdot; p_1, \dots, p_k), \; Q_i(x_T) = q_i \quad (i = 1, \dots, k). \end{gathered} \tag{1.9}$$

Here L, F, ϕ_j, Q_i, q_j are the same as in (1.5)–(1.7),

$$\max_t |\widetilde{f}(t,v)| \to 0, \quad \text{as} \quad \| v \| \to \infty$$

and

$$\| \widetilde{\phi}(\cdot; p_1, \ldots, p_k) \| = o(|p_1| + \ldots + |p_k|) \quad \text{as} \quad |p_1| + \ldots + |p_k| \to \infty.$$

Assume also that

$$\det(Q_i((\mathcal{L}\phi_j)_T)) \neq 0$$

where operator $\mathcal{L}$ is defined in connection with Pr. (1.5)–(1.7). Then Pr. (1.9) has at least one solution.

1.3. *HALANAY'S BOUNDARY VALUE PROBLEM*

A. Halanay [236] has proposed the following additional condition for Eq. (1.1)

$$\mathcal{N}(\phi, x_T) = 0 \tag{1.10}$$

where $\mathcal{N} : (C[-h,0])^2 \to C[-h,0]$ is a given operator and $\phi := x_{t_0}$ is not given in advance. For linear case this condition has a form

$$\mathcal{A}\phi + \mathcal{B}x_T = \psi \tag{1.11}$$

where $\mathcal{A}, \mathcal{B} : C[-h,0] \to C[-h,0]$ are given linear operators. Problem (1.1), (1.10) is reduced to the determination of the function $\phi \in C[-h,0]$ from the equation

$$\mathcal{N}(\phi, x_T(\cdot;\phi)) = 0.$$

Problem (1.5), (1.11) is reduced to the determination ϕ from the equation

$$\mathcal{A}\phi + \mathcal{B}(\mathcal{L}\phi)_T = \psi - (\mathcal{M}f)_T.$$

If we put $\mathcal{L}_t\phi := (\mathcal{L}\phi)_t$ then the latter equation takes the form

$$(\mathcal{A} + \mathcal{B}\mathcal{L}_T)\phi = \psi - (\mathcal{M}f)_T. \tag{1.12}$$

Let the operator $\mathcal{A}$ be inversible, so that $\mathcal{A}^{-1}$ is defined on $C[-h,0]$. Then Eq. (1.12) is equivalent to the equation

$$(I + \mathcal{A}^{-1}\mathcal{B}\mathcal{L}_T)\phi = \mathcal{A}^{-1}(\psi - (\mathcal{M}f)_T).$$

Assume also that $T - t_0 \geq h$. Then $\mathcal{L}_T$ maps any bounded set of initial functions from $C[-h,0]$ into bounded set of functions with uniformly bounded derivatives. Hence by virtue of the Arzelá's theorem this operator is completely continuous. It means that the operator $\mathcal{A}^{-1}\mathcal{B}\mathcal{L}_T$ is completely continuous as well. Using the Riesz – Schauder's theory of the linear equations with completely continuous operators we obtain that if Pr. (1.5), (1.11) with $f = 0, \psi = 0$ has the only trivial solution (or, equivalently, -1 is not eigenvalue of the operator $\mathcal{A}^{-1}\mathcal{B}\mathcal{L}_T$) then corresponding nonhomogeneous problem has only one solution for any functions f and ψ. If the homogeneous problem has nontrivial solutions then for the existence of solution of the nonhomogeneous problem it is necessary and sufficient that the function $\mathcal{A}^{-1}(\psi-(\mathcal{M}f)_T)$ be orthogonal to any solution of the conjugate homogeneous equation.

Now let be $T - t_0 < h$. Then the previously formulated results are valid under additional assumption about operator $\mathcal{A}^{-1}\mathcal{B}$. Thus it is sufficient to assume that there is a number $g \in [0, T - t_0)$ such that if $\| \phi \| \leq 1$ and $\phi|_{[\theta,0]}$ for some $\theta \in [-h, -g)$ satisfies the Lipschitz condition with a constant 1, then $(\mathcal{A}^{-1}\mathcal{B}\phi)|_{[\theta+g,0]}$ satisfies the Lipschitz condition with a constant l_θ, depending only on θ. (Such condition is satisfied by many difference and integral operators, by operators of multiplication on a function etc.) Then it is easy to check that the operator $\mathcal{A}^{-1}\mathcal{B}\mathcal{L}_T$ in some degree $k \geq 2$ will be completely continuous.

In fact let us choose some initial function ϕ with $\| \phi \| \leq 1$. Then $\| \mathcal{L}_T\phi \|$ and the Lipschitz constant for $(\mathcal{L}\phi)|_{[t_0,T]}$ will be less then some constants r and ρ respectively, which depend only on operator $\mathcal{L}$. Consequently $\mathcal{L}_T\phi$ satisfies the Lipschitz condition with a constant ρ on the interval $[t_0 - T, 0]$. Therefore $\| \mathcal{A}^{-1}\mathcal{B}\mathcal{L}_T\phi \| \leq \| \mathcal{A}^{-1}\mathcal{B} \| r$ and $\mathcal{A}^{-1}\mathcal{B}\mathcal{L}_T\phi$ satisfies on the interval $[t_0 - T + g, 0]$ the Lipschitz condition with a constant $\max\{r,\rho\}l_{t_0-T}$. If $2(T - t_0) - g \geq h$ then the function $\mathcal{L}_T(\mathcal{A}^{-1}\mathcal{B}\mathcal{L}_T)\phi$ satisfies the Lipschitz condition with a constant $\rho_1 := \max\{\rho l_{t_0-T}, \| \mathcal{A}^{-1}\mathcal{B} \| r\rho\}$. From here it follows the validity of the assertion for $k = 2$. If $2(T - t_0) - g < h$ then we have $\| (\mathcal{A}^{-1}\mathcal{B}\mathcal{L}_T)^2\phi \| \leq \| \mathcal{A}^{-1}\mathcal{B} \|^2 r^2$ and $(\mathcal{A}^{-1}\mathcal{B}\mathcal{L}_T)^2\phi$ satisfies the Lipschitz condition with a constant $\max\{\| \mathcal{A}^{-1}\mathcal{B} \| r^2, \rho_1\}l_{2(t_0-T)+g}$ on the interval $[2(t_0 - T + g), 0]$. Now if $3(T - t_0) - 2g \geq h$ we obtain our assertion with $k = 3$. If $3(T - t_0) - 2g < h$ then we are continuing in a similar way. As a result we prove the assertion of the previous item with a k equal integral part of $(h - g)(T - t_0 - g)^{-1}$.

Remark that the basic theorems of the Riesz – Schauder's theory are applicable without any changes for the case when the completely continuous operator is replaced on the operator some positive entire degree of which is completely continuous. This leads to the Fredholm type theorems for the considered case.

Using the notion of rotation of the vector field in the Banach space we can obtain the theorem, similar to Thm. 1.1, about existence of at least one solution of the problem obtained by sufficiently small nonlinear perturbation of Eq. (1.5) and boundary condition (1.11) if -1 is not eigenvalue of the operator $\mathcal{A}^{-1}\mathcal{B}\mathcal{L}_T$.

1.4. *PERIODIC PROBLEM*

Let us assume that in Eq. (1.1) we have $f(t+T,v) \equiv f(t,v)$ for some $T = \text{const} > 0$. Then the problem of the searching of T–periodic solution of this equation is equivalent to the problem of searching of the solution of Eq. (1.1) on the interval $[0,T]$ under boundary condition $\phi - x_T = 0$, i.e. this is a particular case of the Halanay problem.

Hence from Subs. 1.3 it follows that if T–periodic equation (1.5) with $f = 0$ has no nontrivial T–periodic solutions, then for any continuous T–periodic function f it has exactly one T–periodic solution. But if Eq. (1.5) with $f = 0$ has nontrivial T–periodic solutions, then for existence of T–periodic solutions with T–periodic function f it is necessary and sufficient the absence of resonance. Criterion of this absence is orthogonality of $f|_{[0,T]}$ to the restriction of any T–periodic solution of homogeneous equation of the form (1.5) with the operator conjugate to L (see 3.5.4, where the conjugate operator is written in the explicit form). Let us note that the conjugate equation to the linear RDE is an equation of advanced type. But in the process of construction of the periodic solution of such equation the unknown function in the righthand side can be shifted on any number of periods. This gives the possibility to pass to the equivalent RDE.

The simplest case is autonomous Eq. (1.5)

$$\dot{x}(t) = Lx_t + f(t) \quad (t \in \mathbf{R}) \tag{1.13}$$

where $L : C[-h,0] \to \mathbf{R}^n$ is a linear functional, and $f \in C\mathbf{R}$ is 2π–periodic function (we can always pass to this period by the change of variable $t = (T/2\pi)t_1$). Let us denote by $f_j \in \mathbf{C}^n (j \in \mathbf{Z})$ the Fourier coefficients of the function f, i.e.

$$f_j := \int_{-\pi}^{\pi} e^{-ijt} f(t)dt.$$

Then if for $f = 0$ Eq. (1.13) has no nontrivial 2π–periodic solutions, then for any function f a unique 2π–periodic solution of Eq. (1.13) is represented as a sum of the absolutely convergent Fourier series

$$x(t) = \sum_{j=-\infty}^{\infty} x_j e^{-ijt}$$

where coefficients x_j are defined by the formulae

$$x_j := (ijI - A'_j)^{-1} f_j \quad (\forall j \in \mathbf{Z}).$$

Here A_j is a matrix whose successive columns are the vectors $L(\theta \mapsto e^{ij\theta}e_k)$, where $e_k (k = 1, \ldots, n)$ are the columns of the matrix J. Necessary and sufficient condition for the absence of above mentioned nontrivial solutions is

$$\det(ijI - A_j') \neq 0 \quad (\forall j \in \mathbf{Z}). \tag{1.14}$$

If this condition is not valid for some $j = \bar{j}$, then for the absence of resonance it is necessary and sufficient the solvability with respect to x_j of the equation

$$(ijI - A_j')x_j = f_j.$$

The relation

$$f_j \in (ijI - A_j')\mathbf{R}^n$$

can serve as a criterion of this. If this relation is valid then the linear manifold of 2π–periodic solutions of Eq. (1.14) has dimension $[n - \mathrm{rk}(ijI - A_j')]$. The conditions of the absence of resonance and dimension of the manifold of 2π–periodic solutions, if $\det(ijI - A_j') = 0$ more than for one j (there can exist only finite number of such j), have a similar form.

2. BOUNDARY VALUE PROBLEMS FOR FDEs OF NONEVOLUTIONARY TYPE

Here we consider briefly the statements of the boundary value problems for FDEs of nonevolutionary type. For such equations the independent variable can not be treated as a time because the deviation of the argument in the differential law can be twosided. In order to underline this we denote the independent variable by x. Equations of such type appear in particular as the Euler equations in the problems of calculus of variations with deviating argument. But the diffusion of these equations out of the framework of the "pure" mathematics is not sufficient up to now.

Moreover, usual subdivision of FDEs on RDEs, NDEs and ADEs for nonevolutionary equations is not entirely adequate to their properties. It seems that more appropriate is the subdivision of these equations on the equations with unique principal term (containing the higher derivative of an unknown function) and the equations with nonunique principal term. Besides, here it is quite natural to consider systems of FDEs of second order, moreover that the Euler equations just mentioned above have for the simple cases the second order.

2.1. *FDES WITH UNIQUE PRINCIPAL TERM*

Let us consider here one of the possible variants of the statement of the boundary value problem for FDE

$$\ddot{u}(x) = F(u, \dot{u})(x) \quad (a \leq x \leq b), \quad \dot{} = d/dx. \tag{2.1}$$

Here $-\infty < a < b < \infty$, and F is an operator (details see below). As additional conditions we assign the values of u on some intervals $[a_1, a]$ and $[b, b_1]$ where $-\infty < a_1 \leq a$,$b \leq b_1 < \infty$:

$$u(x) = \phi(x) \quad (a_1 \leq x \leq q, \; b \leq x \leq b_1).$$

In addition $\phi|_{[a_1,a]} \in C^1$ if $a_1 < a$, and $\phi|_{[b,b_1]} \in C^1$ if $b < b_1$. It is assumed that the operator F maps continuously $C[a_1, b_1] \times C_{[a,b]}[a_1, b_1]$ in $L[a, b]$, where $C_{[a,b]}[a_1, b_1]$ is a space of functions $[a_1, b_1] \to \mathbf{R}^n$ which are continuous on $[a, b]$ and uniformly continuous on $[a_1, b_1] \setminus [a, b]$ with an uniform norm. The solution is interpreted as a function $u \in AC^1[a, b]$, and Eq. (2.1) must be satisfied almost everywhere on $[a, b]$. Hence, when the argument x runs through the interval $[a, b]$ the deviating argument is contained in $[a_1, b_1]$.

The choice of such spaces can be explained as follows. Even the simplest examples show that in general it is possible to achieve the concordance of the values of the solution u and the boundary function ϕ for $x = a$ and $x = b$ for the second order equations. But the values of $\dot{u}$ and $\dot{\phi}$ for these x can not be coordinated; the space $C_{[a,b]}[a_1, b_1]$ obtained from here. This disconcordance can generate discontinuity of $\ddot{u}$ by virtue of Eq. (2.1). Thus the choice $C[a, b]$ as a domain of the values of the operator F could be in general case an extremely strong restriction. This disconcordance makes difficult to write Eq. (2.1) in the seeming more natural functional form

$$\ddot{u}(x) = f(x, u_x, \dot{u}_x) \;\; (v_x(\xi) := v(x + \xi), \;\; -h \leq \xi \leq h_1; \;\; h, h_1 = \text{const} \geq 0).$$

(Some cases when this disconcordance can be avoided will be indicated below.)

It is easy to pass from an arbitrary boundary function to the function $\phi(x) \equiv 0$. For this it is sufficient to choose any function $\phi_0 \in C^2[a, b]$ for which $\phi_0(a) = \phi(a), \phi_0(b) = \phi(b)$ and after that to replace the function u on u_1 using the formulas

$$u(x) = u_1(x) + \phi_0(x) \;\; (a \leq x \leq b),$$

$$u(x) = u_1(x) + \phi(x) \;\; (a_1 \leq x \leq a, \;\; b \leq x \leq b_1).$$

Then for u_1 the boundary function turns out to be zero. Let us assume that this change is already done, hence the boundary condition for Eq. (2.1) has a form

$$u(x) = 0 \;\; (a_1 \leq x \leq a, \;\; b \leq x \leq b_1). \tag{2.2}$$

One of the possible ways to investigate Pr. (2.1), (2.2) is the following. Let us denote $\ddot{u}(x) =: w(x)$. Then by (2.2) we have

$$u(x) = \int_a^b G(x,s)w(s)ds, \quad \dot{u}(x) = \int_a^b G_1(x,s)w(s)ds \ (a \le x \le b). \quad (2.3)$$

Here G is the Green function, and G_1 is its derivative with respect to x:

$$G(x,s) := \begin{cases} -(x-a)(b-s)(b-a)^{-1} & (a \le x \le s), \\ -(s-a)(b-x)(b-a)^{-1} & (s \le x \le b), \end{cases}$$

$$G_1(x,s) = \begin{cases} -(b-s)(b-a)^{-1} & (a \le x < s), \\ (s-a)(b-a)^{-1} & (s < x \le b). \end{cases}$$

Substitution (2.3) into (2.1) gives us the equivalent operator equation in the space $L[a,b]$

$$w = F\left(\int_a^b G(\cdot,s)w(s)ds, \quad \int_a^b G_1(\cdot,s)w(s)ds\right). \quad (2.4)$$

In addition, by virtue of (2.2), both functions representing the arguments of F must be continued by zero outside of $[a,b]$. If w is found we obtain u from the first formula (2.3).

Some fixed point theorem can be applied to Eq. (2.4). Let us apply, e.g., the Banach contraction mappings theorem. In order to do this we assume that the operator F satisfies the Lipschitz condition with respect to its arguments with constants l and l_1 respectively. Then for any $w_1, w_2 \in L[a,b]$ we have:

$$\| F\left(\int_a^b G(\cdot,s)w_2(s)ds, \int_a^b G_1(\cdot,s)w_2(s)ds\right) -$$

$$-F\left(\int_a^b G(\cdot,s)w_1(s)ds, \int_a^b G_1(\cdot,s)w_1(s)ds\right) \|_{L[a,b]} \le$$

$$\le l \| \int_a^b G(\cdot,s)[w_2(s)-w_1(s)]ds \| + l_1 \| \int_a^b G_1(\cdot,s)[w_2(s)-w_1(s)]ds \| \le$$

$$\leq (l/4)(b-a)\int_a^b |w_2(s)-w_1(s)|ds + l_1\int_a^b |w_2(s)-w_1(s)|ds =$$

$$= [(l/4)(b-a)+l_1] \parallel w_2 - w_1 \parallel_{L[a,b]} .$$

From here and the Banach's theorem it follows

Theorem 2.1 Let

$$(l/4)(b-a)+l_1 < 1. \tag{2.5}$$

Then Pr. (2.1), (2.2) has a unique solution which can be obtained by the iteration method applied to Eq. (2.4) with regard to the first equality (2.3).

Let us indicate some cases when the disconcordance, mentioned above, can be overcome, and, hence, we can assume that

$$F : (C[a_1, b_1])^2 \mapsto C[a, b].$$

1. Let the operator F contain $2n$ parameters. Then the statement of the boundary value problem (2.1), (2.2) can be supplemented by the additional condition $\dot{u}(a^+) = \dot{u}(b^-) = 0$ which excludes the disconcordance.

2. Let the operator F be "onesided", i.e. $a_1 < a, b = b_1$ (the case $a_1 = a, b < b_1$ can be considered similarly). Then the boundary condition can be formulated as "onesided", i.e. the condition at $x = b$ can be completely excluded, and, instead of it, at $x = a$ must be put the condition $\dot{u}(a) = 0$ which provides the continuity of $\dot{u}$ at $x = a$. (In addition in (2.4) the function G has a form $G(x,s) = x - s$ for $a \leq x \leq s$; $G(x,s) = 0$ for $s \leq x \leq b$.) Let us remark that though the additional condition is prescribed for $x \leq a$ and the solution is constructed for $x \geq a$, the problem can not be entirely called as an initial value problem because Eq. (2.1) is not assumed to be the evolutionary one.

3. Assume that the deviating argument in Eq. (2.1) belongs $[a, b]$, i. e. $a_1 = a, b_1 = b$. Then the boundary conditions (2.2) become one–point ones, the problem of discontinuity of $\dot{u}(x)$ for $x = a$ and $x = b$ disappears, and instead of (2.2) additional conditions can be prescribed at other points or can be distributed over the interval $[a, b]$.

For linear case Eq. (2.1) has a form

$$\ddot{u}(x) = \mathcal{L}_1(u)(x) + \mathcal{L}_2(\dot{u})(x) + f(x) \quad (a \leq x \leq b), \tag{2.6}$$

where

$$\mathcal{L}_1 : C[a_1, b_1] \to L[a, b] \quad \text{and } \mathcal{L}_2 : C_{[a,b]}[a_1, b_1] \to L[a, b]$$

are linear operators, and $f \in L[a,b]$. Sufficient condition (2.5) of solvability of Pr. (2.6), (2.2) has a form

$$[(b-a)/4] \parallel \mathcal{L}_1 \parallel + \parallel \mathcal{L}_2 \parallel < 1.$$

Equation (2.4) has a form

$$w = \mathcal{L}_1 \left(\int_a^b G(\cdot, s) w(s) ds \right) + \mathcal{L}_2 \left(\int_a^b G_1(\cdot, s) w(s) ds \right) + f. \tag{2.7}$$

Let us require that $\mathcal{L}_2$ maps any bounded set of functions from $AC[a,b]$ with uniformly bounded variation, continued on $[a_1,b_1] \setminus [a,b]$ by zero, into compact set in $L[a,b]$. (It seems that this is not too strong restriction because the identity embedding $C[a,b]$ in $L[a,b]$ has this property.) Then it is easy to check that the operator $L[a,b] \to L[a,b]$ defined by the righthand side of (2.7) is completely continuous. In fact, the map $w \mapsto \int_a^b G(\cdot, s) w(s) ds$ transfer the bounded set of functions in $L[a,b]$ into the compact set in $C[a,b]$. Hence first summand defines completely continuous operator. On the other hand the map

$$w \mapsto \int_a^b G_1(\cdot, s) w(s) ds$$

transfer the bounded set in $L[a,b]$ into the bounded set of absolutely continuous functions with uniformly bounded variation. Therefore the operator defined by the second summand in (2.7) is completely continuous by virtue of the assumption made above.

Complete continuity of the operator defined by the righthand side of Eq. (2.7) means that for this equation (and also for the equivalent problem (2.6), (2.2)) are valid general theorems of the Rietsz – Schauder's theory.

2.2. *FDES WITH NONUNIQUE PRINCIPAL TERM*

Let us consider the vector FDE of the second order in divergent form

$$[F(u, \dot{u})(x)]^{\bullet} + F_1(u, \dot{u})(x) = 0 \quad (a \leq x \leq b) \tag{2.8}$$

with boundary conditions (2.2). Let us assume that in (2.8) the operators $F, F_1 : C[a_1, b_1] \times L[a_1, b_1] \to L[a,b]$. The solution of Pr. (2.8), (2.2) is a function $u \in AC[a,b]$ for which $u(a) = u(b) = 0$ and if this function is continued by zero on $[a_1, b_1] \setminus [a,b]$ then $F(u, \dot{u}) \in AC[a,b]$ and Eq. (2.8) is satisfied almost everywhere on $[a,b]$. The last condition is equivalent to equality

$$F(u,\dot{u})(x)+\int_a^x F_1(u,\dot{u})(s)ds = F(u,\dot{u})(a) \quad (a\le x\le d). \tag{2.9}$$

Since the existence of $\ddot{u}$ is not required, this definition is similar to the definition of the quasisolution for ODEs.

Let be $\eta \in C^1[a,b]$, $\eta(a)=\eta(b)=0$, and u is a solution of Pr. (2.8), (2.2). Let us multiply both parts of (2.8) on $\eta(x)$ and equate the integrals from the both parts over $[a,b]$. Then we have, after integration by parts,

$$\int_a^b \{[-F(u,\dot{u})(x)]'\dot{\eta}(x)+[F_1(u,\dot{u})(x)]'\eta(x)\}dx = 0. \tag{2.10}$$

The opposite conjecture is also true: let be given a function $u\in AC[a,b]$, $u(a)=u(b)=0$, such that equality (2.10) is fulfilled for any function η from the indicated class, if u is continued by zero on $[a_1,b_1]\setminus[a,b]$. Then u is a solution of Pr. (2.8), (2.2). Thus we obtain the notion similar to that of the weak solution which is well–known in the theory of partial differential equation.

Divergent form (2.8) and the notion of the solution given above appear in a natural way in the calculus of variation with deviating argument.

Consider the solvability theorem for the problem under consideration. Let us assume that Eq. (2.9) has a form

$$[\dot{u}(x)-F(u,\dot{u})(x)] = \int_a^x F_1(u,\dot{u})(s)ds + [\dot{u}(a)-F(u,\dot{u})(a)] \quad (a\le x\le b). \tag{2.11}$$

Here operators F, F_1 satisfy the same conditions as in Eq. (2.8). In addition it is assumed that $u\in AC[a,b]$, $u(a)=u(b)=0$, $\dot{u}-F(u,\dot{u})$ must coincide almost everywhere with some AC–function, and square brackets in (2.11) point to the values of this function. Square brackets in the righthand side can not be generally opened, because $\dot{u}(a)$ and $F(u,\dot{u})(a)$ generally are not defined.

In order to overcome this complication let us integrate both parts of (2.11) from a to b. Taking into account boundary conditions for u we obtain

$$0-\int_a^b F(u,\dot{u})(s)ds = \int_a^b (b-s)F_1(u,\dot{u})(s)ds + [\dot{u}(a)-F(u,\dot{u})(a)](b-a).$$

From here excluding the second summand in the righthand side of (2.11) it follows the equation

$$\dot{u}(x) = F(u,\dot{u})(x) + \int_a^x F_1(u,\dot{u})(s)ds -$$
$$-(b-a)^{-1}\left\{\int_a^b [F(u,\dot{u})(s) + (b-s)F_1(u,\dot{u})(s)]ds\right\}. \tag{2.12}$$

This equation, which is equivalent to (2.11) (with regard of boundary condition (2.2)) has to be satisfied almost everywhere on $[a,b]$.

At last let us put

$$\dot{u}(x) =: w(x), \quad u(x) = \int_a^x w(s)ds \tag{2.13}$$

and rewrite Eq. (2.12) in the functional form

$$w = F\left(\int_a^\bullet w(s)ds, w\right) + \int_a^\bullet F_1\left(w(\int_a^\bullet \sigma)d\sigma, w\right)(s)ds -$$
$$-(b-a)^{-1}\{\int_a^b [F\left(\int_a^\bullet w(\sigma)d\sigma, w\right)(s) + \tag{2.14}$$
$$+(b-s)F_1\left(\int_a^\bullet w(\sigma)d\sigma, w\right)(s)]ds\}.$$

It is required to find a solution $w \in L[a,b]$ for which

$$\int_a^b w(x)dx = 0. \tag{2.15}$$

For determination of the values of the operators F and F_1 both their arguments must be continued by zero outside of $[a,b]$.

Let us denote by $L_0[a,b]$ a subspace of the space $L[a,b]$ defined by the condition (2.15). The map $\mathcal{L}$ defined by the righthand side of Eq. (2.14) transform $L_0[a,b]$ into itself. Let us assume that $F(F_1)$ satisfies the Lipschitz condition with respect to the first and second arguments with the constants l and l'(l_1 and l'_1) respectively. Then for $w_1, w_2 \in L_0[a,b]$ we have

$$\| F\left(\int_a^{\bullet} w_2(s)ds, w_2\right) - F\left(\int_a^{\bullet} w_1(s)ds, w_1\right) \|_{L[a,b]} \leq$$

$$\leq l \| \int_a^{\bullet} w_2(s)ds - \int_a^{\bullet} w_1(s)ds \| + l' \| w_2 - w_1 \|_{L[a,b]} \leq (l+l') \| w_2 - w_1 \|_{L[a,b]},$$

$$\| \int_a^{\bullet} F_1\left(\int_a^{\bullet} w_2(s)ds, w_2\right)(s)ds - \int_a^{\bullet} F_1\left(\int_a^{\bullet} w_1(\sigma)d\sigma, w_1\right)(s)ds \|_{L[a,b]} \leq$$

$$\leq \int_a^b (b-x)|F_1\left(\int_a^{\bullet} w_2(\sigma)d\sigma, w_2\right)(x) - F_1\left(\int_a^{\bullet} w_1(\sigma)d\sigma, w_1\right)(x)|dx \leq$$

$$\leq (b-a)(l_1 + l'_1) \| w_2 - w_1 \|_{L[a,b]},$$

$$\| (b-a)^{-1} \int_a^b F\left(\int_a^{\bullet} w_2(\sigma)d\sigma, w_2\right)(s)ds -$$

$$-(b-a)^{-1} \int_a^b F\left(\int_a^{\bullet} w_1(\sigma)d\sigma, w_1\right)(s)ds \|_{L[a,b]} =$$

$$= |\int_a^b F\left(\int_a^{\bullet} w_2(\sigma)d\sigma, w_2\right)(s)ds - \int_a^b F\left(\int_a^{\bullet} w_1(\sigma)d\sigma, w_1\right)(s)ds| \leq$$

$$\leq \| F\left(\int_a^{\bullet} w_2(\sigma)d\sigma, w_2\right)(s)ds - F\left(\int_a^{\bullet} w_1(\sigma)d\sigma, w_1\right)(s)ds \|_{L[a,b]} \leq$$

$$\leq (l+l') \| w_2 - w_1 \|_{L[a,b]},$$

$$\| (b-a)^{-1} \int_a^b (b-s)F_1\left(\int_a^{\bullet} w_2(\sigma)d\sigma, w_2\right)(s)ds -$$

$$-(b-a)^{-1} \int_a^b (b-s)F_1\left(\int_a^{\bullet} w_1(\sigma)d\sigma, w_1\right)(s)ds \|_{L[a,b]} \leq$$

$$\leq \text{(analogously)} \leq (b-a)(l_1+l_1') \parallel w_2 - w_1 \parallel_{L[a,b]} .$$

Hence

$$\parallel \mathcal{L}w_2 - \mathcal{L}w_1 \parallel_{L[a,b]} \leq 2[(l+l') + (b-a)(l_1+l_1')] \parallel w_2 - w_1 \parallel_{L[a,b]} .$$

Applying the Banach's contraction map theorem we obtain

Theorem 2.2 Let

$$2[(l+l') + (b-a)(l_1+l_1')] < 1.$$

Then Pr. (2.11), (2.2) has a unique solution which can be obtained by application of the method of successive approximations to Eq. (2.14) with regard of second equality (2.13).

Solvability theorems of boundary problems for FDE of more special form with nonunique principal term can be found in the paper [264] where further indications can be found.

CHAPTER 12

FREDHOLM ALTERNATIVE FOR PERIODIC SOLUTIONS OF LINEAR FDEs

This chapter is devoted to the conditions of the existence of periodic solutions of linear FDEs and some their properties. Particular case of this problem for RDEs was considered in Subs. 11.1.4 in connection with the Halanay boundary value problem. Here we shall investigate in details the more general case.

1. EXISTENCE OF PERIODIC SOLUTIONS

1.1. *STATEMENT OF THE PROBLEM*

Consider equation

$$\sum_{k=1}^{m} B_k(t)\dot{x}(t-h_k) = \int_{-\infty}^{\infty} (d_s P(t,s))x(t+s) + f(t). \qquad (1.1)$$

Here $x(t), f(t) \in \mathbf{R}^n, B_k(t), P(t,s) \in \mathbf{R}^{n\times n}$; the entries of $f(t), B_k(t)$ and $P(t,s)$ are ω-periodic functions of t ($\omega = \text{const} > 0$); the entries of $P(t,s)$ are functions of bounded variation in $s \in (-\infty, \infty)$ for any fixed t; at last h_k are some constants. All functions are assumed to be measurable. Denote by L_ω the Hilbert space of ω-periodic functions $x : \mathbf{R} \to \mathbf{R}^n$ on $[0,\omega]$ with a finite norm

$$\| x \|_L := \sqrt{(x,x)}, \quad (x,y) := \int_0^\omega x'(t)y(t)dt,$$

where prime is a transposition sign. Let C_ω be a space of continuous ω-periodic functions $x : \mathbf{R} \to \mathbf{R}^n$ on $[0,\omega]$ with a norm

$$\| x \|_C := \max_{0\le t\le\omega} \sqrt{x'(t)x(t)}.$$

At last, W_ω is the Hilbert space of ω-periodic functions $x \in AC(\mathbf{R})$ (i.e. $x(t) \in C_\omega, \dot{x}(t) \in L_\omega$) with a finite norm

$$\| x \|_\omega := [\| x \|_C^2 + \| \dot{x} \|_L^2]^{1/2}.$$

We say that $(n \times n)$ matrix $A(t) \in L_\omega$ (or $A(t) \in C_\omega$) if every column of it is an element of L_ω (or C_ω). In this chapter we use the matrix norm $|A| := (\mathrm{Tr}A'A)^{1/2} = (\sum\limits_{i,j} a_{ij}^2)^{1/2}$ (where $A = (a_{ij})$ and $\mathrm{Tr}A$ is a trace of the matrix A). This norm similar to the Euclidean norm has the properties

$$|cA| = |c||A| \ (c \in \mathbf{R}), \quad |AB| \le |A||B|, \quad |Ax| \le |A||x|.$$

Consider the conjugate equation (see Subs. 3.5.4)

$$\frac{d}{dt}\left[\sum_{k=1}^{m} y'(t+h_k)B_k(t+h_k) + \int_{-\infty}^{\infty} y'(t-s)P(t-s,s)ds\right] = 0. \qquad (1.2)$$

Definition 1.1 The function $x \in W_\omega$, satisfying Eq. (1.1) almost everywhere is called a *ω-periodic solution of (1.1).*

Definition 1.2 The function $y \in L_\omega$ is called a *ω-periodic solution of conjugate equation (1.2)* if exists $c \in \mathbf{R}^n$ such that

$$Y(t) := \sum_{k=1}^{m} y'(t+h_k)B_k(t+h_k) + \int_{-\infty}^{\infty} y'(t-s)P'(t-s,s)ds = c'$$

almost everywhere.

Definition 1.3 Alternative of Fredholm takes place for Eq. (1.1) if one of the following properties is valid:

1) homogeneous equation (1.1) (i.e. Eq. (1.1) with $f(t) \equiv 0$) can have only the zero ω-periodic solution for any $f \in L_\omega$; Eq. (1.2) possesses the same property (the case of impossibility of resonance).

2) homogeneous equation (1.1) has $d \ge 1$ linear independent ω-periodic solutions $x_1(t), \dots, x_d(t)$, equation (1.2) also has d linear independent ω-periodic solutions $y_1(t), \dots, y_d(t)$, and nonhomogeneous Eq. (1.1) has ω-periodic solutions only for those $f \in L_\omega$ which satisfies the conditions

$$(f, y_i) = 0, \quad i = 1, \dots, d \qquad (1.3)$$

(the case of possibility of *resonance* which takes place if conditions (1.3) are broken).

We shall consider in this chapter some conditions for validity of the Fredholm alternative for Eq. (1.1). Below it is always assumed that

$$\sup_{k,t} |B_k(t)| < \infty.$$

Let us define the operator $T : L_\omega \to L_\omega$ as follows

$$Tz(t) := \sum_{k=1}^{m} B_k(t)z(t-h_k), \quad z \in L_\omega. \tag{1.4}$$

Consider in L_ω two operator equations

$$Tz(t) = f, \tag{1.5}$$

$$T^*u(t) = g. \tag{1.6}$$

Here $T^* : L_\omega \to L_\omega$ is an operator conjugated with T in L_ω, i.e. $(x_1, Tx_2) = (T^*x_1, x_2)$ for any $x_1, x_2 \in L_\omega$. Let us find T^*. We have

$$(x_1, Tx_2) = \int_0^\omega x_1'(t) \sum_{k=1}^{m} B_k(t)x_2(t-h_k)dt =$$

$$= \sum_{k=1}^{m} \int_{-h_k}^{\omega-h_k} x_1'(t+h_k)B_k(t+h_k)x_2(t)dt.$$

By virtue of ω–periodicity of the integrals

$$\int_{-h_k}^{\omega-h_k} x_1'(t+h_k)B_k(t+h_k)x_2(t)dt = \int_0^\omega x_1'(t+h_k)B_k(t+h_k)x_2(t)dt.$$

From here it follows

$$(T^*x_1, x_2) = \int_0^\omega \left[\sum_{k=1}^{m} x_1'(t+h_k)B_k(t+h_k)\right] x_2(t)dt.$$

As a result we obtain that

$$T^*u(t) = \sum_{k=1}^{m} B_k'(t+h_k)u(t+h_k). \tag{1.7}$$

1.2. *CONDITIONS OF THE FREDHOLM ALTERNATIVE VALIDITY*

Theorem 1.1 Assume that $\forall B_k \in C_\omega, f(t) \in L_\omega$, *the operator* U, *defined by the equality*

$$Ux(t) = \int_{-\infty}^{\infty} [d_s P(t,s)]x(t+s)$$

maps C_ω *in* L_ω*. (This assumption is valid if* $\int_0^\omega V(t)dt < \infty$ *where* $V(t)$ *is the variation of* $P(t,\cdot)$ *on* $\mathbf{R}$*.) Further the matrix* $P(t,s)$ *satisfies the conditions*

$$\int_{-\infty}^{\infty} P(\cdot,s)ds \in L_\omega, \tag{1.8}$$

$$P(t,-\infty) = P(t,\infty) = 0, \tag{1.9}$$

$$\sum_{i,j=1}^{n} \int_0^\omega dt \int_{-\infty}^{\infty} p_{ij}^2(t,s-t)ds < \infty. \tag{1.10}$$

In (1.8) it is assumed that the integral exists for almost all t. Then the Fredholm alternative for Eq. (1.1) is valid if and only if it is valid for operator equations (1.5).

(The definition of the Fredholm alternative for Eq. (1.5) is just the same as that of for Eq. (1.1).)

The proof of Thm. 1.1 is found on the following three lemmas in which it is assumed that all the assumptions of Thm. 1.1 are fulfilled.

Lemma 1.1 *Let* $x(t)$ *be a* ω*–periodic solution of (1.1). Then the function*

$$t \mapsto z(t) := \frac{1}{\omega}x(0) - \dot{x}(t) \tag{1.11}$$

is a solution in L_ω *of the equation*

$$\sum_{k=1}^{m} B_k(t)z(t-h_k) + \int_{-\infty}^{\infty} P(t,s)z(t+s)ds + \frac{A(t)}{\omega}\int_0^\omega z(s)ds = -f(t). \tag{1.12}$$

Here

$$A(t) := -\sum_{k=1}^{m} B_k(t) - \int_{-\infty}^{\infty} P(t,s)ds. \tag{1.13}$$

Conversely, if $z \in L_\omega$ *is a solution of (1.12), then the function*

$$t \mapsto x(t) := \frac{t}{\omega}\int_0^\omega z(s)ds + \int_t^\omega z(s)ds \tag{1.14}$$

is a ω*–periodic solution of (1.1).*

Proof Let $x(t)$ be a ω–periodic solution of (1.1). Then integrating (1.11) we have

$$x(0) = \int_0^\omega z(t)dt.$$

From here and (1.11)

$$\dot{x}(t) = \frac{1}{\omega}\int_0^\omega z(s)ds - z(t). \tag{1.15}$$

Integrating (1.15) from 0 to t we obtain that

$$x(t) = x(0) + \frac{t}{\omega}\int_0^\omega z(s)ds - \int_0^t z(s)ds =$$

$$= x(0) + \frac{t}{\omega}\int_0^\omega z(s)ds - x(0) + \int_t^\omega z(s)ds = \frac{t}{\omega}\int_0^\omega z(s)ds + \int_t^\omega z(s)ds. \tag{1.16}$$

Substitution (1.15), (1.16) in (1.1) yields that $z(t)$ satisfies the equation

$$\sum_{k=1}^m B_k(t)\left\{\frac{1}{\omega}\int_0^\omega z(s)ds - z(t-h_k)\right\} =$$

$$= \int_{-\infty}^\infty [d_sP(t,s)]\left\{\frac{t+s}{\omega}\int_0^\omega z(\alpha)d\alpha + \int_{t+s}^\omega z(\alpha)d\alpha\right\} + f(t). \tag{1.17}$$

The last integral in (1.17) can be transformed as follows

$$\int_{-\infty}^\infty [d_sP(t,s)]\left\{\frac{t+s}{\omega}\int_0^\omega z(\alpha)d\alpha + \int_{t+s}^\omega z(\alpha)d\alpha\right\} =$$

$$= P(t,s)\left\{\frac{t+s}{\omega}\int_0^\omega z(\alpha)d\alpha + \int_{t+s}^\omega z(\alpha)d\alpha\right\}\Bigg|_{s=-\infty}^{\infty} -$$

$$-\int_{-\infty}^\infty P(t,s)\left\{\frac{1}{\omega}\int_0^\omega z(\alpha)d\alpha - z(t+s)\right\}ds.$$

But

$$x(t) = \frac{t}{\omega} \int_0^\omega z(s)ds + \int_t^\omega z(\alpha)d\alpha$$

is a continuous periodic function, and $P(t, -\infty) = P(t, \infty) = 0$. Hence, Eq. (1.17) takes the form (1.12). First part of the lemma is proven.

Prove the second part. Let $z \in L_\omega$ and satisfy (1.22). Then $x(t)$, defined by (1.14), will be a ω–periodic function. In fact

$$x(t+\omega) = \frac{t+\omega}{\omega} \int_0^\omega z(\alpha)d\alpha + \int_{t+\omega}^\omega z(\alpha)d\alpha =$$

$$= \frac{t}{\omega} \int_0^\omega z(\alpha)d\alpha + \int_0^\omega z(\alpha)d\alpha + \int_{t+\omega}^t z(\alpha)d\alpha + \int_t^\omega z(\alpha)d\alpha = x(t).$$

If we replace z in (1.12) by the righthand side of (1.11) and implement the same transformations in inverse order we obtain that $x(t)$ satisfies (1.1). Lemma 1.1 is proven.□

Let us define the operator $U : L_\omega \to L_\omega$

$$Uz(t) = \int_{-\infty}^{\infty} P(t,s)z(t+s)ds + \frac{A(t)}{\omega} \int_0^\omega z(s)ds. \tag{1.18}$$

Then Eq. (1.12) takes the form

$$Tz + Uz = -f \tag{1.19}$$

where Tz is defined by (1.4).

Lemma 1.2 The operator T is bounded and the operator U is completely continuous in L_ω.

Proof The boundedness of T follows from the assumption $B_k(t) \in C_\omega$. The operator U is completely continuous because it is a sum of the one–dimensional operator

$$z \mapsto \frac{A(\cdot)}{\omega} \int_0^\omega z(s)ds$$

and integral operator which has by virtue of (1.10) a square integrable kernel (see, e.g. [266]). □

Lemma 1.3 Homogeneous equation, conjugate with (1.12) in L_ω, has a form

$$\sum_{k=0}^{m} y'(t+h_k)B_k(t+h_k) + \frac{1}{\omega}\int_0^\omega y'(s)A(s)ds + \\ + \int_{-\infty}^{\infty} y'(t-\alpha)P(t-\alpha,\alpha)d\alpha = 0. \tag{1.20}$$

In addition, each solution $y \in L_\omega$ of (1.20) is a ω-periodic solution of (1.2) and vice versa.

Proof Operator T^* is given by (1.7). Let us find U^*. We have

$$(U^*y, z) = (y, Uz) = \int_0^\omega y'(t)\frac{A(t)}{\omega}dt \int_0^\omega z(\alpha)d\alpha + \\ + \int_0^\omega y'(t)dt \int_{-\infty}^{\infty} P(t,s)z(t+s)ds.$$

Put $\xi = t + s$. Then using periodicity of the integrands and the Fubini theorem we have

$$(U^*y, z) = \int_0^\omega \left\{ \frac{1}{\omega}\int_0^\omega y'(t)A(t)dt + \int_{-\infty}^{\infty} y'(\xi - s)P(\xi - s, s)ds \right\} z(\xi)d\xi.$$

Hence

$$U^*y = \frac{1}{\omega}\int_0^\omega y'(t)A(t)dt + \int_{-\infty}^{\infty} y'(\xi - s)P(\xi - s, s)ds.$$

Taking into account (1.7) we can see that the equation $T^*y + U^*y = 0$ has a form (1.20).

Further, if $y \in L_\omega$ satisfies (1.20) then differentiating (1.20), we obtain that y is a ω-periodic solution of (1.2). Conversely, let y be a ω-periodic solution of (1.2). Denote

$$Y(t) = \sum_{k=1}^{m} y'(t+h_k)B_k(t+h_k) + \int_{-\infty}^{\infty} y'(t-\alpha)P(t-\alpha,\alpha)d\alpha.$$

Then integrating both parts of (1.2)

$$Y(t) = Y(\omega) = \frac{1}{\omega}\int_0^\omega Y(s)ds. \tag{1.21}$$

But

$$\frac{1}{\omega}\int_0^\omega Y(s)ds = -\frac{1}{\omega}\int_0^\omega y'(t)A(t)dt.$$

So, (1.21) coincides with (1.20). Lemma 1.3 is proven. □

Proof of Thm. 1.1 It is known the theorem [266] that the Fredholm alternative for the equation $Bx = f$ in the Banach space is valid if and only if the linear operator B can be represented as a sum $B = W + V$, where W is a linear operator which has two–sided linear bounded inverse operator, and V is a compact operator.

From here we can derive the following conclusion. Let B_1 and B_2 be two operators such that their difference is a completely continuous operator. Then for the equations $B_1x = f$ and $B_2x = f$ the Fredholm alternative is valid (or invalid) simultaneously.

In fact, if the Fredholm alternative is valid for the equation $B_1x = f$, then $B_1 = W_1 + V_1$, where operators W_1 and V_1 satisfy the conditions of theorem from [266] cited above. But then $B_2 = W_1 + V_1 + (B_2 - B_1) = W_1 + V_2$, where V_2 is a compact operator. Hence, the Fredholm alternative is also valid for the equation $B_2y = g$.

By virtue of Lemma 1.1 the finding of periodic solution of Eq. (1.1) is equivalent to the solution in L_ω of the operator equation

$$(T + U)z = f. \tag{1.22}$$

By virtue of Lemma 1.3 any solution of the conjugate equation

$$(T^* + U^*)y = 0$$

is the ω–periodic solution of Eq. (1.2). Hence, the Fredholm alternative for Eq. (1.22) is equivalent to the Fredholm alternative on periodic solutions of (1.1) (see definition 1.3). But the operator U is completely continuous. Hence, validity of the Fredholm alternative for Eq. (1.5) is necessary and sufficient condition for its validity for Eq. (1.1). Thm. 1.1 is proven. □

The following remarks are valid under made assumptions made above.

Remark 1.1 For equation

$$\sum_{k=1}^m B_k(t)\dot{x}(t - h_k) + \sum_{i=1}^m A_i(t)x(t - \tau_i) = f(t)$$

the conjugate equation has a form

$$-\frac{d}{dt}\left\{\sum_{k=1}^m y'(t + h_k)B_k(t + h_k)\right\} + \sum_{i=1}^m y'(t + \tau_i)A_i(t + \tau_i) = 0.$$

Here $A_i(t) \in \mathbf{R}^{n\times n}$ are some measurable matrices.

Remark 1.2 For equation

$$\sum_{k=1}^{m} B_k(t)\dot{x}(t-h_k) = \int_a^b [d_s P(t,s)]x(t+s) + f(t)$$

it is sufficient to assume additionally that $P(t,a) = 0$. In this case the conjugate equation has a form

$$\frac{d}{dt}\left\{\sum_{k=1}^{m} y'(t+h_k)B_k(t+h_k) + \int_a^b y'(t-s)P(t-s,s)ds\right\} +$$

$$+y'(t-b)P(t-b,b) = 0.$$

Remark 1.3 Sufficient condition for validity of the Fredholm alternative for Eq. (1.1) is the existence of the bounded inverse operator T^{-1}. In particular, for the equation

$$\dot{x}(t) = \int_{-\infty}^{\infty} [d_s P(t,s)]x(t+s) + f(t)$$

the operator T is the identity. Hence, for this equation the Fredholm alternative does take place.

Remark 1.4 The proof of the Fredholm alternative for the equation

$$\int_{-\infty}^{\infty} [d_s r(t,s)]\dot{x}(t+s) = \int_{-\infty}^{\infty} [d_s P(t,s)]x(t+s) + f(t)$$

is given in [289.1, 389.3–389.5].

2. CONNECTION BETWEEN BOUNDEDNESS AND PERIODICITY

For linear periodic ODEs from the existence of a bounded solution it follows the existence of a periodic solution. Using the fixed point theorem and the shift operator this result of H. Massera was generalized for RDEs [236] and for NDEs [289.1] of the form

$$\dot{x}(t) = L(t, x_t, \dot{x}_t) + f(t), \quad f \in C_\omega. \tag{2.1}$$

Here $L(t;\cdot,\cdot) : C[-h,0]\times C[-h,0] \to \mathbf{R}^n$ is a linear measurable, ω–periodic in $t \in \mathbf{R}$ operator, satisfying the condition

$$|L(t, x_t, \dot{x}_t)| \le C_1 \parallel x_t \parallel + C_2 \parallel \dot{x}_t \parallel$$

where constants $C_1 > 0, 0 < C_2 < 1$.

3. PERIODIC SOLUTION OF LINEAR DIFFERENCE EQUATIONS (DcEs)

3.1. *STATEMENT OF THE PROBLEM*

By virtue of Thm. 1.1 the Fredholm alternative for Eq. (1.1) is valid under conditions of this theorem if and only if it is valid for the DcE in L_ω

$$Tz = f, \tag{3.1}$$

$$Tz(t) := \sum_{k=1}^{m} B_k(t)z(t-h_k), \quad B_k \in C_\omega. \tag{3.2}$$

Definition 3.1 Bounded operator T is called as the *Fredholm operator* in L_ω if the Fredholm alternative is valid for Eq. (3.1). Sometimes such operators are called as the *Fredholm operator of index zero.*

In this paragraph we discuss the conditions under which the first part of definition 1.3 (nonresonance case) holds for Eq. (3.1), i.e. Eq. (3.1) with $f(t) \equiv 0$ can have only the zero ω–periodic solution and Eq. (3.1) has the only ω–periodic solution for any $f \in L_\omega$.

3.2. *STATIONARY CASE. COMMENSURABLE SHIFTS OF THE ARGUMENT*

In this section we can assume that $\omega = 2\pi$ because if it is not a case we can do a change of variable $t \to (\omega/2\pi)t$. Consider the operator $T : L_{2\pi} \to L_{2\pi}$

$$Tz(t) := \sum_{k=0}^{m-1} B_k z(t-kq), \quad q = \frac{2\pi}{m}, \quad z(t) \in \mathbf{R}^n, \tag{3.3}$$

and the DcE

$$\sum_{k=0}^{m-1} B_k z(t-kq) = f(t), \quad f \in L_\omega \tag{3.4}$$

where B_k are constant matrices. Important feature of (3.3), (3.4) comparing with (1.4), (1.1) is that in the latter delays $h_k = 2\pi k/m$ are commensurable.

Theorem 3.1 The first case of the Fredholm alternative (nonresonance case) for Eq. (3.4) holds if and only if the equation

$$\det\left(\sum_{k=0}^{m-1} B_k \lambda^k\right) = 0 \tag{3.5}$$

has no roots with $\lambda^m = 1$, i.e. for which simultaneously $|\lambda| = 1$ and $\arg\lambda =$ $q_j = \frac{2\pi}{m}j$, *where j is an integer.*

Proof Necessity. Let Eq. (3.5) have a root $\lambda_1 = \exp(i\frac{2\pi}{m}j)$, where $i^2 = -1$ and an integer j. Then Eq. (3.4) has infinitely many linear independent 2π-periodic solutions $z(t) = \exp(i(j+ml)t)$ for any integer l. So the Fredholm alternative for Eq. (3.4) does not hold.

Sufficiency. The quasipolynomial $D(\mu)$ induced by (3.4) is

$$D(\mu) = \det\left(\sum_{k=0}^{m-1} B_k \exp(-kq\mu)\right). \tag{3.6}$$

The zeros μ of (3.6) are given by the formula

$$\mu_{ln} = -q^{-1}\ln|\lambda_l| + iq^{-1}(\arg\lambda_l + 2\pi n), \qquad l = 1,\dots,r; \quad n = 0,\pm1,\pm2,\dots \tag{3.7}$$

where r is a number of different roots of Eq. (3.5). Under conditions of Thm. 3.1 all the numbers $in, n = 0, \pm1, \dots$ are located outside of ε–neighbourhood of the zeros of (3.6), where

$$\varepsilon = \min\{(\min|\ln|\lambda_j||, |\lambda_j| \neq 1);\ \ (\min \mathrm{Fr}(q^{-1}\arg\lambda_j),\ \ |\lambda_j| = 1)\}.$$

Here Fr is a fractional part of α. In this case for all integer n we have $|D(in)| \geq C_1 > 0$ ([42], where C_j are positive constants.

Hence for all integer n the matrices

$$B(in) := \sum_{k=0}^{m-1} B_k \exp(-inqk)$$

are invertible and

$$\| B^{-1}(in) \| \leq C_2. \tag{3.8}$$

Take $f \in L_{2\pi}$ and expand it in the Fourier series

$$f(t) = \sum_{n=-\infty}^{\infty} f_n \exp(int).$$

Then the solution of (3.4) can be represented as follows

$$z(t) = \sum_{n=-\infty}^{\infty} B^{-1}(in)\exp(int). \tag{3.9}$$

By virtue of (3.8) series (3.9) is convergent in $L_{2\pi}$. So, using the Parseval equality

$$\| z \|_L^2 = \sum_{n=-\infty}^{\infty} f_n' \, [B^{-1}(in)]' \, B^{-1}(in) f_n \leq$$

$$\leq C_2^2 \sum_{n=-\infty}^{\infty} f_n' f_n = C_2^2 \, \| f \|_L^2 \, .$$

Therefore, the operator (3.3) has a bounded inverse operator in $L_{2\pi}$, i.e. it is the Fredholm operator. Theorem 3.1 is proven.□

Example 3.1 Consider a scalar homogeneous NDE

$$\dot{x}(t) - \dot{x}(t-\pi) = x(t) - x(t-\pi).$$

It is clear that any π–periodic function $x \in W_\pi$ is a solution of this equation. Hence the space of periodic solutions is infinite dimensional. It means that the Fredholm alternative for this equation is not valid. The same conclusion can be deduced from Thm. 3.1. In fact by virtue of (3.3) the corresponding operator T has a form

$$Tz(t) = z(t) - z(t-\pi). \tag{3.10}$$

Hence, Eq. (3.5) is $1 - \lambda = 0$, i.e. has a root $\lambda = 1$. Using Thm. 3.1 we obtain that operator (3.10) is not the Fredholm operator.

The same arguments show that the Fredholm alternative is not valid for the equation

$$\dot{x}(t) - \dot{x}(t-\pi h) = (\sin t)x(t)$$

for any rational h.

3.3. *ARBITRARY DELAYS*

Consider a stationary operator T with arbitrary h_k

$$Tz := \sum_{k=1}^{m} B_k z(t-h_k). \tag{3.11}$$

The quasipolynomial corresponding (3.11) has a form

$$D(p) = \det\left(\sum_{k=1}^{m} B_k \exp(-h_k p)\right). \tag{3.12}$$

Theorem 3.2 If there exists $\varepsilon > 0$ *such that (3.12) has no roots in the strip* $|\mathrm{Re} p| \le \varepsilon$ *then (3.11) is the Fredholm operator.*

Proof It is known [42] that under conditions of Thm. 3.2 the matrices

$$B(in) := \sum_{k=0}^{m} B_k \exp(-inh_k)$$

are invertible and satisfy (3.8). Hence repeating the end of the proof of Thm. 3.1 we obtain the assertion of Thm. 3.2.□

3.4. *VARIABLE COEFFICIENTS, DELAYS DIVISIBLE BY A PERIOD*

Consider the Fredholm alternative for Eqs. (3.1), (3.2). Denote $A(t) := B_1(t) + \ldots + B_m(t)$. Let in L_ω be given the operator T_1

$$T_1 z(t) := A(t)z(t), \quad z(t) \in \mathbf{R}^n, \quad A(t) \in L_\omega. \tag{3.13}$$

It is quite obvious that if all h_k divisible by ω then the periodic problem for the DcE

$$\sum_{k=0}^{m} B_k(t)z(t - h_k) = f(t)$$

is equivalent to the periodic problem for the equation $A(t)z(t) = f(t)$. It means that it is sufficient to investigate the operator (3.13).

Lemma 3.1 Operator (3.13) is bounded in L_ω if and only if

$$\underset{0 \le t \le \omega}{\text{vrai sup}} \quad |A(t)| < \infty. \tag{3.14}$$

Proof Sufficiency is evident. For the proof of necessity assume the opposite, i.e. that for some a_{lj}

$$\underset{0 \le t \le \omega}{\text{vrai sup}} \quad |a_{lj}(t)| = \infty.$$

Let us introduce the set D_m and the sequence of functions $\phi_m(t)$ as follows

$$D_m := \{t \in [0, \omega] : |a_{lj}(t)| > m\}, \quad \delta_m := \text{mes} D_m > 0.$$

$$\phi_m(t) := \begin{cases} 1, & t \in [0, \omega] \setminus D_m, \\ m^{-1/2}\delta_m^{-1/2}, & t \in D_m. \end{cases}$$

Let e_j be a vector whose j–th component equals 1, and all others equal zero. Then all the functions $\phi_m(t)e_j \in L_\omega$ because of estimates

$$\int_0^\omega e_j' \phi_m^2(t) e_j \, dt = \int_{[0,\omega]\setminus D_m} dt + \int_{D_m} \phi_m^2(t) dt \le \omega + m^{-1} \le \omega + 1.$$

But for the sequence

$$y_m'(t) = [A(t)\phi_m(t)e_j]' = (\phi_m(t)a_{1j}(t), \ldots, \phi_m(t)a_{nj}(t))$$

we have

$$\| y_m(t) \|_L^2 \ge \int_0^\omega \phi_m^2(t) a_{lj}^2(t) dt \ge \int_{D_m} m^{-1}\delta_m^{-1} m^2 dt = m \to \infty.$$

So there is a sequence of the functions $\phi_m e_j$ such that

$$\| \phi_m(t)e_j \|_L^2 \leq \omega + 1, \quad \| T_1(\phi_m(t)e_j) \|_L^2 \to \infty. \tag{3.15}$$

But (3.15) is a contradiction with the boundedness of T_1. Lemma 3.1 is proven.□

Lemma 3.2 The bounded operator (3.13) is the Fredholm operator in L_ω if and only if for some operator norm $|A^{-1}(t)|$ of the matrix $A^{-1}(t)$

$$\operatorname*{vrai\,sup}_{0\leq t\leq\omega} \ |A^{-1}(t)| < \infty \tag{3.16}$$

Proof Sufficiency. Under condition (3.16) the operator $(T_1^{-1}z)(t) = A^{-1}(t)z(t)$ is bounded. Hence, (3.13) is the Fredholm operator.

Necessity. Assume that (3.13) is the Fredholm operator. Then homogeneous equation

$$(Tz)(t) = A(t)z(t) = 0 \tag{3.17}$$

can not have in L_ω nontrivial solutions. In fact, if $z(t)$ is a nontrivial solution of (3.17) then for any conditions scalar function $\psi(t) \neq 0$, $0 \leq t \leq \omega$ the product $\psi(t)z(t)$ also is a nontrivial solution of (3.17). Let us take any denumerable set of linear independent functions $\psi_m(t) \neq 0$, $0 \leq t \leq \omega$. Then we obtain the denumerable set $\{\psi_m(t)z(t)\}$ of nontrivial solutions of Eq. (3.17). But the Fredholm operator can have only finite number of linear independent solutions for the corresponding homogeneous equation.Hence, there does not exist in L_ω a nontrivial solution of Eq. (3.17). From the absence of nontrivial solutions and the definition of the Fredholm operator it follows that the operator (3.13) maps L_ω on the whole space L_ω. By the Banach's theorem about open mapping, the inverse operator $(T_1^{-1}z)(t) = A^{-1}(t)z(t)$ is bounded. Hence by Lemma 3.1 the matrix $A^{-1}(t)$ has to satisfy (3.16).□

Remark 3.1 Sufficient condition for the bounded operator (3.13) to be Fredholm operator is

$$\operatorname*{vrai\,inf}_{0\leq t\leq\omega} \ |\det A(t)| > 0.$$

If $A(t) \in C_\omega$ then (3.13) is the Fredholm operator if

$$\det A(t) \neq 0, \quad 0 \leq t \leq \omega. \tag{3.18}$$

Consider the operator

$$Tz = \sum_{k=0}^{m} B_k(t)z(t-k\omega), \quad B_k(t) \in L_\omega. \tag{3.19}$$

Theorem 3.3 Operator (3.19) be bounded and the Fredholm operator if and only if conditions (3.14) and (3.16) are valid for $A = B_1 + \ldots + B_m$.

Proof We have in L_ω

$$Tz = \sum_{k=1}^{m} B_k(t)z(t-k\omega) = \left(\sum_{k=1}^{m} B_k(t)\right) z(t).$$

It means that operator (3.19) coincides identically with (3.13). Hence the assertion of Thm. 3.3 follows from Lemmas 3.1, 3.2.□

Example 3.2 Consider the homogeneous scalar NDE

$$\dot{x}(t) - (\sin t)\dot{x}(t-2\pi) = x(t) - x(t-2\pi). \tag{3.20}$$

Any 2π–periodic solution $x(t)$ of this equation must satisfy the equality

$$\dot{x}(t)(1-\sin t) = 0.$$

Hence $\dot{x}(t) \equiv 0$, i.e. $x(t) = \text{const}$. So, up to linear dependence, the only 2π periodic solution of Eq. (3.20) is $x(t) \equiv 1$. By virtue of (1.2) the conjugate equation to (3.20) is

$$\frac{d}{dt}[y(t) - (\sin(t+2\pi))y(t+2\pi)] = y(t) - y(t+2\pi).$$

But this equation has no 2π–periodic solutions from $L_{2\pi}$. In fact, any 2π–periodic solution $y(t)$ must be such that

$$y(t)(1-\sin t) = c.$$

Hence

$$y(t) = \frac{c}{1-\sin t}.$$

This solution for any $c \neq 0$ does not belong $L_{2\pi}$. So, the original equation (3.20) has the nontrivial 2π–periodic solution $x(t) \equiv 1$, but the conjugate equation has no nontrivial 2π–periodic solution from $L_{2\pi}$. It means that the Fredholm alternative for Eq. (3.20) is not valid.

The same conclusion can be confirmed by the following arguments. By Thm. 1.1 the Fredholm alternative for Eq. (3.20) is valid if and only if the operator

$$Tz(t) := z(t) - (\sin t)z(t-2\pi)$$

is the Fredholm operator in the space $L_{2\pi}$. But

$$\operatorname*{vrai\ sup}_{0 \le t \le 2\pi} \; [(1-\sin t)^{-1}] = \infty.$$

So by Thm. 3.3, T is not the Fredholm operator.

3.5. *VARIABLE COEFFICIENTS, DELAYS COMMENSURABLE WITH A PERIOD*

Consider the DcE

$$(Tz)(t) := \sum_{k=0}^{N-1} B_k(t)z(t-kq) = f(t), \tag{3.21}$$

where

$$q = \frac{\omega}{N}, \quad f \in L_\omega, \quad B_k \in L_\omega, \qquad \operatorname*{vrai\,sup}_{0\le t\le\omega} |B_k(t)| < \infty, \; k = 0,\dots,N-1.$$

If z is a solution of (3.21) then the following equations hold

$$\left.\begin{array}{c}
\sum_{k=0}^{N-1} B_k(t)z(t-kq) = f(t), \\[2ex]
\sum_{k=0}^{N-1} B_k(t-q)z(t-(k+1)q) = f(t-q), \\[2ex]
\dots\dots \\[2ex]
\sum_{k=0}^{N-1} B_k(t-(N-1)q)z(t-(k+N-1)q) = f(t-(N-1)q).
\end{array}\right\} \tag{3.22}$$

Let us introduce vectors $Y(t), F(t) \in \mathbf{R}^{nN}$ and matrix $D(t) \in \mathbf{R}^{nN\times nN}$:

$$Y'(t) := (y_1'(t),\dots,y_N'(t))', \quad y_i(t) := z(t-(i-1)q), \tag{3.23}$$

$$F'(t) := (f'(t), f'(t-q),\dots,f'(t-(N-1)q))', \tag{3.24}$$

$$D(t) :=$$

$$= \begin{pmatrix}
B_0(t) & B_1(t) & \dots & B_{N-1}(t) \\
B_{N-1}(t-q) & B_0(t-q) & \dots & B_{N-2}(t-q) \\
\dots & \dots & \dots & \dots \\
B_1(t-(N-1)q) & B_2(t-(N-1)q) & \dots & B_0(t-(N-1)q)
\end{pmatrix}. \tag{3.25}$$

Then Eq. (3.22) takes the form

$$D(t)Y(t) = F(t). \tag{3.26}$$

So, any solution $z \in L_\omega$ of Eq. (3.22) generates the solution $Y \in L_\omega$ of Eq. (3.26) defined by (3.23).

Consequently, assume that a function $f \in L_\omega$ exists such that there is a unique solution of (3.26) with $F(t)$ given by (3.24). Then the first n components of $Y(t)$ compose the solution of (3.21). In fact $Y(t)$ satisfies also the equation

$$D(t-q)Y(t-q) = F(t-q). \tag{3.27}$$

By virtue of ω–periodicity of $B_k(t) = B_k(t - Nq)$ and $f(t) = f(t - Nq)$ the last equation in (3.27) is

$$B_1(t)y_1(t-q) + B_2(t)y_2(t-q) + \ldots + B_0(t)y_N(t-q) = f(t). \tag{3.28}$$

Using (3.28) we can represent (3.27) in a form similar (3.26):

$$\left.\begin{aligned} &B_0(t)y_N(t-q) + B_1(t)y_1(t-q) + \ldots + \\ &\quad +B_{N-1}(t)y_0(t-q) = f(t), \\ &B_{N-1}(t-q)y_N(t-q) + B_0(t-q)y_1(t-q) + \ldots + \\ &\quad +B_{N-2}(t-q)y_0(t-q) = f(t-q), \\ &B_1(t-(N-1)q)y_N(t-q) + \ldots + \\ &\quad +B_0(t-(N-1)q)y_0(t-q) = f(t-(N-1)q), \end{aligned}\right\} \tag{3.29}$$

Eq. (3.29) has the same matrix of coefficients and righthand side as (3.26). Assumption about uniqueness of the solution of (3.26) leads us to the equalities $y_N(t-q) = y_1(t), y_1(t-q) = y_2(t), \ldots, y_{N-1}(t-q) = y_N(t)$. Further we have that

$$y_3(t) = y_2(t-q) = y_1(t-2q), \ldots, y_N(t) = y_{N-1}(t-q) =$$

$$= y_{N-2}(t-2q) = \ldots = y_1(t-(N-1)q).$$

Hence, the function $z(t)$ coinciding with the first n components of $Y(t)$ (i.e. $z(t) = y_1(t)$) is a solution of (3.21) in L_ω.

Denote the entries of the matrix $D^{-1}(t)$ as $\delta_{i,j}(t), i,j = 1,\ldots,nN$ and

$$r_{p,s} := \sum_{l=0}^{N-1} \delta_{p,ln+s}(t); \quad p,s = 1,\ldots,n.$$

Theorem 3.4 Assume that $B_k \in L_\omega$ and

$$\operatorname*{vrai\ sup}_{0\le t\le\omega} |B_k(t)| < \infty, \quad k = 0,\dots,N-1. \tag{3.30}$$

Then the operator (3.21) is the Fredholm operator if

$$\operatorname*{vrai\ sup}_{0\le t\le\omega} |\delta_{i,j}(t)| < \infty, \quad i,j = 1,\dots,nN. \tag{3.31}$$

If the operator (3.21) is the Fredholm operator then

$$\operatorname*{vrai\ sup}_{0\le t\le\omega} |r_{p,s}(t)| < \infty, \quad p,s = 1,\dots,n. \tag{3.32}$$

Proof From (3.30) it follows that operator (3.21) is bounded on L_ω. Under condition (3.31) Eq. (3.26) has a solution Y for any $F \in L_\omega$, and $\| Y \|_L \le C \| F \|_L$ ($C = \text{const} > 0$). Moreover, for the function F of the form (3.24) Eq. (3.26) has a unique solution and the first n component of this solution, i.e. the vector function z, is a solution of (3.21). In addition

$$\| z \|_L \le \| Y \|_L \le C \| F \|_L = C \| f \|_L .$$

Hence the operator (3.21) has a bounded inverse operator and because of it is the Fredholm operator. Sufficiently of conditions (3.31) is proven.

Now let us prove the necessity of (3.32). Assume that (3.31) is the bounded Fredholm operator. Then similar to the proof of Lemma 3.2 we obtain that the homogeneous DcE

$$Tz := \sum_{k=0}^{N-1} B_k(t) z(t-kq) = 0$$

can not have nontrivial solutions in L_ω. Moreover, from the Fredholm property of (3.21) and the Banach's theorem on open mapping it follows that the operator (3.21) has a bounded inverse operator. It means that for any $f \in L_\omega$ there exists a solution z of (3.21) and a constant C such that

$$\| z \|_L \le C \| f \|_L . \tag{3.33}$$

Now let be given a periodic function $f(t)$ with a period $q = \omega/N$. From (3.26) it follows that $z(t) = R(t)f(t)$. Further, similar Lemma 3.1, we obtain that (3.32) is a necessary condition for validity of inequality (3.33). □

Example 3.3. Consider the scalar equation

$$Tz := z(t) - (\sin t) z(t-\pi) = f(t), \quad f(t) = f(t+2\pi). \tag{3.34}$$

For (3.34) matrix (3.25) and its inverse has a form

$$D(t) = \begin{pmatrix} 1 & \sin t \\ -\sin t & 1 \end{pmatrix}, \quad D^{-1}(t) = \frac{1}{1+\sin^2 t} D'(t).$$

Hence conditions (3.30) and (3.31) hold. By Thm. 3.4 operator (3.34) is the Fredholm operator in $L_{2\pi}$.

Example 3.4 Consider the scalar equation

$$(Tz)(t) := z(t) - (\sin t) z\left(t - \frac{N-1}{N} 2\pi\right) = f(t), \qquad f(t) = f(t+2\pi), \tag{3.35}$$

where $N \geq 3$ is an integer. Matrix (3.25) for Eq. (3.35) is

$$D(t) = \begin{pmatrix} 1 & 0 & \dots & 0 & \sin t \\ \sin(t-q) & 1 & \dots & 0 & 0 \\ \dots & \dots & \dots & \dots & \dots \\ 0 & 0 & \dots & \sin(t-(N-1)q) & 1 \end{pmatrix}.$$

So, condition (3.30) holds, and condition (3.31) is valid if

$$\min_{0 \leq t \leq 2\pi} |\det D(t)| > 0$$

where

$$\det D(t) = 1 + (-1)^{N+1} \sin t \sin(t-q) \dots \sin(t-(N-1)q).$$

But $q < \pi$ yields this immediately. Hence operator (3.35), satisfying (3.31), is the Fredholm operator.

Example 3.5 Consider the scalar equation

$$z(t) + \left(1 + \frac{1}{2}\sin t\right) z(t-\pi) = f(t), \quad f(t) = f(t+2\pi). \tag{3.36}$$

The matrices $D(t), D^{-1}(t)$ and $R(t)$ for (3.36) are

$$D(t) = \begin{pmatrix} 1 & 1 + \frac{1}{2}\sin t \\ 1 - \frac{1}{2}\sin t & 1 \end{pmatrix},$$

$$D^{-1}(t) = \frac{4}{\sin^2 t}\begin{pmatrix} 1 & -\left(1+\frac{1}{2}\sin t\right) \\ -\left(1-\frac{1}{2}\sin t\right) & 1 \end{pmatrix},$$

$$R(t) = \frac{4}{\sin^2 t}\left(1-\left(1-\frac{1}{2}\sin t\right)\right) = \frac{2}{\sin t}.$$

But $\operatorname{vrai\,sup}_{0\le t<\omega} |R(t)| = \infty$. Hence, necessary condition (3.31) is not valid and operator (3.36) is not the Fredholm operator.

3.6. *VARIABLE COEFFICIENTS, ARBITRARY CONSTANT DELAYS*

Let be given the operator $T : L_\omega \to L_\omega$

$$(Tz)(t) := \sum_{k=0}^{N-1} B_k(t)z(t-kq) + Q(t)z(t-h). \tag{3.37}$$

Let us introduce the matrix $R(t)$ of dimension $nN \times nN$

$$R(t) := \begin{pmatrix} Q(t) & 0 & \dots & 0 \\ 0 & Q(t-q) & 0\dots & 0 \\ \dots & \dots & \dots & \dots \\ 0 & 0 & \dots & Q(t-(N-1)q) \end{pmatrix}.$$

Theorem 3.5 Assume that $B_k(t) \in C_\omega$, $Q(t) \in C_\omega$, $h\omega^{-1}$ is an irrational number and matrix $D(t)$ defined by (3.25) is invertible. Then the operator (3.37) is the Fredholm operator if one of the following two conditions holds

1) $\lim_{\gamma\to 0^-} \int_0^\omega \ln(|D^{-1}(t)R(t)|+\gamma)dt < 0$,

2) matrices $Q(t)$ and $R(t)$ are invertible and

$$\lim_{\gamma\to 0^-} \int_0^\omega \ln(|R^{-1}(t)D(t)|+\gamma)dt < 0.$$

The proof of this theorem see, e.g., [289.1].

4. NDEs WITH SMALL NONLINEARITIES

Consider the equation

$$\sum_{k=1}^{m} B_k(t)\dot{x}(t-h_k) + \int_{-\infty}^{\infty} [d_s P(t,s)]x(t+s) = \tag{4.1}$$

$$= f(t) + \mu T(t, x(\cdot); \mu)(t); \quad x(t), f(t) \in \mathbf{R}^n.$$

Here T is a continuous operator, $T : \mathbf{R} \times W_\omega \times [-\mu_0, \mu_0] \to \mathbf{R}^n$, μ_0 is a given constant, all other objects in (4.1) are the same as in Thm. 1.1.

The generating linear equation for (4.1) is the following one

$$Uz(t) := \sum_{k=1}^{m} B_k(t)\dot{z}(t-h_k) + \int_{-\infty}^{\infty} [d_s P(t,s)]z(t+s) = f(t) \tag{4.2}$$

with the conjugate equation

$$\frac{d}{dt}\left[\sum_{k=1}^{m} y'(t+h_k)B_k(t+h_k) - \int_{-\infty}^{\infty} y'(t-\alpha)P(t-\alpha,\alpha)d\alpha\right] = 0. \tag{4.3}$$

We shall illustrate the method of a small parameter for the nonresonance case (i.e., for the case when Eq. (4.2) with $f(t) \equiv 0$ has the only trivial solution and there exists ω–periodic solution of Eq. (4.2) for any $f \in L_\omega$). The resonance case is considered, e.g., [289.1].

Lemma 4.1 For the nonresonance case a unique periodic solution $\overline{z}$ of Eq. (4.2) satisfies the estimate (C is a positive constant)

$$\| \overline{z} \|_W \leq C \| f \|_L \, . \tag{4.4}$$

Proof Under conditions of Lemma 4.1 the linear operator U given by (4.2) is bounded on W_ω. By the Fredholm alternative for the nonresonance case the operator U maps the space W_ω on the whole space L_ω. Hence by the Banach theorem the operator U^{-1} is also linear and bounded. Lemma 4.1 is proven. □

Assume now that the operator $(\tau, x(\cdot); \mu) \mapsto T(\tau, x(\cdot); \mu)$ in (4.1):

1) is defined on $\mathbf{R} \times G \times [-\mu_0, \mu_0]$, where $\mu_0 > 0$ and

$$G := \{x(\cdot) \in W_\omega : \| x(\cdot) - \overline{z}(\cdot) \|_W \leq \rho; \rho > 0\};$$

Here $\overline{z}$ is a unique periodic solution of Eq. (3.2) with a given function f.

2) the function $t \mapsto T(t, x(\cdot); \mu)(t)$ belongs to L_ω for any fixed $x(\cdot) \in G$, $|\mu| \leq \mu_0$;

3) is continuous in μ uniformly with respect to $x(\cdot) \in G$ and $t \in (-\infty, \infty)$;

4) satisfies the Lipschitz condition:

$$\| T(\tau, x_1(\cdot); \mu) - T(\tau, x_2(\cdot); \mu) \|_L \leq$$

$$\leq C_1 \| x_1(\cdot) - x_2(\cdot) \|_W \quad (C_1 > 0), \tag{4.5}$$

$$-\infty < \tau < \infty, \quad |\mu| \leq \mu_0.$$

Denote

$$p := \max_{|\mu| \leq \mu_0} \| t \mapsto T(t, \overline{z}(\cdot); \mu)(t) \|_L,$$

$$\mu_1 := \frac{\rho}{C(C_1 \rho + p)}, \quad \mu_2 := \min(\mu_0, \mu_1).$$

where the constants C and C_1 are the same as in (4.4), (4.5).

Theorem 4.1 Assume that the matrices $B_k(t), P(t,s)$ satisfy the conditions of Thm. 1.1, the operator T satisfies the conditions 1) – 4). Then in the nonresonance case if $q := \mu_2 C C_1 < 1$ there is a unique periodic solution $x(t;\mu)$ of Eq. (3.1) for all $\mu, |\mu| \leq \mu_2$, which is a continuous function of (t,μ). The function $(t,\mu) \mapsto x(t;\mu)$ is a limit of the uniformly convergent sequence $\{x_j(t;\mu)\}$, where $x_0(t;\mu) = \overline{z}(t) = U^{-1}f(t)$, and $x_j(t;\mu)$ is a periodic solution of the equation

$$\sum_{k=1}^{m} B_k(t)\dot{x}_j(t - h_k, \mu) + \int_{-\infty}^{\infty} [d_s p(t,s)] x_j(t+s;\mu) =$$

$$= f(t) + \mu T(t, x_{j-1}(\cdot;\mu);\mu)(t).$$

Proof By virtue of Lemma 4.1, the problem of existence of the periodic solution of the equation

$$(Ux)(t) = f(t) + \mu T(t, x(\cdot);\mu)(t)$$

is equivalent to the solution in W_ω the operator equation

$$x(\cdot) = U^{-1}f + \mu U^{-1}T(t, x(\cdot);\mu) = \overline{z}(\cdot) + \mu U^{-1}T(t, x(\cdot);\mu).$$

But under conditions of Thm. 4.1 the operator $x(\cdot) \mapsto \overline{z} + \mu U^{-1}T(\cdot; x(\cdot);\mu)$ maps G into itself and is a contraction operator. Hence using a contraction mapping principle we obtain the assertion of Thm. 4.1. □

Example 4.1 Consider the scalar NDE

$$\dot{x}(t) + \frac{1}{2}\dot{x}(t-\pi) = x\left(t - \frac{3\pi}{2}\right) + \sin t + \mu x^2(t). \tag{4.6}$$

The quasipolynomial corresponding to the linear generating equation is

$$\lambda - \frac{1}{2}\lambda \exp(-\pi\lambda) = \exp\left(-\frac{3\pi}{2}\lambda\right). \tag{4.7}$$

Quasipolynomial (4.7) has no roots of the form $\lambda = in,\ n \in \mathbf{Z}$. Moreover from Thms. 1.1, 3.1 it follows that the Fredholm alternative is valid for the generating linear equation. It means that the nonresonance case does take place. The zero approximation $x_0(t) = -2\cos t$. The first approximation is a solution of the equation

$$\dot{x}_1(t) - \frac{1}{2}\dot{x}_1(t-\pi) = x_1\left(t - \frac{3\pi}{2}\right) + \sin t + \mu 4\cos^2 t.$$

Hence

$$x_1(t) = -2\cos t + \mu\left(-2 + \frac{1}{2}\sin 2t\right).$$

Similarly further approximation to the periodic solution of Eq. (4.6) can be found.

5. PERIODIC SOLUTIONS OF AUTONOMOUS FDEs WITH SMALL PARAMETER

Consider the equation

$$\sum_{r=1}^{m} B_r\dot{x}(t-h_r) + \int_{-\infty}^{\infty} [dP(s)]x(t+s) = \mu T(x(\cdot);\mu)(t). \tag{5.1}$$

Though Eq. (5.1) is a particular case of Eq. (4.1) there is the essential distinction between them. Namely, Eq. (5.1) can have a periodic solution which period, generally, does not coincide with the period of the generating solution and depends on μ.

The generating equation for (5.1) is

$$\sum_{r=1}^{m} B_r\dot{z}(t-h_r) + \int_{-\infty}^{\infty} [dP(s)]z(t+s) = 0 \tag{5.2}$$

with a quasipolynomial

$$\det\left[\sum_{r=1}^{m} B_r\lambda\exp(-\lambda h_r) + \int_{-\infty}^{\infty} [dP(s)]\exp(\lambda s)\right] = 0. \tag{5.3}$$

Assume that the quasipolynomial (5.3) has a finite number l of simple imaginary roots $\lambda_j = ib_jd$, $j = 1,\ldots,l$, where $b_j \geq 0$ are integers and $d > 0$. If $\forall b_j > 0$, then these roots correspond $2l$ periodic solutions ϕ_j with a period $\omega = 2\pi/d$ of Eq. (5.2):

$$\phi_1(t) = \sin(b_1dt), \quad \phi_2(t) = \cos(b_1dt),$$

$$\ldots, \phi_{2l-1}(t) = \sin(b_ldt), \; \phi_{2l}(t) = \cos(b_ldt).$$

If $b_1 = 0$ then Eq. (5.2) has $2l - 1$ such solutions. The general ω–periodic solution $\omega = 2\pi/d$ of Eq. (5.2) is

$$z(t, M) = M_1\phi_1(t) + \ldots + M_k\phi_k(t),$$

where $k = 2l$ or $k = 2l - 1$.

Assume that the domain of definition of the operator $T(x(\cdot);\mu)$ is W_ω. The conjugate equation to (5.2) has a form

$$\frac{d}{dt}\left[\sum_{r=1}^{m} y'(t + h_r)B_r - \int_{-\infty}^{\infty} y'(t - s)P(s)ds\right] = 0,$$

with the quasipolynomial

$$\det\left[\lambda\left(\sum_{r=1}^{m} B_r \exp(\lambda h_r) - \int_{-\infty}^{\infty} P(s)\exp(-\lambda s)ds\right)\right] = 0. \tag{5.4}$$

By virtue of (1.9) the quasipolynomial (5.4) can be represented as

$$\det\left[\sum_{r=1}^{m} B_r\lambda\exp(\lambda h_r) - \int_{-\infty}^{\infty} [dP(s)\exp(-\lambda s)\right] = 0. \tag{5.5}$$

Comparing (5.3) and (5.5) we can see that every root λ_j of (5.3) corresponds the root $(-\lambda_j)$ of (5.5). Hence the equation, conjugate with (5.2), has the same general periodic solution as Eq. (5.2).

Let us seek a solution $x(t;\mu)$ of (5.1) with a period $\tau(\mu)$ such that

$$\lim_{\mu\to 0} x(t, \mu) = z_0(t, M) = M_1\phi_1(t) + \ldots + M_{k-1}\phi_{k-1}(t) \tag{5.6}$$

where $M_1, \ldots, M_{k-1}$ are some constants to be defined. Let us remark that formally righthand side of (5.6) must also contain the term $M_k\phi_k(t)$, but it can be shown that it is possible to take $M_k = 0$ due to autonomy of Eq. (5.1). Period $\tau(\mu)$ can differ from ω.

Put $\tau(\mu) = \omega(1+\mu\alpha(\mu))$, where $\alpha(\cdot)$ is a continuous unknown function, and $t = \xi(1+\mu\alpha(\mu))$. Then Eq. (5.1) takes the form

$$\sum_{r=1}^{m} B_r \dot{u}(\xi - \widetilde{h}_r(\mu)) + (1+\mu\alpha(\mu)) \int_{-\infty}^{\infty} [d_\eta P_1(\eta,\mu)] u(\xi+\eta) = \tag{5.7}$$

$$= \mu(1+\mu\alpha(\mu))\widetilde{T}(u(\cdot);\mu)(\xi).$$

Here

$$u(\xi) = x(t), \quad \widetilde{h}_k(\mu) = \frac{h_k}{1+\mu\alpha(\mu)}, \quad \eta = \frac{s}{1+\mu\alpha(\mu)},$$

$$P_1(\eta,\mu) = P(s), \quad \widetilde{T}(u(\cdot);\mu) = T(x(\cdot);\mu).$$

If $x(t;\mu)$ is $\tau(\mu)$–periodic solution of Eq. (5.1), then $u(\xi)$ is ω–periodic solution of Eq. (5.7) and vice versa. Rewrite Eq. (5.7) as follows

$$\sum_{r=1}^{m} B_r \dot{u}(\xi - \widetilde{h}_r(\mu)) + \int_{-\infty}^{\infty} [d_\eta P_1(\eta,\mu)] u(\xi+\eta) = \mu Q(u(\cdot);\mu)(\xi). \tag{5.8}$$

Here

$$Q(u(\cdot);\mu)(\xi) := (1+\mu\alpha(\mu))\widetilde{T}(u(\cdot);\mu)(\xi) - \alpha(\mu) \int_{-\infty}^{\infty} [d_\eta P_1(\eta;\mu)] u(\xi+\eta). \tag{5.9}$$

The operator Q, defined for $u(\cdot) \in \widetilde{G}_2 \subset W_\omega, |\mu| \le \mu_0$, can be expanded: $Q(u(\cdot);\mu) = Q^0(u(\cdot);\mu) + Q^1(u(\cdot);\mu)$, where Q^0 is a projection of Q on the subspace L^0, generated by the functions $\phi_1(t), \dots, \phi_{k-1}(t)$, and Q^1 is a projection of Q on the subspace L^1, which is an orthogonal complement of L^0 with respect to the whole space L_ω.

Consider auxiliary equation

$$\sum_{r=1}^{m} B_r \dot{u}(\xi - \widetilde{h}_r(\mu)) + \int_{-\infty}^{\infty} [d_\eta P_1(\eta,\mu)] u(\xi+\eta) = \mu Q^1(u(\cdot);\mu)(\xi). \tag{5.10}$$

Essential feature of (5.10) is that every μ corresponds its own linear equation

$$\sum_{r=1}^{m} B_r \dot{u}(\xi - \widetilde{h}_k(\mu)) + \int_{-\infty}^{\infty} [d_\eta P_1(\eta,\mu)] u(\xi+\eta) = f(\xi), \tag{5.11}$$

that is not the case for Eq. (4.1). Because of this we have to impose additional restrictions on the generating equation (5.2), comparing with nonautonomous case. Assume that the following two conditions are satisfied:

1) quasipolynomial

$$\det\left(\sum_{r=1}^{m} B_r \exp(-h_r(\mu)\lambda)\right) = 0 \tag{5.12}$$

has no roots in some strip $|\mathrm{Re}\lambda| \le \varepsilon, \varepsilon > 0$ for all modulo sufficiently small μ;

2) quasipolynomial $\det B(\lambda, \mu) = 0$, where

$$B(\lambda,\mu) := \sum_{r=1}^{m} B_r \exp(-\lambda \widetilde{h}_r(\mu)) + \int_{-\infty}^{\infty} [d_\eta P_1(\eta,\mu)] \exp(\lambda\eta),$$

for all modulo sufficiently small μ has no roots of the form $\lambda_j = \pm i\overline{n}d$, $\overline{n} \in \mathbf{Z}$ in the strip $|\mathrm{Re}\lambda| \le \varepsilon$ with the exception of $\lambda_j = \pm ib_j d$ $(j = 1, \dots, l)$.

Consider the linear equation

$$\sum_{r=1}^{m} B_r \dot{u}(\xi - \widetilde{h}_r(\mu)) + \int_{-\infty}^{\infty} [d_\eta P_1(\eta,\mu)] u(\xi + \eta) = f(\xi). \tag{5.13}$$

Lemma 5.1 Assume that the conditions 1), 2) are met. Then for any $f \in L^1$ Eq. (5.13) has a unique solution $u(\cdot;\mu) \in W^1$ such that

$$\| u(\cdot;\mu) \|_W \le R \| f \|_L . \tag{5.14}$$

Here a constant R does not depend on μ for all modulo sufficiently small $\mu, |\mu| \le \mu_0$.

Proof By virtue of Thm. 3.2 and condition 1) the Fredholm alternative is valid for Eq. (5.13) for all $|\mu| \le C$. From here and condition 2) it follows that for any $f \in L^1$ and μ there exists the solution $u(\xi;\mu)$. In this case [42] for all integers $n, n \ne \pm b_j$, $j = 1, \dots, l$, the estimate, uniform with respect to n and $\mu, |\mu| \le \overline{C}$, is valid

$$\| B^{-1}(\pm ind, \mu) \| \le C_1. \tag{5.15}$$

If

$$f(\xi) = \sum_{n \ne b_j} f_n \exp(ind\xi)$$

then

$$u(\xi,\mu) = \sum_{n \ne b_j} u_n \exp(ind\xi) = \sum_{n \ne b_j} B^{-1}(ind;\mu) f_n \exp(ind\xi).$$

From (5.15) and the Parseval equality we obtain

$$\| u(\cdot;\mu) \|_L^2 = \sum_{n\neq b_j} u_n' \overline{u}_n =$$

$$= \sum_{n\neq b_j} f_n' [B^{-1}(ind;\mu)]' \overline{B^{-1}(ind;\mu) f_n} \leq C_1^2 \| f \|_L^2 . \tag{5.16}$$

Let us expand the derivative of the solution $u(\cdot,\mu)$ in the Fourier series

$$\dot{u}(\xi;\mu) = \sum_{n\neq b_j} v_n \exp(ind\xi).$$

Put the Fourier series for $u(\xi,\mu)$ and $f(\xi)$ into (5.11). Then we find the coefficients v_n

$$v_n = \left[\sum_{r=1}^{m} B_r \exp(-ind\widetilde{h}_r(\mu))\right]^{-1} \left(f_n - \int_{-\infty}^{\infty} [d_\eta P_1(\eta;\mu)] \exp(ind\eta) u_n \right). \tag{5.17}$$

Using (5.17), the estimates (5.15), (5.16) and the boundedness with respect to $\mu, |\mu| \leq \overline{C}$, of the integrals

$$\int_{-\infty}^{\infty} \| d_\eta P_1(\eta;\mu) \| \leq p,$$

we obtain

$$\| \dot{u}(\cdot;\mu) \|_L^2 = \sum_{n\neq b_j} v_n' \overline{v}_n \leq$$

$$\leq 2C_1^2 \left(\sum_{n\neq b_j} f_n' \overline{f}_n + p^2 \sum_{n\neq b_j} u_n' \overline{u}_n \right) \leq C_2 \sum_{n\neq b_j} f_n' \overline{f}_n = C_2 \| f \|_L^2$$

where C_1 and C_2 are some positive constants. From here it follows (5.4).□

Denote $S(\mu)$ the contraction of the operator

$$u(\cdot) \mapsto \left(\xi \mapsto \sum_{r=1}^{m} B_r \dot{u}(\xi - \widetilde{h}_r(\mu)) + \int_{-\infty}^{\infty} [d_\eta P_1(\eta;\mu)] u(\xi+\eta) \right)$$

on the subspace W^1. Lemma 5.1 claims that there is the operator $S^{-1}(\mu)$ on L^1 uniformly bounded with respect to $\mu, |\mu| \leq C$ for some positive constant C, i.e.

$$\| S^{-1}(\mu) f \|_W = \| u(\cdot;\mu) \|_W \leq R \| f \|_L .$$

Lemma 5.2 Let quasipolynomials (5.3), (5.12) satisfy conditions 1), 2), matrix $P(s)$ and operator T satisfy the conditions of Thm. 4.1, and $M^0 = (M_1^0, \ldots, M_{k-1}^0)$. Then there are numbers $C > 0, \delta > 0$ such that for any $\mu, |\mu| \leq C, |M - M^0| \leq \delta$ (where $M = (M_1, \ldots, M_{k-1})$) Eq. (5.10) has a unique ω-periodic solution $\xi \mapsto \widetilde{u}(\xi, M; \mu)$ which is continuous in $M_1, \ldots,$ M_{k-1}, μ and also

$$\lim_{\mu \to 0} \widetilde{u}(\xi, M; \mu) = M_1^0 \phi_1(\xi) + \ldots + M_{k-1}^0 \phi_{k-1}(\xi). \tag{5.18}$$

Proof Put

$$u_0(\xi, M; \mu) := M_1 \phi_1(\xi) + \ldots + M_{k-1} \phi_{k-1}(\xi)$$

and take as $u_j(\cdot, M; \mu)$ the ω–periodic solution of the equation

$$\sum_{r=1}^{m} B_r \dot{u}_j(\xi - \widetilde{h}_r(\mu)) + \int_{-\infty}^{\infty} [d_\eta P_1(\eta; \mu)] u_j(\xi + \eta) = \mu Q^1(u_{j-1}(\cdot); \mu)(\xi).$$

It means that

$$u_j(\xi) = \mu S^{-1}(\mu) Q^1(u_{j-1}(\cdot); \mu)(\xi).$$

Now the proof of convergence of the successive approximations, using uniform boundedness of $S^{-1}(\mu)$ (see Lemma 5.1), can be fulfilled in much the same way as in Thm. 4.1.□

Let us introduce the functions

$$p_j(M_1, \ldots, M_{k-1}; \alpha, \mu) := \int_0^\omega \phi_j'(\xi) Q(\widetilde{u}(\cdot, M_1, \ldots, M_{k-1}; \mu)(\xi) d\xi =$$

$$= (1 + \mu\alpha) \int_0^\omega \phi_j'(\xi) \widetilde{T}(\widetilde{u}(\cdot, M_1, \ldots, M_{k-1}; \mu)(\xi) d\xi -$$

$$-\alpha \int_0^\omega \phi_j'(\xi) d\xi \int_{-\infty}^{\infty} [d_\eta P_1(\eta; \mu)] \widetilde{u}(\xi + \eta, M_1, \ldots, M_{k-1}; \mu), \quad j = 1, \ldots, k$$

and also

$$r_j(M_1, \ldots, M_{k-1}; \alpha, \mu) := \mu p_j(M_1, \ldots, M_{k-1}; \alpha, \mu) +$$

$$+ \int_0^\omega \phi_j'(\xi) D_\xi \widetilde{u}(\cdot, M_1, \ldots, M_{k-1}; \mu) d\xi, \quad j = 1, \ldots, k.$$

Here the operator $D : W_\omega \to L_\omega$ is given as follows

$$D(\psi(\cdot))(\xi) := \sum_{r=1}^{m} B_r[\dot{\psi}(\xi - h_r) - \dot{\psi}(\xi - \tilde{h}_r(\mu))]+$$

$$+ \int_{-\infty}^{\infty} [dP(s) - d_s P_1(s;\mu)]\psi(\xi + s).$$

Theorem 5.1 Assume that for Eq. (5.1) are met the assumptions of Lemmas 5.1, 5.2. Then for the existence of $T(\mu)$-periodic solution $x(\cdot;\mu)$ of Eq. (5.1), satisfying (5.6) and continuous in $\mu, |\mu| \le C$, it is sufficient that there are continuous functions $\mu \mapsto M(\mu) = (M_1(\mu), \ldots, M_{k-1}(\mu))$ and $\mu \mapsto \alpha(\mu)$ such that for $|\mu| \le C$

$$p_j(M_1(\mu), \ldots, M_{k-1}(\mu); \alpha(\mu), \mu) = 0, \quad j = 1, \ldots, k, \tag{5.19}$$

$$M(0) = M^0, \quad \alpha(0) = \alpha^0. \tag{5.20}$$

The necessary condition for the existence of $T(\mu)$-periodic solution $x(t,\mu)$ of Eq. (5.1) is the existence of continuous functions $\mu \mapsto M(\mu), \alpha(\mu)$ satisfying (5.20) and also

$$r_j(M_1(\mu), \ldots, M_{k-1}(\mu); \alpha(\mu), \mu) = 0, \quad j = 1, \ldots, k. \tag{5.21}$$

Proof If $M(\mu)$ and $\alpha(\mu)$ satisfy (5.19) then $Q^0(u(\cdot);\mu)(\xi) :\equiv 0$. Hence by Lemma 5.2 the function $\xi \mapsto u(\xi;\mu) := \tilde{u}(\xi, M(\mu);\mu)$ is the ω-periodic solution of (5.10). Let us pass to the variable t by the formula $\xi = (1 + \mu\alpha(\mu))^{-1}t$. Then we obtain that the function $t \mapsto x(t,\mu) = u(\xi;\mu)$ is the $T(\mu)$-periodic solution of Eq. (5.1) which is continuous in μ and satisfies (5.6). Sufficient part of Thm. 5.1 is proven.

Now let $t \mapsto x(t;\mu)$ be some $T(\mu)$-periodic solution of (5.1), which is continuous in μ and satisfies (5.6). Then passing to the variable ξ, we find the function $\xi \mapsto \tilde{u}(\xi, M(\mu);\mu)$ which is the ω-periodic solution of Eq. (5.10), and also the equation

$$\sum_{r=1}^{m} B_r \dot{u}(\xi - h_r) + \int_{-\infty}^{\infty} [dp(s)]u(\xi + s) = \mu Q(u(\cdot);\mu)(\xi) + D(\xi)(u(\cdot))(\xi). \tag{5.22}$$

For Eq. (5.22) the Fredholm alternative is valid. Hence homogeneous Eq. (5.22) has ω-periodic solutions $\phi_j(\xi), j = 1, \ldots, k$. Therefore, by virtue of the Fredholm alternative, the orthogonality conditions must be valid. But these conditions coincide with Eqs. (5.21). So, necessity is proven. □

Example 5.1 Consider the scalar equation

$$\dot{x}(t) - \frac{1}{2}\dot{x}\left(t - \frac{\pi}{2}\right) = x\left(t - \frac{3\pi}{2}\right) - \frac{1}{2}x(t) + \\ +\mu\left[\dot{x}^3\left(t - \frac{\pi}{2}\right) - x\left(t - \frac{3\pi}{2}\right)\right]. \tag{5.23}$$

Let us seek the solution $x(t;\mu)$ of Eq. (5.23) with the period $T = 2\pi(1 + \mu\alpha(\mu))$ such that

$$\lim_{\mu \to 0} x(t;\mu) = M^0 \sin t = \tilde{u}(t, M^0). \tag{5.24}$$

Quasiplynomial of the generating equation

$$\lambda - (1/2)\lambda \exp((-\pi/2)\lambda) = \exp((-3\pi/2)\lambda) - 1/2. \tag{5.25}$$

has only two imaginary roots $\lambda = \pm i$. Eq. (5.23) satisfies all the assumptions of Thm. 5.1. The generating equation for Eq. (5.23) has two linear independent solutions $\sin t$ and $\cos t$. Let us find M^0 and $\alpha(0) = \alpha^0$ such that solution (5.24) does exist. In the case under consideration system (5.19) can be written in a form

$$\left.\begin{aligned} &\int_0^{2\pi}\left[\dot{\tilde{u}}^3\left(t - \frac{\pi}{2}, M^0\right) - \tilde{u}\left(t - \frac{3\pi}{2}, M^0\right) + \right. \\ &\left. + \alpha^0\tilde{u}\left(t - \frac{3\pi}{2}, M^0\right) - \frac{\alpha^0}{2}\tilde{u}(t, M^0)\right] \sin t dt = 0, \\ &\int_0^{2\pi}\left[\dot{\tilde{u}}^3\left(t - \frac{\pi}{2}, M^0\right) - \tilde{u}\left(t - \frac{3\pi}{2}, M^0\right) + \right. \\ &\left. + \alpha^0\tilde{u}\left(t - \frac{3\pi}{2}, M^0\right) - \frac{\alpha^0}{2}\tilde{u}(t, M^0)\right] \cos t dt = 0, \end{aligned}\right\} \tag{5.26}$$

Put $\tilde{u}(t, M^0) = M^0 \sin t$ into (5.26). Then we get

$$\frac{3}{2}(M^0)^2 - \alpha^0 = 0, \quad -1 + \alpha^0 = 0. \tag{5.27}$$

System (5.27) has a solution $\alpha^0 = 1, M^0 = \sqrt{2/3}$. Eq. (5.23) has a unique periodic solution (up to the arbitrary shift the time) whose period in the first apprimixation equals $T = 2\pi(1 + \mu)$. This solution is such that

$$\lim_{\mu \to 0} x(t;\mu) = \sqrt{\frac{2}{3}} \sin t.$$

CHAPTER 13

GENERALIZED PERIODIC SOLUTIONS OF SFDEs

In certain applications it could happen that only some characteristics of the whole set of solutions would be periodic functions, but not the individual trajectory. In this chapter we shall formalize this situation using the concept of a generalized periodic solution as a stochastic periodic process. Preliminaries to this chapter contain in Sec. 3.8 and Ch. 10.

In Sec. 1 basic notions and definitions used further on are given. In Sec. 2 conditions for the existence of periodic solutions are established. In Sec. 3 the relationship between stability and periodicity is considered and the possibility to use the Liapunov's direct method for the proof of conditions for the existence of periodic solutions is established. In Sec. 4 conditions for the existence of periodic solutions of concrete systems are studied, formulated immediately in terms of the coefficients of the equations. Periodic solutions of the Itô's SFDEs are considered in Sec. 5.

1. SOME PREREQUISITES

First of all let us remind some properties of the periodic random processes [152]. The random process $\zeta(t) \in \mathbf{R}^n, t \in (-\infty, \infty)$ is called *periodic* with period $T > 0$ if all its finite–dimensional distributions are periodic with a period T, i.e. for any positive integer m and any moments of time $t_1, \ldots, t_m$ the distributions of the probabilities of the vector $(\zeta(t_1 + kT), \ldots, \zeta(t_m + kT))$ do not depend on k for any integer k. It is clear that the moments of the periodic (with a period T) process $\zeta(t)$ (provided that they exist) are periodic with the same period T. Conversely, if for any positive integer m and moments of time $t_1, \ldots, t_m$ and any continuous bounded function f the mathematical expectation $\mathcal{E} f(\xi(t_1 + t), \ldots, \xi(t_m + t))$ is periodic with respect to t with a period T, then the random process $\xi(t)$ is periodic with the same period T as well (henceforth $\mathcal{E}$ will be the symbol of mathematical expectation).

Let the measurable function $(t, x) \mapsto f(t, x), x \in \mathbf{R}^n$ be periodic with respect to t with period T and let the periodic process $\zeta(t) \in \mathbf{R}^n$ be periodic with period T.

The periodicity of a random process, generally speaking, does not imply the periodicity of its trajectories: there exist processes all the trajectories of which are periodic as well as processes not having with probability 1 periodic trajectories. Thus, e.g., the process $\zeta(t) = \xi \sin t$, where ξ is an arbitrary random variable, is periodic with period 2π and all trajectories of this process are periodic with the same period. On the contrary, the random process $\sin(t + \theta)$, where θ is a random variable uniformly distributed on $[0, 2\pi]$, is periodic with period π. But this process has no periodic trajectories with period π.

Let the two processes $\zeta_1(t)$ and $\zeta_2(t)$ be periodic with one and the same period. Their sum, generally speaking, is not a periodic process. For the periodicity of the sum $\zeta_1(t) + \ldots + \zeta_l(t)$ of any finite number of T–periodic processes $\zeta_i(t)$ it suffices to require that these processes should be *T–periodically related,* i.e. that the vector $(\zeta_1(t), \ldots, \zeta_l(t))$ should be T–periodic.

Now let a sequence of T–periodic processes $\zeta_i(t), i = 1, 2, \ldots$, any finite number of which is T–periodically related, converge by probability to the process $\zeta(t)$. Then the process $\zeta(t)$ will be T–periodic and T–periodically related to the elements of the sequence $\zeta_i(t)$. Here by a *convergence* of the sequence $\zeta_i(t)$ to $\zeta(t)$ *by probability* we mean that for any $\varepsilon > 0, \delta > 0$ we can find a number N such that for $i > N$ and for all $0 \le t \le T$ we have

$$\mathcal{P}(|\zeta_i(t) - \zeta(t)| > \delta) < \varepsilon.$$

Here $\mathcal{P}(\cdot)$ is a probability of the event included in the bracket. If

$$\mathcal{P}(|\zeta_i(t) - \zeta(t)| \to 0) = 1, \quad 0 \le t \le T,$$

then we have *convergence almost everywhere* (called also *convergence with probability* 1).

Let $\zeta(t)$ be a periodic process with period T and let $\alpha(t)$ be a determinate function with the same period. Then for any t_1 the following integral as a function of t

$$\int_{t-t_1}^{t} \alpha(s)\zeta(s)ds = \int_{-t_1}^{0} \alpha(s+t)\zeta(t+s)ds$$

represents a T–periodic and T–periodically related to $\zeta(t)$ process. The same will be true for the process

$$\int_{-\infty}^{t} z(t-s)\zeta(s)ds = \int_{0}^{\infty} z(s)\zeta(t-s)ds$$

for any determinate function $z(t)$ provided that respective integrals are convergent.

2. CONDITIONS OF EXISTENCE OF PERIODIC SOLUTIONS

We shall establish sufficient conditions for the existence of periodic solutions of the SRDE

$$\dot{x}(t) = b(t, x_t, \zeta(t)), \quad t > 0, \quad x(t) \in \mathbf{R}^n. \tag{2.1}$$

As usual initial conditions have a form

$$x(\theta) = \phi(\theta), \quad -h \leq \theta \leq 0. \tag{2.2}$$

Here $\zeta(t) \in \mathbf{R}^{n_1}$ is a continuous with probability 1 random process, periodic with period T. The continuous with respect to all arguments functional $(t, \phi, z) \mapsto b(t, \phi, z)$ in Eq. (2.1) is defined on $[0, \infty) \times C[-h, 0] \times \mathbf{R}^{n_1}$. For ϕ and z fixed the function $t \mapsto b(t, \phi, z)$ is periodic with respect to t with period T and satisfies for some $r > 0$ the conditions

$$\begin{gathered} |b(t,\phi,z) - b(t,\psi,z)| \leq r \parallel \phi - \psi \parallel, \quad \forall \phi, \psi \in C[-h,0], \\ \mathcal{P}\left(\int_0^t |b(s, 0, \zeta(s))| ds < \infty \right) = 1, \quad \forall t \geq 0, \quad \forall z \in \mathbf{R}^{n_1}. \end{gathered} \tag{2.3}$$

Definition 2.1 T-periodic solution of Eq. (2.1) is a T-periodic stochastic process periodically connected with $\zeta(t)$ and satisfying (2.1) with probability 1 for every $t > 0$.

Theorem 2.1 Assume that there exists a solution $x(t) = x(t, \phi)$ of Pr. (2.1), (2.2) satisfying the conditions

1) uniformly on $k = 1, 2, \ldots$

$$\lim_{C \to \infty} \frac{1}{k} \sum_{i=0}^{k-1} \mathcal{P}(|x(iT)| > C) = 0; \tag{2.4}$$

2) there exist positive constants $\varepsilon_1, \varepsilon_2$ and C such that for any two points s_1 and s_2 of the interval $[-h, \infty]$ the following inequality holds

$$\mathcal{E}|x(s_1) - x(s_2)|^{\varepsilon_1} \leq C|s_1 - s_2|^{1+\varepsilon_2}. \tag{2.5}$$

Then there exists a T-periodic solution of Eq. (2.1).

Proof Denote by τ_k a random variable independent of $\zeta(t)$ and taking with equal probabilities the values lT, where $l = 0, 1, \ldots, k-1$, i.e.

$$\mathcal{P}(\tau_k = lT) = \frac{1}{k}, \quad k = 1, 2, \ldots.$$

Introduce the sequences of processes ϕ_k, x_k, ζ_k

$$\phi_k(\theta) = x(\tau_k + \theta), \quad x_k(t) = x(\tau_k + t), \quad \zeta_k(t) = \zeta(\tau_k + t).$$

In view of (2.1), (2.2) the process $x_k(t)$ is a solution of the Cauchy problem

$$\dot{x}_k(t) = b(t, x_{kt}, \zeta_k(t)), \ t > 0; \quad x_k(\theta) = \phi_k(\theta), \quad -h \le \theta \le 0. \tag{2.6}$$

In (2.6) by x_{kt} we have denoted, as usual, the segment of the trajectory $x_k, x_{kt}(\theta) := x_k(t+\theta), \ -h \le \theta \le 0$.

Further on, note that for the proof of the theorem it suffices to establish the existence of an initial process $\xi(\theta)$ for which and for any k, θ_i, A_i, B_i we have

$$\begin{aligned} &\mathcal{P}(\xi(\theta_1) \in A_1, \ldots, \xi(\theta_m) \in A_m, \zeta(t_1) \in B_1, \ldots, \zeta(t_N) \in B_N) = \\ &= \mathcal{P}(x(\theta_1 + kT, \xi) \in A_1, \ldots, x(\theta_m + kT, \xi) \in A_m, \\ &\qquad \zeta(t_1 + kT) \in B_1, \ldots, \zeta(t_N + kT) \in B_N). \end{aligned} \tag{2.7}$$

Here k is an arbitrary positive integer, A_i and B_j are arbitrary Borel subsets of $\mathbf{R}^n$, the points $\theta_i \in [-h, 0]$ for any $i = 1, \ldots, m$ and $t_j \in [0, \infty), j = 1, \ldots, N$. We shall show that the distribution of the probabilities of the process $\xi(\theta)$ in (2.7) is a limit of the distributions of the probabilities of the process $\phi_k(\theta)$. From the definition of the processes x_k, ζ_k it follows that

$$\begin{aligned} &\mathcal{P}(x_k(t) \in A_0, \zeta_k(t_1) \in B_1, \ldots, \zeta_k(t_N) \in B_N) = \\ &= \frac{1}{k} \sum_{l=0}^{k-1} \mathcal{P}(x(t + lT) \in A_0, \zeta(t_1 + lT) \in B_1, \ldots, \zeta(t_N + lT) \in B_N). \end{aligned} \tag{2.8}$$

This relation shows that the distribution of the probabilities of the processes $\zeta_k(t)$ for all k is the same and coincides with the distribution of the probabilities of the process $\zeta(t)$. Whence and from the stochastic continuity of the process $\zeta(t)$ we have

$$\sup_{k, |t_1 - t_2| < \Delta} \mathcal{P}(|\zeta_k(t_1) - \zeta_k(t_2)| > \varepsilon) \to 0, \quad \forall \varepsilon > 0, \ \text{ for } \ \Delta \to 0, \tag{2.9}$$

$$\sup_{k,t} \mathcal{P}(|\zeta_k(t)| > C) \to 0 \quad \text{for} \quad C \to \infty. \tag{2.10}$$

Analogous relations are satisfied by the sequence of processes $\phi_k(\theta)$. In fact, the validity of relations (2.9) for ϕ_k follows from (2.5) and conditions (2.10) follow from (2.4), (2.5), (2.8) since

$$\mathcal{P}(|\phi_k(0)| > C) = \frac{1}{k} \sum_{l=0}^{k-1} \mathcal{P}(|x(lT)| > C) \to 0 \quad \text{for} \quad C \to \infty.$$

Relations (2.9), (2.10) for the sequence (ϕ_k, ζ_k) mean that it is weakly compact (in the sense that the sequence of measures generated by (ϕ_k, ζ_k) in the respective functional space is weakly compact) [152]. Hence it follows that there exists a probability space and a sequence of processes $(\phi_k^{(1)}, \zeta_k^{(1)})$ in this space distributed equally with (ϕ_k, ζ_k) such that some subsequence $(\phi_\alpha^{(1)}, \zeta_\alpha^{(1)})$ of it tends with probability 1 to $\phi_0(\theta), \zeta_0(t)$. Since all $\zeta_k(t)$ are distributed equally with $\zeta(t)$, the distribution of the limit process $\zeta_0(t)$ coincides with the distribution of the process $\zeta(t)$. Then in the initial probability space we can introduce a set of random processes $\xi(\theta), \xi_\alpha(\theta)$ such that the joint distribution of the probabilities of $(\xi(\theta), \xi_\alpha(\theta), \zeta(t))$ coincides with the distribution of $(\phi_0(\theta), \phi_\alpha^{(1)}(\theta), \zeta_0(t))$.

We shall check the validity of formula (2.7) for $x(t,\xi)$. From assumptions (2.3), the Grownwall - Bellman's lemma and Eq. (2.1) it follows that

$$|x(t,\xi) - x(t,\xi_\alpha)| \leq |\xi(\theta) - \xi_\alpha(\theta)| e^{rt}, \quad t \geq 0.$$

Hence $x(t,\xi_\alpha)$ tends in probability to $x(t,\xi)$ as $\alpha \to \infty$ for any $t \geq 0$. Now take an arbitrary continuous bounded function f depending on $n_1 N + mn$ variables. By virtue of the properties of the sequence $(\xi, \xi_\alpha, \zeta(t))$

$$\begin{aligned} J := \mathcal{E} f(x(\theta_1 + kT, \xi), \ldots, x(\theta_m + kT, \xi), \zeta(t_1 + kT), \ldots, \\ \zeta(t_N + kT)) = \lim_{\alpha \to \infty} \mathcal{E} f(x(\theta_1 + kT, \xi_\alpha), \ldots, x(\theta_m + kT, \xi_\alpha), \\ \zeta(t_1 + kT), \ldots, \zeta(t_N + kT)). \end{aligned} \tag{2.11}$$

But the joint distribution of (ξ_α, ζ) coincides with the joint distribution of $(\phi_\alpha^{(1)}, \zeta_0)$, and hence with the joint distribution of (ϕ_α, ζ). Whence and from (2.11), (2.8) it follows that

$$\begin{aligned} J = \lim_{\alpha \to \infty} \mathcal{E} f(x(\theta_1 + kT, \phi_\alpha), \ldots, x(\theta_m + kT, \phi_\alpha), \\ \zeta_\alpha(t_1 + kT), \ldots, \zeta_\alpha(t_N + kT)) = \\ = \lim_{\alpha \to \infty} \frac{1}{\alpha} \sum_{l=0}^{\alpha-1} \mathcal{E} f(x(\theta_1 + kT + lT, \phi), \ldots, x(\theta_m + kT + lT, \phi), \\ \zeta(t_1 + kT + lT), \ldots, \zeta(t_N + kT + lT)) = \\ = \lim_{\alpha \to \infty} \frac{1}{\alpha} \sum_{l=k}^{k+\alpha-1} \mathcal{E} f(x(\theta_1 + lT, \phi), \ldots, x(\theta_m + lT, \phi), \\ \zeta(t_1 + lT), \ldots, \zeta(t_N + lT)). \end{aligned} \tag{2.12}$$

We represent last sum in (2.12) in the form

$$\sum_{l=k}^{k+\alpha-1} = \sum_{l=0}^{\alpha-1} + \sum_{l=\alpha}^{k+\alpha-1} - \sum_{l=0}^{k-1}.$$

Whence and from (2.12), (2.8) we obtain that

$$J = \lim_{\alpha\to\infty} \frac{1}{\alpha} \sum_{l=0}^{\alpha-1} \mathcal{E} f(x(\theta_1 + lT, \phi), \ldots, x(\theta_m + lT, \phi),$$

$$\zeta(t_1 + lT), \ldots, \zeta(t_N + lT)) = \lim_{\alpha\to\infty} \mathcal{E} f(x(\theta_1, \phi_\alpha), \ldots, x(\theta_m, \phi_\alpha), \tag{2.13}$$

$$\zeta_\alpha(t_1), \ldots, \zeta_\alpha(t_N)) = \mathcal{E} f(\xi(\theta_1), \ldots, \xi(\theta_m),\ \zeta(t_1), \ldots, \zeta(t_N)).$$

(2.11)–(2.13) imply (2.7). Thm. 2.1 is proven. □

Remark 2.1 Modifying the results of [152], we can prove that the assertion of Thm. 2.1 still holds if all of its assumptions are satisfied except of (2.4) which is replaced by the following condition: uniformly with respect to $t > 0$

$$\lim_{C\to\infty} \frac{1}{t} \int_0^t \mathcal{P}(|x(s)| > C) ds = 0. \tag{2.14}$$

Remark 2.2 For Eq. (2.1) with a multiplicative perturbation $\zeta(t)$:

$$\dot{x}(t) = a(t, x_t) + b(t, x_t)\zeta(t)$$

the assertion of Thm. 2.1 still holds if the assumptions (2.4), (2.5) are satisfied as well as the following relations which replace conditions (2.3):

$$|a(t,\phi) - a(t,\psi)| + |b(t,\phi) - b(t,\psi)| \le r \parallel \phi - \psi \parallel,$$

$$\mathcal{P}\left(\int_0^t |a(s,0) + b(s,0)\zeta(s)| ds < \infty \right) = 1, \quad \forall t \ge 0. \tag{2.15}$$

3. RELATION BETWEEN STABILITY AND PERIODICITY

3.1. *APPLICATION OF THE DIRECT LIAPUNOV METHOD*

We shall study the conditions for the existence of periodic solutions of SRDE

$$\dot{x}(t) = b(t, x_t) + \sigma(t, x_t)\zeta(t), \quad t > 0, \quad x \in \mathbf{R}^n,$$
$$x_t = x(t+\theta), \quad -h \le \theta \le 0, \quad \zeta(t) \in \mathbf{R}^{n_1}. \tag{3.1}$$

The functionals b and σ are assumed to satisfy the general assumptions (2.15) and for any $x(\cdot)$ fixed the functions $t \mapsto b(t,x(\cdot)),\sigma(t,x(\cdot))$ are periodic on t with period T.

By $\zeta(t)$ we have denoted a T-periodic stochastic process with continuous realizations. The solution of Eq. (3.1) for $t > 0$ is determined by initial condition (2.2). It turns out that the existence of a periodic solution of Eq. (3.1) and some of it properties depend on the stability of the solutions of the deterministic equation

$$y(t) = b(t, y_t), \quad t > 0, \quad y_t(\theta) = y(t+\theta), \quad -h \le \theta \le 0. \tag{3.2}$$

The solution of Eq. (3.2) is determined by initial condition (2.2). Denote by $x(t;s,\phi)$ the solution of Eq. (3.1) for $t \ge s$ with initial condition $x_s(s;s,\phi) = \phi$. The notation of the solution $y(t;s,\phi)$ of Eq. (3.2) has an analogous meaning. Denote by $u \mapsto \omega_i(u)$ non-negative scalar continuous functions defined for scalar $u \ge 0$, non-decreasing and vanishing only for $u = 0$. Denote by C_i some positive constants.

Consider a class of continuous scalar functionals $(t,\phi) \mapsto V(t,\phi(\cdot))$ defined on $[0,\infty) \times C[-h,0]$ and satisfying for some ω_i the conditions

$$\omega_1(|\phi(0)|) \le V(t,\phi) \le \omega_2(\| \phi \|), \tag{3.3}$$

$$|V(t,\phi) - V(t,\phi_1)| \le C_1 \, \| \phi - \phi_1 \| \, . \tag{3.4}$$

Moreover, the total derivative $\dot V^0(t,\phi)$ of the functional V along the trajectories of the determinate system (3.2) satisfies the estimate

$$\dot V^0(t,\phi) \le -\omega_3(\| \phi \|), \tag{3.5}$$

$$\dot V^0(t,\phi) := \limsup_{\Delta \to 0+} \frac{1}{\Delta}[V(t+\Delta, y(t+\Delta;t,\phi)) - V(t,\phi)]. \tag{3.6}$$

In relations (3.3)–(3.5) by ϕ and ϕ_1 we have denoted arbitrary functions of $C[-h,0]$, $\Delta \to 0^+$ means that Δ tends to zero from the right.

Theorem 3.1 Assume that $\mathcal{E}|\zeta(t)|^2 \le C < \infty$, there exists a functional V satisfying estimates (3.3)–(3.5) and the matrix σ is bounded:

$$|\sigma(t,\phi(\cdot))| \le C < \infty \quad \text{for any} \quad t \ge 0, \phi(\cdot) \in C[-h,0].$$

Moreover, let $\omega_1(u) \ge C_3 u^p$, $p > 0$, $\omega_3(u) \ge C_2 \omega_2(u)$. Then there exists a periodic solution of Eq. (3.1).

Proof Denote by $x(\cdot)$ the solution of equation (3.1) with the zero initial condition (i.e. $x(t) = x(t;0,0)$) and prove the boundedness of $\mathcal{E}|x(t)|^2, t \ge 0$. We define the total derivative $\dot V(t,\phi)$ of the functional V along the trajectories of Eq. (3.1) by relation (3.6) in the right–hand side of which $y(t+\Delta;t,\phi)$ is replaced by $x(t+\Delta;t,\phi)$. Take an arbitrary moment of time $s\geq 0$ and estimate $\dot{V}(s,x_s)$. For this purpose we introduce the function $z(\cdot)$ defined by the relations

$$\dot{z}(t)=b(t,z_t),\quad t>s;\quad z(t)=x(t),\quad t\leq s. \tag{3.7}$$

In other words, $z(t)=y(t;s,x_s),t\geq s$.

From (3.3)–(3.7) it follows that

$$\dot{V}(s,x_s)\leq \limsup_{\Delta\to 0+}\frac{1}{\Delta}[V(s+\Delta,z_{s+\Delta})-V(s,x_s)]+$$

$$+\limsup_{\Delta\to 0+}\frac{1}{\Delta}[V(s+\Delta,x_{s+\Delta})-V(s+\Delta,z_{s+\Delta})]\leq$$

$$\leq -C_2V(s,x_s)+C_1|\sigma(s,x_s)\zeta(s)|.$$

Last inequality means that

$$V(s,x_s)\leq C_1\int_0^s|\sigma\zeta(t)|\gamma(t,s)dt,\quad \gamma(t,s):=\exp(C_2(t-s)),\quad \sigma=\sigma(t,x_t).$$

Whence in view of (3.3) we have

$$C_3|x(s)|^p\leq C_1\int_0^s\gamma(t,s)|\sigma\zeta(t)|dt.$$

Hence by the conditions of Thm. 3.1

$$\mathcal{E}|x(s)|^2\leq \mathcal{E}[C_4\int_0^s\gamma(t,s)|\sigma\zeta(t)|dt]^{\frac{2}{p}}\leq$$

$$\leq\left[C_4^2\left(\int_0^s\gamma(t,s)dt\right)^2\left(\int_0^s\gamma(t,s)\mathcal{E}|\sigma\zeta(t)|^2dt\right)\right]^{\frac{1}{p}}\leq C_5. \tag{3.8}$$

Now we shall verify that the process $x(t)$ satisfies assumptions (2.5), (2.14) of Thm. 2.1. First by Chebyshev's inequality for any $t>0$ we obtain

$$\frac{1}{t}\int_0^t\mathcal{P}(|x(s)|>C)ds\leq C_5/C^2\to 0\quad\text{for}\quad C\to\infty. \tag{3.9}$$

Further on, for any $t \geq 0, \Delta \geq 0$, from (2.3), (3.1), (3.8) and the conditions of Thm. 3.1 it follows that

$$\mathcal{E}|x(t+\Delta) - x(t)|^2 = \mathcal{E}\left|\int_t^{t+\Delta} (b(s,x_s) + \sigma(s,x_s)\zeta(s))ds\right|^2 \leq$$

$$\leq 2\Delta \int_t^{t+\Delta} \mathcal{E}(|b(s,x_s)|^2 + |\sigma(s,x_s)\zeta(s)|^2)ds \leq$$

$$\leq 3\Delta \int_t^{t+\Delta} \mathcal{E}(r^2 \parallel x_s \parallel^2 + |b(s,0)|^2 + |\sigma(s,x_s)\zeta(s)|^2)ds \leq C_6\Delta^2.$$

From here and (3.9) by virtue of Thm. 2.1 and remark 2.2 it follows the existence of a T–periodic solution of Eq. (3.1). Thm. 3.1 is proven.□

Remark 3.1 Assume that for $t > 0$ there exists a bounded solution of deterministic equation (3.2). Then in view of Thm. 2.1 there exists also a T–periodic random process satisfying Eq. (3.2). Since, however, the realizations of a periodic random process need not be periodic, this does not provide the existence of a determinate periodic solution of Eq. (3.2).

Remark 3.2 The condition of Thm. 3.1 of existence of the functional V satisfying estimates (3.3)–(3.5) imply the stability of deterministic equation (3.2).

Using the inversion theorem 4.1.3 the condition of Thm. 3.1 mentioned above can be replaced by the assumption of exponential stability of the solutions of Eq. (3.2). Moreover, a number of assertions on the existence of periodic solutions of stochastic equations with aftereffect can be obtained using the theorem of stability by the first approximation or of stability under persistent perturbations. Let us formulate, for instance, some conditions for the existence of periodic solutions of equations of the form

$$\begin{aligned} &\dot{x}(t) = b(t,x_t) + \sigma(t,x_t,\zeta(t)), \quad t > 0, \quad x(t) \in \mathbf{R}^n, \\ &x_t(\theta) := x(t+\theta), \quad -h \leq \theta \leq 0, \quad \zeta(t) \in \mathbf{R}^{n_1}. \end{aligned} \tag{3.10}$$

In (3.10) we have used the same notations as in Eq. (3.1). Assume that $b(t,0) \equiv 0$, the trivial solution of deterministic equation (3.2) is exponentially stable and the functional σ is continuous and bounded. Then Eq. (3.10) has a periodic solution. For the justification of this assertion it suffices to make use of the inversion theorem 4.1.3 (by virtue of which there exists a functional satisfying relations (3.3)–(3.5)) and then repeat the proof of Thm. 3.1 with the appropriate modifications.

3.2. *STABILITY OF PERIODIC SOLUTION*

The functioning of real systems takes place in the presence of various perturbations. That is why we are interested not only in the existence of periodic solutions but also in their stability with respect to these perturbations. We shall study this question in more details for Eq. (3.1) with a linear functional $(t,\phi) \mapsto b(t,\phi)$, i.e. with a functional b satisfying for any constants C_1, C_2 and functions $\phi, \psi \in C[-h,0]$ the condition

$$b(t,0) \equiv 0, \quad b(t, C_1\phi + C_2\psi) = C_1 b(t,\phi) + C_2 b(t,\psi). \tag{3.11}$$

For any ϕ fixed the function $b(\cdot,\phi)$ is periodic with period T. Assume that the trivial solution of linear deterministic equation (3.2), (3.11) is exponentially stable and the functional $(t,\phi) \mapsto \sigma(t,\phi)$ is bounded, periodic with respect to t with period T and satisfies the Lipschitz condition

$$|\sigma(t,\phi) - \sigma(t,\psi)| \leq l \parallel \phi - \psi \parallel . \tag{3.12}$$

We shall find conditions under which Eq. (3.1) has a unique periodic solution $x_0(t)$ and any other solution $x(t)$ of this equation tends to $x_0(t)$ as $t \to \infty$ with probability 1, i.e.

$$\mathcal{P}(\lim_{t\to\infty} |x(t) - x_0(t)| = 0) = 1. \tag{3.13}$$

From the exponential stability of Eq. (3.2) it follows the existence of a functional $V(t,\phi)$ satisfying the estimates (see Thm. 4.3.1)

$$C_3 \parallel \phi \parallel \leq V(t,\phi) \leq C_4 \parallel \phi \parallel, \tag{3.14}$$

$$\dot{V}^0(t,\phi) \leq -C_2 \parallel \phi \parallel, \tag{3.15}$$

$$|V(t,\phi) - V(t,\psi)| \leq C_1 \parallel \phi - \psi \parallel . \tag{3.16}$$

Here $\dot{V}^0$ is defined by formula (3.6) and represents the total derivative of the functional V along the trajectories of deterministic equation (3.2), (3.11).

The assumptions made and theorem 3.1 imply the existence of a periodic solution $x_0(t)$ of Eq. (3.1). For the justification of (3.13) we put $z(t) := x_0(t) - x(t,\phi)$ and write down the equation for $z(t)$. In view of (3.1) and the linearity of b we have

$$\dot{z}(t) = b(t, z_t) + (\sigma(t, x_{0t}) - \sigma(t, x_t))\zeta(t). \tag{3.17}$$

Now let us calculate the total derivative $\dot{V}$ of the functional V along the trajectories of Eq. (3.17), defined by equality (3.6), where y is replaced by z. In view of (3.14)–(3.16) we have

$$\dot{V}(s, z_s) \leq -C_2 V(s, z_s) + C_1 |\sigma(s, x_{0t}) - \sigma(s, x_s))\zeta(s)| \leq$$

$$\leq -C_2 V(s, z_s) + C_1 l \parallel z_s \parallel \cdot |\zeta(s)| \leq (-C_2 + C_1 l C_3^{-1} |\zeta(s)|) V(s, z_s).$$

From here and (3.14) it follows that

$$C_3 \parallel z_t \parallel \leq \parallel z_0 \parallel \exp[-C_2 t + C_1 \mathcal{L} C_3^{-1} \int_0^t |\zeta(s)| ds].$$

Thus following theorem is proven:

Theorem 3.2 Let relations (3.11), (3.12) hold, linear deterministic equation (3.2), (3.11) be exponentially stable, $\mathcal{E}|\zeta(t)|^2 \leq C < \infty$ *and with probability 1*

$$\lim_{t\to\infty} [-C_2 t + C_1 l C_3^{-1} \int_0^t |\zeta(s)| ds] = -\infty. \tag{3.18}$$

Then there exists a unique periodic solution of Eq. (3.1) and any other solution tends to the periodic one in the sense of (3.13).

Remark 3.3 If in Eq. (3.1) the coefficient σ does not depend on the phase variables, i.e. $\sigma(t,\phi) \equiv \sigma(t)$, then the assertion of Thm. 3.2 holds if all of its conditions are satisfied except for assumption (3.18) which in this case is superfluous.

4. PERIODIC SOLUTIONS OF CONCRETE CLASSES OF EQUATIONS

4.1. *THE CASE OF QUASILINEAR DETERMINISTIC EQUATION*

We shall establish conditions for the existence of periodic solutions of scalar equations of the form

$$\dot{x}(t) = -a_1(t, x_t, \zeta(t)) x(t) + a_2(t, x_t, \zeta(t)), \quad t > 0. \tag{4.1}$$

The functionals a_1 and a_2 are defined on $[0,\infty) \times C[-h,0] \times \mathbf{R}^n$, continuous with respect to all arguments and T-periodic with respect to the first argument. Moreover,

$$|a_1(t,\phi,z)\phi(0) - a_1(t,\psi,z)\psi(0)| \leq \int_0^h |\phi(-s) - \psi(-s)| dk_1(s), \tag{4.2}$$

$$|a_2(t,\phi,z) - a_2(t,\psi,z)| \leq \int_0^h |\phi(-s) - \psi(-s)| dk_2(s). \tag{4.3}$$

Here (t,ϕ,z) and (t,ψ,z) are arbitrary elements of $[0,\infty) \times C[-h,0] \times \mathbf{R}^n$, the functions $k_1(\cdot)$ and $k_2(\cdot)$ are non-decreasing and bounded:

$$r_i = \int_0^h dk_i(s) < \infty, \quad i = 1,2.$$

The random process $\zeta(t)$ in (4.1) with continuous realizations is T-periodic. The solution of Eq. (4.1) for $t > 0$ is determined by initial condition (2.2).

Theorem 4.1 Let conditions (4.2), (4.3) hold as well as

$$0 \leq r_2 < \lambda \leq a_1(t,\phi,z), \quad \mathcal{E}a_2(t,0,\zeta(t))^2 \leq C < \infty. \tag{4.4}$$

Then Eq. (4.1) has a T-periodic solution.

Proof Denote by $x(t)$ the solution of Eq. (4.1) with zero initial condition (2.2). We shall prove that $\mathcal{E}x^2(t) \leq C < \infty$. From (4.1) and the Cauchy formula it follows that

$$x(t) = \int_0^t a_2(s,x_s,\zeta(s)) \exp[-\int_s^t a_1(\tau,x_\tau,\zeta(\tau))d\tau]ds. \tag{4.5}$$

In view of (4.3) we have

$$\mathcal{E}a_2(t,x_t,\zeta(t))^2 \leq 2\mathcal{E}a_2(t,0,\zeta(t))^2 + 2\int_0^h \mathcal{E}x^2(t-s)dk_2(s). \tag{4.6}$$

From here, condition (4.4) and (4.5) it follows the estimate

$$\mathcal{E}x^2(t) \leq C_1 \int_0^t e^{-\lambda(t-s)}\mathcal{E}a_2^2(s,0,\zeta(s))ds +$$

$$+C_1 \int_0^t e^{-\lambda(t-\tau)}d\tau \int_0^h \mathcal{E}x^2(\tau-s)dk_2(s).$$

Hence, in view of (4.4),

$$\mathcal{E}x^2(t) \leq C_2 + r_2C_1 \int_0^t \rho(\tau)e^{-\lambda(t-\tau)}d\tau, \quad \rho(t) := \sup_{0\leq s\leq t} \mathcal{E}x^2(s).$$

Note that the function

$$t \mapsto \int_0^t \rho(t_1)e^{-\lambda(t-t_1)}dt_1$$

is non-decreasing with respect to t. That is why for any $s \in [0,t]$ we have

$$\mathcal{E}x^2(s) \leq C_2 + r_2C_1 \int_0^t \rho(t_1)e^{-\lambda(t-t_1)}dt_1.$$

Hence

$$\rho(t) \le C_2 + r_2 C_1 \int_0^t \rho(t_1) e^{-\lambda(t-t_1)} dt_1.$$

Represent last relation in the form

$$\rho(t)e^{\lambda t} \le C_2 e^{\lambda t} + C_1 r_2 \int_0^t \rho(t_1) e^{\lambda t_1} dt_1.$$

Whence and from the lemma of Grownwall – Bellman there follows the estimate

$$\rho(t)e^{\lambda t} \le C_2 e^{\lambda t} + C_1 C_2 r_2 \int_0^t e^{r_2(t-s)+\lambda s} ds.$$

Thus for all $t \ge 0$

$$\rho(t) \le C_2 + r_2 C_1 C_2 \int_0^t e^{-(\lambda - r_2)(t-s)} ds \le C_3 < \infty. \tag{4.7}$$

Now we shall show that the process $x(t)$ satisfies conditions (2.5), (2.14). In view of the Chebyshev inequality for any $t > 0$ we have

$$\frac{1}{t}\int_0^t \mathcal{P}[|x(s)| > C] ds \le C_3 C^{-2} \to 0 \quad \text{for} \quad C \to \infty. \tag{4.8}$$

Further on,

$$\mathcal{E}(x(t+\Delta) - x(t))^2 \le \Delta \mathcal{E} \int_t^{t+\Delta} (a_2(s, x_s, \zeta(s)) -$$

$$-a_1(s, x_s, \zeta(s))x(s))^2 ds \le C_5 \Delta^2.$$

From here and (4.8), in view of Thm. 2.1, there follows the existence of a T–periodic solution of Eq. (4.1). Thm. 4.1 is proven.□

Remark Analogously to the proof of Thm. 4.1 we can establish that sufficient conditions for the existence of periodic solutions of Eq. (4.1) (in the case when x and a_2 have values in $\mathbf{R}^n$, and a_1 is a diagonal matrix of dimension $(n \times n)$) are the assumptions (4.2)–(4.4) and the first of inequalities (4.4) is satisfied by each entry of the matrix a_1.

4.2. *LINEAR EQUATIONS*

Consider linear equation of the form (3.1)

$$\dot{x}(t) = b(t, x_t) + \zeta(t), \quad t \geq 0, \quad x(t) \in \mathbf{R}^n, \quad \zeta(t) \in \mathbf{R}^{n_1}. \tag{4.9}$$

The functional $(t, \phi) \mapsto b(t, \phi)$ satisfies condition (3.11) and the process $\zeta(t)$ is the same as in Eq. (4.1). Thm. 3.2 implies

Theorem 4.2 Assume that the linear deterministic equation (3.2), (3.11) is exponentially stable and $\mathcal{E}|\zeta(t)|^2 \leq C < \infty$. Then there exists a unique periodic solution $x_0(t)$ of Eq. (4.9) and any other solution $x(t, \phi)$ of this equation tends to $x_0(t)$ as $t \to \infty$ with probability 1, i.e. relation (3.13) holds. If, moreover, $b(t, \phi)$ does not depend on the time, i.e. $b(t, \phi) \equiv b(\phi)$, then the following representation of $x_0(t)$ is valid

$$x_0(t) = \int_{-\infty}^{t} z(t-s)\zeta(s)ds = \int_{0}^{\infty} z(s)\zeta(t-s)ds. \tag{4.10}$$

Here $z(t)$ is the fundamental matrix of solutions of the homogeneous equation $\dot{x}(t) = b(x_t)$ defined by the relations

$$\dot{z}(t) = b(z_t), \quad t > 0, \quad z(0) = I, \quad z(\tau) \equiv 0, \quad \tau < 0,$$

where I is the unit matrix (see Sect. 3.5).

Under a number of assumptions some statistical characteristics of the periodic solution $x_0(t)$ of Eq. (4.9) can be obtained by means of temporal means. One of these assumptions related to the character of decrease of the dependence of the past of the process ζ on its future has the form: uniformly with respect to t

$$\lim_{|\tau| \to \infty} R(t, t+\tau) = 0,$$
$$R(t, t+\tau) := \mathcal{E}([\zeta(t) - \mathcal{E}\zeta(t)][\zeta(t+\tau) - \mathcal{E}\zeta(t+\tau)]'). \tag{4.11}$$

Theorem 4.3 Let the linear deterministic equation (3.2), (3.11) with $b(t, \phi) \equiv b(\phi)$ be asymptotically stable uniformly with respect to the initial data, $\mathcal{E}|\zeta(t)|^2 \leq C < \infty$ and condition (4.11) holds. Then there is a mean square convergence of the temporal means of the process $x_0(t)$

$$\lim_{t \to \infty} \mathcal{E}[\frac{1}{t} \int_0^t x_0(s)ds - \frac{1}{T} \int_0^T \mathcal{E}x_0(s)ds]^2 = 0. \tag{4.12}$$

Proof The existence of a periodic solution of Eq. (4.9) follows from Thm. 4.1. In view of representation (4.10)

$$\frac{1}{t}\int_0^t x_0(s)ds = \frac{1}{t}\int_0^t \mathcal{E}x_0(s)ds + \frac{1}{t}\int_0^t [\int_0^\infty z(s)\zeta(\tau - s)ds - \int_0^\infty z(s)\mathcal{E}\zeta(\tau - s)ds]d\tau. \tag{4.13}$$

By virtue of T-periodicity of the process $x_0(t)$

$$\lim_{t\to\infty} \frac{1}{t}\int_0^t \mathcal{E}x_0(s)ds = \frac{1}{T}\int_0^T \mathcal{E}x_0(s)ds.$$

From here and (4.13) it is seen that for the proof of (4.12) it suffices to establish the relation

$$\lim_{t\to\infty} \frac{1}{t^2}\int_0^t d\tau_1 \int_0^t d\tau_2 \int_0^\infty z(s_1)\int_0^\infty z(s_2)R(\tau_1 - s_1, \tau_2 - s_2)ds_1 ds_2 = 0. \tag{4.14}$$

But in view of the exponential stability of Eq. (3.2), (3.11) the following estimate holds

$$|z(t)| \le C_1 e^{-ct}, \quad c > 0, \quad C_1 > 0.$$

Now (4.14) follows from [58], p. 54. Thm. 4.3 is proven. □

5. PERIODIC SOLUTIONS OF THE ITÔ's SFDEs

5.1. *EXISTENCE OF PERIODIC SOLUTIONS*

In this section conditions for the existence of periodic solutions of the SFDEs of Itô are established. The method of the shift and averaging of the initial distribution generating a solution with certain properties which was used in Sec. 2 is well known as the method of construction of invariant measures of dynamical systems. This method was used previously for the construction of stationary solutions of equations with aftereffect perturbed by a stationary process and the SFDEs of Itô. The justification of the general theorems of existence of periodic solutions of stochastic equations with aftereffect is similar to the proof of the theorems of existence of stationary solutions. The modifications arising are due to the fact that the distribution of the probabilities of the periodic solutions must be invariant with respect to shifts of the argument by quantities multiple to the period while for the stationary solutions it is invariant with respect to arbitrary shifts of the argument. The conditions of the general theorems of existence of periodic solutions must contain assumptions sufficient for the compactness of the sequence of measures obtained by shifts and averaging of the initial distribution (see, for instance, conditions (2.4), (2.5) of Thm. 2.1).

The above remarks allow us to formulate a number of general assertions of existence of periodic solutions, the proof of which only slightly differs from the proof of the respective theorems of existence of stationary solutions. That is why here we shall give the formulations only of those theorems which will be used further on the obtaining of coefficient conditions for the existence of periodic solutions.

First consider SRDE of the form

$$\left.\begin{aligned} dx(t) &= a(t,x_t)dt + b(t,x_t)d\xi(t), \quad t \geq 0, \\ x_t(\theta) &:= x(t+\theta), \quad -h \leq \theta \leq 0, \quad x(t) \in \mathbf{R}^n, \quad \xi(t) \in \mathbf{R}^l. \end{aligned}\right\} \tag{5.1}$$

Here $\xi(t)$ is the standard Wiener process, the continuous functionals $(t,\phi) \mapsto a(t,\phi), b(t,\phi)$ are defined on $[0,\infty) \times C[-h,0]$ and are T-periodic with respect to the first argument. The solution of Eq. (5.1) for $t \geq 0$ is determined by initial condition (2.2).

Definition 5.1 T-periodic solution of Eq. (5.1) is a T-periodic stochastic process periodically connected with $\xi(t)$ and satisfying (5.1) with probability 1. The former means that all finite-dimensional distributions of the process $(x(t), -\xi(\tau)+\xi(s))$, $\tau < s$, are invariant with respect to all shifts of the arguments by the quantity kT, where k is an arbitrary integer, i.e., that for any positive integers N, m and any $t_1,\dots,t_N,\tau_1,\dots,\tau_m,s_1,\dots,s_m$ such that $\tau_i < s_i, i = 1,\dots,m$, the distribution of the probabilities of the random variable $(x(t_1),\dots,x(t_N),\xi(s_1)-\xi(\tau_1),\dots,\xi(s_m)-\xi(\tau_m))$ coincides with the distribution of the probabilities of the variable $(x(t_1+kT),\dots,x(t_N+kT),\xi(s_1+kT)-\xi(\tau_1+kT),\dots,\xi(s_m+kT)-\xi(\tau_m+kT))$.

Theorem 5.1 Let the assumptions formulated hold and a solution of Pr. (5.1), (2.2) exist for which conditions (2.4), (2.5) (or conditions (2.14), (2.5)) are satisfied. Then there exists a T-periodic solution of Eq. (5.1).

Proof Since the proof of Thm. 5.1 is similar to the proof of Thm. 2.1, we shall only outline its scheme. As in the proof of theorem 2.1, consider a sequence of random variables τ_k taking with equal probabilities the values lT, where $l = 0,1,\dots.k-1$. Introduce the sequences of random processes

$$\phi_k(\theta) = x(\tau_k+\theta), \quad x_k(t) = x(\tau_k+t), \xi_k(t) = \xi(\tau_k+t)-\xi(\tau_k).$$

Under the assumptions made the sequence (ϕ_k, x_k, ξ_k) is weakly compact. As in Sect. 2 it is verified that this sequence tends with probability 1 to the limit $(\bar{\phi}, \bar{x}, \bar{\xi})$, where $\bar{\xi}$ is a Wiener process and the process $(\bar{x}(t), \bar{\xi}(t))$ is T-periodic. Finally, analogously to [257] it is proved that $(\bar{\phi}, \bar{x})$ is a solution of Pr. (5.1), (2.2) where in Eq. (5.1) the process $\xi(t)$ is replaced by $\bar{\xi}(t)$. The conditions for the validity of Thm. 5.1 can be formulated in terms of existence of a solution of Pr. (5.1), (2.2) with a bounded moment. Namely, modifying slightly the proof of Thm. 3 from [257] of existence of stationary solution, the following theorem can be justified.

Theorem 5.2 Assume that the coefficients of Eq. (5.1) satisfy the assumptions of Thm. 5.1 as well the conditions

$$|a(t,\phi)|^2 \le q_1 + \int_{-h}^{0} |\phi(s)|^2 dk_1(s), \quad |b(t,\phi)|^2 \le q_2 + \int_{-h}^{0} |\phi(s)|^2 dk_2(s), \quad (5.2)$$

where q_i are some positive constants and $k_i(s)$ are scalar non-decreasing functions on $[-h, 0]$. Let, further on, a solution $x(t,\phi)$ of Pr. (5.1), (2.2) exist such that for $t \in [-h, 0]$ (2.5) holds and for $t \ge 0$ the second moment is bounded:

$$\mathcal{E}|x(t,\phi)|^2 \le C < \infty, \quad t \ge 0. \quad (5.3)$$

Then there exists a T-periodic solution of Eq. (5.1).

In an analogous way one can formulate conditions for the existence of periodic solutions of SRDE with an arbitrary aftereffect (bounded or not) of the form

$$\left.\begin{aligned} dx(t) &= a(t,x_t)dt + b(t,x_t)d\xi(t), \quad t \ge 0, \\ x_t(\theta) &:= x(t+\theta), \quad \theta \le 0, \quad x(t) \in \mathbf{R}^n, \quad \xi(t) \in \mathbf{R}^l. \end{aligned}\right\} \quad (5.4)$$

The functionals $(t,\phi) \mapsto a(t,\phi) \in \mathbf{R}^n$ and the matrix–functional $(t,\phi) \mapsto b(t,\phi)$ are defined and continuous on $[0,\infty) \times C(-\infty, 0]$ and periodic with respect to t with period T. By $C(-\infty, 0]$ we have denoted the metric space of continuous functions $\theta \mapsto \phi(\theta) \in \mathbf{R}^n$, $\theta \le 0$ with a metric $\rho(\phi, \psi)$:

$$\rho(\phi,\psi) := \sum_{j=1}^{\infty} 2^{-j} \frac{\| \phi - \psi \|_j}{1 + \| \phi - \psi \|_j}, \quad \| \phi \|_j := \max_{-j \le t \le 0} |\phi(t)|.$$

Theorem 5.3 Assume that beside the restrictions imposed above the following inequalities hold

$$|a(t,\phi)|^4 \le q_1 + \int_{-\infty}^{0} |\phi(s)|^4 dk_1(s), \quad |b(t,\phi)|^4 \le q_2 + \int_{-\infty}^{0} |\phi(s)|^4 dk_2(s), \quad (5.5)$$

where $k_i(s)$ *are scalar non-decreasing bounded functions on* $(-\infty, 0]$. *Let a solution* $x(t;\phi)$ *of Pr. (5.4), (2.2) exist satisfying inequality (2.5) for all* $t \le 0$ *and*

$$E|x(t;\phi)|^4 \le C < \infty, \quad t \ge 0. \tag{5.6}$$

Then there exists a T*-periodic solution of Eq. (5.4).*

The proof of Thm. 5.3 is, in general, similar to the proof of Thm. 2 from [257] devoted to stationary solutions. One has only to take into account (as it was already noted) some modifications in the proof due to the fact that in the case considered the shift and averaging of the initial distribution of the solution $x(t;\phi)$ are realized by a quantity multiple of the period T.

Thms. 5.2, 5.3 show that the boundedness of the moments of some solution implies the existence of a periodic solution. Below sufficient conditions will be formulated for the existence of solutions of Eqs. (5.1), (5.4) with a bounded moment by means of the Liapunov's direct method.

In relation to Thms. 5.2, 5.3 we shall also note that for determinate equations of the form (5.1), (5.4) for $b \equiv 0$ the existence of a bounded solution does not imply, in general, the existence of a periodic solution.

5.2. *SCALAR SRDES*

Let us consider scalar equations with a finite aftereffect (5.1) satisfying (5.2) and find the conditions sufficient for validity of (5.3).

We apply the *method of integral inequalities.* First we shall formulate conditions for the existence of a periodic solution of a scalar equation of the form

$$dx(t) = [-a_0(t)x(t) + a(t,x_t)]dt + b(t,x_t)d\xi(t), \quad t \ge 0. \tag{5.7}$$

Theorem 5.4 Assume that the functionals $(t,\phi) \mapsto a(t,\phi), b(t,\phi)$ *satisfy the assumptions of Thm. 5.2, the function* $a_0(\cdot)$ *is periodic with period* T *and the following inequality holds*

$$\inf_{t\ge 0}[a_0(t) - \sqrt{\alpha_{01}} - \alpha_{02}/2] =: a_1 > 0, \quad \alpha_{0j} := \int_{-h}^{0} |dk_j(s)|. \tag{5.8}$$

Then

$$\mathcal{E}x^2(t) \le C(\| \mathcal{E}\phi^2(\cdot) \| + 1), \quad t \ge 0.$$

Before proving the theorem, we shall prove a lemma which is a generalization of lemma 8.1 from [257].

Lemma 5.1 Let the continuous scalar functions $t \mapsto z(t), f(t), t \geq 0$ *satisfy for some constant* C *and for any* $t_1, t_2 : t_2 \geq t_1 \geq 0$ *the inequality*

$$z(t_2) - z(t_1) \leq -C \int_{t_1}^{t_2} z(s)ds + \int_{t_1}^{t_2} f(s)ds. \tag{5.9}$$

Then

$$z(t) \leq z(0)e^{-Ct} + \int_0^t e^{-C(t-s)} f(s)ds. \tag{5.10}$$

Proof Set

$$y(t) := \int_0^t e^{-C(t-s)} f(s)ds, \quad y_1(t) := z(t) - y(t) - z(0)e^{-Ct}.$$

It is clear that

$$y(t_2) - y(t_1) = -C \int_{t_1}^{t_2} y(s)ds + \int_{t_1}^{t_2} f(s)ds. \tag{5.11}$$

Further on, from the definition $y_1(t)$ it follows that

$$y_1(t_2) - y_1(t_1) = [z(t_2) - z(t_1)] - [y(t_2) - y(t_1)] - z(0)e^{-Ct_2} + z(0)e^{-Ct_1}.$$

Replacing in this relation the increment $z(t_2) - z(t_1)$ by the right–hand side of (5.9) and the increment $y(t_2) - y(t_1)$ by the right–hand side of (5.11), we obtain

$$y_1(t_2) - y_1(t_1) \leq -C \int_{t_1}^{t_2} [z(s) - y(s)]ds - z(0)e^{-Ct_2} + z(0)e^{-Ct_1} =$$

$$= -C \int_{t_1}^{t_2} [z(s) - y(s) - z(0)e^{-Cs}]ds = -C \int_{t_1}^{t_2} y_1(s)ds. \tag{5.12}$$

Note that for the proof of (5.10) it suffices to establish that $y_1(t) \leq 0$ for all $t \geq 0$. Suppose that this is not true, i.e. that at some moment $t_0 > 0$ we should have $y_1(t_0) > 0$. This assumption and the equality $y_1(0) = 0$ imply the existence of such an interval $(t_1, t_2), t_2 > t_1$ that $y_1(t_1) = 0, y_1(s) > 0$ for $t_1 < s \leq t_2$. This, however, in view of (5.12) contradicts the assumption made if $C > 0$. If $C \leq 0$ the contradiction follows from (5.12) and the Grownwall – Bellman's lemma. Lemma 5.1 is proven. □

We shall give an estimate to be used below which is a consequence of the Grownwall – Bellman's lemma.

Lemma 5.2 Let the continuous non-negative function $x(\cdot)$ satisfy for some constants $C_i \geq 0$ the inequality

$$x(t) \leq C_1 + C_2 \int_0^t e^{-C_3(t-s)} x(s)ds, \quad t \geq 0.$$

Then, if $C_3 > C_2$, then $x(t) \leq C_3 C_1 (C_3 - C_2)^{-1}$.

For the proof of lemma 5.2 note that the function $\rho(t) = x(t)\exp(C_3 t)$ satisfies the inequality

$$\rho(t) \leq C_1 \exp(C_3 t) + C_2 \int_0^t \rho(s)ds.$$

This inequality and the Grownwall – Bellman's lemma imply the assertion of lemma 5.2.

Proof of Thm. 5.4 Introduce the kernel $k_3(s) = \alpha_{01}^{-\frac{1}{2}} k_1(s) + k_2(s)$. Consider on the trajectories of Eq. (5.7) the following functional

$$V(t, x_t) = x^2(t) + \int_{-h}^{0} dk_3(s) \int_{t+s}^{t} x^2(\tau)d\tau. \tag{5.13}$$

From the Itô's formula (10.1.5) and Eq. (5.7) it follows that the stochastic differential $dV(t, x_t)$ of the functional (5.13) equals

$$dV(t, x_t) = [b^2(t, x_t) + x^2(t)(\sqrt{\alpha_{01}} + \alpha_{02} - 2a_0(t)) -$$
$$- \int_{-h}^{0} x^2(t+s)dk_3(s) + 2x(t)a(t, x_t)]dt + 2x(t)b(t, x_t)d\xi(t). \tag{5.14}$$

We shall transform the separate addends in (5.14). First, in view of (5.2) we have

$$2|x(t)a(t, x_t)| \leq \sqrt{\alpha_{01}} x^2(t) + \alpha_{01}^{-\frac{1}{2}} a^2(t, x_t) \leq$$
$$\leq \alpha_{01}^{\frac{1}{2}} x^2(t) + \alpha_{01}^{-\frac{1}{2}} (q_1 + \int_{-h}^{0} x^2(t+s)dk_1(s).$$

Using the estimate (5.2) for b^2, in view of (5.14) we obtain

$$dV(t, x_t) \leq -2a_1 x^2(t)dt + C_1 dt + 2x(t)b(t, x_t)d\xi(t), \quad C_1 := q_2 + q_1 \alpha_{01}^{-\frac{1}{2}}. \tag{5.15}$$

We substitute in the righthand side of this inequality the expression (5.13) for V, integrate both sides of the inequality obtained from $t_1 \geq 0$ to $t_2 \geq t_1$ and calculate the mathematical expectation. We have

$$\mathcal{E}x^2(t_2) - \mathcal{E}x^2(t_1) \leq -a_2 \int_{t_1}^{t_2} \mathcal{E}x^2(t)dt + C_1(t_2 - t_1) +$$

$$+ \int_{t_1}^{t_2} dt \int_{-h}^{0} \mathcal{E}x^2(t+s)dk_3(s), \quad a_2 := 2a_1 + \alpha_{03}.$$

From here and lemma 5.1 it follows that

$$\left.\begin{aligned} &\mathcal{E}x^2(t) \leq C_2 + \int_0^t d\tau \int_{-h}^{0} \mathcal{E}x^2(\tau+s)z(t-\tau)dk_3(s), \\ &C_2 := C_1/2a_2 + \mathcal{E}x^2(0), \quad z(t-\tau) := e^{-a_2(t-\tau)}. \end{aligned}\right\} \tag{5.16}$$

We shall transform formula (5.16). Denote by $\kappa(\cdot)$ a function such that $\kappa(t) = 0$ for $t < 0$ and $\kappa(t) = 1$ for $t \geq 0$ and set

$$\gamma(\tau) := \sup_{0 \leq t \leq \tau} \mathcal{E}x^2(t).$$

Note that

$$\int_0^t d\tau \int_{-h}^{0} \mathcal{E}x^2(\tau+s)z(t-\tau)dk_3(s) =$$

$$= \int_0^t d\tau \int_{-h}^{0} \mathcal{E}\phi^2(\tau+s)z(t-\tau)[1-\kappa(s+\tau)]dk_3(s) +$$

$$+ \int_0^t d\tau \int_{-h}^{0} \mathcal{E}x^2(\tau+s)\kappa(\tau+s)dk_3(s) \leq$$

$$\leq \int_0^h d\tau \int_{-h}^{0} \mathcal{E}\phi^2(s+\tau)[1-\kappa(s+\tau)]z(t-\tau)dk_3(s) + \alpha_{03} \int_0^t \gamma(\tau)z(t-\tau)d\tau.$$

Hence

$$\mathcal{E}x^2(t) \le C_3 + \alpha_{03} \int_0^t \gamma(\tau) z(t-\tau) d\tau,$$

$$C_3 \ge C_2 + \int_0^h d\tau \int_{-h}^0 \mathcal{E}\phi^2(s+\tau)[1-\kappa(s+\tau)] z(t-\tau) dk_3(s). \tag{5.17}$$

But the integral in the righthand side of (5.17) as a function of t is non–decreasing. This follows from the non–decreasing of $\gamma(t)$ and the identity

$$\int_0^t \gamma(\tau) z(t-\tau) d\tau = (2a_2)^{-1}(1 - z(t))\gamma(0) +$$

$$+ (2a_2)^{-1} \int_0^t (1 - z(t-\tau)) d\gamma(\tau), \quad t \ge 0.$$

Hence, in view of (5.17)

$$\gamma(t) \le C_3 + \alpha_{03} \int_0^t \gamma(\tau) z(t-\tau) d\tau.$$

Last inequality and lemma 5.2 imply the assertion of Thm. 5.4. □

Remark 5.1 The assertion of Thm. 5.4 still holds for equation

$$dx(t) = [-a_0(t, x_t)x(t) + a(t, x_t)]dt + b(t, x_t)d\xi(t), \quad t \ge 0$$

where $a_0(t, \phi)$ is a periodic with respect to t with period T measurable bounded functional.

The sense of condition (5.8) established for the existence of a periodic solution of Eq. (5.7) is that the addend without aftereffect are prevailing and in some sense suppresses the terms with aftereffect. Below conditions for the existence of periodic solutions when this, in general, is not so. Moreover, it will be shown that periodic solutions of system (5.27) may exist even if the righthand sides of Eqs. (5.27) do not contain at all terms of the form $a(t)x(t)$ and there is no dependence on the current values of the phase variables (i.e. in the system there is no instant feedback).

5.3. *METHOD OF LIAPUNOV FUNCTIONALS*

According to Thms. 5.1, 5.2 the question of existence of periodic solutions of Eqs. (5.1), (5.4) is reduced to the question of existence of solutions of these equations with a bounded second moment. We shall formulate the appropriate conditions in terms of existence of Liapunov's functionals. In Lemmas 5.3, 5.4 r_i are some positive constants.

Lemma 5.3 Let the coefficients of Eq. (5.1) satisfy the assumptions of Thm. 5.2 and let a scalar continuous non-negative functional $V : \mathbf{R}\times C[-h,0] \to \mathbf{R}$ *exist such that*

$$\mathcal{E}V(t,\phi) \le r_0(1+\sup_\theta \mathcal{E}|\phi(\theta)|^2), \tag{5.18}$$

$$\mathcal{E}V(t_2,x_{t_2}) - \mathcal{E}V(t_1,x_{t_1}) \le -r_1 \int_{t_1}^{t_2} \mathcal{E}|x(t)|^2 dt + r_2(t_2-t_1), \quad t_2 \ge t_1, \tag{5.19}$$

$$V(t,\phi) \ge r_3|\phi(0)|^2. \tag{5.20}$$

Here $t_2 \ge t_1 \ge 0$ *are arbitrary moments of time. Then*

$$\mathcal{E}|x(t;0,\phi)|^2 \le r_4(1+\sup_\theta \mathcal{E}|\phi(\theta)|^2), \quad t \ge 0. \tag{5.21}$$

for any continuous initial process $\phi(\theta)$, $-h \le \theta \le 0$.

Lemma 5.4 Let the coefficients of Eq. (5.4) satisfy the assumptions of Thm. 5.3 and let a continuous scalar functional $V : \mathbf{R}\times C(-\infty,0] \to \mathbf{R}$ *exist such that for any* $t \ge 0$ *and* $t_2 \ge t_1 \ge 0$

$$r_3\mathcal{E}|\phi(0)|^4 \le \mathcal{E}V(t,\phi(\cdot)) \le r_0(1+\sup_{\theta\le 0} \mathcal{E}|\phi(\theta)|^4), \tag{5.22}$$

$$\mathcal{E}V(t_2,x_{t_2}) - \mathcal{E}V(t_1,x_{t_1}) \le -r_1 \int_{t_1}^{t_2} \mathcal{E}|x(t)|^4 dt + r_2(t_2-t_1). \tag{5.23}$$

Then

$$\mathcal{E}|x(t;0,\phi)|^4 \le r_4(1+\sup_\theta \mathcal{E}|\phi(\theta)|^4).$$

The proofs of Lemmas 5.3, 5.4, given in [282], are parallel to those of the stability theorems from Ch. 10, and so here they will be omitted.

5.3.1 Scalar case. We shall establish conditions for the existence of a periodic solution of a scalar equation with finite aftereffect of the form

$$dx(t) = [-\int_{-h}^{0} x(t+s)dk_0(s) + a(t,x_t)]dt + b(t,x_t)d\xi(t). \tag{5.24}$$

Theorem 5.5 Assume that the functionals $(t,\phi) \mapsto a(t,\phi)$, $b(t,\phi)$ satisfy the assumptions of Thm. 5.2 and

$$r_1 := (1-\alpha_{10})\beta_{00} - (1+\alpha_{10})\alpha_{01}^{1/2} - \frac{1}{2}\alpha_{02} > 0,$$

$$\alpha_{10} := \int_{-h}^{0} |s dk_0(s)| < \frac{1}{2}, \quad \beta_{00} := \int_{-h}^{0} dk_0(s) > 0.$$

Then there exists a T-periodic solution of Eq. (5.24).

Proof We shall establish the existence of a solution of Eq. (5.24) with a bounded second moment. Consider the following functional V

$$V(x_t) := [x(t) - \int_{-h}^{0} dk_0(s) \int_{t+s}^{t} x(t_1)dt_1]^2 +$$

$$+(\beta_{00} + \alpha_{01}^{1/2}) \int_{-h}^{0} |dk_0(s)| \int_{t+s}^{t} dt_1 \int_{t_1}^{t} x^2(t_2)dt_2 + \tag{5.25}$$

$$+ \int_{-h}^{0} [dk_2(s) + (1+\alpha_{10})\alpha_{01}^{-1/2} dk_1(s)] \int_{t+s}^{t} x^2(t_1)dt_1.$$

Let us calculate the stochastic differential $dV(x_t)$ of the functional (5.25). Analogously to (5.14), (5.15) we have, using the Itô's formula,

$$dV(x_t) \le [-r_1 x^2(t) + r_2]dt + 2[x(t) - \int_{-h}^{0} dk_0(s) \int_{t+s}^{t} x(t_1)dt_1] b(t, x_t) d\xi(t),$$

$$r_2 := q_2 + q_1(\alpha_{01}^{-1/2} + \alpha_{10}). \tag{5.26}$$

Relations (5.25), (5.26) show that conditions (5.18), (5.19) of lemma 5.3 are satisfied. Hence, there exist a solution $\bar{x}(t)$ of Eq. (5.24) and a constant C such that $\mathcal{E}V(\bar{x}_t) \le C$ for all $t \ge 0$. On the other hand, the functional (5.25) is not positively definite. Thus, the boundedness of $\mathcal{E}V(\bar{x}_t)$ does not immediately imply the boundedness of $\mathcal{E}x^2(t)$. Note, however that in view of (5.25)

$$\mathcal{E}[x(t) - \int_{-h}^{0} dk_0(s) \int_{t+s}^{t} \bar{x}(t_1)dt_1]^2 \le C.$$

This implies the inequality

$$(1-\alpha_{10})\mathcal{E}\bar{x}^2(t) \le \int_{-h}^{0} |dk_0(s)| \int_{t+s}^{t} \bar{x}^2(t_1)dt_1 + C, \quad t \ge 0.$$

Hence,

$$(1-\alpha_{10})\mathcal{E}\bar{x}^2(t) \le \alpha_{10}\rho(t) + C_1, \quad t \ge 0,$$

$$C_1 := C + \alpha_{00} \int_{-h}^{0} \bar{\phi}^2(s)ds, \quad \rho(t) := \max_{0 \le \tau \le t} \mathcal{E}x^2(\tau).$$

Therefore,

$$(1-2\alpha_{10})\rho(t) \le C_1.$$

Last estimate implies the existence of a solution of Eq. (5.24) satisfying (5.21). From here and Thm. 5.2 there follows the existence of a T–periodic solution of Eq. (5.24). Thm. 5.5 is proven.□

5.3.2 Second order equations. Consider conditions of existence of T–periodic solution of second order equation which is written in the form

$$\left.\begin{aligned} &\dot{x}(t) = y(t), \\ &dy(t) = [-\int_{-h}^{0} y(t+s)dk_0(s) - a_0(t)y(t) - \int_{-h}^{0} x(t+s)dk_1(s)]dt + \\ &+b(t, y_t)d\xi(t), \quad t \ge 0, \\ &y_t(\theta) := y(t+\theta), \quad -h \le \theta \le 0. \end{aligned}\right\} \tag{5.27}$$

Here k_i are functions of bounded variation on $[-h, 0]$, the functional b and the function a_0 are the same as in Eq. (5.26).

Theorem 5.6 Assume that the coefficients of Eq. (5.27) satisfy the assumptions formulated above and

$$\beta_{01} > 0, \quad \inf_{t \ge 0}[a_0(t) - \int_{-h}^{0} |dk_0(s)| - \alpha_{11} - \alpha_{02}] =: r_1 > 0.$$

Then there exists a T–periodic solution of Eq. (5.27).

Proof In view of Thm. 5.2, it suffices to establish the existence of a solution of system (5.27) with a bounded second moment. Consider the functional V

$$V(x_t, y_t) := 2\beta_{01}x^2(t) + y^2(t) + V_0^2(x_t, y_t) +$$

$$+\int_{-h}^{0} |dk_0(s)| \int_{t+s}^{t} [y^2(\tau) + \beta_{01}x^2(\tau)]d\tau + 2\int_{-h}^{0} dk_2(s) \int_{t+s}^{t} y^2(\tau) +$$

$$+\int_{-h}^{0} |dk_1(s)| \int_{t+s}^{t} dt_1 \int_{t_1}^{t} [y^2(\tau) + \beta_{01}x^2(\tau)]d\tau.$$

Here we have denoted

$$V_0(x_t, y_t) := y(t) + \int_{-h}^{0} x(t+s)dk_0(s) - \int_{-h}^{0} dk_1(s) \int_{t+s}^{t} x(\tau)d\tau. \tag{5.28}$$

By the Itô's formula:

$$dV(x_t, y_t) \le -2x^2(t)\beta_{01}(a_0(t) - \int_{-h}^{0} |dk_0(s)| - \alpha_{11})dt - 2y^2(t)r_1 dt +$$

$$+2b(t, y_t)[V_0(x_t, y_t) + y(t)]d\xi(t) + 2q_2 dt. \tag{5.29}$$

The details of the calculations used to obtain (5.29) can be found in Ch. 10. Thus, in view of lemma 5.3, we have established the existence of a solution of system (5.27) with a bounded second moment, hence, of a T-periodic solution as well.□

Now we turn to conditions for the existence of periodic solutions of system (5.27) without the assumption of a jump of the function $k_0(s)$ for $s = 0$.

Theorem 5.7 Assume that the coefficients of system (5.27) satisfy the general assumptions formulated above and that the following relations hold

$$q_0 := \alpha_{10} + \frac{1}{2}\alpha_{21} < 1, \quad \beta_{01} > 0,$$

$$q := \beta_{00} - \beta_{11} > \max(\beta_{01}, (\beta_{01}q_0 + \alpha_{02})(1 - q_0)^{-1}).$$

Then there exists a T-periodic solution of system (5.27).

Proof Consider the functional V

$$V(x_t, y_t) := V_0^2(x_t, y_t) + V_1^2(y_t) + 2\beta_{01}x^2(t) + 2\int_{-h}^{0} dk_2(s)\int_{t+s}^{t} y^2(t_1)dt_1 +$$

$$+(q + 2\beta_{01})[\int_{-h}^{0} |dk_0(s)| \int_{t+s}^{t} dt_1 \int_{t_1}^{t} y^2(t_2)dt_2 +$$

$$+\int_{-h}^{0} |dk_1(s)| \int_{t+s}^{t} dt_1 \int_{t_1}^{t} dt_2 \int_{t_2}^{t} y^2(t_3)dt_3. \tag{5.30}$$

Here V_0 is determined by (5.28) and

$$V_1(y_t) := y(t) - \int_{-h}^{0} dk_0(s) \int_{t+s}^{t} y(t_1)dt_1 + \int_{-h}^{0} dk_1(s) \int_{t+s}^{t} dt_1 \int_{t_1}^{t} y(t_2)dt_2.$$

The stochastic differential of the functional (5.30) satisfies the estimate

$$dV(x_t, y_t) \le -2x^2(t)\beta_{01}(q - q_0)dt + 2b(t, y_t)[V_1(y_t) + V_0(x_t, y_t)]d\xi(t) -$$

$$-2y^2(t)[-\beta_{01}q_0 + q(1 - q_0) - \alpha_{02}]dt + 2q_2 dt.$$

From here, (5.30) and lemma 5.3 there follows the existence of a solution $(\bar{x}(t), \bar{y}(t))$ of system (5.27) such that

$$\mathcal{E}V(\bar{x}_t, \bar{y}_t) \le C, \quad \mathcal{E}\bar{x}^2(t) \le C, \quad t \ge 0. \tag{5.31}$$

We shall show that $\mathcal{E}y^2(t)$ is also bounded for all $t \ge 0$. From (5.28), (5.30), (5.31) it follows that for any $\varepsilon > 0$ and $t \ge 0$

$$C \ge \mathcal{E}[y(t) + \int_{-h}^{0} x(t+s)dk_0(s) - \int_{-h}^{0} dk_1(s) \int_{t+s}^{t} x(\tau)d\tau]^2 \ge$$

$$\ge (1 - \varepsilon\alpha_{00} - \varepsilon\alpha_{11})\mathcal{E}y^2(t) + \varepsilon^{-1} \int_{-h}^{0} x^2(t+s)|dk_0(s)| +$$

$$+\varepsilon^{-1} \int_{-h}^{0} |dk_1(s)| \int_{t+s}^{t} x^2(\tau)d\tau.$$

From here and from the boundedness of $\mathcal{E}x^2(t)$, choosing $\varepsilon > 0$ so that $1 > \varepsilon(\alpha_{00}+\alpha_{11})$, there follows the estimate $\sup_t \mathcal{E}y^2(t) < \infty$. Hence, by Thm. 5.3 the existence of a T–periodic solution of system (5.27) is established.□

5.3.3 Equations with an arbitrary aftereffect. Conditions for the existence of periodic solutions of concrete systems with an arbitrary aftereffect can be established by means of construction of functionals satisfying the assumptions of lemma 5.4. The principal methods of construction of such functionals are similar to those of used in the construction of the functionals considered above. That is why we shall restrict ourselves to the consideration, as an example, only of one typical situation.

We shall establish conditions for the existence of T–periodic solutions of scalar equation of the form (5.4)

$$dx(t) = [-a_0(t,x_t)x(t) + a(t,x_t)]dt + b(t,x_t)d\xi(t). \tag{5.32}$$

Here $(t,\phi) \mapsto a_0(t,\phi)$ is a continuous, bounded periodic with respect to t with period T functional and the functionals a, b satisfy conditions (5.5).

Theorem 5.8 Assume that for all $\phi \in C(-\infty,0]$

$$\inf_{t\geq 0}[a_0(t,\phi) - (\alpha_{01})^{1/4} - \frac{\sqrt{3}}{2}(\alpha_{02})^{1/2}] =: a_1 > 0. \tag{5.33}$$

Then there exists a T–periodic solution of Eq. (5.32).

Proof Consider the functional V

$$V(x_t) := x^4(t) + \int_{-\infty}^{0} (\delta dk_1(s)\varepsilon^{-1} + \delta_1 dk_2(s)) \int_{t-s}^{t} x^4(\tau)d\tau, \tag{5.34}$$

where $\varepsilon > 0, \delta > 0, \delta_1 > 0$ are constants to be chosen below. Let us calculate the stochastic differential of the functional (5.34):

$$\begin{aligned} dV(x_t) = [&x^4(t)(-4a_0 + \alpha_{01}\delta\varepsilon^{-1} + \delta_1\alpha_{02}) + \\ &+4x^3(t)a(t,x_t) + 6x^2(t)b^2(t,x_t) - \\ -\int_0^{\infty} &x^4(t-s)(\delta dk_1(s)\varepsilon^{-1} + \delta_1 dk_2(s)]dt + 4x^3(t)b(t,x_t)d\xi(t). \end{aligned} \tag{5.35}$$

Here we shall estimate the separate addends. We have

$$4x^3(t)a(t,x_t) \le 2x^4(t)\varepsilon + 2\varepsilon^{-1}x^2(t)a^2(t,x_t) \le$$

$$\le 2\varepsilon x^4(t) + \varepsilon^{-1}(\delta^{-1}x^4(t) + \delta a^4(t,x_t)) \le \tag{5.36}$$

$$\le 2\varepsilon x^4(t) + \varepsilon^{-1}(\delta^{-1}x^4(t) + \delta q_1 + \delta \int\limits_{-\infty}^{0} x(t-s)^4 dk_1(s)).$$

Analogously,

$$6x^2(t)b^2(t,x_t) \le 3x^4(t)\delta_1^{-1} + \delta_1 q_2 + \delta_1 \int\limits_{-\infty}^{0} x^4(t-s)dk_2(s). \tag{5.37}$$

Substituting estimates (5.36), (5.37) in (5.35), we obtain

$$dV(x_t) \le [x^4(t)(-4a_0 + \varepsilon^{-1}\delta\alpha_{01} + \delta_1\alpha_{02} + 2\varepsilon +$$

$$+\varepsilon^{-1}\delta^{-1} + 3\delta_1^{-1}) + \delta q_1 + \delta_1 q_2]dt + 4x^3(t)b(t,x_t)d\xi(t).$$

Now we choose the constants $\delta, \delta_1, \varepsilon$ to minimize the factor at $x^4(t)$. Then

$$\delta_1 := \sqrt{3\alpha_{02}^{-1}}, \quad \delta := (\alpha_{01})^{-1/2}, \quad \varepsilon := \alpha_{01}^{1/4}.$$

Hence

$$dV(t,x_t) \le (-a_1 x^4(t) + \delta q_1 + \delta_1 q_2)dt + 4x^3(t)b(t,x_t)d\xi(t).$$

Let us integrate the both sides of the last inequality from $t_1 \ge 0$ to $t_2 \ge t_1$ and calculate the mathematical expectation. In view of the assumptions of Thm. 5.8, the mathematical expectation of the stochastic integral equals zero. Hence,

$$\mathcal{E}V(t_2,x_{t_2}) - \mathcal{E}V(t_1,x_{t_1}) \le -a_1 \int\limits_{t_1}^{t_2} \mathcal{E}x^4(t)dt +$$

$$+(t_2-t_1)(\delta q_1 + \delta_1 q_2), \quad t_2 \ge t_1.$$

Thus the functional (5.34) satisfies all conditions of Lemma 5.4. Thm. 5.8 is proven. □

5.4. *UNIQUENESS OF PERIODIC SOLUTIONS*

We shall say that the T-periodic solution of Eq. (5.4) *is unique* if for any two T-periodic solutions of this equation $x_1(t)$ and $x_2(t)$ such that

$\mathcal{E}|x_1(t)|^2 \le C < \infty$, $\mathcal{E}|x_2(t)|^2 \le C < \infty$ and for all t the following relations holds

$$P(x_1(t) = x_2(t)) = 1.$$

We shall consider sufficient conditions for uniqueness of the periodic solutions of Eq. (5.4). The coefficients $a(t,\phi)$ and $b(t,\phi)$ of these equations satisfy the general conditions of Subs. 5.1, and also for any continuous $\phi, \psi : (-\infty, 0] \to \mathbf{R}^n$ the assumptions

$$\left.\begin{aligned} |a(t,\phi) - a(t,\psi)|^2 &\le \int_{-\infty}^{0} |\phi(t) - \psi(t)|^2 dk_1(t), \\ |b(t,\phi) - b(t,\psi)|^2 &\le \int_{-\infty}^{0} |\phi(t) - \psi(t)|^2 dk_2(t) \end{aligned}\right\} \tag{5.38}$$

where k_i are scalar non–decreasing bounded on $(-\infty, 0]$ functions. We shall study equations containing linear terms $A(t)x(t)$ of the type of instant feedback, i.e. equations

$$\left.\begin{aligned} dx(t) &= [A(t)x(t) + a(t,x_t)]dt + b(t,x_t)d\xi(t), \quad t \ge 0, \\ x_t(\theta) &:= x(t+\theta), \quad \theta \le 0, \quad x(t) \in \mathbf{R}^n, \quad \xi(t) \in \mathbf{R}^l. \end{aligned}\right\} \tag{5.39}$$

Here $A(t)$ is a matrix with continuous T–periodic entries such that the fundamental matrix $Z(t,s)$ of the solutions of the homogeneous equation $\dot{x}(t) = A(t)x(t)$ satisfies for some constant $\lambda > 0$ the estimate

$$|Z(t,s)| \le e^{-\lambda(t-s)}. \tag{5.40}$$

Now assume that Eq. (5.39) has two T–periodic solutions $x_1(t)$ and $x_2(t)$, where $\mathcal{E}|x_1(t)|^2 \le C < \infty, \mathcal{E}|x_2(t)|^2 \le C < \infty$. In view of the Itô's formula their difference $y(t) = x_2(t) - x_1(t)$ satisfies the relation

$$y(t) = Z(t,s)y(s) + \int_s^t Z(t,\tau)[a(\tau,x_{2\tau}) - a(\tau,x_{1\tau})]d\tau +$$

$$+ \int_s^t Z(t,\tau)[b(\tau,x_{2\tau}) - b(\tau,x_{1\tau})]d\xi(\tau)$$

whence it follows that for any $\varepsilon > 0$ and $\varepsilon_1 > 0$

$$\mathcal{E}|y(t)|^2 \leq (1+\varepsilon^{-1})|Z(t,s)|^2\mathcal{E}|y(s)|^2+$$

$$+(1+\varepsilon)[(1+\varepsilon_1)\int_s^t |Z(t,\tau_1)|d\tau_1 \int_s^t |Z(t,\tau)|\mathcal{E}|a(\tau,x_{2\tau})-$$

$$-a(\tau,x_{1\tau})|^2 d\tau + (1+\varepsilon_1^{-1})\int_s^t |Z(t,\tau)|^2\mathcal{E}|b(\tau,x_{2\tau}) - b(\tau,x_{1\tau})|^2 d\tau].$$

Let $s \to -\infty$ in this inequality. Then in view of (5.38), (5.40) we conclude that

$$\left.\begin{aligned} \mathcal{E}|y(t)|^2 \leq (1+\varepsilon)[(1+\varepsilon_1)\lambda^{-2}\alpha_{01} + (1+\varepsilon_1)^{-1}(2\lambda)^{-1}\alpha_{02}]f(t), \\ f(t) := \sup_{s\leq t} \mathcal{E}|y(s)|^2. \end{aligned}\right\} \qquad (5.41)$$

Let us choose ε_1 from the condition of minimum of the righthand side of (5.41). We obtain $\varepsilon_1 = (\lambda\alpha_{02})^{1/2}(2\alpha_{01})^{-1/2}$.

From here and (5.41) it follows that

$$f(t) \leq (1+\varepsilon)[\lambda^{-1}\alpha_{01}^{1/2} + (2\lambda)^{-1}\alpha_{02}^{1/2}]^2 f(t). \qquad (5.42)$$

Now assume that

$$\gamma := \lambda^{-1}\alpha_{01}^{1/2} + (2\lambda)^{-1}\alpha_{02}^{1/2} < 1. \qquad (5.43)$$

Then we choose $\varepsilon > 0$ in (5.42) so that $\gamma(1+\varepsilon) < 1$. Then in view of (5.42) $P(x_1(t) = x_2(t)) = 1$ for all t. Thus the following theorem is proven:

Theorem 5.9 Assume that the coefficients of Eq. (5.39) satisfy conditions (5.38), (5.40), (5.43). Then the periodic solution of Eq. (5.39) is unique.

Part V

CONTROL AND ESTIMATION IN HEREDITARY SYSTEMS

CHAPTER 14

PROBLEMS OF CONTROL FOR DETERMINISTIC FDEs

Many papers have been devoted to control problems for FDEs (see, e.g., [13, 46, 111, 292.8, 345, 395]). Statements of these problems can be obtained from those of for ODEs with natural modifications induced by the infinite dimensional structure of FDEs. We can mention, e.g., optimal control and estimation, controllability, observability, stabilization, control with incomplete state information, large scale systems, adaptive control and identification, etc.

Control and estimation problems for FDEs are classified according to the type of equation and cost functional, restrictions on the trajectory and control, initial and boundary conditions, etc. A distinguishing feature of these problems is the dependence of the state of the system on the previous values of the trajectory. Using this one can give definitions of controllability in various senses: conditional, relative, full, pointwise, zero–controllability, quasicontrollability, etc.

In this chapter we will give certain optimality conditions, and methods and algorithms for solving control problems.

1. THE DYNAMIC PROGRAMMING METHOD FOR DETERMINISTIC RDEs. BELLMAN'S EQUATION

In this section we consider a general approach to control problems for RDEs using the dynamic programming method. This method allows to build optimal closed–loop controls for certain problems.

1.1. *STATEMENT OF THE PROBLEM*

Consider the equation

$$\dot{x}(t) = f(t, x_t, u(t)), \quad t_0 \leq t \leq T. \tag{1.1}$$

Here, $x(t) \in \mathbf{R}^n, u(t) \in U \subseteq \mathbf{R}^m, x_t(\theta) := x(t+\theta), -h \leq \theta \leq 0, f : [t_0, T] \times C[-h, 0] \times U \to \mathbf{R}^n$. The function f, time moments $t_0, T > t_0$, and set U are prescribed. It is assumed that the functional $(t, \psi, u) \mapsto f(t, \psi, u)$ satisfies the *modified Carathéodory condition* (cf. Subs. 3.2.4): for almost all t this functional is continuous in (ψ, u), for all (ψ, u) it is measurable in t, and for some neighbourhood of an arbitrary point in $[t_0, T] \times C[-h, 0]$ there is a function $\lambda \in L_1([t_0, T], [0, \infty))$ such that for any $q, q_1, q_2 \in U$ we have in this neighbourhood:

$$f(t, \psi, q)| \leq \lambda(t),$$

$$|f(t, \psi_1, q_1) - f(t, \psi_2, q_2)| \leq \lambda(t)(\| \psi_1 - \psi_2 \| + |q_1 - q_2|).$$

The initial condition is

$$x_{t_0} = \phi \in C[-h, 0], \tag{1.2}$$

where ϕ is a given function.

Suppose we are also given a cost functional

$$J(u) = F_0(x_T) + \int_{t_0}^{T} F_1(t, x_t, u(t))dt. \tag{1.3}$$

Here, $F_0 : C[-h, 0] \to \mathbf{R}$ and $F_1 : [t_0, T] \times C[-h, 0] \times U \to \mathbf{R}$ are continuous functionals. Note that the righthand side depends on u directly and also via x (which depends on u).

A piecewise continuous function $u : [t_0, T] \to U$ is called an *admissible control* if for an arbitrary $(t_1, \psi) \in [t_0, T) \times C[-h, 0]$ there is a solution of (1.1) on $[t_1, T]$ satisfying the initial condition $x_{t_1} = \psi$ under this control u. The set of all admissible controls is denoted by W. It is assumed that $W \neq \emptyset$.

The *optimal control problem* under consideration is to find an admissible control minimizing the cost functional $J(u)$.

We define the *Bellman functional* $V : [t_0, T] \times C[-h, 0] \to \mathbf{R}$ as follows. Let $(t \mapsto x(t; t_1, \psi; u)) : [t_1, T] \to \mathbf{R}^n$ $(t_1 \in [t_0, T],\ \psi \in C[-h, 0],\ u \in W)$ be the solution of (1.1) with control u and initial condition $x_{t_1} = \psi$. Put $x_\tau(t, \psi; u) := x(\tau + \cdot; t, \psi; u)$. Then

$$V(t, \psi) := \inf_{u \in W} \left[F_0(x_T(t, \psi; u)) + \int_{t}^{T} F_1(s, x_s(t, \psi; u), u(s))ds \right].$$

Define the functional $L_u V : [t_0, T] \times C[-h, 0] \to \mathbf{R}$ by

$$L_u V(t, \psi) := \limsup_{\Delta t \to 0+} \frac{1}{\Delta t} [V(t + \Delta t, x_{t+\Delta t}(t, \psi; u)) - V(t, \psi)].$$

Note that $L_u V$ is the total derivative of V along the trajectories of the system (1.1) with control u.

If, in addition, the functional $(t,\psi) \mapsto V(t,\psi)$ is locally Lipschitz in ψ, then the value $L_u V(t,\psi)$ at any point (t,ψ) does not depend on the complete control u but only on the value $u(t^+)$.

Remark 1.1 A control u defined as a function of t is called an *open–loop control.* Next to it one can consider controls of the form $t \mapsto u(x(t))$, or more generally, $t \mapsto u(t, x_t)$, which are called *closed–loop controls.* Transition from an open–loop control to a closed–loop control cannot improve the solution of a deterministic control problem, since each function $t \mapsto u(t, x_t)$ is at the same time a function of t. However, such a transition is in certain concrete situations useful in realizing optimal or "quasi–optimal" control. (In particular, such a transition is made in the dynamic programming method, described in Subs. 1.2.) Therefore, in papers on optimal control the initial equation (in our case (1.1)) is usually written with u instead of $u(t)$ or $u(t, x_t)$, assuming that the arguments will be determined more precisely during the further investigation. This representation is useful, but not quite consistent. □

1.2. *OPTIMALITY CONDITIONS*

Theorem 1.1 Suppose there are a continuous functional $V_0 : [t_0, T]\times C[-h, 0] \to \mathbf{R}$ *which is locally Lipschitz in the second argument, and a functional* $u_0 : [t_0, T] \times C[-h, 0] \to \mathbf{R}^m$ *satisfying the Carathéodory condition, such that*

$$\inf_{u\in U}[L_u V_0(t,\psi)+F_1(t,\psi,u)] = L_{u_0(t,\psi)}V_0(t,\psi)+F_1(t,\psi,u_0(t,\psi)) = 0, \quad (1.4)$$

$$V_0(T,\psi) = F_0(\psi), \quad (1.5)$$

for all $t \in [t_0, T), \psi \in C[-h, 0]$. *Then* u_0 *is the control direction at* (t,ψ) *(i.e. the optimal solution and optimal control in the Pr. (1.1)–(1.3) are defined by (1.1) and (1.2) with* $u(t) = u_0(t, x_t)$*) and* V_0 *is the Bellman functional in this problem.*

Remark 1.2 We stress that the infimum in (1.4) must be computed with respect to the vector parameter $u \in U$. □

Proof Substituting in the middle part of (1.4) the trajectory $x(\cdot; t, \psi; u)$ with $u(\tau) := u_0(\tau, x_\tau)(t \le \tau \le T)$, we find

$$L_{u_0}V_0(\tau, x_\tau(t,\psi;u)) + F_1(\tau, x_\tau(t,\psi;u), u_0) = 0, \quad t \le \tau \le T, \quad (1.6)$$

where $u_0(\tau) := u_0(\tau, x_\tau(t,\psi;u_0))$. Here $\tau \mapsto V_0(\tau, x_\tau(t,\psi;u))$ is an absolutely continuous function, and hence, by (1.5), (1.6),

$$\left. \begin{aligned} V_0(t,\psi) = F_0(x_T(t,\psi);u) + \int_t^T F_1(\tau, x_\tau(t,\psi;u), u(\tau))d\tau, \\ u(\tau) = u_0(\tau, x_\tau). \end{aligned} \right\} \quad (1.7)$$

Consider an arbitrary admissible control u_1. By (1.4) we find

$$L_{u_1} V_0(\tau, x_\tau(t,\psi;u_1)) + F_1(\tau, x_\tau(t,\psi;u_1), u_1(\tau)) \geq 0.$$

This implies that

$$V_0(t,\psi) \leq F_0(x_T(t,\psi;u_1)) + \int_t^T F_1(\tau, x_\tau(t,\psi;u_1), u_1(\tau))d\tau.$$

Comparison of this inequality with (1.7) proves Thm. 1.1. □

2. LINEAR QUADRATIC PROBLEMS

2.1. *OPTIMAL CONTROL SYNTHESIS*

Here we will determine the solution of the optimal control problem with quadratic cost functional (1.3) and linear equations

$$\dot{x}(t) = A(t)x(t) + A_1(t)x(t-h) + \int_{-h}^{0} G(t,\theta)x(t+\theta)d\theta + B(t)u(t,x_t). \quad (2.1)$$

The initial condition has the form (1.2). Here $t \in [t_0, T), t_0 < T$ is a given time moment, $x(t) \in \mathbf{R}^n, u : [t_0,T] \times C[-h,0] \to \mathbf{R}^m$, the matrix functions A, A_1, G, B are piecewise continuous, and $U = \mathbf{R}^m$. It is required to choose u so that the quadratic cost functional

$$\begin{aligned} J(u) := x'(T)N_0x(T) + \\ + \int_{t_0}^{T} [u'(t,x_t)N_1(t)u(t,x_t) + x'(t)N_2(t)x(t)]dt \to \min_u . \end{aligned} \quad (2.2)$$

Here, prime denotes transposition, all matrices $N_0, N_{1,2}(t)$ are symmetric nonnegative definite, $N_1(t)$ is positive definite, $N_{1,2}(\cdot)$ are continuous, N_0, $N_2(t) \in \mathbf{R}^{n\times n}, N_1(t) \in \mathbf{R}^{m\times m}$.

To solve this Pr. (2.1), (2.2) we will use Thm. 1.1. The relations (1.4), (1.5) become

$$\left.\begin{aligned} &\inf_{u\in\mathbf{R}^m}[L_uV_0(t,\psi)+u'N_1(t)u+\psi'(0)N_2\psi(0)]=0,\\ &V_0(T,\psi)=\psi'(0)N_0\psi(0). \end{aligned}\right\} \tag{2.3}$$

We assume that the Bellman functional V_0 has the form

$$\begin{aligned} V_0(t,\psi)=\psi'(0)P(t)\psi(0)+\psi'(0)\int_{-h}^{0}Q(t,\theta)\psi(\theta)d\theta+\\ +\int_{-h}^{0}\psi'(\theta)Q'(t,\theta)d\theta\psi(0)+\int_{-h}^{0}\int_{-h}^{0}\psi'(\theta)R(t,\theta,\rho)\psi(\rho)d\theta d\rho, \end{aligned} \tag{2.4}$$

where $P(t)$ is a symmetric nonnegative definite matrix, $R(t,\theta,\rho)=R'(t,\rho,\theta)$, and P,Q,R are Lipschitz matrix functions with piecewise continuous derivatives. Substituting (2.4) into (2.3) and equating to zero the quadratic form in $\psi(t+\theta),-h\le\theta\le 0$, we find that the matrices P,Q,R satisfy the matrix equations

$$\left.\begin{aligned} &\dot P(t)+A'(t)P(t)+P(t)A(t)+Q(t,0)+Q'(t,0)+N_2(t)=\\ &\qquad=P(t)B_1(t)P(t),\\ &A'(t)Q(t,\theta)+P(t)G(t,\theta)+R(t,0,\theta)+\left(\frac{\partial}{\partial t}-\frac{\partial}{\partial\theta}\right)Q(t,\theta)=\\ &\qquad=P(t)B_1(t)Q(t,\theta),\\ &G'(t,\theta)Q(t,\rho)+Q'(t,\theta)G(t,\rho)+\left(\frac{\partial}{\partial t}-\frac{\partial}{\partial\theta}-\frac{\partial}{\partial\rho}\right)R(t,\theta,\rho)=\\ &\qquad=Q'(t,\theta)B_1(t)Q(t,\rho),\\ &\qquad B_1:=BN_1^{-1}B'. \end{aligned}\right\} \tag{2.5}$$

The boundary conditions have the form

$$\left.\begin{aligned} &P(t)=N_0,\quad Q(T,\theta)=0,\quad R(T,\theta,\rho)=0,\quad A_1'(t)P(t)=Q'(t,-h),\\ &A_1'(t)Q(t,\theta)+Q'(t,\theta)A_1(t)-R(t,-h,\theta)-R'(t,\theta,-h)=0,\\ &-h\le\theta,\rho\le 0,\quad 0\le t\le T. \end{aligned}\right\} \tag{2.6}$$

Taking of the lefthand side of (2.3) the gradient with respect to u, we conclude that the minimum of this expression (2.3) is at the point $u(t,\psi)$:

$$u(t,\psi) = -N_1^{-1}(t)B'(t)\left[P(t)\psi(0) + \int_{-h}^{0} Q(t,\theta)\psi(\theta)d\theta\right]. \qquad (2.7)$$

So the optimal control is constructed as a closed–loop control.

Note that under these assumptions the boundary value problem (2.5), (2.6) has a unique solution P, Q, R in the class of piecewise continuously differentiable functions [285, 292.8]. It has also been proved that it is possible to find a Bellman functional of the form (2.4).

By (2.7), the optimal process can be obtained as the solution of the equation

$$\dot{x}(t) = [A(t) - B(t)N_1^{-1}(t)B'(t)P(t)]x(t) + A_1(t)x(t-h) +$$

$$+ \int_{-h}^{0} [G(t,\theta) - B(t)N_1^{-1}(t)B'(t)Q(t,\theta)]x(t+\theta)d\theta$$

$$(t_0 \le t \le T)$$

with initial condition (1.2).

2.2. *EXACT SOLUTION*

Here we will give the exact solution of the Pr. (2.5), (2.6) for the case $G = 0, N_2 = 0$. We introduce matrices Z, B_2, P_1 by

$$\dot{Z}(t) = A(t)Z(t), \quad Z(0) = I,$$

$$\left.\begin{aligned} \dot{B}_2(t) &= -B_2(t+h)Z^{-1}(t+h)A_1(t+h)Z(t), \\ B_2(T) &= I, \quad B_2(t) = 0 \ \ (t > T), \end{aligned}\right\} \qquad (2.8)$$

$$\dot{P}_1(t) = P_1(t)B_2'(t)Z^{-1}(t)B_1(t)[Z'(t)]^{-1}B_2(t)P_1(t),$$

$$P_1(T) = Z'(T)N_0Z(T),$$

where $t_0 \le t \le T$ and I is the identity matrix. Then

$$\left.\begin{aligned} P(t) &= [Z'(t)]^{-1}B_2'(t)P_1(t)B_2(t)Z^{-1}(t), \\ Q(t,\theta) &= -[Z'(t)]^{-1}B_2'(t)P_1(t)\dot{B}_2(t+\theta)Z^{-1}(t+\theta), \\ R(t,\theta,\rho) &= [Z'(t+\theta)]^{-1}\dot{B}_2'(t+\theta)P_1(t)\dot{B}_2(t+\rho)Z^{-1}(t+\rho). \end{aligned}\right\} \qquad (2.9)$$

The Pr. (2.8) for B_2 can be explicitly solved (e.g., by the method of steps) and the equation for P_1 can be reduced to a linear equation by a change of variable $D(t) = P_1^{-1}(t)$, as $P_1(t)$ is a nondegenerate matrix. For this it is necessary and sufficient that $\det N_0 \neq 0$ and

$$\det\left[Z^{-1}(T)N_0^{-1}[Z'(T)]^{-1} + \int_t^T B_2'(\tau)Z^{-1}(\tau)B_1(\tau)[Z'(\tau)]^{-1}B_2(\tau)d\tau\right] \neq 0$$

$$(t_0 \leq t \leq T).$$

In Subs. 15.2.2 a procedure for approximately solving (2.5), (2.6) with arbitrary matrices G and N_2 has been proposed.

2.3. *SYSTEMS WITH DELAYS IN THE CONTROL*

Linear quadratic problems with delays in the control can be studied similarly to (2.1), (2.2). For this reason we restrict ourselves to the final formulas related to the equation

$$\left.\begin{aligned} &\dot{x}(t) = A_1(t)x(t-h) + B(t)u(t) + K(t)u(t-h_1), \\ &t_0 \leq t \leq T, \quad x : [t_0 - h, T] \to \mathbf{R}^n, \quad u : [t_0 - h_1, T] \to \mathbf{R}^m, \\ &h \geq 0, \quad h_1 \geq 0. \end{aligned}\right\} \quad (2.10)$$

Here, A_1, B, K are piecewise continuous matrices. The solution x of (2.10) is defined by the initial conditions (1.2) and $u(t_0+\theta_1) = \phi_1(\theta_1), -h \leq \theta_1 \leq 0$, where ϕ_1 is a given piecewise continuous function. The performance index has the form (2.2). At time t the optimal control depends on t and the previous values $x(t+\theta) = \psi(\theta)(-h \leq \theta \leq 0)$ and $u(t+\theta) = \psi_1(\theta_1)(-h_1 \leq \theta_1 \leq 0)$. The Bellman functional for the Pr. (2.10), (2.2) is $V_0(t,\psi) + V_1(t,\psi,\psi_1)$, with V_0 defined by (2.4) and V_1 equal to

$$\begin{aligned} V_1(t,\psi,\psi_1) &= \psi'(0)\int_{-h_1}^{0} P_1(t,\theta_1)\psi_1(\theta_1)d\theta_1 + \int_{-h_1}^{0} \psi_1'(\theta_1)P_1'(t,\theta_1)\psi(0)d\theta_1 + \\ &+ \int_{-h}^{0} d\theta \int_{-h_1}^{0} \psi'(\theta)P_2'(t,\theta,\theta_1)\psi_1(\theta_1)d\theta_1 + \int_{-h}^{0} d\theta \int_{-h_1}^{0} \psi_1'(\theta_1)P_2'(t,\theta,\theta_1)\psi(\theta)d\theta_1 + \\ &+ \int_{-h_1}^{0}\int_{-h_1}^{0} \psi_1'(\theta_1)P_3(t,\theta_1,\theta_2)\psi_1(\theta_2)d\theta d\theta_1. \end{aligned} \quad (2.11)$$

The optimal control of the Pr. (2.10), (2.2) is given by

$$u_0(t) = -N_1^{-1}(t)B'(t)\left[(P(t) + P_1'(t,0))x(t) + \int_{-h_1}^{0} (P_1(t,\theta_1) + \right.$$

$$\left. +P_3(t,0,\theta_1))u_0(t+\theta_1)d\theta_1 + \int_{-h_1}^{0} (Q(t,\theta) + P_2'(t,\theta,0))x(t+\theta)d\theta\right].$$

But (2.10) implies

$$x(t) = \int_{t_0}^{t} [A_1(\tau)x(\tau-h) + B(\tau)u_0(\tau) + K(\tau)u_0(\tau-h_1)]d\tau,$$

$$t_0 \le t \le T,$$

where the initial conditions for x and u_0 must be taken into consideration. The last two equations can be regarded as Volterra equations in x and u_0. By solving these equations we obtain the expressions for the optimal solution and control.

The matrices P, Q, R satisfy the equations

$$\left.\begin{aligned}
&\dot{P}(t) + Q(t,0) + Q'(t,0) + N_2(t) = \\
&= [P(t)B(t) + P_1(t,0)]N_1^{-1}(t)[B'(t)P(t) + P_1'(t,0)], \\
&\left(\frac{\partial}{\partial t} - \frac{\partial}{\partial \theta}\right) Q(t,\theta) + R(t,0,\theta) = \\
&= [P(t)B(t) + P_1(t,0)]N_1^{-1}(t)[B'(t)Q(t,\theta) + P_2'(t,\theta,0)], \\
&\left(\frac{\partial}{\partial t} - \frac{\partial}{\partial \theta} - \frac{\partial}{\partial \theta_1}\right) R(t,\theta,\theta_1) = \\
&= [Q'(t,\theta)B(t) + P_2(t,\theta,0)]N_1^{-1}(t)[B'(t)Q(t,\theta_1) + P_2'(t,\theta_1,0)], \\
&0 \le t \le T.
\end{aligned}\right\} \tag{2.12}$$

The equations for P_i take the form

$$\left.\begin{aligned}
&\left(\frac{\partial}{\partial t}-\frac{\partial}{\partial\theta_1}\right)P_1(t,\theta_1)+P_2(t,0,\theta_1)=\\
&=[P(t)B(t)+P_1(t,0)]N_1^{-1}(t)[B'(t)P_1(t,\theta_1)+P_3(t,0,\theta_1)],\\
&\left(\frac{\partial}{\partial t}-\frac{\partial}{\partial\theta}-\frac{\partial}{\partial\theta_1}\right)P_2(t,\theta,\theta_1)=\\
&=[Q'(t,\theta)B(t)+P_2(t,\theta,0)]N_1^{-1}(t)[P_1(t,\theta_1)+P_3(t,0,\theta_1)],\\
&\left(\frac{\partial}{\partial t}-\frac{\partial}{\partial\theta_1}-\frac{\partial}{\partial\theta_2}\right)P_3(t,\theta_1,\theta_2)=\\
&=[P_1'(t,\theta_1)B(t)+P_3(t,\theta_2,0)]N_1^{-1}(t)[B'(t)P_1(t,\theta_2)+P_3'(t,\theta_2,0)],\\
&-h_1\le\theta_1,\theta_2\le 0.
\end{aligned}\right\}\tag{2.13}$$

And the boundary conditions have the form

$$\left.\begin{aligned}
&\text{If } -h\le\theta,\bar\theta\le 0,\quad -h_1\le\theta_1,\theta_2\le 0,\quad\text{then}\\
&P(T)=N_0,\quad Q(T,\theta)=R(T,\theta,\bar\theta)=0,\\
&P_1(T,\theta_1)=P_2(T,\theta,\theta_1)=0,\quad P_3(T,\theta_1,\theta_2)=0.
\end{aligned}\right\}\tag{2.14}$$

For $t<T$, we have

$$\left.\begin{aligned}
&P(t)K(t)=P_1(t,-h_1),\quad K'(t)Q(t,\theta)=P_2'(t,\theta,-h_1),\\
&A_1'(t)P(t)=Q_2'(t,-h),\\
&A_1'(t)Q(t,\theta)+Q'(t,\theta)A_1(t)=R(t,\theta,-h)+R'(t,\theta,-h),\\
&A_1'(t)P_1(t,\theta_1)=P_2(t,-h,\theta_1),\quad P_3(t,\theta_1,\theta_2)=P_3'(t,\theta_2,\theta_1),\\
&K'(t)P_1(t,\theta_1)+P_1'(t,\theta_1)K(t)=P_3(t,\theta_1,-h)+P_3'(t,\theta_1,-h).
\end{aligned}\right\}\tag{2.15}$$

Under our assumptions the existence and uniqueness of the solution of the Pr. (2.12)–(2.15) has been proved in [285, 292.8], in which is also given a method for approximately determing it.

2.4. *EFFECTS OF DELAYS IN REGULATORS*

Usually, optimal control synthesis for arbitrary actual systems must include some delays in the control because of the finiteness in the time necessary to perform the measurements, process the results of these, and generate a control action. Sometimes an arbitrary small delay in the feedback regulators radically changes the qualitative behaviour of the system.

E.g., consider the dimensionless Euler – Bernoulli equation of transversal oscillation of a beam:

$$\frac{\partial^2\omega(t,x)}{\partial t^2}+\frac{\partial^4\omega(t,x)}{\partial x^4}=0,\quad 0\le x\le 1,\quad t>0,$$

with boundary conditions

$$\omega(t,0)=\frac{\partial\omega(t,0)}{\partial x}=\frac{\partial^3\omega(t,1)}{\partial x^3}=0$$

and feedback

$$\frac{\partial^2\omega(t,1)}{\partial x^2}=-\frac{\partial^2\omega(t-h,1)}{\partial t\partial x}.$$

Here, $\omega(t,x)$ is the displacement of the beam at point x and time t. It is known that for $h=0$ the Euler – Bernoulli beam is uniformly exponentially stabilized by the given feedback. However (see [135, 136]), the system becomes unstable if an arbitrary small delay $h>0$ is introduced into the feedback, and arbitrary large divergence rates are generated. E.g., for a delay h of order 10^{-182} there is a nontrivial solution which diverges at the rate of approximately $\exp(2.23\cdot 10^{181}t)$. Clearly, this means that despite its superficial innocence this equation cannot adequately describe the actual process.

The same phenomenon takes place for the one–dimensional wave equation

$$\frac{\partial^2\omega(t,x)}{\partial t^2}=\frac{\partial^2\omega(t,x)}{\partial x^2},\quad 0\le x\le 1,\quad t>0,$$

with boundary condition

$$\omega(t,0)=0$$

and velocity feedback

$$\frac{\partial\omega(t,1)}{\partial x}=-k\frac{\partial\omega(t-h,1)}{\partial t},$$

where $k>0$ and $h\ge 0$ are given and fixed. For $h=0$ this system is exponentially stable, but for arbitrarily small $h>0$ it becomes unstable

[135]. The similar problem of boundary feedback stabilization of a one–dimensional viscoelastic beam has been considered in [241], in which a comparison of solutions of the elastic and viscoelastic cases has been given.

On the other hand we should not believe that delay has only a harmful effect. On the contrary, sometimes the performance of a system can be improved by a judicious use of delay. This has already been shown by the simplest example in Subs. 3.6.3.

In [359, 360] it has been shown that a regulator with delay can eliminate overcontrol and damp the oscillation. The possibility to replace a PD regulator by a regulator with delay has been proved in [472]. The use of regulators with discrete or distributed delay to stabilize a system and to improve the performance index has been demonstrated in [305, 448].

2.5. *NDE*

Here we will consider certain formulas for the optimal control of a NDE

$$\left.\begin{aligned} &\dot{x}(t) = A_1(t)x(t-h) + A_2(t)\dot{x}(t-h_1) + B(t)u(t), \\ &\qquad t_0 \le t \le T, \quad h \ge 0, \quad h_1 > 0, \\ &x : [t_0 - \max\{h, h_1\}, T] \to \mathbf{R}^n, \quad u : [t_0, T] \to \mathbf{R}^m, \end{aligned}\right\} \tag{2.16}$$

with cost functional (2.2) and initial condition (1.2). We assume that the initial function $\phi \in C^1[-h, 0]$, the matrix function $A_2 \in C[t_0, T]$; the remaining parameters of the Pr. (2.16), (2.2) being as in (2.1), (2.2). The Bellman functional for the Pr. (2.16), (2.2) is equal to $V_0(t, \psi) + V_1(t, \psi, \dot{\psi})$. Here, V_0 is defined by (2.4) and V_1 by (2.11), with ψ_1 replaced by $\dot{\psi}$. The equations and boundary conditions for P, Q, R have the form (2.5)–(2.6). The matrices P_i satisfy the equations ($B_1 = BN_1^{-1}B'$):

$$\left(\frac{\partial}{\partial t} - \frac{\partial}{\partial \theta_1}\right) P_1(t, \theta) + P_2(t, 0, \theta_1) + P(t, \theta_1, 0) =$$

$$= -P(t)B_1(t)P_1(t, \theta),$$

$$\left(\frac{\partial}{\partial t} - \frac{\partial}{\partial \theta} - \frac{\partial}{\partial \theta_1}\right) P_2(t, \theta, \theta_1) = -Q'(t, \theta)B_1(t)P_1(t, \theta_1),$$

$$\left(\frac{\partial}{\partial t} - \frac{\partial}{\partial \theta_1} - \frac{\partial}{\partial \theta_2}\right) P_3(t, \theta_1, \theta_2) = -P'(t, \theta_1)B_1(t)P_1(t, \theta_2).$$

The boundary conditions for $t = T$ are given by (2.14), and for $t < T$ by

$$P(t)A_2(t) = P_1(t,-h_1), \quad P_1(t,0) = 0,$$

$$A_2'(t)Q(t,\theta) = P_2'(t,\theta,-h_1), \quad A_2'(t)P_1(t,\theta) = P_2(t,-h,\theta),$$

$$P_2(t,-h,\theta_1) = P_2(t,\theta,-h_1) = 0,$$

$$-h \leq \theta \leq 0, \quad -h_1 \leq \theta_1 \leq 0.$$

3. OPTIMAL CONTROL OF BILINEAR HEREDITARY SYSTEMS

In this section we propose a method for constructing optimal controls for bilinear hereditary systems.

3.1. *OPTIMALITY CONDITIONS*

Consider the bilinear equation

$$\left.\begin{aligned} &\dot{x}(t) = A_1(t)x(t-h) + [A(t)x(t) + B(t)]u(t,x_t), \\ &h \geq 0,\ x : [t_0 - h, T] \to \mathbf{R}^n,\ u : [t_0, T] \times C[-h,0] \to \mathbf{R}^m,\ t_0 \leq t \leq T \end{aligned}\right\} \tag{3.1}$$

with performance index

$$J(u) + \int_0^T f(t,x_t)dt \to \min_u . \tag{3.2}$$

Here, $J(u)$ is given by (2.2), the matrix functions A, A_1, B are piecewise continuous, and the nonnegative continuous functional f will be chosen in the sequel. The initial condition has the form (1.2). By (1.4), (1.5), the Bellman functional $V(t,\psi)$ for (3.1), (3.2) satisfies

$$\left.\begin{aligned} &\inf_{u \in \mathbf{R}} [L_u V(t,\psi) + \psi'(0)N_2(t)\psi(0) + N_1(t)u^2 + f(t,\psi)] = 0, \\ &V(T,\psi) = \psi'(0)N_0\psi(0). \end{aligned}\right\} \tag{3.3}$$

Let us look for the solution of (3.3) in the form (2.4). We compute $L_u V$, substitute it in (3.3), replace $\dot{x}$ according to (3.1), and finally determine the control u_0 for which the lefthand side of (3.2) is maximal. We obtain

$$u_0(t,\psi) = -N_1^{-1}(t)[A(t)\psi(0) + B(t)]' \left[P(t)\psi(0) + \int_{-h}^0 Q(t,\theta)\psi(\theta)d\theta\right]. \tag{3.4}$$

Now we define $f(t,\psi)$ as follows:

$$f(t,\psi) = N_1^{-1}(t)\left[(A(t)\psi(0)+B(t))'\left(P(t)\psi(0)+\int_{-h}^{0}Q(t,\theta)\psi(\theta)d\theta\right)\right]^2. \tag{3.5}$$

Then the nonlinear terms in (3.3) containing P, Q, R will vanish. Substituting (2.4), (3.4), (3.5) into (3.3) and equating to zero the quadratic forms in $\psi(0)$ and ψ, we obtain a system of linear partial differential equations for the matrices $P(t), Q(t,\theta), P(t,\theta,\rho)$. These equations, defining the optimal cost value and the optimal control synthesis, are:

$$\left.\begin{aligned}
&\dot P(t)+Q(t,0)+Q'(t,0)+N_2(t)=0,\\
&\left(\frac{\partial}{\partial t}-\frac{\partial}{\partial\theta}\right)Q(t,\theta)+R(t,\theta,0)=0,\\
&\left(\frac{\partial}{\partial t}-\frac{\partial}{\partial\theta}-\frac{\partial}{\partial\rho}\right)R(t,\theta,\rho)=0,\\
&0\le t\le T,\ \ -h\le\theta,\rho\le 0.
\end{aligned}\right\} \tag{3.6}$$

By equating to zero the quadratic form depending on $\psi(-h)$ we obtain the boundary conditions (2.6).

Under the conditions imposed there exists a unique solution of the boundary value problem (3.6), (2.6), as well as a unique solution of the Pr. (3.1), (1.2), (3.4).

3.2. *CONSTRUCTION OF THE OPTIMAL CONTROL SYNTHESIS*

We represent the solution of the boundary value problem (3.6), (2.6) as a sum of two terms: the first for $N_2 = 0$ and arbitrary N_0, and the second for $N_0 = 0$ and arbitrary N_2.

By (2.9), the first solution is

$$P(t) = B_2'(t)N_0B_2(t),\quad Q(t,\theta) = -B_2'(t)N_0\dot B_2(t+\theta),$$

$$R(t,\theta,\rho) = \dot B_2'(t+\theta)N_0\dot B_2(t+\rho),$$

$$0\le t\le T,\quad -h\le\theta,\rho\le 0.$$

Here, $B_2 \in \mathbf{R}^{n\times n}$ satisfies the equations

$$\dot B_2(t) = -B_2(t+h)A_1(t+h),\quad 0\le t\le T,$$

$$B_2(s)=0,\quad s>T,\quad B_2(T)=I.$$

At the discontinuity point $T-h$ of the derivative $\dot{B}_2(t)$ we take it to be lefthand continuous; I is the identity matrix.

The second solution is given by recurrence formulas. Let $t_i = T - ih$, where $i = 0, 1, \ldots$. For $t_1 \le t \le T$ we find

$$P(t) = \int_t^T N_2(r)dr, \quad Q(t,\theta) = \int_{t+\theta+h}^T N_2(r)drA_2(t+\theta+h),$$

$$R(t,\theta,\rho) = A_2'(t+\theta+h) \int_{t+h+\max\{\theta,\rho\}}^T N_2(r)drA_2(t+\rho+h).$$

Here, the matrix A_2 is defined for $t_{i+1} \le t \le t_i$ and $-h \le \theta \le 0$ by

$$A_2(t+\theta+h) := \begin{cases} A_1(t+\theta+h), & -h \le \theta \le -t + t_{i+1}, \\ 0, & -t + t_{i+1} < \theta \le 0. \end{cases}$$

We also put

$$Q_1(t_i, \theta + t - t_i) := \begin{cases} Q(t_i, \theta + t - t_i), & -t + t_{i+1} \le \theta \le 0, \\ 0, & -h \le \theta < -t + t_{i+1}, \end{cases}$$

$$R_1(t_i, \theta + t - t_i, \rho) := \begin{cases} R(t_i, \theta + t - t_i, \rho), & -t + t_{i+1} \le \theta \le 0, \\ 0, & -h \le \theta < -t + t_{i+1}. \end{cases}$$

The matrix R_2 is defined for $t_{i+1} \le t \le t_i$, $-t + t_{i+1} \le \theta, \rho \le 0$ by

$$R_2(t_i, \theta + t - t_i, \rho + t - t_i) := R_1(t_i, \theta + t - t_i, \rho + t - t_i).$$

If one of the arguments θ, ρ does not belong to the interval $[-t + t_{i+1}, 0]$, then the function $R_2(t_i, \theta + t - t_i, \rho + t - t_i)$ is put equal to zero. The solution of the boundary value problem (3.6), (2.6) for $t_{i+1} \le t \le t_i, N_0 = 0$, is

$$P(t) = \int_t^{t_i} N_2(r)dr + P(t_i) +$$

$$+ \int_{t-t_i}^0 [Q(t_i, r) + Q'(t_i, r)]dr + \int_{t-t_i}^0 \int_{t-t_i}^0 R(t_i, r, \xi)drd\xi.$$

The matrix Q satisfies the relations

$$Q(t,\theta) = \left[\int\limits_{t+\theta+h}^{t_i} N_2(r)dr + P(t_i) + \int\limits_{t-t_i}^{0} Q(t_i,r)dr + \int\limits_{t-t_{i+1}+\theta}^{0} Q'(t_i,r)dr + \right.$$

$$\left. + \int\limits_{t-t_i}^{0} d\xi \int\limits_{t-t_{i+1}+\theta}^{0} R(t_i,r,\xi)dr \right] A_2(t+\theta+h) +$$

$$+ Q_1(t_i, t+\theta-t_i) + \int\limits_{t-t_i}^{0} R_1(t_i, t+\theta-t_i, r)dr,$$

$$t_{i+1} \leq t \leq t_i, \quad -h \leq \theta \leq 0.$$

Finally, for R we have

$$R(t,\theta,\rho) = A_2'(t+\theta+h) \left[\int\limits_{t+h+\max\{\theta,\rho\}}^{t_i} N_2(r)dr + P(t_i) + \right.$$

$$+ \int\limits_{t-t_{i+1}+\rho}^{0} Q'(t_i,r)dr + \int\limits_{t-t_{i+1}+\theta}^{0} Q(t_i,r)dr +$$

$$\left. + \int\limits_{t-t_{i+1}+\theta}^{0} d\xi \int\limits_{t-t_{i+1}+\rho}^{0} R(t_i,r,\xi)dr \right] A_2(t+\rho+h) +$$

$$+ A_2'(t+\theta+h) \left[Q_1'(t_i, t+\rho-t_i) + \int\limits_{t-t_{i+1}+\theta}^{0} R_1(t_i, r, t+\rho-t_i)dr \right] +$$

$$+ \left[Q_1(t_i, t+\theta-t_i) + \int\limits_{t-t_{i+1}+\rho}^{0} R_1(t_i, t+\theta-t_i, r)dr \right] A_2(t+\rho+h) +$$

$$+ R_2(t_i, t+\theta-t_i, t+\rho-t_i),$$

$$t_{i+1} \leq t \leq t_i, \quad -h \leq \theta, \rho \leq 0.$$

Remark 3.1 It is not difficult to surmise from the above said that similar expansions for the solution are valid for more general bilinear equations [292.8]. □

3.3. *MODEL OF OPTIMAL FEEDBACK CONTROL FOR MICROBIAL GROWTH*

The results of this section can be applied to certain practical problems. E.g., consider the process of microbial growth described by (1.5.4). The initial conditions for this equation are

$$x_1(t_0) = x_{10}, \quad x_1(t_0+\theta) = 0, \quad -h \le \theta < 0, \quad x_2(t_0) = 0.$$

The problem is to achieve a desired value of resulting product $\overline{x}_2$ at time T, and to minimize the spenditure of nourishing substances. We take the cost functional to be

$$C_1[x_2(T) - \overline{x}_2]^2 + C_2 x_1(T) +$$

$$+ \int_{t_0}^{T} [(x_2(t) - \overline{x}_2)^2 + C_3 u^2(t) + f(t, x_1(t), x_2(t))]dt \to \min_u, \tag{3.7}$$

where C_i are nonnegative constants.

The method described in this section was applied to the Pr. (1.5.4), (3.7), and a numerical solution was obtained for the following parameter values:

$$t_0 = 0, \quad T = 3, \quad h = 1, \quad b = 3.5, \quad \overline{x}_2 = 1.5, \quad \alpha = 52,$$

$$\gamma(t) \equiv 0.5, \quad C_i = 1, \quad x_{10} = 4.$$

Fig. 14.3.1 shows the optimal trajectories $x_1(t)$ and $x_2(t)$. It can be seen that $x_2(T)$ is close to $\overline{x}_2$, and $x_1(T)$ is close to zero.

4. CONTROL PROBLEMS WITH PHASE CONSTRAINT FORMULA

4.1. *GENERAL OPTIMALITY CONDITIONS*

We will consider the Pr. (1.1)–(1.3) with the additional constraints

$$x(t) \in X(t), \quad t_0 \le t \le T, \tag{4.1}$$

where $X(t) \subseteq \mathbf{R}^n$ is a given set. In this connection the definition of the set W of admissible controls must include the requirement (4.1). In this section we let $V : [t_0, T] \times C[-h, 0] \to \mathbf{R}$ be a functional such that for any $u \in W, \tau \in [t_0, T], t \in [\tau, T]$ we have

$$V(t, x_t(u)) - V(\tau, x_\tau(u)) = \int_{\tau}^{t} L_u V(s, x_s(u))ds, \tag{4.2}$$

where $x_t(u)(\theta) = x(t+\theta; t_0, \phi; u), \ -h \le \theta \le 0.$

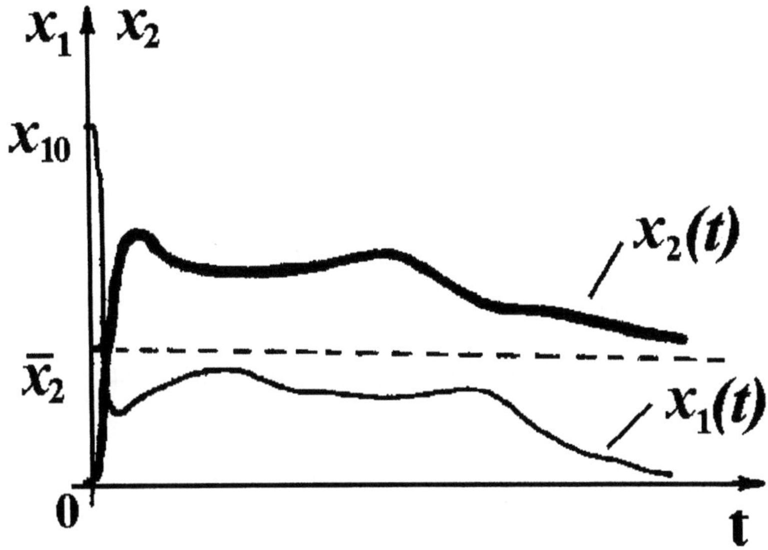

Figure 14.3.1. Optimal trajectories for microbial growth

Theorem 4.1 Suppose there are a functional $(t,\psi) \mapsto V(t,\psi)$ *satisfying (4.2) and an admissible control* u_0 *such that*

$$\int_{t_0}^{T} [F_1(t,x_t(u),u) - L_u V(t,x_t(u))]dt \geq 0, \quad \forall u \in W,$$

$$F_1(t,x_t(u_0),u_0) - L_{u_0} V(t,x_t(u_0)) = 0,$$

$$\alpha := F_0(x_T(u_0)) + V(T,x_T(u_0)) = \inf_{\psi \in X(T+\theta)} [F_0(\psi) + V(T,\psi)].$$

Then u_0 *is the optimal control, and*

$$J(u_0) = \alpha - V(t_0,\phi).$$

The assertion of the theorem follows from the formula

$$J(u) = \int_{t_0}^{T} [F_1(t,x_t(u),u) - L_u V(t,x_t(u))]dt +$$

$$+F_0(x_T(u)) + V(T,x_T(u)) - V(t_0,\phi).$$

4.2. *EQUATIONS WITH DISCRETE DELAYS*

For simplicity we will first consider the problem of optimal control for equations with a single delay:

$$\left.\begin{aligned} &\dot{x}(t) = f(t,x(t),x(t-h),u(t)), \\ &t_0 \leq t \leq T, \quad x(t) \in X(t) \subseteq \mathbf{R}^n, \quad u(t) \in U(t) \subseteq \mathbf{R}^m, \end{aligned}\right\} \quad (4.3)$$

$$J(u) := \min F_0(x(T)) + \int_0^T F_1(t,x(t),u(t))dt \to \min_{u \in W}, \quad (4.4)$$

under the conditions (1.2), (4.1).

Consider functions $g_i : [t_0,T] \times \mathbf{R}^n \to \mathbf{R}$ $(i = 0,1)$ such that $g_i \in C^1$. Introduce functions $H : [t_0,T] \times (\mathbf{R}^n)^2 \times \mathbf{R}^m \to \mathbf{R}$ and $\overline{H}, g : [t_0,T] \times (\mathbf{R}^n)^2 \to \mathbf{R}$ by

$$H(t,x,y,u) := -F_1(t,x,u) + f'(t,x,y,u)\frac{\partial}{\partial x}[g_0(t,x) + g_1(t+h,x)]$$

$$(g_1(t+h,x) := 0 \text{ if } t+h > T),$$

$$\overline{H}(t,x,y) := \sup_{u \in U(t)} H(t,x,y,u), \quad g(t,x,y) := g_0(t,x) + g_1(t,y).$$

Note that

$$\frac{d}{dt}g(t,x(t),x(t-h)) = \frac{\partial g}{\partial t} + \frac{\partial g}{\partial x(t)}\dot{x}(t) + \frac{\partial g}{\partial x(t-h)}\dot{x}(t-h).$$

Integrating both parts of this equality from t_0 to T, we obtain by (4.3), (4.4)

$$J(u) = \int_{t_0}^{T}[\overline{H}(t) - H(t)]dt - \int_{t_0}^{T}[g_t(t) + \overline{H}(t)]dt + C + F_0(x(T)) + \tag{4.5}$$

$$+g(T,x(T),x(T-h)) - g(t_0,\phi(t_0),\phi(t_0-h)).$$

Here, $H(t) := H(t,x(t),x(t-h),u(t))$, $\overline{H}(t) := \overline{H}(t,x(t),x(t-h))$,

$$C := \int_{-h}^{0}\dot{\phi}'(\theta)\frac{\partial g_1(\theta+h,\phi(\theta))}{\partial\phi(\theta)}d\theta,$$

$$g_t(t) := \frac{\partial g(t,x(t),x(t-h))}{\partial t}.$$

Let u_0 be the optimal control of the Pr. (4.3)–(4.4), (1.2), (4.1), and let u be an admissible control. The inequality $J(u) - J(u_0) \geq 0$ and the representation (4.5) imply the following theorem.

Theorem 4.2 Suppose there exist functions g_i and a process $t \mapsto (u(t),x(t))$, $u \in W$, such that

$$\overline{H}(t,x(t),x(t-h)) = H(t,x(t),x(t-h),u(t)),$$

$$C_1 := F_0(x(T)) + g(T,x(T),x(T-h)) =$$

$$= \min_{\substack{x\in X(t)\\ y\in X(T-h)}} [F_0(x) + g(T,x,y)],$$

$$0 = \frac{\partial g(t,x(t),x(t-h))}{\partial t} + \overline{H}(t,x(t),x(t-h)) =$$

$$= \max_{\substack{x\in X(t)\\ y\in X(T-h)}} \left[\frac{\partial g(t,x,y)}{\partial t} + \overline{H}(t,x,y)\right],$$

$$t_0 \leq t \leq T.$$

Then the process $t \mapsto (u(t), x(t))$ *is optimal, and*

$$J(u) = C + C_1 - g(t_0, \phi(t_0), \phi(t_0 - h)).$$

Remark 4.1 The same optimality criterion can be stated for equation with several discrete delays $h_i, 0 < h_1 < \ldots < h_k = h$:

$$\dot{x}(t) = f(t, x(t), x(t-h_1), \ldots, x(t-h_k), u(t)), \quad t_0 \leq t \leq T.$$

If we are also given a cost functional (4.4), constrains (4.1), and an initial condition (1.2), the functions $H, \overline{H}, g$ must be introduced as follows:

$$H(t, x, y_1, \ldots, y_k, u) := -F_1(t, x, u) + f'(t, x, y_1, \ldots, y_k, u)\frac{\partial}{\partial x}\sum_{i=0}^{k} g_i(t + h_i, x)$$

$$(h_0 := 0),$$

$$\overline{H}(t, x, y_1, \ldots, y_k) := \sup_{u \in U(t)} H(t, x, y_1, \ldots, y_k, u),$$

$$g(t, x, y_1, \ldots, y_k) := g_0(t, x) + \sum_{i=1}^{k} g_i(t, y_i). \quad \square$$

Theorem 4.3 Suppose there exist functions $g_i : [t_0, T] \times \mathbf{R}^n \to \mathbf{R}$ $(i = 0, \ldots, k)$ *and a process* $t \mapsto (u(t), x(t)), u \in W$, *such that*

$$\overline{H}(t, x(t), x(t-h_1), \ldots, x(t-h_k)) = H(t, x(t), x(t-h_1), \ldots, x(t-h_k), u(t)),$$

$$C_k := F_0(x(t)) + g(T, x(T), x(T-h_1), \ldots, x(T-h_k) =$$

$$= \min_{\substack{x \in X(T) \\ \cdots \\ y_k \in X(T-h_k)}} [F_0(x) + g(T, x, \ldots, y_k)],$$

$$0 = \frac{\partial}{\partial t} g(t, x(t), x(t-h_1), \ldots, x(t-h_k)) + \overline{H}(t, x(t), x(t-h_1), \ldots, x(t-h_k)) =$$

$$= \min_{\substack{x \in X(T) \\ \cdots \\ y_k \in X(T-h_k)}} \left[\frac{\partial}{\partial t} g(t, x, \ldots, y_k) + \overline{H}(t, x, \ldots, y_k)\right],$$

$$t_0 \leq t \leq T.$$

Then the process $t \mapsto (u(t), x(t))$ *is optimal, and*

$$J(u) = C_k - g(t_0, \phi(t_0), \phi(t_0 - h_1), \ldots, \phi(t_0 - h_k)) +$$

$$+ \sum_{i=1}^{k} \int_{-h_k}^{0} \dot{\phi}'(\theta) \frac{\partial g_i(\theta + h_i, \phi(\theta))}{\partial \phi(\theta)} d\theta.$$

5. NECESSARY OPTIMALITY CONDITIONS

In this section we will give some necessary optimality conditions in the form of the Pontrjagin's principle, also called the maximum principle. A rigourous proof of this principle for hereditary systems is quite long, and can be found in [418]. According to this principle, the determination of the optimal control is reduced to the solution of a boundary value problem for a set of differential equations with both retarded and advanced arguments, together with a maximization condition of the Hamiltonian.

5.1. *SYSTEMS WITH STATE DELAYS*

Consider the optimal control problem

$$\left.\begin{aligned} \dot{x}(t) = f(t, x(t), x(t - h(t)), u(t)), \quad t_0 \le t \le T, \\ x(t) \in \mathbf{R}^n, \quad u(t) \in U(t) \subseteq \mathbf{R}^m, \end{aligned}\right\} \tag{5.1}$$

$$x(t_0 + \theta) = \phi(\theta), \quad \theta \le 0, \tag{5.2}$$

$$J(u) := \int_{t_0}^{T} F_1(t, x(t), u(t)) dt + F_0(x(T)) \to \min_u. \tag{5.3}$$

Here, $h(t)$ is a positive continuously differentiable function, and $\dot{h}(t) < 1$. On the functions f, F_0, u, F_1, ϕ we impose the same restrictions as in Sec. 1. Let H be the scalar function

$$\left.\begin{aligned} H(t, x, y, u, \psi) := -F_1(t, x, y, u) + \psi' f(t, x, y, u), \\ t_0 \le t \le T. \end{aligned}\right\} \tag{5.4}$$

For $t > T$ we put $H(t, x, y, u, \psi) \equiv 0$ for all x, y, u, ψ.

Theorem 5.1 If a function $t \mapsto (u_0(t), x_0(t))$ *is an optimal control and optimal trajectory, then there is a function* $\psi : [t_0, T] \to \mathbf{R}^n$ *such that*

$$H(t, x_0(t), x_0(t - h(t)), u_0(t), \psi(t)) =$$
$$= \max_{u \in U(t)} H(t, x_0(t), x_0(t - h(t)), u, \psi(t)), \tag{5.5}$$

$$\dot{\psi}(t) = -H_x(t, x_0(t), x_0(t - h(t)), u_0(t), \psi(t)) -$$
$$-H_y(r(t), x_0(r(t)), x_0(t), u_0(r(t)), \psi(r(t)))\dot{r}(t), \tag{5.6}$$
$$t_0 \le t \le T,$$

$$\psi(T) = -F_{0x}(x_0(T)), \quad \psi(s) := 0, \quad s > T. \tag{5.7}$$

Here, $H_x(t, x, y, u, \psi) = \partial H(t, x, y, u, \psi)/\partial x$, *the functions* H_y, F_{0x} *are similarly defined, and the function* $t \mapsto r(t)$ *is defined by the equation* $t = r(t) - h(r(t))$.

If in the Pr. (5.1)–(5.3) the delay $h(t)$ is constant, then $r(t) = t+h, \dot{r}(t) = 1$, and (5.6) takes the form

$$\dot{\psi}(t) = -H_x(t, x_0(t), x_0(t - h), u_0(t), \psi(t)) -$$
$$-H_y(t + h, x_0(t + h), x_0(t), u_0(t + h), \psi(t + h)).$$

Hence the solution of the optimal control problem (5.1)–(5.3) is reduced to the solution of the boundary value problem (5.1), (5.6), under the conditions (5.2), (5.4), (5.5), (5.7).

We can similarly state necessary optimality conditions for systems with several delays h_i:

$$\left.\begin{aligned} &\dot{x}(t) = f(t, x(t), x(t - h_1), \ldots, x(t - h_k), u(t)), \\ &t_0 \le t \le T, \quad x(t) \in \mathbf{R}^n, \quad u(t) \in U(t) \subset \mathbf{R}^m, \\ &0 < h_1 < \ldots < h_k = h. \end{aligned}\right\} \tag{5.8}$$

Here, h_i are certain constants, and the Carathéodory function $(t, x, y_1, \ldots, y_k, u) \mapsto f(t, x, y_1, \ldots, y_k, u)$ is differentiable with respect to x, y_i, $i = 1, \ldots, k$. We will consider the Pr. (5.8), (5.2), (5.3). Put

$$H(t, x, y_1, \ldots, y_k, u, \psi) := -F_1(t, x, u) + \psi' f(t, x, y_1, \ldots, y_k).$$

Theorem 5.2 Suppose there is an optimal control $t \mapsto u_0(t)$ and corresponding optimal trajectory $x_0(t)$ for the Pr. (5.8), (5.2), (5.3). Then the conditions (5.7) and the equations

$$H(t, x_0(t), x_0(t-h_1), \dots, x_0(t-h_k), u_0(t), \psi(t)) =$$

$$= \max_{u \in U(t)} H(t, x_0(t), x_0(t-h_1), \dots, x_0(t-h_k), u, \psi(t)),$$

$$\dot{\psi}(t) = -H_x(t, x_0(t), x_0(t-h_1), \dots, x_0(t-h_k), u_0(t), \psi(t)) -$$

$$-\sum_{i=1}^{k} H_{y_i}(t+h_i, x_0(t+h_i), x_0(t+h_i-h_1),$$

$$\dots, x_0(t+h_i-h_k), u_0(t+h_i), \psi(t+h_i)),$$

$$t_0 \le t \le T,$$

hold for the function ψ.

5.2. *SYSTEMS WITH DELAYS IN THE CONTROL*

Suppose the unknown vector function $t \mapsto x(t) \in \mathbf{R}^n$ is the solution of

$$\dot{x}(t) = f(t, x(t), u(t), u(t-h)), \quad t_0 \le t \le T, \tag{5.9}$$

with initial conditions

$$x(t_0) = x_0 (\in \mathbf{R}^n), \quad u(t_0+\theta) = \phi_1(\theta), \quad -h \le \theta \le 0. \tag{5.10}$$

Suppose the cost functional is given by (5.3), there is a constraint $u(t) \in U(t) \subseteq \mathbf{R}^m$, and T is a fixed time moment. Suppose also that the continuous functions $(t, x, u, v) \mapsto f(t, x, u, v), F_1(t, x, u, v), x \mapsto F_0(x)$ are continuously differentiable with respect to x. We take the class of admissible controls to consist of the piecewise continuous functions u on an interval $[t_0 - h, T]$ satisfying (5.10) and $u(t) \in U(t)$.

Theorem 5.3 If there is an optimal process $t \mapsto (u_0(t), x_0(t))$ for the Pr. (5.9), (5.10), (5.3), then there is a function $\psi : [t_0, T+h] \to \mathbf{R}^n$ satisfying

(5.7) and

$$\dot{\psi}(t) = -H_x(t, x_0(t), u_0(t), v_0(t), \psi(t)),$$

$$\Delta_{u_0(t)} H(t, x_0(t), u_0(t), v_0(t), \psi(t)) +$$

$$+\Delta_{v_0(t+h)} H(t+h, x_0(t+h), u_0(t+h), v_0(t+h), \psi(t+h)) \le 0,$$

$$t_0 \le t \le T.$$

Here,

$$H(t, x, u, v, \psi) := -F_1(t, x, u, v) + \psi' f(t, x, u, v), \quad v_0(t) := u_0(t-h),$$

$$\Delta_{u_0(t)} H(t, x_0(t), u_0(t), v_0(t), \psi(t)) :=$$

$$:= H(t, x_0(t), u, v_0(t), \psi(t)) - H(t, x_0(t), u_0(t), v_0(t), \psi(t))),$$

and $\Delta_{v_0(t+h)}$ is similarly defined.

5.3. *SYSTEMS WITH DISTRIBUTED DELAYS*

Suppose we are given a controlled system

$$\left.\begin{aligned} \dot{x}(t) = f(t, x(t), u(t)) + \int_{t_0}^{t} K(t, s, x(s), u(s)) ds, \\ t_0 \le t \le T, \end{aligned}\right\} \tag{5.11}$$

with cost functional (5.3) and conditions

$$x(t) \in \mathbf{R}^n, \quad x(t_0) = x_0, \quad u(t) \in U(t) \subseteq \mathbf{R}^m. \tag{5.12}$$

Put

$$H(t, x, u, \psi) := -F_1(t, x, u) + \psi' f(t, x, u) + \int_t^T \psi'(s) K(s, t, x, u) ds.$$

We assume that the continuous functions $(t, x, u) \mapsto f(t, x, u)$, $F_1(t, x, u)$, $(t, s, x, u) \mapsto K(t, s, x, u)$, $x \mapsto F_0(x)$ are continuously differentiable in x.

Theorem 5.4 *Suppose there is an optimal process $t \mapsto (u_0(t), x_0(t))$ for the Pr. (5.11), (5.12), (5.3). Then there is a function $\psi : [t_0, T] \to \mathbf{R}^n$ such that*

$$\dot{\psi}(t) = -H_x(t, x_0(t), u_0(t), \psi(t)), \quad \psi(T) = -F_{0x}(x_0(T)),$$

$$H(t, x_0(t), u_0(t), \psi(t)) = \max_{u \in U(t)} H(t, x_0(t), u, \psi(t)).$$

Now we will derive optimality conditions for a problem which is linear in x:

$$\left.\begin{aligned} &\dot{x}(t) = A(t)x(t) + \beta(t,u(t)) + \int_{t_0}^{t} [K(t,s)x(s) + \alpha(t,s,u(s))]ds, \\ &t_0 \le t \le T, \quad x(t) \in \mathbf{R}^n, \quad x(t_0) = x_0, \quad u(t) \in U(t) \subseteq \mathbf{R}^m, \end{aligned}\right\} \tag{5.13}$$

$$J(u) := \int_{t_0}^{T} [a_0'(t)x(t) + \alpha_0(t,u(t))]dt + b'x(T) \to \min_u . \tag{5.14}$$

Here, $\alpha_0(t,u) \in \mathbf{R}$, $a_0(t), \beta(t,u), \alpha(t,s,u) \in \mathbf{R}^n$, $A(t), K(t,s) \in \mathbf{R}^{n\times n}$ are continuous functions, and the vector $b \in \mathbf{R}^n$. We define the function $\psi : [t_0, T] \to \mathbf{R}^n$ by

$$\dot{\psi}(t) = a_0(t) - A'(t)\psi(t) - \int_t^T K(s,t)\psi(s)ds, \quad \psi(T) = -b.$$

If an optimal control u_0 exists for the Pr. (5.13), (5.14), then

$$-\alpha_0(t,u_0(t)) + \psi'(t)\beta(t,u_0(t)) + \int_t^T \psi'(s)\alpha(s,t,u_0(t))ds =$$

$$= \max_{u\in U(t)} \left\{ -\alpha_0(t,u) + \psi'(t)\beta(t,u) + \int_t^T \psi'(s)\alpha(s,t,u)ds \right\}.$$

5.4. *LINEAR SYSTEMS WITH DISCRETE AND DISTRIBUTED DELAYS*

Suppose we are given a vector $b \in \mathbf{R}^n$ and two continuous functions $t \mapsto a_0(t) \in \mathbf{R}^n$, $(t,u) \mapsto \alpha_0(t,u) \in \mathbf{R}$. The problem is to find the minimum value of the functional (5.14), where $t \mapsto x(t)$ denotes the solution of the linear equation

$$\dot{x}(t) = \sum_{i=1}^{N} A_i(t)x(t-h_i) + \int_{-h}^{0} A(t,\theta)x(t+\theta)d\theta + B(t)u(t), \tag{5.15}$$

under the conditions (5.2). Here, A_i, A, B are continuous matrix functions, and the constants $h_i \geq 0, h \geq 0$.

We define the function $\psi : [t_0, \infty) \to \mathbf{R}^n$ as the solution of the problem

$$\dot{\psi}(t) = a_0(t) - \sum_{i=1}^{N} A_i'(t+h_i)\psi(t+h_i) - \int_t^{t+h} A'(\theta, t-\theta)\psi(\theta)d\theta,$$

$$t_0 \le t \le T, \quad \psi(T) = -b, \quad \psi(s) \equiv 0, \quad s > T.$$

Because of the identity

$$\psi'(T)x(T) - \psi'(t_0)x(t_0) = \int_{t_0}^{T} d[\psi'(t)x(t)],$$

we find

$$J(u) = \int_0^T \Bigg\{ a_0'(t)x(t) + \alpha_0(t, u(t)) - \dot{\psi}'(t)x(t) -$$

$$-\psi'(t)\left[Bu(t) + \sum_{i=1}^{N} A_i(t)x(t-h_i) + \int_{-h}^{0} A(t,\theta)x(t+\theta)d\theta\right]\Bigg\} dt-$$

$$-\psi'(t_0)x(t_0).$$

Set $\delta(s) := 0$ for $s < 0$ and $\delta(s) := 1$ for $s > 0$ ($\delta(s) + \delta(-s) = 1$), and let C be equal to

$$C := -\int_{t_0}^{T} \left[\psi'(t)\int_{-h}^{0} \delta(-\delta - t + t_0)A(t,\theta)\phi(t+\theta-t_0)d\theta\right] dt.$$

So, $J(u)$ can be written as

$$J(u) = \int_{t_0}^{T} [a_0'x(t) + \alpha_0(t, u(t)) - \psi'(t)Bu(t)-$$

$$-\dot{\psi}'(t)x(t) - x'(t)\int_t^{t+h} A'(s, t-s)\psi(s)ds\Bigg] dt-$$

$$-\sum_{i=1}^{N}\int_{t_0}^{T} \psi'(t)A_i(t)x(t-h_i)dt - \psi'(t_0)x(t_0) + C =$$

$$= \int_{t_0}^{T} [\alpha_0(t, u(t)) - \psi'(t)Bu(t)]dt - \psi'(t_0)x(t_0) +$$

$$+C - \sum_{i=1}^{N} \int_{-h_i}^{0} \psi'(t_0 + \theta + h_i)A_i(t_0 + \theta + h_i)\phi(\theta)\delta(T - h_i - \theta)d\theta.$$

This implies the following theorem.

Theorem 5.5 A control $t \mapsto u_0(t)$ in the Pr. (5.14), (5.15) is optimal if and only if

$$-\alpha_0(t, u_0(t)) + \psi'(t)Bu_0(t) = \max_{u \in U(t)} \{-\alpha_0(t, u) + \psi'(t)Bu\},$$

$$t_0 \leq t \leq T.$$

5.5. *NEUTRAL TYPE SYSTEMS*

It is well known that in general the maximum principle for NDEs is not valid without additional assumptions. Here we will consider necessary optimality conditions for the NDE

$$\left.\begin{aligned} \dot{x}(t) = f(t, x(t), x(t-h), \dot{x}(t-h), u(t)), \\ t_0 \leq t \leq T, \end{aligned}\right\} \tag{5.16}$$

with initial condition (5.2). We assume that the functions $(t, x, y, z, u) \mapsto f(t, x, y, z, u)$ and $\theta \mapsto \phi(\theta)$ are continuous, and continuously differentiable in x, y, z, u, respectively, θ. The cost functional has the form

$$F(x(T)) \to \min_{u(t) \in U}, \quad U \subseteq \mathbf{R}^m, \tag{5.17}$$

where U is a given convex set. We define the adjoint function $\psi : [t_0, T + h] \to \mathbf{R}^n$ by

$$\begin{aligned} \psi(t) := \psi(T) + \int_{t}^{T} [f_1'(s, x(s), x(s-h), \dot{x}(s-h), u(s))\psi(s) + \\ + f_2'(s+h, x(s+h), x(s), \dot{x}(s), u(s+h))\psi(s+h)]ds + \\ + f_3'(t+h, x(t+h), x(t), \dot{x}(t), u(t+h))\psi(t+h), \end{aligned} \tag{5.18}$$

where

$$\left. \begin{aligned} f_i(t, x_1, x_2, x_3, u) &:= \frac{\partial f(t, x_1, x_2, x_3, u)}{\partial x_i}, \\ \psi(T) = -\frac{\partial F(x(T))}{\partial x}&, \quad \psi(t) \equiv 0, \quad t > T. \end{aligned} \right\} \tag{5.19}$$

Theorem 5.6 If a process $t \mapsto (u_0(t), x_0(t))$ *is optimal for the Pr. (5.16)–(5.17), (5.2), then there is a function* $\psi : [t_0, T + h] \to \mathbf{R}^n$ *such that (5.18)–(5.19) hold, as well as*

$$\psi'(t)\left[\frac{\partial}{\partial u} f(t, x_0(t), x_0(t-h), \dot{x}_0(t-h), u_0(t))\right] u_0(t) =$$

$$= \max_{u \in U}\left\{\psi'(t)\left[\frac{\partial}{\partial u} f(t, x_0(t), x_0(t-h), \dot{x}_0(t-h), u)\right] u\right\}.$$

Remark 5.1 The papers [384, 433, 524] are concerned with the control of parabolic systems with delays in either the systems equations or the boundary conditions. A maximum principle has been formulated for systems with delays in the state variables. In [524] the author has considered a scalar linear diffusion system as an example, and have obtained an analytic solution using the maximum principle. □

6. ADAPTIVE CONTROL OF FUNCTIONAL DIFFERENTIAL EQUATIONS

Adaptive controls and, in particular, model reference adaptive controls (MRAC) are used to control objects with unknown dynamical parameters and/or under influence of considerable perturbations. The main goal of the MRAC design is to construct such adaption device that a phase error between system and model outputs tends to zero. Basic methods of the MRAC theory are strongly connected with those of the theory of stability. We shall illustrate these methods for some classes of FDEs.

6.1. *SCALAR EQUATIONS*

Consider RDE

$$\begin{aligned} \dot{x}(t) = -(a + \Delta a + \delta a(t))x(t) + (b + \Delta b + \delta b(t))\times \\ \times x(t - h + \Delta h + \delta h(t)) + f(t), \quad t \geq t_0. \end{aligned} \tag{6.1}$$

Here $\Delta a, \Delta b, \Delta h$ are unknown parameters such that $|\Delta a| \leq A_0, |\Delta b| \leq B_0, |\Delta h| \leq H_0 < h$. Functions $\delta a(t), \delta b(t), \delta h(t)$ can be changed by adaptation circuit and $f(t)$ is an input, $|f(t)| \leq F_0$. Along with RDE (6.1) consider reference model

$$\dot{y}(t) = -ay(t) + by(t-h) + f(t), \quad t \geq t_0. \tag{6.2}$$

Initial conditions for (6.1) and (6.2) are zero

$$x(t_0+\theta) = y(t_0+\theta) = 0, \quad \theta \leq 0.$$

The constants a, b, A_0, B_0, H_0, F_0 are known.

Firstly assume that $\Delta h + \delta h \equiv 0$. Then the error $\varepsilon(t) := x(t) - y(t)$ satisfies the equation

$$\begin{gathered}\varepsilon(t) = -a\varepsilon(t) + b\varepsilon(t-h) + \alpha(t)x(t) + \beta(t)x(t-h), \\ t \geq t_0, \;\; \alpha(t) := \Delta a + \delta a(t), \;\; \beta(t) := \Delta b + \delta b(t).\end{gathered} \tag{6.3}$$

Using direct Liapunov's method find the conditions under which $\varepsilon(t) \to 0$ as $t \to \infty$. Let us introduce the functional

$$V(t, \varepsilon_t, \alpha, \beta) := \gamma(t)\varepsilon^2(t) + \alpha^2(t) + \beta^2(t) + (a\gamma(t) - \frac{1}{2}\dot{\gamma}(t)) \int\limits_{t-h}^{t} \varepsilon^2(s)ds. \tag{6.4}$$

Here $\gamma(t)$ is a positive function, $\gamma \in C^2[t_0; \infty)$.

Then

$$\dot{V} = (\dot{\gamma} - 2a\gamma)\varepsilon^2(t) + 2\gamma b\varepsilon(t)\varepsilon(t-h) +$$

$$+(a\gamma - \dot{\gamma}/2)\varepsilon^2(t) - (a\gamma - \dot{\gamma}/2)\varepsilon^2(t-h) + 2\gamma x(t)\varepsilon(t)\alpha(t) +$$

$$+2\gamma x(t-h)\varepsilon(t)\beta(t) + 2\alpha\dot{\alpha} + 2\beta\dot{\beta} + (a\dot{\gamma} - \ddot{\gamma}/2) \int\limits_{t-h}^{t} \varepsilon^2(s)ds.$$

This expression for $\dot{V}$ prompts us to define $\alpha(t)$ and $\beta(t)$ by the equations

$$\dot{\alpha}(t) = \frac{d}{dt}(\delta a(t)) = -\gamma(t)x(t)\varepsilon(t), \quad \alpha(t_0) = \Delta a, \tag{6.5}$$

$$\dot{\beta}(t) = \frac{d}{dt}(\delta b(t)) = -\gamma(t)x(t-h)\varepsilon(t), \quad \beta(t_0) = \Delta b. \tag{6.6}$$

Taking into account (6.3)–(6.6) we have

$$\begin{gathered}\dot{V} = (\dot{\gamma} - 2a\gamma)\varepsilon^2(t) + 2\gamma b\varepsilon(t)\varepsilon(t-h) + \\ +(a\gamma - \dot{\gamma}/2)\varepsilon^2(t) - (a\gamma - \dot{\gamma}/2)\varepsilon^2(t-h) + (a\dot{\gamma} - \ddot{\gamma}/2) \int\limits_{t-h}^{t} \varepsilon^2(s)ds.\end{gathered} \tag{6.7}$$

Denote by C_i some positive constants.

Theorem 6.1 1) Let for all $t \geq 0$ be

$$0 < C_1 \leq \gamma(t) \leq C_2, \quad -a + |b| < 0,$$
$$-C_3 \leq -a\gamma + |b|\gamma + \dot{\gamma}/2 \leq -C_4 < 0, \quad a\dot{\gamma} - \ddot{\gamma}/2 \leq 0. \tag{6.8}$$

Then $\varepsilon(t) \to 0$ as $t \to \infty$.

2) If besides for any numbers $p, q, p^2 + q^2 \geq \mu^2 > 0$ and $B > 0$ there exists a number $T(B)$ such that

$$\int_t^{t+T(B)} (py(t) + qy(t-h))^2 dt \geq 3B \tag{6.9}$$

then the trivial solution of Eqs. (6.3), (6.5), (6.6) is globally uniformly asymptotically stable and stable under steady acting perturbations.

Remind that the trivial solution of RDE $\dot{x}(t) = f(t, x_t) + R(t, x_t)$ is called *stable under steady–acting perturbations* if for any $\varepsilon > 0$ there are $\delta > 0$ and $\eta > 0$ such that all solutions $x(t, t_0, \phi)$ satisfy inequality $|x(t, t_0, \phi)| \leq \varepsilon$ for $\| \phi \| \leq \delta$ and $R(t, \psi)| \leq \eta$, where $\| \psi \| \leq \varepsilon$. Similarly other stability definitions under steady–acting perturbations can be introduced.

Proof Assumptions of Thm. 6.1 imply that

$$C_5(\varepsilon^2(t) + \alpha^2(t) + \beta^2(t)) \leq V(t, \varepsilon_t, \alpha, \beta) \leq$$
$$\leq C_6 \max_{t-h \leq s \leq t} (\varepsilon^2(s) + \alpha^2(s) + \beta^2(s)).$$

Relation (6.7) yields

$$\dot{V} \leq (-a\gamma + |b|\gamma + \dot{\gamma}/2)\varepsilon^2(t) + (-a\gamma + (|b|\gamma +$$
$$+ \dot{\gamma}/2)\varepsilon^2(t-h) \leq -C_4(\varepsilon^2(t) + \varepsilon^2(t-h)) \leq 0.$$

Assume now that $\varepsilon(t)$ does not tend to zero as $t \to \infty$. Then the integral from $\dot{V}$ is unbounded, but the difference $V(t) - V(0)$ is bounded. Hence $\varepsilon(t) \to 0$, $t \to \infty$. For the proof of the second part of Thm. 6.1 let us use Thm. 4.1.10. Equality $\dot{V} = 0$ is possible if and only if $\varepsilon(t) = \varepsilon(t-h) = 0$. So functional (6.4) satisfies to the first two conditions of that theorem. As a second functional let us take

$$W(t, \varepsilon_t, \alpha, \beta) := (\alpha(t)x(t) + \beta(t)x(t-h))\varepsilon(t) \tag{6.10}$$

For $\mu, \rho > 0$ denote $E(\mu, \rho)$ the domain

$$E(\mu, \rho) := \{(\varepsilon, \alpha, \beta) \in C[-h, 0] \times C[t_0, \infty)^2 :$$
$$\sup |\varepsilon(t)| \leq \rho, \quad \mu \leq \sup[\alpha^2(t) + \beta^2(t)] < a^2\}.$$

By virtue of (6.3), (6.5), (6.6)

$$\dot{W} = (\alpha(t)x(t) + \beta(t)x(t-h))[-a\varepsilon(t) + b\varepsilon(t-h) + \alpha(t)x(t)+$$
$$+\beta(t)x(t-h)] + \varepsilon(t)\frac{d}{dt}(\alpha(t)x(t) + \beta(t)x(t-h)) = \tag{6.11}$$
$$= (\alpha(t)y(t) + \beta(t)y(t-h))^2 + \varepsilon(t)f(\alpha, \beta, x, \dot{x}, \varepsilon(t), \varepsilon(t-h)).$$

For $\dot{W}$ in the domain $E(\mu, \rho)$ the estimate holds

$$\int_s^{s+T(B)} \dot{W}(t)dt \geq \int_s^{s+T(B)} (py(t) + qy(t-h))^2 dt - B/2 \geq$$
$$\geq 2.5B, \ p = \alpha(s), \ q = \beta(s), \ p^2 + q^2 \geq \mu > 0.$$

Statement of Thm. 6.1 follows now from Thm. 4.1.10. □

6.2. *DELAY IDENTIFICATION*

Let the reference model be described by (6.2) and the plant by the equation

$$\dot{x}(t) = -ax(t) + bx(t - h + \Delta h + \delta h(t)) + f(t), \quad t \geq t_0. \tag{6.12}$$

If $x(t)$ is twice continuously differentiable function and $|\Delta h + \delta h(t)| < h$ for all $t > t_0$ then $\varepsilon(t) := x(t) - y(t)$ satisfies RDE

$$\dot{\varepsilon}(t) = -a\varepsilon(t) + b\varepsilon(t-h) + b[x(t-h+\tau(t)) - x(t-h)] =$$
$$= -a\varepsilon(t) + b\varepsilon(t-h) + b\tau(t)\dot{x}(t-h) + b\frac{\tau^2(t)}{2}\ddot{x}(t-h+\theta\tau(t)), \tag{6.13}$$

where $\tau(t) := \Delta h + \delta h(t), \ 0 < \theta < 1$.

Consider the following algorithm of delay identification

$$\dot{\tau}(t) = \frac{d}{dt}(\delta h(t)) = -b\gamma(t, x(t))\dot{x}(t-h)\varepsilon(t) - \eta(t, x(t))\varepsilon(t)/u(t). \tag{6.14}$$

Here $u(t)$ is a sensitivity function defined by the equations

$$\dot{u}(t) = -au(t) + bu(t-h) + b\dot{x}(t-h), \quad u(t) = 0, \quad t \leq t_0$$

and η is some continuous function. Algorithm (6.14) is more complicated than (6.5) since for its realization we must know not only $x(t)$ or $x(t-h)$ but also $\dot{x}(t-h)$. Similarly to the proof of Thm. 6.1 using appropriate functional V can be proven

Theorem 6.2 Let condition 1) Theorem 6.1 be fulfilled, function $x(t)$ be twice continuously differentiable and

$$0 < C_4 \le \eta(t,x) \le C_5, \quad |\ddot{x}(t)| \le C, \quad |u(t)) > 0. \tag{6.15}$$

Then the trivial solution of (6.13), (6.14) is uniformly asymptotically stable and stable under steady-acting perturbations for all sufficiently small Δh such that $|\tau(t)| < h$ for $t \ge t_0$.

6.3. *MULTIDIMENSIONAL SYSTEMS*

Let the reference model have the form

$$\dot{y}(t) = A(q^0)y(t) + B(q^0)y(t-h) + f(t), \quad y(t) \in \mathbf{R}^n. \tag{6.16}$$

Equation of the system is

$$\dot{x}(t) = A(q)x(t) + B(q)x(t-h) + \Phi(t) + f(t), \quad x \in \mathbf{R}^n. \tag{6.17}$$

Here matrices $A(q)$ and $B(q)$ are linear functions of $q \in \mathbf{R}^n$, $f(t)$ is an input and $\Phi(t)$ is the action on the system by the adaptation circuit.

Given

$$q = q^0 + \Delta q, \quad \Delta q = (\Delta q_1, \ldots, \Delta q_m). \tag{6.18}$$

Because of (6.18) Eq. (6.17) has a form

$$\dot{x}(t) = A(q^0)x(t) + B(q^0)x(t-h) + \sum_{i=1}^{m} \left[\frac{\partial A}{\partial q_i} \Delta q_i x(t) + \right.$$
$$\left. + \frac{\partial B}{\partial q_i} \Delta q_i x(t-h) \right] + \Phi(t) + f(t). \tag{6.19}$$

Put $\varepsilon := x - y$. The problem is to choose a function $\Phi(t)$ such that $\varepsilon(t) \to 0$ as $t \to \infty$.

Put

$$\Phi(t) = \sum_{i=1}^{m} \left[\frac{\partial A}{\partial q_i} \delta q_i(t) x(t) + \frac{\partial B}{\partial q_i} \delta q_i(t) x(t-h) \right]. \tag{6.20}$$

In Eq. (6.20) the functions $\delta q_i(t), i = 1, \ldots, m$, must be determined. Represent (6.19) as follows

$$\dot{\varepsilon}(t) = A(q^0)\varepsilon(t) + B(q^0)\varepsilon(t-h) +$$
$$+ \sum_{i=1}^{m} \left[\frac{\partial A}{\partial q_i} x(t) + \frac{\partial B}{\partial q_i} x(t-h) \right] \alpha_i(t), \tag{6.21}$$
$$\alpha_i(t) := \Delta q_i + \delta q_i(t).$$

Define $\alpha_i(t)$ by the equations

$$\dot{\alpha}_i(t) = \phi_i(t, \varepsilon_t, x_t). \tag{6.22}$$

Let us find the functionals ϕ_i using the direct Liapunov's method. Assume that there is a functional

$$V_1(t, \varepsilon(t), \varepsilon_t) := \varepsilon'(t)\Gamma(t)\varepsilon(t) + V_2(\varepsilon_t), \tag{6.23}$$

such that the derivative of (6.23) along the solution of the system

$$\dot{\varepsilon}(t) = A(q^0)\varepsilon(t) + B(q^0)\varepsilon(t-h) \tag{6.24}$$

is negative definite: $\dot{V}_1 \le -\omega_3(|\varepsilon(t)|)$. In the righthand side of (6.23) Γ is a symmetric positive definite smooth matrix-valued function and V_2 is a continuous functional which does not depend on $\varepsilon(t)$. Consider now the functional V:

$$V(t, \varepsilon(t), \varepsilon_t, \alpha(t)) := V_1(t, \varepsilon(t), \varepsilon_t) + \sum_{i=1}^{m} \alpha_i^2(t). \tag{6.25}$$

The derivative $\dot{V}$ of functional (6.25) along the trajectories of (6.21), (6.22) satisfies the inequality

$$\dot{V} \le -\omega_3(\varepsilon(t)) + \varepsilon'(t)\Gamma(t)\left[\sum_{i=1}^{m}(\frac{\partial A}{\partial q_i}x(t) + \frac{\partial B}{\partial q_i}x(t-h))\alpha_i(t)\right] +$$

$$+\left[\sum_{i=1}^{m}(\frac{\partial A}{\partial q_i}x(t) + \frac{\partial B}{\partial q_i}x(t-h))\alpha_i(t)\right]' \Gamma(t)\varepsilon(t) + 2\sum_{i=1}^{m}\alpha_i(t)\phi_i(t, \varepsilon_t, x_t).$$

Further, define ϕ_i as follows:

$$\phi_i(t, \varepsilon_t, x_t) := -\varepsilon'(t)\Gamma(t)\left(\frac{\partial A}{\partial q_i}x(t) + \frac{\partial B}{\partial q_i}x(t-h)\right).$$

Then

$$\dot{V} \le -\omega_3(\varepsilon(t)) \le 0.$$

As a result we get the equations of the adaptation circuit

$$\dot{\alpha}_i(t) = -\varepsilon'(t)\Gamma(t)\left[\frac{\partial A}{\partial q_i}x(t) + \frac{\partial B}{\partial q_i}x(t-h)\right].$$

At last similarly to the proof of Thm. 6.1 it can be proven that under assumptions made above $\varepsilon(t) \to 0$ as $t \to \infty$.

CHAPTER 15

OPTIMAL CONTROL OF STOCHASTIC DELAY SYSTEMS

In practical applications the behaviour of many dynamical systems depends not only on their previous history, but also on unknown disturbances.

This phenomenon can be brought about by the presence of delays and stochastic factors in the equations under consideration. In view of the intrinsic difficulties in solving corresponding problems, progress in this is slow, especially so when stochastic processes are involved. In this chapter we will give some methods and results of the optimal control of stochastic delay systems.

1. DYNAMIC PROGRAMMING METHOD FOR CONTROLLED STOCHASTIC HEREDITARY PROCESSES

In this section we will derive the Bellman equation for controlled SRDEs. In the sequel we will use some notation mentioned in Sec. 6.1.

Suppose we are dealing with a controlled stochastic equation of Itô type

$$\left.\begin{aligned} &dx(t) = f(t, x_t, u)dt + \sigma(t, x_t, u)d\xi(t), \quad t_0 \leq t \leq T, \\ &x_t(\theta) := x(t+\theta), \quad -h \leq \theta \leq 0, \quad x(t) \in \mathbf{R}^n, \quad u(t) \in U \subseteq \mathbf{R}^m. \end{aligned}\right\} \tag{1.1}$$

Here, $t \mapsto \xi(t) \in \mathbf{R}^l$ is the standard Wiener process, and the continuous functionals f, σ are defined on $[t_0, T] \times C[-h, 0] \times U$. The control u chosen from the set U, at any time t. In this situation, a feedback control is a function $u : [t_0, T] \times C[-h, 0] \to U$. The initial condition for (1.1) has the form

$$x_{t_0} = \phi \in C[-h, 0]. \tag{1.2}$$

The criterion $J(u)$ to be defined is

$$J(u) = \mathcal{E}\left[F_0(x_T) + \int_{t_0}^{T} F_1(t, x_t, u(t, x_t))dt\right] \to \min, \tag{1.3}$$

where $\mathcal{E}$ is expectation, and F_0, F_1 are continuous scalar functions. The observations of the state $x(t)$ are assumed to be complete. This means that for any time moment $t \in [t_0, T]$ the realization of $x(t)$ can be exactly measured. A Carathéodory feedback control $(t, \psi) \mapsto u(t, \psi)$ is admissible if for any $t_1 \in [t_0, T), \psi \in C[-h, 0]$ there is in $[t_0, T]$ a solution of (1.1) with initial condition $x_{t_1} = \psi$, while the value of the cost functional (1.3) (with t_1 instead of t_0) is finite under this control. Let W be the class of admissible controls. The problem is to find an admissible control minimizing (1.3) among all such controls.

The Bellman functional $V : [t_0, T] \times C[-h, 0] \to \mathbf{R}$ is defined as follows:

$$V(t, \psi) := \inf_{u \in W} \left\{ \mathcal{E} \left[F_0(x_T(t, \psi; u) + \int_t^T F_1(s, x_s(t, \psi; u), u(s, x_s))ds \right] \right\}.$$

Here, $x_s(t, \psi; u)(\theta) := x(s + \theta; t, \psi; u)$, $-h \leq \theta \leq 0$, and $x(\cdot; t_1, \psi; u)$ is the solution of (1.1) with control $u \in W$ and initial condition $x(t_1 + \cdot; t_1, \psi; u) = \psi$. The operator $L_u V(t, \psi)$ is defined by the relation

$$L_u V(t, \psi) := \limsup_{\Delta t \to 0+} \frac{1}{\Delta t} \mathcal{E}[V(t + \Delta t, x_{t+\Delta t}(t, \psi; u)) - V(t, \psi)]. \quad (1.4)$$

Our optimality conditions are given in the following theorem, whose proof is completely analogous to the proof of Thm. 6.1.1.

Theorem 1.1 Suppose there are a functional $(t, \psi) \mapsto V_0(t, \psi)$ and an admissible control $(t, \psi) \mapsto u_0(t, \psi)$ satisfying

$$\inf_{u \in U} [L_u V_0(t, \psi) + F_1(t, \psi, u)] = L_{u_0(t,\psi)} V_0(t, \psi) + F_1(t, \psi, u_0(t, \psi)) = 0, \quad (1.5)$$

$$V_0(T, \psi) = F_0(\psi). \quad (1.6)$$

Then $(t, \psi) \mapsto u_0(t, \psi)$ is the optimal control, and V_0 is the Bellman function, for the Pr. (1.1)–(1.3).

Formally, the control u_0 defined by (1.5) depends on V_0, so we will sometimes denote it by $u_0(t, \psi; V_0)$ to emphasize this dependence.

2. THE LINEAR QUADRATIC PROBLEM

2.1. *BELLMAN FUNCTIONAL AND OPTIMAL CONTROL*

We encounter substantial difficulties when trying to use (1.5), (1.6) to solve the optimal control problem (1.1)–(1.3). Up till now, mainly linear and quasilinear systems with quadratic criterion have been treated in this manner.

Here we will consider the Itô type stochastic equation

$$\left.\begin{aligned} dx(t) = \Bigg[A(t)x(t) + A_1(t)x(t-h) + \int_{-h}^{0} G(t,\theta)x(t+\theta)d\theta + \\ +B(t)u(t,x_t)\Bigg] dt + \sigma(t)d\xi(t), \\ t_0 \le t \le T, \quad x(t) \in \mathbf{R}^n, \quad u(t,x_t) \in \mathbf{R}^m. \end{aligned}\right\} \tag{2.1}$$

Here, T is arbitrary but fixed, $\sigma(t)$ is a matrix continuous entries, $t \mapsto \xi(t)$ is a Wiener process, and the remaining terms are the same as in (14.2.1). The expression to be minimized is the expectation of (14.2.2), i.e.

$$J(u) := \mathcal{E}\left[x'(T)N_0x(T) + \int_{t_0}^{T}(u'(t,x_t)N_1(t)u(t,x_t) + x'(t)N_2(t)x(t))dt\right] \to$$

$$\to \min_u, \tag{2.2}$$

where the symmetric matrices $N_0 \ge 0,\ N_1(t) > 0,\ N_2(t) \ge 0$, and N_1, N_1 are continuous.

The Bellman equation (1.5) for the Pr. (2.1), (2.2) has the same form (14.2.3) as in the deterministic case:

$$\left.\begin{aligned} \inf_u[L_uV_0(t,\psi) + u'N_1(t)u + \psi'(0)N_2(t)\psi(0)] = 0, \\ V_0(T,\psi) = \psi'(0)N_0\psi(0), \end{aligned}\right\} \tag{2.3}$$

where the operator L_u is defined by (1.4).

We try to find $(t,\psi) \mapsto V(t,\psi)$ of (2.3) in the form

$$V(t,\psi) = V_0(t,\psi) + g(t). \tag{2.4}$$

Here, V_0 is given by (14.2.4) and g is a scalar function. Since we have the representation (2.4) for V, the functional L_uV can be calculated by Itô's formula (10.1.5). As before, a direct calculation shows that the optimal control $(t,\psi) \mapsto u(t,\psi)$ is defined by (14.2.7):

$$u(t,\psi) = -N_1^{-1}(t)B'(t)\left[P(t)\psi(0) + \int_{-h}^{0} Q(t,\theta)\psi(\theta)d\theta\right].$$

Substitute (2.4) and (14.2.7) into (2.3). Equating to zero the quadratic form in $\psi(\theta), -h \leq \theta \leq 0$, we conclude that P, Q, R satisfy (14.2.5) with boundary conditions (14.2.6). The function g is equal to

$$g(t) = \int_t^T \mathrm{Tr}[\sigma'(s)P(s)\sigma(s)]ds. \tag{2.5}$$

Sufficient conditions for the existence of the matrices P, Q, R, as well as certain exact formulas for them, are given in Sec. 14.2.

2.2. *APPROXIMATE SOLUTION*

2.2.1 Method of successive approximation. Here we will describe the method of successive approximation for solving (2.3), and, consequently, (14.2.5), (14.2.6). Let u_0 be a control, defined on $[t_0, T] \times C[-h, 0]$:

$$u_0(t,\psi) = -N_1^{-1}(t)B'(t)\left[P_0(t)\psi(0) + \int_{-h}^{0} Q_0(t,\theta)\psi(\theta)d\theta\right],$$

where P_0, Q_0 are arbitrary continuous matrix functions, and P_0 is symmetric and nonnegative definite. We find the solution $(t,\psi) \mapsto V_{k+1}(t,\psi)(k = 0, 1, \ldots)$ of the problem

$$L_{u_k}V_{k+1}(t,\psi) + u_k'N_1(t)u_k + \psi'(0)N_2(t)\psi(0) = 0,$$

$$V_{k+1}(T,\psi) = \psi'(0)N_0\psi(0),$$

assuming that the functional u_k is known. Then we will determine a functional u_{k+1} minimizing the expression

$$\min_u\{L_uV_{k+1}(t,\psi) + u'N_1(t)u + \psi'(0)N_2(t)\psi(0)\}.$$

By this procedure we construct sequences of controls $(t, P) \mapsto u_k(t,\psi)$ and functionals $(t,\psi) \mapsto V_k(t,\psi)(k = 0, 1, \ldots)$. Moreover, for all $k \geq 1$,

$$u_k(t,\psi) = -N_1^{-1}(t)B'(t)\left[P_k(t)\psi(0) + \int_{-h}^{0} Q_k(t,\theta)\psi(\theta)d\theta\right],$$

$$V_k(t,\psi) = \psi'(0)P_k(t)\psi(0) + \psi'(0)\int_{-h}^{0} Q_k(t,\theta)\psi(\theta)d\theta +$$

$$+g_k(t) + \int_{-h}^{0} \psi'(\theta)Q_k'(t,\theta)\psi(0)d\theta + \int_{-h}^{0}\int_{-h}^{0} \psi'(\theta)R_k(t,\theta,\theta_1)\psi(\theta_1)d\theta d\theta_1.$$

The matrices P_k, Q_k, R_k are the solutions of the linear equations (14.2.5), (14.2.6), where the nonlinear terms are linearized by the rule

$$PB_1P \to P_kB_1P_{k-1} + P_{k-1}B_1P_k - P_{k-1}B_1P_{k-1},$$

$$PB_1Q \to P_kB_1Q_{k-1} + P_{k-1}B_1Q_k - P_{k-1}B_1Q_{k-1},$$

$$Q'B_1Q \to Q'_kB_1Q_{k-1} + Q'_{k-1}B_1Q_k - Q'_{k-1}B_1Q_{k-1},$$

Using the linear equations for P_k, Q_k, R_k it has been proved [292.8] that $V_k(t,\psi) \geq V_{k+1}(t,\psi) \geq 0$, that the sequences $\{u_k\},\{V_k\}$ tend to u, V as $k \to \infty$, where $(t,\psi) \mapsto u(t,\psi), V(t,\psi)$ are the optimal control and Bellman functional for the problem (1.1), (1.3), and that

$$\max_{t,\theta,\theta_1}\{|P_k(t) - P(t)| + |Q_k(t,\theta) - Q(t,\theta)|+$$

$$+|R_k(t,\theta,\theta_1) - R(t,\theta,\theta_1)|\} = O\left(\frac{1}{k!}\right),$$

$$t_0 \leq t \leq T, \quad -h \leq \theta, \theta_1 \leq 0.$$

Here, $|\cdot|$ is the euclidean matrix norm. Analytic formulas have been derived for the matrices P_k, Q_k, R_k, and each $g_k(t)$ is given by (2.5) with $P(t)$ replaced by $P_k(t)$.

Previous results show that obtaining of the optimal control refers to considerable difficulties. Therefore it is highly desirable to simplify the control algorithm, even losing in optimality. Below we describe two of them and evaluate the nonoptimality level for the approximate solution (see 2) and 3)) of Pr. (14.2.5), (14.2.6).

2.2.2 Small delays. In this case we can neglect the delays to calculate P, Q and the optimal control. Let us denote by $P^a(t), u^a(t)$ and $x^a(t)$ the approximate values of $P(t)$, optimal control and trajectories. Then the simplified equations for P^a, u^a and x^a are proposed to take in the form

$$\dot{P}^a(t) + (A'(t) + A'_1(t))P^a(t) + P^a(t)(A(t) + A_1(t))-$$

$$-P^a(t)B(t)N_1^{-1}(t)B'(t)P^a(t) + N_2(t) = 0, \quad P^a(T) = N_0.$$

$$u^a(t) = -N_1^{-1}(t)B'(t)P^a(t)x^a(t), \tag{2.6}$$

$$\dot{x}^a(t) = (A(t) - B(t)N_1^{-1}(t)B'(t)P^a(t))x^a(t) + A_1(t)x^a(t-h)+$$

$$+\int_{-h}^{0} G(t,\tau)x^a(t+\tau)d\tau, \quad x^a(\tau) = \phi(\tau), \quad -h \leq \tau \leq 0. \tag{2.7}$$

2.2.3 Linearization. Let us neglect in the second equation (14.2.5) the terms $R(t,0,\theta)$ and PB_1Q. Then $Q(t,\theta)$ satisfy the ODE with a parameter θ

$$\frac{dQ(t,\theta-t)}{dt}+A'(t)Q(t,\theta-t)+P(t)G(t,\theta-t)=0$$

with boundary condition (14.2.6).

Denote P^a, Q^a, u^a and x^a the approximate value of the corresponding functions. Then

$$\dot{P}^a(t)+A'(t)P^a(t)+A_1'(t+h)P^a(t+h)+\int_0^h G'(s+t,-s)P^a(t+s)ds+$$

$$+P^a(t)A(t)+P^a(t+h)A_1(t+h)+\int_0^h P^a(t+s)G(t+s,-s)ds-$$

$$-P^a(t)B(t)N_1^{-1}(t)B'(t)P^a(t)+N_2(t)=0,$$

$$P^a(T)=N_0,\quad P^a(\tau)=0,\quad \tau>T$$

$$Q^a(t,\tau)=\int_0^{h+\tau} P^a(t+s)G(s+t,\tau-s)ds+P^a(t+\tau+h)A_1(t+\tau+h).$$

Therefore

$$u^a(t)=-N_1^{-1}(t)B'(t)\left[P^a(t)x^a(t)+\int_{-h}^{0} Q^a(t,\tau)x^a(t+\tau)d\tau\right], \tag{2.8}$$

$$\dot{x}^a(t)=(A(t)-B(t)N_1^{-1}(t)B'(t)P^a(t))x^a(t)+A_1(t)x^a(t-h)+$$

$$+\int_{-h}^{0}(G(t,\tau)-B(t)N_1^{-1}(t)B'(t)Q^a(t,\tau))x^a(t+\tau)d\tau, \tag{2.9}$$

$$x^a(\tau)=\phi(\tau),\quad -h\le\tau\le 0.$$

2.2.4 Estimation of the nonoptimality level. Let us define the nonoptimality level of the control u^a by

$$\Delta=\frac{J(u^a)}{J(u^0)},$$

where $u^a(t)$ and $u^0(t)$ are the approximate and optimal controls in Pr. (14.2.1), (14.2.2).

Since the optimal control $u^0(t)$ is unknown, the ratio Δ is also unknown. Then it is reasonable to construct an upper bound for Δ that can be efficiently computed. The efforts needed for this computation must be substantionally less than those of for the computation of the optimal control.

Theorem 2.1 Suppose the approximate control $u^a(t)$ is defined by (2.6) or (2.8).

$$1 \leq \Delta \leq \Delta^0,$$

where

$$\Delta^0 = \frac{J(u^a)\mu}{|\mathcal{Q}|},$$

$$\mathcal{Q} = \phi'(0)\eta(0) + \int_{-h}^{0} \phi'(s)A_1'(s+h)\eta(s+h)ds + \tag{2.10}$$

$$+ \int_0^h \int_{-\theta}^{0} \phi'(s)G'(s+\theta,-\theta)\eta(s+\theta)dsd\theta,$$

$$\dot{\eta}(t) = -A'(t)\eta(t) - A_1'(t+h)\eta(t+h) - \int_0^h G'(t+\theta,-\theta)\eta(t+\theta)d\theta - p(t), \tag{2.11}$$

$$t \in [0,T], \quad \eta(T) = \lambda, \quad \eta(\tau) = 0, \quad \tau > T,$$

$$\lambda = N_0 x^a(T), \quad p(t) = N_2(t)x^a(t), \tag{2.12}$$

$$\mu = \left\{ x^{a\prime}(T)N_0 x^a(T) + \int_0^T \eta'(t)B(t)N_1^{-1}(t)B'(t)\eta(t)dt + \right. \tag{2.13}$$

$$\left. + \int_0^T x^{a\prime}(t)N_2(t)x^a(t)dt \right\}^{1/2}.$$

The proof of Thm. 2.1 is in [286.3, 364]. Thm. 2.1 provides the following algorithm for Δ^0 calculation: (1) in accordance with one of the two schemes, the coefficients $P^a(t), Q^a(t,\tau)$ are computed; (2) the process $x^a(t)$ is computed by formulas (2.7) or (2.9); (3) the value of λ and the function $P(t)$ are determined from (2.12); (4) the function $\eta^a(t)$ is obtained from (2.11); (5) finally, the values $J(u^a), \mathcal{Q}$ and μ are obtained from (14.2.2), (2.10), and (2.13).

So, simplified algorithms for the control of hereditary dynamic systems are constructed. Moreover, the estimate of these simplified algorithms non-optimality level is obtained. This estimate can be computed by relatively simple numerical means, without accurate solving the initial optimal problem.

2.3. *SOME GENERALIZATIONS*

The above mentioned results have been generalized in [292.8] to linear equation with several discrete delays and under disturbance

$$\sigma(t)d\xi(t) + \int_{y\in\mathbf{R}^n} f(t,y)q(dt,dy),$$

where q is a Poisson measure with independent increments.

Another generalization is that to equation

$$dx(t) = \left[A(t)x(t) + A_1(t)x(t-h) + \int_{-h}^{0} G(t,\theta)x(t+\theta)d\theta + B(t)u(t)\right] dt+$$

$$+\sigma(t)d\xi(t) + \sum_{i=1}^{l} \sigma_i(t)x(t)d\xi_i(t), \quad t_0 \le t \le T, \tag{2.14}$$

where $t \mapsto \xi(t), \xi_i(t)$ are mutually independent scalar Wiener processes. The solution of (2.14), (1.3) is given by (2.4), (14.2.7). The functions P, Q, R, g satisfy the same equations (14.2.5), (14.2.6), (2.5), but on the lefthand side of the first equation of (14.2.5) we must add

$$\sum_{i=1}^{l} \sigma_i'(t)P(t)\sigma_i(t)$$

(see [292.8]).

3. APPROXIMATE OPTIMAL CONTROL FOR EQUATIONS WITH SMALL PARAMETERS

3.1. *FORMAL ALGORITHM*

Here we will consider an optimal control problem of the type (1.1)–(1.3) depending on a small parameter:

$$dx(t) = f(t,x_t,u(t,x_t);\varepsilon)dt + \sigma(t,x_t,u(t,x_t);\varepsilon)d\xi(t), \quad t_0 \le t \le T, \tag{3.1}$$

$$J(u;\varepsilon) := \mathcal{E}\left[F_0(x_T;\varepsilon) + \int_0^T F_1(t,x_t,u(t,x_t);\varepsilon)dt\right] \to \inf_{u(t,x_t)\in U}. \qquad (3.2)$$

Then the optimal control u_0, Bellman functional V, and operator L_u also depend on the parameter ε. We assume that the dependence of u_0 on t, ψ, V, ε given by the Bellman equation for (3.1), (3.2) is known.

We represent u_0, L_u, V, F_0, F_1 by series in ε:

$$\left.\begin{aligned} u_0(t,\psi;V,\varepsilon) &= u^0(t,\psi) + \varepsilon u^1(t,\psi) + \ldots, \\ V(t,\psi;\varepsilon) &= V^0(t,\psi) + \varepsilon V^1(t,\psi) + \ldots, \\ L_{u_0}(\varepsilon) &= L^0 + \varepsilon L^1 + \ldots, \\ F_0(\psi;\varepsilon) &= F_0^0(\psi) + \varepsilon F_0^1(\psi) + \ldots, \\ F_1(t,\psi,u_0;\varepsilon) &= F_1^0(t,\psi) + \varepsilon F_1^1(t,\psi) + \ldots. \end{aligned}\right\} \qquad (3.3)$$

The coefficients of all the expansions on the righthand side of (3.3) are independent on ε.

Substituting the expansions (3.3) into (1.5), (1.6) and equating to zero the coefficients at the successive powers of ε, we obtain

$$\sum_{j=0}^{i} L^{i-j}V^j(t,\psi) + F_1^i(t,\psi) = 0, \quad V^i(T,\psi) = F_0^i(\psi), \quad i = 0,1,\ldots. \qquad (3.4)$$

By solving (3.4) for $i = 0,\ldots,k$, we obtain the functionals V^i.

Consider the sequence of controls $(t,\psi) \mapsto u_0(t,\psi;Q_k;\varepsilon)$, where $Q_k := V^0 + \ldots + \varepsilon^k V^k$. We will denote the controls obtained in this way by

$$v_k(t,\psi;\varepsilon) := u_0(t,\psi;Q_k;\varepsilon).$$

Let $(t,\psi) \mapsto \delta_k(t,\psi;\varepsilon)$ be the functional

$$\begin{aligned} \delta_k(t,\psi;\varepsilon) := \sum_{i=0}^{k} \varepsilon^i \left[\sum_{j=0}^{i} L^{i-j}V^j(t,\psi) + F_1^i(t,\psi)\right] - \\ -L_{v_k}Q_k(t,\psi;\varepsilon) - F_1(t,\psi,v_k;\varepsilon). \end{aligned} \qquad (3.5)$$

Equations (3.4), (3.5) imply that the boundary value problem

$$\left.\begin{aligned} \inf_{u\in U}[L_uV(t,\psi;\varepsilon) + F_1(t,\psi,u;\varepsilon)] + \delta_k(t,\psi;\varepsilon) = 0, \\ V(T,\psi;\varepsilon) = Q_k(T,\psi;\varepsilon), \end{aligned}\right\}$$

has the solution $Q_k(t,\psi;\varepsilon)$, since

$$\left.\begin{aligned} L_{v_k}Q_k(t,\psi;\varepsilon)+F_1(t,\psi,v_k;\varepsilon)+\delta_k(t,\psi;\varepsilon)=0,\\ Q_k(T,\psi;\varepsilon)=F_0^0(\psi)+\ldots+\varepsilon^k F_0^k(\psi).\end{aligned}\right\} \tag{3.6}$$

Comparing (3.6) with (1.5), (1.6), we conclude that v_k and Q_k are the optimal control and Bellman functional for the problem (3.1) with minimization criterion $J_k(u;\varepsilon)$:

$$\left.\begin{aligned} J_k(u;\varepsilon):=J(u;\varepsilon)+r_k,\\ r_k:=\mathcal{E}\left[-F_0(x_T;\varepsilon)+\sum_{j=0}^{k}\varepsilon^j F_0^j(x_T)+\int_{t_0}^{T}\delta_k(s,x_s;\varepsilon)ds\right].\end{aligned}\right\} \tag{3.7}$$

If we could show that $r_k=O(\varepsilon^{k+1})$, then v_k could be shown to be the kth approximation to the optimal control. (The latter means that

$$0\le J(v_k;\varepsilon)-J(u_0;\varepsilon)=O(\varepsilon^{k+1}).)$$

This estimate has been proved in [292.8] for quasilinear systems with quadratic cost functional.

3.2. *QUASILINEAR EQUATION WITH QUADRATIC COST*

We will consider in some detail the zeroth approximation of the optimal control for a quasilinear equation of the type (1.1):

$$\begin{aligned} dx(t)=[\varepsilon f(t,x_t)+B(t)u(t,x_t)]dt+\sigma(t)d\xi(t),\\ t_0\le t\le T,\quad x(t)\in\mathbf{R}^n,\quad u(t,x_t)\in\mathbf{R}^m.\end{aligned} \tag{3.8}$$

The initial condition for (3.8) is (1.2), and the cost functional $J(u)$ is given by (2.2). We assume f to be a continuous functional, and

$$|f(t,\psi)|\le a_0+\int_{-h}^{0}|\psi(\theta)|^2dr(\theta), \tag{3.9}$$

where r is a nondecreasing function on $[-h,0]$. The matrices B and σ are taken measurable and bounded.

In the expansions (3.3) we find

$$\left.\begin{aligned} L^0 &= \frac{\partial}{\partial t} + \frac{1}{2}\mathrm{Tr}\left(\sigma\sigma'\frac{\partial^2}{\partial x^2}\right), \\ V^0(t,\psi) &= \psi'(0)P(t)\psi(0) + \int_t^T \mathrm{Tr}[\sigma(s)\sigma'(s)P(s)]ds, \\ u^0(t,\psi) &= -N_1^{-1}B'(t)P(t)\psi(0). \end{aligned}\right\} \tag{3.10}$$

Here, the matrix $P(t)$ satisfies the conditions

$$\dot{P}(t) + N_2(t) = P(t)B_1(t)P(t), \quad P(T) = N_0, \quad B_1 := BN_1^{-1}B'.$$

Taking into account (3.7), we see that the control $v_0 := u^0(t,\psi)$ is optimal for Eq. (3.8) with cost functional

$$\left.\begin{aligned} J_0(u;\varepsilon) &:= J(u) + \mathcal{E}\int_{t_0}^T \delta_0(s,x_s;\varepsilon)ds \to \min_u, \\ \delta_0(t,\psi;\varepsilon) &:= -2\varepsilon f'(t,\psi)P(t)\psi(0). \end{aligned}\right\} \tag{3.11}$$

The Bellman functional for (3.8), (3.11) is V^0. By (3.9), (3.10),

$$|\delta_0(t,\psi;\varepsilon)| \leq \varepsilon C(1 + \|\psi\|^2). \tag{3.12}$$

In the sequel C will denote various constants not depending on t, ψ, ε. Let u be the optimal control in the Pr. (3.8), (2.2). Then

$$J(u) \leq J(u^0) = J_0(u^0;\varepsilon) - \mathcal{E}\int_{t_0}^T \delta_0(s, x_s^{u^0};\varepsilon)ds, \tag{3.13}$$

where x_s^u is the solution of the Pr. (3.8), (1.2) with control u. However, the second moment $\mathcal{E}|x_s^{u^0}|^2$ is bounded, because of (3.8)–(3.10). This, (3.12), and (3.13) imply that

$$-J(u) + J_0(u^0;\varepsilon) \leq \varepsilon C(1 + \|\phi\|^2). \tag{3.14}$$

On the other hand,

$$J_0(u^0;\varepsilon) \leq J_0(u;\varepsilon) = J(u) + \mathcal{E}\int_{t_0}^T \delta_0(s, x_s^u;\varepsilon)ds. \tag{3.15}$$

But $J(u) \leq J(0)$. This inequality and the positive definiteness of $N_1(t)$ imply that

$$\mathcal{E}\int_{t_0}^{T} |u(s,x_s)|^2 ds < \infty.$$

Consequently we may conclude, as above, that the second moment $\mathcal{E}|x_s^u|^2$ is bounded. So, by (3.15),

$$J(u) - J_0(u^0;\varepsilon) \leq \varepsilon C(1+ \parallel \phi \parallel^2). \tag{3.16}$$

Using (3.14), (3.16) we find

$$0 \leq J(u^0) - J(u) \leq |J(u) - J_0(u^0;\varepsilon)| + |J(u^0) - J_0(u^0;\varepsilon)| \leq$$

$$\leq \varepsilon C(1+ \parallel \phi \parallel^2).$$

Hence u^0 is the zeroth approximation of the optimal control.

4. ANOTHER APPROACH TO THE PROBLEM OF OPTIMAL SYNTHESIS CONTROL

One of the main difficulties in applying the dynamic programming equation for hereditary systems is that for a general functional $(t,\psi) \mapsto V(t,\psi)$ the form of the operator L_u is unknown. However, if we suitably restrict the set of admissible functionals V, then L_u can be given explicitly. For such classes of hereditary systems we can sometimes obtain analytic formulas for the successive approximations to the optimal control. This approach has been developed in [292.8], and we will give it here.

4.1. *ADMISSIBLE FUNCTIONALS*

Let us once again consider the optimal control problem (1.1)–(1.3). Let $PC[-h,0]$ be the space of piecewise continuous functions $\psi : [-h,0] \to \mathbf{R}^n$. For a functional $V : [t_0,T]\times PC[-h,0] \to \mathbf{R}$ we define the function $((t,x) \mapsto V_\psi(t,x)) : [t_0,T] \times \mathbf{R}^n \to \mathbf{R}$ as $V(t,\psi^x)$, where $\psi^x(\theta) := \psi(\theta)(-h \leq \theta < 0)$, $\psi^x(0) := x$. We assume that

$$L_u V(t,\psi) = L_u V_\psi(t,x) = \frac{\partial}{\partial t} V_\psi(t,x) + f'(t,\phi,u) \bigtriangledown V_\psi(t,x) + \\ + \frac{1}{2}\mathrm{Tr}[\sigma'(t,\psi,u) \bigtriangledown^2 V_\psi(t,x)\sigma(t,\psi,u)]. \tag{4.1}$$

Here, L_u is defined by (1.4), and $\bigtriangledown V_\psi(t,x) := \partial V_\psi(t,x)/\partial x$. We will denote the class of such functional by D; it is sometimes used in control problems.

Of course, all previous assertions in this chapter can be reformulated, with minor modifications, in terms of the functionals $V \in D$ ([292.8]).

E.g., the modification of Thm. 1.1 is

Theorem 4.1 Suppose there are a functional $V_0 \in D$ satisfying (1.6) and an admissible u_0 such that

$$\inf_{u \in U}\{L_u V_0(t,\psi) + F_1(t,\psi,u)\} =$$

$$= \frac{\partial}{\partial t} V_{0\psi}(t,x) + f'(t,\psi,u_0) \nabla V_{0\psi}(t,x) + \tag{4.2}$$

$$+\frac{1}{2}\mathrm{Tr}[\sigma'(t,\psi,u_0) \nabla^2 V_{0\psi}(t,x)\sigma(t,\psi,u_0)] = 0.$$

Then u_0 is the optimal control and V^0 the Bellman function of the Pr. (1.1)–(1.3).

For the linear quadratic problem (2.1), (2.2) the solution of (4.2), (1.6) is given by (2.4). Note that the functional V defined by (2.4) belongs to D.

4.2. *QUASILINEAR QUADRATIC PROBLEMS*

We will apply the above described approach to the Pr. (3.8), (2.2). Then

$$L_u V_\psi(t,x) = L^0 V_\psi(t,x) + [\varepsilon f(t,\psi) + B(t)u]' \nabla V_\psi(t,x),$$

where L^0 is defined by (3.10). The solution of (3.8), (2.2) is, by (4.1), related to the investigation of the equation

$$\left.\begin{aligned} &L^0 V_\psi(t,x) + \varepsilon f'(t,\psi) \nabla V_\psi(t,x) + x' N_2(t) x = \\ &= \frac{1}{4} \nabla V_\psi'(t,x) B_1(t) \nabla V_\psi(t,x), \quad t_0 \le t \le T, \\ &V(T,\psi) = x' N_0 x, \quad x = \psi(0), \quad B_1 = B N_1^{-1} B'. \end{aligned}\right\} \tag{4.3}$$

Under certain restrictions, the optimal control u_0 can be written as

$$u_0(t,\psi) = -\tfrac{1}{2} N_1^{-1}(t) B'(t) \nabla V_\psi(t,x). \tag{4.4}$$

Using the general scheme (3.3)–(3.5) we find that an approximate solution of (4.3) may be looked for as the cut–off expansions

$$\left.\begin{aligned} V(t,\psi;\varepsilon) &= V^0(t,\psi) + \varepsilon V^1(t,\psi) + \ldots, \\ u_0(t,\psi;\varepsilon) &= u^0(t,\psi) + \varepsilon u^1(t,\psi) + \ldots. \end{aligned}\right\} \tag{4.5}$$

Here, V^0 and u^0 are defined by (3.10), (4.4).

By (4.3), ((4.5), for $i \geq 1$,

$$\left.\begin{aligned} &L^0 V_\psi(t,x) + f'(t,\psi) \nabla V_\psi^{i-1}(t,x) = \\ &= \frac{1}{4}\sum_{j=0}^{i-1} \nabla V_\psi^{j\prime}(t,x) B_1(t) \nabla V_\psi^{i-j}(t,x), \\ &V_\psi^i(T,x) = 0, \quad x = \psi(0). \end{aligned}\right\} \tag{4.6}$$

Solving (4.6) for $i = 1, \ldots, k$, we find the approximate optimal control v_k:

$$v_k(t,\psi) = -\frac{1}{2} N_1^{-1}(t) B'(t) \nabla [V_\psi^0(t,x) + \ldots + \varepsilon^k V_\psi^k(t,x)]. \tag{4.7}$$

Equation (4.6) implies that the representation

$$V^i(t,\psi) = \mathcal{E} \int_t^T S_i(t, y_s) ds \tag{4.8}$$

holds. Here

$$S_i(t,P) := f'(t,\psi) \nabla V_\psi^{i-1}(t,x) - \frac{1}{4}\sum_{j=0}^{i-1} \nabla V_\psi^{j\prime}(t,x) B_1(t) \nabla V_\psi^{i-j}(t,x),$$

$$x := \psi(0).$$

The process $s \mapsto y(s)$ satisfies the linear stochastic equation

$$\left.\begin{aligned} &dy(s) = -B_1(s)P(s)y(s)ds + \sigma(s)d\xi(s), \quad t \leq s \leq T, \\ &y_t = \psi \in H[-h, 0]. \end{aligned}\right\} \tag{4.9}$$

If $\sigma\sigma'$ is nondegenerate, then $s \mapsto y(s)$ is the Gaussian process with expectation $m(s)$ and convariance matrix $D(s)$, where

$$\left.\begin{aligned} &m(s) := z(s,t)x, \quad x = \psi(0), \\ &D(s) := \int_t^s z(s,\tau)\sigma(\tau)\sigma'(\tau)z'(s,\tau)d\tau. \end{aligned}\right\} \tag{4.10}$$

Here, the Cauchy matrix $z(s,t)$ is defined by

$$\frac{\partial z(s,t)}{\partial s} = -B_1(s)P(s)z(s,t), \quad z(t,t) = I,$$

where I is the identity matrix.

In certain cases the righthand side of (4.8) may be reduced by quadratures. E.g., suppose $f(t,x_t) = f(t,x(t-h))$. Then

$$V^1(t,\psi) = 2\int_t^{t+h} ds \int_{\mathbf{R}^n} f'(s,\psi(s-t-h))P(s)yp(t,\psi(0),s,y)dy +$$

$$+2\int_{t+h}^{T} ds \int_{\mathbf{R}^n} dy \int_{\mathbf{R}^n} f'(s,y^1)P(s)yp(t,\psi(0),s-h,y^1)p(s-h,y^1,s,y)dy^1,$$

$$0 \le t \le T-h,$$

$$V^1(t,\psi) = 2\int_t^{T} ds \int_{\mathbf{R}^n} f'(s,\psi(s-t-h))P(s)yp(t,\psi(0),s,y)dy,$$

$$T-h \le t \le T.$$

Here, the transition function p of the process (4.9) is

$$p(t,x,s,y) = [(2\pi)^n\Delta(s)]^{-1/2}\exp\left[-\frac{1}{2}(y-m(s))'D^{-1}(s)(x-m(s))\right],$$

where $m(s), D(s)$ are given by (4.10), and $\Delta(s) := \det D(s)$.

We will finally consider an auxiliary problem for which the control (4.7) is optimal. Eq. (3.6) implies that the functionals $(t,\psi) \mapsto Q_k(t,\psi;\varepsilon) = V^0 + \ldots + \varepsilon^k V^k$ satisfy

$$L^0 Q_{k\psi}(t,x) + \varepsilon f'(t,\psi)\nabla Q_{k\psi}(t,x) + x'N_2(t)x + \delta_k(t,\psi;\varepsilon) =$$

$$= \frac{1}{4}\nabla Q'_{k\psi}(t,x)B_1(t)\nabla Q_{k\psi}(t,x),$$

$$Q_k(T,\psi) = x'N_0x, \quad x = \psi(0),$$

where

$$\delta_k(t,\psi;\varepsilon) :=$$

$$= \left[\frac{1}{4}\sum_{j=1}^{k}\sum_{i=0}^{j-1}\varepsilon^i \nabla V_\psi^{j\prime}(t,x)B_1(t)\nabla V_\psi^{k+1+i-j}(t,x) - f'(t,x)\nabla V_\psi^k(t,x)\right]\varepsilon^{k+1}.$$

Suppose we are given Eq. (3.8). The cost functional to be minimized is taken to be $u \mapsto J_k(u;\varepsilon)$, which is according to (3.7) defined by

$$J_k(u;\varepsilon) = J(u) + \mathcal{E}\int_{t_0}^{T} \delta_k(t, x_t; \varepsilon)dt \to \inf_u . \tag{4.11}$$

We assume that

$$|\nabla V_\psi^i(t,x)|^2 \leq C(1+\|\psi\|^2), \quad i = 2,\dots,k.$$

Then the control v_k in (4.7) and the functional Q_k are the optimal control and Bellman functional for the Pr. (3.8), (4.11); moreover, $0 \leq J(v_k) - J(u) = O(\varepsilon^{k+1})$. The details of the proof are the same as for the zeroth approximation, and can be found in [292.8].

CHAPTER 16

STATE ESTIMATES OF SYSTEMS WITH DELAY

In this chapter we will consider filtering problems for Gaussian unobservable processes and linear observations with delays. We will investigate the dependence of optimal estimates on the delays. We use a least–squares method to estimate the coordinates of the system at each time moment, based on input data and corresponding measurements with noise.

1. FILTERING OF GAUSSIAN PROCESSES

1.1. *PROBLEM STATEMENT*

We are given a basic probability space $\{\Omega, \sigma, P\}$, a family of σ–algebras $\{f_t\}, t \geq t_0$, and an f_t–adapted stochastic process $t \mapsto (x(t), y(t))$, where $t \mapsto x(t) \in \mathbf{R}^n$ is the Gaussian unobservable component and $t \mapsto y(t) \in \mathbf{R}^l$ the observable component, with bounded second moments. The filtering problem is to find the optimal (in the mean square sense) estimate $m_0(T)$ for a value $x(T)$ under observations $y(t), t_0 \leq t \leq T$. It is well known that this optimal estimate is the conditional expectation

$$m_0(T) = \mathcal{E}(x(T)|f_T^y), \tag{1.1}$$

where f_t^y is the minimal σ–algebra generated by the observations $y(t), t_0 \leq t \leq T$. We assume that y satisfies the equation

$$\left.\begin{aligned} dy(t) &= A(t)x(t-h)dt + d\xi(t), \quad t_0 \leq t \leq T, \\ y(t_0) &= 0, \quad x(s) = 0, \quad t_0 - h \leq s \leq t_0. \end{aligned}\right\} \tag{1.2}$$

Here, $t \mapsto \xi(t) \in \mathbf{R}^l$ is a Gaussian stochastic process with independent increments,

$$\left.\begin{aligned} &\mathcal{E}\xi(t) = 0, \quad \mathcal{E}x(t) = 0, \quad \mathcal{E}(x(t)x'(s)) = R(t,s), \\ &\mathcal{E}(x(t)d\xi'(s)) = Q(t,s)ds, \quad \mathcal{E}(d\xi(s)d\xi'(s)) = N(s)ds. \end{aligned}\right\} \tag{1.3}$$

The matrices A, R, Q, N are piecewise continuous, $Q(t,s) \equiv 0$ for $t < s$, and $N(s)$ is uniformly positive definite with respect to s.

The delay h in the observation equation (1.2) is due to the finiteness of the time necessary to perform the observation process and process its results.

1.2. *INTEGRAL REPRESENTATION FOR THE OPTIMAL ESTIMATE*

Lemma 1.1 There is a deterministic (non–stochastic) matrix function u : $[t_0, T] \to \mathbf{R}^{n\times l}$ such that under the conditions of Subs. 1.1 we find for the function (1.1)

$$m_0(T) = \int_{t_0}^{T} u_0(t)dy(t), \quad \int_{t_0}^{T} \| u_0(t) \|^2 \, dt < \infty. \tag{1.4}$$

Proof. Without loss of generality we may assume that $t_0 = 0$. Let

$$t_k^{(i)} = 2^{-i}kT, \quad k = 0, 1, \ldots, 2^i, \quad i = 1, 2, \ldots.$$

Let $f_{t,i}^y$ be the minimal σ–algebra generated by the quantities $y(t_0^{(i)}), y(t_1^{(i)}) - y(t_0^{(i)}), \ldots, y(t_{2^i}^{(i)} - y(t_{2^i-1}^{(i)})$. Then, by [292.5]

$$\lim_{i\to\infty} \mathcal{E}|\mathcal{E}(x(T)|f_{t,i}^y) - \mathcal{E}(x(T)|f_t^y)|^2 = 0. \tag{1.5}$$

Moreover, for any integer $i = 1, 2, \ldots$ we find

$$\mathcal{E}(x(t)|f_{t,i}^y) = \int_0^T u_i(t)dy(t), \tag{1.6}$$

where $\{u_i\}$ is a sequence of deterministic piecewise constant functions. We will prove that $\{u_i\}$ is a fundamental sequence in $L_2[0, T]$. Put

$$\left.\begin{aligned}
x_0(t) &:= \int_0^t \alpha(t,s)d\xi(s), \quad \alpha(t,s) := Q(t,s)N^{-1}(s), \\
\beta_i(\tau) &:= u_i(\tau) + \int_{\tau+h}^{T} u_i(t)A(t)\alpha(t-h,\tau)dt, \\
\Delta_{ij}u(t) &:= u_i(t) - u_j(t), \quad \Delta_{ij}\beta(t) := \beta_i(t) - \beta_j(t).
\end{aligned}\right\} \tag{1.7}$$

Then

$$\mathcal{E}([x(t) - x_0(t)]d\xi'(s)) = \mathcal{E}(x(t)d\xi'(s)) - \int_0^t \alpha(t,\tau)\mathcal{E}(d\xi(\tau)d\xi'(s)) =$$

$$= Q(t,s)ds - Q(t,s)N^{-1}(s)N(s)ds = 0.$$

Consequently,

$$\mathcal{E}|\mathcal{E}(x(T)|f_{t,i}^{y}) - \mathcal{E}(x(T)|f_{t,j}^{y})|^2 = \left|\int_0^T \Delta_{ij} u(t)A(t)[x(t-h) - x_0(t-h)]dt\right|^2 +$$

$$+\mathrm{Tr}\left[\int_0^T \Delta_{ij}\beta(t)N(t)\Delta_{ij}\beta'(t)dt\right]. \tag{1.8}$$

Relation (1.5) implies that the lefthand side of (1.8) tends to zero as $i, j \to \infty$. However, the righthand side of (1.8) consists of two nonnegative terms, and so both these terms tend to zero as $i, j \to \infty$. Since the matrix $N(t)$ is positive definite,

$$\lim_{i,j\to\infty} \int_0^T |\Delta_{ij}\beta(t)|^2 dt = 0.$$

Hence the sequence $\{\beta_i\}_{i=1}^{\infty}$ is fundamental in $L_2[0,T]$.

But the u_i are solutions of the linear Volterra equation (1.7), hence the values of a linear functional on β_i. So the sequence $\{u_i\}$ is also fundamental in $L_2[0,T]$. This implies the existence of a function $u_0 \in L_2[0,T]$ such that

$$\lim_{i\to\infty} \int_0^T |u_i(t) - u_0(t)|^2 dt = 0. \tag{1.9}$$

Now (1.4) follows from (1.5)–(1.6), (1.8)–(1.9), using the same arguments as in [292.5] □

1.3. *THE FUNDAMENTAL FILTERING EQUATION*

Let us determine the kernel u_0 of the optimal estimate (1.4).

Theorem 1.1 Under the assumptions of Subs. 1.1 the kernel u_0 is the unique solution of the integral equation

$$u_0(s)N(s) + \int_{t_0}^T u_0(t)Z(t,s)dt = P(s), \tag{1.10}$$

where

$$\left.\begin{aligned} P(s) &:= R(T, s-h)A'(s) + Q(T,s), \\ Z(t,s) &:= A(t)R(t-h, s-h)A'(s) + A(t)Q(t-h,s) + Q'(s-h,t)A'(s). \end{aligned}\right\} \tag{1.11}$$

Proof Let $m_0(T)$ be the mean–square optimal estimate of $x(T)$, and let $m(T)$ be an arbitrary f_T^y–adapted random variable for which $\mathcal{E}|m(T)|^2 < \infty$. Then

$$\mathcal{E}([x(T) - m_0(T)]m'(T)) = 0. \tag{1.12}$$

We assume that $m(T)$ has the form (1.4) with kernel $u(s)$. Then (1.12) implies

$$\int_{t_0}^{T} \left[\mathcal{E}(x(T)dy'(s)) - \int_{t_0}^{T} u_0(t)\mathcal{E}(dy(t)dy'(s)) \right] u'(s) = 0.$$

Taking into account that here $u(s)$ is arbitrary, we obtain an equation for u_0:

$$\mathcal{E}(x(T)dy'(s)) = \int_{t_0}^{T} u_0(t)\mathcal{E}(dy(t)dy'(s)). \tag{1.13}$$

Using (1.2)–(1.3) we find

$$\mathcal{E}(x(T)dy'(s)) = \mathcal{E}(x(T)x'(s-h))A'(s)ds + \mathcal{E}(x(T)d\xi'(s)) =$$

$$= [R(T, s-h)A'(s) + Q(T,s)]ds = P(s)ds,$$

$$\int_{t_0}^{T} u_0(t)\mathcal{E}(dy(t)dy'(s)) =$$

$$= \int_{t_0}^{T} u_0(t)A(t)\mathcal{E}(x(t-h)dy'(s))dt + \int_{t_0}^{T} u_0(t)\mathcal{E}(d\xi(t)dy'(s)) =$$

$$= \int_{t_0}^{T} u_0(t)A(t)[R(t-h, s-h)A'(s) + Q(t-h,s)]dsdt+$$

$$+ \int_{t_0}^{T} u_0(t)Q'(s-h,t)A'(s)dsdt + u_0(s)N(s)ds =$$

$$= \int_{t_0}^{T} u_0(t)[A(t)R(t-h, s-h)A'(s) + A(t)Q(t-h,s)+$$

$$+Q'(s-h,t)A'(s)]dtds + u_0(s)N(s)ds =$$

$$= \left[\int_{t_0}^{T} u_0(t)Z(t,s)dt + u_0(s)N(s)\right] ds.$$

This implies (1.10), (1.11), as well as the existence of a solution of (1.10).

Now we prove uniqueness. Let $s \mapsto \Delta u(s)$ be the difference between two solutions of (1.10). Then by (1.13) (which is equivalent to (1.10))

$$\int_{t_0}^{T} \Delta u(t)\mathcal{E}(dy(t)dy'(s)) = 0.$$

Multiplying both sides of this equation by $\Delta u'(s)$ and integrating with respect to $s \in [t_0, T]$, we find

$$S := \int_{t_0}^{T}\int_{t_0}^{T} \Delta y(t)\mathcal{E}(dy(t)dy'(s))\Delta u'(s) = 0.$$

Hence $\mathrm{Tr}S = 0$, i.e.

$$\mathcal{E}\left|\int_{t_0}^{T} \Delta u(t)dy(t)\right|^2 = 0.$$

We put

$$x_0(t) := \int_{t_0}^{t} \alpha(t,s)d\xi(s), \quad \alpha(t,s) := Q(t,s)N^{-1}(s),$$
$$\beta(s) := \Delta u(s) + \int_{s+h}^{T} \Delta u(t)A(t)\alpha(t-h,s)dt. \tag{1.14}$$

But

$$\mathcal{E}([x(t) - x_0(t)]d\xi'(s)) = Q(t,s)ds - \alpha(t,s)N(s)ds = 0.$$

Consequently,

$$\mathcal{E}\left|\int_{t_0}^{T} \Delta u(t)dy(t)\right|^2 =$$

$$= \mathcal{E}\left|\int_{t_0}^{T} \Delta u(t)[A(t)(x(t-h) - x_0(t-h))dt + A(t)x_0(t-h)dt + d\xi(t)]\right|^2 =$$

$$= \mathcal{E}\left| \int_{t_0+h}^{T} \Delta u(t)A(t)[x(t-h) - x_0(t-h)]dt + \right.$$

$$\left. + \int_{t_0}^{T} \Delta u(t)A(t) \int_{t_0}^{t-h} \alpha(t-h,s)d\xi(s)dt + \int_{t_0}^{T} \Delta u(s)d\xi(s) \right|^2 =$$

$$= \mathcal{E}\left| \int_{t_0+h}^{T} \Delta u(t)A(t)[x(t-h) - x_0(t-h)]dt + \int_{t_0}^{T} \beta(s)d\xi(s) \right|^2 =$$

$$= \mathcal{E}\left| \int_{t_0}^{T} \Delta u(t)A(t)[x(t-h) - x_0(t-h)]dt \right|^2 +$$

$$+ \mathrm{Tr} \int_{t_0}^{T} \beta(s)N(s)\beta'(s)ds = 0.$$

Since $N(s)$ is positive definite, we obtain $\beta(s) = 0$ almost everywhere on $[t_0, T]$. The relations (1.14) show that the equation $\beta(s) = 0$ is a linear homogeneous Volterra equation in Δu. Therefore $\Delta u(s) = 0$ almost everywhere. □

1.4. *DUAL OPTIMAL CONTROL PROBLEM*

We will show that the solution of the filtering problem can be obtained as the solution of an auxiliary optimal control problem.

Consider the controlled process $t \mapsto m_u(t)$ governed by the relation

$$m_u(t) := \int_{t_0}^{t} u(t,s)dy(s), \tag{1.15}$$

with cost functional to be minimized:

$$J(u) := \mathcal{E}\left[z_u'(T)N_0 z_u(T) + \int_{t_0}^{T} z_u'(s)N_1(s)z_u(s)ds \right], \tag{1.16}$$

$$z_u(s) := x(s) - m_u(s).$$

Here N_0 and the values of the piecewise continuous function N_1 are symmetric nonnegative definite matrices, x is a stochastic process, described by

(1.2), and u is a control used to minimized (1.16). The set of admissible controls consists of the measurable functions u such that

$$\sup_{t_0 \le t \le T} \int_{t_0}^{t} |u(t,s)|^2 ds < \infty, \quad u : [t_0, T] \times [t_0, T] \to \mathbf{R}^{n \times m}.$$

Let u_0 be an optimal control for (1.15), (1.16), and $u_\varepsilon := u_0 + \varepsilon u$, $\varepsilon > 0$, an admissible one. It is clear that the optimal control u_0 must satisfy the condition

$$J_0(u_0) := \limsup_{\varepsilon \to 0+} \frac{1}{\varepsilon}[J(u_\varepsilon) - J(u_0)] \ge 0.$$

Computing $J_0(u_0)$ we find

$$J_0(u_0) =$$

$$= -2\mathrm{Tr}\left(N_0 \int_{t_0}^{T} \left[\mathcal{E}(x(T)dy'(\tau)) - \int_{t_0}^{T} u_0(T,s)\mathcal{E}(dy(s)dy'(\tau)) \right] u(T,\tau) \right) -$$

$$-2\mathrm{Tr} \int_{t_0}^{T} \left(N_1(t) \int_{t_0}^{t} \left[\mathcal{E}(x(t)dy'(\tau)) - \int_{t_0}^{t} u_0(t,s)\mathcal{E}(dy(s)dy'(\tau)) \right] u(t,\tau) \right) dt. \tag{1.17}$$

For (1.17) to be nonnegative for any admissible control u, the expression in square brackets must vanish. Hence there follows, similarly to (1.10), an equation for u_0:

$$u_0(t,s)N(s) + \int_{t_0}^{t} u_0(t,\tau)Z(\tau,s)d\tau = P(t,s),$$

where

$$P(t,s) := R(t,s-h)A'(s) + Q(t,s),$$

and the matrix $Z(t,s)$ is defined by (1.11).

If the optimal control u_0 for the Pr. (1.15), (1.16) has been found, then the optimal estimate is defined by (1.4) with $u_0(t) = u_0(T,t)$. In particular, if $N_0 = I$ and $N_1 = 0$, then the error estimate is

$$J(u_0) = \mathrm{Tr}\left[R(T,T) - \int_{t_0}^{T} Q(T,s)N^{-1}(s)Q'(T,s)ds \right].$$

1.5. *PARTICULAR CASES*

In certain cases the method of optimal estimation described above can be simplified, and the solution can be obtained in analytical form. We will consider some of these cases.

1.5.1 Suppose $h \geq T - t_0$. Then

$$u_0(s) = Q(T,s)N^{-1}(s),$$

$$J(u_0) = \operatorname{Tr}\left[R(T,T) - \int_{t_0}^{T} Q(T,s)N^{-1}(s)Q'(T,s)ds\right].$$

1.5.2 Suppose $h \leq T - t_0$. We assume that $x(t) = \chi(t)x_0$, where χ is a piecewise continuous deterministic matrix function on $[t_0, T], \chi(t_0) = I$, the gaussian variable x_0 is independent of $\xi(t)$, and $\mathcal{E}x_0 = 0$, $\mathcal{E}(x_0 x_0') =: D_0$, with I the identity matrix. Then $Q(t,s) = 0$ and $R(t,s) = \chi(t)D_0\chi'(s)$. In this case $u_0(s)$ for $t_0 \leq s \leq T$ satisfies the equation

$$u_0(s) = \left[\chi(T) - \int_{t_0+h}^{T} u_0(t)A(t)\chi(t-h)dt\right] D_0\chi'(s-h)A'(s)N^{-1}(s). \tag{1.18}$$

We try to find a solution of this equation in the form

$$u_0(s) = \chi(T)F\chi'(s-h)A'(s)N^{-1}(s), \tag{1.19}$$

where the matrix F is to be determined. Substituting (1.19) into (1.18) we obtain

$$\left.\begin{aligned} F = (D_0^{-1} + G)^{-1},\ \ G &:= \int_{t_0}^{T-h} \chi'(t)A_1(t+h)\chi(t)dt, \\ A_1 &:= A'N^{-1}A. \end{aligned}\right\} \tag{1.20}$$

Therefore, under assumptions stated earlier, the optimal estimate is defined by (1.4), (1.19), (1.20). By (1.19), (1.20) the error of this estimate is

$$J(u_0) = \mathrm{Tr}\Bigg[\chi(T)D_0\chi'(T)+$$

$$+\int_{t_0+h}^{T}\chi(T)F\chi'(s-h)A'(s)N^{-1}(s)A(s)\chi(s-h)F\chi'(T)ds-$$

$$-2\int_{t_0+h}^{T}\chi(T)D_0\chi'(s-h)A'(s)N^{-1}(s)A(s)\chi(s-h)F\chi'(T)ds+$$

$$+\int_{t_0+h}^{T}\int_{t_0+h}^{T}\chi(T)F\chi'(t-h)A'(t)N^{-1}(t)A(t)\chi(t-h)\times$$

$$\times D_0\chi'(s-h)A'(s)N^{-1}(s)A(s)\chi(s-h)F\chi'(T)dsdt\Bigg] =$$

$$= \mathrm{Tr}[\chi(T)(D_0 + FGF - 2D_0GF + FGD_0GF)\chi'(T)].$$

But $F = D_0(I - GF) = F' = D_0 - FGD_0$, so

$$D_0 + FGF - 2D_0GF + FGD_0GF = D_0 + (F - D_0)GF-$$

$$-D_0GF + FGD_0GF = D_0 - FGD_0GF - D_0GF + FGD_0GF = F.$$

Hence the error of the optimal estimate is

$$J(u_0) = \mathrm{Tr}[\chi(T)F\chi'(T)]. \tag{1.21}$$

1.6. *DEPENDENCE OF THE ERROR OF THE OPTIMAL ESTIMATE ON THE DELAY*

Here we will study the interesting problem of the dependence of the error of the optimal estimate on the value of the delay h obtained in observations. Since $x(s) = 0$ for $s < t_0$, and $x(t)$ and $\xi(t)$ are independent, the observations (1.2) on the interval $[t_0, t_0 + h]$ do not contain information on the process x, and the general observation time for x is equal to $T - t_0 - h$, which decreases as h grows. Hence it is natural to assume that, as a function of h, the error of the estimate is nondecreasing as $h \in [0, T - t_0]$ increases, and is constant for $h \geq T - t_0$.

In particular, if A_1 is a constant matrix, then this assumption holds, since (1.21) means that

$$\frac{d}{dh} J(u_0) = \mathrm{Tr}\left[\chi(T)\left(\frac{dF}{dh}\right)\chi'(T)\right].$$

But

$$\frac{dF}{dh} = -F\left(\frac{d}{dh}F^{-1}\right)F = -F\frac{dG}{dh} = F\chi'(T-h)A_1\chi(T-h)F.$$

Consequently,

$$\frac{d}{dh} J(u_0) = \mathrm{Tr}[\chi(T)F\chi'(T-h)A_1\chi(T-h)F\chi'(T)] \geq 0.$$

Thus, in this case the function $J(u_0)$, regarded as a function of h, does not decrease.

However, there are situations in which increase of h leads to decrease of $J(u_0)$. As an example we will consider a scalar process. Suppose the unobservable process $t \mapsto x(t)$ and the observations $y(t)$ are scalar, and such that

$$\left.\begin{aligned} &x(t) = \chi(t)x_0, \quad \chi(t_0) = 1, \quad x(s) = 0, \quad s \in [t_0 - h, t_0), \\ &dy(t) = A(t)x(t-h)dt + d\xi(t), \ t_0 \leq t \leq T, \quad y(t_0) = 0. \end{aligned}\right\} \tag{1.22}$$

Here, x_0 is a gaussian variable with expectation zero and variance σ_0^2, independent of the gaussian process $t \mapsto \xi(t)$, which has independent increments, $\mathcal{E}\xi(t) = 0$, $\mathcal{E}(d\xi(t))^2 = \sigma^2(t)dt$, $\sigma(t) > 0$. In this case the solution of the filtering problem is

$$m(T) = \int_{t_0+h}^{T} u_0(s)dy(s), \quad u_0(s) = \chi(T)F(h)\chi(s-h)A(s)\sigma^{-2}(s),$$

$$F(h) = [\sigma_0^{-2} + G(h)]^{-1}, \quad G(h) = \int_{t_0}^{T-h} \chi^2(t)A_1(t+h)dt,$$

$$A_1(t) = A^2(t)\sigma^{-2}(t).$$

The error of the estimate equals

$$J(h) = \chi^2(T)F(h).$$

So, minimization of $J(h)$ with respect to h is equivalent to minimization of the function $h \mapsto F(h)$, or maximization of $G(h), h \in [0, T - h_0]$. Note that

$$\max G(h) = G(0), \quad 0 \leq h \leq T - t_0,$$

if $A_1(t) = \text{const}$ or if $x(t) \equiv x_0$, i.e. $\chi(t) \equiv 1$. The interior extremal points of G satisfy the equation

$$\frac{d}{dt}G(h) = -\chi^2(T-h)A_1(t) + \int_{t_0}^{T-h} \chi^2(t)dA_1(t+h) =$$

$$= -\chi^2(T-h)A_1(T) + \chi^2(t)A_1(t+h)|_{t_0}^{T-h} - \int_{t_0}^{T-h} A_1(t+h)d\chi^2(t) =$$

$$= -A_1(t) - \int_{t_0}^{T-h} A_1(t+h)d\chi^2(t) = 0. \tag{1.23}$$

We will give some examples in which (1.23) has a unique solution $h_0 > 0$, defining an interior maximum of G, i.e. an interior minimum of J.

Example 1.1 Let $t_0 = 0, \chi(t) = e^{-\alpha t}, A(t) = e^{\beta t}, \sigma^2(t) = \sigma_0^2 = 1, \beta \geq \alpha > 1/(2T)$. Then (1.23) takes the form

$$e^{2\beta h} = 2\alpha \int_0^{T-h} e^{2\beta(t+h)-2\alpha t}dt.$$

By solving this equation we find for h_0:

$$h_0 = T - \frac{1}{2\alpha} g\left(\frac{\beta}{\alpha}\right),$$

$$g(r) := \begin{cases} (r-1)^{-1}\ln r, & r > 1, \\ 1, & r = 1. \end{cases}$$

The corresponding minimum value of the error is

$$J(h_0) = e^{-2\alpha T}[1 + G(h_0)]^{-1},$$

$$G(h) := \frac{e^{2\beta h}}{2(\beta - \alpha)}[e^{2(\beta-\alpha)(T-h)} - 1].$$

Fig. 16.1.1 shows the graph of the function $h \mapsto J(h)$ for $T = 1$ and various values of α, β:

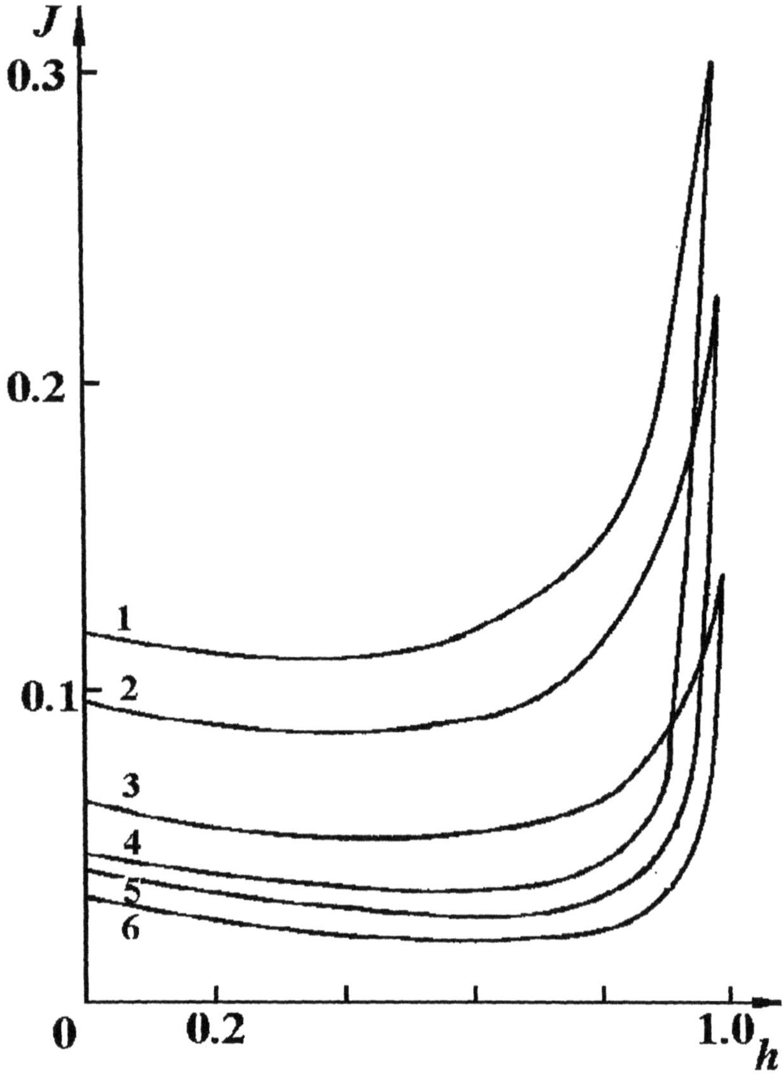

Figure 16.1.1. Dependence of the error of the optimal estimate on the delay h obtained by observation in example 1.1

1) $\alpha = 0.6, \beta = 1, h_0 = 0.361$;
2) $\alpha = 0.75, \beta = 1, h_0 = 0.425$;
3) $\alpha = \beta = 1, h_0 = 0.5$;
4) $\alpha = 0.6, \beta = 2, h_0 = 0.570$;
5) $\alpha = 0.75, \beta = 2, h_0 = 0.608$;
6) $\alpha = 1, \beta = 2, h_0 = 0.653$. □

Example 1.2 Consider the filtering problem (1.22) with $t_0 = 0, \chi(t) = \sqrt{1 - \alpha t/T}, A(t) = \sqrt{t}, \sigma^2(t) = \sigma_0^2 = 1, \alpha < 1$. Then (1.23) takes the form

$$h + \int_0^{T-h} (t+h)\left(-\frac{\alpha}{T}\right) dt = 0.$$

Hence

$$\frac{Th}{\alpha} = \int_0^{T-h} (t+h) dt = \frac{T^2 - h^2}{2}.$$

By solving this equation we find

$$h_0 = T\left(\sqrt{1 + \frac{1}{\alpha^2}} - \frac{1}{\alpha}\right),$$

which correspond to the minimum of the function

$$J(h) = (1-\alpha)[1 + G(h)]^{-1},$$

$$G(h) := \int_0^{T-h} \left(1 - \alpha\frac{t}{T}\right)(t+h) dt =$$

$$= (T-h)\left[h + \frac{T-h}{2}\left(1 - \frac{\alpha h}{T}\right) - \frac{\alpha}{3T}(T-h)^2\right].$$

Fig. 16.1.2 gives the graph of the function $h \mapsto J(h)$ for $T = 1$ and
1) $\alpha = 0.1, h_0 = 0.0498$;
2) $\alpha = 0.3, h_0 = 0.147$;
3) $\alpha = 0.5, h_0 = 0.236$;
4) $\alpha = 0.7, h_0 = 0.315$;
5) $\alpha = 0.9, h_0 = 0.384$.□

Example 1.3 Consider the filtering problem (1.22) with $t_0 = 0, A(t) = \sqrt{t}, \sigma^2(t) = \sigma_0^2 = 1, \chi(t) = 1$ for $t \leq \alpha T, \chi(t) = \sqrt{\beta}$ for $t > \alpha T$, and with $\alpha \leq \beta < 1$. For $h > T(1-\alpha)$ Eq. (1.23) does not have solutions.

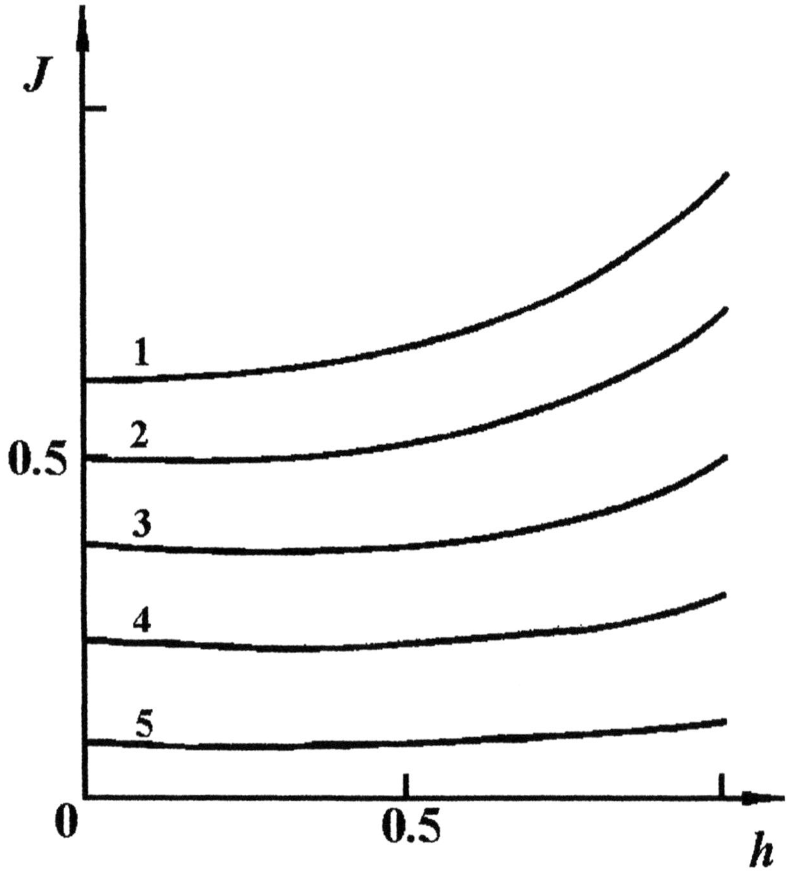

Figure 16.1.2. Influence of various parameters on the error in example 1.2.

Let $h \le T(1-\alpha)$. Then (1.23) becomes $h = (1-\beta)(\alpha T + h)$. Hence $h_0 = (1-\beta)\alpha T\beta^{-1}$ is the minimum of $J(h) = \beta[1+G(h)]^{-1}$, where

$$2G(t) := \begin{cases} \beta T^2 - h^2 + (1-\beta)(\alpha T + h)^2, & h \le T(1-\alpha), \\ T^2 - h^2, & h > T(1-\alpha). \end{cases}$$

Fig. 16.1.3 gives the graph of $J(h)$ for $T = 1$ and
1) $\alpha = 0, \beta = 0.9, h_0 = 0$;
2) $\alpha = 0.5, \beta = 0.8, h_0 = 0.125$;
3) $\alpha = 0.4, \beta = 0.6, h_0 = 0.267$;
4) $\alpha = 0.2, \beta = 0.5, h_0 = 0.2$;
5) $\alpha = \beta = 0.5, h_0 = 0.5$;
6) $\alpha = 0.1, \beta = 0.3, h_0 = 0.233$;
7) $\alpha = \beta = 0.2, h_0 = 0.8$.□

1.7. *SOME GENERALIZATIONS*

1.7.1. Suppose that the observation equations involve many delays, i.e.

$$dy(t) = \sum_{i=1}^{r} A_i(t)x(t-h_i)dt + d\xi(t), \quad y(t_0) = 0, \tag{1.24}$$

where $h_i \ge 0$, the processes $t \mapsto x(t), \xi(t)$ are as in Subs. 1.1, and the matrix functions A_i are piecewise continuous.

Then the optimal estimate is given by (1.4), where the matrix function u_0 is the solution of (1.10) with

$$P(s) = \sum_{i=1}^{r} R(T, s-h_i)A_i(s) + Q(T,s),$$

$$Z(t,s) = \sum_{i,j=1}^{r} A_i(t)R(t-h_i, s-h_i)A_j'(s)+$$

$$+\sum_{i=1}^{r} A_i(t)Q(t-h_i, s) + \sum_{j=1}^{r} Q'(s-h_j, t)A_j'(s).$$

1.7.2 One can show that smoothing and interpolation problems can be reduced to filtering problems. As an example we consider a smoothing problem. The process $t \mapsto (x(t), y(t))$ is supposed to be as in Subs. 1.1. The smoothing problem consists of constructing a mean–square optimal estimate $x(T_1), T_1 > T$, under the observation (1.2) on the interval $t_0 \le t \le T$. This problem can be reduced to the filtering problem. In fact, putting

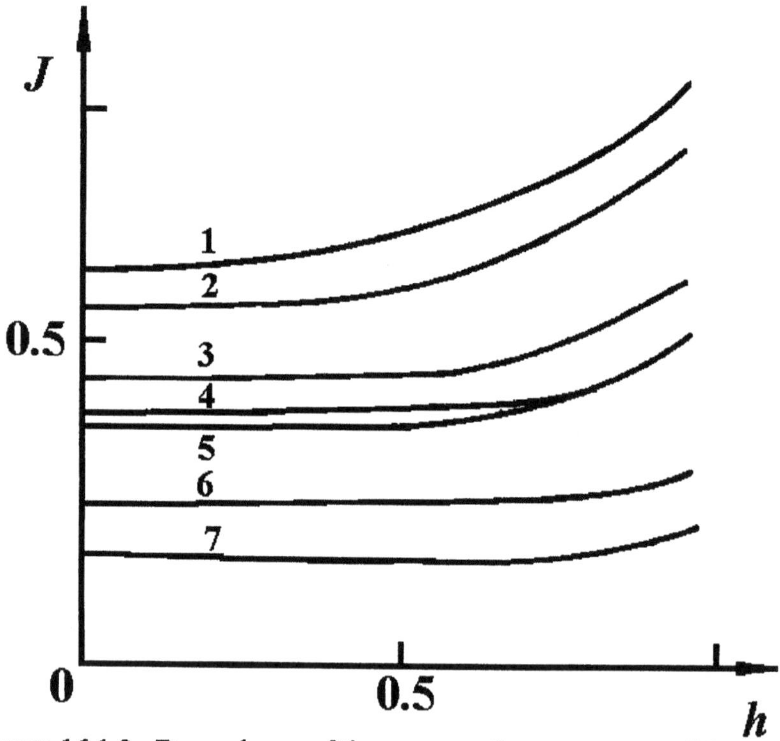

Figure 16.1.3. Dependence of the error on the parameters of the filtering problem in example 1.3.

$\overline{A}(t) = A(t)$ for $t_0 \le t \le T, \overline{A}(t) = 0$ for $t > T$, and $\overline{Q}(t,s) = Q(t,s)$ for $t_0 \le s \le t \le T, \overline{Q}(t,s) = 0$ otherwise. Consider the auxiliary filtering problem of estimating $x(T_1)$ under the observations (1.2) on $[t_0, T_1]$, with $A(t)$ replaced by $\overline{A}(t)$ and $\mathcal{E}(x(t)d\xi'(s)) = \overline{Q}(t,s)ds$. For $t \ge s \ge T$ the processes x and ξ are uncorrelated and gaussian, and therefore are independent. Hence the solution of the auxiliary filtering problem is at the same time the solution of the initial smoothing problem. Therefore the solution of the smoothing problem is given by

$$m_0(T_1) = \int_{t_0}^{T} u_0(t)dy(t).$$

Here u_0 is the solution of (1.10) with $T_1, \overline{A}, \overline{Q}$ instead of T, A, Q.

2. FILTERING OF SOLUTIONS OF ITÔ's EQUATIONS WITH DELAY

In this section we will consider the form of the filtering equations for the particular case when the unobservable process $t \mapsto x(t)$ is described by a linear stochastic equation. We confine ourselves to stating the problem and the resulting filtering equations. The reader is referred to [292.1] for a comprehensive coverage of the subject.

2.1. *PROBLEM STATEMENT*

The unobservable process $t \mapsto x(t)$ is such that

$$\left.\begin{aligned} &dx(t) = \sum_{i=1}^{k} g_i(t)x(t-\tau_i)dt + \sigma_1(t)d\xi_1(t), \quad t_0 \le t \le T, \\ &0 \le \tau_1 < \ldots < \tau_k, \quad x(s) = 0, \quad s < t_0, \quad s(t_0) = x_0. \end{aligned}\right\} \tag{2.1}$$

The observation process $t \mapsto y(t)$ satisfies (1.24), i.e.

$$dy(t) = \sum_{i=1}^{r} A_i(t)x(t-h_i)dt + d\xi(t), \quad y(t_0) = 0.$$

The processes ξ_1, ξ, x are mutually independent, ξ_1 is the standard Wiener process, x_0 is a gaussian vector, $\mathcal{E}x_0 = 0, \mathcal{E}(x_0x_0') = D_0$, the matrix functions g_i and σ_1 are piecewise continuous. The problem is to find the optimal mean–square estimate $m_0(T)$ for $x(T)$ under the observations $y(t), t_0 \le t \le T$.

2.2. *DUAL CONTROL PROBLEM*

The optimal estimate of the vector $x(T)$ under the observations $y(t), t_0 \leq t \leq T$, is defined by (1.4), where the matrix function u_0 is the optimal control of the deterministic system

$$\left.\begin{aligned} \dot{\alpha}(t) &= -\sum_{i=1}^{k} g_i'(t+\tau_i)\alpha(t+\tau_i) + \sum_{j=1}^{r} A_j'(t+h_j)u(t+h_j), \\ \alpha(T) &= I, \quad \alpha(s) = 0, \quad s > T. \end{aligned}\right\} \tag{2.2}$$

Here, $t \mapsto \alpha(t) \in \mathbf{R}^{n \times n}$ is a square matrix function. The cost functional is

$$J(u) = \operatorname{Tr}\left[\alpha'(t_0)D_0\alpha(t_0) + \int_{t_0}^{T} (\alpha'(t)N_1(t)\alpha(t) + u'(t)N(t)u(t))dt\right] \to \min_u, \quad N_1 := \sigma_1\sigma_1'. \tag{2.3}$$

Linear quadratic problems of the type (2.2), (2.3) were investigated in Ch. 14, where it has been proved that the solution of (2.2), (2.3) can be reduced to the study of a boundary value problem for nonlinear differential equations. Here we will illustrate the application of the results in Ch. 14 for $k = r = 1, h_1 = h < T - t_0$. In this case the optimal control $u_0(t)$ is

$$u_0(t) = \left\{\begin{array}{ll} 0, & t_0 \leq t \leq t_0 + h, \\ N^{-1}(t)A_1(t)[P(T-h)\alpha(t-h) + \int_{-\tau_1}^{0} Q_1(T-t,\tau)\alpha(t-\tau-h)d\tau], & \\ & t_0 + h < t \leq T. \end{array}\right\} \tag{2.4}$$

The covariance matrix $D(T)$ is equal to

$$D(T) = \mathcal{E}([x(T) - m(T)][x(T) - m(T)]') = P(T). \tag{2.5}$$

The matrices $P(t), Q_1(t,\tau)$ and $R_1(t,\tau,\rho)$ are the solutions of the equations

$$\left.\begin{aligned} &\dot{P}(t) + P(t)A_0(t+h)P(t) - Q_1(t,0) - Q_1'(t,0) - N_1(t) = 0, \\ &\left(\frac{\partial}{\partial t} + \frac{\partial}{\partial \tau}\right) Q_1(t,\tau) + P(t)A_0(t+h)Q_1(t,\tau) - R_1(t,0,\tau) = 0, \\ &\left(\frac{\partial}{\partial t} + \frac{\partial}{\partial \tau} + \frac{\partial}{\partial \rho}\right) R_1(t,\tau,\rho) + Q_1'(t,\tau)A_0(t+\tau)Q_1(t,\rho) = 0, \end{aligned}\right\} \tag{2.6}$$

$$t_0 \le t \le T, \quad -\tau_1 \le \tau, \rho \le 0, \quad A_0 := A_1' N^{-1} A_1, \quad A_0(s) \equiv 0, \quad s > T.$$

The boundary conditions for (2.6) are

$$\left.\begin{aligned} P(t_0) = D_0, \quad Q_1(t_0, \tau) = R_1(t_0, \tau, \rho) = 0, \\ Q_1'(t, -\tau_1) = g(t + \tau_1) P(t), \\ R_1(t, -\tau_1, \tau) + R_1'(t, \tau, -\tau_1) = 2A(t + \tau_1) Q_1(t, T_1), \\ t_0 \le t \le T, \quad -\tau_1 \le \tau, \rho \le 0. \end{aligned}\right\} \quad (2.7)$$

Hence the construction of the optimal estimate $m(T)$ and covariance matrix $D(T)$ consists of the following steps:

1) solve the boundary value problem (2.6), (2.7), and determine the matrices $P(t), Q_1(t, \tau)$;

2) find the solution α of (2.2), with control $t \mapsto u_0(t)$ given by (2.4);

3) substitute the matrices P, Q_1, α in (2.4), and determine the kernel $u_0(t)$ of the optimal estimate (1.4) and error (2.5).

In certain cases this algorithm can be simplified. We will consider some such cases.

2.2.1 If $k = r = 1, \sigma_1 = 0$, then by Cauchy's formula (Subs. 3.5.3) the solution of Pr. (2.1) can be represented as $x(t) = \chi(t) x_0$. In this case the solution of the filtering problem has been obtained in Subs. 1.5. Moreover, for $\sigma_1 = 0$ the exact analytical solution of (2.6), (2.7) is given by (14.2.9).

2.2.2 Suppose $k = r = 1, \tau_1 = 0$, and (2.1) holds for all $t \ge t_0 - h$. Then we obtain for the covariance matrix $D(t)$ the equation

$$\dot{D} - DA' - AD = -Dz_1 D + Dz_1 \overline{\sigma}_1 + \overline{\sigma}_1 z_1 D + \sigma_1 \sigma_1 - \overline{\sigma}_1 z_1 \overline{\sigma}_1,$$

$$D(t_0) = D_0.$$

Here, the matrix functions $z_1, \overline{\sigma}_1$ are given by

$$z_1(t) = z'(t - h, t) A_1'(t) N_1^{-1}(t) A_1(t) z(t - h, t),$$

$$\overline{\sigma}_1(t) = \int_{t-h}^{t} z(t, s) N_1(s) z'(t, s) ds,$$

where the matrix function z is defined by

$$\frac{\partial z(t, s)}{\partial t} = g_1(t) z(t, s), \quad z(s, s) = I.$$

For the optimal estimate $m(t)$ we have the equation

$$dm(t) - g_1(t)m(t)dt = \left[D(t) - \int_{t-h}^{t} \sigma_1(s)ds\right] z'(t,t-h)A_1'(t)N^{-1}(t)\times$$

$$\times[dy(t) - A_1(t)z(t-h,t)m(t)dt],$$

$$m(t_0) = m_0.$$

Remark 2.1 The results in this section have been obtained by reducing the initial filtering problem to the dual optimal control problem. A similar procedure has been applied in [292.8] for solving a filtering problem under both continuous and discrete observations. The same filtering algorithms can also be obtained by the maximum likelihood method. The latter was used in [13] to evaluate the parameters of equation (2.5.2), which describes the Volterra population model in a stochastic environment. □

3. MINIMAX FILTERING IN SYSTEMS WITH DELAY

Minimax estimation problems arise if we know only domains to which the perturbations can belong. The exact solution of minimax estimation problems can be obtained quite rarely. Because of this it is advisable to do away with the search of the optimal estimator. Moreover, keeping in mind a simplicity of the process, very easily to use ready algorithms, etc. it is often sufficient to find a simplified solution which is not far from the optimal one.

One such simplified solution is proposed here. It is based on the interpretation of all perturbations as standard Wiener process and using the procedure from Sec. 2 as an approximate solution for the original problem. Detailed exposition of minimax filtering problems is given in the book [364].

3.1. *STATEMENT OF THE PROBLEM*

We shall treat the linear dynamical system with delay

$$\dot{x}(t) = A(t)x(t) + B(t)x(t-\overline{h}) + C(t)u(t), \quad t \in [0,T],$$

$$x(\tau) = 0, \quad -\overline{h} \le \tau < 0, \quad \overline{h} = \text{const}, \quad \overline{h} < T$$

where $x(t) = (x_1(t),\dots,x_n(t))' \in \mathbf{R}^n$ is the system state vector, $A(t), B(t) \in \mathbf{R}^{n\times n}, C(t) \in \mathbf{R}^{n\times r}, u(t) = (u_1(t),\dots,u_r(t))' \in \mathbf{R}^r$ is a disturbance vector, T is a prescribed moment.

We assume

$$|u_j(t)| \le \gamma_j(t), \quad j = 1, \dots, r$$

where $\gamma_j(t)$ are given positive functions.

Let the measurements of the state vector be available on $[0, T]$:

$$z(t) = H'(t)x(t-h) + \rho(t), \quad z(t), \rho(t) \in \mathbf{R}^m,$$

$$H(t) \in \mathbf{R}^{n \times m}, \quad h = \text{const}, \quad h < T.$$

It is supposed that the measurement errors $\rho(t) = (\rho_1(t), \dots, \rho_m(t))'$ satisfy the restrictions

$$|\rho_l(t)| \le \sigma_l(t), \quad t \in [0, T], \quad l = 1, \dots, m$$

where $\sigma_l(t), l = 1, \dots, m$ are known positive functions.

Besides, we assume that the initial state belongs to a given parallelepiped:

$$|x_d(0)| \le \sigma_{-d}, \quad d = 1, \dots, n$$

where $\sigma_{-d} > 0, d = 1, \dots, n$ are known.

The entries of the matrices $A(t), B(t), C(t), H(t)$, and the functions $u_j(t), \gamma_j(t), \rho_l(t), \sigma_l(t)$ are assumed to be piecewise–continuous. Further we introduce the notation $x(0) \in \Pi, u \in U, \rho \in \sigma$ for the elements of the sets defined by inequalities mentioned above.

It is required to find an estimate for a scalar value $Lx = a'x(T)$, where $a \in \mathbf{R}^n$ is a given vector, by linear functional

$$\hat{L}z = \int_0^T \Phi'(t)z(t)dt, \quad \Phi(t) = (\Phi_1(t), \dots, \Phi_m(t))' \in \mathbf{R}^m,$$

$$\Phi(t) = \Phi_{pc}(t) + \sum_{s=1}^{M} \overline{\Phi}^s \delta(t - t_s). \tag{3.1}$$

Here $\Phi_{pc}(t)$ is a piecewise–continuous function, $\overline{\Phi}^s \in \mathbf{R}^m$, $t_s \in [0, T]$, $s = 1, \dots, M, M$ is an arbitrary positive integer, $\delta(t)$ is the Dirac delta–function. We shall denote the set (3.1) as $\mathcal{F}$.

The optimal guaranteed estimation problem is to find the estimator Φ^0 that minimizes the guaranteed estimation error

$$D(\Phi^0) = D = \inf_{\Phi \in \mathcal{F}} D(\Phi), \quad D(\Phi) := \sup_{x(0) \in \Pi, u \in U, \rho \in \sigma} |\hat{L}(z) - L(x)|.$$

Remark 3.1 It can be shown that the use of nonlinear estimates does not improve the estimation quality (see, e.g., [364] for the case $h = \overline{h} = 0$).

In order to construct a mathematical programming problem that corresponds to the guaranteed estimation problem let us introduce the following quantities:

1) the function $\xi_\Phi(t)$: with a given $\Phi(t), \Phi(\tau) = 0,\ T < \tau \le T + h$

$$\dot{\xi}_\Phi(t) = -A'(t)\xi_\Phi(t) - B'(t+\overline{h})\xi_\Phi(t+\overline{h}) + H(t+h)\Phi(t+h),\ 0 \le t \le T,$$

$$\xi_\Phi(T) = a,\ \xi_\Phi(\tau) = 0,\ B(\tau) = 0,\ H(\tau) = 0,\ \tau > T;$$

2) the functional $I(\Phi_-, \Phi, w) : \mathbf{R}^n \times \mathcal{F} \times L_1^r \to \mathbf{R}$,

$$I(\Phi_-, \Phi, w) := \sum_{d=1}^{n} \sigma_{-d}|\Phi_{-d}| + \sum_{l=1}^{m} \int_h^T \sigma_l(t)|\Phi_l(t)|dt + \sum_{j=1}^{r} \int_0^{T-h} \gamma_j(t)|w_j(t)|dt \tag{3.2}$$

and the number

$$\kappa := \sum_{j=1}^{r} \int_{T-h}^{T} \gamma_j(t)|C_j'(t)\xi_0(t)|dt,\quad C(t) = (C_1(t), \dots, C_r(t))'.$$

One can verify that the problem to find an optimal estimate of L reduces to the solving of the following nonsmooth mathematical programming problem:

$$I_0 := \inf_{\Phi_- \in \mathbf{R}^n, \Phi \in \mathcal{F}, w \in L_0^r} I(\Phi_-, \Phi, w) \tag{3.3}$$

under restrictions on (Φ_-, Φ, w):

$$0 = \Phi_- - \xi_\Phi(0),\quad 0 = w(t) - C'(t)\xi_\Phi(t),\quad 0 \le t \le T - h. \tag{3.4}$$

If (Φ_-^0, Φ^0, w^0) is the solution of (3.3), (3.4), then Φ^0 is the optimal estimator and $D = I_0 + k$.

In order to get the appropriate solution for (3.3), (3.4) we change the nonsmooth functional I by a quadratic one. The obtained quadratic problem appears when solving the stochastic mean–square filtering problem. Its solution determines the estimator ϕ, obtained in Sec. 2. Let us use the estimator ϕ instead of the optimal algorithm Φ^0. We define the approximation quality by the ratio

$$\Delta = \frac{D(\phi)}{D}.$$

The main problem is to construct the upper bound for the value of Δ, which can be calculated without knowledge of the accurate solution to (3.3), (3.4).

3.2. *APPROXIMATE SOLUTION*

Let us change the nonsmooth problem (3.3), (3.4) by a simpler linear–quadratic one

$$J_1 = \inf_{\Phi_- \in \mathbf{R}^n, \Phi \in L_2^m, w \in L_2^r} J(\Phi_-, \Phi, w),$$

$$J(\Phi_-, \Phi, w) := \Phi_-' P_0 \Phi_- + \int_h^T \Phi'(t) R(t) \Phi(t) dt + \int_0^{T-h} w'(t) Q(t) w(t) dt \tag{3.5}$$

under conditions (3.4).

Let us imagine that $u(t), \rho(t)$ are Gaussian white noises, and $x(0)$ is a stochastic value, which are mutually independent with characteristics

$$Ex(0) = 0,\ Ex(0)x'(0) = P_0, \quad Eu(t) = 0,\ Eu(t)u'(s) = Q(t)\delta(t-s),$$

$$E\rho(t) = 0,\ E\rho(t)\rho'(s) = R(t)\delta(t-s).$$

Then we obtain the optimal stochastic filtering problem, considered in §2, which reduces to (3.5). Hence for our deterministic case we can introduce formally the following weight matrices

$$P_0 = \mathrm{diag}\{\sigma_{-1}^2, \ldots, \sigma_{-n}^2\},$$

$$R(t) = \beta_1 \mathrm{diag}\{\sigma_1^2(t), \ldots, \sigma_m^2(t)\}, \quad Q(t) = \beta_2 \mathrm{diag}\{\gamma_1^2(t), \ldots, \gamma_r^2(t)\}, \tag{3.6}$$

where $\beta_1, \beta_2 > 0$ are specified constants.

The estimator ϕ (that is obtained by the solving of (3.4), (3.5)) is determined by the relations (see Sec. 2)

$$\phi(t) = \begin{cases} R^{-1}(t) H'(t) \left[P(t-h)\xi(t-h) + \int_{-h}^{0} P_1(t-h, \tau)\, \xi(t-h-\tau) d\tau \right], \\ \qquad \text{if} \quad h \le t \le T, \\ 0 \qquad \text{if} \quad 0 \le t < h \end{cases} \tag{3.7}$$

where

$$\dot{\xi}(t) = -A'(t)\xi(t) - B'(t+\overline{h})\xi(t+\overline{h}) +$$

$$+A_0(t+h)\left[P(t)\xi(t) + \int_{-h}^{0} P_1(t,\tau)\xi(t-\tau)d\tau\right], \tag{3.8}$$

$$0 \leq t \leq T,\ \xi(T) = a,\ \xi(\tau) = 0,\ \tau > T.$$

(moreover, $\xi(t) = \xi_\phi(t),\ \xi(0) = \xi_\phi(0) = \phi_-$),

$$\dot{P}(t) = A(t)P(t) + P(t)A'(t) - P(t)A_0(t+h)P(t) +$$

$$+C(t)Q(t)C'(t) + P_1(t,0) + P_1'(t,0),$$

$$\left(\frac{\partial}{\partial t} + \frac{\partial}{\partial \tau}\right) P_1(t,\tau) = A(t)P_1(t,\tau) - P(t)A_0(t+h)P_1(t,\tau) + P_2(t,0,\tau),$$

$$\left(\frac{\partial}{\partial t} + \frac{\partial}{\partial \tau} + \frac{\partial}{\partial s}\right) P_2(t,\tau,s) = -P_1'(t,\tau)A_0(t+h)P_1(t,s),$$

$$0 \leq t \leq T,\quad -\overline{h} < \tau, s \leq 0,$$

$$A_0(\mu) := H(\mu)R^{-1}(\mu)H'(\mu),\ 0 \leq \mu \leq T; \tag{3.9}$$

$$A_0(\mu) := 0,\ B(\mu) := 0,\ \mu > T.$$

The boundary conditions for (3.9) have the form

$$P(0) = P_0,\ P_1(0,\tau) = P_2(0,\tau,s) = 0,$$

$$P_1'(t,-\overline{h}) = B(t+\overline{h})P(t),$$

$$B(t+\overline{h})P_1(t,\tau) = P_2(t,-\overline{h},\tau) = P_2'(t,\tau-\overline{h}), \tag{3.10}$$

$$0 \leq t \leq T,\quad -\overline{h} < \tau, s \leq 0;$$

Relations (3.7)–(3.10) will give us the estimator $\phi(t)$ (which is optimal for the approximating problem) and the function $\xi(t)$.

The degree Δ of approximation is estimated in the following theorem [286]:

Theorem 3.1 The nonoptimality degree Δ satisfies the inequalities:

$$1 \leq \Delta \leq \Delta^0,$$

$$\Delta^0 := \frac{\mathcal{N}_1(\Psi_-,\Psi,\chi) + \kappa}{\mathcal{N}_2^2(\Psi_-,\Psi,\chi)\mathcal{N}_\infty^{-1}(\Psi_-,\Psi,\chi) + \kappa} \tag{3.11}$$

where

$$\mathcal{N}_1 := \sum_{d=1}^{n} |\Psi_{-d}| + \sum_{l=1}^{m} \beta_1^{-1} \int_h^T |\Psi_l(t)|dt + \sum_{j=1}^{r} \beta_2^{-1} \int_h^{T-h} |\chi_j(t)|dt,$$

$$\mathcal{N}_2 := \left\{ \sum_{d=1}^{n} |\Psi_{-d}^2| + \sum_{l=1}^{m} \beta_1^{-1} \int_h^T \Psi_l^2(t)|dt + \sum_{j=1}^{r} \beta_2^{-1} \int_h^{T-h} \chi_j^2(t)|dt \right\}^{\frac{1}{2}}, \tag{3.12}$$

$$\mathcal{N}_\infty := \max \left\{ \max_{1 \le d \le n} |\Psi_{-d}|, \sup_{\substack{1 \le l \le m \\ t \in [h,T]}} |\Psi_l(t)|, \sup_{\substack{1 \le j \le r \\ t \in [0,T-h]}} |\chi_j(t)| \right\},$$

$$\Psi_{-d} := \sigma_{-d}\phi_{-d}, \; \Psi_l(t) := \beta_1 \sigma_l(t)\phi_l(t), \; \chi_j(t) := \beta_2 \gamma_j(t) C_j'(t) \xi_\phi(t), \tag{3.13}$$

The proof given in [286] is based on the duality theory for mathematical programming problems.

Remark 3.2 In the scalar case ($n = m = r = 1$) the upper bound Δ^0 can coincide with Δ. It occurs, e.g., when there is no noise in the object: $C(t) = 0$, or when we consider the stationary system with zero delays and with no *a priori* information about the initial state: $A(t) = 0, B(t) = 0, C(t) = 1, H(t) = 1, \gamma(t) = \gamma, \sigma(t) = \sigma, h = \overline{h} = 0, \sigma_- = \infty; \beta_1 = \beta_2 = 1$.

The algorithm for the estimate (3.11) calculation consists of the following steps:

- *Step 1.* Solving of the differential equation system (3.9) with the boundary conditions (3.10), where the matrices $P_0, Q(t)$, and $R(t)$ are given by (3.6);
- *Step 2.* Solving of the FDE (3.8) (in backward time) for determination of the function $\xi(t) = \xi_\phi(t)$ and the vector $\xi(0) = \xi_\phi(0) = \phi_-$;
- *Step 3.* Calculation of the linear–quadratic problem estimator ϕ using the formula (3.7);
- *Step 4.* Calculation of the auxiliary functions (3.13) and computation of the quadratures (3.12);
- *Step 5.* Calculation of the value (3.11).

Thus, the steps 1–3 consist of the corresponding linear–quadratic problem solving, and the steps 4,5 consist of the calculation of quadratures and of elementary operations over them.

Assume $n = m = 1, A(t) = 0, B(t) = B = \text{const}, C(t) = 0, H(t) = 1, a = 1, \sigma(t) = \sigma = \text{const}$. It means that the system and the measurements have the form

$$\dot{x}(t) = Bx(t-\bar{h}), x(\tau) = 0, \ -\bar{h} \le \tau < 0, \ |x(0)| \le \sigma_-,$$

$$z(t) = x(t-h) + \rho(t), \ \ |\rho(t)| \le \sigma, \ \ t \in [0,T],$$

and we must evaluate $x(T)$.

Assume in addition that there is no prior information about $x(0)$, i.e., $\sigma_- \to \infty$.

Then one can verify that the estimate Δ^0 does not depend on β_1 and takes the simplest form:

$$\Delta = \Delta^0 = \frac{\int_h^T |\phi(t)|dt \cdot \max_{h \le t \le T} |\phi(t)|}{\int_h^T \phi^2(t)dt}$$

where the approximating estimator ϕ is determined by formulas (3.7)–(3.10). One can calculate that

$$\phi(t) = \alpha Y(t-h),$$

$$\alpha := Y(T)\left(\int_h^T Y^2(t-h)dt\right)^{-1} = \text{const},$$

where

$$\dot{Y}(t) = BY(t-\bar{h}), \ Y(\tau) = 0, \ -\bar{h} \le \tau < 0, \ Y(0) = 1,$$

i.e.,

$$Y(t) = 1 + \sum_{j=1}^{p} \frac{B^j(t-j\bar{h})^j}{j!}, \ \ p\bar{h} \le t < (p+1)\bar{h}, \ \ p = 0, 1, \ldots.$$

Some numerical results (with $h = \bar{h}$) are presented in Table 16.3.1. It shows that our estimate provides a good result: the use of the linear-quadratic estimator does not influence sharply in the estimation accuracy; moreover, in a number of cases it does not practically worsen the estimation quality. It should be noted that in the case in question the growth of delay is beneficial for the decrease of Δ^0.

BT	$hT^{-1} = \overline{h}T^{-1}$					
	0.0	0.05	0.1	0.2	0.25	0.5
− 1	1.46	1.39	1.32	1.21	1.15	1.00
− 2	1.76	1.62	1.48	1.28	1.22	1.00
− 5	1.98	1.55	1.29	1.36	1.22	1.00

Table 16.3.1 The values of $\Delta = \Delta^0$

BIBLIOGRAPHY

[1] C. Abdallah, P. Dorato, J. Benitez–Read, Delayed positive feedback can stabilize oscillatory systems, *Proc. of the American Control Conf.*, (1993), pp. 3106–3107.

[2] D.L. Abrahamson, E.F. Infante, A Liapunov functional for linear Volterra integrodifferential equations, *Quarterly of Appl. Math.* **11: 1** (1983), pp. 35–44.

[3] G.E. Adomian, R. Rach,

1) Nonlinear stochastic differential delay equations, *J. Math. Anal. & Appl.* **91: 1** (1983), pp. 94–101.
2) A nonlinear differential delay equation, *J. Math. Anal & Appl.* **91: 2** (1983) pp. 301–304.

[4] V.N. Afanasiev, V.B. Kolmanovskii, V.R. Nosov, *Mathematical Theory of Control System Design,* Kluwer Academic Publishers, Dordrecht, 1996, 700pp.

[5] M. Agarval, C. Canudas, On–line estimation of time delay and continuous–time progress parameters, *Int. J. Control,* **46: 1** (1987), pp. 295–311.

[6] P. Agathoklis, S. Foda, Stability and the matrix Liapunov equations for delay systems, *Int. J. Control* **49: 2** (1989), pp. 417–432.

[7] Z. Agur, Population dynamics in harshly varying environments, *Int. Centre Theoret. Physics,* World Scientific, Italy, 1986, pp. 440–454.

[8] M. Ait Babram, K. Ezzinbi, R. Benkhalti, Periodic solutions of Functional Differential Equations of Neutral Type, *J. of Math. Anal. Applic.* 1996, p. 204, pp. 898–909.

[9] G. Alain, R. Colly, Mathematical study of a muscle construction model in which the fibre is a continuum of elements, *Adv. Appl. Math.* **9: 1** (1998), pp. 104–126.

[10] V. Alexiades, Almost periodic solutions of an integrodifferential system with infinite delay, *Nonlinear Anal., Th., Methods & Applic.* **5: 4** (1981), pp. 401–410.

[11] T. Amemiya, Arazawak H. Tokumaru, Delay independent stabilization and decay rate assignability of linear systems with limited measurable state variables, *Int. J. Control* **47: 1** (1988), pp 65–73.

[12] J.M. Ananjevski, V.B. Kolmanovskii, Stabilization of some nonlinear hereditary mechanical systems, *Nonlinear Anal., Th., Methods & Applic.* **15: 2** (1990), pp. 101–114.

[13] E.A. Andreeva, V.B. Kolmanovskii, L.E. Shaikhet, *Control of Systems with Delay,* Nauka, Moscow, 1992, 400pp. (In Russian.)

[14] D. Andrencci, A. Fasano, M. Primicerio, On a mathematical model for the crystallization of polymers, *Proc. 4 th Europ. Conf. Math. in Industry,* Kluwer AP, 1991, pp. 3–16.

[15] A.A. Andronov, A.G. Mayer, The simplest linear system with delay, *Autom. Remote Control* **7: 213** (1946), pp. 95–106. (In Russian.)

[16] O. Arino, I. Györi, M. Pituk, Asymptotically Diagonal Delay Differential systems, *J. Math. Anal. Applic.* **204: 3**, (1996), pp. 701–728.

[17] O. Arino, V. Nosov, On stability of a class of neutral type functional differential equations, *J. of Math. and Computer in Simulation*, 1997, Special Issue "Delay systems".

[18] Z. Artstein, G. Zarakostas, Convergence in the delay population equation, *SIAM J. Appl. Math.* **38: 2** (1980), pp. 261–272.

[19] N.H. Arutunian, V.B. Kolmanovskii, *Greep Theory of Nongomogeneous Bodies,* Nauka, Moscow, 1983, 380pp. (In Russian.)

[20] N. Azbelev, V. Maksimov, L. Rakhmatullina, *Introduction to the Theory of Linear Functional Differential Equations,* World Federation Publishers, 1995, 176p.

[21] V.G. Babskii, A.D. Myshkis, Mathematical models in biology connected with regard of delays, Appendix to J.D. Murray, *Lectures on Non-linear Differential Equations,* Models in Biology, Mir, Moscow, 1983, pp. 383–394. (In Russian.)

[22] A. Bagchi, Approximations of Itô integral arising stochastic time–delay systems, *SIAM J. Control Optim* **22: 6**, (1984), pp. 878–888.

[23] D.D. Bainov, V.B. Kolmanovskii, Periodic solutions of stochastic functional differential equations, *J. Math. of Toyama Univ.* **14**, (1991), pp. 1–39.

[24] D. Bainov, D. Mishev, *Oscillation Theory for Neutral Differential Equations with Delay,* A. Hilger, N–Y, 1991, 280p.

[25] C.T.H. Baker, V.B. Kolmanovskii, Stability of chemostat equations for two species competing for renewable resource, *J. Dynamic systems and Applications,* **4: 4**, (1995), pp. 601–611.

[26] C.T.H. Baker, A. Tang, Stability of grazing systems with infinite delay, *Mathematics and Computers in Simulation,* **N 12** (1997), pp. 149–163.

[27] K. Balachandran, Nonlinear perturbations of quasilinear delay control systems, *Kibernetika* **25: 1**, (1989).

[28] H.T. Banks, Delay systems in biological models: approximation techniques, *Nonlinear Systems & Appl.*, Acad. Press, New York, 1977, pp. 21–28.

[29] H.T. Banks, J.M. Mahaffy, Global asymptotic stability of certain models for protein synthesis and repression, *Quarterly Appl. Math.* **36: 3**, (1978), pp. 209–221.

[30] J. Bansal, K.S. Chang, Optimal control of dispertion type distributed–parameter systems with time–delay in interacting two–point boundary conditions, *Int. J. Control* **16: 3**, (1972), pp. 481–500.

[31] M. Bardi, Exchange of stability along a branch of periodic solutions of a single species model, *Lect. Notes in Math.* **1017**, (1983), pp. 73–82.

[32] B.R. Barmish, *New Tools for Robustness of Linear Systems*, Macmillan, N–Y, 1994.

[33] B.R. Barmish, Z. Shi, Robust stability of perturbed systems with time delay, *Automatica* **25: 3**, (1989), pp. 371–381.

[34] H.A.M. Basher, R. Mukundan, Model reference control of uncertain systems with delay in plant state and control, *Int. J. Systems Sci.* **18: 9**, (1987), pp. 1609–1626.

[35] L. Bass, A.I. Bracken, K. Holmaker, B.R. Jefferies, Integro–differential equations for the selforganization of liver zones by competitive exclusion of cell types, *J. Austral. Math. Soc.* **29: 2** (1987), pp. 156–194.

[36] J.R. Beddington, R.M. May, Time delay are not necessarily distabilizing, *Math. Biosc.* **27: 112** (1975), pp. 108–117.

[37] J. Belair, Stability in a model of a delayed neural network, *J. Dynamics Diff. Eqs.*, **5** (1993), pp. 607–623.

[38] J. Belair, S. Dufour, Stability in a three–dimensional system of delay–differential equations, *Canadian applied math. quartely* **4: 2** (1996), pp. 135–156.

[39] I. Belair, M.C. Mackey, Consumer memory and price fluctuations in commodity markets: an integrodifferential model, *J. Dinamics & Diff. Eq.* **1: 3** (1989), pp. 299–325.

[40] D. Bell, S. Mohammed, On the solutions of stochastic ordinary differential equations with small delays, *Stochastics* **28** (1989), pp. 293–299.

[41] A. Bellen, V.B. Kolmanovskii, L. Torelli, R. Vermiglio, About stability of some functional differential equations of neutral type, *J. of Math. Anal. and Applic.* **189** (1995), pp. 59–84.

[42] R. Bellman, K.L. Cooke, *Differential–Difference Equation*, Acad. Press, New York – London, 1963.

[43] P.A. Belmisov, A.D. Drozdov, V.B. Kolmanovskii, *Stability of Viscoelastic Systems*, Saratov Univ. Press, Saratov, 1991, 280pp. (In Russian.)

[44] E.A. Bencer, L.P. Neuwirth, Traffic flow: Laplace transform, *Amer. Math. Monthly* **80: 4** (1973), pp. 417–423.

[45] M.H. Bender, A counterintuitive result in Bell's model of delayed maturity, *Math. Biosci.*, **131: 1** (1996), pp. 103–109.

[46] A. Bensoussan, G. Da Prato, M.C. Delfour, S.K. Mitter, Representation and control of infinite dimensional systems, *Systems & Control: foundation and applications,* **1& 2**, Birkhäuser, Boston, 1993.

[47] J. Bentsman, J. Fakhfakh, R.B. Lehman, Vibrational stabilization of linear time delay systems and their robustness with respect to delay size, *Systems, Control Letters,* **12: 3** (1989), pp. 267–272.

[48] E. Beretta, A. Fasano, Y. Hosono, V.B. Kolmanovskii, Stability analysis of the Phytoplankton Vertical Steady States in a Laboratory Test tube, *Math. Methods in the Applied Sciences,* **17** (1994), pp. 551–575.

[49] E. Beretta, V.B. Kolmanovskii, L. Shaikhet, Stability of Epidemic model with time delay influenced by stochastic perturbations, *Mathematics and Computers in Simulation,* **N 6** (1997).

[50] L. Berezansky, E. Braverman, Explicit conditions of exponential stability for a linear impulsive Differential Equation, *J. of Math. Anal. and Applic.,* **214: 2** (1997), pp. 439–458.

[51] Y. Biran, B. Innis, Optimal control of bilinear systems: time–varying effects of cancer drugs, *Automatica* (1979).

[52] S.P. Blythl, R.M. Anderson, Distributed incubation and infection periods in models of transmission dynamics of the human immunodeficiency virus (HIV), (IMA), *J. Math. Appl. Med. & Biol.* **5: 1** (1988), pp. 1–19.

[53] S.P. Blythe, R.M. Nisbet, W.S.C. Gurney, Instability and complex dynamic behaviour in population models with long time delays, *Theor. Pop. Biol.* **2** (1982), pp. 147–176.

[54] G.A. Bocharov, G.I. Marchuk, A.A. Romanyukha, Numerical solution by LMMs of stiff delay differential systems modelling an immune response, *Num. Math.* **73** (1996), pp. 131–148.

[55] G.A. Bocharov, A.A. Romanyukha, Mathematical model of antiviral immune response. Influenza A virus infection, *J. Theor. Biol.* **167** (1994), pp. 323–360.

[56] F.G. Boese, Delay independent stability of a special sequence of neutral difference–differential equations with one delay, *J. Diff. Eq.* **90** (1991), pp. 397–407.

[57] V.M. Bogdan, Existence of solutions to differential equations of relativistic mechanics involving Lorentzian time delays, *J. Math. Anal. & Appl.* **118: 2** (1986), pp. 561–573.

[58] V.L. Bogomolov, Control of power of hydrostation, *Autom. & Remote Control* **4/5** (1941), pp. 103–129, (In Russian.)

[59] G.N. Bojadziev, R.W. Lardner, Asymptotic solutions of partial differential equations with damping and delay, *Quart. Appl. Math.* **33: 3** (1975), pp. 205–214.

[60] R.D. Braddock, P. van den Driessche,

1) A population model with time delay, *Math. Sc.* **5: 1** (1980), pp. 55–66,

2) On a two lag differential delay equation, *J. Austral. Math. Soc.*

B **24: 3** (1983), pp. 292–317.

[61] P. Brandi, R. Ceppitelli, Existence, uiqueness and continuous dependence for hereditary differential equations, *J. Diff. Equations* **81** (1989), pp. 317–339.

[62] Y. Brian, B. Innis, Optimal control of bilinear systems: time varying effects of cancer drugs, *Automatica* **5** (1979).

[63] F. Braner, Characteristic return times for harvested population models with time lag, *Math. Biosc.* **45: 3/4** (1979), pp. 295–311.

[64] R.K. Brayton, Small–signal stability criterion for electrical networks containing lossless transmission lines, *IBM J. Res. & Developm.* **12: 8** (1968), pp. 431–440.

[65] N. Britten, Travelling wave front solutions of a differential–difference equation arising in the modelling of myelinated nerve axon, *Lect. Notes in Math.* **1151** (1985), pp. 77–89.

[66] W.E. Brumbey, On the asymptotic behaviour of solutions of differential–difference equations of neutral type, *J. Diff. Eq.* **7** (1970), pp. 175–188.

[67] H. Brunner, P.J. van der Houwen, *The Numerical Solution of Volterra Equations*, N–Y, North Holland, 1986.

[68] H. Bunke, *Gewöhnliche, Differential gleichungen mit Zufälligen Parametern*, Academic–Verlag, Berlin, 1972, 207s.

[69] D.N. Burghes, Population dynamics. An introduction to differential equations, *Int. J. Math. Educ. Sci. & Techn.* **6: 3** (1975), pp. 265–276.

[70] J.A. Burns, P.D. Hirsch, A difference equation approach to parameter estimation for differential delay equations, *Appl. Math. & Comp.* **7: 4** (1980), pp. 281–311.

[71] J.A. Burns, K. Ito, G. Propst, On nonconvergence of adjoint semigroups for control systems with delays, *SIAM J. Control & Optim.* **26: 6** (1988).

[72] T.A. Burton, *Volterra Integral and Differential Equations*, Acad. Press, New York, 1983.

[73] T. Burton, L. Hatvani, Stability theorems for nonautonomous functional differential equations by Liapunov functionals, *Tdioku Math. J.* **41** (1989), pp. 65–104.

[74] T. Burton, V. Hudson, Repellers in systems with infinite delay, *J. Math. Anal. & Appl.* **137: 1** (1989), pp. 240–263.

[75] S. Busenberg, K.L. Cooke, Periodic solutions of a periodic nonlinear delay differential equation, *SIAM J. Appl. Math.* **35: 4** (1978), pp. 704–721.

[76] S.N. Busenberg, M. Martelli (eds.), *Delay Differential Equations and Dynamical Systems*, Springer, Berlin, 1991.

[77] S.N. Busenberg, C.G. Trevis, On the use of reducible–functional dif-

ferential equations in biological models, *J. Math. Anal. & Appl.* **89: 1** (1982), pp. 46–66.

[78] S.N. Busenberg, M.A. Iannelli, Class of nonlinear diffusion problems in age–dependent population dynamics, *Nonlinear Anal.: Theory Methods & Appl.* **7: 5** (1983), pp. 501–529.

[79] A.G. Buyvolova, V.B. Kolmanovskii, Time optimal control of chemostat, (In Russian.) *Automatic and Remote Control* 1 (1994), pp. 41–49.

[80] D. Bylarsky, Lag matrix in Leontief model formed by means of random functionals theory, *Ekon. Mat. Obz.* **13: 4** (1977), pp. 405–419.

[81] B.F. Bylov, R.E. Vinograd, D.N. Grobman, V.V. Nemytskii, *Theory of Liapunov Exponents,* (In Russian.) Nauka, Moscow, 1966.

[82] B. Cahlon, J. Herband, A confidence headway for safe car following, *J. Math. Anal. & Appl.* **69: 2** (1979), pp. 511–530.

[83] Y. Cao, Multiexistence of slowly oscillating periodic solutions for diffrential delay equations, *SIAM J. Math. Anal.* **26: 2** (1995), pp. 436–445.

[84] V. Capasso, Asymptotic stability for an integro–differential reaction–diffusion system, *J. Math. Anal. & Appl.* **103** (1984), pp. 575–588.

[85] D.A. Carlson, An elementary proof of the maximum principle for optimal control problems governed by a Volterra integral equation, *J. Optim. Theory & Appl.* **54: 1** (1987), pp. 43–61.

[86] G.C. Carter, Time–delay estimation, *IEEE Truns. Acoustics. Speech & Signal Processing* **29** (1981), special issue.

[87] A. Casal, M. Freedman, A Poincará–Lindstedt approach to bifurcation problems for differential–delay equations, *IEEE Truns. Autom. Control* **25: 5** (1980), pp. 967–973.

[88] P.A. Casal, A. Somolinos, Forced oscilations for the sun–flower equation entrainment, *Nonlinear Anal. Theory, Methods & Appl.* **6: 4** (1982), pp. 397–414.

[89] J.S. Cassell, Z. Hou, Initial value problem for mixed–type differential equations, *Monatshefte für Mathematik* **124: 2** (1997), pp. 133–145.

[90] A.M. Castelfranco, H.W. Stech, Periodic solution in a model of recurrent neutral feedback, *SIAM J. Appl. Math.* **47: 3** (1987), pp. 573–588.

[91] C. Castillo–Chavez, Nonlinear character dependent models with constant time delay in population dynamics, *J. Math. Anal. & Appl.* **128** (1987), pp. 1–29.

[92] C. Castillo–Chavez, H.R. Thieme, On the role of variable infectivity in the dynamics of the human immunodeficiency virus epidemic, In: C. Castillo–Chavez (ed.), *Math. and Statistical Approaches to AIDS Epidemology,* Springer, Berlin, 1989, pp. 157–176.

[93] C. Castello–Chavez, K. Cooke, W. Huang, S.A. Levin, On the role of long incubation periods in the dynamics of acquired immunodeficiency

syndrome (AIDS). Part1. Single population models, *J. Math. Biol.* **27: 4** (1989), pp. 373–398.

[94] L. Cesari, *Asymptotic Behaviour and Stability Problems in Ordinary Differential Equations*, Springer, N–Y, 1959.

[95] J.M.F. Chamayon, A probabilistic approach to a differential–difference equation arising in analytic number theory, *Math. Comp.* **27** (1973), pp. 197–203.

[96] W.L. Chan, Guo Bao Z. Hu, Overtaking optimal control problem of the age–dependent population with infinite horizon, *J. Math. Anal. & Appl.* **150: 1** (1990), pp. 41–53.

[97] M.H. Chang, Discrete approximation of nonlinear filtering for stochastic delay equations, *Stoch. Anal. & Appl.* **5: 3** (1987), pp. 267–298.

[98] P.H. Chang, J.W. Lee, A model reference observer for time–delay control and its application to robot trajectory control, *IEEE Trans. on control systems technology* **4: 1** (1996).

[99] N.G. Chebotarev, N.N. Meimen, Problem of Routh–Hurwitz for polynomials and entire functions, *Tr. Mat. Inst. Steklov.* **26** (1949), (in Russian.)

[100] M. Chen, Differential–difference equations for the radial distribution function of hard rods and hard spheres under Percus–Jevick approximation, *J. Math. Anal. & Appl.* **64: 3** (1978), pp. 629–650.

[101] E. Cheres, Hypersurface attractivity for systems with delays driven by min–max controller, *Int. J. Systems Sc.* **19** (1988), p.10.

[102] F.L. Chernousko, V.B. Kolmanovskii,

1) *Optimal Control Under Stochastic Disturbances*, Nauka, Moscow, 1978. (In Russian.)

2) Computational and approximate methods of optimal control, *J. Soviet Math.* **12** (1979), pp. 310–353.

[103] Y. Cherrnault, *Mathematical Modelling in Biomedicine; Optimal Control of Biomedical Systems*, Reidel, Dordrecht, 1986.

[104] N.G. Chetaev, *Stability of Motion*, Nauka, Moscow, 1965. (In Russian.)

[105] I.D. Cheuskov, About one system of equations with delay appearing in aircontrol, *Funct. Anal., Theory & Appl.* **54** (1990), pp. 123–130, (in Russian).

[106] O.Yu. Cherkasov, A.G. Yakushev, Optimal evasion from an object using pure pursuit method, *Vestnik Moscow University* **N 1** (1996), pp. 50–55. (In Russian.)

[107] Henjin Chi, J. Bell, B. Hassard, Numerical solution of nonlinear advance delay–differential equations from nerve conduction theory, *J. Math. Biol.* **24: 3** (1986), pp. 583–601.

[108] U. Chiasson, A method of computing the interval of delay values for which a differential delay system is stable, *IEEE Trans. Autom. Control* **33: 12** (1988), pp. 1176–1178.

[109] Lyn–Horng Chou, Ing–Rong–Horng, Bor–Sen Chen, Dynamical feedback compensator for uncertain time delay systems containing saturating actuator, *Int. J. Control* **49: 3** (1989), pp. 961–968.

[110] Shui–Nee Chow, Existence of periodic solutions of autonomic functional differential equations, *J. Diff. Eq.* **15: 2** (1974), pp. 350–378.

[111] E.N. Chukwu, *Stability and Time Optimal Control of Hereditary Systems,* N–Y, Academic Press, 1992.

[112] D.H. Chyung, E.B. Lee, Linear optimal control problems with delay, *SIAM J. Control* **4: 1** (1966).

[113] M.D. Ciletti, Differential games with an information time lag, *Topics in Differential Games,* North–Holland, Amsterdam, 1973, pp. 179–268.

[114] D.S. Cohen, P.S. Hagen, H.C. Simpson, Spatial structures in predator–prey communites with hereditary effects and diffusion, *Math. Biosc.* **44: 314** (1979), pp. 167–177.

[115] D.S. Cohen, S. Rosenblatt, A delay logistic equation with variable growth rate, *SIAM J. Appl. Math.* **42: 3** (1982), pp. 608–624.

[116] B.D. Coleman, M.E. Gurtin, Waves in materials with memory: II. On the growth and decay of one–dimensional acceleration waves, *Arch. Rat. Mech. & Anal.* **19: 4** (1965), pp. 239–265; 266–298.

[117] B.D. Coleman, M.E. Gurtin, R.I. Herrera, Waves in materials with memory: I. The velocity of one–dimensional shock and acceleration waves, *Arch. Rat. Mech. & Anal.* **19:1** (1965), pp. 1–19.

[118] B.D. Coleman, V.J. Mizel, Stability of functional differential equations, *Arch. Rat. Mech. & Anal.* **30: 3** (1968), pp. 173–196.

[119] B.D. Coleman, D.R. Owen, On the initial value problem for a class of functional–differential equations, *Arch. Rat. Mech. & Anal.* **55: 4** (1974), pp. 275–299.

[120] K.L. Cooke, P. van der Driessche, Analysis of an SEIRS epidemic model with two delays, *J. Math. Biol.* **35** (1996), pp. 240–260.

[121] K.L. Cooke, Z. Grossman, Discrete delay, distributed delay and stability switches, *J. Math. Anal. Appl.* **86: 2** (1982), pp. 592–627.

[122] K.L. Cooke, J.L. Kaplan, M. Sorg, Stability of functional differential equation for the motion of a radiating charged particle, *Nonlinear Anal. Theory Methods & Appl.* **10: 5** (1981), pp. 1133–1139.

[123] C. Corduneanu, *Integral Equations and Stability of Feedback Systems,* Acad. Press, New York–London, 1973.

[124] C. Corduneanu, V. Lakshmikantham, Equations with infinite delay, *Nonlinear Analysis, TMA* **4** (1980), pp. 831–877.

[125] F. Corpet, C. Chevalet, M. Gillois, A. Micali, Dynamics and stability

of induction of the lactose operon of Escherischia Coli, *Lect. Notes Biomath.* **49** (1983), pp. 123–143.

[126] M. Courtemanche, P.J. Keener, L. Glass, A delay equation representation of pulse circulation on a ring in excitable media, *SIAM J. Appl. Math.* **56: 1** (1996), pp. 119–142.

[127] M.R. Crisci, V.B. Kolmanovskii, E. Russo, A. Vecchio,

1) Stability of continuous and discrete Volterra integro–differential equations by Liapunov approach, *J. of Integral Equations and Appl.* **7: 4** (1995), pp. 393–411.

2) Boundedness of discrete Volterra equations, *J. of Math. Anal. & Appl.* **211** (1997), pp. 106–130.

[128] J.M. Cushing, Integro–differential equations and delay models in population dynamics, *Lect. Notes in Biomath.* **20** 1977, Springer, Berlin.

[129] E. Ait Dads, K. Ezzinbi, Pseudo–almost periodic solutions of some delay differential equations, *J. Math. Anal. & Applic.* **201: 3** (1996), pp. 840–850.

[130] Lo Sheng Dai, Nonconstant periodic solutions in predator prey systems with continuous time delay, *Math. Biosc.* **53: 1/2** (1981), pp. 149–157.

[131] M. Dambrine, J.P. Richard, Stability and stability domains analysis for nonlinear differential–difference equations, *J. Dynamic Systems and Applic.* **3** (1994), pp. 318–396.

[132] M. Dambrine, J.P. Richard, Comparison principle for functional Differential Equations, *Proceedings 15-th IMACS World Congress,* 1 1997, Berlin, pp. 221–226.

[133] M. Dambrine, J.P. Richard, P. Borne, Feedback control of time–delay systems with bounded control and state, *J. Math. Problems in Engineering* **1: 1** (1995), pp. 77–87.

[134] P. Das, N. Misra, A necessary and sufficient conditions for the solutions of a functional differential equation to be oscillatory or tend to zero, *J. Math. Anal. & Applic.* **205: 1** (1997), pp. 78–87.

[135] R. Datko, Two examples of ill posedness with respect to small time delays in stabilized elastic systems, *IEEE Trans. AC* **38: 1** (1993), pp. 163–166.

[136] R. Datko, J. Lagnese, M.P. Polis, An example of the effect of time delays in boundary feedback stabilization of wave equations, *SIAM J. Control Optim.* **24** (1986), pp. 152–156.

[137] P.L. Davis, On the hyperbolicity of the equations of the linear theory of heat conduction for materials with memory, *SIAM J. Appl. Math.* **30: 1** (1976), pp. 75–80.

[138] E.J. Davison, B.R. Copeland, Gain margin and time lag tolerance con-

strains applied to the stabilization problem and robust servo–mechanism problem, *IEEE Trans. Autom. Control* **30: 3** (1985), pp. 229–239.

[139] M.C. Delfour, A. Manitius, Control systems with delays; areas of applications and present status of the linear theory, *Anal. Syst. et Orientat. Nouv. Coll. TRTA Versailles–Rocquencourt 1976,* pp. 345–392.

[140] M. Delpech, M. Maurette, Feasibility of time delay compensation for a space teleoperation task, in: J. Chretien (ed.), *Autom. Control in Space,* Toulouse, 1985, pp. 279–286.

[141] B.P. Demidovich, On the dissipativity of a nonlinear system of differential equations, *Trans. of Moscow State Univ.* **1** 1962, pp. 3–8. (In Russian.)

[142] M.W. Derstine, H.M. Gibbs, F.A. Hopf, L.D. Sanders, Distinguishing chaos from noise in an optically bistable system, *IEEE, J. Quantum Electronics* **QE-21: 9** (1985), pp. 1419–1422.

[143] W. Desch, R. Grimmer, W. Schappacher, Wellposedness and wave propagation for a class of integrodifferential equations in Banach space, *J. Diff. Eq.* **74: 2** (1988), pp. 391–411.

[144] G. Diblasio, K. Kunish, E. Sinestrari, Stability for abstract linear functional differential equations, *Israel J. Math.* **50** (1985), pp. 231–263.

[145] J. Diblik,

1) Stability of the trivial solution of retarded functional differential equations, *J. Diff. Eq.* **26: 2** (1990), pp. 219–223.

2) Applications of Vajevskii's method to the differential equations with delay, *Dokl. Akad. Nauk SSSR* **310: 2** (1990), pp. 285–287. (In Russian.)

[146] B.F. Dibrov, M.A. Livshits, M.V. Volkenstein, The effect of a time lag in the immune reaction, *Lect. Notes Control & Inform. Sc.* **18** (1979), pp. 87–94.

[147] O. Diekmann, The cell size distribution and semigroups of linear operators, *Lect. Notes Biomath.* **68** (1986), pp. 46–77.

[148] O. Diekmann, S.A. van Gils, S.M. Verduyn Lunel, H. -O. Waltner, *Delay Equations,* Springer, N–Y, 1995, 534pp.

[149] J.A. Dixon, A nonlinear weakly singular Volterra integrodifferential equation arising from a reaction–diffusion on study in a small cell, *J. Comb. & Appl. Math.* **18: 3** (1987), pp. 289–305.

[150] R.K. Dodd, Generalized Backlund transformation for some nonlinear partial differential–difference equations, *J. Phys. A* **11: 1** (1978), pp. 81–92.

[151] R.K. Dodd, Periodic orbits arising from Hopf bifurcations in a Volterra prey–predator model, *J. Math. Biol.* **35** (1997), pp. 432–452.

[152] A.Ja. Dorogovzev, *Periodic and Stationary Solutions of Infinite Dimensional Deterministic and Stochastic Dynamical Systems,* Kiev, Visha shk., 1992, 319pp. (In Russian.)

[153] B. Dorizzi, B. Grammaticos, M. Le Berre, I. Pomean, E. Ressayre, A. Tallet, Statistics and dimension of chaos in differential delay systems, *Phys. Rev. A* **35: 1** (1987), pp. 328–339.

[154] P. Dormayer, The stability of special symmetric solution of $\dot{x}(t) = af(x(t-1))$ with small amplitudes, *Nonlinear Anal. TMA* **14: 8** (1990), pp. 701–715.

[155] R.D. Driver,

1) The 'backwards' problem for a delay differential system arising in classical electrodynamics, Proc. 5th Conf. Nonlinear Oscillation, Kiev 2 (1970), pp. 137–143.
2) *Ordinary and Delay Differential Equations,* Springer, New York, etc., 1977.
3) A neutral system with state–dependent delay, *J. Diff. Eq.* **54: 1** (1984), pp. 73–86.

[156] A.D. Drozdov, V.B. Kolmanovskii,

1) Stability of equations with aftereffect describing dynamics of viscoelastic bodies, *Diff. Eq.* **28: 2** (1992), pp. 187–198. (In Russian.)
2) Stochastic stability of viscoelastic bars, *J. Stochastic Analysis and Appl.* **10: 3** (1992), pp. 265–277.
3) *Stability in Viscoelasticity,* North Holland, 1994, 616pp.

[157] A.D. Drozdov, V.B. Kolmanovskii, D. Trigiante, Stability of predator–prey systems, *Automatic and Remote Control* **11** (1992), pp. 57–64. (In Russian.)

[158] A.D. Drozdov, V.B. Kolmanovskii, P. Nistri, P. Zecca, Stability of nonhomogeneous aging viscoelastic bodies under dynamic loading, *Nonlinear Analysis, TMA* **24: 9** (1995), pp. 1361–1377.

[159] L. Dugart, E.I. Verriest, (eds.) *Stability and Control of Time–Delay Systems,* N–Y, Springer, 1997, 317pp.

[160] G. Dunkel, Single–species model for population growth depending on past history, *Lect. Notes in Math.* **60** (1968), pp. 92–99.

[161] M. Ďurišová, L. Dedík, M. Balan, Building a structured model of a complex pharmacokinetic system with time delays, *Bulletin of Mathematical Biology* **57: 6** (1995), pp. 787–808.

[162] C.J. Duyn, Regularity properties of the solution of an equation arising in the theory of turbulence, *J. Diff. Eq.* **33: 2** (1979), pp. 226–238.

[163] L.C. Edie, Car–following and steady state theory for noncongested traffic, *Oper. Res.* **9: 1** (1961), pp. 66–76.

[164] S. Elaydi, *An Introduction to Difference Equations,* N–Y, Springer, 1996.

[165] R.H. Elderkin, Nonlinear, globally age–dependent population models: some basic theory, *J. Math. Anal. & Appl.* **108: 2** (1985), pp. 546–562.

[166] M.El–Doma, Analysis of nonlinear integro–differential equation arising in age–dependent epidemic models, *Nonlinear Anal. Theory, Methods & Appl.* **11: 8** (1987), pp. 913–937.

[167] M.E. El'jasberg, About stability of metals cutting, Izv. Akad. Nauk SSSR, OTN 9 (1958). (In Russian.)

[168] H. El–Owaidy, A.A. Ammar, Stable oscillations in a predator–prey model with time lag, *J. Math. Anal. & Appl.* **139: 1** (1988), pp. 191–199.

[169] S.A. Elsanousi, Filtering for a signal given by a linear stochastic retarded differential equation, *J. Math. Anal. Applic.* **212** (1997), pp. 75–87.

[170] L.E. El'sgoltz, S.B. Norkin, *Introduction to the Theory and Application of Differential Equations with Deviating Arguments,* N–Y, Academic Press, 1973.

[171] I.B. Emmerth, On the existence of solution of nonlinear differential equation with delay, *Arch. Rat. Mech. Anal.* **68: 2** (1978), pp. 155–169.

[172] I.R. Epstein, Delay effects and differential delay equations in chemical kinetics, *Intern Reviews in Physical Chemistry* **11** (1992), pp. 135–160.

[173] L.H. Erbe, H.J. Freedman, V.S.H. Rao, Three–species food–chain models with mutual interference and time delay, *Math. Biosc.* **80: 1** (1986), pp. 57–80.

[174] L. Faina, A. Salvadori, Existence theorems for optimal problems in hereditary structures, *J. Nonlinear Anal.* **22** (1995), TMA pp. 643–654.

[175] M. Farkas, *Periodic Motions,* Springer, 1994, 577pp.

[176] C.D. Farmer, Chaotic attractors of an infinite–dimensional dynamical system, *Physica D* **4: 3** (1982), pp. 366–393.

[177] Wei Feng, Xin Lu, Harmless delays for permanence in a class of population models with diffusion effects, *J. of Math Anal. & Appl.* **206: 2** (1997), pp. 547–566.

[178] M.L. Fernandez, F. Zanolin, Existence and uniqueness results for periodic solutions of differential systems with application to the competing species problem, *J. Math. Anal. & Appl.* **137: 2** (1979), pp. 335–348.

[179] J.M. Ferreira, On the stability of a distributed network, *SIAM J. Math. Anal.* **17: 1** (1986), pp. 38–45.

[180] Y.A. Fiagbedzi, A.E. Pearson, Feedback stabilization of linear au-

tonomous time lag systems, *IEEE AC* **31: 9** (1986), pp. 847–855.

[181] G.M. Fichtengolz, *Course of Differential and Integral Calculus,* Vol 3, Fizmatgiz, Moscow, 1960. (In Russian.)

[182] W. Fickett, A mathematical problem from detonation theory, *Quart. Appl.* **46: 3** (1988), pp. 459–471.

[183] A.C. Fowler,

1) Approximate solution of a model of biological immune responses incorporating delay, *J. Math. Biol.* **13: 1** (1981), pp. 23–45.

2) Asymptotic analysis of the delayed logistic equation when the delay is large, *IMA J. Appl. Math.* **28: 1** (1982), pp. 41–49.

[184] L. Fox, D.F. Mayers, I.R. Ockenden, A.B. Tauler, On a functional differential equation, *J. Inst. Math. Anal. & Appl.* **8: 3** (1971), pp. 271–307.

[185] A. Freedman, L.A. Romero, Functional differential equation for the determination of the viscosity function in a rheometer, *Arch. Rat. Mech. Anal.* **107: 1** (1989), pp. 85–97.

[186] H.I. Freedman, K. Gopalsamy, Nonoccurence of stability in systems with discrete delays, *Canad. Math. Bull.* **31: 1** (1988), pp. 52–58.

[187] H.I. Freedman, J.W.H. So, P. Waltman, Chemostat competition with time delays, *IMACS Ann. Comput. & Appl. Math.* **5: 1/4** (1989), pp. 171–173. (12th IMACS World Congr. Sc.: Com. Paris, July 1988).

[188] H.I. Freedman, V. Ito, Hari Sree, The trade–off between mutual interference and time lags in predator–prey systems, *Bull. Math. Biol.* **45: 6** (1983), pp. 991–1004.

[189] E. Fridman, L. Fridman, E. Shustin, Steady modes in a discontinuous control systems with time delay, *Pure Math. and its Applic.* **4** (1993), pp. 59–73.

[190] L. Fridman, E. Shustin, E. Fridman, Steady modes in the relay control systems with time delay, *Proceedings 15th IMACS World Congress,* Vol 1, Berlin, 1997, pp. 239–244.

[191] G. Friesecke, Exponentially growing solutions for a delay–diffusion equation with negative feedback, **98: 1** (1992), pp. 1–18.

[192] M. Fu, A.W. Olbort, M.P. Polis, Robust stability for time–delay systems: the edge theorem and graphical tests, *IEEE Trans. Autom. Control* **34** (1989), pp. 813–820.

[193] K. Furuta, M. Yamakita, Y. Sato, Computation of optimal control for linear systems with delay, *Int. J. Control* **48: 2**, pp. 577–589.

[194] G. Fusco, M. Ianneli, L. Salvadori (Eds.) *Advanced Topic in the Theory of Dynamical Systems,* Academic Press, London, 1989.

[195] G. Gabisch, H.W. Lorenz, Business cycle theory, *Lect. Notes in Econ. & Math. Systems* **283** (1987), 229pp.

[196] I.V. Gaishun, Discrete equations with variable structure and stability of their solutions, *J. Diff. Eqs.* **33: 12** (1997), pp. 1607–1614.

[197] I.V. Gaishun, L.B. Knyashiche, Nonmonotone Liapunov functionals stability conditions of equations with delay, *Diff. Eqs.* **30: 8** (1994). (In Russian.)

[198] J.I. Gao, L.M. Narducci, H. Sadiky, M. Squicciarini, J.M. Yuan, Higher–order bifurcations in a bistable system with delay, *Phys. Rev. A* **30: 2** (1984), pp. 901–904.

[199] L. Gastaldi, F. Tomarelli, A uniqueness result for a nonlinear hyperbolic equation, *Ann. Mat. Pura Appl.* **137** (1984), pp. 175–205.

[200] J.A. Gatica, P. Waltman

1) A singular functional differential equation arising in an immunological model *Lect. Notes in Math.* **564** (1976), pp. 114–124.

2) A system of functional differential equations modelling threshold phenomena, *Applicable Anal.* **28** (1989), pp. 39–50.

[201] V.G. Gerasimov, *Theoretical Foundation of Control of Heat Processes,* Gosenergoizdat, Moscow, 1949. (In Russian.)

[202] T.R. Gicev, A.L. Dontchev, Convergence of the solution of convex optimal control problem with delay, *Rend. Math.* **11: 4** (1978), pp. 635–643.

[203] M.I. Gil

1) *Norm Estimates for Operator–Valued Functions and Applications,* Marcel Dekker, N–Y.,1995.

2) On absolute stability of neutral type functional differential systems, *Nonlinear Studies* **3: 1** (1996), pp. 59–72.

3) On exponential stability of nonlinear systems with delaying argument, *Nonlinear Analysis TMA* **31: 5/6** (1998), pp. 755–764.

[204] L.P. Gilbert, A.M. Johnson, An application of the Jordan canonical form to the epidemic problem, *J. Appl. Probab.* **17: 2** (1980), pp. 313–323.

[205] L. Glass, M.C. Mackey, *From Clocks to Chaos,* Princeton Univ. Press, 1988, 248pp.

[206] E.A. Godin, V.B. Kolmanovskii, E.Sh. Stengold, *Hypertension Disease as a Result of Altered–Stress–Strained Vascular State,* Nauka, Moscow, 1990, 360pp. (In Russian.)

[207] K. Gopalsamy, *Equations of Mathematical Ecology,* Kluwer A.P., Dordrecht, 1992, 341pp.

[208] H. Gorecki, S. Fuksa, P. Grabomski, A. Korytowski, *Analysis and Synthesis of Time Delay Systems,* John Wiley, 1989.

[209] V.D. Gorjachenko, S.L. Zolotarev, V.A. Kolchin, *Qualitative Methods in Nuclear Reactor Dynamics,* Energoatomizdat, Moscow, 1988. (In Russian.)

[210] A. Goubet-Bartholoméüs, H. Dambrine, J.P. Richard, Stability of perturbed systems with time-varying delays, *System, Control Letters* **31** (1997), pp. 155-163.

[211] R.C. Graef, M.K. Grammatikopoulos, P.W. Spikes, On the asymptotic behaviour of solutions of second order nonlinear neutral delay differential equation, *J. Math. Anal. & Appl.* **156** (1991), pp. 23-39.

[212] M.K. Grammatikopoulos, S.A. Tersian, On the periodic solutions of nonlinear neutral differential equations, *J. Dynamic Systems and Applic.* **6** (1997), pp. 197-206.

[213] G. Gripenberg, S. -O. Londen, O. Staffans, *Volterra Integral and Functional Equations,* Cambridge University Press, 1990, 880pp.

[214] F. Grodins, J. Buell, A. Bart, Mathematical analysis and digital simulation of the respiratory control system, *J. Appl. Physiology* **22: 2** (1967), pp. 260-276.

[215] P.S. Gromova, A.M. Zverkin, About trigonometric series the sum of which is a continuous unbounded function-solutions of an equation with delay, *Diff. Uravn.* **4: 10** (1968), pp. 1174-1784. (In Russian.)

[216] S. Grossberg,

1) Global ratio limit theorems for some nonlinear functional-differential equations, *Bull. Amer. Math. Soc.* **74: 1** (1968), pp. 95-99; 100-105.

2) A prediction theory for some nonlinear functional-differential equations 1. Learning of lists, *J. Math. Anal. & Appl.* **21: 3** (1968), pp. 643-694.

3) A prediction theory for some nonlinear functional-differential equations 2. Learning of parameters, *J. Math. Anal. & Appl.* **22: 2** (1968), pp. 490-522.

4) On varitional of some nonlinear difference-differential equations, *J. Diff. Eq.* **6: 3** (1969), pp. 534-577.

5) On the global limits and oscillations of a system of nonlinear differential equations describing a flow on a probabilistic network, *J. Diff. Eq.* **5: 3** (1969), pp. 531-563.

6) Learning and energy entropy dependence in some nonlinear functional differential systems, *Bull. Amer. Math. Soc.* **75: 6** (1969), pp. 1238-1242.

[217] Z. Grossman, I. Gumowski, Self-sustained oscillations in the Jacob-Monod mode of generegulation, *Lect. Notes Comp. Sc.* **40** (1976), pp. 145-154.

[218] F. Grupp, On difference–differential equations in the theory of series, *J. Number Th.* **24** (1984), pp. 154–173.

[219] I. Guintchev, On the stability of zero solution of retarded functional–differential systems with an application to nuclear reactor dynamics, *Zag. drgannielin* **22** (1984), pp. 107–126.

[220] V.V. Guljaev, A.S. Dmitriev, V.E. Kislov, Strange attractors in the circle: selfoscillating systems, *Dokl. Akad. Nauk SSSR* **282: 1** (1985), pp. 53–66. (In Russian.)

[221] I. Gumovski,

1) Sur le calcul des solutions periodiques de l'equation de Cherwell–Wright, *C.R. Acad. Sci.* **268: 3** (1969), pp. A157–A159.

2) Periodic solution of differential equations with pure delay arising in a biological problem, *Simul. Syst. Proc. 8th AICA Congr., Delft* 1976 (1977), Amsterdam, pp. 633–639.

3) Qualitative properties of some dynamical systems with pure delay, *Lect. Notes Biomath.* **41** (1981), pp. 139–152.

[222] I. Gumowski, P. Sonrias, Analysis of the blood platelet control mechanism, *Sim. Syst.'79,* Amsterdam, 1980, pp. 1071–1076.

[223] Guo–Bao–Shu, Guang–Tian Zhu, The linear invariant population evolution equation with time delay, *J. Syst. Sc. & Math. Sc.* **8: 4** (1988), pp. 346–350.

[224] Dajun Guo, V. Lakshmikantham, Positive solution of nonlinear integral equations arising in infections diseases, *J. Math. Anal. & Appl.* **134: 1** (1988), pp. 1–8.

[225] Nazesh N. Gupta, J. Houndeshell, A differential–difference equations model of a dynamic aquatic eco–system, *Int. J. Syst. Sc.* **7: 5** (1976), pp. 481–492.

[226] H. Guretsky, *Analysis and Synthesis of Control Systems with Delay,* Mashinostroenie, Moscow, 1974. (In Russian.)

[227] W.S.C. Gurney, R.M. Nisbet, J.H. Lawton, Nicholson's blowflies revisited, *Nature* **187** (1980), pp. 17–21.

[228] P. Guttorp, R. Kulperger, Statistical inference for some Volterra population process in a random environment, *Canad. J. Stat.* **12: 4** (1984), pp. 289–302.

[229] M. Gyllenberg, H.J.A.M. Heijmans, An abstract delay–differential equation modelling size dependent cell growth and division, *SIAM J. Math. Anal.* **18: 1** (1987), pp. 74–88.

[230] I. Gyori,

1) Connections between compartmental systems with pipes and integro–differential equations, *Math. Modell.* **7: 9/12** (1986), pp. 1215–1238.

2) Remarks about differential equations of the population growth, *Alkalm. Mat. Lapok.* **1: 1/2** (1975), pp. 139–150.

[231] I. Gyori, J. Eller,

1) Compartmental systems with pipes, *Math. Biosc.* **53: 3/4** (1981), pp. 223–247.

2) Analytic properties and indentifiability problems of compartmental models with time lags, *Math. & Comp. Meth. Physiol. Satell. Symp. 28th Int. Congr. Physiol. Sci.*, Budapest, 1980 (1981), pp. 61–74, Oxford.

[232] I. Gyori, G. Ladas, *Oscillation Theory of Delay Differential Equations,* Clarendon Press, Oxford, 1991.

[233] S.D. Haddad, Combustion and heat release in disel engines, *Principles and Performance in Diesel Engineering*, ed. by S.D. Haddad, and N. Watson, Ellis Horwood Limited, 1984, p. 28.

[234] J.R. Haddock, T. Krisztin, J. Terjéki, Invariance principles for autonomous functional differential equations, *J. Integral Eqns.* **10** (1989), pp. 123–136.

[235] K.P. Hadeler, Delay equations in biology, *Lect. Notes in Math* **730** (1979), Berlin, pp. 139–156.

[236] A. Halanay, *Differential Equations: Stability, Oscillations, Time–Lag,* N–Y, Acad. Press, 1966.

[237] J.B.S. Haldane, A contribution to the theory of price fluctuations, *Rev. Econ. Stud.* **1** (1933), pp. 186–195.

[238] J.K. Hale, S.M. Verduyn Lunel, Introduction to functional differential equations, N–Y, Springer, 1993.

[239] Y. Halevi, Reduced–order models with delay. *International Journal of Control*, **64: 4** (1996), pp. 733–744.

[240] A.J. Hall, G.C. Wake, A functional differential equation arising in modelling of cell growth, *J. Austral. Math. Soc. B* **30: 4** (1989), pp. 424–435.

[241] K.B. Hannsgen, I. Renardy, R.L. Wheeler, Effectivness and robustness with respect to time delays of boundary feedback stabilization in one–dimensional viscoelasticity, *SIAM J. Contr. Optim.* **26: 5** (1988), pp. 1200–1234.

[242] F.B. Hanson, Analysis of a singular functional differential equation that arises from stochastic modelling of population growth, *J. Math. Anal. & Appl.* **96: 2** (1983), pp. 405–419.

[243] A. Haraux, *Nonlinear Evolution Equations. Global Behavior of Solutions,* Springer, Berlin, 1981.

[244] J. Harband, The existence of monotonic solutions of a nonlinear car–following equation, *J. Math. Anal. & Appl.* **57: 2** (1977), pp. 257–272.

[245] U. van der Heiden, H.–O. Walther, Existence of chaos in control systems with delayed feedback, *J. Diff. Eqns.* **47** (1983), pp. 273–295.

[246] R.H. Hering, Stabilities of functional differential equations in fading memory spaces, *Dynamic Systems and Applic.* **4** (1995), pp. 497–512.

[247] H.W. Hethcote, M.A. Lewis, P. van der Driessche, An epidemiological model with a delay and a nonlinear incidence rate, *J. Math. Biol.* **27: 1** (1989), pp. 49–64.

[248] H.W. Hethcote, H.W. Stech, P. van der Driessche, Nonlinear oscillations in epidemic models, *SIAM J. Appl. Math.* **40: 1** (1981), pp. 1–9.

[249] Y. Hino, S. Murakami, T. Naito, *Functional-Differential Equations with Infinite Delay,* Springer, 1991.

[250] J.T. Hoag, R.D. Driver, A delayed–advanced model for the electrodynamic two–body problem, *Nonlinear Anal. Theory Methods & Appl.* **15: 2** (1990), pp. 165–184.

[251] M. Niezgódka, Large–time solutions of the two–phase Stephan problem with delay, *J. of Math Anal. & Appl.* **170** (1992), pp. 433–448.

[252] K.H. Hoffman, J. Sprekels, Automatic delay–control in a two–phase Stefan problem, In: L. Collatz (ed.), *Differential–Difference Equations,* Birkhäuser, Basel, 1983, pp. 119–136.

[253] W.J. Hrusa, A nonlinear functional differential equation in Banach space with applications to material with fading memory, *Arch. Rat. Mech. & Anal.* **84: 2** (1983), pp. 99–137.

[254] D. Husainov, A. Shaturko, *Liapunov Method in Stability of Functional Differential Systems,* Naukova Dumka, Kiev, 1997, 236pp. (In Russian.)

[255] G.E. Hutchinson, Circular causal systems in ecology, *Ann. N.Y. Acad. of Sci.* **50** (1948), pp. 221–246.

[256] A.J. Isaksson, M. Hagberg, and L.E. Johnsson, Benchmarking for paper machine MD–control: simulation results, *Control Eng. Practice* **3: 10** (1995), pp. 1491–1498.

[257] K. Ito, M. Nisio, On stationary solutions of stochastic differential equation, *J. Math. Sci. Kyoto Univ.* **4** (1954), pp. 1–75.

[258] I. Ivanova, On almost periodic solution of nonlinear system of neutral type equations with a small delay, *Abstracts of the IVth International Seminar "Stability and nonlinear control systems oscillations"*, Moscow, June, 1996, pp. 83–84. (In Russian.)

[259] A.F. Ize, *Functional Differential Equations and Bifurcations*, Springer, New York, 1980.

[260] O.L.R. Jacobs, Designing feedback controllers to regulate deposited mass in hot–dip galvanizing, *Control Eng. Practice* **3: 11** (1995), pp. 1529–1542.

[261] C.S. Kahane, On the stability of solutions of linear differential systems with slowly varying coefficients, *Czech. Math. Journ.* **42: 117** (1992), pp. 715–726.

[262] S. Kakutani, L. Marcus, On the nonlinear difference–diferential equation $y'(t) = (A - By(t-1))y(t)$, *Contr. to the Theory of Nonlinear Oscillations* **4** (1958), pp. 1–18.

[263] G.A. Kamenskii, A.D. Myshkis,

1) Boundary value problem for second order quasilinear differential equations with of divergent kind with deviating argument, *Diff. Eqs.* **10: 12** (1974), pp. 2137–2146. (In Russian.)

2) Statement of boundary value problems for differential equations with deviating arguments and some leading terms, *Diff. Eqs.* **10: 3** (1974), pp. 409–418. (In Russian.)

[264] G.A. Kamenskii, A.D. Myshkis, A.L. Skubachevskii, Generalized and smooth solutions of boundary value problems for functional differential equations with many senior members, *Časopis propěstování matematiky* **111** (1986), pp.254–266.

[265] D. Kannan, Evolution operators and equations of an interacted prey–predator population incorporating historical actions, *Nonlinear Anal., Theory, Meth. & Appl.* **2: 1** (1978), pp. 59–68.

[266] L.V. Kantorovich, G.P. Akilov, *Functional Analysis,* Nauka, Moscow, 1977.

[267] F. Kappel, Differenzen–Differetialgleichungen, *Ber. Math. Stat. Seh. Forschungszentrum Gras* **1** (1973), pp. 1–77.

[268] F. Kappel, K. Kunisch, Invariance result for delay and Volterra equations in fractional order Sobolev spaces, *Trans. Amer. Math. Soc.* **304: 1** (1987), pp. 1–51.

[269] P. Kappel, D. Salamon, Spline approximations for retarded systems and Riccati equation, *SIAM J. Control Opt.* **25: 4** (1987), pp. 1082–1114.

[270] F. Kappel, W. Schappacher,

1) Some considerations to the fundamental theory of infinite delay equations, *J. Diff. Eq.* **37** (1980), pp. 141–183.

2) *Evolution Equations and Their Applications,* Pitman, Boston–London, 1982.

[271] J.N. Kapur,

1) Predator–prey models with discrete time lags, *Nat. Acad. Sci. Letters* **2: 7** (1979), pp. 273–275.

2) Bifurcation theory in population models with time delays, *Indian J. Pure & Appl. Math.* **11: 11** (1980), pp. 1397–1410.

3) Stability analysis of a difference equation population model with delayed recruitment, *Indian J. Pure & Appl. Math.* **12: 1** (1981), pp. 7–21.

[272] G. Karakostas,

1) The effect of seasonal variations to the delay population equation, *Nonlinear Anal. Theory. Methods & Appl.* **6: 11** (1982), pp. 1143–1154.

2) Asymptotic equilibrium in prey–predator interaction with nonlinear "per capita" growth rates, *Funkc. Ekv.* **30: 2/3** (1987), pp. 383–397.

[273] I. Kashiwagi, I. Flugge–Lotz, Periodic solutions of differential–difference equations with delay control, *Int. J. Control* **6: 1** (1967), pp. 1–25.

[274] J. Kato,

1) On Liapunov–Razumikhin type theorems for functional differential equations, *Funkc. Ekvacioj* **16** (1973), pp. 225–239.

2) Liapunov second method in functional differential equations, *Tohoku Math. J.* **32** (1980), pp. 487–497.

[275] B.Z. Khan, B. Lehman, Setpoint PI controllers for systems with large normalized dead time, *IEEE Transactions on Control Systems Technology* **4: 4** (1996), pp. 459–466.

[276] V.L. Kharitonov, A.P. Zhabko, Robust stability of time delay systems, *IEEE Trans. AC* **39: 12** (1994), pp. 2388–2397.

[277] M. Khoo, R. Kronauer, K. Strohl, A. Slutsky, Factors inducing periodic breathing in humans: a general models, *J.Appl. Physiol.* **53: 3** (1982), pp. 644–659.

[278] A.V. Kim, *Direct Liapunov Method in Stability Theory of Hereditary Systems,* Ural Univ. Press, Sverdlovsk, 1992. (In Russian.)

[279] K.D. Kim, N.K. Bose, Vertex implications of stability for a class of delay differential interval systems, *IEEE Trans. Circuits & Systems* **37: 7** (1990), pp. 969–972.

[280] J. Kogan, *Robust Stability and Convexity,* N–Y, Springer, 1995, 176pp.

[281] Yu.S. Kolesov, D.I. Shvitra, *Autovibrations in Systems with Delay,* Mokslas, Vilnius, 1979. (In Russian.)

[282] V.B. Kolmanovskii

1) Periodic solutions of stochastic equation with delay, *Diff. Uravn.* **24: 6** (1988), pp. 965–972. (In Russian.)

2) On chemostat stability, *Differential Eqs.* **11** (1992), pp. 1565–1568. (In Russian.)

3) Stability of some systems with arbitrary delays, *Reports of Russian Academy of Science* **331: 4** (1993), pp. 421–424. (In Russian.)

4) Stability of some nonlinear functional differential equations, *J. Nonlinear Diff. Eqs. and Appl.* **2** (1995), pp. 185–198.

5) The stability of certain retarded systems with variable coefficients, *J. Appl. Maths. Mechs.* **59: 1** (1995), pp. 65–74. (In Russian.)

6) Applications of differential inequalities for stability of some functional differential equations, *Nonlinear Analysis, TMA* **25: 9–10** (1995), pp. 1017–1028.

7) Stability in probability of Lotka–Volterra system, *J. Diff. Eqs.* **32: 11** (1996), pp. 1480–1488.

8) Stability of discrete Volterra equations, *Reports of Russian Acad. of Sci.* **349: 5** (1996), pp. 610–614. (In Russian.)

9) Stability of neutral type functional differential equations, *Appl. Math. and Mech.* **60: 2** (1996), pp. 210–222. (In Russian.)

10) Boundedness of some discrete systems, *Autom. and Remote Control* 4 (1997), pp. 121–132. (In Russian.)

[283] V.B. Kolmanovskii, R.Z. Hasminskii, Stability of linear systems with retardation, *Ivz. Vyssh. Uchebn. Zav. Mat.* **53: 4** (1966), pp. 58–65. (In Russian.)

[284] V.B. Kolmanovskii, N.I. Koroleva,

1) Optimal control of bilinear hereditary systems and their applications, *Appl. Math. Letters* **4: 2** (1991), pp. 1–3.

2) Synthesis of bilinear systems with delay in control, *Appl. Math. and Mech.* **57: 1** (1993), pp. 32–40. (In Russian.)

[285] V.B. Kolmanovskii, T.L. Maizenberg, Optimal estimation of system states and problems of control of systems with delay, *Applied Math. and Mech.* **41: 3** (1977), pp. 446–456. (In Russian.)

[286] V.B. Kolmanovskii, A.I. Matasov,

1) Accuracy estimate of approximate solution of minimax filtering problem for systems with delay, *Doklady Physics* **39: 11** (1994), pp. 758–760.

2) Minimax filtering problems in systems with delay, *Autom. and Remote Control* **6: 11** (1996), pp. 125–147. (In Russian.)

3) An approximate method for solving optimal control problem in hereditary systems, *Doklady Mathematics* **55: 3** (1997), pp. 475–478.

[287] V.B. Kolmanovskii, A.D. Myshkis

1) *Applied Theory of Functional Differential Equations*, Kluwer Academic Press Publishers, Dordrecht, 1992.

2) Stability in the first approximation of some Volterra difference equations, *J. of Difference Equations and Applications*, **3** (1998), pp. 563–569.

[288] V.B. Kolmanovskii, A.D. Myshkis, V.R. Nosov,, Modern theory of equations with delay from the point of view of its application, *Sovr. Probl. Mat. Fiz.*, Tbilisi, 1987, pp. 280–290.

[289] V.B. Kolmanovskii, V.R. Nosov,

1) *Stability and Periodic Solutions of Regulated Systems with Delay*, Nauka, Moscow, 1981. (In Russian.)

2) Stability of neutral type functional differential equations, *Nonlinear Anal. Theory, Methods. & Appl.* **6: 9** (1982), pp. 873–910.

3) *Stability of Functional Differential Equations*, Acad. Press, New York–London, 1986.

[290] V.B. Kolmanovskii, J. –P. Richard (eds.),

1) Time–delay systems, *Special Issue of the journal "Mathematics and Computers in simulation"*, *Elsevier Sc. Publ.* **6** (1997).

2) Stability of some systems with distributed delay, *JESA*, *European Journal of Automatic Control* **31: 6** (1997), pp. 971–981.

[291] V.B. Kolmanovskii, A.M. Rodionov, Stability of some discrete processes, *Autom. and Remote Control* **2** (1995), pp. 3–13. (In Russian.)

[292] V.B. Kolmanovskii, L.E. Shaichet,

1) Filtering of Gaussian stochastic processes by observations with delays, *Lect. Notes in Economics and Math. Systems, Advances in Optimization*, Springer, **382** (1992), pp. 330–345.

2) Successive approximations to the optimal control of stochastic systems with aftereffect, In the book: *Random Operators and Stoch. Eq–ns*, Netherlands, Part 1: 1992, 1,2, pp. 245–258, Part 2: 1993, 1,3, pp. 375–391.

3) On a method of Liapunov functionals construction for stochastic delay equations, *J. Diff. Eqs.* **29: 11** (1993), pp. 1909–1920.

4) Stability of stochastic equations with delay, *Autom. and Remote Control* **7** (1993), pp. 66–85. (In Russian.)

5) Some problems of optimal control and estimation for system with delay, in the book: *Differential Equations and Control Theory*, eds. Deng, Liang, Lu, Marcel Dekker, Ruan, N–Y, 1995, pp. 129–141.

6) General method of Liapunov functionals construction for stability of stochastic difference equations, *World Scientific series in Applicable Analysis* **4**, 1995, pp. 397–439.

7) Numerical solution of some optimal control problems, *Stochastic Analysis and Appl.* **14: 1** (1996), pp. 101–130.

8) Control of systems with aftereffect, *Amer. Math. Soc. RI* **157** (1996), 359pp.

9) Stability of some stochastic Volterra Equations, *J. Diff. Equat.* **33: 11** (1997), pp. 1495–1502.

[293] V.B. Kolmanovskii, P.A. Velmisov, Stability of equations for viscoelastic plates in liquid, *J. Diff. Eqs.* **30: 11** (1997), pp. 1966–1982.

[294] V.B. Kolmanovskii, L. Torelli, R. Vermiglio, Stability of some test equations with delay, *SIAM J. Math. Anal.* **25** (1994), pp. 948–961.

[295] G.E. Kolosov, *Optimal Design of Control Systems: Stochastic and Deterministic Problems,* N–Y, Marcel Dekker, 1998, 315p.

[296] S.A.L.M. Kooijman, B.W. Kooi, M.P. Boer, Rotifers do it with delay: the behaviour of reproducery vs dividers in chemostat, *Nonlin. World* **3** (1996), pp. 107–128.

[297] M. Kot, *Delay Differential Models and the Effects of Economic Delays on the Stability of Open–Access Fisheries,* M.S. Thesis, Cornell Univ., 1979.

[298] N.N. Krasovskii, *Stability of Motion,* Stanford Univ. Press, 1963.

[299] L. Krivy,

1) Some models of predator–prey with time deviation, *Sb. Pr. Ped. Ostrave A* **17: 77** (1982), pp. 15–24.

2) Retardation in models of isolated population reproduction, *Sb. Pr. Ped. Ostrave A* **18: 82** (1983), pp. 3–12.

3) The Maya's model of predator–prey system with time delay, *Sb. Pr. Ped. Ostrave A* **20: 92** (1985), pp. 3–15.

[300] I. Krolik, M. Eizenman, S. Pasupathy, Time delay estimate of signals with uncertain spectra, *IEEE Trans. Acoustics, Speech & Signal Proc.* **36: 12** (1988).

[301] Y. Kuang, *Delay Differential Equations with Applications in Population Dynamics,* Academic Press, London, 1993.

[302] Y. Kuang, H.L. Smith

1) Global stability in diffusive delay Lotka–Volterra Systems, *Diff. & Integral Eq.* **4: 1** (1991), pp. 117–128.

2) Convergence in Lotka–Volterra type diffusive delay systems without dominating instantaneous negative feedback, *J. Austral. Math. Soc. B* **34: 4** (1993), pp. 471–494.

[303] M.R.S. Kulenovic, G. Ladas, Oscillations of the sunflower equation, *Quart. Appl. Math.* **46: 2** (1988), pp. 23–29.

[304] V.G. Kurbatov, *Linear Differential–Difference Equations,* Univ. Press, Voronezh, 1990. (In Russian.)

[305] W.H. Kwon, C.W. Lee, S.W. Kim, Performance improvement using a small delay, *Autom. & Remote Control* **2** (1986), pp. 31–38.

[306] G. Ladas, I.P. Stavroulakis, On delay differential inequalities of first order, *Funkcialaj* **25** (1982), pp. 105–111.

[307] G.S. Ladde,

1) Differential inequalities and stochastic functional differential equations, *J. Math. Phys.* **1: 6** (1974), pp. 738–743.

2) Stochastic stability analysis of model ecosystems with time delay, *Link. Sci. & Appl. Autom. Control*, Proc. 7th World Congr. Int. Fed. Autom. Control., Helsinki 1978, **2** (1979), pp. 1149–1154.

[308] V. Lakshmikatham, S. Leela, A technique in stability theory delay–differential equations, *Nonlin. Anal. Th. Meth. & Appl.* **3: 3** (1979), pp. 317–323.

[309] V. Lakshmikatham, S. Leela, A. Martynyuk, Stability Analysis of Nonlinear Systems, M. Dekker, N–Y, 1989.

[310] V. Lakshmikatham, H.R.M. Rao, Theory of integro–differential equations, Gordon and Breach, 1995.

[311] M.R. Langlais, Stabilization in nonlinear age dependent population model, Biomath. & Relat. Comput. Probl., Dordrecht, 1988, pp.217–248.

[312] A.B. Larson, The hog cycle as harmonic motion, *J. Farm. Econ.* **46** (1946), pp. 375–386.

[313] A. Lasota, Ergodic problems in biology, *Asterisque* **50** (1977), pp. 239–250.

[314] R. Laugalis, D. Shvitra,

1) Red blood production mathematical modelling, *Diff. Eqs. & Appl.*, Vilnius, **34** (1983), pp. 29–39.

2) Control of white blood cell production problem, *Lit. Mat. Sb.* **26: 1** (1986), pp. 70–80.

[315] H.A. Lauwerier, Mathematical models of epidemics, Math Centre Tracks, **138** (1981).

[316] M.A. Lavrentjev, B.V. Shabat, Methods of Theory of Functions of complex variable, Fizmatgiz, Moscow, 1958.

[317] A.Y. Lee, Design of stability augmentation systems for automotive vehicles, *J. of Dynamics Systems, Measurm., and Control* **112** (1990), pp. 489–495.

[318] E.B. Lee, Variational problems for systems having delay in the control action, *IEEE Trans. Autom. Control AC* **13** (1968), pp. 387–399.

[319] C.S. Lee, G. Leitmann, Continuous feedback guaranteeing uniform ultimate boundedness for uncertain linear delay systems: an application to river pollution control, *Comput. math. applic.* **16: 10/11** (1988), pp. 929–938.

[320] E.B. Lee, Y. You, Optimal synthesis for infinite–dimensional linear delayed state–output systems: a semicausality approach, *Appl. Math. Optim.* **19** (1989), pp. 113–136.

[321] B. Lehman, J. Bentsman, Vibrational control of linear time lag systems with arbitrary large but bounded delays, *IEEE AC* **37: 10** (1992), pp. 1576–1582.

[322] B. Lehman, V.B. Kolmanovskii, Extensions of classical averaging techniques to delay differential equations, in Proc. IEEE 1994 Conf. Decision Contr. **1** pp. 411–416.

[323] B. Lehman, K. Shujaee, Delay independent stability conditions and decay estimates for time–varying functional differential equations, *IEEE Trans. AC* **39: 8** (1994), pp. 1673–1676.

[324] S.M. Lenhart, C.C. Travis, Global stability of a biological model with time delay, *Proc. Amer. Math. Soc.* **96: 1** (1986), pp. 75–78.

[325] A. Leung,

1) Conditions for global stability concerning a prey–predator model with delay effects, *SIAM J. Appl. Math.* **36: 2** (1979), pp. 281–286.

2) Periodic solutions for a prey–predator differential delay equation, *J. Diff. Eq.* **26: 3** (1977), pp. 391–403.

[326] D. Levi, Nonlinear differential–difference equations as Backlund transformations, *J. Phys. A.* **14: 5** (1981), pp. 1083–1098.

[327] D. Levi, O. Ragnisso, Nonlinear differential–difference equations with N–dependent coefficients 1–2, *J. Phys. A* **12: 7** (1979), pp. L157–162; L163–167.

[328] J.J. Levin, Boundedness and oscillation of some Volterra and delay equations, *Diff. Uravn.* **5: 2** (1969), pp. 369–398.

[329] S. Levin, L. Segal, Pattern generation in space and aspects, *SIAM Rev.* **17** (1985), pp. 45–67.

[330] R.M. Lewis, B.D. O. Anderson, Necessary and sufficient conditions for delay–independent stability of linear autonomous system, *IEEE AC* **25: 4** (1980), pp. 735–739.

[331] X.Y. Li, S.N. Chow, Maximum principle of optimal control for functional differential systems, *J. Optim. Theory & Appl.* **54: 2** (1987).

[332] Juan Lin, P.B. Kahn, Averaging methods in the delayed–logistic equation, *J. Math. Biol.* **10: 1** (1980), pp. 89–96.

[333] Laifn Lin, J.E. Cihen, Equilibrium and local stability in a logistic matrix model for age–structured populations, *J. Math. Biol.* **25: 1** (1987), pp. 73–88.

[334] D.A. Limkins, R.I. Vitney, Mode analysis of physiological oscillations intercoupled via pure time delays, *Bull. Math. Biol.* **44: 1** (1982), pp. 57–74.

[335] Ching–Tien Lion, Y. Shyong Chou, Maximum energy control of time delay systems via piecewise linear polynomial functions, *Int. J. Control* **46: 5** (1987).

[336] Y. Lin, B. Anderson, Model reduction with time delay, *Proc. IEEE* **134: 6** (1987), pp. 349–367.

[337] E. Liz, J.J. Nieto, Periodic boundary value problems for a class of functional differential equations, *J. Math. Anal. Applic.* **200: 3** (1986), pp. 680–686.

[338] M. Lizana, Asymptotic expansion for the initial value problem of the sunflower equation, Proc. 11th Int. Conf. Nonlin. Oscill., Budapest Aug. 1987 17–23 (1987), pp. 297–300.

[339] A.S. Lodge, J.B. McLeod, J.A. Nohel, A nonlinear singular perturbed Volterra integrodifferential equation occurring in polymer rheology, *Proc. R. Soc. Edinburgh A* **80** (1978), pp. 99–137.

[340] S.O. Londen, I.A. Nohel, Nonlinear Volterra integrodifferential equation occurring in heat flow, *J. Int. Eqs.* **6** (1984), pp. 11–50.

[341] S.T. Lunt, G. Tolman, Some effects of delays in control systems, *Meas. & Control* **13: 5** (1980), pp. 184–185.

[342] N. Macdonald,

1) Time lags in biological models, *Lect. Notes Biomath.* **27** (1978), 112pp.

2) An activation–inhibition model of cyclic granulopoiesis in chronic granulocytic leukomia, *Math. Biosc.* **54: 112** (1981), pp. 61–70.

[343] M.C. Mackey, J.G. Milton, Feedback, delays and the origins of blood cell dynamics, *Comm. Theor. Biol.* **1** (1990), pp. 299–327.

[344] J.M. Mahaffy, Stability of periodic solutions for a model of genetic repression with delays, *J. Math. Biol.* **22: 2** (1985), pp. 137–144.

[345] Yu. L. Maistrenko, E. Yu. Romanenko, A.N. Sharkovskii, Integrodifferential equations and numerical population dynamics, *Diff. Eqs. & Appl.*, Vilnius **29** (1981), pp. 69–78.

[346] Zavarei M. Malek, M. Jamshidi, Time–Delay Systems; Analysis, Optimization & Application, North Holland, Amsterdam, 1987.

[347] J. Mallet–Paret, Morse decomposition for delay differential equations, *J. Diff. Eqs.* **62** (1986), pp. 266–274.

[348] J. Mallet–Paret, R.D. Nussbaum, Global continuation and complicated trajectories for periodic solutions of a differential–delay equation, Proc. of Symposia "Nonlinear Functional Analysis and Applications", **45: 2** (1986), pp. 155–167.

[349] A. Manitius, R. Triggiani, Function space controllability of linear retarded systems, a derivation from abstract operator conditions, *SIAM J. Control & Optim.* **16: 4** (1978), pp. 599–645.

[350] X. Mao, Razumikhin–type theorems on exponential stability of neutral stochastic functional differential equations, *SIAM J. Math. Anal.* **28: 2** (1997), pp. 389–401.

[351] P. Marcati, Some considerations on the mathematical approach to nonlinear age dependent population dynamics, *Comput. & Math.* **9: 3** (1983), pp. 361–370.

[352] C. Marcelli, Comparison theorems in hereditary structures and applications, *J. Math. Anal. Appl.* **196** (1995), pp. 1073–1092.

[353] C. Marcelli, P. Rubbioni, A new extension of classical Müller's theorem, *Nonlin. Anal. TMA* **28: 11** (1997), pp. 1759–1767.

[354] C. Marcelli, A. Salvadori, On the equivalence of various functional differential settings, Proceedings 15th IMACS world congress, **1**, Berlin, 1997, pp. 209–213.

[355] G.I. Marchuk, Mathematical Modelling of Immune response in infectious diseases, Kluwer AP, 1997, 358pp.

[356] M. Marcus, V.J. Mizel,

1) Semilinear hereditary hyperbolic systems with nonlocal boundary conditions, *J. Math. Anal. & Appl.* **77: 1** (1980), pp. 1–19.

2) Stochastic functional differential equations modelling materials with selective recall, *Stochastics* **25** (1988), pp. 195–230.

[357] C.M. Marcus, R.M. Westervelt, Stability of analog networks with delay, *Phys. Rev. A* **39** (1989), pp. 347–359.

[358] M. Marek, I. Schreiber, Chaotic behaviour of deterministic dissipative systems, Academia, Praha, 1991.

[359] J.E. Marshall, Control of time delay systems, *IEEE Control Eng. Series* **10** (1978).

[360] J.E. Marshall, S.V. Salehi, Improvement of system performance by the use of time–delay elements, *IEEE Proc. D. Control Th. & Appl.* **129: 5** (1982), pp. 177–181.

[361] M. Martelli, K. Schmitt, H. Smith, Periodic solutions of some nonlinear delay–differential equations, *J. Math. Anal. & Appl.* **74: 2** (1980), pp. 494–503.

[362] R.H. Martin, H.L. Smith, Convergence in Lotka–Volterra systems with diffusion and delay, Proc. Workshop. Diff. Eqs. & Appl. Retzhof, Austria, M. Dekker, 1989.

[363] R.H. Martin, H.L. Smith, Reaction–diffusion systems with time delays: monotonicity, invariance, comparison, and convergence, *J. Reine Angew. Math.* **413** (1991), pp. 1–35.

[364] A.I. Matasov, Estimators for uncertain dynamic systems, Kluwer Academic Publishers, 1988.

[365] S. Maynard, Models in ecology, Cambridge Univ. Press, 1974.

[366] M.C. Memory, Stable and unstable mainfolds for partial functional differential equations, *Nonlinear Anal., Th., Meth. & Appl.* **16: 2** (1991), pp. 131–142.

[367] V.N. Michelkevich, Yu. A. Chabanov, Identification and synthesis of control system of infeed grinding process, *Elektromash. & Electrooborudov.* **31** (1980), pp. 42–52.

[368] M.M. Milman,

1) Repersentation for the optimal controll on for stable hereditary systems, *Lect. Notes in Control & Inform. Sc.* **102** (1987), pp. 253–271.

2) Approximation of the linear quadratic optimal control low for hereditary systems with delays in the control, *SIAM J. Control* **26: 2** (1988), pp. 291–320.

[369] J. Milota, Stability in models with long memories, Biomath. & Relat. Comput. Probl., Dordrecht, 1988, pp. 523–527.

[370] N. Minorsky, Control problems, *J. Franklin Inst.* **232: 6** (1941), pp. 519–551.

[371] Yu. A. Mitropolski, A.Yu. Shvets,

1) Delay effect influence on stability of a pendulum with vinrating central point, Anal. Meth. of Nonlin. Vibration Investigation, Kiev, 1980, pp. 115–120.

2) Vibration of a pendulum with vibrating central point regarding delay effect, Anal. Meth. of Vibrating Investigation, Kiev, 1980, pp. 120–128.

[372] S.E.A. Mohammed,

1) Retarded functional differential equations: A global point of view, Pitman Research Notes in Math. **21** (1980).

2) Stochastic functional differential equations, Pitman Research Notes in Math. **99** (1983).

3) Unstable invariant distributions for a class of stochastic delay equations, Proc. Edinburgh Math. Soc. **31: 1** (1988), pp. 1–23.

[373] R.R. Mohler, Z. Farooqi, T. Heiling, Lymphocyte distribution and lymphatic dynamics, Immun. Volume Ded. to G.I. Marchuk on the Occas. of his 60th birthday, New York, 1986, pp. 317–333.

[374] T. Mori, H. Kokame, Stability of $\dot{x} = Ax + Bx(t - \tau)$, *IEEE Trans. AC* **34: 4** (1989), pp. 460–462.

[375] H.C. Morris, Approximate periodic solutions of a delay differential model of interacting populations, *Int. J. Systems Sc.* **9: 1** (1978), pp. 111–119.

[376] H.C. Morris, R.K. Dodd, Pattern formation in a delay equation with diffusion, *Int. J. Systems Sc.* **11: 6** (1980), pp. 787–792.

[377] H.C. Morris, E.E. Ryan, R.K. Dodd, Periodic solutions and chaos in a delay–differential equation modelling haematopoesis, *Nonlin. Anal. Th. Methods & Appl.* **7: 6** (1983), pp. 623–660.

[378] I.H. Mufti, On the stability of a single–species population model with time lag, *Int. J. Systems Sc.* **10: 10** (1979), pp. 1149–1154.

[379] D. Mukherjee, Global stability of an S–I. Epidemic model with maturation delay, *Cybernetica* **15: 3** (1997), pp. 179–188.

[380] J.D. Murray, Mathematical Biology, Springer, New York, 1989.

[381] A.D. Myshkis,

1) General theory of differential equations with delay, *Usp. Mat. Nauk* **4: 5** (1949), pp. 99–141. (In Russian). (Engl. transl.: *Transl. AMS* **55** (1951), pp. 1–62. Also *Transl. AMS* **1: 4** (1962), pp. 207–267).

2) General theorems of the theory of ordinary differential equations in nonclassical cases, 1st Math. Summer School, Nauk. Dumka, Kiev, 1964, pp. 45–116 (in Russian).

3) Linear Differentiat Equations with Delayed Argument, Nauka, Moscow, 1972. (In Russian). (First ed.: 1951, German transl.: Lineare Differentialyleichungen mit nacheilendem Argument, VEB Deutsch, Verlag, Berlin, 1955).

4) On some problems of the theory of differential equations with deviating argument, *Usp. Mat. Nauk* **32: 2** (1977), pp. 174–202. (In Russian). (Engl. transl.: Russian Math. Survegs **32** (1977), pp. 181–210).

[382] A.D. Myshkis, L.E. El'sgol'tz, The status and problems of the theory of differential equations with deviating arguments, *Usp. Mat. Nauk* **22: 2** (1967), pp. 21–57.

[383] A.D. Myshkis, S.N. Shimanov, L.E. El'sgol'tz, Stability and oscillations of systems with time lag, Tr. ICNO–1, Kiev **2** (1963), pp. 241–267.

[384] W. Nababan, K.L. Teo, Necessary conditions for optimal control for systems governed by parabolic partial delay–differential equations in

divergence form with boundary conditions, *J. Optim. Theory & Appl.* **36** (1982), pp. 565–573.

[385] B. Neta, J.O. Igwe, Finite differences versus finite elements for solving nonlinear integro–differential equations, *J. Math. Anal. & Appl.* **112: 2** (1985), pp. 607–618.

[386] S.I. Niculescu, Systèmes a rètard. Aspects qualitatifs sur stabilité et la stabilization, Paris, Diderot, 1997, 340pp.

[387] G. Nitsche, U. Dressler, Controlling chaotic dynamical systems using time delay coordinates, *Physica D* **58** (1992), pp. 153–164.

[388] V.W. Noonlurg, Competing species model with behavioral adaptation, *J. Math. Biol.* **24: 5** (1986), pp. 543–555.

[389] V.R. Nosov,

1) Linear boundary–value problems with small lags, *Diff. Uravn.* **3** (1967), pp. 1025–1028.

2) Sur l'arternative de Fredholm pour les systèmes linéaires généraux *à* coefficients périodiques et *à* argument déplacé, *C.R. Acad. Sci., Paris* **270: 17**(1970), pp. 1097–1100.

3) Solutions periodiques des systems differentiels fonctionnels quasilineaires, *C.R. Acad. Sci., Paris* **270: 18** (1970), pp. 1170–1173.

4) Periodic solutions of linear equations of general form with deviating arguments, *Diff. Uravn.* **7** (1971), pp. 639–650.

5) Periodic solutions of quasilinear functional differential equations, *Izv. Vysš Učebn. Zaved. Matematica* **5** (1973), pp. 55–62.

6) On stability of some non stationary equations, *Automatic and Remote Control* **9** (1997), pp. 31–42.

[390] V.R. Nosov, M.M. Pushkin, Adaptive control for a linear–quadratic problem with a priori incertitudes, *Izv. AN SSSR, Techn. Kybernetika* **4** (1987), pp. 153–158.

[391] L.I. Novakovskaya, Construction of asymptotic solutions for first order differential equations with random delay, *Ukr. Mat. Zh.* **41: 11** (1989), pp. 1569–1573.

[392] C. Nowak, A maximum principle for optimal control problems with neutral functional differential systems, In : Cybernetics and Systems (Ed. K. Trappl.), Kluwer AP, 1988, pp. 23–30.

[393] L. Nunney, The effect of long time delays in predator–prey systems, *Theoret. Popul. Biol.* **27: 2** (1985), pp. 202–221.

[394] R.D. Nussbaum,

1) Periodic solutions of analytic functional differential equations are analytic, *Michigan Math. J.* **20** (1973), pp. 249–255.

2) Differential–delay equations with two time lags, *Memories AMS* **6: 205** (1978), 62pp.

3) Asymptotic analysis of functional differential equations and solutions of long period, *Arch. Rat. Mech. & Anal.* **81: 4** (1983), pp. 373–397.

[395] M.N. Oguztoreli, Time–Lag Control Systems, Acad. Press, New York, 1966.

[396] M.N. Oguztoreli, T.M. Caelli, An inverse problem in neutral processing, *Biol. Cybern.* **53: 4** (1986), pp. 239–245.

[397] A.W. Olbrot, L. Pandolfi, Null controllability of a class of functional differential systems, *Int. J. Control* **47: 1** (1988), pp. 193–208.

[398] A.P. Oskolkov, M.M. Ahmatov, Asymptotic difference scheme for filtering liquid equation with delay, *Dokl. Akad. Nauk SSSR* **163** (1987), pp. 138–142. (In Russian).

[399] G. Oster, Internal variables in population dynamics, *Lect. Math. Life Sci.* **8** (1976), pp. 37–68.

[400] K. Pakdaman, C.P. Malta, C. Grotta–Ragazzo, O. Arino, J.-F. Vibert, Transient oscillations in continuous–time excitatory ring neural networks with delay, *Phys. Rev. E* **55: 3** (1997).

[401] K.R. Palanisamy, G.P. Rao, Minimum energy control of time–delayed systems via Walsh functions, *Opt. Control Appl. Methods* 4 (1983).

[402] L. Pandolfi, Dynamic stabilization of systems with input delays, *Automatica* **27: 6** (1991), pp. 1047–1050.

[403] C.V. Pao, Systems of parabolic equations with continuous and discrete delays, *J. Math. Anal. and Appl.* **205** (1997), pp. 157–185.

[404] C.V. Pao, J.M. Mahaffy, Qualitative analysis of a coupled reaction–diffusion model in biology with time delays, *J. Math. Anal. & Appl.* **109: 2** (1985), pp. 355–371.

[405] W. Pauwels, Optimal dynamic advertising policies in the presence of a continuously distributed time lags, *JOTA* **22** (1977), pp. 79–89.

[406] M.L. Pena, Asymptotic expansion for the initial value problem of the sunflower equation, *J. Math. Anal. & Appl.* **143: 2** (1989), pp. 471–479.

[407] N.V. Perzev, Asymptotic behaviour of linear differential equations with delay, *Izv. Vuzov Math.* **9: 412** (1966), pp. 48–52.

[408] E. Pinney, Ordinary Difference–Differential Equations, Univ. Calif. Press, Berkley, 1958.

[409] A.B. Piunovskiy, Optimal control of random sequences in problems with constraints, Kluwer AP, Dordrecht, 1997, 345pp.

[410] W.M. Post, C.C. Travis, Global stability in ecologic models with continuous time delay, In Integral and Functional Diff. Eqs.(ed.: T. Herdman), M. Dekker, N–Y, 1981.

[411] H.M. Portella, F.M.O. Castro, N. Arata, Exact solution of the diffusion equation of cosmic–ray nucleons with rising cross section, *J. Phys. G* **14: 8** (1988), pp. 1157–1165.

[412] M. Peschel, W. Mende, The Predator–Prey Model: Do We Live in Volterra World? Akad. Verlag, Berlin, 1966.

[413] V.G. Pisarenko, Problems of Relativistic Dynamics of Many Bodies and Nonlinear Field Theory, Naukova Dumka, Kiev, 1974.

[414] R.E. Plant, A. Kitzhugh, Differential–difference equation modelling recurrent neutral feedback, *SIAM J. Appl. Math.* **40: 1** (1981), pp. 150–162.

[415] S.R. Pliska, On a functional differential equation that arises in Markov control problems, *J. Diff. Eqs* **28: 1** (1978), pp. 390–405.

[416] M. Podovski, Analysis of a class of differential equations with delay effect and its application to nuclear reactor kinetics, *God. Vissh. Uchebn. Zav. Pril. Mat.* **11: 3** (1977), pp. 83–93.

[417] B.T. Polyak, P.S. Cherbakov, Probability approach to robust stability of systems with delay, *Automatic and Remote Control* **12** (1996), pp. 97–108.

[418] L.S. Pontryagin, V.G. Boltyanskii, R.V. Gamkrelidze, E.F. Mishchenko, The Mathematical Theory of Optimal Processes, Interscience, 1962.

[419] B.K. Powell, Nonlinear low frequency phenomenological engine modeling and analysis, Proceedings of the 1987 American Control Conference, Minneapolis, MN, June 10–12, (1987), pp. 332–340.

[420] M.A. Pozio, Behavior of solutions of some abstract functional differential equations and applications to predator–prey dynamics, *Nonlinear Anal. Th. Methods & Appl.* **4: 5** (1980), pp. 917–938.

[421] K. Pyragas, A. Tamaševičius, Experimental control of chaos by delayed self–controlling feedback, *Phys. Lett.A* **180** (1993), pp. 99–102.

[422] M.R.M. Rao, V.N. Pal, Asymptotic stability of grazing systems with unbounded delay, *J. Math. Anal. Applic.* **163** (1992), pp. 60–72.

[423] B. Rashap, P. Khargonekar, J. Grizzle, M. Elta, J. Freudenberg, F. Terry, Real–time control of reactive ion etching: Identification and disturbance rejection, Proceeding of 32nd IEEE Conference on Decision and Control, 1993, pp. 3379–3386.

[424] B.S. Razumikhin, Stability of Hereditary systems, Nauka, Moscow, 1988.

[425] J. Real, Stochastic partial differential equations with delay, *Stochastics* **8: 2** (1982), pp. 81–102.

[426] D.C. Reber, A finite difference technique for solving optimization problems governed by linear functional differential equations, *J. Diff. Eqs.* **32: 2** (1979), pp. 193–232.

[427] M. Renardy,

1) A class of quasilinear parabolic equations with infinite delay and application to a problem of viscoelasticity, *J. Diff. Eqs.* **48: 2** (1983), pp. 280–292.

2) Singular perturbed hyperbolic evolution problems with infinite delay and application to a polymer rheology, *SIAM J. Math. Anal.* **15: 2** (1984), pp. 333–349.

[428] V. Rezvan, Absolute stability of automatic systems with delay, Nauka, Moscow, 1983.

[429] A.M. Rodionov, Stability of some differential and discrete equations with delay, *Automatic and Remote Control* **12** (1996), pp. 38–47.

[430] A.E. Rodkina,

1) On regularity of solutions of stochastic equations for hereditary systems, *Automatic and Remote control* **5** (184), pp. 87–96.

2) On stabilization of stochastic system with retarded argument, *J. Functional Diff. Eqs.* **3: 1–2** (1995), pp. 207–214.

3) On rate control of n–link manipulator robot, *Mathematics and Computers in Simulation* **6** (1997).

[431] G. Rosen, Time delay produced by essential nonlinearity in population growth models, *Bull. Math. Biol.* **49** (1987), pp. 253–255.

[432] N. Rouche, P. Abetz, M. Laloy, Liapunov's direct method in stability theory, Mir, Moscow, 1980.

[433] I.S. Sadek, Optimal control of time–delay systems with distributed parameters, *J. Optim. Th. & Appl.* **67: 3** (1990), pp. 567–585.

[434] D. Salamon, Control and observation of neutral systems, *Pitman Research Notes in Math.* **91** (1982).

[435] M. Saleem, S.U. Siddiqui, V. Gupta, A mathematical model with young predation, *J. Math. Biol.* **25: 1** (1987), pp. 89–101.

[436] N.I. Sancho, C. Mitchell, Economic optimization in controlled fisheries, *Math. Biosc.* **27: 1/2** (1975), pp. 1–7.

[437] T. Sasagava, Optimal control of stochastic systems dependent on the past, *Trans. Soc. Instrum. & Control Eng.* **12: 3** (1976), pp. 257–263.

[438] F.G. Schinskey, Control of pH, The Control Handbook, CRC Press, ed. by William S. Levine, pp. 1205–1217, 1996.

[439] W.A. Schlup, The retarded Josephson equation, *J. Math. Phys.* **19: 12** (1978), pp. 2469–2475.

[440] K. Schmidt (ed.), Delay and Functional Differential Equations and their Applications, Acad. Press, NY, 1972.

[441] D. Schneider, Control of processes with time delays, *IEEE Transactions on Industrial Applications IA* **24** (1988), pp. 186–219.

[442] G. Seifert,

1) A temperature equation for a right heat conductor with memory, *Quart. Appl. Math.* **38: 2** (1980), pp. 246–252.

2) Uniform stability of almost periodic solutions of delay–differential equations, *Lect. Notes in Math.* **1017** (1983), pp. 563–567.

3) On a delay–differential equation for single species population variations, *Nonlin. Anal. Th., Methods & Appl.* **11: 9** (1987), pp. 1051–1059.

[443] S.P. Sethi, T.W. McGuire, Optimal skill mix: An application of the maximum principle for systems with retarded controls, *J. Optim. Th. & Appl.* **23: 2** (1977), pp. 245–275.

[444] S.P. Sethi, G.L. Thompson, Optimal control theory (application to management science), Martinus Nijhoff Publ., Boston, 1981.

[445] N.C. Severo, A recurssion theorem on solving differential–difference equations and applications to some stochastic processes, *J. Appl. Probab.* **6: 3** (1969), pp. 673–681.

[446] D. Sforza, Infinite delay parabolic integro–differential equation with singular kernels, *Nonlinear Anal., Th., Methods & Appl.* **16: 6** (1990), pp. 559–596.

[447] L.E. Shaikhet,

1) On an optimal control problem of partly observable stochastic Volterra's process, *Probl. Control & Inform. Th.* **16: 6** (1987), pp. 439–448.

2) Necessary and sufficient conditions of asymptotic mean square stability for stochastic linear difference equations, *Applied Math. Lett.* **10: 3** (1997), pp. 111–115.

3) Stability in probability of nonlinear stochastic hereditary systems, *Dynamic systems and applic.* **4: 2** (1995), pp. 199–204.

4) Modern state and development perspectives of Liapunov functionals method in the stability theory of stochastic hereditary systems, *Theory of stochastic Processes* **2: 18** (1996), pp. 248–259.

[448] N. Shanmugthasan, R.D. Johnson, Exploitation of time delays for improved process control, *Int. J. Control.* **48: 3** (1988), pp. 1137–1152.

[449] A.N. Sharkovsky, Yu.L. Maistrenko, E.Yu. Romanenko, Difference Equations and their Applications, Naulova Dumka, Kiev, 1986.

[450] T.B. Sheridan, Space teleoperation through time delay: Review and prognosis, *IEEE Trans. on Robotics and Autom., Special Issue on Teleoperation in Space* **9: 5** October 1993.

[451] Akiza Shilata, Nobuhiko Saito, Time delays and chaos in two competing species, *Math. Biosc.* **51: 3/4** (1980), pp. 199–211.

[452] A. Shimbell, Contributions to the mathematical biophysics of the control nervous system with special reference to learning, *Bull. Math. Biophys.* **12** (1950), pp. 241–275.

[453] W.M. Shong, T.J. Tarn, On the spectral controllability of delay differential equations, *IEEE Trans. Autom. Control. AC* **26: 2** (1981), pp. 527–528.

[454] K. Shrestha, The lag relationship between producer and consumer prices. An unobservable variable approach, *Econ. Letters* **22: 2/3** (1986), pp. 173–179.

[455] V.P. Shukla, Conditions for global stability of two–species population models with discrete time delay, *Bull. Math. Biol.* **45: 5** (1983), pp. 793–805.

[456] A.Yu. Shvets,

1) Influence of delay on vibrational–rotational movements of double pendulum, *Methods of Nonlin. Mech. & Appl.* (1982), pp. 141–149.

2) Resonance pendulum vibrations regarding nonconstant delay effects, *Asympt. Integration of Diff. Eqs.*, Kiev (1985), pp. 173–179.

[457] C. Simo, Strungle attractor for differential delay equations, *Lect. Notes in Phys.* **179** (1983), pp. 188–191.

[458] A.S.C. Sinha,

1) Further stability results of delay–differential systems for digital networks, *Int. J. Systems Sc.* **5: 4** (1974), pp. 317–321.

2) Solution of a certain nonlinear functional differential equation arising from articular cartilage, *Int. J. Systems Sc.* **15: 10** (1984), pp. 1049–1053.

[459] G.L. Slatek, W.R. Wells, Optimal evasive tactics against a proportional navigation missile with time delay, *J. Of Spacecraft and Rockets* **10**, May (1973), pp. 309–313.

[460] G.M. Smith, Models in Ecology, Cambridge Univ. Press, Cambridge, 1974.

[461] R.E. Smith, S.K. Sarrafin, Effect of time–delay on flying qualities: An update, *J. of Guidance, Control and Dynamics* **9** (1986), pp. 578–584.

[462] H.L. Smith, P. Waltman, The theory of the chemostat: Dynamics of Microbial Competition, Cambridge UP, 1995.

[463] W. Snow, Existence, uniqueness and stability for nonlinear differential-difference equations in the neutral case, Preprint NY University, N–Y., 1965.

[464] M.A. Soliman, W. Ray, The optimal control of processes containing pure time delays. Transfer functions models, Proc. 13th Joint Autom.

Control Conf. Amer. Autom. Control Councel, Stanford, Calif, 1972 (1972), pp. 294–298.

[465] F. Solimano, E. Bereta, Existence of a globally asymptotically stable equilibrium in Volterra models with continuous time delay, *J. Math. Biol.* **18: 2** (1983), pp. 93–102.

[466] A. Solominos, Periodic solutions of the sunflower equation, *Quart. Appl. Math.* **35: 4** (1978), pp. 465–478.

[467] A. Sparks, J. Buffington, S. Banda, Fighter aircraft lateral directional axes full envelope control law design, *Int. J. Control* **59: 4** (1994), pp. 893–924.

[468] H. Stats, G.A. de Mars, D.T. Wilson, C.L. Tang, Problem of spike eleimination in lasers, *J. Appl. Phys.* **36: 1** (1965), p. 1510.

[469] H.W. Stech, Hopf bifurcation calculations for functional differential equations, *J. Math. Anal. & Appl.* **109: 2** (1985), pp. 472–491.

[470] H. Stech, M.Williams, Stability in a class of cyclic epidemic models with delay, *J. Math. Biol.* **11: 1** (1981), pp. 95–103.

[471] G. Stepan, Retarded Dynamical Systems: Stability and Characteristic Functions, Pitman, NY, 1990.

[472] I.H. Suh, Z. Bien, Use of time delay action in the controller design, *IEEE Trans. Autom. Control* **25** (1980), pp. 600–603.

[473] H.H. Swart, Viable controls in age–dependent population dynamics, *J. Math. Biol.* **27: 3** (1989), pp. 297–308.

[474] G.M. Swisher, S. Tenqchen, Design of proportional minus delay action feedback controller for second and third order systems, Proc. Amer. Control Conf. (1988), pp. 254–260.

[475] G. Szabo, A remark of M. Farkas: stable oscillations in a predator–prey model with time lag, *J. Math. Anal. & Appl.* **128: 1** (1987), pp. 205–206.

[476] L. Szekelyhidi, Stability properties of FDE describing the scientific laws, *J. Math. Anal. & Appl.* **150: 1** (1990), pp. 151–158.

[477] G. Tadmor, Optimal controls and their discontinuities in quadratic problems of delay systems, *IEEE Trans. Autom. Control AC* **30: 7** (1985).

[478] H. Tamura, A discrete dynamical model with distributed transport delays and its hierarchial optimization to preserve stream quality, *IEEE Trans. SMC* **41** (1974), pp. 424–429.

[479] H.R. Thieme,

1) Asymptotic proportionality (weak ergodicity) and conditional asymptotic equality of solutions to time–heterogeneous sublinear difference and differential equations, *J. Diff. Eqs.* **73: 2** (1988), pp. 237–268.

2) Global asymptotic stability in epidemic model, *Lect. Notes in*

Math. **1017** (1983), pp. 608–615.

[480] R. Torrejon, J. Young, On a quasilinear wave equation with memeory, *Nonlinear Anal., Th., Methods & Appl.* **16: 1** (1991), pp. 61–78.

[481] T. Trombetti, A model approach to nonlinear stability in nuclear reactor space dynamics, *Int. J. Engin. Sc.* **21: 9** (1983), pp. 1069–1079.

[482] S. Tuljapurkar, Population dynamics in variable environments, *Lect. Notes in Bio. Math.* **85** (1980).

[483] M. Tylikowski, On stability of the stochastic partial differential equation with deviated argument describing transversal oscillation of a beam squeezed by the axial force, *Zec. Nauk. PSI* **465** (1979), pp. 217–224.

[484] E.F. Tsar'kov, Stochastic Disturbances of Functional Differential Equations, Zinante, Riga, 1989.

[485] Ya. Z. Tsypkin, M. Fu, Robust stability of time–delay systems with an uncertain time–delay constant, *Int. J. Cont.* **57: 4** (1993), pp. 865–879.

[486] R.G. Underwood, A functional differential equation approach to solving infinite games, *SIAM J. Control & Optim.* **14: 2** (1976), pp. 251–267.

[487] P. van den Driessche, Some epidemiological models with delays, in Diff. Eqs. and Applic. to biology and industry, M. Martelli (eds.), World Scientific, 1996, pp. 507–520.

[488] H.R. van der Vaart, A comparative investigation of certain difference equations and related differential equations. Implications for model building, *Bull. Math. Biol.* **35: 1/2** (1973), pp. 195–201.

[489] R. Vallee, C. Delisle, Periodicity windows in a dynamical system with a delayed feedback, *Phys. Rev. A* **34: 1** (1986), pp. 309–318.

[490] R. Vallee, C. Delisle, J. Chrostowski, Noise versus chaos in acousto–optic bistability, *Phys. Rev. A* **30: 1** (1984), pp. 336–342.

[491] J.M.E. van de Vegte, P.Milgram, R.H. Kwong, Teleoperator control models: effect of time delay and imperfect system knowledge, *IEEE Trans. Systems, Math. & Cybern.* **20: 6** (1990), 1258–1272.

[492] G.C. Varley, G.R. Gradwell, M.R. Hassel, Insect population ecology, London, Blackwell Scientific, 1973.

[493] C. Varsan, On necessary conditions for stochastic control problems, *Revue Roumaine de mathematique pures et appliques* **28: 7** (1983), pp. 641–663.

[494] V.A. Vasiliev, V.M. Romanovskii, V.G. Yachno, Autowave Processes, Nauka, Moscow, 1987.

[495] A.B. Vasilieva, Neutral type equations with small delays, *Reports of Academy of Sci. of USSR* **145: 3** (1962), pp. 495–497.

[496] A.B. Vasilieva, V.F. Butuzov, Asymptotic expansions of solutions of singular perturbed equations, Nauka, Moscow, 272p.

[497] Y.V. Venkatesh, Energy Methods in Time–varying system stability and Instability Analyses, Springer, N–Y, 1977.

[498] A.F. Verlan, V.S. Sizikov, Integral equations: methods, algorithms, and programs, Naukova Dumka, Kiev, 1986.

[499] E.I. Verriest, A.F. Ivanov, Robust stability of systems with delayed feedback, Circuits, Systems and Signal Processing, **13: 2** / **13: 3** (1994).

[500] V.R. Vinokurov, The optimal control of systems described by integral equations, *Izv. Vuzov* **62: 7** (1967), pp. 21–33.

[501] W. Voigt, On finite–difference methods for parabolic functional differential equations on unbounded domains, Num. Methods and Applic. Proc. 3rd Intern. Conf., Sofia, 1989, pp. 559–567.

[502] V. Volterra, Théorie mathématique de la lutte pour la vie, Gauthier–Villars, Paris, 1931.

[503] I.N. Voznesenskiy, To Choose Control of Power and Heat Supply Turbines, Soviet Energooborudov., Moscow, 1934. (In Russian).

[504] H.O. Walther,

1) Asymptotic stability for some functional differential equations, *Proc. Roy. Soc. Edinburgh A* **74** (1976), pp. 253–255.
2) One transcendental equation in the stability analysis of a population growth model, *J. Math. Biol.* **3: 2** (1976), pp. 187–195.
3) Hyperbolic periodic solutions, heteroclinic connections and transversal homoclinic points in autonomous differential delay equation, *Mem. AMS* **79: 402** (1989), 104pp.

[505] P. Waltman, Competition models in population biology, Ph., SIAM, 1983.

[506] P.L. Wangersky, J. W. Cunningham, Time–lag in prey–predator population models, *Ecology* **38** (1957), pp. 136–139.

[507] A. Wikan, Dynamic consequences of reproductive delay in Leslie matrix. Models with nonlinear survival probabilities, *J. Math. Biosci.* **146: 1** (1997), pp. 37–62.

[508] J. Wu, X. Zou, Patterns of sustained oscillations in neutral networks with delayed interactions, *Appl. Math. and Computation* **73** (1995), pp. 55–75.

[509] Xue–Ping Wang, Phase–space description of time delay in scattering theory, *Comm. Partial Diff. Eqs.* **13: 2** (1988), pp. 236–259.

[510] V. Watanabe, Finite spectrum assignment and observer or multivariable systems with commensurable delays, *IEEE Trans. Autom. Control* **31: 6** (1986), pp. 543–550.

[511] M. Wazewska–Czyzevsia, A. Lasota, Mathematical problems of dynamics of red blood cell system, *Ann. Polish Math. Soc. Ser 3. Appl. Math.* **17** (1988), pp. 23–40.

[512] G.F. Webb

1) Nonlinear age–dependent population dynamics in L1, *J. Integral Eqs.* **5: 4** (1983), pp. 309–328.

2) Dynamics of populations structured by internal variables, *Math. Z.* **185: 3** (1985), pp. 319–335.

[513] C. Wen, Z. Huaguang, Y. Chaowan, An output tracking control for nonlinear systems with uncertainties and disturbances using time delay control, *Cybernetica* **15: 3** (1997), pp. 229–240.

[514] J. Weyer, A mathematical model for chemical mass recruitment of ants, *J. Math. Biol.* **21: 3** (1985), pp. 307–315.

[515] F.S. Wheeler, Two differential difference equations arising in number theory, *Trans. AMS* **318: 2** (1989), pp. 491–523.

[516] J.A. Wheeler, R.P. Feyman, Classical electrodynamics in terms of direct in terpartical action, *Rev. Modern. Phys.* **21** (1949), pp. 425–433.

[517] B.S. White, The effect of a rapidly–fluctuating random environment on systems of interacting species, *SIAM J. Appl. Math.* **32: 3** (1977), pp. 666–693.

[518] A. Wikan, E. Mjolhus, Overcompensatory recruitment and generation delay in discrete age–structured population models, *J. of Math. Biol.* **35** (1996), pp. 195–239.

[519] D.L. William, Multivariable singularly perturbed feedback systems with time delay, *IEEE Trans. Autom. Control* **32: 11** (1987), pp. 990–994.

[520] L.R. Williams, R.W. Legyetti, Nonzero solutions of nonlinear integral equations modelling infection disease, *SIAM J. Math. Anal.* **13: 1** (1982), pp.112–121.

[521] S.I. Wilson, K.H. Wong, Convergence of a feasible direction algorithm for relaxed controls in time lag systems, *J. Optim. Yh. & Appl.* **53: 3** (1987), pp. 461–479.

[522] A. Winfree, 24 hard problems about mathematics of 24 hours rythms, Nonlin. Oscillations Biol., (1979), pp. 93–126.

[523] B. Wittenmark, Sampling of a system with a time delay, *IEEE Trans. Autom. Control* **30: 5** (1985), pp. 507–510.

[524] K.H. Wong, Optimal control computation for parabolic systems with boundary conditions involving time delays, *J. Optim. Th. & Appl.* **53** (1987), pp. 475–507.

[525] K.H. Wong, D. Clements, K.L. Teo,

1) Optimal control computation for linear time lag–systems with linear terminal constrains, *J. Optim. Th. & Appl.* **44** (1984), pp. 509–526.

2) Optimal control computation for nonlinear time lag systems, *J. Optim. Th. & Appl.* **47** (1985), pp. 91–107.

[526] A. Wozz–Busekros, Global stability in ecological systems with continuous time delay, *SIAM J. Appl. Math.* **35: 1** (1978), pp. 123–134.

[527] E.M. Wright, A functional equation in the heuristic theory of primes, *Math. Gazette* **45** (1961), pp. 15–16.

[528] Jiong–yu Wu, On a class of prey–predator population models with time lag, *Acta Math. Appl. Sin.* **11: 1** (1988), pp. 12–21.

[529] K. Yamahaka, E. Shimemura, Effects of mismashed Smith controller on stability in systems with time delays, *Automatica* **23** (1987), pp. 787–791.

[530] R.T. Yanushevsky, Control of Objects with Delay, Nauka, Moscow, 1979.

[531] T. Yonoyama, J. Sugie, Perturbing uniformly stable nonlinear scalar delay–differential equations, *Nonlinear Analysis, TMA* **12** (1988), pp. 303–311.

[532] J. Young, A. Arapostathis, Stabilization of discrete–time linear systems with a time delay in feedback loop, *Int. J. Control* **48: 4** (1988), pp. 1475–1485.

[533] Li Yong, Lin Zhenghua, Li Zhaoxing, A Massera type criterion for linear functional differential equations with advance and delay, *J. Math. Anal. Appl.* **200: 3** (1996), pp. 717–725.

[534] Kiyoshi Yoshida, The Hopf bifurcation and its stability for semilinear diffusion equations with time delay arising in ecology, *Hiroshima Math. J.* **12: 2** (1982), pp. 321–348.

[535] K. Yoshida, Functional Analysis, Berlin, Springer, 1965.

[536] G.E. Young, S. Rao, Robust sliding–mode control of a nonlinear process with uncertainty and delay, *J. of Dynamic Systems, Measurement, and Control* **109** (1987), pp. 203–208.

[537] R.L. Zandamela, Dortman on economic inerpretation of optimal control theory: an extention, *Riv. Mat. Sc. Econ. Soc.* **8: 2** (1985), pp. 115–122.

[538] B. Zhang, Necessary and sufficient conditions for boundedness and oscillation in the retarded Liénard equation, *J. Math. Anal. Applic.* **200** (1996), pp. 453–473.

[539] B.G. Zhang, K. Gopalsamy, Global attractivity and oscillation in a periodic delay logistic equation, *J. Math. Anal. & Appl.* **150: 1** (1990), pp. 274–283.

[540] A. Zhao, J. Yan, Asymptotic behaviour of solutions of impulsive delay differential equations, *J. Math Anal. Appl.* **201: 3** (1996), pp. 943–954.

[541] V.I. Zhdanov,

1) About doublesided solutions of charged particles movement equation with delay, *Izv. Vuz. Ser. Mat.* **5** (1983), pp. 20–25. (In Russian).

2) Finite parameter decision's family of the problem of two bodies in dynamics with delay, *Dokl. Akad. Nauk SSSR A* **7** (1983), pp. 13–16. (In Russian).

[542] S.L. Zolotarev, To the investigation of numerical species on diffusional models regarding delay effect, *Dyn. Biol. Populations*, Gorki, (1987), pp. 55–60. (In Russian).

[543] A.M. Zsykunov, Adaptive Control of Objects with Delay, Nauka, Moscow, 1984.

[544] A.M. Zverkin,

1) Dependence of stability of solutions of linear differential equations with lag upon the choice of the initial moment, *Vestnik Moskov. Univ. Ser. Mat. Mekh. Astr.* **5** (1959), pp. 15–20.

2) Theorems of existence and unicity for equations with deviating argument in critical case, *Tr. Sem. Theor. Diff. Uravn. Otkl. Argum.* **1** (1962), pp. 37–46.

INDEX

MIX
Papier aus verantwortungsvollen Quellen
Paper from responsible sources
FSC® C105338

If you have any concerns about our products,
you can contact us on
ProductSafety@springernature.com

In case Publisher is established outside the EU,
the EU authorized representative is:
Springer Nature Customer Service Center GmbH
Europaplatz 3, 69115 Heidelberg, Germany

Printed by Libri Plureos GmbH
in Hamburg, Germany